Graduate Texts in Mathematics 73

Springer
New York
Berlin
Heidelberg
Barcelona
Hong Kong
London
Milan
Paris
Singapore
Tokyo

Graduate Texts in Mathematics

(continued after index)

Thomas W. Hungerford

ALGEBRA

Springer

Thomas W. Hungerford
Department of Mathematics
Cleveland State University
Cleveland, Ohio 44115
USA

Mathematics Subject Classification (2000): 12-01, 13-01, 15-01, 16-01, 18-01, 20-01

Library of Congress Cataloging in Publication Data
Hungerford, Thomas W.
Algebra
Bibliography: p.
1. Algebra. I. Title.
QA155.H83 512 73-15693

Printed on acid-free paper

Originally published by Holt, Rinehart and Winston, Inc.
Printed and bound by R.R. Donnelley & Sons, Harrisonburg, VA.
Printed in the United States of America.

19 18 17 16 15 14 13 12 11

ISBN 0-387-90518-9
ISBN 3-540-90518-9 SPIN 10792683

Springer-Verlag New York Berlin Heidelberg
A member of BertelsmannSpringer Science+Business Media GmbH

To Mary

Preface to the Springer Edition

The reception given to the first edition of *Algebra* indicates that is has filled a definite need: to provide a self-contained, one-volume, graduate level algebra text that is readable by the average graduate student and flexible enough to accomodate a wide variety of instructors and course contents. Since it has been so well received, an extensive revision at this time does not seem warranted. Therefore, no substantial changes have been made in the text for this revised printing. However, all known misprints and errors have been corrected and several proofs have been rewritten.

I am grateful to Paul Halmos and F. W. Gehring, and the Springer staff, for their encouragement and assistance in bringing out this edition. It is gratifying to know that *Algebra* will continue to be available to the mathematical community. Springer-Verlag is to be commended for its willingness to continue to produce high quality mathematics texts at a time when many other publishers are looking to less elegant but more lucrative ventures.

Seattle, Washington
June, 1980

THOMAS W. HUNGERFORD

Note on the eighth printing (1996): A number of corrrections were incorporated in the fifth printing, thanks to the sharp-eyed diligence of George Bergman and his students at Berkeley and Keqin Feng of the Chinese University of Science and Technology. Additional corrections appear in this printing, thanks to Victor Boyko, Bob Cacioppo, and Joe L. Mott.

Preface

Note: A complete discussion of possible ways of using this text, including suggested course outlines, is given on page xv.

This book is intended to serve as a basic text for an algebra course at the beginning graduate level. Its writing was begun several years ago when I was unable to find a one-volume text which I considered suitable for such a course. My criteria for "suitability," which I hope are met in the present book, are as follows.

(i) A conscious effort has been made to produce a text which an average (but reasonably prepared) graduate student might read by himself without undue difficulty. The stress is on clarity rather than brevity.

(ii) For the reader's convenience the book is essentially self-contained. Consequently it includes much undergraduate level material which may be easily omitted by the better prepared reader.

(iii) Since there is no universal agreement on the content of a first year graduate algebra course we have included more material than could reasonably be covered in a single year. The major areas covered are treated in sufficient breadth and depth for the first year graduate level. Unfortunately reasons of space and economics have forced the omission of certain topics, such as valuation theory. For the most part these omitted subjects are those which seem to be least likely to be covered in a one year course.

(iv) The text is arranged to provide the instructor with maximum flexibility in the choice, order and degree of coverage of topics, without sacrificing readability for the student.

(v) There is an unusually large number of exercises.

There are, in theory, no formal prerequisites other than some elementary facts about sets, functions, the integers, and the real numbers, and a certain amount of "mathematical maturity." In actual practice, however, an undergraduate course in modern algebra is probably a necessity for most students. Indeed the book is written on this assumption, so that a number of concepts with which the typical graduate student may be assumed to be acquainted (for example, matrices) are presented in examples, exercises, and occasional proofs before they are formally treated in the text.

The guiding philosophical principle throughout the book is that the material should be presented in the *maximum useable generality consistent with good pedagogy*. The principle is relatively easy to apply to various technical questions. It is more difficult to apply to broader questions of conceptual organization. On the one hand, for example, the student must be made aware of relatively recent insights into the nature of algebra: the heart of the matter is the study of morphisms (maps); many deep and important concepts are best viewed as universal mapping properties. On the other hand, a high level of abstraction and generality is best appreciated and fully understood only by those who have a firm grounding in the special situations which motivated these abstractions. Consequently, concepts which can be characterized by a universal mapping property are not *defined* via this property if there is available a definition which is more familiar to or comprehensible by the student. In such cases the universal mapping property is then given in a theorem.

Categories are introduced early and some *terminology* of category theory is used frequently thereafter. However, the language of categories is employed chiefly as a useful convenience. A reader who is unfamiliar with categories should have little difficulty reading most of the book, even as a casual reference. Nevertheless, an instructor who so desires may give a substantial categorical flavor to the entire course without difficulty by treating Chapter X (Categories) at an early stage. Since it is essentially independent of the rest of the book it may be read at any time.

Other features of the mathematical exposition are as follows.

Infinite sets, infinite cardinal numbers, and transfinite arguments are used routinely. All of the necessary set theoretic prerequisites, including complete proofs of the relevant facts of cardinal arithmetic, are given in the Introduction.

The proof of the Sylow Theorems suggested by R. J. Nunke seems to clarify an area which is frequently confusing to many students.

Our treatment of Galois theory is based on that of Irving Kaplansky, who has successfully extended certain ideas of Emil Artin. The Galois group and the basic connection between subgroups and subfields are defined in the context of an absolutely general pair of fields. Among other things this permits easy generalization of various results to the infinite dimensional case. The Fundamental Theorem is proved at the beginning, before splitting fields, normality, separability, etc. have been introduced. Consequently the very real danger in many presentations, namely that student will lose sight of the forest for the trees, is minimized and perhaps avoided entirely.

In dealing with separable field extensions we distinguish the algebraic and the transcendental cases. This seems to be far better from a pedogogical standpoint than the Bourbaki method of presenting both cases simultaneously.

If one assumes that all rings have identities, all homomorphisms preserve identities and all modules are unitary, then a very quick treatment of semisimple rings and modules is possible. Unfortunately such an approach does not adequately prepare a student to read much of the literature in the theory of noncommutative rings. Consequently the structure theory of rings (in particular, semisimple left Artinian rings) is presented in a more general context. This treatment includes the situation mentioned above, but also deals fully with rings without identity, the Jacobson radical and related topics. In addition the prime radical and Goldie's Theorem on semiprime rings are discussed.

There are a large number of exercises of varying scope and difficulty. My experience in attempting to "star" the more difficult ones has thoroughly convinced me of the truth of the old adage: one man's meat is another's poison. Consequently no exercises are starred. The exercises are important in that a student is unlikely to appreciate or to master the material fully if he does not do a reasonable number of exercises. But the exercises are not an integral part of the text in the sense that non-trivial proofs of certain needed results are left entirely to the reader as exercises.

Nevertheless, most students are quite capable of proving nontrivial propositions provided that they are given appropriate guidance. Consequently, some theorems in the text are followed by a "sketch of proof" rather than a complete proof. Sometimes such a sketch is no more than a reference to appropriate theorems. On other occasions it may present the more difficult parts of a proof or a necessary "trick" in full detail and omit the rest. Frequently all the major steps of a proof will be stated, with the reasons or the routine calculational details left to the reader. Some of these latter "sketches" would be considered complete proofs by many people. In such cases the word "sketch" serves to warn the student that the proof in question is somewhat more concise than and possibly not as easy to follow as some of the "complete" proofs given elsewhere in the text.

Seattle, Washington
September, 1973

THOMAS W. HUNGERFORD

Acknowledgments

A large number of people have influenced the writing of this book either directly or indirectly. My first thanks go to Charles Conway, Vincent McBrien, Raymond Swords, S.J., and Paul Halmos. Without their advice, encouragement, and assistance at various stages of my educational career I would not have become a mathematician. I also wish to thank my thesis advisor Saunders Mac Lane, who was my first guide in the art of mathematical exposition. I can only hope that this book approaches the high standard of excellence exemplified by his own books.

My colleagues at the University of Washington have offered advice on various parts of the manuscript. In particular I am grateful to R. J. Nunke, G. S. Monk, R. Warfield, and D. Knudson. Thanks are also due to the students who have used preliminary versions of the manuscript during the past four years. Their comments have substantially improved the final product.

It is a pleasure to acknowledge the help of the secretarial staff at the University of Washington. Two preliminary versions were typed by Donna Thompson, sometimes assisted by Jan Nigh, Pat Watanabe, Pam Brink, and Sandra Evans. The final version was typed by Sonja Ogle, Kay Kolodziej Martin, and Vicki Caryl, with occasional assistance from Lois Bond, Geri Button, and Jan Schille.

Mary, my wife, deserves an accolade for her patience during the (seemingly interminable) time the book was being written. The final word belongs to our daughter Anne, age three, and our son Tom, age two, whose somewhat unexpected arrival after eleven years of marriage substantially prolonged the writing of this book: a small price to pay for such a progeny.

Suggestions on the Use of this Book

GENERAL INFORMATION

Within a given section all definitions, lemmas, theorems, propositions and corollaries are numbered consecutively (for example, in section 3 of some chapter the fourth numbered item is Item 3.4). The exercises in each section are numbered in a separate system. Cross references are given in accordance with the following scheme.

(i) Section 3 of Chapter V is referred to as section 3 throughout Chapter V and as section V.3 elsewhere.
(ii) Exercise 2 of section 3 of Chapter V is referred to as Exercise 2 throughout section V.3, as Exercise 3.2 throughout the other sections of Chapter V, and as Exercise V.3.2 elsewhere.
(iii) The fourth numbered item (Definition, Theorem, Corollary, Proposition, or Lemma) of section 3 of Chapter V is referred to as Item 3.4 throughout Chapter V and as Item V.3.4 elsewhere.

The symbol ■ is used to denote the end of a proof. A complete list of mathematical symbols precedes the index.

For those whose Latin is a bit rusty, the phrase *mutatis mutandis* may be roughly translated: "by changing the things which (obviously) must be changed (in order that the argument will carry over and make sense in the present situation)."

The title "proposition" is applied in this book only to those results which are *not* used in the sequel (except possibly in occasional exercises or in the proof of other "propositions"). Consequently **a reader who wishes to follow only the main line of the development may omit all propositions (and their lemmas and corollaries) without hindering his progress.** Results labeled as lemmas or theorems are almost always used at some point in the sequel. When a theorem is only needed in one or two places after its initial appearance, this fact is usually noted. The few minor exceptions to this labeling scheme should cause little difficulty.

INTERDEPENDENCE OF CHAPTERS

The table on the next page shows chapter interdependence and should be read in conjunction with the Table of Contents and the notes below (indicated by superscripts). In addition the reader should consult the introduction to each chapter for information on the interdependence of the various sections of the chapter.

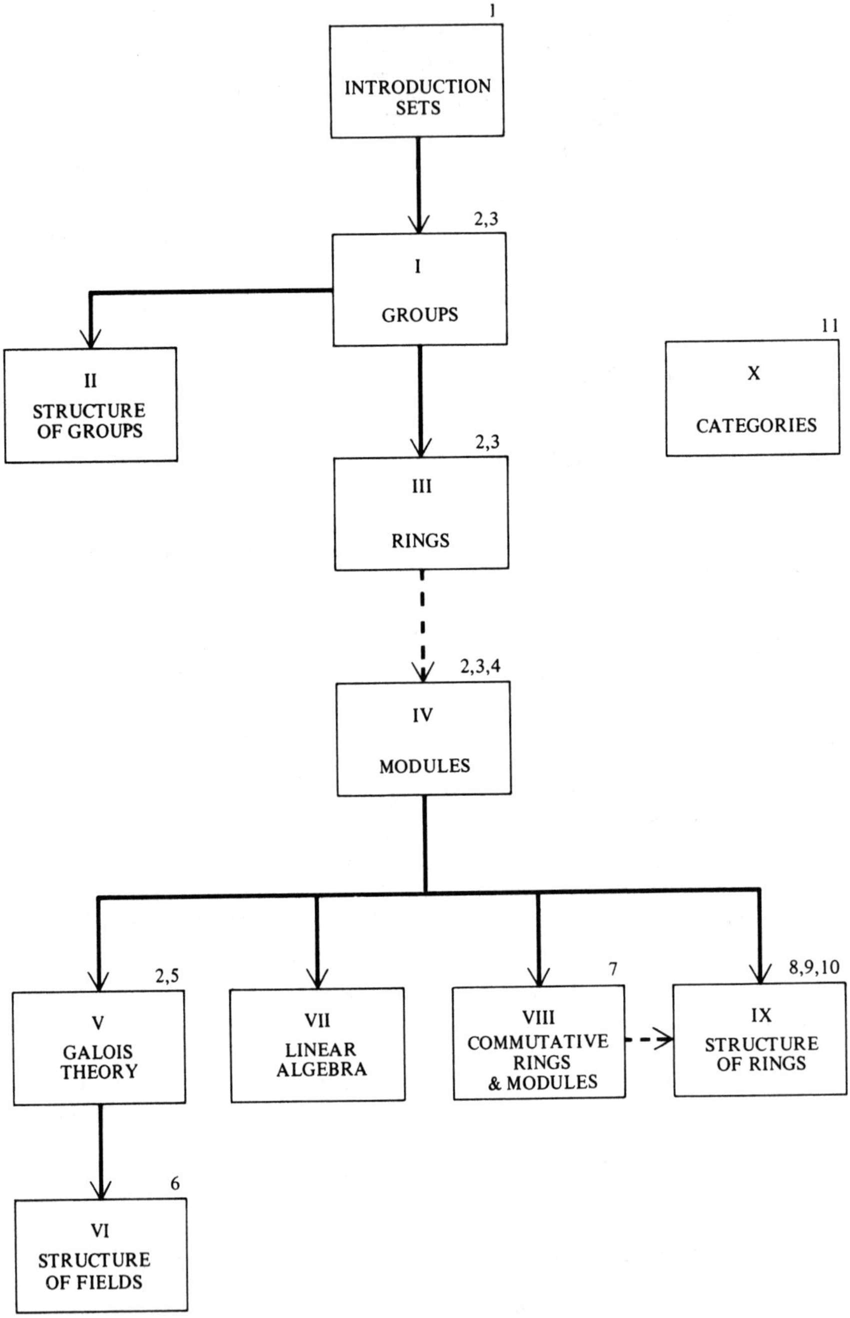
1
INTRODUCTION
SETS
2,3
I
GROUPS
II
STRUCTURE
OF GROUPS
11
X
CATEGORIES
2,3
III
RINGS
2,3,4
IV
MODULES
2,5
V
GALOIS
THEORY
VII
LINEAR
ALGEBRA
7
VIII
COMMUTATIVE
RINGS
& MODULES
8,9,10
IX
STRUCTURE
OF RINGS
6
VI
STRUCTURE
OF FIELDS

NOTES

1. Sections 1-7 of the Introduction are essential and are used frequently in the sequel. Except for Section 7 (Zorn's Lemma) this material is almost all elementary. The student should also know a definition of cardinal number (Section 8, through Definition 8.4). The rest of Section 8 is needed only five times. (Theorems II.1.2 and IV.2.6; Lemma V.3.5; Theorems V.3.6 and VI.1.9). Unless one wants to spend a considerable amount of time on cardinal arithmetic, this material may well be postponed until needed or assigned as outside reading for those interested.

2. A student who has had an undergraduate modern algebra course (or its equivalent) and is familiar with the contents of the Introduction can probably begin reading immediately any one of Chapters I, III, IV, or V.

3. A reader who wishes to skip Chapter I is strongly advised to scan Section I.7 to insure that he is familiar with the language of category theory introduced there.

4. With one exception, the only things from Chapter III needed in Chapter IV are the basic definitions of Section III.1. However Section III.3 is a prerequisite for Section IV.6.

5. Some knowledge of solvable groups (Sections II.7, II.8) is needed for the study of radical field extensions (Section V.9).

6. Chapter VI requires only the first six sections of Chapter V.

7. The proof of the Hilbert Nullstellensatz (Section VIII.7) requires some knowledge of transcendence degrees (Section VI.1) as well as material from Section V.3.

8. Section VIII.1 (Chain Conditions) is used extensively in Chapter IX, but Chapter IX is independent of the rest of Chapter VIII.

9. The basic connection between matrices and endomorphisms of free modules (Section VII.1, through Theorem VII.1.4) is used in studying the structure of rings (Chapter IX).

10. Section V.3 is a prerequisite for Section IX.6.

11. Sections I.7, IV.4, and IV.5 are prerequisites for Chapter X; otherwise Chapter X is essentially independent of the rest of the book.

SUGGESTED COURSE OUTLINES

The information given above, together with the introductions to the various chapters, is sufficient for designing a wide variety of courses of varying content and length. Here are some of the possible one quarter courses (30 class meetings) on specific topics.

These descriptions are somewhat elastic depending on how much is assumed, the level of the class, etc. Under the heading *Review* we list background material (often of an elementary nature) which is frequently used in the course. This material may be assumed or covered briefly or assigned as outside reading or treated in detail if necessary, depending on the background of the class. It is assumed without explicit mention that the student is familiar with the appropriate parts of the Introduction (see note 1, p. xvii). Almost all of these courses can be shortened by omitting all *Propostions* and their associated *Lemmas* and *Corollaries* (see page xv).

GROUP THEORY

Review: Introduction, omitting most of Section 8 (see note 1, p. xvii). **Basic Course:** Chapters I and II, with the possible omission of Sections I.9, II.3 and the last half of II.7. It is also possible to omit Sections II.1 and II.2 or at least postpone them until after the Sylow Theorems (Section II.5).

MODULES AND THE STRUCTURE OF RINGS

Review: Sections III.1 and III.2 (through Theorem III.2.13). **Basic Course:** the rest of Section III.2; Sections 1-5 of Chapter IV[1]; Section VII.1 (through Theorem VII.1.4); Section VIII.1; Sections 1-4 of Chapter IX. **Additional Topics:** Sections III.4, IV.6, IV.7, IX.5; Section IV.5 if not covered earlier; Section IX.6; material from Chapter VIII.

FIELDS AND GALOIS THEORY

Review: polynomials, modules, vector spaces (Sections III.5, III.6, IV.1, IV.2). Solvable groups (Sections II.7, II.8) are used in Section V.9. **Basic Course[2]:** Sections 1-3 of Chapter V, omitting the appendices; Definition V.4.1 and Theorems V.4.2 and V.4.12; Section V.5 (through Theorem 5.3); Theorem V.6.2; Section V.7, omitting Proposition V.7.7—Corollary V.7.9; Theorem V.8.1; Section V.9 (through Corollary V.9.5); Section VI.1. **Additional Topics:** the rest of Sections V.5 and V.6 (at least through Definition V.6.10); the appendices to Sections V.1-V.3; the rest of Sections V.4, V.9, and V.7; Section V.8; Section VI.2.

LINEAR ALGEBRA

Review: Sections 3-6 of Chapter III and Section IV.1; selected parts of Section IV.2 (finite dimensional vector spaces). **Basic Course:** structure of torsion modules over a PID (Section IV.6, omitting material on free modules); Sections 1-5 of Chapter VII, omitting appendices and possibly the *Propositions*.

[1]If the stress is primarily on rings, one may omit most of Chapter IV. Specifically, one need only cover Section IV.1; Section IV.2 (through Theorem IV.2.4); Definition IV.2.8; and Section IV.3 (through Definition IV.3.6).

[2]The outline given here is designed so that the solvability of polynomial equations can be discussed quickly after the Fundamental Theorem and splitting fields are presented; it requires using Theorem V.7.2 as a definition, in place of Definition V.7.1. The discussion may be further shortened if one considers only finite dimensional extensions and omits algebraic closures, as indicated in the note preceding Theorem V.3.3.

COMMUTATIVE ALGEBRA

Review: Sections III.1, III.2 (through Theorem III.2.13). **Basic Course:** the rest of Section III.2; Sections III.3 and III.4; Section IV.1; Section IV.2 (through Corollary IV.2.2); Section IV.3 (through Proposition IV.3.5); Sections 1-6 of Chapter VIII, with the possible omission of *Propositions*. **Additional topics:** Section VIII.7 (which also requires background from Sections V.3 and VI.1).

Table of Contents

INTRODUCTION

PREREQUISITES AND PRELIMINARIES

In Sections 1–6 we summarize for the reader's convenience some basic material with which he is assumed to be thoroughly familiar (with the possible exception of the distinction between sets and proper classes (Section 2), the characterization of the Cartesian product by a universal mapping property (Theorem 5.2) and the Recursion Theorem 6.2). The definition of cardinal number (first part of Section 8) will be used frequently. The Axiom of Choice and its equivalents (Section 7) and cardinal arithmetic (last part of Section 8) may be postponed until this information is actually used. Finally the reader is presumed to have some familiarity with the fields **Q**, **R**, and **C** of rational, real, and complex numbers respectively.

1. LOGIC

We adopt the usual logical conventions, and consider only statements that have a truth value of either true or false (not both). If P and Q are statements, then the statement "P and Q" is true if both P and Q are true and false otherwise. The statement "P or Q" is true in all cases except when both P and Q are false. An implication is a statement of the form "P implies Q" or "if P, then Q" (written symbolically as $P \Rightarrow Q$). An implication is false if P is true and Q is false; it is true in all other cases. In particular, *an implication with a false premise is always a true implication.* An equivalence or biconditional is a statement of the form "P implies Q and Q implies P." This is generally abbreviated to "P if and only if Q" (symbolically $P \Leftrightarrow Q$). The biconditional "$P \Leftrightarrow Q$" is true exactly when P and Q are both true or both false; otherwise it is false. The negation of the statement P is the statement "it is not the case that P." It is true if and only if P is false.

2. SETS AND CLASSES

Our approach to the theory of sets will be quite informal. Nevertheless in order to define adequately both cardinal numbers (Section 8) and categories (Section I.7) it will be necessary to introduce at least the rudiments of a formal axiomatization of set theory. In fact the entire discussion may, if desired, be made rigorously precise; see Eisenberg [8] or Suppes [10]. An axiomatic approach to set theory is also useful in order to avoid certain paradoxes that are apt to cause difficulty in a purely intuitive treatment of the subject. A paradox occurs in an axiom system when both a statement and its negation are deducible from the axioms. This in turn implies (by an exercise in elementary logic) that *every* statement in the system is true, which is hardly a very desirable state of affairs.

In the Gödel-Bernays form of axiomatic set theory, which we shall follow, the primitive (undefined) notions are **class, membership,** and **equality.** Intuitively we consider a class to be a collection A of objects (elements) such that given any object x it is possible to determine whether or not x is a member (or element) of A. We write $x \in A$ for "x is an element of A" and $x \notin A$ for "x is not an element of A." The axioms are formulated in terms of these primitive notions and the first-order predicate calculus (that is, the language of sentences built up by using the connectives *and*, *or*, *not*, *implies* and the quantifiers *there exists* and *for all*). For instance, equality is assumed to have the following properties for all classes $A, B, C : A = A$; $A = B \Rightarrow B = A$; $A = B$ and $B = C \Rightarrow A = C$; $A = B$ and $x \in A \Rightarrow x \in B$. The **axiom of extensionality** asserts that two classes with the same elements are equal (formally, $[x \in A \Leftrightarrow x \in B] \Rightarrow A = B$).

A class A is defined to be a **set** if and only if there exists a class B such that $A \in B$. Thus a set is a particular kind of class. A class that is not a set is called a **proper class.** Intuitively the distinction between sets and proper classes is not too clear. Roughly speaking a set is a "small" class and a proper class is exceptionally "large." The **axiom of class formation** asserts that for any statement $P(y)$ in the first-order predicate calculus involving a variable y, there exists a class A such that $x \in A$ if and only if x is a set and the statement $P(x)$ is true. We denote this class A by $\{x \mid P(x)\}$, and refer to "the class of all x such that $P(x)$." Sometimes a class is described simply by listing its elements in brackets, for example, $\{a,b,c\}$.

EXAMPLE.[1] Consider the class $M = \{X \mid X$ is a set and $X \notin X\}$. The statement $X \notin X$ is not unreasonable since many sets satisfy it (for example, the set of all books is not a book). M is a proper class. For if M were a set, then either $M \in M$ or $M \notin M$. But by the definition of M, $M \in M$ implies $M \notin M$ and $M \notin M$ implies $M \in M$. Thus in either case the assumption that M is a set leads to an untenable paradox: $M \in M$ and $M \notin M$.

We shall now review a number of familiar topics (unions, intersections, functions, relations, Cartesian products, etc.). The presentation will be informal with the mention of axioms omitted for the most part. However, it is also to be understood that there are sufficient axioms to guarantee that when one of these constructions is performed on *sets*, the result is also a set (for example, the union of sets is a set; a subclass of a set is a set). The usual way of proving that a given class is a set is to show that it may be obtained from a set by a sequence of these admissible constructions.

A class A is a **subclass** of a class B (written $A \subset B$) provided:

$$\text{for all } x \in A, \quad x \in A \quad \Rightarrow \quad x \in B. \tag{1}$$

[1]This was first propounded (in somewhat different form) by Bertrand Russell in 1902 as a paradox that indicated the necessity of a formal axiomatization of set theory.

By the axioms of extensionality and the properties of equality:

$$A = B \quad \Leftrightarrow \quad A \subset B \quad \text{and} \quad B \subset A.$$

A subclass A of a class B that is itself a set is called a **subset** of B. There are axioms to insure that a subclass of a set is a subset.

The **empty set** or **null set** (denoted $\varnothing$) is the set with no elements (that is, given any x, $x \notin \varnothing$). Since the statement "$x \in \varnothing$" is always false, the implication (1) is always true when $A = \varnothing$. Therefore $\varnothing \subset B$ for every class B. A is said to be a proper subclass of B if $A \subset B$ but $A \neq \varnothing$ and $A \neq B$.

The **power axiom** asserts that for every *set* A the class $P(A)$ of all subsets of A is itself a set. $P(A)$ is called the **power set** of A; it is also denoted 2^A.

A **family of sets** indexed by (the nonempty class) I is a collection of *sets* A_i, one for each $i \in I$ (denoted $\{A_i \mid i \in I\}$). Given such a family, its **union** and **intersection** are defined to be respectively the classes

$$\bigcup_{i \in I} A_i = \{x \mid x \in A_i \quad \text{for some} \quad i \in I\}; \text{ and}$$

$$\bigcap_{i \in I} A_i = \{x \mid x \in A_i \quad \text{for every} \quad i \in I\}.$$

If I is a set, then suitable axioms insure that $\bigcup_{i \in I} A_i$ and $\bigcap_{i \in I} A_i$ are actually sets. If $I = \{1,2,\ldots,n\}$ one frequently writes $A_1 \cup A_2 \cup \cdots \cup A_n$ in place of $\bigcup_{i \in I} A_i$ and similarly for intersections. If $A \cap B = \varnothing$, A and B are said to be **disjoint.**

If A and B are classes, the **relative complement** of A in B is the following subclass of B:

$$B - A = \{x \mid x \in B \quad \text{and} \quad x \notin A\}.$$

If all the classes under discussion are subsets of some fixed set U (called the universe of discussion), then $U - A$ is denoted A' and called simply the **complement** of A. The reader should verify the following statements.

$$A \cap (\bigcup_{i \in I} B_i) = \bigcup_{i \in I} (A \cap B_i) \quad \text{and} \tag{2}$$

$$A \cup (\bigcap_{i \in I} B_i) = \bigcap_{i \in I} (A \cup B_i).$$

$$(\bigcup_{i \in I} A_i)' = \bigcap_{i \in I} A_i' \quad \text{and} \quad (\bigcap_{i \in I} A_i)' = \bigcup_{i \in I} A_i' \text{ (DeMorgan's Laws).} \tag{3}$$

$$A \cup B = B \quad \Leftrightarrow \quad A \subset B \quad \Leftrightarrow \quad A \cap B = A. \tag{4}$$

3. FUNCTIONS

Given classes A and B, a **function** (or **map** or **mapping**) f from A to B (written $f: A \to B$) assigns to each $a \in A$ exactly one element $b \in B$; b is called the value of the function at a or the image of a and is usually written $f(a)$. A is the **domain** of the function (sometimes written Dom f) and B is the **range** or **codomain.** Sometimes it is convenient to denote the effect of the function f on an element of A by $a \mapsto f(a)$. Two functions are **equal** if they have the same domain and range and have the same value for each element of their common domain.

If $f: A \to B$ is a function and $S \subset A$, the function from S to B given by

$$a \mapsto f(a), \quad \text{for} \quad a \in S$$

is called the **restriction** of f to S and is denoted $f \mid S : S \to B$. If A is any class, the **identity function** on A (denoted $1_A : A \to A$) is the function given by $a \mapsto a$. If $S \subset A$, the function $1_A \mid S : S \to A$ is called the **inclusion map** of S into A.

Let $f: A \to B$ and $g : B \to C$ be functions. The **composite** of f and g is the function $A \to C$ given by

$$a \mapsto g(f(a)), \quad a \in A.$$

The composite function is denoted $g \circ f$ or simply gf. If $h : C \to D$ is a third function, it is easy to verify that $h(gf) = (hg)f$. If $f: A \to B$, then $f \circ 1_A = f = 1_B \circ f: A \to B$.

A **diagram** of functions:

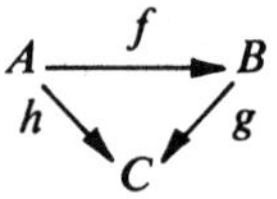

is said to be **commutative** if $gf = h$. Similarly, the diagram:

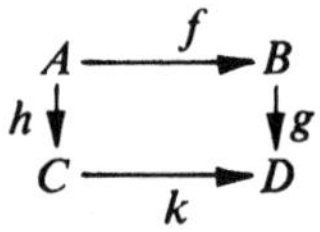

is **commutative** if $kh = gf$. Frequently we shall deal with more complicated diagrams composed of a number of triangles and squares as above. Such a diagram is said to be commutative if every triangle and square in it is commutative.

Let $f: A \to B$ be a function. If $S \subset A$, the **image of S under f** (denoted $f(S)$) is the class

$$\{b \in B \mid b = f(a) \quad \text{for some} \quad a \in S\}.$$

The class $f(A)$ is called the **image of f** and is sometimes denoted $\operatorname{Im} f$. If $T \subset B$, the **inverse image of T** under f (denoted $f^{-1}(T)$) is the class

$$\{a \in A \mid f(a) \in T\}.$$

If T consists of a single element, $T = \{b\}$, we write $f^{-1}(b)$ in place of $f^{-1}(T)$. The following facts can be easily verified:

$$\text{for } S \subset A,\ f^{-1}(f(S)) \supset S; \tag{5}$$

$$\text{for } T \subset B,\ f(f^{-1}(T)) \subset T. \tag{6}$$

For any family $\{T_i \mid i \in I\}$ of subsets of B,

$$f^{-1}(\bigcup_{i \in I} T_i) = \bigcup_{i \in I} f^{-1}(T_i); \tag{7}$$

$$f^{-1}(\bigcap_{i \in I} T_i) = \bigcap_{i \in I} f^{-1}(T_i). \tag{8}$$

A function $f: A \to B$ is said to be **injective** (or **one-to-one**) provided

$$\text{for all } a,a' \in A, \quad a \neq a' \quad \Rightarrow \quad f(a) \neq f(a');$$

alternatively, f is injective if and only if

$$\text{for all } a,a' \in A, \quad f(a) = f(a') \quad \Rightarrow \quad a = a'.$$

A function f is **surjective** (or **onto**) provided $f(A) = B$; in other words,

$$\text{for each } b \in B,\ b = f(a) \quad \text{for some} \quad a \in A.$$

A function f is said to be **bijective** (or a **bijection** or a **one-to-one correspondence**) if it is both injective and surjective. It follows immediately from these definitions that for any class A, the identity map $1_A : A \to A$ is bijective. The reader should verify that for maps $f: A \to B$ and $g : B \to C$,

$$f \text{ and } g \text{ injective} \quad \Rightarrow \quad gf \text{ is injective;} \tag{9}$$

$$f \text{ and } g \text{ surjective} \quad \Rightarrow \quad gf \text{ is surjective;} \tag{10}$$

$$gf \text{ injective} \quad \Rightarrow \quad f \text{ is injective;} \tag{11}$$

$$gf \text{ surjective} \quad \Rightarrow \quad g \text{ is surjective.} \tag{12}$$

Theorem 3.1. *Let* $f: A \to B$ *be a function, with* A *nonempty.*

(i) f *is injective if and only if there is a map* $g : B \to A$ *such that* $gf = 1_A$.
(ii) *If* A *is a set, then* f *is surjective if and only if there is a map* $h : B \to A$ *such that* $fh = 1_B$.

PROOF. Since every identity map is bijective, (11) and (12) prove the implications ($\Leftarrow$) in (i) and (ii). Conversely if f is injective, then for each $b \in f(A)$ there is a unique $a \in A$ with $f(a) = b$. Choose a fixed $a_0 \in A$ and verify that the map $g : B \to A$ defined by

$$g(b) = \begin{cases} a & \text{if} \quad b \in f(A) \quad \text{and} \quad f(a) = b \\ a_0 & \text{if} \quad b \notin f(A) \end{cases}$$

is such that $gf = 1_A$. For the converse of (ii) suppose f is surjective. Then $f^{-1}(b) \subset A$ is a nonempty set for every $b \in B$. For each $b \in B$ choose $a_b \in f^{-1}(b)$ (Note: this requires the Axiom of Choice; see Section 7). Verify that the map $h : B \to A$ defined by $h(b) = a_b$ is such that $fh = 1_B$. ■

The map g as in Theorem 3.1 is called a **left inverse** of f and h is called a **right inverse** of f. If a map $f: A \to B$ has both a left inverse g and a right inverse h, then

$$g = g1_B = g(fh) = (gf)h = 1_A h = h$$

and the map $g = h$ is called a **two-sided inverse** of f. This argument also shows that the two-sided inverse of a map (if it has one) is unique. By Theorem 3.1 if A is a set and $f: A \to B$ a function, then

$$f \text{ is bijective} \quad \Leftrightarrow \quad f \text{ has a two-sided inverse.}^2 \tag{13}$$

The unique two-sided inverse of a bijection f is denoted f^{-1}; clearly f is a two-sided inverse of f^{-1} so that f^{-1} is also a bijection.

[2](13) is actually true even when A is a proper class; see Eisenberg [8; p. 146].

4. RELATIONS AND PARTITIONS

The **axiom of pair formation** states that for any two sets [elements] a,b there is a set $P = \{a,b\}$ such that $x \in P$ if and only if $x = a$ or $x = b$; if $a = b$ then P is the **singleton** $\{a\}$. The **ordered pair** (a,b) is defined to be the set $\{\{a\}, \{a,b\}\}$; its **first component** is a and its **second component** is b. It is easy to verify that $(a,b) = (a',b')$ if and only if $a = a'$ and $b = b'$. The **Cartesian product** of classes A and B is the class

$$A \times B = \{(a,b) \mid a \in A, b \in B\}.$$

Note that $A \times \varnothing = \varnothing = \varnothing \times B$.

A subclass R of $A \times B$ is called a **relation** on $A \times B$. For example, if $f: A \rightarrow B$ is a function, the **graph** of f is the relation $R = \{(a,f(a)) \mid a \in A\}$. Since f is a function, R has the special property:

every element of A is the first component of one and only one ordered pair in R. (14)

Conversely any relation R on $A \times B$ that satisfies (14), determines a unique function $f: A \rightarrow B$ whose graph is R (simply define $f(a) = b$, where (a,b) is the unique ordered pair in R with first component a). For this reason it is customary in a formal axiomatic presentation of set theory to identify a function with its graph, that is, to *define* a function to be a relation satisfying (14). This is necessary, for example, in order to prove from the axioms that the image of a set under a function is in fact a set.

Another advantage of this approach is that it permits us to define functions with empty domain. For since $\varnothing \times B = \varnothing$ is the unique subset of $\varnothing \times B$ and vacuously satisfies (14), there is a unique function $\varnothing \rightarrow B$. It is also clear from (14) that there can be a function with empty range only if the domain is also empty. Whenever convenient we shall think of a function as a relation satisfying (14).

A relation R on $A \times A$ is an **equivalence relation** on A provided R is:

$$\textbf{reflexive: } (a,a) \in R \quad \text{for all} \quad a \in A; \tag{15}$$

$$\textbf{symmetric: } (a,b) \in R \quad \Rightarrow \quad (b,a) \in R; \tag{16}$$

$$\textbf{transitive: } (a,b) \in R \quad \text{and} \quad (b,c) \in R \quad \Rightarrow \quad (a,c) \in R. \tag{17}$$

If R is an equivalence relation on A and $(a,b) \in R$, we say that a is equivalent to b under R and write $a \sim b$ or aRb; in this notation (15)–(17) become:

$$a \sim a; \tag{15$'$}$$

$$a \sim b \quad \Rightarrow \quad b \sim a; \tag{16$'$}$$

$$a \sim b \quad \text{and} \quad b \sim c \quad \Rightarrow \quad a \sim c. \tag{17$'$}$$

Let R ($\sim$) be an equivalence relation on A. If $a \in A$, the **equivalence class** of a (denoted $\bar{a}$) is the class of all those elements of A that are equivalent to a; that is, $\bar{a} = \{b \in A \mid b \sim a\}$. The class of all equivalence classes in A is denoted A/R and called the **quotient class** of A by R. Since R is reflexive, $a \in \bar{a}$ for every $a \in A$; hence

$$\bar{a} \neq \varnothing, \quad \text{for every} \quad a \in A; \quad \text{and} \quad \text{if } A \text{ is a set} \tag{18}$$

$$\bigcup_{a \in A} \bar{a} = A = \bigcup_{\bar{a} \in A/R} \bar{a}. \tag{19}$$

Also observe that

$$\bar{a} = \bar{b} \Leftrightarrow \cdot\, a \sim b; \tag{20}$$

for if $\bar{a} = \bar{b}$, then $a \in \bar{a} \Rightarrow a \in \bar{b} \Rightarrow a \sim b$. Conversely, if $a \sim b$ and $c \in \bar{a}$, then $c \sim a$ and $a \sim b \Rightarrow c \sim b \Rightarrow c \in \bar{b}$. Thus $\bar{a} \subset \bar{b}$; a symmetric argument shows that $\bar{b} \subset \bar{a}$ and therefore $\bar{a} = \bar{b}$. Next we prove:

$$\text{for } a,b \in A, \quad \text{either} \quad \bar{a} \cap \bar{b} = \varnothing \quad \text{or} \quad \bar{a} = \bar{b}. \tag{21}$$

If $\bar{a} \cap \bar{b} \neq \varnothing$, then there is an element $c \in \bar{a} \cap \bar{b}$. Hence $c \sim a$ and $c \sim b$. Using symmetry, transitivity and (20) we have: $a \sim c$ and $c \sim b \Rightarrow a \sim b \Rightarrow \bar{a} = \bar{b}$.

Let A be a nonempty class and $\{A_i \mid i \in I\}$ a family of subsets of A such that:

$$A_i \neq \varnothing, \quad \text{for each} \quad i \in I;$$
$$\bigcup_{i \in I} A_i = A;$$
$$A_i \cap A_j = \varnothing \quad \text{for all} \quad i \neq j \in I;$$

then $\{A_i \mid i \in I\}$ is said to be a **partition** of A.

Theorem 4.1. *If* A *is a nonempty set, then the assignment* R $\mapsto$ A/R *defines a bijection from the set* E(A) *of all equivalence relations on* A *onto the set* Q(A) *of all partitions of* A.

SKETCH OF PROOF. If R is an equivalence relation on A, then the set A/R of equivalence classes is a partition of A by (18), (19), and (21) so that $R \mapsto A/R$ defines a function $f: E(A) \to Q(A)$. Define a function $g: Q(A) \to E(A)$ as follows. If $S = \{A_i \mid i \in I\}$ is a partition of A, let $g(S)$ be the equivalence relation on A given by:

$$a \sim b \quad \Leftrightarrow \quad a \in A_i \quad \text{and} \quad b \in A_i \quad \text{for some (unique) } i \in I. \tag{22}$$

Verify that $g(S)$ is in fact an equivalence relation such that $\bar{a} = A_i$ for $a \in A_i$. Complete the proof by verifying that $fg = 1_{Q(A)}$ and $gf = 1_{E(A)}$. Then f is bijective by (13). ■

5. PRODUCTS

Note. In this section we deal only with *sets*. No proper classes are involved.

Consider the Cartesian product of two sets $A_1 \times A_2$. An element of $A_1 \times A_2$ is a pair (a_1,a_2) with $a_i \in A_i$, $i = 1,2$. Thus the pair (a_1,a_2) determines a function $f: \{1,2\} \to A_1 \cup A_2$ by: $f(1) = a_1$, $f(2) = a_2$. Conversely, every function $f: \{1,2\} \to A_1 \cup A_2$ with the property that $f(1) \in A_1$ and $f(2) \in A_2$ determines an element $(a_1,a_2) = (f(1),f(2))$ of $A_1 \times A_2$. Therefore it is not difficult to see that there is a one-to-one correspondence between the set of all functions of this kind and the set $A_1 \times A_2$. This fact leads us to generalize the notion of Cartesian product as follows.

Definition 5.1. *Let* $\{A_i \mid i \in I\}$ *be a family of sets indexed by a (nonempty) set* I. *The* **(Cartesian) product** *of the sets* A_i *is the set of all functions* $f: I \to \bigcup_{i \in I} A_i$ *such that* $f(i) \in A_i$ *for all* $i \in I$. *It is denoted* $\prod_{i \in I} A_i$.

If $I = \{1,2, \ldots, n\}$, the product $\prod_{i\varepsilon I} A_i$ is often denoted by $A_1 \times A_2 \times \cdots \times A_n$ and is identified with the set of all ordered n-tuples $(a_1,a_2, \ldots, a_n)$, where $a_i \varepsilon A_i$ for $i = 1,2, \ldots, n$ just as in the case mentioned above, where $I = \{1,2\}$. A similar notation is often convenient when I is infinite. We shall sometimes denote the function $f \varepsilon \prod_{i\varepsilon I} A_i$ by $\{a_i\}_{i\varepsilon I}$ or simply $\{a_i\}$, where $f(i) = a_i \varepsilon A_i$ for each $i \varepsilon I$.

If some $A_j = \emptyset$, then $\prod_{i\varepsilon I} A_i = \emptyset$ since there can be no function $f: I \to \bigcup A_i$ such that $f(j) \varepsilon A_j$.

If $\{A_i \mid i \varepsilon I\}$ and $\{B_i \mid i \varepsilon I\}$ are families of sets such that $B_i \subset A_i$ for each $i \varepsilon I$, then every function $I \to \bigcup_{i\varepsilon I} B_i$ may be considered to be a function $I \to \bigcup_{i\varepsilon I} A_i$. Therefore we consider $\prod_{i\varepsilon I} B_i$ to be a subset of $\prod_{i\varepsilon I} A_i$.

Let $\prod_{i\varepsilon I} A_i$ be a Cartesian product. For each $k \varepsilon I$ define a map $\pi_k : \prod_{i\varepsilon I} A_i \to A_k$ by $f \mapsto f(k)$, or in the other notation, $\{a_i\} \mapsto a_k$. π_k is called the **(canonical) projection** of the product onto its kth component (or factor). If every A_i is nonempty, then each π_k is surjective (see Exercise 7.6).

The product $\prod_{i\varepsilon I} A_i$ and its projections are precisely what we need in order to prove:

Theorem 5.2. *Let* $\{A_i \mid i \varepsilon I\}$ *be a family of sets indexed by* I. *Then there exists a set* D, *together with a family of maps* $\{\pi_i : D \to A_i \mid i \varepsilon I\}$ *with the following property: for any set* C *and family of maps* $\{\varphi_i : C \to A_i \mid i \varepsilon I\}$, *there exists a unique map* $\varphi : C \to D$ *such that* $\pi_i\varphi = \varphi_i$ *for all* $i \varepsilon I$. *Furthermore,* D *is uniquely determined up to a bijection.*

The last sentence means that if D' is a set and $\{\pi_i' : D' \to A_i \mid i \varepsilon I\}$ a family of maps, which have the same property as D and $\{\pi_i\}$, then there is a bijection $D \to D'$.

PROOF OF 5.2. (Existence) Let $D = \prod_{i\varepsilon I} A_i$ and let the maps π_i be the projections onto the ith components. Given C and the maps φ_i, define $\varphi : C \to \prod_{i\varepsilon I} A_i$ by $c \mapsto f_c$, where $f_c(i) = \varphi_i(c) \varepsilon A_i$. It follows immediately that $\pi_i\varphi = \varphi_i$ for all $i \varepsilon I$. To show that φ is unique we assume that $\varphi' : C \to \prod_{i\varepsilon I} A_i$ is another map such that $\pi_i\varphi' = \varphi_i$ for all $i \varepsilon I$ and prove that $\varphi = \varphi'$. To do this we must show that for each $c \varepsilon C$, $\varphi(c)$, and $\varphi'(c)$ are the same element of $\prod_{i\varepsilon I} A_i$ — that is, $\varphi(c)$ and $\varphi'(c)$ agree as functions on I: $(\varphi(c))(i) = (\varphi'(c))(i)$ for all $i \varepsilon I$. But by hypothesis and the definition of π_i we have for every $i \varepsilon I$:

$$(\varphi'(c))(i) = \pi_i\varphi'(c) = \varphi_i(c) = f_c(i) = (\varphi(c))(i).$$

(Uniqueness) Suppose D' (with maps $\pi_i' : D' \to A_i$) has the same property as $D = \prod_{i\varepsilon I} A_i$. If we apply this property (for D) to the family of maps $\{\pi_i' : D' \to A_i\}$ and also apply it (for D') to the family $\{\pi_i : D \to A_i\}$, we obtain (unique) maps

$\varphi : D' \to D$ and $\psi : D \to D'$ such that the following diagrams are commutative for each $i \in I$:

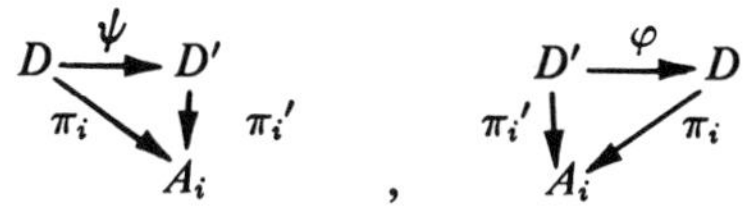

Combining these gives for each $i \in I$ a commutative diagram

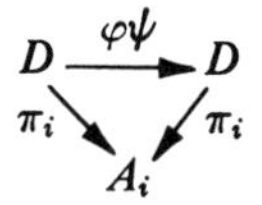

Thus $\varphi\psi : D \to D$ is a map such that $\pi_i(\varphi\psi) = \pi_i$ for all $i \in I$. But by the proof above, there is a unique map with this property. Since the map $1_D : D \to D$ is also such that $\pi_i 1_D = \pi_i$ for all $i \in I$, we must have $\varphi\psi = 1_D$ by uniqueness. A similar argument shows that $\psi\varphi = 1_{D'}$. Therefore, φ is a bijection by (13) and $D = \prod_{i \in I} A_i$ is uniquely determined up to a bijection. ■

Observe that the statement of Theorem 5.2 does not mention elements; it involves only sets and maps. It says, in effect, that the product $\prod_{i \in I} A_i$ is characterized by a certain **universal mapping property.** We shall discuss this concept with more precision when we deal with categories and functors below.

6. THE INTEGERS

We do not intend to give an axiomatic development of the integers. Instead we assume that the reader is thoroughly familiar with the set $\mathbf{Z}$ of integers, the set $\mathbf{N} = \{0,1,2,\ldots\}$ of nonnegative integers (or natural numbers) the set $\mathbf{N}^* = \{1,2,\ldots\}$ of positive integers and the elementary properties of addition, multiplication, and order. In particular, for all $a,b,c \in \mathbf{Z}$:

$$(a + b) + c = a + (b + c) \quad \text{and} \quad (ab)c = a(bc) \quad \text{(associative laws);} \tag{23}$$

$$a + b = b + a \quad \text{and} \quad ab = ba \quad \text{(commutative laws);} \tag{24}$$

$$a(b + c) = ab + ac \quad \text{and} \quad (a + b)c = ac + bc \quad \text{(distributive laws);} \tag{25}$$

$$a + 0 = a \quad \text{and} \quad a1 = a \quad \text{(identity elements);} \tag{26}$$

for each $a \in \mathbf{Z}$ there exists $-a \in \mathbf{Z}$ such that $a + (-a) = 0$ (additive inverse); we write $a - b$ for $a + (-b)$. (27)

$$ab = 0 \iff a = 0 \quad \text{or} \quad b = 0; \tag{28}$$

$$a < b \implies a + c < b + c \quad \text{for all} \quad c \in \mathbf{Z}; \tag{29}$$

$$a < b \implies ad < bd \quad \text{for all} \quad d \in \mathbf{N}^*. \tag{30}$$

We write $a < b$ and $b > a$ interchangeably and write $a \leq b$ if $a < b$ or $a = b$. The absolute value $|a|$ of $a \in \mathbf{Z}$ is defined to be a if $a \geq 0$ and $-a$ if $a < 0$. Finally we assume as a basic axiom the

Law of Well Ordering. *Every nonempty subset* S *of* **N** *contains a least element* (*that is, an element* $b \varepsilon S$ *such that* $b \leq c$ *for all* $c \varepsilon S$).

In particular, 0 is the least element of **N**.

In addition to the above we require certain facts from elementary number theory, some of which are briefly reviewed here.

Theorem 6.1. (*Principle of Mathematical Induction*) *If* S *is a subset of the set* **N** *of natural numbers such that* $0 \varepsilon S$ *and either*

(i) $n \varepsilon S \Rightarrow n + 1 \varepsilon S$ *for all* $n \varepsilon \mathbf{N}$;

or

(ii) $m \varepsilon S$ *for all* $0 \leq m < n \Rightarrow n \varepsilon S$ *for all* $n \varepsilon \mathbf{N}$;

then $S = \mathbf{N}$.

PROOF. If $\mathbf{N} - S \neq \varnothing$, let $n \neq 0$ be its least element. Then for every $m < n$, we must have $m \not\varepsilon \mathbf{N} - S$ and hence $m \varepsilon S$. Consequently either (i) or (ii) implies $n \varepsilon S$, which is a contradiction. Therefore $\mathbf{N} - S = \varnothing$ and $\mathbf{N} = S$. ■

REMARK. Theorem 6.1 also holds with 0, **N** replaced by c, $M_c = \{x \varepsilon \mathbf{Z} \mid x \geq c\}$ for any $c \varepsilon \mathbf{Z}$.

In order to insure that various recursive or inductive definitions and proofs in the sequel (for example, Theorems 8.8 and III.3.7 below) are valid, we need a technical result:

Theorem 6.2. (*Recursion Theorem*) *If* S *is a set,* $a \varepsilon S$ *and for each* $n \varepsilon \mathbf{N}$, $f_n : S \rightarrow S$ *is a function, then there is a unique function* $\varphi : \mathbf{N} \rightarrow S$ *such that* $\varphi(0) = a$ *and* $\varphi(n + 1) = f_n(\varphi(n))$ *for every* $n \varepsilon \mathbf{N}$.

SKETCH OF PROOF. We shall construct a relation R on $\mathbf{N} \times S$ that is the graph of a function $\varphi : \mathbf{N} \rightarrow S$ with the desired properties. Let $\mathcal{G}$ be the set of all subsets Y of $\mathbf{N} \times S$ such that

$$(0,a) \varepsilon Y; \quad \text{and} \quad (n,x) \varepsilon Y \Rightarrow (n + 1, f_n(x)) \varepsilon Y \quad \text{for all} \quad n \varepsilon \mathbf{N}.$$

Then $\mathcal{G} \neq \varnothing$ since $\mathbf{N} \times S \varepsilon \mathcal{G}$. Let $R = \bigcap_{Y \varepsilon \mathcal{G}} Y$; then $R \varepsilon \mathcal{G}$. Let M be the subset of **N** consisting of all those $n \varepsilon \mathbf{N}$ for which there exists a *unique* $x_n \varepsilon S$ such that $(n,x_n) \varepsilon R$. We shall prove $M = \mathbf{N}$ by induction. If $0 \not\varepsilon M$, then there exists $(0,b) \varepsilon R$ with $b \neq a$ and the set $R - \{(0,b)\} \subset \mathbf{N} \times S$ is in $\mathcal{G}$. Consequently $R = \bigcap_{Y \varepsilon \mathcal{G}} Y \subset R - \{(0,b)\}$, which is a contradiction. Therefore, $0 \varepsilon M$. Suppose inductively that $n \varepsilon M$ (that is, $(n,x_n) \varepsilon R$ for a unique $x_n \varepsilon S$). Then $(n + 1, f_n(x_n)) \varepsilon R$ also. If $(n + 1, c) \varepsilon R$ with $c \neq f_n(x_n)$ then $R - \{(n + 1, c)\} \varepsilon \mathcal{G}$ (verify!), which leads to a contradiction as above. Therefore, $x_{n+1} = f_n(x_n)$ is the unique element of S such that $(n + 1, x_{n+1}) \varepsilon R$. Therefore by induction (Theorem 6.1) $\mathbf{N} = M$, whence the assignment $n \mapsto x_n$ defines a function $\varphi : \mathbf{N} \to S$ with graph R. Since $(0,a) \varepsilon R$ we must have $\varphi(0) = a$. For each $n \varepsilon \mathbf{N}$, $(n,x_n) = (n,\varphi(n)) \varepsilon R$ and hence $(n+1,f_n(\varphi(n))) \varepsilon R$ since $R \varepsilon \mathcal{G}$. But $(n+1,x_{n+1}) \varepsilon R$ and the uniqueness of x_{n+1} imply that $\varphi(n+1) = x_{n+1} = f_n(\varphi(n))$. ■

If A is a nonempty set, then a **sequence** in A is a function $\mathbf{N} \to A$. A sequence is usually denoted $\{a_0,a_1,\ldots\}$ or $\{a_i\}_{i\varepsilon N}$ or $\{a_i\}$, where $a_i \varepsilon A$ is the image of $i \varepsilon \mathbf{N}$. Similarly a function $\mathbf{N}^* \to A$ is also called a sequence and denoted $\{a_1,a_2,\ldots\}$ or $\{a_i\}_{i\varepsilon \mathbf{N}^*}$ or $\{a_i\}$; this will cause no confusion in context.

Theorem 6.3. (*Division Algorithm*) *If* a,b, ε **Z** *and* $a \neq 0$, *then there exists unique integers* q *and* r *such that* $b = aq + r$, *and* $0 \leq r < |a|$.

SKETCH OF PROOF. Show that the set $S = \{b - ax \mid x \varepsilon \mathbf{Z}, b - ax \geq 0\}$ is a nonempty subset of **N** and therefore contains a least element $r = b - aq$ (for some $q \varepsilon \mathbf{Z}$). Thus $b = aq + r$. Use the fact that r is the least element in S to show $0 \leq r < |a|$ and the uniqueness of q,r. ■

We say that an integer $a \neq 0$ **divides** an integer b (written $a \mid b$) if there is an integer k such that $ak = b$. If a does not divide b we write $a \nmid b$.

Definition 6.4. *The positive integer* c *is said to be the* **greatest common divisor** *of the integers* $a_1,a_2,\ldots,a_n$ *if:*

(1) $c \mid a_i$ *for* $1 \leq i \leq n$;
(2) $d \varepsilon \mathbf{Z}$ *and* $d \mid a_i$ *for* $1 \leq i \leq n \Rightarrow d \mid c$.

c *is denoted* $(a_1,a_2,\ldots,a_n)$.

Theorem 6.5. *If* $a_1,a_2,\ldots,a_n$ *are integers, not all* 0, *then* $(a_1,a_2,\ldots,a_n)$ *exists. Furthermore there are integers* $k_1,k_2,\ldots,k_n$ *such that*

$$(a_1,a_2,\ldots,a_n) = k_1a_1 + k_2a_2 + \cdots + k_na_n.$$

SKETCH OF PROOF. Use the Division Algorithm to show that the least positive element of the nonempty set $S = \{x_1a_1 + x_2a_2 + \cdots + x_na_n \mid x_i \varepsilon \mathbf{Z}, \sum_i x_ia_i > 0\}$ is the greatest common divisor of $a_1,\ldots,a_n$. For details see Shockley [51,p.10]. ■

The integers $a_1,a_2,\ldots,a_n$ are said to be **relatively prime** if $(a_1,a_2,\ldots,a_n) = 1$. A positive integer $p > 1$ is said to be **prime** if its only divisors are ± 1 and $\pm p$. Thus if p is prime and $a \varepsilon \mathbf{Z}$, either $(a,p) = p$ (if $p \mid a$) or $(a,p) = 1$ (if $p \nmid a$).

Theorem 6.6. *If* a *and* b *are relatively prime integers and* $a \mid bc$, *then* $a \mid c$. *If* p *is prime and* $p \mid a_1a_2\cdots a_n$, *then* $p \mid a_i$ *for some* i.

SKETCH OF PROOF. By Theorem 6.5 $1 = ra + sb$, whence $c = rac + sbc$. Therefore $a \mid c$. The second statement now follows by induction on n. ■

Theorem 6.7. (*Fundamental Theorem of Arithmetic*) *Any positive integer* $n > 1$ *may be written uniquely in the form* $n = p_1^{t_1}p_2^{t_2}\cdots p_k^{t_k}$, *where* $p_1 < p_2 < \cdots < p_k$ *are primes and* $t_i > 0$ *for all* i.

The proof, which proceeds by induction, may be found in Shockley [51, p.17].

Let $m > 0$ be a fixed integer. If $a,b \in \mathbf{Z}$ and $m \mid (a - b)$ then a is said to be **congruent** to b **modulo m**. This is denoted by $a \equiv b \pmod{m}$.

Theorem 6.8. *Let* $m > 0$ *be an integer and* $a,b,c,d \in \mathbf{Z}$.

(i) *Congruence modulo* m *is an equivalence relation on the set of integers* $\mathbf{Z}$, *which has precisely* m *equivalence classes.*

(ii) *If* $a \equiv b \pmod{m}$ *and* $c \equiv d \pmod{m}$, *then* $a + c \equiv b + d \pmod{m}$ *and* $ac \equiv bd \pmod{m}$.

(iii) *If* $ab \equiv ac \pmod{m}$ *and* a *and* m *are relatively prime, then* $b \equiv c \pmod{m}$.

PROOF. (i) The fact that congruence modulo m is an equivalence relation is an easy consequence of the appropriate definitions. Denote the equivalence class of an integer a by $\bar{a}$ and recall property (20), which can be stated in this context as:

$$\bar{a} = \bar{b} \quad \Leftrightarrow \quad a \equiv b \pmod{m}. \tag{20'}$$

Given any $a \in \mathbf{Z}$, there are integers q and r, with $0 \leq r < m$, such that $a = mq + r$. Hence $a - r = mq$ and $a \equiv r \pmod{m}$; therefore, $\bar{a} = \bar{r}$ by (20′). Since a was arbitrary and $0 \leq r < m$, it follows that every equivalence class must be one of $\bar{0},\bar{1},\bar{2},\bar{3},\ldots,\overline{(m-1)}$. However, these m equivalence classes are distinct: for if $0 \leq i < j < m$, then $0 < (j - i) < m$ and $m \nmid (j - i)$. Thus $i \not\equiv j \pmod{m}$ and hence $\bar{i} \neq \bar{j}$ by (20′). Therefore, there are exactly m equivalence classes.

(ii) We are given $m \mid a - b$ and $m \mid c - d$. Hence m divides $(a - b) + (c - d) = (a + c) - (b + d)$ and therefore $a + c \equiv b + d \pmod{m}$. Likewise, m divides $(a - b)c + (c - d)b$ and therefore divides $ac - bc + cb - db = ac - bd$; thus $ac \equiv bd \pmod{m}$.

(iii) Since $ab \equiv ac \pmod{m}$, $m \mid a(b - c)$. Since $(m,a) = 1$, $m \mid b - c$ by Theorem 6.6, and thus $b \equiv c \pmod{m}$. ■

7. THE AXIOM OF CHOICE, ORDER, AND ZORN'S LEMMA

Note. In this section we deal only with *sets*. No proper classes are involved.

If $I \neq \emptyset$ and $\{A_i \mid i \in I\}$ is a family of sets such that $A_i \neq \emptyset$ for all $i \in I$, then we would like to know that $\prod_{i \in I} A_i \neq \emptyset$. It has been proved that this apparently innocuous conclusion cannot be deduced from the usual axioms of set theory (although it is not inconsistent with them — see P. J. Cohen [59]). Consequently we shall assume

The Axiom of Choice. *The product of a family of nonempty sets indexed by a nonempty set is nonempty.*

See Exercise 4 for another version of the Axiom of Choice. There are two propositions equivalent to the Axiom of Choice that are essential in the proofs of a number of important theorems. In order to state these equivalent propositions we must introduce some additional concepts.

A **partially ordered set** is a nonempty set A together with a relation R on $A \times A$ (called a **partial ordering** of A) which is reflexive and transitive (see (15), (17) in section 4) and

$$\textbf{antisymmetric}: (a,b) \in R \quad \text{and} \quad (b,a) \in R \quad \Rightarrow \quad a = b. \tag{31}$$

If R is a partial ordering of A, then we usually write $a \leq b$ in place of $(a,b) \in R$. In this notation the conditions (15), (17), and (31) become (for all $a,b,c \in A$):

$$a \leq a;$$
$$a \leq b \quad \text{and} \quad b \leq c \quad \Rightarrow \quad a \leq c;$$
$$a \leq b \quad \text{and} \quad b \leq a \quad \Rightarrow \quad a = b.$$

We write $a < b$ if $a \leq b$ and $a \neq b$.

Elements $a,b \in A$ are said to be **comparable,** provided $a \leq b$ or $b \leq a$. However, two given elements of a partially ordered set need not be comparable. A partial ordering of a set A such that any two elements are comparable is called a **linear** (or **total** or **simple**) **ordering.**

EXAMPLE. Let A be the power set (set of all subsets) of $\{1,2,3,4,5\}$. Define $C \leq D$ if and only if $C \subset D$. Then A is partially ordered, but not linearly ordered (for example, $\{1,2\}$ and $\{3,4\}$ are not comparable).

Let $(A,\leq)$ be a partially ordered set. An element $a \in A$ is **maximal** in A if for every *$c \in A$ which is comparable to a, $c \leq a$;* in other words, for all $c \in A$, $a \leq c \Rightarrow a = c$. Note that if a is maximal, it need not be the case that $c \leq a$ for *all* $c \in A$ (there may exist $c \in A$ that are not comparable to a). Furthermore, a given set may have many maximal elements (Exercise 5) or none at all (for example, **Z** with its usual ordering). An **upper bound** of a nonempty subset B of A is an element $d \in A$ such that $b \leq d$ for *every* $b \in B$. A nonempty subset B of A that is linearly ordered by $\leq$ is called a **chain** in A.

Zorn's Lemma. *If* A *is a nonempty partially ordered set such that every chain in* A *has an upper bound in* A, *then* A *contains a maximal element.*

Assuming that all the other usual axioms of set theory hold, it can be proved that Zorn's Lemma is true if and only if the Axiom of Choice holds; that is, the two are equivalent — see E. Hewitt and K. Stromberg [57; p. 14]. Zorn's Lemma is a powerful tool and will be used frequently in the sequel.

Let B be a nonempty subset of a partially ordered set $(A,\leq)$. An element $c \in B$ is a **least** (or **minimum**) **element** of B provided $c \leq b$ for *every* $b \in B$. If every nonempty subset of A has a least element, then A is said to be **well ordered.** Every well-ordered set is linearly ordered (but not vice versa) since for all $a,b \in A$ the subset $\{a,b\}$ must have a least element; that is, $a \leq b$ or $b \leq a$. Here is another statement that can be proved to be equivalent to the Axiom of Choice (see E. Hewitt and K. Stromberg [57; p.14]).

The Well Ordering Principle. *If* A *is a nonempty set, then there exists a linear ordering* $\leq$ *of* A *such that* (A,$\leq$) *is well ordered.*

EXAMPLES. We have already assumed (Section 6) that the set **N** of natural numbers is well ordered. The set **Z** of all integers with the usual ordering by magnitude is linearly ordered but *not* well ordered (for example, the subset of negative integers has no least element). However, each of the following is a well ordering of **Z** (where by definition $a < b \Leftrightarrow a$ is to the left of b):

(i) $0,1,-1,2,-2,3,-3,\ldots,n,-n,\ldots$;
(ii) $0,1,3,5,7,\ldots,2,4,6,8,\ldots,-1,-2,-3,-4,\ldots$;
(iii) $0,3,4,5,6,\ldots,-1,-2,-3,-4,\ldots,1,2$.

These orderings are quite different from one another. Every nonzero element a in ordering (i) has an immediate predecessor (that is an element c such that a is the least element in the subset $\{x \mid c < x\}$). But the elements -1 and 2 in ordering (ii) and -1 and 1 in ordering (iii) have no immediate predecessors. There are no maximal elements in orderings (i) and (ii), but 2 is a maximal element in ordering (iii). The element 0 is the least element in all three orderings.

The chief advantage of the well-ordering principle is that it enables us to extend the principle of mathematical induction for positive integers (Theorem 6.1) to any well ordered set.

Theorem 7.1. (*Principle of Transfinite Induction*) *If* B *is a subset of a well-ordered set* (A,$\leq$) *such that for every* a ε A,

$$\{c \,\varepsilon\, A \mid c < a\} \subset B \quad \Rightarrow \quad a \,\varepsilon\, B,$$

then B = A.

PROOF. If $A - B \neq \emptyset$, then there is a least element $a \,\varepsilon\, A - B$. By the definitions of *least element* and $A - B$ we must have $\{c \,\varepsilon\, A \mid c < a\} \subset B$. By hypothesis then, $a \,\varepsilon\, B$ so that $a \,\varepsilon\, B \cap (A - B) = \emptyset$, which is a contradiction. Therefore, $A - B = \emptyset$ and $A = B$. ■

EXERCISES

1. Let $(A,\leq)$ be a partially ordered set and B a nonempty subset. A **lower bound** of B is an element $d \,\varepsilon\, A$ such that $d \leq b$ for *every* $b \,\varepsilon\, B$. A **greatest lower bound (g.l.b.)** of B is a lower bound d_0 of B such that $d \leq d_0$ for every other lower bound d of B. A **least upper bound (l.u.b.)** of B is an upper bound t_0 of B such that $t_0 \leq t$ for every other upper bound t of B. $(A,\leq)$ is a **lattice** if for all $a,b \,\varepsilon\, A$ the set $\{a,b\}$ has both a greatest lower bound and a least upper bound.

(a) If $S \neq \emptyset$, then the power set $P(S)$ ordered by set-theoretic inclusion is a lattice, which has a unique maximal element.

(b) Give an example of a partially ordered set which is *not* a lattice.

(c) Give an example of a lattice with no maximal element and an example of a partially ordered set with two maximal elements.

2. A lattice $(A,\leq)$ (see Exercise 1) is said to be **complete** if every nonempty subset of A has both a least upper bound and a greatest lower bound. A map of partially ordered sets $f: A \to B$ is said to preserve order if $a \leq a'$ in A implies $f(a) \leq f(a')$ in B. Prove that an order-preserving map f of a complete lattice A into itself has at least one fixed element (that is, an $a \in A$ such that $f(a) = a$).

3. Exhibit a well ordering of the set $\mathbf{Q}$ of rational numbers.

4. Let S be a set. A **choice function** for S is a function f from the set of all nonempty subsets of S to S such that $f(A) \in A$ for all $A \neq \emptyset, A \subset S$. Show that the Axiom of Choice is equivalent to the statement that every set S has a choice function.

5. Let S be the set of all points (x,y) in the plane with $y \leq 0$. Define an ordering by $(x_1,y_1) \leq (x_2,y_2) \Leftrightarrow x_1 = x_2$ and $y_1 \leq y_2$. Show that this is a partial ordering of S, and that S has infinitely many maximal elements.

6. Prove that if all the sets in the family $\{A_i \mid i \in I \neq \emptyset\}$ are nonempty, then each of the projections $\pi_k : \prod_{i \in I} A_i \to A_k$ is surjective.

7. Let $(A,\leq)$ be a linearly ordered set. The **immediate successor** of $a \in A$ (if it exists) is the least element in the set $\{x \in A \mid a < x\}$. Prove that if A is well ordered by $\leq$, then at most one element of A has no immediate successor. Give an example of a linearly ordered set in which precisely two elements have no immediate successor.

8. CARDINAL NUMBERS

The definition and elementary properties of cardinal numbers will be needed frequently in the sequel. The remainder of this section (beginning with Theorem 8.5), however, will be used only occasionally (Theorems II.1.2 and IV.2.6; Lemma V.3.5; Theorems V.3.6 and VI.1.9). It may be omitted for the present, if desired.

Two sets, A and B, are said to be **equipollent,** if there exists a bijective map $A \to B$; in this case we write $A \sim B$.

Theorem 8.1. *Equipollence is an equivalence relation on the class $\mathcal{S}$ of all sets.*

PROOF. Exercise; note that $\emptyset \sim \emptyset$ since $\emptyset \subset \emptyset \times \emptyset$ is a relation that is (vacuously) a bijective function.[3] ■

Let $I_0 = \emptyset$ and for each $n \in \mathbf{N}^*$ let $I_n = \{1,2,3,\ldots,n\}$. It is not difficult to prove that I_m and I_n are equipollent if and only if $m = n$ (Exercise 1). To say that a set A

[3]See page 6.

has precisely n elements means that A and I_n are equipollent, that is, that A and I_n are in the same equivalence class under the relation of equipollence. Such a set A (with $A \sim I_n$ for some unique $n \geq 0$) is said to be **finite**; a set that is not finite is **infinite.** Thus, for a finite set A, the equivalence class of A under equipollence provides an answer to the question: how many elements are contained in A? These considerations motivate

Definition 8.2. *The* **cardinal number** (*or* **cardinality**) *of a set* A, *denoted* $|A|$, *is the equivalence class of* A *under the equivalence relation of equipollence.* $|A|$ *is an infinite or finite cardinal according as* A *is an infinite or finite set.*

Cardinal numbers will also be denoted by lower case Greek letters: α,β,γ, etc. For the reasons indicated in the preceding paragraph we shall *identify* the integer $n \geq 0$ with the cardinal number $|I_n|$ and write $|I_n| = n$, so that the cardinal number of a finite set is precisely the number of elements in the set.

Cardinal numbers are frequently defined somewhat differently than we have done so that a cardinal number is in fact a set (instead of a proper class as in Definition 8.2). We have chosen this definition both to save time and because it better reflects the intuitive notion of "the number of elements in a set." No matter what definition of cardinality is used, cardinal numbers possess the following properties (the first two of which are, in our case, immediate consequences of Theorem 8.1 and Definition 8.2).

(i) *Every set has a unique cardinal number;*

(ii) *two sets have the same cardinal number if and only if they are equipollent* ($|A| = |B| \Leftrightarrow A \sim B$);

(iii) *the cardinal number of a finite set is the number of elements in the set.*

Therefore statements about cardinal numbers are simply statements about equipollence of sets.

EXAMPLE. The cardinal number of the set **N** of natural numbers is customarily denoted $\aleph_0$ (read "aleph-naught"). A set A of cardinality $\aleph_0$ (that is, one which is equipollent to **N**) is said to be **denumerable.** The set **N***, the set **Z** of integers, and the set **Q** of rational numbers are denumerable (Exercise 3), but the set **R** of real numbers is *not* denumerable (Exercise 9).

Definition 8.3. *Let* α *and* β *be cardinal numbers. The* **sum** $\alpha + \beta$ *is defined to be the cardinal number* $|A \cup B|$, *where* A *and* B *are disjoint sets such that* $|A| = \alpha$ *and* $|B| = \beta$. *The* **product** $\alpha\beta$ *is defined to be the cardinal number* $|A \times B|$.

It is not actually necessary for A and B to be disjoint in the definition of the product $\alpha\beta$ (Exercise 4). By the definition of a cardinal number α there always exists a set A such that $|A| = \alpha$. It is easy to verify that *disjoint* sets, as required for the definition of $\alpha + \beta$, always exist and that the sum $\alpha + \beta$ and product $\alpha\beta$ are independent of the choice of the sets A,B (Exercise 4). Addition and multiplication of cardinals are associative and commutative, and the distributive laws hold (Exercise 5). Furthermore, addition and multiplication of finite cardinals agree with addition and multiplication of the nonnegative integers with which they are identified; for if A has m elements, B has n elements and $A \cap B = \emptyset$, then $A \cup B$ has $m + n$ elements and $A \times B$ has mn elements (for more precision, see Exercise 6).

Definition 8.4. *Let α,β be cardinal numbers and* A,B *sets such that* $|A| = \alpha$, $|B| = \beta$. *α is* **less than or equal to** *β, denoted $\alpha \leq \beta$ or $\beta \geq \alpha$, if* A *is equipollent with a subset of* B (*that is, there is an injective map* $A \to B$). *α is* **strictly less than** *β, denoted $\alpha < \beta$ or $\beta > \alpha$, if $\alpha \leq \beta$ and $\alpha \neq \beta$.*

It is easy to verify that the definition of $\leq$ does not depend on the choice of A and B (Exercise 7). It is shown in Theorem 8.7 that the class of all cardinal numbers is linearly ordered by $\leq$. For finite cardinals $\leq$ agrees with the usual ordering of the nonnegative integers (Exercise 1). The fact that there is no largest cardinal number is an immediate consequence of

Theorem 8.5. *If* A *is a set and* P(A) *its power set, then* $|A| < |P(A)|$.

SKETCH OF PROOF. The assignment $a \mapsto \{a\}$ defines an injective map $A \to P(A)$ so that $|A| \leq |P(A)|$. If there were a bijective map $f: A \to P(A)$, then for some $a_0 \in A$, $f(a_0) = B$, where $B = \{a \in A \mid a \notin f(a)\} \subset A$. But this yields a contradiction: $a_0 \in B$ and $a_0 \notin B$. Therefore $|A| \neq |P(A)|$ and hence $|A| < |P(A)|$. ■

REMARK. By Theorem 8.5, $\aleph_0 = |\mathbf{N}| < |P(\mathbf{N})|$. It can be shown that $|P(\mathbf{N})| = |\mathbf{R}|$, where $\mathbf{R}$ is the set of real numbers. The conjecture that there is no cardinal number β such that $\aleph_0 < \beta < |P(\mathbf{N})| = |\mathbf{R}|$ is called the **Continuum Hypothesis.** It has been proved to be independent of the Axiom of Choice and of the other basic axioms of set theory; see P. J. Cohen [59].

The remainder of this section is devoted to developing certain facts that will be needed at several points in the sequel (see the first paragraph of this section).

Theorem 8.6. (*Schroeder-Bernstein*) *If* A *and* B *are sets such that* $|A| \leq |B|$ *and* $|B| \leq |A|$, *then* $|A| = |B|$.

SKETCH OF PROOF. By hypothesis there are injective maps $f: A \to B$ and $g: B \to A$. We shall use f and g to construct a bijection $h: A \to B$. This will imply that $A \sim B$ and hence $|A| = |B|$. If $a \in A$, then since g is injective the set $g^{-1}(a)$ is either empty (in which case we say that a is *parentless*) or consists of exactly one element $b \in B$ (in which case we write $g^{-1}(a) = b$ and say that b is the *parent* of a). Similarly for $b \in B$, we have either $f^{-1}(b) = \emptyset$ (b is *parentless*) or $f^{-1}(b) = a' \in A$ (a' is the *parent* of b). If we continue to trace back the "ancestry" of an element $a \in A$ in this manner, one of three things must happen. Either we reach a parentless element in A (an *ancestor* of $a \in A$), or we reach a parentless element in B (an *ancestor* of a), or the ancestry of $a \varepsilon A$ can be traced back forever (*infinite ancestry*). Now define three subsets of A [resp. B] as follows:

$$\begin{aligned} A_1 &= \{a \varepsilon A \mid a \text{ has a parentless ancestor in } A\}; \\ A_2 &= \{a \varepsilon A \mid a \text{ has a parentless ancestor in } B\}; \\ A_3 &= \{a \varepsilon A \mid a \text{ has infinite ancestry}\}; \\ B_1 &= \{b \varepsilon B \mid b \text{ has a parentless ancestor in } A\}; \\ B_2 &= \{b \varepsilon B \mid b \text{ has a parentless ancestor in } B\}; \\ B_3 &= \{b \varepsilon B \mid b \text{ has infinite ancestry}\}. \end{aligned}$$

Verify that the A_i [resp. B_i] are pairwise disjoint, that their union is A [resp. B]; that $f \mid A_i$ is a bijection $A_i \to B_i$ for $i = 1, 3$; and that $g \mid B_2$ is a bijection $B_2 \to A_2$. Consequently the map $h : A \to B$ given as follows is a well-defined bijection:

$$h(a) = \begin{cases} f(a) & \text{if } a \varepsilon A_1 \cup A_3; \\ g^{-1}(a) & \text{if } a \varepsilon A_2. \end{cases} \qquad \blacksquare$$

Theorem 8.7. *The class of all cardinal numbers is linearly ordered by* $\leq$. *If* α *and* β *are cardinal numbers, then exactly one of the following is true:*

$$\alpha < \beta; \quad \alpha = \beta; \quad \beta < \alpha \quad (\textit{Trichotomy Law}).$$

SKETCH OF PROOF. It is easy to verify that $\leq$ is a *partial* ordering. Let α,β be cardinals and A,B be sets such that $|A| = \alpha$, $|B| = \beta$. We shall show that $\leq$ is a linear ordering (that is, either $\alpha \leq \beta$ or $\beta \leq \alpha$) by applying Zorn's Lemma to the set $\mathfrak{F}$ of all pairs (f,X), where $X \subset A$ and $f : X \to B$ is an injective map. Verify that $\mathfrak{F} \neq \varnothing$ and that the ordering of $\mathfrak{F}$ given by $(f_1,X_1) \leq (f_2,X_2)$ if and only if $X_1 \subset X_2$ and $f_2 \mid X_1 = f_1$ is a partial ordering of $\mathfrak{F}$. If $\{(f_i,X_i) \mid i \varepsilon I\}$ is a chain in $\mathfrak{F}$, let $X = \bigcup_{i \varepsilon I} X_i$ and define $f : X \to B$ by $f(x) = f_i(x)$ for $x \varepsilon X_i$. Show that f is a well-defined injective map, and that (f,X) is an upper bound in $\mathfrak{F}$ of the given chain. Therefore by Zorn's Lemma there is a maximal element (g,X) of $\mathfrak{F}$. We claim that either $X = A$ or $\operatorname{Im} g = B$. For if both of these statements were false we could find $a \varepsilon A - X$ and $b \varepsilon B - \operatorname{Im} g$ and define an injective map $h : X \cup \{a\} \to B$ by $h(x) = g(x)$ for $x \varepsilon X$ and $h(a) = b$. Then $(h, X \cup \{a\}) \varepsilon \mathfrak{F}$ and $(g,X) < (h, X \cup \{a\})$, which contradicts the maximality of (g,X). Therefore either $X = A$ so that $|A| \leq |B|$ or $\operatorname{Im} g = B$ in which case the injective map $B \xrightarrow{g^{-1}} X \subset A$ shows that $|B| \leq |A|$. Use these facts, the Schroeder-Bernstein Theorem 8.6 and Definition 8.4 to prove the Trichotomy Law. $\blacksquare$

REMARKS. A family of functions partially ordered as in the proof of Theorem 8.7 is said to be **ordered by extension.** The proof of the theorem is a typical example of the use of Zorn's Lemma. The details of similar arguments in the sequel will frequently be abbreviated.

Theorem 8.8. *Every infinite set has a denumerable subset. In particular,* $\aleph_0 \leq \alpha$ *for every infinite cardinal number* α.

SKETCH OF PROOF. If B is a finite subset of the infinite set A, then $A - B$ is nonempty. For each finite subset B of A, choose an element $x_B \in A - B$ (Axiom of Choice). Let F be the set of all finite subsets of A and define a map $f: F \to F$ by $f(B) = B \cup \{x_B\}$. Choose $a \in A$. By the Recursion Theorem 6.2 (with $f_n = f$ for all n) there exists a function $\varphi : \mathbf{N} \to F$ such that

$$\varphi(0) = \{a\} \quad \text{and} \quad \varphi(n+1) = f(\varphi(n)) = \varphi(n) \cup \{x_{\varphi(n)}\} \; (n \geq 0).$$

Let $g : \mathbf{N} \to A$ be the function defined by

$$g(0) = a; \, g(1) = x_{\varphi(0)} = x_{\{a\}}; \ldots ; g(n+1) = x_{\varphi(n)}; \ldots .$$

Use the order properties of $\mathbf{N}$ and the following facts to verify that g is injective:

(i) $g(n) \in \varphi(n)$ for all $n \geq 0$;
(ii) $g(n) \notin \varphi(n-1)$ for all $n \geq 1$;
(iii) $g(n) \notin \varphi(m)$ for all $m < n$.

Therefore Im g is a subset of A such that $|\text{Im } g| = |\mathbf{N}| = \aleph_0$. ■

Lemma 8.9. *If* A *is an infinite set and* F *a finite set then* $|A \cup F| = |A|$. *In particular,* $\alpha + n = \alpha$ *for every infinite cardinal number* α *and every natural number (finite cardinal)* n.

SKETCH OF PROOF. It suffices to assume $A \cap F = \emptyset$ (replace F by $F - A$ if necessary). If $F = \{b_1, b_2, \ldots, b_n\}$ and $D = \{x_i \mid i \in \mathbf{N}^*\}$ is a denumerable subset of A (Theorem 8.8), verify that $f: A \to A \cup F$ is a bijection, where f is given by

$$f(x) = \begin{cases} b_i & \text{for } x = x_i, \ 1 \leq i \leq n; \\ x_{i-n} & \text{for } x = x_i, \ i > n; \\ x & \text{for } x \in A - D. \end{cases}$$ ■

Theorem 8.10. *If* α *and* β *are cardinal numbers such that* $\beta \leq \alpha$ *and* α *is infinite, then* $\alpha + \beta = \alpha$.

SKETCH OF PROOF. It suffices to prove $\alpha + \alpha = \alpha$ (simply verify that $\alpha \leq \alpha + \beta \leq \alpha + \alpha = \alpha$ and apply the Schroeder-Bernstein Theorem to conclude $\alpha + \beta = \alpha$). Let A be a set with $|A| = \alpha$ and let $\mathfrak{F}$ be the set of all pairs (f,X), where $X \subset A$ and $f: X \times \{0,1\} \to X$ is a bijection. Partially order $\mathfrak{F}$ by extension (as in the proof of Theorem 8.7) and verify that the hypotheses of Zorn's Lemma are satisfied. The only difficulty is showing that $\mathfrak{F} \neq \emptyset$. To do this note that the map $\mathbf{N} \times \{0,1\} \to \mathbf{N}$ given by $(n,0) \mapsto 2n$ and $(n,1) \mapsto 2n+1$ is a bijection. Use this fact to construct a bijection $f: D \times \{0,1\} \to D$, where D is a denumerable subset of A (that is, $|D| = |\mathbf{N}|$; see Theorem 8.8). Therefore by Zorn's Lemma there is a maximal element $(g,C) \in \mathfrak{F}$.

Clearly $C_0 = \{(c,0) \mid c \in C\}$ and $C_1 = \{(c,1) \mid c \in C\}$ are disjoint sets such that $|C_0| = |C| = |C_1|$ and $C \times \{0,1\} = C_0 \cup C_1$. The map $g : C \times \{0,1\} \to C$ is a bijection. Therefore by Definition 8.3,

$$|C| = |C \times \{0,1\}| = |C_0 \cup C_1| = |C_0| + |C_1| = |C| + |C|.$$

To complete the proof we shall show that $|C| = \alpha$. If $A - C$ were infinite, it would contain a denumerable subset B by Theorem 8.8, and as above, there would be a bijection $\zeta : B \times \{0,1\} \to B$. By combining ζ with g, we could then construct a bijection $h : (C \cup B) \times \{0,1\} \to C \cup B$ so that $(g,C) < (h,C \cup B) \in \mathfrak{F}$, which would contradict the maximality of (g,C). Therefore $A - C$ must be finite. Since A is infinite and $A = C \cup (A - C)$, C must also be infinite. Thus by Lemma 8.9, $|C| = |C \cup (A - C)| = |A| = \alpha$. ■

Theorem 8.11. *If α and β are cardinal numbers such that $0 \neq \beta \leq \alpha$ and α is infinite, then $\alpha\beta = \alpha$; in particular, $\alpha\aleph_0 = \alpha$ and if β is finite $\aleph_0\beta = \aleph_0$.*

SKETCH OF PROOF. Since $\alpha \leq \alpha\beta \leq \alpha\alpha$ it suffices (as in the proof of Theorem 8.10) to prove $\alpha\alpha = \alpha$. Let A be an infinite set with $|A| = \alpha$ and let $\mathfrak{F}$ be the set of all bijections $f: X \times X \to X$, where X is an infinite subset of A. To show that $\mathfrak{F} \neq \varnothing$, use the facts that A has a denumerable subset D (so that $|D| = |\mathbf{N}| = |\mathbf{N}^*|$) and that the map $\mathbf{N}^* \times \mathbf{N}^* \to \mathbf{N}^*$ given by $(m,n) \mapsto 2^{m-1}(2n - 1)$ is a bijection. Partially order $\mathfrak{F}$ by extension and use Zorn's Lemma to obtain a maximal element $g : B \times B \to B$. By the definition of g, $|B||B| = |B \times B| = |B|$. To complete the proof we shall show that $|B| = |A| = \alpha$.

Suppose $|A - B| > |B|$. Then by Definition 8.4 there is a subset C of $A - B$ such that $|C| = |B|$. Verify that $|C| = |B| = |B \times B| = |B \times C| = |C \times B| = |C \times C|$ and that these sets are mutually disjoint. Consequently by Definition 8.3 and Theorem 8.10 $|(B \cup C) \times (B \cup C)| = |(B \times B) \cup (B \times C) \cup (C \times B) \cup (C \times C)| = |B \times B| + |B \times C| + |C \times B| + |C \times C| = (|B| + |B|) + (|C| + |C|) = |B| + |C| = |B \cup C|$ and there is a bijection $(B \cup C) \times (B \cup C) \to (B \cup C)$, which contradicts the maximality of g in $\mathfrak{F}$. Therefore, by Theorems 8.7 and 8.10 $|A - B| \leqslant |B|$ and $|B| = |A - B| + |B| = |(A - B) \cup B| = |A| = \alpha$. ■

Theorem 8.12. *Let* A *be a set and for each integer* $n \geq 1$ *let* $A^n = A \times A \times \cdots \times A$ (n *factors*).

(i) *If* A *is finite, then* $|A^n| = |A|^n$, *and if* A *is infinite, then* $|A^n| = |A|$.
(ii) $|\bigcup_{n \in \mathbf{N}^*} A^n| = \aleph_0|A|$.

SKETCH OF PROOF. (i) is trivial if $|A|$ is finite and may be proved by induction on n if $|A|$ is infinite (the case $n = 2$ is given by Theorem 8.11). (ii) The sets A^n ($n \geq 1$) are mutually disjoint. If A is infinite, then by (i) there is for each n a bijection $f_n : A^n \to A$. The map $\bigcup_{n \in \mathbf{N}^*} A^n \to \mathbf{N}^* \times A$, which sends $u \in A^n$ onto $(n, f_n(u))$, is a bijection. Therefore $|\bigcup_{n \in \mathbf{N}^*} A^n| = |\mathbf{N}^* \times A| = |\mathbf{N}^*||A| = \aleph_0|A|$. (ii) is obviously true if $A = \varnothing$. Suppose, therefore, that A is nonempty and finite. Then each A^n is nonempty and it is easy to show that $\aleph_0 = |\mathbf{N}^*| \leq |\bigcup_{n \in \mathbf{N}^*} A^n|$. Furthermore each A^n is finite and there is for each n an injective map $g_n : A^n \to \mathbf{N}^*$. The map $\bigcup_{n \in \mathbf{N}^*} A^n \to \mathbf{N}^* \times \mathbf{N}^*$, which sends $u \in A^n$ onto $(n, g_n(u))$ is injective so that $|\bigcup_{n \in \mathbf{N}^*} A^n| \leq |\mathbf{N}^* \times \mathbf{N}^*| = |\mathbf{N}^*| = \aleph_0$ by Theorem 8.11. Therefore by the Schroeder-Bernstein Theorem $|\bigcup_{n \in \mathbf{N}^*} A^n| = \aleph_0$. But $\aleph_0 = \aleph_0|A|$ since A is finite (Theorem 8.11). ■

Corollary 8.13. *If* A *is an infinite set and* F(A) *the set of all finite subsets of* A, *then* $|F(A)| = |A|$.

PROOF. The map $A \to F(A)$ given by $a \mapsto \{a\}$ is injective so that $|A| \leq |F(A)|$. For each n-element subset S of A, choose $(a_1, \ldots, a_n) \in A^n$ such that $S = \{a_1, \ldots, a_n\}$. This defines an injective map $F(A) \to \bigcup_{n \in \mathbf{N}^*} A^n$ so that $|F(A)| \leq |\bigcup_{n \in \mathbf{N}^*} A^n| = \aleph_0|A| = |A|$ by Theorems 8.11 and 8.12. Therefore, $|A| = |F(A)|$ by the Schroeder-Bernstein Theorem 8.6. ■

EXERCISES

1. Let $I_0 = \emptyset$ and for each $n \in \mathbf{N}^*$ let $I_n = \{1,2,3, \ldots, n\}$.
   (a) I_n is not equipollent to any of its proper subsets [*Hint:* induction].
   (b) I_m and I_n are equipollent if and only if $m = n$.
   (c) I_m is equipollent to a subset of I_n but I_n is not equipollent to any subset of I_m if and only if $m < n$.

2. (a) Every infinite set is equipollent to one of its proper subsets.
   (b) A set is finite if and only if it is not equipollent to one of its proper subsets [see Exercise 1].

3. (a) $\mathbf{Z}$ is a denumerable set.
   (b) The set $\mathbf{Q}$ of rational numbers is denumerable. [*Hint:* show that $|\mathbf{Z}| \leq |\mathbf{Q}| \leq |\mathbf{Z} \times \mathbf{Z}| = |\mathbf{Z}|$.]

4. If A,A',B,B' are sets such that $|A| = |A'|$ and $|B| = |B'|$, then $|A \times B| = |A' \times B'|$. If in addition $A \cap B = \emptyset = A' \cap B'$, then $|A \cup B| = |A' \cup B'|$. Therefore multiplication and addition of cardinals is well defined.

5. For all cardinal numbers α,β,γ:
   (a) $\alpha + \beta = \beta + \alpha$ and $\alpha\beta = \beta\alpha$ (commutative laws).
   (b) $(\alpha + \beta) + \gamma = \alpha + (\beta + \gamma)$ and $(\alpha\beta)\gamma = \alpha(\beta\gamma)$ (associative laws).
   (c) $\alpha(\beta + \gamma) = \alpha\beta + \alpha\gamma$ and $(\alpha + \beta)\gamma = \alpha\gamma + \beta\gamma$ (distributive laws).
   (d) $\alpha + 0 = \alpha$ and $\alpha 1 = \alpha$.
   (e) If $\alpha \neq 0$, then there is no β such that $\alpha + \beta = 0$ and if $\alpha \neq 1$, then there is no β such that $\alpha\beta = 1$. Therefore subtraction and division of cardinal numbers cannot be defined.

6. Let I_n be as in Exercise 1. If $A \sim I_m$ and $B \sim I_n$ and $A \cap B = \emptyset$, then $(A \cup B) \sim I_{m+n}$ and $A \times B \sim I_{mn}$. Thus if we identify $|A|$ with m and $|B|$ with n, then $|A| + |B| = m + n$ and $|A||B| = mn$.

7. If $A \sim A'$, $B \sim B'$ and $f: A \to B$ is injective, then there is an injective map $A' \to B'$. Therefore the relation $\leq$ on cardinal numbers is well defined.

8. An infinite subset of a denumerable set is denumerable.

9. The infinite set of real numbers $\mathbf{R}$ is not denumerable (that is, $\aleph_0 < |\mathbf{R}|$). [*Hint:* it suffices to show that the open interval (0,1) is not denumerable by Exercise 8. You may assume each real number can be written as an infinite decimal. If (0,1) is denumerable there is a bijection $f: \mathbf{N}^* \to (0,1)$. Construct an infinite decimal (real number) $.a_1a_2\cdots$ in (0,1) such that a_n is not the nth digit in the decimal expansion of $f(n)$. This number cannot be in Im f.]

10. If α,β are cardinals, define α^β to be the cardinal number of the set of all functions $B \to A$, where A,B are sets such that $|A| = \alpha$, $|B| = \beta$.
    (a) α^β is independent of the choice of A,B.
    (b) $\alpha^{\beta+\gamma} = (\alpha^\beta)(\alpha^\gamma)$; $(\alpha\beta)^\gamma = (\alpha^\gamma)(\beta^\gamma)$; $\alpha^{\beta\gamma} = (\alpha^\beta)^\gamma$.
    (c) If $\alpha \leq \beta$, then $\alpha^\gamma \leq \beta^\gamma$.
    (d) If α,β are finite with $\alpha > 1$, $\beta > 1$ and γ is infinite, then $\alpha^\gamma = \beta^\gamma$.
    (e) For every finite cardinal n, $\alpha^n = \alpha\alpha\cdots\alpha$ (n factors). Hence $\alpha^n = \alpha$ if α is infinite.
    (f) If $P(A)$ is the power set of a set A, then $|P(A)| = 2^{|A|}$.

11. If I is an infinite set, and for each $i \in I$ A_i is a finite set, then $|\bigcup_{i\in I} A_i| \leq |I|$.

12. Let α be a fixed cardinal number and suppose that for every $i \in I$, A_i is a set with $|A_i| = \alpha$. Then $|\bigcup_{i\in I} A_i| \leq |I|\alpha$.

CHAPTER I

GROUPS

The concept of a group is of fundamental importance in the study of algebra. Groups which are, from the point of view of algebraic structure, essentially the same are said to be isomorphic. Ideally the goal in studying groups is to classify all groups up to isomorphism, which in practice means finding necessary and sufficient conditions for two groups to be isomorphic. At present there is little hope of classifying arbitrary groups. But it is possible to obtain complete structure theorems for various restricted classes of groups, such as cyclic groups (Section 3), finitely generated abelian groups (Section II.2), groups satisfying chain conditions (Section II.3) and finite groups of small order (Section II.6). In order to prove even these limited structure theorems, it is necessary to develop a large amount of miscellaneous information about the structure of (more or less) arbitrary groups (Sections 1, 2, 4, 5, and 8 of Chapter I and Sections 4 and 5 of Chapter II). In addition we shall study some classes of groups whose structure is known in large part and which have useful applications in other areas of mathematics, such as symmetric groups (Section 6), free [abelian] groups (Sections 9 and II.1), nilpotent and solvable groups (Sections II.7 and II.8).

There is a basic truth that applies not only to groups but also to many other algebraic objects (for example, rings, modules, vector spaces, fields): in order to study effectively an object with a given algebraic structure, it is necessary to study as well the functions that preserve the given algebraic structure (such functions are called homomorphisms). Indeed a number of concepts that are common to the theory of groups, rings, modules, etc. may be described completely in terms of objects and homomorphisms. In order to provide a convenient language and a useful conceptual framework in which to view these common concepts, the notion of a category is introduced in Section 7 and used frequently thereafter. Of course it is quite possible to study groups, rings, etc. without ever mentioning categories. However, the small amount of effort needed to comprehend this notion now will pay large dividends later in terms of increased understanding of the fundamental relationships among the various algebraic structures to be encountered.

With occasional exceptions such as Section 7, each section in this chapter depends on the sections preceding it.

1. SEMIGROUPS, MONOIDS AND GROUPS

If G is a nonempty set, a **binary operation** on G is a function $G \times G \to G$. There are several commonly used notations for the image of (a,b) under a binary operation: ab (multiplicative notation), $a + b$ (additive notation), $a \cdot b$, $a * b$, etc. For convenience we shall generally use the multiplicative notation throughout this chapter and refer to ab as the **product** of a and b. A set may have several binary operations defined on it (for example, ordinary addition and multiplication on $\mathbf{Z}$ given by $(a,b) \mapsto a + b$ and $(a,b) \mapsto ab$ respectively).

Definition 1.1. *A* **semigroup** *is a nonempty set* G *together with a binary operation on* G *which is*

(i) *associative:* $a(bc) = (ab)c$ *for all* $a, b, c \in G$;

a **monoid** *is a semigroup* G *which contains a*

(ii) *(two-sided) identity element* $e \in G$ *such that* $ae = ea = a$ *for all* $a \in G$.

A **group** *is a monoid* G *such that*

(iii) *for every* $a \in G$ *there exists a (two-sided) inverse element* $a^{-1} \in G$ *such that* $a^{-1}a = aa^{-1} = e$.

A semigroup G *is said to be* **abelian** *or* **commutative** *if its binary operation is*

(iv) *commutative:* $ab = ba$ *for all* $a,b \in G$.

Our principal interest is in groups. However, semigroups and monoids are convenient for stating certain theorems in the greatest generality. Examples are given below. The **order** of a group G is the cardinal number $|G|$. G is said to be finite [resp. infinite] if $|G|$ is finite [resp. infinite].

Theorem 1.2. *If* G *is a monoid, then the identity element* e *is unique. If* G *is a group, then*

(i) $c \in G$ *and* $cc = c \Rightarrow c = e$;
(ii) *for all* $a, b, c \in G$ $ab = ac \Rightarrow b = c$ *and* $ba = ca \Rightarrow b = c$ *(left and right cancellation)*;
(iii) *for each* $a \in G$, *the inverse element* a^{-1} *is unique;*
(iv) *for each* $a \in G$, $(a^{-1})^{-1} = a$;
(v) *for* $a, b \in G$, $(ab)^{-1} = b^{-1}a^{-1}$;
(vi) *for* $a, b \in G$ *the equations* $ax = b$ *and* $ya = b$ *have unique solutions in* $G : x = a^{-1}b$ *and* $y = ba^{-1}$.

SKETCH OF PROOF. If e' is also a two-sided identity, then $e = ee' = e'$. (i) $cc = c \Rightarrow c^{-1}(cc) = c^{-1}c \Rightarrow (c^{-1}c)c = c^{-1}c \Rightarrow ec = e \Rightarrow c = e$; (ii), (iii) and (vi) are proved similarly. (v) $(ab)(b^{-1}a^{-1}) = a(bb^{-1})a^{-1} = (ae)a^{-1} = aa^{-1} = e \Rightarrow (ab)^{-1} = b^{-1}a^{-1}$ by (iii); (iv) is proved similarly. ■

If G is a monoid and the binary operation is written multiplicatively, then the identity element of G will always be denoted e. If the binary operation is written additively, then $a + b$ $(a, b \in G)$ is called the **sum** of a and b, and the identity element is denoted 0; if G is a group the inverse of $a \in G$ is denoted by $-a$. We write $a - b$ for $a + (-b)$. Abelian groups are frequently written additively.

The axioms used in Definition 1.1 to define a group can actually be weakened considerably.

Proposition 1.3. *Let* G *be a semigroup. Then* G *is a group if and only if the following conditions hold:*

(i) *there exists an element* $e \in G$ *such that* $ea = a$ *for all* $a \in G$ *(left identity element)*;
(ii) *for each* $a \in G$, *there exists an element* $a^{-1} \in G$ *such that* $a^{-1}a = e$ *(left inverse)*.

REMARK. An analogous result holds for "right inverses" and a "right identity."

SKETCH OF PROOF OF 1.3. ($\Rightarrow$) Trivial. ($\Leftarrow$) Note that Theorem 1.2(i) is true under these hypotheses. $G \neq \emptyset$ since $e \in G$. If $a \in G$, then by (ii) $(aa^{-1})(aa^{-1}) = a(a^{-1}a)a^{-1} = a(ea^{-1}) = aa^{-1}$ and hence $aa^{-1} = e$ by Theorem 1.2(i). Thus a^{-1} is a two-sided inverse of a. Since $ae = a(a^{-1}a) = (aa^{-1})a = ea = a$ for every $a \in G$, e is a two-sided identity. Therefore G is a group by Definition 1.1. ■

Proposition 1.4. *Let* G *be a semigroup. Then* G *is a group if and only if for all* $a, b \in G$ *the equations* $ax = b$ *and* $ya = b$ *have solutions in* G.

PROOF. Exercise; use Proposition 1.3. ■

EXAMPLES. The integers **Z**, the rational numbers **Q**, and the real numbers **R** are each infinite abelian groups under ordinary addition. Each is a monoid under ordinary multiplication, but not a group (0 has no inverse). However, the nonzero elements of **Q** and **R** respectively form infinite abelian groups under multiplication. The even integers under multiplication form a semigroup that is not a monoid.

EXAMPLE. Consider the square with vertices consecutively numbered 1,2,3,4, center at the origin of the x-y plane, and sides parallel to the axes.

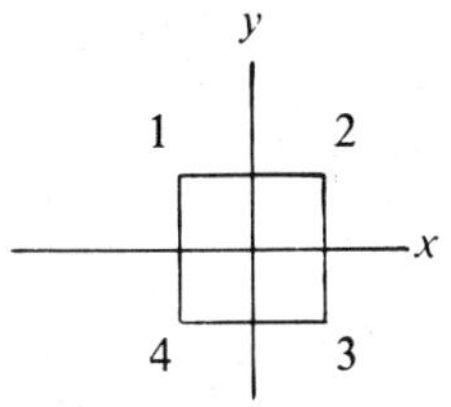

Let $D_4{}^*$ be the following set of "transformations" of the square. $D_4{}^* = \{R, R^2, R^3, I, T_x, T_y, T_{1,3}, T_{2,4}\}$, where R is a counterclockwise rotation about the center of 90°, R^2 a counterclockwise rotation of 180°, R^3 a counterclockwise rotation of 270°

and I a rotation of 360° (= 0°); T_x is a reflection about the x axis, $T_{1,3}$ a reflection about the diagonal through vertices 1 and 3; similarly for T_y and $T_{2,4}$. Note that each $U \in D_4^*$ is a bijection of the square onto itself. Define the binary operation in D_4^* to be composition of functions: for $U,V \in D_4^*$, $U \circ V$ is the transformation V followed by the transformation U. D_4^* is a nonabelian group of order 8 called the **group of symmetries of the square.** Notice that each symmetry (element of D_4^*) is completely determined by its action on the vertices.

EXAMPLE. Let S be a nonempty set and $A(S)$ the set of all bijections $S \to S$. Under the operation of composition of functions, $f \circ g$, $A(S)$ is a group, since composition is associative, composition of bijections is a bijection, 1_S is a bijection, and every bijection has an inverse (see (13) of Introduction, Section 3). The elements of $A(S)$ are called **permutations** and $A(S)$ is called the group of permutations on the set S. If $S = \{1,2,3,\ldots,n\}$, then $A(S)$ is called the **symmetric group on n letters** and denoted S_n. Verify that $|S_n| = n!$ (Exercise 5). The groups S_n play an important role in the theory of finite groups.

Since an element σ of S_n is a function on the finite set $S = \{1,2,\ldots,n\}$, it can be described by listing the elements of S on a line and the image of each element under σ directly below it: $\begin{pmatrix} 1 & 2 & 3 & \cdots & n \\ i_1 & i_2 & i_3 & & i_n \end{pmatrix}$. The product $\sigma\tau$ of two elements of S_n is the composition function τ *followed by* σ; that is, the function on S given by $k \mapsto \sigma(\tau(k))$.[1] For instance, let $\sigma = \begin{pmatrix} 1 & 2 & 3 & 4 \\ 3 & 1 & 2 & 4 \end{pmatrix}$ and $\tau = \begin{pmatrix} 1 & 2 & 3 & 4 \\ 4 & 1 & 2 & 3 \end{pmatrix}$ be elements of S_4. Then under $\sigma\tau$, $1 \mapsto \sigma(\tau(1)) = \sigma(4) = 4$, etc.; thus $\sigma\tau = \begin{pmatrix} 1 & 2 & 3 & 4 \\ 3 & 1 & 2 & 4 \end{pmatrix}\begin{pmatrix} 1 & 2 & 3 & 4 \\ 4 & 1 & 2 & 3 \end{pmatrix}$ $= \begin{pmatrix} 1 & 2 & 3 & 4 \\ 4 & 3 & 1 & 2 \end{pmatrix}$; similarly, $\tau\sigma = \begin{pmatrix} 1 & 2 & 3 & 4 \\ 4 & 1 & 2 & 3 \end{pmatrix}\begin{pmatrix} 1 & 2 & 3 & 4 \\ 3 & 1 & 2 & 4 \end{pmatrix} = \begin{pmatrix} 1 & 2 & 3 & 4 \\ 2 & 4 & 1 & 3 \end{pmatrix}$ This example also shows that S_n need not be abelian.

Another source of examples is the following method of constructing new groups from old. Let G and H be groups with identities e_G, e_H respectively, and define the **direct product** of G and H to be the group whose underlying set is $G \times H$ and whose binary operation is given by:

$$(a,b)(a',b') = (aa',bb'), \quad \text{where} \quad a,a' \in G;\ b,b' \in H.$$

Observe that there are three different operations in G, H and $G \times H$ involved in this statement. It is easy to verify that $G \times H$ is, in fact, a group that is abelian if both G and H are; (e_G,e_H) is the identity and (a^{-1},b^{-1}) the inverse of (a,b). Clearly $|G \times H| = |G||H|$ (Introduction, Definition 8.3). If G and H are written additively, then we write $G \oplus H$ in place of $G \times H$.

Theorem 1.5. *Let* R ($\sim$) *be an equivalence relation on a monoid* G *such that* $a_1 \sim a_2$ *and* $b_1 \sim b_2$ *imply* $a_1b_1 \sim a_2b_2$ *for all* $a_i,b_i \in G$. *Then the set* G/R *of all equivalence classes of* G *under* R *is a monoid under the binary operation defined by* $(\bar{a})(\bar{b}) = \overline{ab}$, *where* $\bar{x}$ *denotes the equivalence class of* $x \in G$. *If* G *is an* [*abelian*] *group, then so is* G/R.

[1]In many books, however, the product $\sigma\tau$ is defined to be "σ followed by τ."

An equivalence relation on a monoid G that satisfies the hypothesis of the theorem is called a **congruence relation** on G.

PROOF OF 1.5. If $\bar{a}_1 = \bar{a}_2$ and $\bar{b}_1 = \bar{b}_2$ $(a_i, b_i \in G)$, then $a_1 \sim a_2$ and $b_1 \sim b_2$ by (20) of Introduction, Section 4. Then by hypothesis $a_1b_1 \sim a_2b_2$ so that $\overline{a_1b_1} = \overline{a_2b_2}$ by (20) again. Therefore the binary operation in G/R is well defined (that is, independent of the choice of equivalence class representatives). It is associative since $\bar{a}(\bar{b}\,\bar{c}) = \bar{a}(\overline{bc}) = \overline{a(bc)} = \overline{(ab)c} = (\overline{ab})\bar{c} = (\bar{a}\,\bar{b})\bar{c}$. $\bar{e}$ is the identity element since $(\bar{a})(\bar{e}) = \overline{ae} = \bar{a} = \overline{ea} = (\bar{e})(\bar{a})$. Therefore G/R is a monoid. If G is a group, then $\bar{a} \in G/R$ clearly has inverse $\overline{a^{-1}}$ so that G/R is also a group. Similarly, G abelian implies G/R abelian. ■

EXAMPLE. Let m be a fixed integer. Congruence modulo m is a congruence relation on the additive group $\mathbf{Z}$ by Introduction, Theorem 6.8. Let Z_m denote the set of equivalence classes of $\mathbf{Z}$ under congruence modulo m. By Theorem 1.5 (with additive notation) Z_m is an abelian group, with addition given by $\bar{a} + \bar{b} = \overline{a+b}$ $(a,b \in \mathbf{Z})$. The proof of Introduction, Theorem 6.8 shows that $Z_m = \{\bar{0},\bar{1},\ldots,\overline{m-1}\}$ so that Z_m is a finite group of order m under addition. Z_m is called the (additive) group of **integers modulo m.** Similarly since $\mathbf{Z}$ is a commutative monoid under multiplication, and congruence modulo m is also a congruence relation with respect to multiplication (Introduction, Theorem 6.8), Z_m is a commutative monoid, with multiplication given by $(\bar{a})(\bar{b}) = \overline{ab}$ $(a,b \in \mathbf{Z})$. Verify that for all $\bar{a}, \bar{b}, \bar{c} \in Z_m$:

$$\bar{a}(\bar{b} + \bar{c}) = \bar{a}\bar{b} + \bar{a}\bar{c} \quad \text{and} \quad (\bar{a} + \bar{b})\bar{c} = \bar{a}\bar{c} + \bar{b}\bar{c} \text{ (distributivity).}$$

Furthermore if p is prime, then the nonzero elements of Z_p form a multiplicative group of order $p - 1$ (Exercise 7). It is customary to denote the elements of Z_m as $0,1,\ldots,m-1$ rather than $\bar{0},\bar{1},\ldots,\overline{m-1}$. In context this ambiguous notation will cause no difficulty and will be used whenever convenient.

EXAMPLE. The following relation on the additive group $\mathbf{Q}$ of rational numbers is a congruence relation (Exercise 8):

$$a \sim b \Leftrightarrow a - b \in \mathbf{Z}.$$

By Theorem 1.5 the set of equivalence classes (denoted $\mathbf{Q}/\mathbf{Z}$) is an (infinite) abelian group, with addition given by $\bar{a} + \bar{b} = \overline{a+b}$. $\mathbf{Q}/\mathbf{Z}$ is called the **group of rationals modulo one.**

Given $a_1, \ldots, a_n \in G$ $(n \geq 3)$ it is intuitively plausible that there are many ways of inserting parentheses in the expression $a_1a_2\cdots a_n$ so as to yield a "meaningful" product in G of these n elements in this order. Furthermore it is plausible that any two such products can be proved equal by repeated use of the associative law. A necessary prerequisite for further study of groups and rings is a precise statement and proof of these conjectures and related ones.

Given any sequence of elements of a semigroup G, $\{a_1,a_2\ldots\}$ define inductively a **meaningful product** of $a_1, \ldots, a_n$ (in this order) as follows. If $n = 1$, the only meaningful product is a_1. If $n > 1$, then a meaningful product is defined to be any product of the form $(a_1\cdots a_m)(a_{m+1}\cdots a_n)$ where $m < n$ and $(a_1\cdots a_m)$ and $(a_{m+1}\cdots a_n)$ are meaningful products of m and $n - m$ elements respectively.[2] Note that for each

[2]To show that this definition is in fact well defined requires a stronger version of the Recursion Theorem 6.2 of the Introduction; see C. W. Burrill [56; p. 57].

$n \geq 3$ there may be many meaningful products of $a_1, \ldots, a_n$. For each $n \in \mathbf{N}^*$ we single out a particular meaningful product by defining inductively the **standard n product** $\prod_{i=1}^{n} a_i$ of $a_1, \ldots, a_n$ as follows:

$$\prod_{i=1}^{1} a_i = a_1; \quad \text{and for } n > 1, \prod_{i=1}^{n} a_i = \left(\prod_{i=1}^{n-1} a_i\right)a_n.$$

The fact that this definition defines for each $n \in \mathbf{N}^*$ a unique element of G (which is clearly a meaningful product) is a consequence of the Recursion Theorem 6.2 of the Introduction (Exercise 16).

Theorem 1.6. (*Generalized Associative Law*) *If* G *is a semigroup and* $a_1, \ldots, a_n \in G$, *then any two meaningful products of* $a_1, \ldots, a_n$ *in this order are equal.*

PROOF. We use induction to show that for every n any meaningful product $a_1 \cdots a_n$ is equal to the standard n product $\prod_{i=1}^{n} a_i$. This is certainly true for $n = 1, 2$. If $n > 2$, then by definition $(a_1 \cdots a_n) = (a_1 \cdots a_m)(a_{m+1} \cdots a_n)$ for some $m < n$. Therefore, by induction and associativity:

$$(a_1 \cdots a_n) = (a_1 \cdots a_m)(a_{m+1} \cdots a_n) = \left(\prod_{i=1}^{m} a_i\right)\left(\prod_{i=1}^{n-m} a_{m+i}\right)$$

$$= \left(\prod_{i=1}^{m} a_i\right)\left(\left(\prod_{i=1}^{n-m-1} a_{m+i}\right)a_n\right) = \left(\left(\prod_{i=1}^{m} a_i\right)\left(\prod_{i=1}^{n-m-1} a_{m+i}\right)\right)a_n$$

$$= \left(\prod_{i=1}^{n-1} a_i\right)a_n = \prod_{i=1}^{n} a_i. \quad \blacksquare$$

In view of Theorem 1.6 we can and do write any meaningful product of $a_1, \ldots, a_n \in G$ (G a semigroup) as $a_1a_2 \cdots a_n$ without parentheses or ambiguity.

Corollary 1.7. (*Generalized Commutative Law*) *If* G *is a commutative semigroup and* $a_1, \ldots, a_n \in G$, *then for any permutation* $i_1, \ldots, i_n$ *of* $1, 2, \ldots n$, $a_1a_2 \cdots a_n = a_{i_1}a_{i_2} \cdots a_{i_n}$.

PROOF. Exercise. $\blacksquare$

Definition 1.8. *Let* G *be a semigroup,* $a \in G$ *and* $n \in \mathbf{N}^*$. *The element* $a^n \in G$ *is defined to be the standard* n *product* $\prod_{i=1}^{n} a_i$ *with* $a_i = a$ *for* $1 \leq i \leq n$. *If* G *is a monoid,* a^0 *is defined to be the identity element* e. *If* G *is a group, then for each* $n \in \mathbf{N}^*$, a^{-n} *is defined to be* $(a^{-1})^n \in G$.

The remarks preceding Theorem 1.6 and Exercise 16 show that exponentiation is well defined. By definition, then, $a^1 = a$, $a^2 = aa$, $a^3 = (aa)a = aaa, \ldots, a^n = a^{n-1}a$

$= aa\cdots a$ (n factors). Note that we may have $a^m = a^n$ with $m \neq n$ (for example, in **C**, $-1 = i^2 = i^6$).

ADDITIVE NOTATION. If the binary operation in G is written additively, then we write na in place of a^n. Thus $0a = 0$, $1a = a$, $na = (n-1)a + a$, etc.

Theorem 1.9. *If* G *is a group [resp. semigroup, monoid] and* $a \in G$, *then for all* $m, n \in \mathbf{Z}$ [*resp.* $\mathbf{N}^*$, $\mathbf{N}$]:

(i) $a^m a^n = a^{m+n}$ (*additive notation:* $ma + na = (m+n)a$);
(ii) $(a^m)^n = a^{mn}$ (*additive notation:* $n(ma) = mna$).

SKETCH OF PROOF. Verify that $(a^n)^{-1} = (a^{-1})^n$ for all $n \in \mathbf{N}$ and that $a^{-n} = (a^{-1})^n$ for all $n \in \mathbf{Z}$. (i) is true for $m > 0$ and $n > 0$ since the product of a standard n product and a standard m product is a meaningful product equal to the standard $(m+n)$ product by Theorem 1.6. For $m < 0$, and $n < 0$ replace a, m, n by $a^{-1}, -m, -n$ and use the preceding argument. The case $m = 0$ or $n = 0$ is trivial and the cases $m \geq 0, n < 0$ and $m < 0, n \geq 0$ are handled by induction on m and n respectively. (ii) is trivial if $m = 0$. The case when $m > 0$ and $n \in \mathbf{Z}$ is proved by induction on m. Use this result to prove the case $m < 0$ and $n \in \mathbf{Z}$. ■

EXERCISES

1. Give examples other than those in the text of semigroups and monoids that are not groups.

2. Let G be a group (written additively), S a nonempty set, and $M(S,G)$ the set of all functions $f: S \to G$. Define addition in $M(S,G)$ as follows: $(f+g): S \to G$ is given by $s \mapsto f(s) + g(s) \in G$. Prove that $M(S,G)$ is a group, which is abelian if G is.

3. Is it true that a semigroup which has a *left* identity element and in which every element has a *right* inverse (see Proposition 1.3) is a group?

4. Write out a multiplication table for the group D_4^*.

5. Prove that the symmetric group on n letters, S_n, has order $n!$.

6. Write out an addition table for $Z_2 \oplus Z_2$. $Z_2 \oplus Z_2$ is called the **Klein four group**.

7. If p is prime, then the nonzero elements of Z_p form a group of order $p - 1$ under multiplication. [*Hint:* $\bar{a} \neq \bar{0} \Rightarrow (a,p) = 1$; use Introduction, Theorem 6.5.] Show that this statement is false if p is not prime.

8. (a) The relation given by $a \sim b \Leftrightarrow a - b \in \mathbf{Z}$ is a congruence relation on the additive group **Q** [see Theorem 1.5].
(b) The set $\mathbf{Q}/\mathbf{Z}$ of equivalence classes is an infinite abelian group.

9. Let p be a fixed prime. Let R_p be the set of all those rational numbers whose denominator is relatively prime to p. Let R^p be the set of rationals whose denominator is a power of p (p^i, $i \geq 0$). Prove that both R_p and R^p are abelian groups under ordinary addition of rationals.

10. Let p be a prime and let $Z(p^\infty)$ be the following subset of the group $\mathbf{Q}/\mathbf{Z}$ (see pg. 27):

$$Z(p^\infty) = \{\overline{a/b} \in \mathbf{Q}/\mathbf{Z} \mid a,b \in \mathbf{Z} \text{ and } b = p^i \text{ for some } i \geq 0\}.$$

Show that $Z(p^\infty)$ is an infinite group under the addition operation of $\mathbf{Q}/\mathbf{Z}$.

11. The following conditions on a group G are equivalent: (i) G is abelian; (ii) $(ab)^2 = a^2b^2$ for all $a,b \in G$; (iii) $(ab)^{-1} = a^{-1}b^{-1}$ for all $a,b \in G$; (iv) $(ab)^n = a^nb^n$ for all $n \in \mathbf{Z}$ and all $a,b \in G$; (v) $(ab)^n = a^nb^n$ for three consecutive integers n and all $a,b \in G$. Show that (v) $\Rightarrow$ (i) is false if "three" is replaced by "two."

12. If G is a group, $a,b \in G$ and $bab^{-1} = a^r$ for some $r \in \mathbf{N}$, then $b^jab^{-j} = a^{r^j}$ for all $j \in \mathbf{N}$.

13. If $a^2 = e$ for all elements a of a group G, then G is abelian.

14. If G is a finite group of even order, then G contains an element $a \neq e$ such that $a^2 = e$.

15. Let G be a nonempty finite set with an associative binary operation such that for all $a,b,c \in G$ $ab = ac \Rightarrow b = c$ and $ba = ca \Rightarrow b = c$. Then G is a group. Show that this conclusion may be false if G is infinite.

16. Let $a_1,a_2,\ldots$ be a sequence of elements in a semigroup G. Then there exists a unique function $\psi : \mathbf{N}^* \to G$ such that $\psi(1) = a_1$, $\psi(2) = a_1a_2$, $\psi(3) = (a_1a_2)a_3$ and for $n \geq 1$, $\psi(n+1) = (\psi(n))a_{n+1}$. Note that $\psi(n)$ is precisely the standard n product $\prod_{i=1}^{n} a_i$. [*Hint:* Applying the Recursion Theorem 6.2 of the Introduction with $a = a_1$, $S = G$ and $f_n : G \to G$ given by $x \mapsto xa_{n+2}$ yields a function $\varphi : \mathbf{N} \to G$. Let $\psi = \varphi\theta$, where $\theta : \mathbf{N}^* \to \mathbf{N}$ is given by $k \mapsto k - 1$.]

2. HOMOMORPHISMS AND SUBGROUPS

Essential to the study of any class of algebraic objects are the functions that preserve the given algebraic structure in the following sense.

Definition 2.1. *Let* G *and* H *be semigroups. A function* $f : G \to H$ *is a* **homomorphism** *provided*

$$f(ab) = f(a)f(b) \quad \textit{for all} \quad a,b \in G.$$

If f *is injective as a map of sets,* f *is said to be a* **monomorphism.** *If* f *is surjective,* f *is called an* **epimorphism.** *If* f *is bijective,* f *is called an* **isomorphism.** *In this case* G *and* H *are said to be* **isomorphic** (*written* $G \cong H$). *A homomorphism* $f : G \to G$ *is called an* **endomorphism** *of* G *and an isomorphism* $f : G \to G$ *is called an* **automorphism** *of* G.

If $f : G \to H$ and $g : H \to K$ are homomorphisms of semigroups, it is easy to see that $gf : G \to K$ is also a homomorphism. Likewise the composition of monomorphisms is a monomorphism; similarly for epimorphisms, isomorphisms and automorphisms. If G and H are groups with identities e_G and e_H respectively and

$f: G \to H$ is a homomorphism, then $f(e_G) = e_H$; however, this is not true for monoids (Exercise 1). Furthermore $f(a^{-1}) = f(a)^{-1}$ for all $a \in G$ (Exercise 1).

EXAMPLE. The map $f: \mathbf{Z} \to Z_m$ given by $x \mapsto \bar{x}$ (that is, each integer is mapped onto its equivalence class in Z_m) is an epimorphism of additive groups. f is called the canonical epimorphism of $\mathbf{Z}$ onto Z_m. Similarly, the map $g: \mathbf{Q} \to \mathbf{Q}/\mathbf{Z}$ given by $r \mapsto \bar{r}$ is also an epimorphism of additive groups.

EXAMPLE. If A is an abelian group, then the map given by $a \mapsto a^{-1}$ is an automorphism of A. The map given by $a \mapsto a^2$ is an endomorphism of A.

EXAMPLE. Let $1 < m$, $k \in \mathbf{N}^*$. The map $g: Z_m \to Z_{mk}$ given by $\bar{x} \mapsto \overline{kx}$ is a monomorphism.

EXAMPLE. Given groups G and H, there are four homomorphisms: $G \underset{\pi_1}{\overset{\iota_1}{\rightleftarrows}} G \times H \underset{\pi_2}{\overset{\iota_2}{\leftrightarrows}} H$, given by $\iota_1(g) = (g,e)$; $\iota_2(h) = (e,h)$; $\pi_1(g,h) = g$; $\pi_2(g,h) = h$. ι_i is a monomorphism and π_j is an epimorphism $(i,j = 1,2)$.

Definition 2.2. *Let* $f: G \to H$ *be a homomorphism of groups. The* **kernel** *of* f *(denoted* $Ker\, f$) *is* $\{a \in G \mid f(a) = e \in H\}$. *If* A *is a subset of* G, *then* $f(A) = \{b \in H \mid b = f(a)$ *for some* $a \in A\}$ *is the* **image of A**. $f(G)$ *is called the* **image of f** *and denoted* $Im\, f$. *If* B *is a subset of* H, $f^{-1}(B) = \{a \in G \mid f(a) \in B\}$ *is the* **inverse image** *of* B.

Theorem 2.3. *Let* $f: G \to H$ *be a homomorphism of groups. Then*

(i) f *is a monomorphism if and only if* $Ker\, f = \{e\}$;

(ii) f *is an isomorphism if and only if there is a homomorphism* $f^{-1}: H \to G$ *such that* $ff^{-1} = 1_H$ *and* $f^{-1}f = 1_G$.

PROOF. (i) If f is a monomorphism and $a \in \text{Ker } f$, then $f(a) = e_H = f(e)$, whence $a = e$ and $\text{Ker } f = \{e\}$. If $\text{Ker } f = \{e\}$ and $f(a) = f(b)$, then $e_H = f(a)f(b)^{-1} = f(a)f(b^{-1}) = f(ab^{-1})$ so that $ab^{-1} \in \text{Ker } f$. Therefore, $ab^{-1} = e$ (that is, $a = b$) and f is a monomorphism.

(ii) If f is an isomorphism, then by (13) of Introduction, Section 3 there is a map of sets $f^{-1}: H \to G$ such that $f^{-1}f = 1_G$ and $ff^{-1} = 1_H$. f^{-1} is easily seen to be a homomorphism. The converse is an immediate consequence of (13) of Introduction, Section 3 and Definition 2.1. ■

Let G be a semigroup and H a nonempty subset of G. If for every $a,b \in H$ we have $ab \in H$, we say that H is **closed** under the product in G. This amounts to saying that the binary operation on G, when restricted to H, is in fact a binary operation on H.

Definition 2.4. *Let* G *be a group and* H *a nonempty subset that is closed under the product in* G. *If* H *is itself a group under the product in* G, *then* H *is said to be a* **subgroup** *of* G. *This is denoted by* $H < G$.

Two examples of subgroups of a group G are G itself and the **trivial subgroup** $\langle e \rangle$ consisting only of the identity element. A subgroup H such that $H \neq G$, $H \neq \langle e \rangle$ is called a **proper subgroup.**

EXAMPLE. The set of all multiples of some fixed integer n is a subgroup of $\mathbf{Z}$, which is isomorphic to $\mathbf{Z}$ (Exercise 7).

EXAMPLE. In S_n, the group of all permutations of $\{1,2,\ldots,n\}$, the set of all permutations that leave n fixed forms a subgroup isomorphic to S_{n-1} (Exercise 8).

EXAMPLE. In $Z_6 = \{0,1,2,3,4,5\}$, both $\{0,3\}$ and $\{0,2,4\}$ are subgroups under addition. If p is prime, $(Z_p,+)$ has no proper subgroups.

EXAMPLE. If $f: G \to H$ is a homomorphism of groups, then Ker f is a subgroup of G. If A is a subgroup of G, $f(A)$ is a subgroup of H; in particular Im f is a subgroup of H. If B is a subgroup of H, $f^{-1}(B)$ is a subgroup of G (Exercise 9).

EXAMPLE. If G is a group, then the set Aut G of all automorphisms of G is a group, with composition of functions as binary operation (Exercise 15).

By Theorem 1.2 the identity element of any subgroup H is the identity element of G and the inverse of $a \in H$ is the inverse a^{-1} of a in G.

Theorem 2.5. *Let* H *be a nonempty subset of a group* G. *Then* H *is a subgroup of* G *if and only if* $ab^{-1} \in H$ *for all* $a,b \in H$.

PROOF. ($\Leftarrow$) There exists $a \in H$ and hence $e = aa^{-1} \in H$. Thus for any $b \in H$, $b^{-1} = eb^{-1} \in H$. If $a,b \in H$, then $b^{-1} \in H$ and hence $ab = a(b^{-1})^{-1} \in H$. The product in H is associative since G is a group. Therefore H is a (sub)group. The converse is trivial. ■

Corollary 2.6. *If* G *is a group and* $\{H_i \mid i \in I\}$ *is a nonempty family of subgroups, then* $\bigcap_{i \in I} H_i$ is a subgroup of G.

PROOF. Exercise. ■

Definition 2.7. *Let* G *be a group and* X *a subset of* G. *Let* $\{H_i \mid i \in I\}$ *be the family of all subgroups of* G *which contain* X. *Then* $\bigcap_{i \in I} H_i$ *is called the* **subgroup** *of* G **generated by the set** X *and denoted* $\langle X \rangle$.

The elements of X are the **generators** of the subgroup $\langle X \rangle$, which may also be generated by other subsets (that is, we may have $\langle X \rangle = \langle Y \rangle$ with $X \neq Y$). If $X = \{a_1,\ldots,a_n\}$, we write $\langle a_1,\ldots,a_n \rangle$ in place of $\langle X \rangle$. If $G = \langle a_1,\ldots,a_n \rangle$, $(a_i \in G)$, G is said to be **finitely generated.** If $a \in G$, the subgroup $\langle a \rangle$ is called the **cyclic (sub)-group** generated by a.

Theorem 2.8. *If* G *is a group and* X *is a nonempty subset of* G, *then the subgroup* $\langle X\rangle$ *generated by* X *consists of all finite products* $a_1^{n_1}a_2^{n_2}\cdots a_t^{n_t}$ $(a_i \in X; n_i \in \mathbf{Z})$. *In particular for every* $a \in G$, $\langle a\rangle = \{a^n \mid n \in \mathbf{Z}\}$.

SKETCH OF PROOF. Show that the set H of all such products is a subgroup of G that contains X and is contained in every subgroup containing X. Therefore $H < \langle X\rangle < H$. ■

EXAMPLES. The additive group **Z** is an infinite cyclic group with generator 1, since by Definition 1.8 (additive notation), $m1 = m$ for all $m \in \mathbf{Z}$. Of course *the "powers" of the generating element need not all be distinct* as they are in **Z**. The trivial subgroup $\langle e\rangle$ of any group is cyclic; the multiplicative subgroup $\langle i\rangle$ in **C** is cyclic of order 4 and for each m the additive group Z_m is cyclic of order m with generator $1 \in Z_m$. In Section 3 we shall prove that every cyclic subgroup is isomorphic either to **Z** or Z_m for some m. Also, see Exercise 12.

If $\{H_i \mid i \in I\}$ is a family of subgroups of a group G, then $\bigcup_{i\in I} H_i$ is not a subgroup of G in general. The subgroup $\langle \bigcup_{i\in I} H_i\rangle$ generated by the set $\bigcup_{i\in I} H_i$ is called the **subgroup generated by the groups** $\{\mathbf{H_i} \mid \mathbf{i} \in \mathbf{I}\}$. If H and K are subgroups, the subgroup $\langle H \cup K\rangle$ generated by H and K is called the **join** of H and K and is denoted $H \vee K$ (additive notation: $H + K$).

EXERCISES

1. If $f: G \to H$ is a homomorphism of groups, then $f(e_G) = e_H$ and $f(a^{-1}) = f(a)^{-1}$ for all $a \in G$. Show by example that the first conclusion may be false if G, H are monoids that are not groups.

2. A group G is abelian if and only if the map $G \to G$ given by $x \mapsto x^{-1}$ is an automorphism.

3. Let Q_8 be the group (under ordinary matrix multiplication) generated by the complex matrices $A = \begin{pmatrix} 0 & 1 \\ -1 & 0 \end{pmatrix}$ and $B = \begin{pmatrix} 0 & i \\ i & 0 \end{pmatrix}$, where $i^2 = -1$. Show that Q_8 is a nonabelian group of order 8. Q_8 is called the **quaternion group.** [*Hint:* Observe that $BA = A^3B$, whence every element of Q_8 is of the form A^iB^j. Note also that $A^4 = B^4 = I$, where $I = \begin{pmatrix} 1 & 0 \\ 0 & 1 \end{pmatrix}$ is the identity element of Q_8.]

4. Let H be the group (under matrix multiplication) of real matrices generated by $C = \begin{pmatrix} 0 & 1 \\ -1 & 0 \end{pmatrix}$ and $D = \begin{pmatrix} 0 & 1 \\ 1 & 0 \end{pmatrix}$. Show that H is a nonabelian group of order 8 which is *not* isomorphic to the quaternion group of Exercise 3, but is isomorphic to the group D_4^*.

5. Let S be a nonempty subset of a group G and define a relation on G by $a \sim b$ if and only if $ab^{-1} \in S$. Show that $\sim$ is an equivalence relation if and only if S is a subgroup of G.

6. A nonempty finite subset of a group is a subgroup if and only if it is closed under the product in G.

7. If n is a fixed integer, then $\{kn \mid k \in \mathbf{Z}\} \subset \mathbf{Z}$ is an additive subgroup of $\mathbf{Z}$, which is isomorphic to $\mathbf{Z}$.

8. The set $\{\sigma \in S_n \mid \sigma(n) = n\}$ is a subgroup of S_n which is isomorphic to S_{n-1}.

9. Let $f: G \to H$ be a homomorphism of groups, A a subgroup of G, and B a subgroup of H.
(a) Ker f and $f^{-1}(B)$ are subgroups of G.
(b) $f(A)$ is a subgroup of H.

10. List all subgroups of $Z_2 \oplus Z_2$. Is $Z_2 \oplus Z_2$ isomorphic to Z_4?

11. If G is a group, then $C = \{a \in G \mid ax = xa \text{ for all } x \in G\}$ is an abelian subgroup of G. C is called the **center** of G.

12. The group D_4^* is not cyclic, but can be generated by two elements. The same is true of S_n (nontrivial). What is the minimal number of generators of the additive group $\mathbf{Z} \oplus \mathbf{Z}$?

13. If $G = \langle a \rangle$ is a cyclic group and H is any group, then every homomorphism $f: G \to H$ is completely determined by the element $f(a) \in H$.

14. The following cyclic subgroups are all isomorphic: the multiplicative group $\langle i \rangle$ in $\mathbf{C}$, the additive group Z_4 and the subgroup $\left\langle \begin{pmatrix} 1 & 2 & 3 & 4 \\ 2 & 3 & 4 & 1 \end{pmatrix} \right\rangle$ of S_4.

15. Let G be a group and Aut G the set of all automorphisms of G.
(a) Aut G is a group with composition of functions as binary operation. [*Hint:* $1_G \in$ Aut G is an identity; inverses exist by Theorem 2.3.]
(b) Aut $\mathbf{Z} \cong Z_2$ and Aut $Z_6 \cong Z_2$; Aut $Z_8 \cong Z_2 \oplus Z_2$; Aut $Z_p \cong Z_{p-1}$ (p prime).
(c) What is Aut Z_n for arbitrary $n \in \mathbf{N}^*$?

16. For each prime p the additive subgroup $Z(p^\infty)$ of $\mathbf{Q}/\mathbf{Z}$ (Exercise 1.10) is generated by the set $\{\overline{1/p^n} \mid n \in \mathbf{N}^*\}$.

17. Let G be an abelian group and let H,K be subgroups of G. Show that the join $H \vee K$ is the set $\{ab \mid a \in H, b \in K\}$. Extend this result to any finite number of subgroups of G.

18. (a) Let G be a group and $\{H_i \mid i \in I\}$ a family of subgroups. State and prove a condition that will imply that $\bigcup_{i \in I} H_i$ is a subgroup, that is, that $\bigcup_{i \in I} H_i = \langle \bigcup_{i \in I} H_i \rangle$.
(b) Give an example of a group G and a family of subgroups $\{H_i \mid i \in I\}$ such that $\bigcup_{i \in I} H_i \neq \langle \bigcup_{i \in I} H_i \rangle$.

19. (a) The set of all subgroups of a group G, partially ordered by set theoretic inclusion, forms a complete lattice (Introduction, Exercises 7.1 and 7.2) in which the g.l.b. of $\{H_i \mid i \in I\}$ is $\bigcap_{i \in I} H_i$ and the l.u.b. is $\langle \bigcup_{i \in I} H_i \rangle$.
(b) Exhibit the lattice of subgroups of the groups S_3, D_4^*, Z_6, Z_{27}, and Z_{36}.

3. CYCLIC GROUPS

The structure of cyclic groups is relatively simple. We shall completely characterize all cyclic groups (up to isomorphism).

Theorem 3.1. *Every subgroup* H *of the additive group* **Z** *is cyclic. Either* $H = \langle 0 \rangle$ *or* $H = \langle m \rangle$, *where* m *is the least positive integer in* H. *If* $H \neq \langle 0 \rangle$, *then* H *is infinite.*

PROOF. Either $H = \langle 0 \rangle$ or H contains a least positive integer m. Clearly $\langle m \rangle = \{km \mid k \in \mathbf{Z}\} \subset H$. Conversely if $h \in H$, then $h = qm + r$ with $q, r \in \mathbf{Z}$ and $0 \leq r < m$ (division algorithm). Since $r = h - qm \in H$ the minimality of m implies $r = 0$ and $h = qm$. Hence $H \subset \langle m \rangle$. If $H \neq \langle 0 \rangle$, it is clear that $H = \langle m \rangle$ is infinite. ■

Theorem 3.2. *Every infinite cyclic group is isomorphic to the additive group* **Z** *and every finite cyclic group of order* m *is isomorphic to the additive group* $\mathbf{Z}_m$.

PROOF. If $G = \langle a \rangle$ is a cyclic group then the map $\alpha : \mathbf{Z} \to G$ given by $k \mapsto a^k$ is an epimorphism by Theorems 1.9 and 2.8. If Ker $\alpha = 0$, then $\mathbf{Z} \cong G$ by Theorem 2.3 (i). Otherwise Ker α is a nontrivial subgroup of **Z** (Exercise 2.9) and hence Ker $\alpha = \langle m \rangle$, where m is the least positive integer such that $a^m = e$ (Theorem 3.1). For all $r, s \in \mathbf{Z}$,

$$\begin{aligned} a^r = a^s \quad &\Leftrightarrow \quad a^{r-s} = e \quad \Leftrightarrow \quad r - s \in \text{Ker } \alpha = \langle m \rangle \\ &\Leftrightarrow \quad m \mid (r - s) \quad \Leftrightarrow \quad \bar{r} = \bar{s} \text{ in } Z_m, \end{aligned}$$

(where $\bar{k}$ is the congruence class of $k \in \mathbf{Z}$). Therefore the map $\beta : Z_m \to G$ given by $\bar{k} \mapsto a^k$ is a well-defined epimorphism. Since

$$\beta(\bar{k}) = e \quad \Leftrightarrow \quad a^k = e = a^0 \quad \Leftrightarrow \quad \bar{k} = \bar{0} \text{ in } Z_m,$$

β is a monomorphism (Theorem 2.3(i)), and hence an isomorphism $Z_m \cong G$. ■

Definition 3.3. *Let* G *be a group and* $a \in G$. *The* **order** *of* a *is the order of the cyclic subgroup* $\langle a \rangle$ *and is denoted* $|a|$.

Theorem 3.4. *Let* G *be a group and* $a \in G$. *If* a *has infinite order, then*

(i) $a^k = e$ *if and only if* $k = 0$;
(ii) *the elements* a^k $(k \in \mathbf{Z})$ *are all distinct.*

If a *has finite order* $m > 0$, *then*

(iii) m *is the least positive integer such that* $a^m = e$;
(iv) $a^k = e$ *if and only if* $m \mid k$;
(v) $a^r = a^s$ *if and only if* $r \equiv s$ (*mod* m);
(vi) $\langle a \rangle$ *consists of the distinct elements* $a, a^2, \ldots, a^{m-1}, a^m = e$;
(vii) *for each* k *such that* $k \mid m$, $|a^k| = m/k$.

SKETCH OF PROOF. (i)–(vi) are immediate consequences of the proof of Theorem 3.2. (vii) $(a^k)^{m/k} = a^m = e$ and $(a^k)^r \neq e$ for all $0 < r < m/k$ since otherwise $a^{kr} = e$ with $kr < k(m/k) = m$ contradicting (iii). Therefore, $|a^k| = m/k$ by (iii). ■

Theorem 3.5. *Every homomorphic image and every subgroup of a cyclic group* G *is cyclic. In particular, if* H *is a nontrivial subgroup of* $G = \langle a \rangle$ *and* m *is the least positive integer such that* $a^m \in H$, *then* $H = \langle a^m \rangle$.

SKETCH OF PROOF. If $f: G \to K$ is a homomorphism of groups, then $\operatorname{Im} f = \langle f(a) \rangle$. To prove the second statement simply translate the proof of Theorem 3.1 into multiplicative notation (that is, replace every $t \in \mathbf{Z}$ by a^t throughout). This proof works even if G is finite. ■

Recall that two distinct elements in a group may generate the same cyclic subgroup.

Theorem 3.6. *Let* $G = \langle a \rangle$ *be a cyclic group. If* G *is infinite, then* a *and* a^{-1} *are the only generators of* G. *If* G *is finite of order* m, *then* a^k *is a generator of* G *if and only if* $(k,m) = 1$.

SKETCH OF PROOF. It suffices to assume either that $G = \mathbf{Z}$, in which case the conclusion is easy to prove, or that $G = Z_m$. If $(k,m) = 1$, there are $c,d \in \mathbf{Z}$ such that $ck + dm = 1$; use this fact to show that $\bar{k}$ generates Z_m. If $(k,m) = r > 1$, show that for $n = m/r < m$, $n\bar{k} = \overline{nk} = \bar{0}$ and hence $\bar{k}$ cannot generate Z_m. ■

A naive hope might be that the techniques used above could be extended to groups with two generators and eventually to all finitely generated groups, and thus provide a description of the structure of such groups. Unfortunately, however, even groups with only two generators may have a very complex structure. (They need not be abelian for one thing; see Exercises 2.3 and 2.4.) Eventually we shall be able to characterize all finitely generated abelian groups, but even this will require a great deal more machinery.

EXERCISES

1. Let a,b be elements of group G. Show that $|a| = |a^{-1}|$; $|ab| = |ba|$, and $|a| = |cac^{-1}|$ for all $c \in G$.

2. Let G be an abelian group containing elements a and b of orders m and n respectively. Show that G contains an element whose order is the least common multiple of m and n. [*Hint:* first try the case when $(m,n) = 1$.]

3. Let G be an abelian group of order pq, with $(p,q) = 1$. Assume there exist $a,b \in G$ such that $|a| = p$, $|b| = q$ and show that G is cyclic.

4. If $f: G \to H$ is a homomorphism, $a \in G$, and $f(a)$ has finite order in H, then $|a|$ is infinite or $|f(a)|$ divides $|a|$.

5. Let G be the multiplicative group of all nonsingular 2×2 matrices with rational entries. Show that $a = \begin{pmatrix} 0 & -1 \\ 1 & 0 \end{pmatrix}$ has order 4 and $b = \begin{pmatrix} 0 & 1 \\ -1 & -1 \end{pmatrix}$ has order 3, but ab has infinite order. Conversely, show that the additive group $Z_2 \oplus \mathbf{Z}$ contains nonzero elements a,b of infinite order such that $a + b$ has finite order.

6. If G is a cyclic group of order n and $k \mid n$, then G has exactly one subgroup of order k.

7. Let p be prime and H a subgroup of $Z(p^\infty)$ (Exercise 1.10).
(a) Every element of $Z(p^\infty)$ has finite order p^n for some $n \geq 0$.
(b) If at least one element of H has order p^k and no element of H has order greater than p^k, then H is the cyclic subgroup generated by $\overline{1/p^k}$, whence $H \cong Z_{p^k}$.
(c) If there is no upper bound on the orders of elements of H, then $H = Z(p^\infty)$; [see Exercise 2.16].
(d) The only proper subgroups of $Z(p^\infty)$ are the finite cyclic groups $C_n = \langle \overline{1/p^n} \rangle$ $(n = 1,2, \ldots)$. Furthermore, $\langle 0 \rangle = C_0 < C_1 < C_2 < C_3 < \cdots$.
(e) Let $x_1, x_2, \ldots$ be elements of an abelian group G such that $|x_1| = p$, $px_2 = x_1$, $px_3 = x_2, \ldots, px_{n+1} = x_n, \ldots$. The subgroup generated by the x_i $(i \geq 1)$ is isomorphic to $Z(p^\infty)$. [*Hint:* Verify that the map induced by $x_i \mapsto \overline{1/p^i}$ is a well-defined isomorphism.]

8. A group that has only a finite number of subgroups must be finite.

9. If G is an abelian group, then the set T of all elements of G with finite order is a subgroup of G. [Compare Exercise 5.]

10. An infinite group is cyclic if and only if it is isomorphic to each of its proper subgroups.

4. COSETS AND COUNTING

In this section we obtain the first significant theorems relating the structure of a finite group G with the number theoretic properties of its order $|G|$. We begin by extending the concept of congruence modulo m in the group $\mathbf{Z}$. By definition $a \equiv b$ (mod m) if and only if $m \mid a - b$, that is, if and only if $a - b$ is an element of the subgroup $\langle m \rangle = \{mk \mid k \in \mathbf{Z}\}$. More generally (and in multiplicative notation) we have

Definition 4.1. *Let* H *be a subgroup of a group* G *and* a,b ε G. a *is* **right congruent** *to* b **modulo** H, *denoted* $a \equiv_r b$ (*mod* H) *if* $ab^{-1} \in H$. a *is* **left congruent** *to* b **modulo** H, *denoted* $a \equiv_l b$ (*mod* H), *if* $a^{-1}b \in H$.

If G is abelian, then right and left congruence modulo H coincide (since $ab^{-1} \in H \Leftrightarrow (ab^{-1})^{-1} \in H$ and $(ab^{-1})^{-1} = ba^{-1} = a^{-1}b$). There also exist nonabelian groups G and subgroups H such that right and left congruence coincide (Section 5), but this is not true in general.

Theorem 4.2. *Let* H *be a subgroup of a group* G.

(i) *Right [resp. left] congruence modulo* H *is an equivalence relation on* G.

(ii) *The equivalence class of* a ε G *under right [resp. left] congruence modulo* H *is the set* $Ha = \{ha \mid h \varepsilon H\}$ [*resp.* $aH = \{ah \mid h \varepsilon H\}$].

(iii) $|Ha| = |H| = |aH|$ *for all* $a \varepsilon G$.

The set Ha is called a **right coset** of H in G and aH is called a **left coset** of H in G. In general it is *not* the case that a right coset is also a left coset (Exercise 2).

PROOF OF 4.2. We write $a \equiv b$ for $a \equiv_r b \pmod{H}$ and prove the theorem for right congruence and right cosets. Analogous arguments apply to left congruence.

(i) Let $a,b,c \varepsilon G$. Then $a \equiv a$ since $aa^{-1} = e \varepsilon H$; hence $\equiv$ is reflexive. $\equiv$ is clearly symmetric ($a \equiv b \Rightarrow ab^{-1} \varepsilon H \Rightarrow (ab^{-1})^{-1} \varepsilon H \Rightarrow ba^{-1} \varepsilon H \Rightarrow b \equiv a$). Finally $a \equiv b$ and $b \equiv c$ imply $ab^{-1} \varepsilon H$ and $bc^{-1} \varepsilon H$. Thus $ac^{-1} = (ab^{-1})(bc^{-1}) \varepsilon H$ and $a \equiv c$; hence $\equiv$ is transitive. Therefore, right congruence modulo H is an equivalence relation.

(ii) The equivalence class of $a \varepsilon G$ under right congruence is $\{x \varepsilon G \mid x \equiv a\}$ $= \{x \varepsilon G \mid xa^{-1} \varepsilon H\} = \{x \varepsilon G \mid xa^{-1} = h \varepsilon H\} = \{x \varepsilon G \mid x = ha; h \varepsilon H\}$ $= \{ha \mid h \varepsilon H\} = Ha$.

(iii) The map $Ha \to H$ given by $ha \mapsto h$ is easily seen to be a bijection. ∎

Corollary 4.3. *Let H be a subgroup of a group* G.

(i) G *is the union of the right [resp. left] cosets of* H *in* G.

(ii) *Two right [resp. left] cosets of* H *in* G *are either disjoint or equal.*

(iii) *For all* $a,b \varepsilon G$, $Ha = Hb \Leftrightarrow ab^{-1} \varepsilon H$ *and* $aH = bH \Leftrightarrow a^{-1}b \varepsilon H$.

(iv) *If* $\mathcal{R}$ *is the set of distinct right cosets of* H *in* G *and* $\mathcal{L}$ *is the set of distinct left cosets of* H *in* G, *then* $|\mathcal{R}| = |\mathcal{L}|$.

PROOF. (i)–(iii) are immediate consequences of the theorem and statements (19)–(21) of Introduction, Section 4. (iv) The map $\mathcal{R} \to \mathcal{L}$ given by $Ha \mapsto a^{-1}H$ is a bijection since $Ha = Hb \Leftrightarrow ab^{-1} \varepsilon H \Leftrightarrow (a^{-1})^{-1}b^{-1} \varepsilon H \Leftrightarrow a^{-1}H = b^{-1}H$. ∎

ADDITIVE NOTATION. If H is a subgroup of an additive group, then right congruence modulo H is defined by: $a \equiv_r b \pmod{H} \Leftrightarrow a - b \varepsilon H$. The equivalence class of $a \varepsilon G$ is the right coset $H + a = \{h + a \mid h \varepsilon H\}$; similarly for left congruence and left cosets.

Definition 4.4. *Let* H *be a subgroup of a group* G. *The* **index of H in G,** *denoted* [G : H], *is the cardinal number of the set of distinct right [resp. left] cosets of* H *in* G.

In view of Corollary 4.3 (iv), $[G : H]$ does not depend on whether right or left cosets are used in the definition. Our principal interest is in the case when $[G : H]$ is finite, which can occur even when G and H are infinite groups (for example, $[\mathbf{Z} : \langle m \rangle] = m$ by Introduction, Theorem 6.8(i)). Note that if $H = \langle e \rangle$, then $Ha = \{a\}$ for every $a \varepsilon G$ and $[G : H] = |G|$.

A **complete set of right coset representatives** of a subgroup H in a group G is a set $\{a_i\}$ consisting of precisely one element from each right coset of H in G. Clearly the set $\{a_i\}$ has cardinality $[G : H]$. Note that such a set contains exactly one element of H since $H = He$ is itself a right coset. Analogous statements apply to left cosets.

Theorem 4.5. *If* K,H,G *are groups with* $K < H < G$, *then* $[G : K] = [G : H][H : K]$. *If any two of these indices are finite, then so is the third.*

PROOF. By Corollary 4.3 $G = \bigcup_{i\varepsilon I} Ha_i$ with $a_i \varepsilon G$, $|I| = [G : H]$ and the cosets Ha_i mutually disjoint (that is, $Ha_i = Ha_j \Leftrightarrow i = j$). Similarly $H = \bigcup_{j\varepsilon J} Kb_j$ with $b_j \varepsilon H$, $|J| = [H : K]$ and the cosets Kb_j are mutually disjoint. Therefore $G = \bigcup_{i\varepsilon I} Ha_i = \bigcup_{i\varepsilon I} (\bigcup_{j\varepsilon J} Kb_j)a_i = \bigcup_{(i,j)\varepsilon I\times J} Kb_ja_i$. It suffices to show that the cosets Kb_ja_i are mutually disjoint. For then by Corollary 4.3. we must have $[G : K] = |I \times J|$, whence $[G : K] = |I \times J| = |I||J| = [G : H][H : K]$. If $Kb_ja_i = Kb_ra_t$, then $b_ja_i = kb_ra_t$ $(k \varepsilon K)$. Since $b_j,b_r,k \varepsilon H$ we have $Ha_i = Hb_ja_i = Hkb_ra_t = Ha_t$; hence $i = t$ and $b_j = kb_r$. Thus $Kb_j = Kkb_r = Kb_r$ and $j = r$. Therefore, the cosets Kb_ja_i are mutually disjoint. The last statement of the theorem is obvious. ■

Corollary 4.6. (*Lagrange*). *If* H *is a subgroup of a group* G, *then* $|G| = [G : H]|H|$. *In particular if* G *is finite, the order* $|a|$ *of* $a \varepsilon G$ *divides* $|G|$.

PROOF. Apply the theorem with $K = \langle e\rangle$ for the first statement. The second is a special case of the first with $H = \langle a\rangle$. ■

A number of proofs in the theory of (finite) groups rely on various "counting" techniques, some of which we now introduce. If G is a group and H,K are subsets of G, we denote by HK the set $\{ab \mid a \varepsilon H, b \varepsilon K\}$; a right or left coset of a subgroup is a special case. If H,K are subgroups, HK may *not* be a subgroup (Exercise 7).

Theorem 4.7. *Let* H *and* K *be finite subgroups of a group* G. *Then* $|HK| = |H||K|/|H \cap K|$.

SKETCH OF PROOF. $C = H \cap K$ is a subgroup of K of index $n = |K|/|H \cap K|$ and K is the disjoint union of right cosets $Ck_1 \cup Ck_2 \cup \cdots \cup Ck_n$ for some $k_i \varepsilon K$. Since $HC = H$, this implies that HK is the disjoint union $Hk_1 \cup Hk_2 \cup \cdots \cup Hk_n$. Therefore, $|HK| = |H|\cdot n = |H||K|/|H \cap K|$. ■

Proposition 4.8. *If* H *and* K *are subgroups of a group* G, *then* $[H : H \cap K] \leq [G : K]$. *If* $[G : K]$ *is finite, then* $[H : H \cap K] = [G : K]$ *if and only if* $G = KH$.

SKETCH OF PROOF. Let A be the set of all right cosets of $H \cap K$ in H and B the set of all right cosets of K in G. The map $\varphi : A \to B$ given by $(H \cap K)h \mapsto Kh$

$(h \in H)$ is well defined since $(H \cap K)h' = (H \cap K)h$ implies $h'h^{-1} \in H \cap K \subset K$ and hence $Kh' = Kh$. Show that φ is injective. Then $[H : H \cap K] = |A| \leq |B| = [G : K]$. If $[G : K]$ is finite, then show that $[H : H \cap K] = [G : K]$ if and only if φ is surjective and that φ is surjective if and only if $G = KH$. Note that for $h \in H$, $k \in K$, $Kkh = Kh$ since $(kh)h^{-1} = k \in K$. ■

Proposition 4.9. *Let* H *and* K *be subgroups of finite index of a group* G. *Then* $[G : H \cap K]$ *is finite and* $[G : H \cap K] \leq [G : H][G : K]$. *Furthermore,* $[G : H \cap K] = [G : H][G : K]$ *if and only if* $G = HK$.

PROOF. Exercise; use Theorem 4.5 and Proposition 4.8. ■

EXERCISES

1. Let G be a group and $\{H_i \mid i \in I\}$ a family of subgroups. Then for any $a \in G$, $(\bigcap_i H_i)a = \bigcap_i H_ia$.

2. (a) Let H be the cyclic subgroup (of order 2) of S_3 generated by $\begin{pmatrix} 1 & 2 & 3 \\ 2 & 1 & 3 \end{pmatrix}$. Then no left coset of H (except H itself) is also a right coset. There exists $a \in S_3$ such that $aH \cap Ha = \{a\}$.
   (b) If K is the cyclic subgroup (of order 3) of S_3 generated by $\begin{pmatrix} 1 & 2 & 3 \\ 2 & 3 & 1 \end{pmatrix}$, then every left coset of K is also a right coset of K.

3. The following conditions on a finite group G are equivalent.
   (i) $|G|$ is prime.
   (ii) $G \neq \langle e \rangle$ and G has no proper subgroups.
   (iii) $G \cong Z_p$ for some prime p.

4. (Euler-Fermat) Let a be an integer and p a prime such that $p \nmid a$. Then $a^{p-1} \equiv 1 \pmod{p}$. [*Hint:* Consider $\bar{a} \in Z_p$ and the multiplicative group of nonzero elements of Z_p; see Exercise 1.7.] It follows that $a^p \equiv a \pmod{p}$ for any integer a.

5. Prove that there are only two distinct groups of order 4 (up to isomorphism), namely Z_4 and $Z_2 \oplus Z_2$. [*Hint:* By Lagrange's Theorem 4.6 a group of order 4 that is not cyclic must consist of an identity and three elements of order 2.]

6. Let H,K be subgroups of a group G. Then HK is a subgroup of G if and only if $HK = KH$.

7. Let G be a group of order p^km, with p prime and $(p,m) = 1$. Let H be a subgroup of order p^k and K a subgroup of order p^d, with $0 < d \leq k$ and $K \not\subset H$. Show that HK is not a subgroup of G.

8. If H and K are subgroups of finite index of a group G such that $[G : H]$ and $[G : K]$ are relatively prime, then $G = HK$.

9. If H,K and N are subgroups of a group G such that $H < N$, then $HK \cap N = H(K \cap N)$.

10. Let H,K,N be subgroups of a group G such that $H < K$, $H \cap N = K \cap N$, and $HN = KN$. Show that $H = K$.

11. Let G be a group of order $2n$; then G contains an element of order 2. If n is odd and G abelian, there is only one element of order 2.

12. If H and K are subgroups of a group G, then $[H \vee K : H] \geq [K : H \cap K]$.

13. If $p > q$ are primes, a group of order pq has at most one subgroup of order p. [*Hint:* Suppose H,K are distinct subgroups of order p. Show $H \cap K = \langle e \rangle$; use Exercise 12 to get a contradiction.]

14. Let G be a group and $a,b \in G$ such that (i) $|a| = 4 = |b|$; (ii) $a^2 = b^2$; (iii) $ba = a^3b = a^{-1}b$; (iv) $a \neq b$; (v) $G = \langle a,b \rangle$. Show that $|G| = 8$ and $G \cong Q_8$. (See Exercise 2.3; observe that the generators A,B of Q_8 also satisfy (i)–(v).)

5. NORMALITY, QUOTIENT GROUPS, AND HOMOMORPHISMS

We shall study those subgroups N of a group G such that left and right congruence modulo N coincide. Such subgroups play an important role in determining both the structure of a group G and the nature of homomorphisms with domain G.

Theorem 5.1. *If* N *is a subgroup of a group* G, *then the following conditions are equivalent.*

(i) *Left and right congruence modulo* N *coincide* (*that is, define the same equivalence relation on* G);
(ii) *every left coset of* N *in* G *is a right coset of* N *in* G;
(iii) $aN = Na$ *for all* $a \in G$;
(iv) *for all* $a \in G$, $aNa^{-1} \subset N$, *where* $aNa^{-1} = \{ana^{-1} \mid n \in N\}$;
(v) *for all* $a \in G$, $aNa^{-1} = N$.

PROOF. (i) $\Leftrightarrow$ (iii) Two equivalence relations R and S are identical if and only if the equivalence class of each element under R is equal to its equivalence class under S. In this case the equivalence classes are the left and right cosets respectively of N. (ii) $\Rightarrow$ (iii) If $aN = Nb$ for some $b \in G$, then $a \in Nb \cap Na$, which implies $Nb = Na$ since two *right* cosets are either disjoint or equal. (iii) $\Rightarrow$ (iv) is trivial. (iv) $\Rightarrow$ (v) We have $aNa^{-1} \subset N$. Since (iv) also holds for $a^{-1} \in G$, $a^{-1}Na \subset N$. Therefore for every $n \in N$, $n = a(a^{-1}na)a^{-1} \in aNa^{-1}$ and $N \subset aNa^{-1}$. (v) $\Rightarrow$ (ii) is immediate. ∎

Definition 5.2. *A subgroup* N *of a group* G *which satisfies the equivalent conditions of Theorem 5.1 is said to be* **normal** *in* G (*or a* **normal subgroup** *of* G); *we write* $N \lhd G$ *if* N *is normal in* G.

In view of Theorem 5.1 we may omit the subscripts "r" and "l" when denoting congruence modulo a normal subgroup.

EXAMPLES. Every subgroup of an abelian group is trivially normal. The subgroup H generated by $\begin{pmatrix} 1 & 2 & 3 \\ 2 & 3 & 1 \end{pmatrix}$ in S_3 is normal (Exercise 4.2). More generally any subgroup N of index 2 in a group G is normal (Exercise 1). The intersection of any family of normal subgroups is a normal subgroup (Exercise 2).

If G is a group with subgroups N and M such that $N \lhd M$ and $M \lhd G$, it does not follow that $N \lhd G$ (Exercise 10). However, it is easy to see that if N is normal in G, then N is normal in every subgroup of G containing N.

Recall that the join $H \vee K$ of two subgroups is the subgroup $\langle H \cup K\rangle$ generated by H and K.

Theorem 5.3. *Let* K *and* N *be subgroups of a group* G *with* N *normal in* G. *Then*

(i) $N \cap K$ *is a normal subgroup of* K;
(ii) N *is a normal subgroup of* $N \vee K$;
(iii) $NK = N \vee K = KN$;
(iv) *if* K *is normal in* G *and* $K \cap N = \langle e\rangle$, *then* $nk = kn$ *for all* $k \in K$ *and* $n \in N$.

PROOF. (i) If $n \in N \cap K$ and $a \in K$, then $ana^{-1} \in N$ since $N \lhd G$ and $ana^{-1} \in K$ since $K < G$. Thus $a(N \cap K)a^{-1} \subset N \cap K$ and $N \cap K \lhd K$. (ii) is trivial since $N < N \vee K$. (iii) Clearly $NK \subset N \vee K$. An element x of $N \vee K$ is a product of the form $n_1k_1n_2k_2 \cdots n_rk_r$, with $n_i \in N$, $k_i \in K$ (Theorem 2.8). Since $N \lhd G$, $n_ik_j = k_jn_i'$, $n_i' \in N$ and therefore x can be written in the form $n(k_1 \cdots k_r)$, $n \in N$. Thus $N \vee K \subset NK$. Similarly $KN = N \vee K$. (iv) Let $k \in K$ and $n \in N$. Then $nkn^{-1} \in K$ since $K \lhd G$ and $kn^{-1}k^{-1} \in N$ since $N \lhd G$. Hence $(nkn^{-1})k^{-1} = n(kn^{-1}k^{-1}) \in N \cap K = \langle e\rangle$, which implies $kn = nk$. ■

Theorem 5.4. *If* N *is a normal subgroup of a group* G *and* G/N *is the set of all (left) cosets of* N *in* G, *then* G/N *is a group of order* [G : N] *under the binary operation given by* $(aN)(bN) = abN$.

PROOF. Since the coset aN [resp. bN, abN] is simply the equivalence class of $a \in G$ [resp. $b \in G$, $ab \in G$] under the equivalence relation of congruence modulo N, it suffices by Theorem 1.5 to show that congruence modulo N is a congruence relation, that is, that $a_1 \equiv a \pmod N$ and $b_1 \equiv b \pmod N$ imply $a_1b_1 \equiv ab \pmod N$. By assumption $a_1a^{-1} = n_1 \in N$ and $b_1b^{-1} = n_2 \in N$. Hence $(a_1b_1)(ab)^{-1} = a_1b_1b^{-1}a^{-1} = (a_1n_2)a^{-1}$. But since N is normal, $a_1N = Na_1$ which implies that $a_1n_2 = n_3a_1$ for some $n_3 \in N$. Consequently $(a_1b_1)(ab)^{-1} = (a_1n_2)a^{-1} = n_3a_1a^{-1} = n_3n_1 \in N$, whence $a_1b_1 \equiv ab \pmod N$. ■

If N is a normal subgroup of a group G, then the group G/N, as in Theorem 5.4, is called the **quotient group** or **factor group** of G by N. If G is written additively, then the group operation in G/N is given by $(a + N) + (b + N) = (a + b) + N$.

REMARK. If $m > 1$ is a (fixed) integer and $k \in \mathbf{Z}$, then the remarks preceding Definition 4.1 show that the equivalence class of k under congruence modulo m is precisely the coset of $\langle m \rangle$ in $\mathbf{Z}$ which contains k; that is, as sets, $Z_m = \mathbf{Z}/\langle m \rangle$. Theorems 1.5 and 5.4 show that the group operations coincide, whence $Z_m = \mathbf{Z}/\langle m \rangle$ as groups.

We now explore the relationships between normal subgroups, quotient groups, and homomorphisms.

Theorem 5.5. *If* $f: G \to H$ *is a homomorphism of groups, then the kernel of* f *is a normal subgroup of* G. *Conversely, if* N *is a normal subgroup of* G, *then the map* $\pi : G \to G/N$ *given by* $\pi(a) = aN$ *is an epimorphism with kernel* N.

PROOF. If $x \in \operatorname{Ker} f$ and $a \in G$, then

$$f(axa^{-1}) = f(a)f(x)f(a^{-1}) = f(a)ef(a)^{-1} = e$$

and $axa^{-1} \in \operatorname{Ker} f$. Therefore $a(\operatorname{Ker} f)a^{-1} \subset \operatorname{Ker} f$ and $\operatorname{Ker} f \lhd G$. The map $\pi : G \to G/N$ is clearly surjective and since $\pi(ab) = abN = aNbN = \pi(a)\pi(b)$, π is an epimorphism. $\operatorname{Ker} \pi = \{a \in G \mid \pi(a) = eN = N\} = \{a \in G \mid aN = N\}$ $= \{a \in G \mid a \in N\} = N$. ■

The map $\pi : G \to G/N$ is called the **canonical epimorphism** or **projection.** Hereafter unless stated otherwise $G \to G/N$ $(N \lhd G)$ always denotes the canonical epimorphism.

Theorem 5.6. *If* $f : G \to H$ *is a homomorphism of groups and* N *is a normal subgroup of* G *contained in the kernel of* f, *then there is a unique homomorphism* $\bar{f} : G/N \to H$ *such that* $\bar{f}(aN) = f(a)$ *for all* $a \in G$. *Im* f = *Im* $\bar{f}$ *and Ker* $\bar{f}$ = (*Ker* f)/N. $\bar{f}$ *is an isomorphism if and only if* f *is an epimorphism and* N = *Ker* f.

The essential part of the conclusion may be rephrased: there exists a unique homomorphism $\bar{f}: G/N \to H$ such that the diagram:

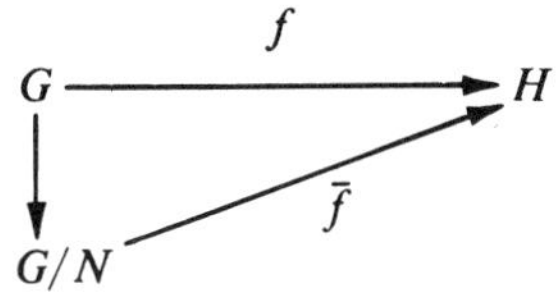

is commutative. Corollary 5.8 below may also be stated in terms of commutative diagrams.

PROOF OF 5.6. If $b \in aN$, then $b = an$, $n \in N$, and $f(b) = f(an) = f(a)f(n)$ $= f(a)e = f(a)$, since $N < \operatorname{Ker} f$. Therefore, f has the same effect on every element of the coset aN and the map $\bar{f}: G/N \to H$ given by $\bar{f}(aN) = f(a)$ is a well-defined function. Since $\bar{f}(aNbN) = \bar{f}(abN) = f(ab) = f(a)f(b) = \bar{f}(aN)\bar{f}(bN)$, $\bar{f}$ is a homomorphism. Clearly $\operatorname{Im} \bar{f} = \operatorname{Im} f$ and

$$aN \in \text{Ker } \bar{f} \Leftrightarrow f(a) = e \Leftrightarrow a \in \text{Ker } f,$$

whence Ker $\bar{f} = \{aN \mid a \in \text{Ker } f\} = (\text{Ker } f)/N$. $\bar{f}$ is unique since it is completely determined by f. Finally it is clear that $\bar{f}$ is an epimorphism if and only if f is. By Theorem 2.3 $\bar{f}$ is a monomorphism if and only if Ker $\bar{f} = (\text{Ker } f)/N$ is the trivial subgroup of G/N, which occurs if and only if Ker $f = N$. ■

Corollary 5.7. (*First Isomorphism Theorem*) *If* $f : G \to H$ *is a homomorphism of groups, then* f *induces an isomorphism* $G/\text{Ker } f \cong \text{Im } f$.

PROOF. $f : G \to \text{Im } f$ is an epimorphism. Apply Theorem 5.6 with $N = \text{Ker } f$. ■

Corollary 5.8. *If* $f : G \to H$ *is a homomorphism of groups,* $N \lhd G$, $M \lhd H$, *and* $f(N) < M$, *then* f *induces a homomorphism* $\bar{f} : G/N \to H/M$, *given by* $aN \mapsto f(a)M$. $\bar{f}$ *is an isomorphism if and only if* $\text{Im } f \vee M = H$ *and* $f^{-1}(M) \subset N$. *In particular if* f *is an epimorphism such that* $f(N) = M$ *and* $\text{Ker } f \subset N$, *then* $\bar{f}$ *is an isomorphism.*

SKETCH OF PROOF. Consider the composition $G \xrightarrow{f} H \xrightarrow{\pi} H/M$ and verify that $N \subset f^{-1}(M) = \text{Ker } \pi f$. By Theorem 5.6 (applied to πf) the map $G/N \to H/M$ given by $aN \mapsto (\pi f)(a) = f(a)M$ is a homomorphism that is an isomorphism if and only if πf is an epimorphism and $N = \text{Ker } \pi f$. But the latter conditions hold if and only if $\text{Im } f \vee M = H$ and $f^{-1}(M) \subset N$. If f is an epimorphism, then $H = \text{Im } f = \text{Im } f \vee M$. If $f(N) = M$ and $\text{Ker } f \subset N$, then $f^{-1}(M) \subset N$, whence $\bar{f}$ is an isomorphism. ■

Corollary 5.9. (*Second Isomorphism Theorem*) *If* K *and* N *are subgroups of a group* G, *with* N *normal in* G, *then* $K/(N \cap K) \cong NK/N$.

PROOF. $N \lhd NK = N \vee K$ by Theorem 5.3. The composition $K \xrightarrow{\subset} NK \xrightarrow{\pi} NK/N$ is a homomorphism f with kernel $K \cap N$, whence $\bar{f}: K/K \cap N \cong \text{Im } f$ by Corollary 5.7. Every element in NK/N is of the form nkN $(n \in N, k \in K)$. The normality of N implies that $nk = kn_1$ $(n_1 \in N)$, whence $nkN = kn_1N = kN = f(k)$. Therefore f is an epimorphism and hence $\text{Im } f = NK/N$. ■

Corollary 5.10. (*Third Isomorphism Theorem*). *If* H *and* K *are normal subgroups of a group* G *such that* $K < H$, *then* H/K *is a normal subgroup of* G/K *and* $(G/K)/(H/K) \cong G/H$.

PROOF. The identity map $1_G : G \to G$ has $1_G(K) < H$ and therefore induces an epimorphism $I : G/K \to G/H$, with $I(aK) = aH$. Since $H = I(aK)$ if and only if $a \in H$, $\text{Ker } I = \{aK \mid a \in H\} = H/K$. Hence $H/K \lhd G/K$ by Theorem 5.5 and $G/H = \text{Im } I \cong (G/K)/\text{Ker } I = (G/K)/(H/K)$ by Corollary 5.7. ■

Theorem 5.11. *If* $f : G \to H$ *is an epimorphism of groups, then the assignment* $K \mapsto f(K)$ *defines a one-to-one correspondence between the set* $S_f(G)$ *of all subgroups* K *of* G *which contain Ker* f *and the set* $S(H)$ *of all subgroups of* H*. Under this correspondence normal subgroups correspond to normal subgroups.*

SKETCH OF PROOF. By Exercise 2.9 the assignment $K \mapsto f(K)$ defines a function $\varphi : S_f(G) \to S(H)$ and $f^{-1}(J)$ is a subgroup of G for every subgroup J of H. Since $J < H$ implies $\text{Ker } f < f^{-1}(J)$ and $f(f^{-1}(J)) = J$, φ is surjective. Exercise 18 shows that $f^{-1}(f(K)) = K$ if and only if $\text{Ker } f < K$. It follows that φ is injective. To prove the last statement verify that $K \lhd G$ implies $f(K) \lhd H$ and $J \lhd H$ implies $f^{-1}(J) \lhd G$. ■

Corollary 5.12. *If* N *is a normal subgroup of a group* G*, then every subgroup of* G/N *is of the form* K/N*, where* K *is a subgroup of* G *that contains* N*. Furthermore,* K/N *is normal in* G/N *if and only if* K *is normal in* G.

PROOF. Apply Theorem 5.11 to the canonical epimorphism $\pi : G \to G/N$. If $N < K < G$, then $\pi(K) = K/N$. ■

EXERCISES

1. If N is a subgroup of index 2 in a group G, then N is normal in G.
2. If $\{N_i \mid i \in I\}$ is a family of normal subgroups of a group G, then $\bigcap_{i \in I} N_i$ is a normal subgroup of G.
3. Let N be a subgroup of a group G. N is normal in G if and only if (right) congruence modulo N is a congruence relation on G.
4. Let $\sim$ be an equivalence relation on a group G and let $N = \{a \in G \mid a \sim e\}$. Then $\sim$ is a congruence relation on G if and only if N is a normal subgroup of G and $\sim$ is congruence modulo N.
5. Let $N < S_4$ consist of all those permutations σ such that $\sigma(4) = 4$. Is N normal in S_4?
6. Let $H < G$; then the set aHa^{-1} is a subgroup for each $a \in G$, and $H \cong aHa^{-1}$.
7. Let G be a finite group and H a subgroup of G of order n. If H is the only subgroup of G of order n, then H is normal in G.
8. All subgroups of the quaternion group are normal (see Exercises 2.3 and 4.14).
9. (a) If G is a group, then the center of G is a normal subgroup of G (see Exercise 2.11);
   (b) the center of S_n is the identity subgroup for all $n > 2$.
10. Find subgroups H and K of D_4^* such that $H \lhd K$ and $K \lhd D_4^*$, but H is not normal in D_4^*.
11. If H is a cyclic subgroup of a group G and H is normal in G, then every subgroup of H is normal in G. [Compare Exercise 10.]

12. If H is a normal subgroup of a group G such that H and G/H are finitely generated, then so is G.

13. (a) Let $H \lhd G$, $K \lhd G$. Show that $H \vee K$ is normal in G.
(b) Prove that the set of all normal subgroups of G forms a complete lattice under inclusion (Introduction, Exercise 7.2).

14. If $N_1 \lhd G_1$, $N_2 \lhd G_2$ then $(N_1 \times N_2) \lhd (G_1 \times G_2)$ and $(G_1 \times G_2)/(N_1 \times N_2) \cong (G_1/N_1) \times (G_2/N_2)$.

15. Let $N \lhd G$ and $K \lhd G$. If $N \cap K = \langle e \rangle$ and $N \vee K = G$, then $G/N \cong K$.

16. If $f: G \to H$ is a homomorphism, H is abelian and N is a subgroup of G containing Ker f, then N is normal in G.

17. (a) Consider the subgroups $\langle 6 \rangle$ and $\langle 30 \rangle$ of $\mathbf{Z}$ and show that $\langle 6 \rangle/\langle 30 \rangle \cong Z_5$.
(b) For any $k,m > 0$, $\langle k \rangle/\langle km \rangle \cong Z_m$; in particular, $\mathbf{Z}/\langle m \rangle = \langle 1 \rangle/\langle m \rangle \cong Z_m$.

18. If $f: G \to H$ is a homomorphism with kernel N and $K < G$, then prove that $f^{-1}(f(K)) = KN$. Hence $f^{-1}(f(K)) = K$ if and only if $N < K$.

19. If $N \lhd G$, $[G : N]$ finite, $H < G$, $|H|$ finite, and $[G : N]$ and $|H|$ are relatively prime, then $H < N$.

20. If $N \lhd G$, $|N|$ finite, $H < G$, $[G : H]$ finite, and $[G : H]$ and $|N|$ are relatively prime, then $N < H$.

21. If H is a subgroup of $Z(p^\infty)$ and $H \neq Z(p^\infty)$, then $Z(p^\infty)/H \cong Z(p^\infty)$. [*Hint:* if $H = \langle \overline{1/p^n} \rangle$, let $x_i = \overline{1/p^{n+i}} + H$ and apply Exercise 3.7(e).]

6. SYMMETRIC, ALTERNATING, AND DIHEDRAL GROUPS

In this section we shall study in some detail the symmetric group S_n and certain of its subgroups. By definition S_n is the group of all bijections $I_n \to I_n$, where $I_n = \{1,2,\ldots,n\}$. The elements of S_n are called permutations. In addition to the notation given on page 26 for permutations in S_n there is another standard notation:

Definition 6.1. *Let* $i_1,i_2,\ldots,i_r$, $(r \leq n)$ *be distinct elements of* $I_n = \{1,2,\ldots n\}$. *Then* $(i_1i_2i_3\cdots i_r)$ *denotes the permutation that maps* $i_1 \mapsto i_2$, $i_2 \mapsto i_3$, $i_3 \mapsto i_4, \ldots,$ $i_{r-1} \mapsto i_r$, *and* $i_r \mapsto i_1$, *and maps every other element of* I_n *onto itself.* $(i_1i_2\cdots i_r)$ *is called a* **cycle** *of length* r *or an* r*-cycle; a 2-cycle is called a* **transposition**.

The cycle notation is not unique (see below); indeed, strictly speaking, the cycle notation is ambiguous since $(i_1\cdots i_r)$ may be an element of any S_n, $n \geq r$. In context, however, this will cause no confusion. A 1-cycle (k) is the identity permutation. Clearly, an r-cycle is an element of order r in S_n. Also observe that if τ is a cycle and $\tau(x) \neq x$ for some $x \in I_n$, then $\tau = (x\tau(x)\tau^2(x)\cdots\tau^d(x))$ for some $d \geq 1$. The inverse of the cycle $(i_1i_2\cdots i_r)$ is the cycle $(i_ri_{r-1}i_{r-2}\cdots i_2i_1) = (i_1i_ri_{r-1}i_{r-2}\cdots i_2)$ (verify!).

EXAMPLES. The permutation $\tau = \begin{pmatrix} 1 & 2 & 3 & 4 \\ 4 & 1 & 2 & 3 \end{pmatrix}$ is a 4-cycle: $\tau = (1432)$ $= (4321) = (3214) = (2143)$. If σ is the 3-cycle (125), then $\sigma\tau = (125)(1432) = (1435)$

(remember: permutations are functions and $\sigma\tau$ means τ followed by σ); similarly $\tau\sigma = (1432)(125) = (2543)$ so that $\sigma\tau \neq \tau\sigma$. There is one case, however, when two permutations do commute.

Definition 6.2. *The permutations $\sigma_1,\sigma_2,\dots,\sigma_r$ of S_n are said to be* **disjoint** *provided that for each $1 \leq i \leq r$, and every $k \in I_n$, $\sigma_i(k) \neq k$ implies $\sigma_j(k) = k$ for all $j \neq i$.*

In other words $\sigma_1,\sigma_2,\dots,\sigma_r$ are disjoint if and only if no element of I_n is moved by more than one of $\sigma_1,\dots,\sigma_r$. It is easy to see that $\tau\sigma = \sigma\tau$ whenever σ and τ are disjoint.

Theorem 6.3. *Every nonidentity permutation in S_n is uniquely (up to the order of the factors) a product of disjoint cycles, each of which has length at least 2.*

SKETCH OF PROOF. Let $\sigma \in S_n$, $\sigma \neq (1)$. Verify that the following is an equivalence relation on I_n: for $x,y \in I_n$, $x \sim y$ if and only if $y = \sigma^m(x)$ for some $m \in \mathbf{Z}$. The equivalence classes $\{B_i \mid 1 \leq i \leq s\}$ of this equivalence relation are called the *orbits* of σ and form a partition of I_n (Introduction, Theorem 4.1). Note that if $x \in B_i$, then $B_i = \{u \mid x \sim u\} = \{\sigma^m(x) \mid m \in \mathbf{Z}\}$. Let $B_1,B_2,\dots,B_r$ $(1 \leq r \leq s)$ be those orbits that contain more than one element each ($r \geq 1$ since $\sigma \neq (1)$). For each $i \leq r$ define $\sigma_i \in S_n$ by:

$$\sigma_i(x) = \begin{cases} \sigma(x) & \text{if } x \in B_i; \\ x & \text{if } x \notin B_i. \end{cases}$$

Each σ_i is a well-defined nonidentity permutation of I_n since $\sigma \mid B_i$ is a bijection $B_i \to B_i$. $\sigma_1,\sigma_2,\dots,\sigma_r$ are disjoint permutations since the sets $B_1,\dots,B_r$ are mutually disjoint. Finally verify that $\sigma = \sigma_1\sigma_2\cdots\sigma_r$; (note that $x \in B_i$ implies $\sigma(x) = \sigma_i(x)$ if $i \leq r$ and $\sigma(x) = x$ if $i > r$; use disjointness). We must show that each σ_i is a cycle.

If $x \in B_i$ $(i \leq r)$, then since B_i is finite there is a least positive integer d such that $\sigma^d(x) = \sigma^j(x)$ for some j $(0 \leq j < d)$. Since $\sigma^{d-j}(x) = x$ and $0 < d - j \leq d$, we must have $j = 0$ and $\sigma^d(x) = x$. Hence $(x\sigma(x)\sigma^2(x)\cdots\sigma^{d-1}(x))$ is a well-defined cycle of length at least 2. If $\sigma^m(x) \in B_i$, then $m = ad + b$ for some $a,b \in \mathbf{Z}$ such that $0 \leq b < d$. Hence $\sigma^m(x) = \sigma^{b+ad}(x) = \sigma^b\sigma^{ad}(x) = \sigma^b(x) \in \{x,\sigma(x),\sigma^2(x),\dots,\sigma^{d-1}(x)\}$. Therefore $B_i = \{x,\sigma(x),\sigma^2(x),\dots,\sigma^{d-1}(x)\}$ and it follows that σ_i is the cycle

$$(x\sigma(x)\sigma^2(x)\cdots\sigma^{d-1}(x)).$$

Suppose $\tau_1,\dots,\tau_t$ are disjoint cycles such that $\sigma = \tau_1\tau_2\cdots\tau_t$. Let $x \in I_n$ be such that $\sigma(x) \neq x$. By disjointness there exists a unique j $(1 \leq j \leq t)$ with $\sigma(x) = \tau_j(x)$. Since $\sigma\tau_j = \tau_j\sigma$, we have $\sigma^k(x) = \tau_j^k(x)$ for all $k \in \mathbf{Z}$. Therefore, the orbit of x under τ_j is precisely the orbit of x under σ, say B_i. Consequently, $\tau_j(y) = \sigma(y)$ for every $y \in B_i$ (since $y = \sigma^n(x) = \tau_j^n(x)$ for some $n \in \mathbf{Z}$). Since τ_j is a cycle it has only one nontrivial orbit (verify!), which must be B_i since $x \neq \sigma(x) = \tau_j(x)$. Therefore $\tau_j(y) = y$ for all $y \notin B_i$, whence $\tau_j = \sigma_i$. A suitable inductive argument shows that $r = t$ and (after reindexing) $\sigma_i = \tau_i$ for each $i = 1,2,\dots,r$. ■

Corollary 6.4. *The order of a permutation $\sigma \in S_n$ is the least common multiple of the orders of its disjoint cycles.*

PROOF. Let $\sigma = \sigma_1 \cdots \sigma_r$, with $\{\sigma_i\}$ disjoint cycles. Since disjoint cycles commute, $\sigma^m = \sigma_1^m \cdots \sigma_r^m$ for all $m \in \mathbf{Z}$ and $\sigma^m = (1)$ if and only if $\sigma_i^m = (1)$ for all i. Therefore $\sigma^m = (1)$ if and only if $|\sigma_i|$ divides m for all i (Theorem 3.4). Since $|\sigma|$ is the least such m, the conclusion follows. ■

Corollary 6.5. *Every permutation in S_n can be written as a product of (not necessarily disjoint) transpositions.*

PROOF. It suffices by Theorem 6.3 to show that every cycle is a product of transpositions. This is easy: $(x_1) = (x_1x_2)(x_1x_2)$ and for $r > 1$, $(x_1x_2x_3\cdots x_r) = (x_1x_r)(x_1x_{r-1})\cdots(x_1x_3)(x_1x_2)$. ■

Definition 6.6. *A permutation $\tau \in S_n$ is said to be* **even** *[resp.* **odd***] if τ can be written as a product of an even [resp. odd] number of transpositions.*

The **sign of a permutation** τ, denoted sgn τ, is 1 or -1 according as τ is even or odd. The fact that sgn τ is well defined is an immediate consequence of

Theorem 6.7. *A permutation in S_n $(n \geq 2)$ cannot be both even and odd.*

PROOF. Let $i_1,i_2,\ldots,i_n$ be the integers $1,2,\ldots,n$ in some order and define $\Delta(i_1,\ldots,i_n)$ to be the integer $\prod (i_j - i_k)$, where the product is taken over all pairs (j,k) such that $1 \leq j < k \leq n$. Note that $\Delta(i_1,\ldots,i_n) \neq 0$. We first compute $\Delta(\sigma(i_1),\ldots,\sigma(i_n))$ when $\sigma \in S_n$ is a transposition, say $\sigma = (i_ci_d)$ with $c < d$. We have $\Delta(i_1,\ldots,i_n) = (i_c - i_d)ABCDEFG$, where

$$A = \prod_{\substack{j<k \\ j,k \neq c,d}} (i_j - i_k); \quad B = \prod_{j<c} (i_j - i_c); \quad C = \prod_{j<c} (i_j - i_d);$$

$$D = \prod_{c<j<d} (i_j - i_d); \quad E = \prod_{c<k<d} (i_c - i_k); \quad F = \prod_{d<k} (i_c - i_k);$$

$$G = \prod_{d<k} (i_d - i_k).$$

We write $\sigma(A)$ for $\prod\limits_{\substack{j<k \\ j,k \neq c,d}} (\sigma(i_j) - \sigma(i_k))$ and similarly for $\sigma(B)$, $\sigma(C)$, etc. Verify that $\sigma(A) = A$; $\sigma(B) = C$ and $\sigma(C) = B$; $\sigma(D) = (-1)^{d-c-1}E$ and $\sigma(E) = (-1)^{d-c-1}D$; $\sigma(F) = G$, and $\sigma(G) = F$. Finally, $\sigma(i_c - i_d) = \sigma(i_c) - \sigma(i_d) = i_d - i_c = -(i_c - i_d)$. Consequently,

$$\begin{aligned} \Delta(\sigma(i_1),\ldots,\sigma(i_n)) &= \sigma(i_c - i_d)\sigma(A)\sigma(B)\cdots\sigma(G) = (-1)^{1+2(d-c-1)}(i_c - i_d)ABCDEFG \\ &= -\Delta(i_1,\ldots,i_n). \end{aligned}$$

Suppose for some $\tau \in S_n$, $\tau = \tau_1\cdots\tau_r$ and $\tau = \sigma_1\cdots\sigma_s$ with τ_i, σ_j transpositions, r even and s odd. Then for $(i_1,\ldots,i_n) = (1,2,\ldots,n)$ the previous paragraph implies $\Delta(\tau(1),\ldots,\tau(n)) = \Delta(\tau_1\cdots\tau_r(1),\ldots,\tau_1\cdots\tau_r(n)) = -\Delta(\tau_2\cdots\tau_r(1),\ldots,$

$\tau_2 \cdots \tau_r(n)) = \cdots = (-1)^r\Delta(1,2,\ldots,n) = \Delta(1,2,\ldots,n)$. Similarly $\Delta(\tau(1),\ldots,\tau(n)) = (-1)^s\Delta(1,2,\ldots,n) = -\Delta(1,2,\ldots,n)$, whence $\Delta(1,2,\ldots,n) = -\Delta(1,2,\ldots,n)$. This is a contradiction since $\Delta(1,2,\ldots,n) \neq 0$. ■

Theorem 6.8. *For each* $n \geq 2$, *let* A_n *be the set of all even permutations of* S_n. *Then* A_n *is a normal subgroup of* S_n *of index* 2 *and order* $|S_n|/2 = n!/2$. *Furthermore* A_n *is the only subgroup of* S_n *of index* 2.

The group A_n is called the **alternating group on n letters** or the **alternating group of degree n.**

SKETCH OF PROOF OF 6.8. Let C be the multiplicative subgroup $\{1,-1\}$ of the integers. Define a map $f: S_n \to C$ by $\sigma \mapsto \text{sgn}\,\sigma$ and verify that f is an epimorphism of groups. Since the kernel of f is clearly A_n, A_n is normal in S_n. By the First Isomorphism Theorem $S_n/A_n \cong C$, which implies $[S_n : A_n] = 2$ and $|A_n| = |S_n|/2$. A_n is the unique subgroup of S_n of index 2 by Exercise 6. ■

Definition 6.9. *A group* G *is said to be* **simple** *if* G *has no proper normal subgroups.*

The only simple abelian groups are the Z_p with p prime (Exercise 4.3). There are a number of nonabelian simple groups; in particular, we have

Theorem 6.10. *The alternating group* A_n *is simple if and only if* $n \neq 4$.

The proof we shall give is quite elementary. It will be preceded by two lemmas. Recall that if τ is a 2-cycle, $\tau^2 = (1)$ and hence $\tau = \tau^{-1}$.

Lemma 6.11. *Let* r,s *be distinct elements of* $\{1,2,\ldots,n\}$. *Then* A_n $(n \geq 3)$ *is generated by the 3-cycles* $\{(rsk) \mid 1 \leq k \leq n, k \neq r,s\}$.

PROOF. Assume $n > 3$ (the case $n = 3$ is trivial). Every element of A_n is a product of terms of the form $(ab)(cd)$ or $(ab)(ac)$, where a,b,c,d are distinct elements of $\{1,2,\ldots,n\}$. Since $(ab)(cd) = (acb)(acd)$ and $(ab)(ac) = (acb)$, A_n is generated by the set of all 3-cycles. Any 3-cycle is of the form (rsa), (ras), (rab), (sab), or (abc), where a,b,c are distinct and $a,b,c \neq r,s$. Since $(ras) = (rsa)^2$, $(rab) = (rsb)(rsa)^2$, $(sab) = (rsb)^2(rsa)$, and $(abc) = (rsa)^2(rsc)(rsb)^2(rsa)$, A_n is generated by

$$\{(rsk) \mid 1 \leq k \leq n, k \neq r,s\}. \quad ■$$

Lemma 6.12. *If* N *is a normal subgroup of* A_n $(n \geq 3)$ *and* N *contains a 3-cycle, then* $N = A_n$.

PROOF. If $(rsc) \in N$, then for any $k \neq r,s,c$, $(rsk) = (rs)(ck)(rsc)^2(ck)(rs) = [(rs)(ck)](rsc)^2[(rs)(ck)]^{-1} \in N$. Hence $N = A_n$ by Lemma 6.11. ■

PROOF OF THEOREM 6.10. $A_2 = (1)$ and A_3 is the simple cyclic group of order 3. It is easy to verify that $\{(1),(12)(34),(13)(24),(14)(23)\}$ is a normal subgroup of A_4 (Exercise 7). If $n \geq 5$ and N is a nontrivial normal subgroup of A_n we shall show $N = A_n$ by considering the possible cases.

CASE 1. N contains a 3-cycle; hence $N = A_n$ by Lemma 6.12.

CASE 2. N contains an element σ, the product of disjoint cycles, at least one of which has length $r \geq 4$. Thus $\sigma = (a_1a_2\cdots a_r)\tau$ (disjoint). Let $\delta = (a_1a_2a_3) \in A_n$. Then $\sigma^{-1}(\delta\sigma\delta^{-1}) \in N$ by normality. But

$$\sigma^{-1}(\delta\sigma\delta^{-1}) = \tau^{-1}(a_1a_ra_{r-1}\cdots a_2)(a_1a_2a_3)(a_1a_2\cdots a_r)\tau(a_1a_3a_2) = (a_1a_3a_r) \in N.$$

Hence $N = A_n$ by Lemma 6.12.

CASE 3. N contains an element σ, the product of disjoint cycles, at least two of which have length 3, so that $\sigma = (a_1a_2a_3)(a_4a_5a_6)\tau$ (disjoint). Let $\delta = (a_1a_2a_4) \in A_n$. Then as above, N contains $\sigma^{-1}(\delta\sigma\delta^{-1}) = \tau^{-1}(a_4a_6a_5)(a_1a_3a_2)(a_1a_2a_4)(a_1a_2a_3)(a_4a_5a_6)\tau(a_1a_4a_2) = (a_1a_4a_2a_6a_3)$. Hence $N = A_n$ by case 2.

CASE 4. N contains an element σ that is the product of one 3-cycle and some 2-cycles, say $\sigma = (a_1a_2a_3)\tau$ (disjoint), with τ a product of disjoint 2-cycles. Then $\sigma^2 \in N$ and $\sigma^2 = (a_1a_2a_3)\tau(a_1a_2a_3)\tau = (a_1a_2a_3)^2\tau^2 = (a_1a_2a_3)^2 = (a_1a_3a_2)$, whence $N = A_n$ by Lemma 6.12.

CASE 5. Every element of N is the product of (an even number of) disjoint 2-cycles. Let $\sigma \in N$, with $\sigma = (a_1a_2)(a_3a_4)\tau$ (disjoint). Let $\delta = (a_1a_2a_3) \in A_n$; then $\sigma^{-1}(\delta\sigma\delta^{-1}) \in N$ as above. Now $\sigma^{-1}(\delta\sigma\delta^{-1}) = \tau^{-1}(a_3a_4)(a_1a_2)(a_1a_2a_3)(a_1a_2)(a_3a_4)\tau(a_1a_3a_2) = (a_1a_3)(a_2a_4)$. Since $n \geq 5$, there is an element $b \in \{1,2,\ldots,n\}$ distinct from a_1,a_2,a_3,a_4. Since $\xi = (a_1a_3b) \in A_n$ and $\zeta = (a_1a_3)(a_2a_4) \in N$, $\zeta(\xi\zeta\xi^{-1}) \in N$. But $\zeta(\xi\zeta\xi^{-1}) = (a_1a_3)(a_2a_4)(a_1a_3b)(a_1a_3)(a_2a_4)(a_1ba_3) = (a_1a_3b) \in N$. Hence $N = A_n$ by Lemma 6.12.

Since the cases listed cover all the possibilities, A_n has no proper normal subgroups and hence is simple. ■

Another important subgroup of S_n ($n \geq 3$) is the subgroup D_n generated by $a = (123\cdots n)$ and

$$b = \begin{pmatrix} 1 & 2 & 3 & 4 & 5 & \cdots & i & \cdots & n-1 & n \\ 1 & n & n-1 & n-2 & n-3 & \cdots & n+2-i & \cdots & 3 & 2 \end{pmatrix}$$

$= \prod_{2 \leq i < n+2-i} (i\ \ n+2-i)$. D_n is called the **dihedral group of degree n.** The group D_n is isomorphic to and usually identified with the group of all symmetries of a regular polygon with n sides (Exercise 13). In particular D_4 is (isomorphic to) the group D_4^* of symmetries of the square (see pages 25–26).

Theorem 6.13. *For each* $n \geq 3$ *the dihedral group* D_n *is a group of order* $2n$ *whose generators* a *and* b *satisfy:*

(i) $a^n = (1)$; $b^2 = (1)$; $a^k \neq (1)$ if $0 < k < n$;
(ii) $ba = a^{-1}b$.

Any group G *which is generated by elements* $a,b \in G$ *satisfying (i) and (ii) for some* $n \geq 3$ (*with* $e \in G$ *in place of* (1)) *is isomorphic to* D_n.

SKETCH OF PROOF. Verify that $a,b \in D_n$ as defined above satisfy (i) and (ii), whence $D_n = \langle a,b \rangle = \{a^ib^j \mid 0 \leq i < n; j = 0,1\}$ (see Theorem 2.8). Then verify that the $2n$ elements a^ib^j $(0 \leq i < n; j = 0,1)$ are all distinct (just check their action on 1 and 2), whence $|D_n| = 2n$.

Suppose G is a group generated by $a,b \in G$ and a,b satisfy (i) and (ii) for some $n \geq 3$. By Theorem 2.8 every element of G is a finite product $a^{m_1}b^{m_2}a^{m_3}b^{m_4} \cdots b^{m_k}$ $(m_i \in \mathbf{Z})$. By repeated use of (i) and (ii) any such product may be written in the form a^ib^j with $0 \leq i < n$ and $j = 0,1$ (in particular note that $b^2 = e$ and (ii) imply $b = b^{-1}$ and $ab = ba^{-1}$). Denote the generators of D_n by a_1,b_1 to avoid confusion and verify that the map $f: D_n \to G$ given by $a_1{}^ib_1{}^j \to a^ib^j$ is an epimorphism of groups. To complete the proof we show that f is a monomorphism. Suppose $f(a_1{}^ib_1{}^j) = a^ib^j = e \in G$ with $0 \leq i < n$ and $j = 0,1$. If $j = 1$, then $a^i = b$ and by (ii) $a^{i+1} = a^ia = ba = a^{-1}b = a^{-1}a^i = a^{i-1}$, which implies $a^2 = e$. This contradicts (i) since $n \geq 3$. Therefore $j = 0$ and $e = a^ib^0 = a^i$ with $0 \leq i < n$, which implies $i = 0$ by (i). Thus $f(a_1{}^ib_1{}^j) = e$ implies $a_1{}^ib_1{}^j = a_1{}^0b_1{}^0 = (1)$. Therefore f is a monomorphism by Theorem 2.3. ■

This theorem is an example of a characterization of a group in terms of "generators and relations." A detailed discussion of this idea will be given in Section 9.

EXERCISES

1. Find four different subgroups of S_4 that are isomorphic to S_3 and nine isomorphic to S_2.

2. (a) S_n is generated by the $n - 1$ transpositions $(12), (13), (14), \ldots, (1n)$. [*Hint:* $(1i)(1j)(1i) = (ij)$.]
(b) S_n is generated by the $n - 1$ transpositions $(12), (23), (34), \ldots, (n-1\ n)$. [*Hint:* $(1j) = (1\ j-1)(j-1\ j)(1\ j-1)$; use (a).]

3. If $\sigma = (i_1i_2 \cdots i_r) \in S_n$ and $\tau \in S_n$, then $\tau\sigma\tau^{-1}$ is the r-cycle $(\tau(i_1)\tau(i_2) \cdots \tau(i_r))$.

4. (a) S_n is generated by $\sigma_1 = (12)$ and $\tau = (123 \cdots n)$. [*Hint:* Apply Exercise 3 to σ_1, $\sigma_2 = \tau\sigma_1\tau^{-1}$, $\sigma_3 = \tau\sigma_2\tau^{-1}, \ldots, \sigma_{n-1} = \tau\sigma_{n-2}\tau^{-1}$ and use Exercise 2(b).]
(b) S_n is generated by (12) and $(23 \cdots n)$.

5. Let $\sigma,\tau \in S_n$. If σ is even (odd), then so is $\tau\sigma\tau^{-1}$.

6. A_n is the only subgroup of S_n of index 2. [*Hint:* Show that a subgroup of index 2 must contain all 3-cycles of S_n and apply Lemma 6.11.]

7. Show that $N = \{(1),(12)(34),(13)(24),(14)(23)\}$ is a normal subgroup of S_4 contained in A_4 such that $S_4/N \cong S_3$ and $A_4/N \cong Z_3$.

8. The group A_4 has no subgroup of order 6.

9. For $n \geq 3$ let G_n be the multiplicative group of complex matrices generated by $x = \begin{pmatrix} 0 & 1 \\ 1 & 0 \end{pmatrix}$ and $y = \begin{pmatrix} e^{2\pi i/n} & 0 \\ 0 & e^{-2\pi i/n} \end{pmatrix}$, where $i^2 = -1$. Show that $G_n \cong D_n$. (*Hint:* recall that $e^{2\pi i} = 1$ and $e^{k2\pi i} \neq 1$, where k is real, unless $k \in \mathbf{Z}$.)

10. Let a be the generator of order n of D_n. Show that $\langle a \rangle \lhd D_n$ and $D_n/\langle a \rangle \cong Z_2$.

11. Find all normal subgroups of D_n.

12. The center (Exercise 2.11) of the group D_n is $\langle e \rangle$ if n is odd and isomorphic to Z_2 if n is even.

13. For each $n \geq 3$ let P_n be a regular polygon of n sides (for $n = 3$, P_n is an equilateral triangle; for $n = 4$, a square). A *symmetry* of P_n is a bijection $P_n \to P_n$ that preserves distances and maps adjacent vertices onto adjacent vertices.
(a) The set D_n^* of all symmetries of P_n is a group under the binary operation of composition of functions.
(b) Every $f \varepsilon D_n^*$ is completely determined by its action on the vertices of P_n. Number the vertices consecutively $1,2,\ldots,n$; then each $f \varepsilon D_n^*$ determines a unique permutation σ_f of $\{1,2,\ldots,n\}$. The assignment $f \mapsto \sigma_f$ defines a monomorphism of groups $\varphi : D_n^* \to S_n$.
(c) D_n^* is generated by f and g, where f is a rotation of $2\pi/n$ degrees about the center of P_n and g is a reflection about the "diameter" through the center and vertex 1.
(d) $\sigma_f = (123\cdots n)$ and $\sigma_g = \begin{pmatrix} 1 & 2 & 3 & \cdots & n-1 & n \\ 1 & n & n-1 & \cdots & 3 & 2 \end{pmatrix}$, whence $Im\ \varphi = D_n$ and $D_n^* \cong D_n$.

7. CATEGORIES: PRODUCTS, COPRODUCTS, AND FREE OBJECTS

Since we now have several examples at hand, this is an appropriate time to introduce the concept of a category. Categories will serve as a useful language and provide a general context for dealing with a number of different mathematical situations. They are studied in more detail in Chapter X.

The intuitive idea underlying the definition of a category is that several of the mathematical objects already introduced (sets, groups, monoids) or to be introduced (rings, modules) together with the appropriate maps of these objects (functions for sets; homomorphisms for groups, etc.) have a number of formal properties in common. For example, in each case composition of maps (when defined) is associative; each object A has an identity map $1_A : A \to A$ with certain properties. These notions are formalized in

Definition 7.1. *A* **category** *is a class* $\mathcal{C}$ *of objects (denoted* A,B,C, . . .) *together with*

(i) *a class of disjoint sets, denoted hom*(A,B), *one for each pair of objects in* $\mathcal{C}$;*(an element* f *of hom*(A,B) *is called a* **morphism** *from* A *to* B *and is denoted* f : A → B);
(ii) *for each triple* (A,B,C) *of objects of* $\mathcal{C}$ *a function*

$$hom(B,C) \times hom(A,B) \to hom(A,C);$$

(for morphisms f : A → B, g : B → C, *this function is written* (g,f) ↦ g ∘ f *and* g ∘ f : A → C *is called the* **composite** *of* f *and* g); *all subject to the two axioms:*

(I) Associativity. *If* f : A → B g : B → C, h : C → D *are morphisms of* $\mathcal{C}$, *then* h ∘ (g ∘ f) = (h ∘ g) ∘ f.

(II) Identity. *For each object* B *of* $\mathcal{C}$ *there exists a morphism* $1_B : B \to B$ *such that for any* $f : A \to B$, $g : B \to C$,

$$1_B \circ f = f \quad \text{and} \quad g \circ 1_B = g.$$

In a category $\mathcal{C}$ a morphism $f : A \to B$ is called an **equivalence** if there is in $\mathcal{C}$ a morphism $g : B \to A$ such that $g \circ f = 1_A$ and $f \circ g = 1_B$. The composite of two equivalences, when defined, is an equivalence. If $f : A \to B$ is an equivalence, A and B are said to be **equivalent.**

EXAMPLE. Let $\mathcal{S}$ be the class of all sets; for $A,B \in \mathcal{S}$, hom(A,B) is the set of all functions $f : A \to B$. Then $\mathcal{S}$ is easily seen to be a category. By (13) of Introduction, Section 3, a morphism f of $\mathcal{S}$ is an equivalence if and only if f is a bijection.

EXAMPLE. Let $\mathcal{G}$ be the category whose objects are all groups; hom(A,B) is the set of all group homomorphisms $f : A \to B$. By Theorem 2.3, a morphism f is an equivalence if and only if f is an isomorphism. The category $\mathcal{A}$ of all abelian groups is defined similarly.

EXAMPLE. A (multiplicative) group G can be considered as a category with one object, G. Let hom(G,G) be the set of elements of G; composition of morphisms a,b is simply the composition ab given by the binary operation in G. Every morphism is an equivalence (since every element of G has an inverse). 1_G is the identity element e of G.

EXAMPLE. Let the objects be all partially ordered sets $(S,\leq)$. A morphism $(S,\leq) \to (T,\leq)$ is a function $f : S \to T$ such that for $x,y \in S$, $x \leq y \Rightarrow f(x) \leq f(y)$.

EXAMPLE. Let $\mathcal{C}$ be any category and define the category $\mathcal{D}$ whose objects are all morphisms of $\mathcal{C}$. If $f : A \to B$ and $g : C \to D$ are morphisms of $\mathcal{C}$, then hom(f,g) consists of all pairs (α,β), where $\alpha : A \to C$, $\beta : B \to D$ are morphisms of $\mathcal{C}$ such that the following diagram is commutative:

$$\begin{array}{ccc} A & \xrightarrow{f} & B \\ \alpha \downarrow & & \downarrow \beta \\ C & \xrightarrow{g} & D \end{array}$$

Definition 7.2. *Let* $\mathcal{C}$ *be a category and* $\{A_i \mid i \in I\}$ *a family of objects of* $\mathcal{C}$. *A* **product** *for the family* $\{A_i \mid i \in I\}$ *is an object* P *of* $\mathcal{C}$ *together with a family of morphisms* $\{\pi_i : P \to A_i \mid i \in I\}$ *such that for any object* B *and family of morphisms* $\{\varphi_i : B \to A_i \mid i \in I\}$, *there is a unique morphism* $\varphi : B \to P$ *such that* $\pi_i \circ \varphi = \varphi_i$ *for all* $i \in I$.

A product P of $\{A_i \mid i \in I\}$ is usually denoted $\prod_{i \in I} A_i$. It is sometimes helpful to describe a product in terms of commutative diagrams, especially in the case $I = \{1,2\}$. A product for $\{A_1,A_2\}$ is a diagram (of objects and morphisms) $A_1 \xleftarrow{\pi_1} P \xrightarrow{\pi_2} A_2$ such that: for any other diagram of the form $A_1 \xleftarrow{\varphi_1} B \xrightarrow{\varphi_2} A_2$, there is a unique morphism $\varphi : B \to P$ such that the following diagram is commutative:

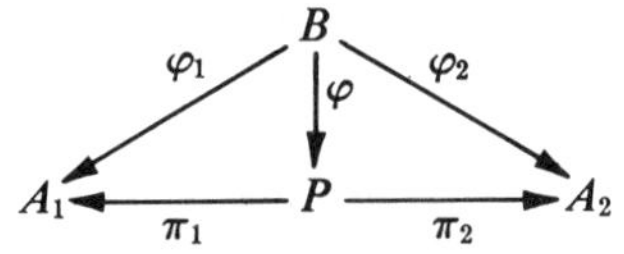

A family of objects in a category need not have a product. In several familiar categories, however, products always exist. For example, in the category of sets the Cartesian product $\prod_{i \varepsilon I} A_i$ is a product of the family $\{A_i \mid i \varepsilon I\}$ by Introduction, Theorem 5.2. In the next section we shall show that products exist in the category of groups.

Theorem 7.3. *If* $(P,\{\pi_i\})$ *and* $(Q,\{\psi_i\})$ *are both products of the family* $\{A_i \mid i \varepsilon I\}$ *of objects of a category* $\mathcal{C}$, *then* P *and* Q *are equivalent.*

PROOF. Since P and Q are both products, there exist morphisms $f: P \to Q$ and $g: Q \to P$ such that the following diagrams are commutative for each $i \varepsilon I$:

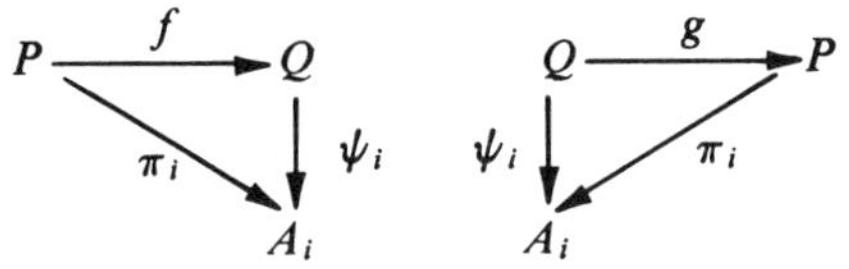

Composing these gives for each $i \varepsilon I$ a commutative diagram:

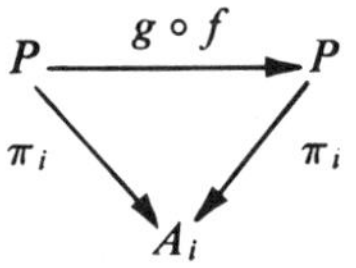

Thus $g \circ f: P \to P$ is a morphism such that $\pi_i \circ (g \circ f) = \pi_i$ for all $i \varepsilon I$. But by the definition of product there is a unique morphism with this property. Since the map $1_P: P \to P$ is also such that $\pi_i \circ 1_P = \pi_i$ for all $i \varepsilon I$, we must have $g \circ f = 1_P$ by uniqueness. Similarly, using the fact that Q is a product, one shows that $f \circ g = 1_Q$. Hence $f: P \to Q$ is an equivalence. ■

Since abstract categories involve only objects and morphisms (no elements), every statement about them has a dual statement, obtained by reversing all the arrows (morphisms) in the original statement. For example, the dual of Definition 7.2 is

Definition 7.4. *A* **coproduct** (*or* **sum**) *for the family* $\{A_i \mid i \varepsilon I\}$ *of objects in a category* $\mathcal{C}$ *is an object* S *of* $\mathcal{C}$, *together with a family of morphisms* $\{\iota_i : A_i \to S \mid i \varepsilon I\}$ *such that for any object* B *and family of morphisms* $\{\psi_i : A_i \to B \mid i \varepsilon I\}$, *there is a unique morphism* $\psi: S \to B$ *such that* $\psi \circ \iota_i = \psi_i$ *for all* $i \varepsilon I$.

There is no uniform notation for coproducts, although $\coprod_{i \in I} A_i$ is sometimes used. In the next two sections we shall discuss coproducts in the category $\mathcal{G}$ of groups and the category $\mathcal{A}$ of abelian groups. The following theorem may be proved by using the "dual argument" to the one used to prove Theorem 7.3 (do it!).

Theorem 7.5. *If* $(S, \{\iota_i\})$ *and* $(S', \{\lambda_i\})$ *are both coproducts for the family* $\{A_i \mid i \in I\}$ *of objects of a category* $\mathcal{C}$, *then* S *and* S′ *are equivalent.*

In several of the categories mentioned above (for example, groups), every object in the category is in fact a set (usually with some additional structure) and every morphism $f : A \to B$ in the category is a function on the "underlying sets" (usually with some other properties as well). We formalize this idea in

Definition 7.6. *A* **concrete category** *is a category* $\mathcal{C}$ *together with a function* σ *that assigns to each object* A *of* $\mathcal{C}$ *a set* $\sigma(A)$ *(called the underlying set of* A*) in such a way that:*

(i) *every morphism* $A \to B$ *of* $\mathcal{C}$ *is a function on the underlying sets* $\sigma(A) \to \sigma(B)$;
(ii) *the identity morphism of each object* A *of* $\mathcal{C}$ *is the identity function on the underlying set* $\sigma(A)$;
(iii) *composition of morphisms in* $\mathcal{C}$ *agrees with composition of functions on the underlying sets.*

EXAMPLES. The category of groups, equipped with the function that assigns to each group its underlying set in the usual sense, is a concrete category. Similarly the categories of abelian groups and partially ordered sets, with the obvious underlying sets, are concrete categories. However, in the third example after Definition 7.1, if the function σ assigns to the group G the usual underlying set G, then the category in question is not a concrete category (since the morphisms are not functions on the set G).

Concrete categories are frequently useful since one has available not only the properties of a category, but also certain properties of sets, subsets, etc. Since in virtually every concrete category we are interested in, the function σ assigns to an object its underlying set in the usual sense (as in the examples above), we shall denote both the object and its underlying set by the same symbol and omit any explicit reference to σ. There is little chance of confusion since we shall be careful in a concrete category $\mathcal{C}$ to distinguish *morphisms* of $\mathcal{C}$ (which are by definition also functions on the underlying sets) and *maps* (functions on the underlying sets, which may not be morphisms of $\mathcal{C}$).

Definition 7.7. *Let* F *be an object in a concrete category* $\mathcal{C}$, X *a nonempty set, and* $i : X \to F$ *a map (of sets).* F *is* **free on the set** X *provided that for any object* A *of* $\mathcal{C}$ *and map (of sets)* $f : X \to A$, *there exists a unique morphism of* $\mathcal{C}$, $\bar{f} : F \to A$, *such that* $\bar{f}i = f$ *(as a map of sets* $X \to A$*).*

The essential fact about a free object F is that *in order to define a morphism with domain F, it suffices to specify the image of the subset $i(X)$ as is seen in the following examples.*

EXAMPLES. Let G be any group and $g \in G$. Then the map $\bar{f} : \mathbf{Z} \to G$ defined by $\bar{f}(n) = g^n$ is easily seen to be the unique homomorphism $\mathbf{Z} \to G$ such that $1 \mapsto g$. Consequently, if $X = \{1\}$ and $i : X \to \mathbf{Z}$ is the inclusion map, then $\mathbf{Z}$ is free on X in the category of groups; (given $f : X \to G$, let $g = f(1)$ and define $\bar{f}$ as above). In other words, to determine a unique homomorphism from $\mathbf{Z}$ to G we need only specify the image of $1 \in \mathbf{Z}$ (that is, the image of $i(X)$). The (additive) group $\mathbf{Q}$ of rational numbers does *not* have this property. It is not difficult to show that there is no nontrivial homomorphism $\mathbf{Q} \to S_3$. Thus for any set X, function $i : X \to \mathbf{Q}$ and function $f : X \to S_3$ with $f(x_1) \neq (1)$ for some $x_1 \in X$, there is no homomorphism $\bar{f} : \mathbf{Q} \to S_3$ with $\bar{f}i = f$.

Theorem 7.8. *If $\mathcal{C}$ is a concrete category, F and F' are objects of $\mathcal{C}$ such that F is free on the set X and F' is free on the set X' and $|X| = |X'|$, then F is equivalent to F'.*

Note that the hypotheses are satisfied when F and F' are both free on the same set X.

PROOF OF 7.8. Since F, F' are free and $|X| = |X'|$, there is a bijection $f : X \to X'$ and maps $i : X \to F$ and $j : X' \to F'$. Consider the map $jf : X \to F'$. Since F is free, there is a morphism $\varphi : F \to F'$ such that the diagram:

$$\begin{array}{ccc} F & \xrightarrow{\varphi} & F' \\ \uparrow{\scriptstyle i} & & \uparrow{\scriptstyle j} \\ X & \xrightarrow[f]{} & X' \end{array}$$

is commutative. Similarly, since the bijection f has an inverse $f^{-1} : X' \to X$ and F' is free, there is a morphism $\psi : F' \to F$ such that:

$$\begin{array}{ccc} F' & \xrightarrow{\psi} & F \\ \uparrow{\scriptstyle j} & & \uparrow{\scriptstyle i} \\ X' & \xrightarrow[f^{-1}]{} & X \end{array}$$

is commutative. Combining these gives a commutative diagram:

$$\begin{array}{ccc} F & \xrightarrow{\psi \circ \varphi} & F \\ \uparrow{\scriptstyle i} & & \uparrow{\scriptstyle i} \\ X & \xrightarrow[f^{-1}f = 1_X]{} & X \end{array}$$

Hence $(\psi \circ \varphi)i = i1_X = i$. But $1_F i = i$. Thus by the uniqueness property of free objects we must have $\psi \circ \varphi = 1_F$. A similar argument shows that $\varphi \circ \psi = 1_{F'}$. Therefore F is equivalent to F'. ■

Products, coproducts, and free objects are all defined via **universal mapping properties** (that is, in terms of the existence of certain uniquely determined morphisms). We have also seen that any two products (or coproducts) for a given family of objects are actually equivalent (Theorems 7.3 and 7.5). Likewise two free objects on the same set are equivalent (Theorem 7.8). Furthermore there is a distinct similarity between the proofs of Theorems 7.3 and 7.8. Consequently it is not surprising that all of the notions just mentioned are in fact special cases of a single concept.

Definition 7.9. *An object* I *in a category* $\mathcal{C}$ *is said to be* **universal** (*or* **initial**) *if for each object* C *of* $\mathcal{C}$ *there exists one and only one morphism* I → C. *An object* T *of* $\mathcal{C}$ *is said to be* **couniversal** (*or* **terminal**) *if for each object* C *of* $\mathcal{C}$ *there exists one and only one morphism* C → T.

We shall show below that products, coproducts, and free objects may be considered as (co)universal objects in suitably chosen categories. However, this characterization is not needed in the sequel. Since universal objects will not be mentioned again (except in occasional exercises) until Sections III.4, III.5, and IV.5, the reader may wish to omit the following material for the present.

Theorem 7.10. *Any two universal [resp. couniversal] objects in a category* $\mathcal{C}$ *are equivalent.*

PROOF. Let I and J be universal objects in $\mathcal{C}$. Since I is universal, there is a unique morphism $f: I \to J$. Similarly, since J is universal, there is a unique morphism $g: J \to I$. The composition $g \circ f: I \to I$ is a morphism of $\mathcal{C}$. But $1_I: I \to I$ is also a morphism of $\mathcal{C}$. The universality of I implies that there is a unique morphism $I \to I$, whence $g \circ f = 1_I$. Similarly the universality of J implies that $f \circ g = 1_J$. Therefore $f: I \to J$ is an equivalence. The proof for couniversal objects is analogous. ■

EXAMPLE. The trivial group $\langle e \rangle$ is both universal and couniversal in the category of groups.

EXAMPLE. Let F be a free object on the set X (with $i: X \to F$) in a concrete category $\mathcal{C}$. Define a new category $\mathcal{D}$ as follows. The objects of $\mathcal{D}$ are all maps of sets $f: X \to A$, where A is (the underlying set of) an object of $\mathcal{C}$. A morphism in $\mathcal{D}$ from $f: X \to A$ to $g: X \to B$ is defined to be a morphism $h: A \to B$ of $\mathcal{C}$ such that the diagram:

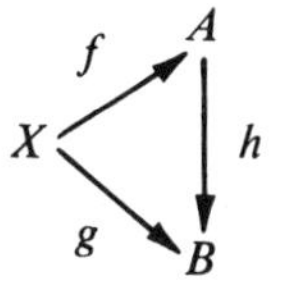

is commutative (that is, $hf = g$). Verify that $1_A : A \to A$ is the identity morphism from f to f in $\mathfrak{D}$ and that h is an equivalence in $\mathfrak{D}$ if and only if h is an equivalence in $\mathcal{C}$. Since F is free on the set X, there is for each map $f : X \to A$ a unique morphism $\bar{f} : F \to A$ such that $\bar{f}i = f$. This is precisely the statement that $i : X \to F$ is a universal object in the category $\mathfrak{D}$.

EXAMPLE. Let $\{A_i \mid i \in I\}$ be a family of objects in a category $\mathcal{C}$. Define a category $\mathcal{E}$ whose objects are all pairs $(B,\{f_i \mid i \in I\})$, where B is an object of $\mathcal{C}$ and for each i, $f_i : B \to A_i$ is a morphism of $\mathcal{C}$. A morphism in $\mathcal{E}$ from $(B,\{f_i \mid i \in I\})$ to $(D,\{g_i \mid i \in I\})$ is defined to be a morphism $h : B \to D$ of $\mathcal{C}$ such that $g_i \circ h = f_i$ for every $i \in I$. Verify that 1_B is the identity morphism from $(B, \{f_i\})$ to $(B, \{f_i\})$ in $\mathcal{E}$ and that h is an equivalence in $\mathcal{E}$ if and only if h is an equivalence in $\mathcal{C}$. If a product exists in $\mathcal{C}$ for the family $\{A_i \mid i \in I\}$ (with maps $\pi_k : \prod A_i \to A_k$ for each $k \in I$), then for every $(B,\{f_i\})$ in $\mathcal{E}$ there exists a unique morphism $f : B \to \prod A_i$ such that $\pi_i \circ f = f_i$ for every $i \in I$. But this says that $(\prod A_i,\{\pi_i \mid i \in I\})$ is a couniversal object in the category $\mathcal{E}$. Similarly the coproduct of a family of objects in $\mathcal{C}$ may be considered as a universal object in an appropriately constructed category.

Since a product $\prod A_i$ of a family $\{A_i \mid i \in I\}$ in a category may be considered as a couniversal object in a suitable category, it follows immediately from Theorem 7.10 that $\prod A_i$ is uniquely determined up to equivalence. Analogous results hold for coproducts and free objects.

EXERCISES

1. A *pointed set* is a pair (S,x) with S a set and $x \in S$. A morphism of pointed sets $(S,x) \to (S',x')$ is a triple (f,x,x'), where $f : S \to S'$ is a function such that $f(x) = x'$. Show that pointed sets form a category.

2. If $f : A \to B$ is an equivalence in a category $\mathcal{C}$ and $g : B \to A$ is the morphism such that $g \circ f = 1_A$, $f \circ g = 1_B$, show that g is unique.

3. In the category $\mathcal{G}$ of groups, show that the group $G_1 \times G_2$ together with the homomorphisms $\pi_1 : G_1 \times G_2 \to G_1$ and $\pi_2 : G_1 \times G_2 \to G_2$ (as in the Example preceding Definition 2.2) is a product for $\{G_1,G_2\}$.

4. In the category $\mathcal{A}$ of abelian groups, show that the group $A_1 \times A_2$, together with the homomorphisms $\iota_1 : A_1 \to A_1 \times A_2$ and $\iota_2 : A_2 \to A_1 \times A_2$ (as in the Example preceding Definition 2.2) is a coproduct for $\{A_1,A_2\}$.

5. Every family $\{A_i \mid i \in I\}$ in the category of sets has a coproduct. [*Hint:* consider $\dot{\bigcup} A_i = \{(a,i) \in (\bigcup A_i) \times I \mid a \in A_i\}$ with $A_i \to \dot{\bigcup} A_i$ given by $a \mapsto (a,i)$. $\dot{\bigcup} A_i$ is called the **disjoint union** of the sets A_i.]

6. (a) Show that in the category $\mathcal{S}_*$ of pointed sets (see Exercise 1) products always exist; describe them.
(b) Show that in $\mathcal{S}_*$ every family of objects has a coproduct (often called a "wedge product"); describe this coproduct.

7. Let F be a free object on a set X $(i : X \to F)$ in a concrete category $\mathcal{C}$. If $\mathcal{C}$ contains an object whose underlying set has at least two elements in it, then i is an injective map of sets.

8. Suppose X is a set and F is a free object on X (with $i: X \to F$) in the category of groups (the existence of F is proved in Section 9). Prove that $i(X)$ is a set of generators for the group F. [*Hint:* If G is the subgroup of F generated by $i(X)$, then there is a homomorphism $\varphi: F \to G$ such that $\varphi i = i$. Show that $F \xrightarrow{\varphi} G \xrightarrow{\subset} F$ is the identity map.]

8. DIRECT PRODUCTS AND DIRECT SUMS

In this section we study products in the category of groups and coproducts in the category of abelian groups. These products and coproducts are important not only as a means of constructing new groups from old, but also for describing the structure of certain groups in terms of particular subgroups (whose structure, for instance, may already be known).

We begin by extending the definition of the direct product $G \times H$ of groups G and H (see page 26) to an arbitrary (possibly infinite) family of groups $\{G_i \mid i \varepsilon I\}$. Define a binary operation on the Cartesian product (of sets) $\prod_{i \varepsilon I} G_i$ as follows. If $f,g \varepsilon \prod_{i \varepsilon I} G_i$ (that is, $f,g: I \to \bigcup_{i \varepsilon I} G_i$ and $f(i),g(i) \varepsilon G_i$ for each i), then $fg: I \to \bigcup_{i \varepsilon I} G_i$ is the function given by $i \to f(i)g(i)$. Since each G_i is a group, $f(i)g(i) \varepsilon G_i$ for every i, whence $fg \varepsilon \prod_{i \varepsilon I} G_i$ by Introduction, Definition 5.1. If we identify $f \varepsilon \prod_{i \varepsilon I} G_i$ with its image $\{a_i\}(a_i = f(i)$ for each $i \varepsilon I)$ as is usually done in the case when I is finite, then the binary operation in $\prod_{i \varepsilon I} G_i$ is the familiar component-wise multiplication: $\{a_i\}\{b_i\} = \{a_ib_i\}$. $\prod_{i \varepsilon I} G_i$, together with this binary operation, is called the **direct product** (or **complete direct sum**) of the family of groups $\{G_i \mid i \varepsilon I\}$. If $I = \{1,2,\dots,n\}$, $\prod_{i \varepsilon I} G_i$ is usually denoted $G_1 \times G_2 \times \cdots \times G_n$ (or in additive notation, $G_1 \oplus G_2 \oplus \cdots \oplus G_n$).

Theorem 8.1. *If* $\{G_i \mid i \varepsilon I\}$ *is a family of groups, then*

(i) *the direct product* $\prod_{i \varepsilon I} G_i$ *is a group;*

(ii) *for each* $k \varepsilon I$, *the map* $\pi_k : \prod_{i \varepsilon I} G_i \to G_k$ *given by* $f \mapsto f(k)$ [*or* $\{a_i\} \mapsto a_k$] *is an epimorphism of groups.*

PROOF. Exercise. ∎

The maps π_k in Theorem 8.1 are called the **canonical projections** of the direct product.

Theorem 8.2. *Let* $\{G_i \mid i \varepsilon I\}$ *be a family of groups and* $\{\varphi_i : H \to G_i \mid i \varepsilon I\}$ *a family of group homomorphisms. Then there is a unique homomorphism* $\varphi : H \to \prod_{i \varepsilon I} G_i$ *such that* $\pi_i\varphi = \varphi_i$ *for all* $i \varepsilon I$ *and this property determines* $\prod_{i \varepsilon I} G_i$ *uniquely up to isomorphism. In other words,* $\prod_{i \varepsilon I} G_i$ *is a product in the category of groups.*

PROOF. By Introduction, Theorem 5.2, the map of sets $\varphi : H \to \prod_{i \in I} G_i$ given by $\varphi(a) = \{\varphi_i(a)\}_{i \in I} \in \prod_{i \in I} G_i$ is the unique function such that $\pi_i\varphi = \varphi_i$ for all $i \in I$. It is easy to verify that φ is a homomorphism. Hence $\prod_{i \in I} G_i$ is a product (in the categorical sense) and therefore determined up to isomorphism (equivalence) by Theorem 7.3. ■

Since the direct product of abelian groups is clearly abelian, it follows that the direct product of abelian groups is a product in the category of abelian groups also.

Definition 8.3. *The* **(external) weak direct product** *of a family of groups* $\{G_i \mid i \in I\}$, *denoted* $\prod_{i \in I}^{w} G_i$, *is the set of all* $f \in \prod_{i \in I} G_i$ *such that* $f(i) = e_i$, *the identity in* G_i, *for all but a finite number of* $i \in I$. *If all the groups* G_i *are (additive) abelian,* $\prod_{i \in I}^{w} G_i$ *is usually called the* **(external) direct sum** *and is denoted* $\sum_{i \in I} G_i$.

If I is finite, the weak direct product coincides with the direct product. In any case, we have

Theorem 8.4. *If* $\{G_i \mid i \in I\}$ *is a family of groups, then*

(i) $\prod_{i \in I}^{w} G_i$ *is a normal subgroup of* $\prod_{i \in I} G_i$;

(ii) *for each* $k \in I$, *the map* $\iota_k : G_k \to \prod_{i \in I}^{w} G_i$ *given by* $\iota_k(a) = \{a_i\}_{i \in I}$, *where* $a_i = e$ *for* $i \neq k$ *and* $a_k = a$, *is a monomorphism of groups;*

(iii) *for each* $i \in I$, $\iota_i(G_i)$ *is a normal subgroup of* $\prod_{i \in I} G_i$.

PROOF. Exercise. ■

The maps ι_k in Theorem 8.4 are called the **canonical injections.**

Theorem 8.5. *Let* $\{A_i \mid i \in I\}$ *be a family of abelian groups (written additively). If* B *is an abelian group and* $\{\psi_i : A_i \to B \mid i \in I\}$ *a family of homomorphisms, then there is a unique homomorphism* $\psi : \sum_{i \in I} A_i \to B$ *such that* $\psi\iota_i = \psi_i$ *for all* $i \in I$ *and this property determines* $\sum_{i \in I} A_i$ *uniquely up to isomorphism. In other words,* $\sum_{i \in I} A_i$ *is a coproduct in the category of abelian groups.*

REMARK. The theorem is false if the word abelian is omitted. The external weak direct product is *not* a coproduct in the category of all groups (Exercise 4).

PROOF OF 8.5. Throughout this proof all groups will be written additively. If $0 \neq \{a_i\} \in \sum A_i$, then only finitely many of the a_i are nonzero, say $a_{i_1}, a_{i_2}, \ldots, a_{i_r}$. Define $\psi : \sum A_i \to B$ by $\psi\{0\} = 0$ and $\psi(\{a_i\}) = \psi_{i_1}(a_{i_1}) + \psi_{i_2}(a_{i_2}) + \cdots + \psi_{i_r}(a_{i_r}) = \sum_{i \in I_0} \psi_i(a_i)$, where I_0 is the set $\{i_1, i_2, \ldots, i_r\} = \{i \in I \mid a_i \neq 0\}$. Since B is abelian, it is readily verified that ψ is a homomorphism and that $\psi\iota_i = \psi_i$ for all $i \in I$. For each $\{a_i\} \in \sum A_i$, $\{a_i\} = \sum_{i \in I_0} \iota_i(a_i)$, I_0 finite as above. If $\xi : \sum A_i \to B$ is a homomorphism such that $\xi\iota_i = \psi_i$ for all i then $\xi(\{a_i\}) = \xi(\sum_{I_0} \iota_i(a)) = \sum_{I_0} \xi\iota_i(a_i) = \sum_{I_0} \psi_i(a_i) = \sum_{I_0} \psi\iota_i(a_i) = \psi(\sum_{I_0} \iota_i(a_i)) = \psi(\{a_i\})$; hence $\xi = \psi$ and ψ is unique. Therefore $\sum A_i$ is a coproduct in the category of abelian groups and hence is determined up to isomorphism (equivalence) by Theorem 7.5. ■

Next we investigate conditions under which a group G is isomorphic to the weak direct product of a family of its subgroups.

Theorem 8.6. *Let* $\{N_i \mid i \in I\}$ *be a family of normal subgroups of a group* G *such that*

(i) $G = \langle \bigcup_{i \in I} N_i \rangle$;

(ii) *for each* $k \in I$, $N_k \cap \langle \bigcup_{i \neq k} N_i \rangle = \langle e \rangle$.

Then $G \cong \prod_{i \in I}^{w} N_i$.

Before proving the theorem we note a special case that is frequently used. Observe that for normal subgroups $N_1, N_2, \ldots, N_r$ of a group G, $\langle N_1 \cup N_2 \cup \cdots \cup N_r \rangle = N_1N_2\cdots N_r = \{n_1n_2\cdots n_r \mid n_i \in N_i\}$ by an easily proved generalization of Theorem 5.3. In additive notation $N_1N_2\cdots N_r$ is written $N_1 + N_2 + \cdots + N_r$. It may be helpful for the reader to keep the following corollary in mind since the proof of the general case is essentially the same.

Corollary 8.7. *If* $N_1, N_2, \ldots, N_r$ *are normal subgroups of a group* G *such that* $G = N_1N_2\cdots N_r$ *and for each* $1 \leq k \leq r$, $N_k \cap (N_1\cdots N_{k-1}N_{k+1}\cdots N_r) = \langle e \rangle$, *then* $G \cong N_1 \times N_2 \times \cdots \times N_r$. ■

PROOF OF THEOREM 8.6. If $\{a_i\} \in \prod^w N_i$, then $a_i = e$ for all but a finite number of $i \in I$. Let I_0 be the finite set $\{i \in I \mid a_i \neq e\}$. Then $\prod_{i \in I_0} a_i$ is a well-defined element of G, since for $a \in N_i$ and $b \in N_j$, $(i \neq j)$, $ab = ba$ by Theorem 5.3(iv). Consequently the map $\varphi : \prod^w N_i \to G$, given by $\{a_i\} \mapsto \prod_{i \in I_0} a_i \in G$ (and $\{e\} \mapsto e$), is a homomorphism such that $\varphi\iota_i(a_i) = a_i$ for $a_i \in N_i$.

Since G is generated by the subgroups N_i, every element a of G is a finite product of elements from various N_i. Since elements of N_i and N_j commute (for $i \neq j$), a can be written as a product $\prod_{i \in I_0} a_i$, where $a_i \in N_i$ and I_0 is some finite subset of I. Hence $\prod_{i \in I_0} \iota_i(a_i) \in \prod^w N_i$ and $\varphi(\prod_{i \in I_0} \iota_i(a_i)) = \prod_{i \in I_0} \varphi\iota_i(a_i) = \prod_{i \in I_0} a_i = a$. Therefore, φ is an epimorphism.

Suppose $\varphi(\{a_i\}) = \prod_{i \in I_0} a_i = e \in G$. Clearly we may assume for convenience of notation that $I_0 = \{1,2,\ldots,n\}$. Then $\prod_{i \in I_0} a_i = a_1a_2\cdots a_n = e$, with $a_i \in N_i$. Hence $a_1^{-1} = a_2\cdots a_n \in N_1 \cap \langle \bigcup_{i \neq 1} N_i \rangle = \langle e \rangle$ and therefore $a_1 = e$. Repetition of this argument shows that $a_i = e$ for all $i \in I$. Hence φ is a monomorphism. ■

Theorem 8.6 motivates

Definition 8.8. *Let* $\{N_i \mid i \in I\}$ *be a family of normal subgroups of a group* G *such that* $G = \langle \bigcup_{i \in I} N_i \rangle$ *and for each* $k \in I$, $N_k \cap \langle \bigcup_{i \neq k} N_i \rangle = \langle e \rangle$. *Then* G *is said to be the* **internal weak direct product** *of the family* $\{N_i \mid i \in I\}$ (*or the* **internal direct sum** *if* G *is (additive) abelian*).

As an easy corollary of Theorem 8.6 we have the following characterization of internal weak direct products.

Theorem 8.9. *Let* $\{N_i \mid i \in I\}$ *be a family of normal subgroups of a group* G. G *is the internal weak direct product of the family* $\{N_i \mid i \in I\}$ *if and only if every nonidentity element of* G *is a unique product* $a_{i_1}a_{i_2}\cdots a_{i_n}$ *with* $i_1, \ldots, i_n$ *distinct elements of* I *and* $e \neq a_{i_k} \in N_{i_k}$ *for each* $k = 1,2,\ldots,n$.

PROOF. Exercise. ■

There is a distinction between internal and external weak direct products. If a group G is the *internal* weak direct product of groups N_i, then by definition each N_i is actually a subgroup of G and G is *isomorphic* to the *external* weak direct product $\prod^w_{i \in I} N_i$. However, the external weak direct product $\prod^w_{i \in I} N_i$ does *not* actually contain the groups N_i, but only isomorphic copies of them (namely the $\iota_i(N_i)$ — see Theorem 8.4 and Exercise 10). Practically speaking, this distinction is not very important and the adjectives "internal" and "external" will be omitted whenever no confusion is possible. In fact we shall use the following notation.

NOTATION. We write $G = \prod^w_{i \in I} N_i$ to indicate that the group G is the internal weak direct product of the family of its subgroups $\{N_i \mid i \in I\}$.

Theorem 8.10. *Let* $\{f_i : G_i \to H_i \mid i \in I\}$ *be a family of homomorphisms of groups and let* $f = \prod f_i$ *be the map* $\prod_{i \in I} G_i \to \prod_{i \in I} H_i$, *given by* $\{a_i\} \mapsto \{f_i(a_i)\}$. *Then* f *is a homomorphism of groups such that* $f(\prod^w_{i \in I} G_i) \subset \prod^w_{i \in I} H_i$, $\text{Ker } f = \prod_{i \in I} \text{Ker } f_i$ *and* $\text{Im } f = \prod_{i \in I} \text{Im } f_i$. *Consequently* f *is a monomorphism [resp. epimorphism] if and only if each* f_i *is.*

PROOF. Exercise. ■

Corollary 8.11. *Let* $\{G_i \mid i \in I\}$ *and* $\{N_i \mid i \in I\}$ *be families of groups such that* N_i *is a normal subgroup of* G_i *for each* $i \in I$.

(i) $\prod_{i \in I} N_i$ *is a normal subgroup of* $\prod_{i \in I} G_i$ *and* $\prod_{i \in I} G_i / \prod_{i \in I} N_i \cong \prod_{i \in I} G_i/N_i$.

(ii) $\prod^w_{i \in I} N_i$ *is a normal subgroup of* $\prod^w_{i \in I} G_i$ *and* $\prod^w_{i \in I} G_i / \prod^w_{i \in I} N_i \cong \prod^w_{i \in I} G_i/N_i$.

PROOF. (i) For each i, let $\pi_i : G_i \to G_i/N_i$ be the canonical epimorphism. By Theorem 8.10, the map $\prod \pi_i : \prod_{i \in I} G_i \to \prod_{i \in I} G_i/N_i$ is an epimorphism with kernel $\prod_{i \in I} N_i$. Therefore $\prod G_i / \prod N_i \cong \prod G_i/N_i$ by the First Isomorphism Theorem. (ii) is similar. ■

EXERCISES

1. S_3 is *not* the direct product of any family of its proper subgroups. The same is true of Z_{p^n} (p prime, $n \geq 1$) and $\mathbf{Z}$.

2. Give an example of groups H_i, K_j such that $H_1 \times H_2 \cong K_1 \times K_2$ and no H_i is isomorphic to any K_j.

3. Let G be an (additive) abelian group with subgroups H and K. Show that $G \cong H \oplus K$ if and only if there are homomorphisms $H \underset{\iota_1}{\overset{\pi_1}{\leftrightarrows}} G \underset{\iota_2}{\overset{\pi_2}{\rightleftarrows}} K$ such that $\pi_1\iota_1 = 1_H$, $\pi_2\iota_2 = 1_K$, $\pi_1\iota_2 = 0$ and $\pi_2\iota_1 = 0$, where 0 is the map sending every element onto the zero (identity) element, and $\iota_1\pi_1(x) + \iota_2\pi_2(x) = x$ for all $x \in G$.

4. Give an example to show that the weak direct product is not a coproduct in the category of all groups. (*Hint:* it suffices to consider the case of two factors $G \times H$.)

5. Let G, H be finite cyclic groups. Then $G \times H$ is cyclic if and only if $(|G|,|H|) = 1$.

6. Every finitely generated abelian group $G \neq \langle e \rangle$ in which every element (except e) has order p (p prime) is isomorphic to $Z_p \oplus Z_p \oplus \cdots \oplus Z_p$ (n summands) for some $n \geq 1$. [*Hint:* Let $A = \{a_1, \ldots, a_n\}$ be a set of generators such that no proper subset of A generates G. Show that $\langle a_i \rangle \cong Z_p$ and $G = \langle a_1 \rangle \times \langle a_2 \rangle \times \cdots \times \langle a_n \rangle$.]

7. Let H,K,N be nontrivial normal subgroups of a group G and suppose $G = H \times K$. Prove that N is in the center of G or N intersects one of H,K nontrivially. Give examples to show that both possibilities can actually occur when G is nonabelian.

8. Corollary 8.7 is false if one of the N_i is not normal.

9. If a group G is the (internal) direct product of its subgroups H,K, then $H \cong G/K$ and $G/H \cong K$.

10. If $\{G_i \mid i \in I\}$ is a family of groups, then $\prod^w G_i$ is the internal weak direct product its subgroups $\{\iota_i(G_i) \mid i \in I\}$.

11. Let $\{N_i \mid i \in I\}$ be a family of subgroups of a group G. Then G is the internal weak direct product of $\{N_i \mid i \in I\}$ if and only if: (i) $a_ia_j = a_ja_i$ for all $i \neq j$ and $a_i \in N_i$, $a_j \in N_j$; (ii) every nonidentity element of G is uniquely a product $a_{i_1} \cdots a_{i_n}$, where $i_1, \ldots, i_n$ are distinct elements of I and $e \neq a_{i_k} \in N_{i_k}$ for each k. [Compare Theorem 8.9.]

12. A normal subgroup H of a group G is said to be a **direct factor** (**direct summand** if G is additive abelian) if there exists a (normal) subgroup K of G such that $G = H \times K$.

(a) If H is a direct factor of K and K is a direct factor of G, then H is normal in G. [Compare Exercise 5.10.]

(b) If H is a direct factor of G, then every homomorphism $H \to G$ may be extended to an endomorphism $G \to G$. However, a monomorphism $H \to G$ need not be extendible to an automorphism $G \to G$.

13. Let $\{G_i \mid i \in I\}$ be a family of groups and $J \subset I$. The map $\alpha : \prod_{j\in J} G_j \to \prod_{i \in I} G_i$ given by $\{a_j\} \mapsto \{b_i\}$, where $b_j = a_j$ for $j \in J$ and $b_i = e_i$ (identity of G_i) for $i \notin J$, is a monomorphism of groups and $\prod_{i\in I} G_i/\alpha(\prod_{j\in J} G_j) \cong \prod_{i \in I-J} G_i$.

14. For $i = 1,2$ let $H_i \lhd G_i$ and give examples to show that each of the following statements may be false: (a) $G_1 \cong G_2$ and $H_1 \cong H_2 \Rightarrow G_1/H_1 \cong G_2/H_2$. (b) $G_1 \cong G_2$ and $G_1/H_1 \cong G_2/H_2 \Rightarrow H_1 \cong H_2$. (c) $H_1 \cong H_2$ and $G_1/H_1 \cong G_2/H_2 \Rightarrow G_1 \cong G_2$.

9. FREE GROUPS, FREE PRODUCTS, AND GENERATORS AND RELATIONS

We shall show that free objects (free groups) exist in the (concrete) category of groups, and we shall use these to develop a method of describing groups in terms of "generators and relations." In addition, we indicate how to construct coproducts (free products) in the category of groups.

Given a set X we shall construct a group F that is free on the set X in the sense of Definition 7.7. If $X = \varnothing$, F is the trivial group $\langle e \rangle$. If $X \neq \varnothing$, let X^{-1} be a set disjoint from X such that $|X| = |X^{-1}|$. Choose a bijection $X \to X^{-1}$ and denote the image of $x \in X$ by x^{-1}. Finally choose a set that is disjoint from $X \cup X^{-1}$ and has exactly one element; denote this element by 1. A **word** on X is a sequence $(a_1,a_2,\ldots)$ with $a_i \in X \cup X^{-1} \cup \{1\}$ such that for some $n \in \mathbf{N}^*$, $a_k = 1$ for all $k \geq n$. The constant sequence $(1,1,\ldots)$ is called the **empty word** and is denoted 1. (This ambiguous notation will cause no confusion.) A word $(a_1,a_2,\ldots)$ on X is said to be **reduced** provided that

(i) for all $x \in X$, x and x^{-1} are not adjacent (that is, $a_i = x \Rightarrow a_{i+1} \neq x^{-1}$ and $a_i = x^{-1} \Rightarrow a_{i+1} \neq x$ for all $i \in \mathbf{N}^*$, $x \in X$) and

(ii) $a_k = 1$ implies $a_i = 1$ for all $i \geq k$.

In particular, the empty word 1 is reduced.

Every nonempty reduced word is of the form $(x_1^{\lambda_1},x_2^{\lambda_2},\ldots,x_n^{\lambda_n},1,1,\ldots)$, where $n \in \mathbf{N}^*$, $x_i \in X$ and $\lambda_i = \pm 1$ (and by convention x^1 denotes x for all $x \in X$). Hereafter we shall denote this word by $x_1^{\lambda_1}x_2^{\lambda_2}\cdots x_n^{\lambda_n}$. This new notation is both more tractable and more suggestive. Observe that the definition of equality of sequences shows that two reduced words $x_1^{\lambda_1}\cdots x_m^{\lambda_m}$ and $y_1^{\delta_1}\cdots y_n^{\delta_n}$ $(x_i,y_j \in X; \lambda_i,\delta_j = \pm 1)$ are equal if and only if both are 1 or $m = n$ and $x_i = y_i$, $\lambda_i = \delta_i$ for each $i = 1,2,\ldots,n$. Consequently the map from X into the set $F(X)$ of all reduced words on X given by $x \mapsto x^1 = x$ is injective. We shall identify X with its image and consider X to be a subset of $F(X)$.

Next we define a binary operation on the set $F = F(X)$ of all reduced words on X. The empty word 1 is to act as an identity element ($w1 = 1w = w$ for all $w \in F$). Informally, we would like to have the product of nonempty reduced words to be given by juxtaposition, that is,

$$(x_1^{\lambda_1}\cdots x_m^{\lambda_m})(y_1^{\delta_1}\cdots y_n^{\delta_n}) = x_1^{\lambda_1}\cdots x_m^{\lambda_m}y_1^{\delta_1}\cdots y_n^{\delta_n}.$$

Unfortunately the word on the right side of the equation may not be reduced (for example, if $x_m^{\lambda_m} = y_1^{-\delta_1}$). Therefore, we define the product to be given by juxtaposition and (if necessary) cancellation of adjacent terms of the form xx^{-1} or $x^{-1}x$; for example $(x_1^1x_2^{-1}x_3^1)(x_3^{-1}x_2^1x_4^1) = x_1^1x_4^1$. More precisely, if $x_1^{\lambda_1}\cdots x_m^{\lambda_m}$ and $y_1^{\delta_1}\cdots y_n^{\delta_n}$ are nonempty reduced words on X with $m \leq n$, let k be the largest integer $(0 \leq k \leq m)$ such that $x_{m-j}^{\lambda_{m-j}} = y_{j+1}^{-\delta_{j+1}}$ for $j = 0,1,\ldots,k-1$. Then define

$$(x_1^{\lambda_1}\cdots x_m^{\lambda_m})(y_1^{\delta_1}\cdots y_n^{\delta_n}) = \begin{cases} x_1^{\lambda_1}\cdots x_{m-k}^{\lambda_{m-k}}y_{k+1}^{\delta_{k+1}}\cdots y_n^{\delta_n} \text{ if } k < m; \\ y_{m+1}^{\delta_{m+1}}\cdots y_n^{\delta_n} \text{ if } k = m < n; \\ 1 \text{ if } k = m = n. \end{cases}$$

If $m > n$, the product is defined analogously. The definition insures that the product of reduced words is a reduced word.

Theorem 9.1. *If X is a nonempty set and $F = F(X)$ is the set of all reduced words on X, then F is a group under the binary operation defined above and $F = \langle X\rangle$.*

The group $F = F(X)$ is called the **free group on the set** X. (The terminology "free" is explained by Theorem 9.2 below.)

SKETCH OF PROOF OF 9.1. Since 1 is an identity element and $x_1^{\delta_1}\cdots x_n^{\delta_n}$ has inverse $x_n^{-\delta_n}\cdots x_1^{-\delta_1}$, we need only verify associativity. This may be done by induction and a tedious examination of cases or by the following more elegant device. For each $x \in X$ and $\delta = \pm 1$ let $|x^\delta|$ be the map $F \to F$ given by $1 \mapsto x^\delta$ and

$$x_1^{\delta_1}\cdots x_n^{\delta_n} \mapsto \begin{cases} x^\delta x_1^{\delta_1}\cdots x_n^{\delta_n} & \text{if } x^\delta \neq x_1^{-\delta_1}; \\ x_2^{\delta_2}\cdots x_n^{\delta_n} & \text{if } x^\delta = x_1^{-\delta_1} \ (= 1 \text{ if } n = 1). \end{cases}$$

Since $|x||x^{-1}| = 1_F = |x^{-1}||x|$, every $|x^\delta|$ is a permutation (bijection) of F (with inverse $|x^{-\delta}|$) by (13) of Introduction, Section 3. Let $A(F)$ be the group of all permutations of F (see page 26) and F_0 the subgroup generated by $\{|x| \mid x \in X\}$. The map $\varphi : F \to F_0$ given by $1 \mapsto 1_F$ and $x_1^{\delta_1}\cdots x_n^{\delta_n} \mapsto |x_1^{\delta_1}|\cdots|x_n^{\delta_n}|$ is clearly a surjection such that $\varphi(w_1w_2) = \varphi(w_1)\varphi(w_2)$ for all $w_i \in F$. Since $1 \mapsto x_1^{\delta_1}\cdots x_n^{\delta_n}$ under the map $|x_1^{\delta_1}|\cdots|x_n^{\delta_n}|$, it follows that φ is injective. The fact that F_0 is a group implies that associativity holds in F and that φ is an isomorphism of groups. Obviously $F = \langle X\rangle$. ∎

Certain properties of free groups are easily derived. For instance if $|X| \geq 2$, then the free group on X is nonabelian ($x,y \in X$ and $x \neq y \Rightarrow x^{-1}y^{-1}xy$ is reduced $\Rightarrow x^{-1}y^{-1}xy \neq 1 \Rightarrow xy \neq yx$). Similarly every element (except 1) in a free group has infinite order (Exercise 1). If $X = \{a\}$, then the free group on X is the infinite cyclic group $\langle a\rangle$ (Exercise 2). A decidedly nontrivial fact is that every subgroup of a free group is itself a free group on some set (see J. Rotman [19]).

Theorem 9.2. *Let F be the free group on a set X and $\iota : X \to F$ the inclusion map. If G is a group and $f : X \to G$ a map of sets, then there exists a unique homomorphism of groups $\bar{f} : F \to G$ such that $\bar{f}\iota = f$. In other words, F is a free object on the set X in the category of groups.*

REMARK. If F' is another free object on the set X in the category of groups (with $\lambda : X \to F'$), then Theorems 7.8 and 9.2 imply that there is an isomorphism $\varphi : F \cong F'$ such that $\varphi\iota = \lambda$. In particular $\lambda(X)$ is a set of generators of F'; this fact may also be proved directly from the definition of a free object.

SKETCH OF PROOF OF 9.2. Define $\bar{f}(1) = e$ and if $x_1^{\delta_1}\cdots x_n^{\delta_n}$ is a nonempty reduced word on X, define $\bar{f}(x_1^{\delta_1}\cdots x_n^{\delta_n}) = f(x_1)^{\delta_1}f(x_2)^{\delta_2}\cdots f(x_n)^{\delta_n}$. Since G is a group and $\delta_i = \pm 1$, the product $f(x_1)^{\delta_1}\cdots f(x_n)^{\delta_n}$ is a well-defined element of G. Verify that $\bar{f}$ is a homomorphism such that $\bar{f}\iota = f$. If $g : F \to G$ is any *homomorphism* such that $g\iota = f$, then $g(x_1^{\delta_1}\cdots x_n^{\delta_n}) = g(x_1^{\delta_1})\cdots g(x_n^{\delta_n}) = g(x_1)^{\delta_1}\cdots g(x_n)^{\delta_n} = g\iota(x_1)^{\delta_1}\cdots g\iota(x_n)^{\delta_n} = f(x_1)^{\delta_1}\cdots f(x_n)^{\delta_n} = \bar{f}(x_1^{\delta_1}\cdots x_n^{\delta_n})$. Therefore $\bar{f}$ is unique. ■

Corollary 9.3. *Every group G is the homomorphic image of a free group.*

PROOF. Let X be a set of generators of G and let F be the free group on the set X. By Theorem 9.2 the inclusion map $X \to G$ induces a homomorphism $\bar{f} : F \to G$ such that $x \mapsto x \in G$. Since $G = \langle X \rangle$, the proof of Theorem 9.2 shows that $\bar{f}$ is an epimorphism. ■

An immediate consequence of Corollary 9.3 and the First Isomorphism Theorem is that any group G is isomorphic to a quotient group F/N, where $G = \langle X \rangle$, F is the free group on X and N is the kernel of the epimorphism $F \to G$ of Corollary 9.3. Therefore, in order to describe G up to isomorphism we need only specify X, F, and N. But F is determined up to isomorphism by X (Theorem 7.8) and N is determined by any subset that generates it as a subgroup of F. Now if $w = x_1^{\delta_1}\cdots x_n^{\delta_n} \in F$ is a generator of N, then under the epimorphism $F \to G$, $w \mapsto x_1^{\delta_1}\cdots x_n^{\delta_n} = e \in G$. The equation $x_1^{\delta_1}\cdots x_n^{\delta_n} = e$ in G is called a **relation** on the generators x_i. Clearly a given group G may be completely described by specifying a set X of generators of G and a suitable set R of relations on these generators. This description is not unique since there are many possible choices of both X and R for a given group G (see Exercises 6 and 9).

Conversely, suppose we are given a set X and a set Y of (reduced) words on the elements of X. Question: does there exist a group G such that G is generated by X and all the relations $w = e$ ($w \in Y$) are valid (where $w = x_1^{\delta_1}\cdots x_n^{\delta_n}$ now denotes a product in G)? We shall see that the answer is yes, providing one allows for the possibility that in the group G the elements of X may not all be distinct. For instance, if $a,b \in X$ and a^1b^{-1} is a (reduced) word in Y, then any group containing a,b and satisfying $a^1b^{-1} = e$ must have $a = b$.

Given a set of "generators" X and a set Y of (reduced) words on the elements of X, we construct such a group as follows. Let F be the free group on X and N the *normal* subgroup of F generated by Y.[3] Let G be the quotient group F/N and identify X with its image in F/N under the map $X \subset F \to F/N$; as noted above, this may involve identifying some elements of X with one another. Then G is a group generated by X (subject to identifications) and by construction all the relations $w = e$ ($w \in Y$) are satisfied ($w = x_1^{\delta_1}\cdots x_n^{\delta_n} \in Y \Rightarrow x_1^{\delta_1}\cdots x_n^{\delta_n} \in N \Rightarrow x_1^{\delta_1}N\cdots x_1^{\delta_n}N = N$; that is, $x_1^{\delta_1}\cdots x_n^{\delta_n} = e$ in $G = F/N$).

[3]The normal subgroup generated by a set $S \subset F$ is the intersection of all normal subgroups of F that contain S; see Exercise 5.2.

Definition 9.4. *Let* X *be a set and* Y *a set of (reduced) words on* X. *A group* G *is said to be the* **group defined by** *the* **generators** x ε X **and relations** w = e (w ε Y) *provided* G ≅ F/N, *where* F *is the free group on* X *and* N *the normal subgroup of* F *generated* by Y. *One says that* (X | Y) *is a* **presentation** *of* G.

The preceding discussion shows that the group defined by given generators and relations always exists. Furthermore it is the largest possible such group in the following sense.

Theorem 9.5. *(Van Dyck) Let* X *be a set,* Y *a set of (reduced) words on* X *and* G *the group defined by the generators* x ε X *and relations* w = e (w ε Y). *If* H *is any group such that* H = ⟨X⟩ *and* H *satisfies all the relations* w = e (w ε Y), *then there is an epimorphism* G → H.

REMARK. The elements of Y are being interpreted as words on X, products in G, and products in H as the context indicates.

PROOF OF 9.5. If F is the free group on X then the inclusion map $X \to H$ induces an epimorphism $\varphi : F \to H$ by Corollary 9.3. Since H satisfies the relations $w = e$ $(w \varepsilon Y)$, $Y \subset \text{Ker } \varphi$. Consequently, the normal subgroup N generated by Y in F is contained in Ker φ. By Corollary 5.8 φ induces an epimorphism $F/N \to H/0$. Therefore the composition $G \cong F/N \to H/0 \cong H$ is an epimorphism. ■

The following examples of groups defined by generators and relations illustrate the sort of *ad hoc* arguments that are often the only way of investigating a given presentation. When convenient, we shall use exponential notation for words (for example, x^2y^{-3} in place of $x^1x^1y^{-1}y^{-1}y^{-1}$).

EXAMPLE. Let G be the group defined by generators a,b and relations $a^4 = e$, $a^2b^{-2} = e$ and $abab^{-1} = e$. Since Q_8, the quaternion group of order 8, is generated by elements a,b satisfying these relations (Exercise 4.14), there is an epimorphism $\varphi : G \to Q_8$ by Theorem 9.5. Hence $|G| \geq |Q_8| = 8$. Let F be the free group on $\{a,b\}$ and N the normal subgroup generated by $\{a^4,a^2b^{-2},abab^{-1}\}$. It is not difficult to show that every element of F/N is of the form a^ib^jN with $0 \leq i \leq 3$ and $j = 0,1$, whence $|G| = |F/N| \leq 8$. Therefore $|G| = 8$ and φ is an isomorphism. Thus the group defined by the given generators and relations is (isomorphic to) Q_8.

EXAMPLE. The group defined by the generators a,b and the relations $a^n = e$ $(3 \leq n \varepsilon \mathbf{N}^*)$, $b^2 = e$ and $abab = e$ (or $ba = a^{-1}b$) is the dihedral group D_n (Exercise 8).

EXAMPLE. The group defined by one generator b and the single relation $b^m = e$ $(m \varepsilon \mathbf{N}^*)$ is Z_m (Exercise 9).

EXAMPLE. The free group F on a set X is the group defined by the generators $x \varepsilon X$ and no relations (recall that $\langle \varnothing \rangle = \langle e \rangle$ by Definition 2.7). The terminology "free" arises from the fact that F is relation-free.

We close this section with a brief discussion of coproducts (free products) in the category of groups. Most of the details are left to the reader since the process is quite similar to the construction of free groups.

Given a family of groups $\{G_i \mid i \in I\}$ we may assume (by relabeling if necessary) that the G_i are mutually disjoint sets. Let $X = \bigcup_{i \in I} G_i$ and let $\{1\}$ be a one-element set disjoint from X. A *word* on X is any sequence $(a_1,a_2,\ldots)$ such that $a_i \in X \cup \{1\}$ and for some $n \in \mathbf{N}^*$, $a_i = 1$ for all $i \geq n$. A word $(a_1,a_2,\ldots)$ is *reduced* provided:

(i) no $a_i \in X$ is the identity element in its group G_j;

(ii) for all $i,j \geq 1$, a_i and a_{i+1} are *not* in the same group G_j;

(iii) $a_k = 1$ implies $a_i = 1$ for all $i \geq k$.

In particular $1 = (1,1,\ldots)$ is reduced. Every reduced word ($\neq 1$) may be written uniquely as $a_1a_2\cdots a_n = (a_1,a_2,\ldots,a_n,1,1,\ldots)$, where $a_i \in X$.

Let $\prod^*_{i \in I} G_i$ (or $G_1 * G_2 * \cdots * G_n$ if I is finite) be the set of all reduced words on X. $\prod^*_{i \in I} G_i$ forms a group, called the **free product** of the family $\{G_i \mid i \in I\}$, under the binary operation defined as follows. 1 is the identity element and the product of two reduced words ($\neq 1$) essentially is to be given by juxtaposition. Since the juxtaposed product of two reduced words may not be reduced, one must make the necessary cancellations and contractions. For example, if $a_i,b_i \in G_i$ for $i = 1,2,3$, then $(a_1a_2a_3)(a_3^{-1}b_2b_1b_3) = a_1c_2b_1b_3 = (a_1,c_2,b_1,b_3,1,1,\ldots)$, where $c_2 = a_2b_2 \in G_2$. Finally, for each $k \in I$ the map $\iota_k : G_k \to \prod^*_{i \in I} G_i$ given by $e \mapsto 1$ and $a \mapsto a = (a,1,1,\ldots)$ is a monomorphism of groups. Consequently, we sometimes identify G_k with its isomorphic image in $\prod^*_{i \in I} G_i$ (for example Exercise 15).

Theorem 9.6. *Let* $\{G_i \mid i \in I\}$ *be a family of groups and* $\prod^*_{i \in I} G_i$ *their free product. If* $\{\psi_i : G_i \to H \mid i \in I\}$ *is a family of group homomorphisms, then there exists a unique homomorphism* $\psi : \prod^*_{i \in I} G_i \to H$ *such that* $\psi\iota_i = \psi_i$ *for all* $i \in I$ *and this property determines* $\prod^*_{i \in I} G_i$ *uniquely up to isomorphism. In other words,* $\prod^*_{i \in I} G_i$ *is a coproduct in the category of groups.*

SKETCH OF PROOF. If $a_1a_2\cdots a_n$ is a reduced word in $\prod^*_{i \in I} G_i$ with $a_k \in G_{i_k}$, define $\psi(a_1\cdots a_n)$ to be $\psi_{i_1}(a_1)\psi_{i_2}(a_2)\cdots\psi_{i_n}(a_n) \in H$. ■

EXERCISES

1. Every nonidentity element in a free group F has infinite order.

2. Show that the free group on the set $\{a\}$ is an infinite cyclic group, and hence isomorphic to $\mathbf{Z}$.

3. Let F be a free group and let N be the subgroup generated by the set $\{x^n \mid x \in F$, n a fixed integer$\}$. Show that $N \lhd F$.

4. Let F be the free group on the set X, and let $Y \subset X$. If H is the smallest normal subgroup of F containing Y, then F/H is a free group.

5. The group defined by generators a,b and relations $a^8 = b^2a^4 = ab^{-1}ab = e$ has order at most 16.

6. The cyclic group of order 6 is the group defined by generators a,b and relations $a^2 = b^3 = a^{-1}b^{-1}ab = e$.

7. Show that the group defined by generators a,b and relations $a^2 = e$, $b^3 = e$ is infinite and nonabelian.

8. The group defined by generators a,b and relations $a^n = e$ $(3 \leq n \in N^*)$, $b^2 = e$ and $abab = e$ is the dihedral group D_n. [See Theorem 6.13.]

9. The group defined by the generator b and the relation $b^m = e$ $(m \in N^*)$ is the cyclic group Z_m.

10. The operation of free product is commutative and associative: for any groups A,B,C, $A * B \cong B * A$ and $A * (B * C) \cong (A * B) * C$.

11. If N is the normal subgroup of $A * B$ generated by A, then $(A * B)/N \cong B$.

12. If G and H each have more than one element, then $G * H$ is an infinite group with center $\langle e \rangle$.

13. A free group is a free product of infinite cyclic groups.

14. If G is the group defined by generators a,b and relations $a^2 = e$, $b^3 = e$, then $G \cong Z_2 * Z_3$. [See Exercise 12 and compare Exercise 6.]

15. If $f : G_1 \to G_2$ and $g : H_1 \to H_2$ are homomorphisms of groups, then there is a unique homomorphism $h : G_1 * H_1 \to G_2 * H_2$ such that $h \mid G_1 = f$ and $h \mid H_1 = g$.

CHAPTER II

THE STRUCTURE OF GROUPS

We continue our study of groups according to the plan outlined in the introduction of Chapter I. The chief emphasis will be on obtaining structure theorems of some depth for certain classes of abelian groups and for various classes of (possibly non-abelian) groups that share some desirable properties with abelian groups. The chapter has three main divisions which are essentially independent of one another, except that results from one may be used as examples or motivation in the others. The interdependence of the sections is as follows.

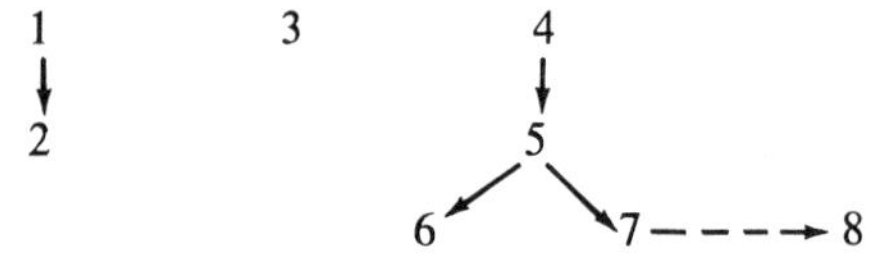

Most of Section 8 is independent of the rest of the chapter.

1. FREE ABELIAN GROUPS

We shall investigate free objects in the category of *abelian* groups. As is the usual custom when dealing with abelian groups **additive notation is used throughout this section.** The following dictionary may be helpful.

ab	$a + b$
a^{-1}	$-a$
e	0
a^n	na
ab^{-1}	$a - b$
HK	$H + K$
aH	$a + H$

$G \times H$ .... $G \oplus H$
$H \vee K$.......................... $H + K$
$\prod_{i \in I}^{w} G_i$ .... $\sum_{i \in I} G_i$
weak direct product .... direct sum

For any group G in additive notation, $(m + n)a = ma + na$ $(a \in G;\ m,n \in \mathbf{Z})$. If the group is abelian, then $m(a + b) = ma + mb$. If X is a nonempty subset of G, then by Theorem I.2.8 the subgroup $\langle X \rangle$ generated by X in additive notation consists of all **linear combinations** $n_1x_1 + n_2x_2 + \cdots + n_kx_k$ $(n_i \in \mathbf{Z},\ x_i \in X)$. In particular, the cyclic group $\langle x \rangle$ is $\{nx \mid n \in \mathbf{Z}\}$.

A **basis** of an abelian group F is a subset X of F such that (i) $F = \langle X \rangle$; and (ii) for distinct $x_1,x_2,\ldots,x_k \in X$ and $n_i \in \mathbf{Z}$,

$$n_1x_1 + n_2x_2 + \cdots + n_kx_k = 0 \quad \Rightarrow \quad n_i = 0 \text{ for every } i.$$

The reader should not be misled by the tempting analogy with bases of vector spaces (Exercise 2).

Theorem 1.1. *The following conditions on an abelian group* F *are equivalent.*

(i) F *has a nonempty basis.*
(ii) F *is the (internal) direct sum of a family of infinite cyclic subgroups.*
(iii) F *is (isomorphic to) a direct sum of copies of the additive group* **Z** *of integers.*
(iv) *There exists a nonempty set* X *and a function* $\iota : X \to F$ *with the following property: given an abelian group* G *and function* $f : X \to G$, *there exists a unique homomorphism of groups* $\bar{f} : F \to G$ *such that* $\bar{f}\iota = f$. *In other words,* F *is a free object in the category of abelian groups.*

An abelian group F that satisfies the conditions of Theorem 1.1 is called a **free abelian group** (on the set X). By definition the trivial group 0 is the free abelian group on the null set $\emptyset$.

SKETCH OF PROOF OF 1.1. (i) $\Rightarrow$ (ii) If X is a basis of F, then for each $x \in X$, $nx = 0$ if and only if $n = 0$. Hence each subgroup $\langle x \rangle$ $(x \in X)$ is infinite cyclic (and normal since F is abelian). Since $F = \langle X \rangle$, we also have $F = \langle \bigcup_{x \in X} \langle x \rangle \rangle$. If for some $z \in X$, $\langle z \rangle \cap \langle \bigcup_{\substack{x \in X \\ x \neq z}} \langle x \rangle \rangle \neq 0$, then for some nonzero $n \in \mathbf{Z}$, $nz = n_1x_1 + \cdots + n_kx_k$ with $z,x_1,\ldots,x_k$ distinct elements of X, which contradicts the fact that X is a basis. Therefore $\langle z \rangle \cap \langle \bigcup_{\substack{x \in X \\ x \neq z}} \langle x \rangle \rangle = 0$ and hence $F = \sum_{x \in X} \langle x \rangle$ by Definition I.8.8.

(ii) $\Rightarrow$ (iii) Theorems I.3.2, I.8.6, and I.8.10.

(iii) $\Rightarrow$ (i) Suppose $F \cong \sum \mathbf{Z}$ and the copies of $\mathbf{Z}$ are indexed by a set X. For each $x \in X$, let θ_x be the element $\{u_i\}$ of $\sum \mathbf{Z}$, where $u_i = 0$ for $i \neq x$, and $u_x = 1$. Verify that $\{\theta_x \mid x \in X\}$ is a basis of $\sum \mathbf{Z}$ and use the isomorphism $F \cong \sum \mathbf{Z}$ to obtain a basis of F.

(i) $\Rightarrow$ (iv) Let X be a basis of F and $\iota : X \to F$ the inclusion map. Suppose we are given a map $f : X \to G$. If $u \in F$, then $u = n_1x_1 + \cdots + n_kx_k$ $(n_i \in \mathbf{Z};\ x_i \in X)$ since X generates F. If $u = m_1x_1 + \cdots + m_kx_k$, $(m_k \in \mathbf{Z})$, then $\sum_{i=1}^{k} (n_i - m_i)x_i = 0$, whence

$n_i = m_i$ for every i since X is a basis. Consequently the map $\bar{f}: F \to G$, given by $\bar{f}(u) = \bar{f}\left(\sum_{i=1}^{k} n_i x_i\right) = n_1 f(x_1) + \cdots + n_k f(x_k)$, is a well-defined function such that $\bar{f}\iota = f$. Since G is abelian $\bar{f}$ is easily seen to be a homomorphism. Since X generates F, any homomorphism $F \to G$ is completely detemined by its action on X. Thus if $g: F \to G$ is a homomorphism such that $g\iota = f$, then for any $x \in X$ $g(x) = g(\iota(x)) = f(x) = \bar{f}(x)$, whence $g = \bar{f}$ and $\bar{f}$ is unique. Therefore, by Definition I.7.7 F is a free object on the set X in the category of abelian groups.

(iv) $\Rightarrow$ (iii). Given $\iota: X \to F$, construct the direct sum $\sum \mathbf{Z}$ with the copies of $\mathbf{Z}$ indexed by X. Let $Y = \{\theta_x \mid x \in X\}$ be a basis of $\sum \mathbf{Z}$ as in the proof of (iii) $\Rightarrow$ (i). The proof of (iii) $\Rightarrow$ (i) $\Rightarrow$ (iv) shows that $\sum \mathbf{Z}$ is a free object on the set Y. Since we clearly have $|X| = |Y|$, $F \cong \sum \mathbf{Z}$ by Theorem I.7.8. ■

Given any set X, the proof of Theorem 1.1 indicates how to construct a free abelian group F with basis X. Simply let F be the direct sum $\sum \mathbf{Z}$, with the copies of $\mathbf{Z}$ indexed by X. As in the proof of (iii) $\Rightarrow$ (i), $\{\theta_x \mid x \in X\}$ is a basis of $F = \sum \mathbf{Z}$, and F is free on the set $\{\theta_x \mid x \in X\}$. Since the map $\iota: X \to F$ given by $x \mapsto \theta_x$ is injective it follows easily that F is free on X in the sense of condition (iv) of Theorem 1.1. In this situation we shall identify X with its image under ι so that $X \subset F$ and the cyclic subgroup $\langle \theta_x \rangle = \{n\theta_x \mid n \in \mathbf{Z}\} = \mathbf{Z}\theta_x$ is written $\langle x \rangle = \mathbf{Z}x$. In this notation $F = \sum_{x \in X} \langle \theta_x \rangle$ is written $F = \sum_{x \in X} \mathbf{Z}x$, and a typical element of F has the form $n_1x_1 + \cdots + n_kx_k$ ($n_i \in \mathbf{Z}$, $x_i \in X$). In particular, $X = \iota(X)$ is a basis of F.

Theorem 1.2. *Any two bases of a free abelian group* F *have the same cardinality.*

The cardinal number of any basis X of the free abelian group F is thus an invariant of F; $|X|$ is called the **rank** of F.

SKETCH OF PROOF OF 1.2. First suppose F has a basis X of finite cardinality n so that $F \cong \mathbf{Z} \oplus \cdots \oplus \mathbf{Z}$ (n summands). For any subgroup G of F verify that $2G = \{2u \mid u \in G\}$ is a subgroup of G. Verify that the restriction of the isomorphism $F \cong \mathbf{Z} \oplus \cdots \oplus \mathbf{Z}$ to $2F$ is an isomorphism $2F \cong 2\mathbf{Z} \oplus \cdots \oplus 2\mathbf{Z}$, whence $F/2F \cong \mathbf{Z}/2\mathbf{Z} \oplus \cdots \oplus \mathbf{Z}/2\mathbf{Z} \cong Z_2 \oplus \cdots \oplus Z_2$ (n summands) by Corollary I.8.11. Therefore $|F/2F| = 2^n$. If Y is another basis of F and r any integer such that $|Y| \geq r$, then a similar argument shows that $|F/2F| \geq 2^r$, whence $2^r \leq 2^n$ and $r \leq n$. It follows that $|Y| = m \leq n$ and $|F/2F| = 2^m$. Therefore $2^m = 2^n$ and $|X| = n = m = |Y|$.

If one basis of F is infinite, then all bases are infinite by the previous paragraph. Consequently, in order to complete the proof it suffices to show that $|X| = |F|$, if X is *any* infinite basis of F. Clearly $|X| \leq |F|$. Let $S = \bigcup_{n \in \mathbf{N}^*} X^n$, where $X^n = X \times \cdots \times X$ (n factors). For each $s = (x_1, \ldots, x_n) \in S$ let G_s be the subgroup $\langle x_1, \ldots, x_n \rangle$. Then $G_s \cong \mathbf{Z}y_1 \oplus \cdots \oplus \mathbf{Z}y_t$ where $y_1, \ldots, y_t$ ($t \leq n$) are the distinct elements of $\{x_1, \ldots, x_n\}$. Therefore, $|G_s| = |\mathbf{Z}^t| = |\mathbf{Z}| = \aleph_0$ by Introduction, Theorem 8.12. Since $F = \bigcup_{s \in S} G_s$, we have $|F| = |\bigcup_{s \in S} G_s| \leq |S|\aleph_0$ by Introduction, Exercise 8.12. But by Introduction, Theorems 8.11 and 8.12, $|S| = |X|$, whence $|F| \leq |X|\aleph_0 = |X|$. Therefore $|F| = |X|$ by the Schroeder-Bernstein Theorem. ■

Proposition 1.3. *Let* F_1 *be the free abelian group on the set* X_1 *and* F_2 *the free abelian group on the set* X_2. *Then* $F_1 \cong F_2$ *if and only if* F_1 *and* F_2 *have the same rank* (*that is,* $|X_1| = |X_2|$).

REMARK. Proposition 1.3 is also true for arbitrary nonabelian free groups (as in Section I.9); see Exercise 12.

SKETCH OF PROOF OF 1.3. If $\alpha : F_1 \cong F_2$, then $\alpha(X_1)$ is a basis of F_2, whence $|X_1| = |\alpha(X_1)| = |X_2|$ by Theorem 1.2. The converse is Theorem I.7.8. ∎

Theorem 1.4. *Every abelian group* G *is the homomorphic image of a free abelian group of rank* $|X|$, *where* X *is a set of generators of* G.

PROOF. Let F be the free abelian group on the set X. Then $F = \sum_{x \in X} \mathbf{Z}x$ and rank $F = |X|$. By Theorem 1.1 the inclusion map $X \to G$ induces a homomorphism $\bar{f} : F \to G$ such that $1x \mapsto x \in G$, whence $X \subset \operatorname{Im} \bar{f}$. Since X generates G we must have $\operatorname{Im} \bar{f} = G$. ∎

We now prove a theorem that will be extremely useful in analyzing the structure of finitely generated abelian groups (Section 2). We shall need

Lemma 1.5. *If* $\{x_1, \ldots, x_n\}$ *is a basis of a free abelian group* F *and* $a \in \mathbf{Z}$, *then for all* $i \neq j$ $\{x_1, \ldots, x_{j-1}, x_j + ax_i, x_{j+1}, \ldots, x_n\}$ *is also a basis of* F.

PROOF. Since $x_j = -ax_i + (x_j + ax_i)$, it follows that $F = \langle x_1, \ldots, x_{j-1}, x_j + ax_i, x_{j+1}, \ldots, x_n \rangle$. If $k_1x_1 + \cdots + k_j(x_j + ax_i) + \cdots + k_nx_n = 0$ $(k_i \in \mathbf{Z})$, then $k_1x_1 + \cdots + (k_i + k_ja)x_i + \cdots + k_jx_j + \cdots + k_nx_n = 0$, which implies that $k_t = 0$ for all t. ∎

Theorem 1.6. *If* F *is a free abelian group of finite rank* n *and* G *is a nonzero subgroup of* F, *then there exists a basis* $\{x_1, \ldots, x_n\}$ *of* F, *an integer* r $(1 \leq r \leq n)$ *and positive integers* $d_1, \ldots, d_r$ *such that* $d_1 \mid d_2 \mid \cdots \mid d_r$ *and* G *is free abelian with basis* $\{d_1x_1, \ldots, d_rx_r\}$.

REMARKS. Every subgroup of a free abelian group of (possibly infinite) rank α is free of rank at most α; see Theorem IV.6.1. The notation "$d_1 \mid d_2 \mid \ldots \mid d_r$" means "$d_1$ divides d_2, d_2 divides d_3, etc."

PROOF OF 1.6. If $n = 1$, then $F = \langle x_1 \rangle \cong \mathbf{Z}$ and $G = \langle d_1x_1 \rangle \cong \mathbf{Z}$ $(d_i \in \mathbf{N}^*)$ by Theorems I.3.5, I.3.1, and I.3.2. Proceeding inductively, assume the theorem is true for all free abelian groups of rank less than n. Let S be the set of all those integers s such that there exists a basis $\{y_1, \ldots, y_n\}$ of F and an element in G of the form $sy_1 + k_2y_2 + \cdots + k_ny_n$ $(k_i \in \mathbf{Z})$. Note that in this case $\{y_2, y_1, y_3, \ldots, y_n\}$ is also a basis of F, whence $k_2 \in S$; similarly $k_j \in S$ for $j = 3,4, \ldots, n$. Since $G \neq 0$, we have $S \neq \varnothing$. Hence S contains a least positive integer d_1 and for some basis $\{y_1, \ldots, y_n\}$

of F there exists $v \in G$ such that $v = d_1y_1 + k_2y_2 + \cdots + k_ny_n$. By the division algorithm for each $i = 2, \ldots, n$, $k_i = d_1q_i + r_i$ with $0 \leq r_i < d_1$, whence $v = d_1(y_1 + q_2y_2 + \cdots + q_ny_n) + r_2y_2 + \cdots + r_ny_n$. Let $x_1 = y_1 + q_2y_2 + \cdots + q_ny_n$; then by Lemma 1.5 $W = \{x_1,y_2, \ldots, y_n\}$ is a basis of F. Since $v \in G$, $r_i < d_1$ and W in *any* order is a basis of F, the minimality of d_1 in S implies that $0 = r_2 = r_3 = \cdots = r_n$ so that $d_1x_1 = v \in G$.

Let $H = \langle y_2,y_3, \ldots, y_n\rangle$. Then H is a free abelian group of rank $n - 1$ such that $F = \langle x_1\rangle \oplus H$. Furthermore we claim that $G = \langle v\rangle \oplus (G \cap H) = \langle d_1x_1\rangle \oplus (G \cap H)$. Since $\{x_1,y_2, \ldots, y_n\}$ is a basis of F, $\langle v\rangle \cap (G \cap H) = 0$. If $u = t_1x_1 + t_2y_2 + \cdots + t_ny_n \in G$ ($t_i \in \mathbf{Z}$), then by the division algorithm $t_1 = d_1q_1 + r_1$ with $0 \leq r_1 < d_1$. Thus G contains $u - q_1v = r_1x_1 + t_2y_2 + \cdots + t_ny_n$. The minimality of d_1 in S implies that $r_1 = 0$, whence $t_2y_2 + \cdots + t_ny_n \in G \cap H$ and $u = q_1v + (t_2y_2 + \cdots + t_ny_n)$. Hence $G = \langle v\rangle + (G \cap H)$, which proves our assertion (Definition I.8.8).

Either $G \cap H = 0$, in which case $G = \langle d_1x_1\rangle$ and the theorem is true or $G \cap H \neq 0$. Then by the inductive assumption there is a basis $\{x_2,x_3, \ldots, x_n\}$ of H and positive integers $r,d_2,d_3, \ldots, d_r$ such that $d_2 \mid d_3 \mid \cdots \mid d_r$ and $G \cap H$ is free abelian with basis $\{d_2x_2, \ldots, d_rx_r\}$. Since $F = \langle x_1\rangle \oplus H$ and $G = \langle d_1x_1\rangle \oplus (G \cap H)$, it follows easily that $\{x_1,x_2, \ldots, x_n\}$ is a basis of F and $\{d_1x_1, \ldots, d_rx_r\}$ is a basis of G. To complete the inductive step of the proof we need only show that $d_1 \mid d_2$. By the division algorithm $d_2 = qd_1 + r_0$ with $0 \leq r_0 < d_1$. Since $\{x_2,x_1 + qx_2,x_3, \ldots, x_n\}$ is a basis of F by Lemma 1.5 and $r_0x_2 + d_1(x_1 + qx_2) = d_1x_1 + d_2x_2 \in G$, the minimality of d_1 in S implies that $r_0 = 0$, whence $d_1 \mid d_2$. ■

Corollary 1.7. *If* G *is a finitely generated abelian group generated by* n *elements, then every subgroup* H *of* G *may be generated by* m *elements with* m $\leq$ n.

The corollary is false if the word *abelian* is omitted (Exercise 8).

PROOF OF 1.7. By Theorem 1.4 there is a free abelian group F of rank n and an epimorphism $\pi : F \to G$. $\pi^{-1}(H)$ is a subgroup of F, and therefore, free of rank $m \leq n$ by Theorem 1.6. The image under π of any basis of $\pi^{-1}(H)$ is a set of at most m elements that generates $\pi(\pi^{-1}(H)) = H$. ■

EXERCISES

1. (a) If G is an abelian group and $m \in \mathbf{Z}$, then $mG = \{mu \mid u \in G\}$ is a subgroup of G.
   (b) If $G \cong \sum_{i \in I} G_i$, then $mG \cong \sum_{i \in I} mG_i$ and $G/mG \cong \sum_{i \in I} G_i/mG_i$.

2. A subset X of an abelian group F is said to be **linearly independent** if $n_1x_1 + \cdots + n_kx_k = 0$ always implies $n_i = 0$ for all i (where $n_i \in \mathbf{Z}$ and $x_1, \ldots, x_k$ are distinct elements of X).
   (a) X is linearly independent if and only if every nonzero element of the subgroup $\langle X\rangle$ may be written uniquely in the form $n_1x_1 + \cdots + n_kx_k$ ($n_i \in \mathbf{Z}, n_i \neq 0$, $x_1, \ldots, x_k$ distinct elements of X).
   (b) If F is free abelian of finite rank n, it is *not* true that every linearly independent subset of n elements is a basis [*Hint:* consider $F = \mathbf{Z}$].

(c) If F is free abelian, it is *not* true that every linearly independent subset of F may be extended to a basis of F.

(d) If F is free abelian, it is *not* true that every generating set of F contains a basis of F. However, if F is also finitely generated by n elements, F has rank $m \leq n$.

3. Let $X = \{a_i \mid i \in I\}$ be a set. Then the free abelian group on X is (isomorphic to) the group defined by the generators X and the relations (in multiplicative notation) $\{a_i a_j a_i^{-1} a_j^{-1} = e \mid i,j \in I\}$.

4. A free abelian group is a free group (Section I.9) if and only if it is cyclic.

5. The direct sum of a family of free abelian groups is a free abelian group. (A direct product of free abelian groups need not be free abelian; see L. Fuchs [13, p. 168].)

6. If $F = \sum_{x \in X} \mathbf{Z}x$ is a free abelian group, and G is the subgroup with basis $X' = X - \{x_0\}$ for some $x_0 \in X$, then $F/G \cong \mathbf{Z}x_0$. Generalize this result to arbitrary subsets X' of X.

7. A nonzero free abelian group has a subgroup of index n for every positive integer n.

8. Let G be the multiplicative group generated by the real matrices $a = \begin{pmatrix} 2 & 0 \\ 0 & 1 \end{pmatrix}$ and $b = \begin{pmatrix} 1 & 1 \\ 0 & 1 \end{pmatrix}$. If H is the set of all matrices in G whose (main) diagonal entries are 1, then H is a subgroup that is *not* finitely generated.

9. Let G be a finitely generated abelian group in which no element (except 0) has finite order. Then G is a free abelian group. [*Hint:* Theorem 1.6.]

10. (a) Show that the additive group of rationals $\mathbf{Q}$ is not finitely generated.

(b) Show that $\mathbf{Q}$ is not free.

(c) Conclude that Exercise 9 is false if the hypothesis "finitely generated" is omitted.

11. (a) Let G be the additive group of all polynomials in x with integer coefficients. Show that G is isomorphic to the group Q^* of all positive rationals (under multiplication). [*Hint:* Use the Fundamental Theorem of Arithmetic to construct an isomorphism.]

(b) The group Q^* is free abelian with basis $\{p \mid p \text{ is prime in } \mathbf{Z}\}$.

12. Let F be the free (not necessarily abelian) group on a set X (as in Section I.9) and G the free group on a set Y. Let F' be the subgroup of F *generated* by $\{aba^{-1}b^{-1} \mid a,b \in F\}$ and similarly for G'.

(a) $F' \lhd F$, $G' \lhd G$ and F/F', G/G' are abelian [see Theorem 7.8 below].

(b) F/F' [resp. G/G'] is a free abelian group of rank $|X|$ [resp. $|Y|$]. [*Hint:* $\{xF' \mid x \in X\}$ is a basis of F/F'.]

(c) $F \cong G$ if and only if $|X| = |Y|$. [*Hint:* if $\varphi : F \cong G$, then φ induces an isomorphism $F/F' \cong G/G'$. Apply Proposition 1.3 and (b). The converse is Theorem I.7.8.]

2. FINITELY GENERATED ABELIAN GROUPS

We begin by proving two different structure theorems for finitely generated abelian groups. A uniqueness theorem (2.6) then shows that each structure theorem provides a set of numerical invariants for a given group (that is, two groups have the same invariants if and only if they are isomorphic). Thus each structure theorem leads to a complete classification (up to isomorphism) of all finitely generated abelian groups. As in Section 1, **all groups are written additively.** Many of the results (though not the proofs) in this section may be extended to certain abelian groups that are not finitely generated; see L. Fuchs [13] or I. Kaplansky [17].

All of the structure theorems to be proved here are special cases of corresponding theorems for finitely generated modules over a principal ideal domain (Section IV.6). Some readers may prefer the method of proof used in Section IV.6 to the one used here, which depends heavily on Theorem 1.6.

Theorem 2.1. *Every finitely generated abelian group* G *is* (*isomorphic to*) *a finite direct sum of cyclic groups in which the finite cyclic summands* (*if any*) *are of orders* $m_1, \ldots, m_t$, *where* $m_1 > 1$ *and* $m_1 \mid m_2 \mid \cdots \mid m_t$.

PROOF. If $G \neq 0$ and G is generated by n elements, then there is a free abelian group F of rank n and an epimorphism $\pi : F \to G$ by Theorem 1.4. If π is an isomorphism, then $G \cong F \cong \mathbf{Z} \oplus \cdots \oplus \mathbf{Z}$ (n summands). If not, then by Theorem 1.6 there is a basis $\{x_1, \ldots, x_n\}$ of F and positive integers $d_1, \ldots, d_r$ such that $1 \leq r \leq n$, $d_1 \mid d_2 \mid \cdots \mid d_r$ and $\{d_1x_1, \ldots, d_rx_r\}$ is a basis of $K = \text{Ker } \pi$. Now $F = \sum_{i=1}^{n} \langle x_i \rangle$ and $K = \sum_{i=1}^{r} \langle d_ix_i \rangle$, where $\langle x_i \rangle \cong \mathbf{Z}$ and under the same isomorphism $\langle d_ix_i \rangle \cong d_i\mathbf{Z} = \{d_iu \mid u \in \mathbf{Z}\}$. For $i = r+1, r+2, \ldots, n$ let $d_i = 0$ so that $K = \sum_{i=1}^{n} \langle d_ix_i \rangle$. Then by Corollaries I.5.7, I.5.8, and I.8.11

$$G \cong F/K = \sum_{i=1}^{n} \langle x_i \rangle \Big/ \sum_{i=1}^{n} \langle d_ix_i \rangle \cong \sum_{i=1}^{n} \langle x_i \rangle / \langle d_ix_i \rangle \cong \sum_{i=1}^{n} \mathbf{Z}/d_i\mathbf{Z}.$$

If $d_i = 1$, then $\mathbf{Z}/d_i\mathbf{Z} = \mathbf{Z}/\mathbf{Z} = 0$; if $d_i > 1$, then $\mathbf{Z}/d_i\mathbf{Z} \cong Z_{d_i}$; if $d_i = 0$, then $\mathbf{Z}/d_i\mathbf{Z} = \mathbf{Z}/0 \cong \mathbf{Z}$. Let $m_1, \ldots, m_t$ be those d_i (in order) such that $d_i \neq 0, 1$ and let s be the number of d_i such that $d_i = 0$. Then

$$G \cong Z_{m_1} \oplus \cdots \oplus Z_{m_t} \oplus (\mathbf{Z} \oplus \cdots \oplus \mathbf{Z}),$$

where $m_1 > 1$, $m_1 \mid m_2 \mid \cdots \mid m_t$ and $(\mathbf{Z} \oplus \cdots \oplus \mathbf{Z})$ has rank s. ■

Theorem 2.2. *Every finitely generated abelian group* G *is* (*isomorphic to*) *a finite direct sum of cyclic groups, each of which is either infinite or of order a power of a prime.*

SKETCH OF PROOF. The theorem is an immediate consequence of Theorem 2.1 and the following lemma. Another proof is sketched in Exercise 4. ■

Lemma 2.3. *If* m *is a positive integer and* $m = p_1^{n_1}p_2^{n_2}\cdots p_t^{n_t}$ ($p_1, \ldots, p_t$ *distinct primes and each* $n_i > 0$), *then* $Z_m \cong Z_{p_1^{n_1}} \oplus Z_{p_2^{n_2}} \oplus \cdots \oplus Z_{p_t^{n_t}}$.

SKETCH OF PROOF. Use induction on the number t of primes in the prime decomposition of m and the fact that

$$Z_{rn} \cong Z_r \oplus Z_n \quad \text{whenever} \quad (r,n) = 1,$$

which we now prove. The element $n = n1 \in Z_{rn}$ has order r (Theorem I.3.4 (vii)), whence $Z_r \cong \langle n1 \rangle < Z_{rn}$ and the map $\psi_1 : Z_r \to Z_{rn}$ given by $k \mapsto nk$ is a monomorphism. Similarly the map $\psi_2 : Z_n \to Z_{rn}$ given by $k \mapsto rk$ is a monomorphism. By the proof of Theorem I.8.5 the map $\psi : Z_r \oplus Z_n \to Z_{rn}$ given by $(x,y) \mapsto \psi_1(x) + \psi_2(y) = nx + ry$ is a well-defined homomorphism. Since $(r,n) = 1$, $ra + nb = 1$ for some $a,b \in \mathbf{Z}$ (Introduction, Theorem 6.5). Hence $k = rak + nbk = \psi(bk,ak)$ for all $k \in Z_{rn}$ and ψ is an epimorphism. Since $|Z_r \oplus Z_n| = rn = |Z_{rn}|$, ψ must also be a monomorphism. ■

Corollary 2.4. *If* G *is a finite abelian group of order* n, *then* G *has a subgroup of order* m *for every positive integer* m *that divides* n.

SKETCH OF PROOF. Use Theorem 2.2 and observe that $G \cong \sum_{i=1}^{k} G_i$ implies that $|G| = |G_1||G_2|\cdots|G_k|$ and for $i \le r, p^{r-i}Z_{p^r} \cong Z_{p^i}$ by Lemma 2.5 (v) below. ■

REMARK. Corollary 2.4 may be false if G is not abelian (Exercise I.6.8).

In Theorem 2.6 below we shall show that the orders of the cyclic summands in the decompositions of Theorems 2.1 and 2.2 are in fact uniquely determined by the group G. First we collect a number of miscellaneous facts about abelian groups that will be used in the proof.

Lemma 2.5. *Let* G *be an abelian group,* m *an integer and* p *a prime integer. Then each of the following is a subgroup of* G:

(i) $mG = \{mu \mid u \in G\}$;
(ii) $G[m] = \{u \in G \mid mu = 0\}$;
(iii) $G(p) = \{u \in G \mid |u| = p^n$ *for some* $n \ge 0\}$;
(iv) $G_t = \{u \in G \mid |u|$ *is finite*$\}$.

In particular there are isomorphisms

(v) $Z_{p^n}[p] \cong Z_p$ $(n \ge 1)$ *and* $p^mZ_{p^n} \cong Z_{p^{n-m}}$ $(m < n)$.

Let H *and* G_i ($i \in I$) *be abelian groups.*

(vi) *If* $g : G \to \sum_{i \in I} G_i$ *is an isomorphism, then the restrictions of* g *to* mG *and* G[m] *respectively are isomorphisms* $mG \cong \sum_{i \in I} mG_i$ *and* $G[m] \cong \sum_{i \in I} G_i[m]$.

(vii) *If* $f : G \to H$ *is an isomorphism, then the restrictions of* f *to* G_t *and* G(p) *respectively are isomorphisms* $G_t \cong H_t$ *and* $G(p) \cong H(p)$.

SKETCH OF PROOF. (i)–(iv) are exercises; the hypothesis that G is abelian is essential (S_3 provides counterexamples for (i)–(iii) and Exercise I.3.5 for (iv)). (v) $p^{n-1} \in Z_{p^n}$ has order p by Theorem I.3.4 (vii), whence $\langle p^{n-1}\rangle \cong Z_p$ and $\langle p^{n-1}\rangle < Z_{p^n}[p]$. If $u \in Z_{p^n}[p]$, then $pu = 0$ in Z_{p^n} so that $pu \equiv 0 \pmod{p^n}$ in **Z**. But $p^n \mid pu$ implies $p^{n-1} \mid u$. Therefore, in Z_{p^n}, $u \in \langle p^{n-1}\rangle$ and $Z_{p^n}[p] < \langle p^{n-1}\rangle$. For the second statement note that $p^m \in Z_{p^n}$ has order p^{n-m} by Theorem I.3.4 (vii). Therefore $p^mZ_{p^n} = \langle p^m\rangle \cong Z_{p^{n-m}}$. (vi) is an exercise. (vii) If $f:G \to H$ is a homomorphism and $x \in G(p)$ has order p^n, then $p^nf(x) = f(p^nx) = f(0) = 0$. Therefore $f(x) \in H(p)$. Hence $f : G(p) \to H(p)$. If f is an isomorphism then the same argument shows that $f^{-1} : H(p) \to G(p)$. Since $ff^{-1} = 1_{H(p)}$ and $f^{-1}f = 1_{G(p)}$, $G(p) \cong H(p)$. The other conclusion of (vii) is proved similarly. ■

If G is an abelian group, then the subgroup G_t defined in Lemma 2.5 is called the **torsion subgroup** of G. If $G = G_t$, then G is said to be a **torsion group.** If $G_t = 0$, then G is said to be **torsion-free.** For a complete classification of all *denumerable* torsion groups, see I. Kaplansky [17].

Theorem 2.6. *Let* G *be a finitely generated abelian group.*

(i) *There is a unique nonnegative integer* s *such that the number of infinite cyclic summands in any decomposition of* G *as a direct sum of cyclic groups is precisely* s;

(ii) *either* G *is free abelian or there is a unique list of (not necessarily distinct) positive integers* $m_1, \ldots, m_t$ *such that* $m_1 > 1$, $m_1 \mid m_2 \mid \cdots \mid m_t$ *and*

$$G \cong Z_{m_1} \oplus \cdots \oplus Z_{m_t} \oplus F$$

with F *free abelian;*

(iii) *either* G *is free abelian or there is a list of positive integers* $p_1^{s_1}, \ldots, p_k^{s_k}$, *which is unique except for the order of its members, such that* $p_1, \ldots, p_k$ *are (not necessarily distinct) primes,* $s_1, \ldots, s_k$ *are (not necessarily distinct) positive integers and*

$$G \cong Z_{p_1^{s_1}} \oplus \cdots \oplus Z_{p_k^{s_k}} \oplus F$$

with F *free abelian.*

PROOF. (i) Any decomposition of G as a direct sum of cyclic groups (and there is at least one by Theorem 2.1) yields an isomorphism $G \cong H \oplus F$, where H is a direct sum of finite cyclic groups (possibly 0) and F is a free abelian group whose rank is precisely the number s of infinite cyclic summands in the decomposition. If $\iota : H \to H \oplus F$ is the canonical injection $(h \mapsto (h,0))$, then clearly $\iota(H)$ is the torsion subgroup of $H \oplus F$. By Lemma 2.5, $G_t \cong \iota(H)$ under the isomorphism $G \cong H \oplus F$. Consequently by Corollary I.5.8, $G/G_t \cong (F \oplus H)/\iota(H) \cong F$. Therefore, any decomposition of G leads to the conclusion that G/G_t is a free abelian group whose rank is the number s of infinite cyclic summands in the decomposition. Since G/G_t does not depend on the particular decomposition and the rank of G/G_t is an invariant by Theorem 1.2, s is uniquely determined.

(iii) Suppose G has two decompositions, say

$$G \cong \sum_{i=1}^{r} Z_{n_i} \oplus F \qquad \text{and} \qquad G = \sum_{j=1}^{d} Z_{k_j} \oplus F',$$

with each n_i, k_j a power of a prime (different primes may occur) and F, F' free abelian; (there is at least one such decomposition by Theorem 2.2). We must show that $r = d$ and (after reordering) $n_i = k_i$ for every i. It is easy to see that the torsion subgroup of $\sum Z_{n_i} \oplus F$ is (isomorphic to) $\sum Z_{n_i}$ and similarly for the other decomposition. Hence $\sum_{i=1}^{r} Z_{n_i} \cong G_t \cong \sum_{j=1}^{d} Z_{k_j}$ by Lemma 2.5. For each prime p, $(\sum Z_{n_i})(p)$ is obviously (isomorphic to) the direct sum of those Z_{n_i} such that n_i is a power of p and similarly for the other decomposition. Since $(\sum Z_{n_i})(p) \cong (\sum Z_{k_i})(p)$ for each prime p by Lemma 2.5, it suffices to assume that $G = G_t$ and each n_i, k_j is a power of a fixed prime p (so that $G = G(p)$). Hence we have

$$\sum_{i=1}^{r} Z_{p^{a_i}} \cong G \cong \sum_{j=1}^{d} Z_{p^{c_j}} (1 \leq a_1 \leq a_2 \leq \cdots \leq a_r;\ 1 \leq c_1 \leq c_2 \leq \cdots \leq c_d).$$

We first show that in any two such decompositions of a group we must have $r = d$. Lemma 2.5 and the first decomposition of G show that

$$G[p] \cong \sum_{i=1}^{r} Z_{p^{a_i}}[p] \cong Z_p \oplus \cdots \oplus Z_p \text{ (r summands)},$$

whence $|G[p]| = p^r$. A similar argument with the second decomposition shows that $|G[p]| = p^d$. Therefore, $p^r = p^d$ and $r = d$.

Let v $(1 \leq v \leq r)$ be the first integer such that $a_i = c_i$ for all $i < v$ and $a_v \neq c_v$. We may assume that $a_v < c_v$. Since $p^{a_v} Z_{p^{a_i}} = 0$ for $a_i \leq a_v$, the first decomposition and Lemma 2.5 imply that

$$p^{a_v} G \cong \sum_{i=1}^{r} p^{a_v} Z_{p^{a_i}} \cong \sum_{i=v+1}^{r} Z_{p^{a_i - a_v}},$$

with $a_{v+1} - a_v \leq a_{v+2} - a_v \leq \cdots \leq a_r - a_v$. Clearly, there are at most $r - (v+1) + 1 = r - v$ nonzero summands. Similarly since $a_i = c_i$ for $i < v$ and $a_v < c_v$ the second decomposition implies that

$$p^{a_v} G \cong \sum_{i=v}^{r} Z_{p^{c_i - a_v}},$$

with $1 \leq c_v - a_v \leq c_{v+1} - a_v \leq \cdots \leq c_r - a_v$. Obviously there are at least $r - v + 1$ nonzero summands. Therefore, we have two decompositions of the group $p^{a_v} G$ as a direct sum of cyclic groups of prime power order and the number of summands in the first decomposition is less than the number of summands in the second. This contradicts the part of the Theorem proved in the previous paragraph (and applied here to $p^{a_v} G$). Hence we must have $a_i = c_i$ for all i.

(ii) Suppose G has two decompositions, say

$$G \cong Z_{m_1} \oplus \cdots \oplus Z_{m_t} \oplus F \text{ and } G \cong Z_{k_1} \oplus \cdots \oplus Z_{k_d} \oplus F'$$

with $m_1 > 1$, $m_1 \mid m_2 \mid \cdots \mid m_t$, $k_1 > 1$, $k_1 \mid k_2 \mid \cdots \mid k_d$ and F, F' free abelian; (one such decomposition exists by Theorem 2.1). Each m_i, k_j has a prime decomposition and by inserting factors of the form p^0 we may assume that the same (distinct) primes $p_1, \ldots, p_r$ occur in all the factorizations, say

$$\begin{array}{ll} m_1 = p_1^{a_{11}}p_2^{a_{12}}\cdots p_r^{a_{1r}} & k_1 = p_1^{c_{11}}p_2^{c_{12}}\cdots p_r^{c_{1r}} \\ m_2 = p_1^{a_{21}}p_2^{a_{22}}\cdots p_r^{a_{2r}} & k_2 = p_1^{c_{21}}p_2^{c_{22}}\cdots p_r^{c_{2r}} \\ \cdot & \cdot \\ \cdot & \cdot \\ \cdot & \cdot \\ m_t = p_1^{a_{t1}}p_2^{a_{t2}}\cdots p_r^{a_{tr}} & k_d = p_1^{c_{d1}}p_2^{c_{d2}}\cdots p_r^{c_{dr}}. \end{array}$$

Since $m_1 \mid m_2 \mid \cdots \mid m_t$, we must have for each j, $0 \leq a_{1j} \leq a_{2j} \leq \cdots \leq a_{tj}$. Similarly $0 \leq c_{1j} \leq c_{2j} \leq \cdots \leq c_{dj}$ for each j. By Lemmas 2.3 and 2.5

$$\sum_{i,j} Z_{p_j^{a_{ij}}} \cong \sum_{i=1}^{t} Z_{m_i} \cong G_t \cong \sum_{i=1}^{d} Z_{k_i} \cong \sum_{i,j} Z_{p_j^{c_{ij}}},$$

where some summands may be zero. It follows that for each $j = 1,2, \ldots, r$

$$\sum_{i=1}^{t} Z_{p_j^{a_{ij}}} \cong G(p_j) \cong \sum_{i=1}^{d} Z_{p_j^{c_{ij}}}.$$

Since $m_1 > 1$, there is some p_j such that $1 \leq a_{1j} \leq \cdots \leq a_{tj}$, whence $\sum_{i=1}^{t} Z_{p_j^{a_{ij}}}$ has t nonzero summands. By (iii) $\sum_{i=1}^{d} Z_{p_j^{c_{ij}}}$ has exactly t nonzero summands, whence $t \leq d$. Similarly $k_1 > 1$ implies that $d \leq t$ and hence $d = t$. By (iii) we now must have $a_{ij} = c_{ij}$ for all i,j, which implies that $m_i = k_i$ for $i = 1,2, \ldots, t$. ■

If G is a finitely generated abelian group, then the uniquely determined integers $m_1, \ldots, m_t$ as in Theorem 2.6 (ii) are called the **invariant factors** of G. The uniquely determined prime powers as in Theorem 2.6 (iii) are called the **elementary divisors** of G.

Corollary 2.7. *Two finitely generated abelian groups G and H are isomorphic if and only if G/G_t and H/H_t have the same rank and G and H have the same invariant factors [resp. elementary divisors].*

PROOF. Exercise. ■

EXAMPLE. All finite abelian groups of order 1500 may be determined up to isomorphism as follows. Since the product of the elementary divisors of a finite group G must be $|G|$ and $1500 = 2^2 \cdot 3 \cdot 5^3$, the only possible families of elementary divisors are $\{2,2,3,5^3\}$, $\{2,2,3,5,5^2\}$, $\{2,2,3,5,5,5\}$, $\{2^2,3,5^3\}$, $\{2^2,3,5,5^2\}$ and $\{2^2,3,5,5,5\}$. Each of these six families determines an abelian group of order 1500 (for example, $\{2,2,3,5^3\}$ determines $Z_2 \oplus Z_2 \oplus Z_3 \oplus Z_{125}$). By Theorem 2.2 every abelian group of order 1500 is isomorphic to one of these six groups and no two of the six are isomorphic by Corollary 2.7.

If the invariant factors $m_1, \ldots, m_t$ of a finitely generated abelian group G are known, then the proof of Theorem 2.6 shows that the elementary divisors of G are the prime powers p^n $(n > 0)$ which appear in the prime factorizations of $m_1, \ldots, m_t$. Conversely if the elementary divisors of G are known, they may be arranged in the following way (after the insertion of some terms of the form p^0 if necessary):

$$\begin{array}{l} p_1^{n_{11}}, p_2^{n_{12}}, \ldots, p_r^{n_{1r}} \\ p_1^{n_{21}}, p_2^{n_{22}}, \ldots, p_r^{n_{2r}} \\ \vdots \qquad\qquad \vdots \\ p_1^{n_{t1}}, p_2^{n_{t2}}, \ldots, p_r^{n_{tr}}. \end{array}$$

where $p_1, \ldots, p_r$ are distinct primes; for each $j = 1,2,\ldots,r$, $0 \leq n_{1j} \leq n_{2j} \leq \cdots \leq n_{tj}$ with some $n_{ij} \neq 0$; and finally $n_{1j} \neq 0$ for some j. By the definition of elementary divisors (Theorem 2.6 (iii)), $G \cong \sum_{i=1}^{t} \sum_{j=1}^{r} Z_{p_j{}^{n_{ij}}} \oplus F$ where F is free abelian (and some finite summands are 0, namely those with $p_j^{n_{ij}} = p_j{}^0 = 1$). For each $i = 1,2,\ldots,t$ let $m_i = p_1^{n_{i1}} p_2^{n_{i2}} \cdots p_r^{n_{ir}}$ (that is, m_i is the product of the ith row in the array above). Since some $n_{1j} \neq 0$, $m_1 > 1$ and by construction $m_1 \mid m_2 \mid \cdots \mid m_t$. By Lemma 2.3 $G \cong \sum_{i=1}^{t} \left(\sum_{j=1}^{r} Z_{p_j{}^{n_{ij}}} \right) \oplus F \cong \sum_{i=1}^{t} Z_{m_i} \oplus F$. Therefore, $m_1, \ldots, m_t$ are the invariant factors of G by Theorem 2.6 (ii).

EXAMPLE. If G is the group $Z_5 \oplus Z_{15} \oplus Z_{25} \oplus Z_{36} \oplus Z_{54}$, then by Lemma 2.3 $G \cong Z_5 \oplus (Z_5 \oplus Z_3) \oplus Z_{25} \oplus (Z_9 \oplus Z_4) \oplus (Z_{27} \oplus Z_2)$. Hence the elementary divisors of G are $2,2^2,3,3^2,3^3,5,5,5^2$ which may be arranged as explained above:

$$\begin{array}{lll} 2^0, & 3, & 5 \\ 2, & 3^2, & 5 \\ 2^2, & 3^3, & 5^2. \end{array}$$

Consequently the invariant factors of G are $1 \cdot 3 \cdot 5 = 15$, $2 \cdot 3^2 \cdot 5 = 90$, and $2^2 \cdot 3^3 \cdot 5^2 = 2700$ so that $G \cong Z_{15} \oplus Z_{90} \oplus Z_{2700}$.

A topic that would fit naturally into this section is the determination of the structure of a finitely generated abelian group which is described by generators and relations. However, since certain matrix techniques are probably the best way to handle this question, it will be treated in the Appendix to Section VII.2. The interested reader should have little or no difficulty in reading that material at the present time.

EXERCISES

1. Show that a finite abelian group that is not cyclic contains a subgroup which is isomorphic to $Z_p \oplus Z_p$ for some prime p.

2. Let G be a finite abelian group and x an element of maximal order. Show that $\langle x \rangle$ is a direct summand of G. Use this to obtain another proof of Theorem 2.1.

3. Suppose G is a finite abelian p-group (Exercise 7) and $x \in G$ has maximal order. If $\bar{y} \in G/\langle x \rangle$ has order p^r, then there is a representative $y \in G$ of the coset $\bar{y}$ such that $|y| = p^r$. [Note that if $|x| = p^t$, then $p^t G = 0$.]

4. Use Exercises 3 and 7 to obtain a proof of Theorem 2.2 which is independent of Theorem 2.1. [*Hint:* If G is a p-group, let $x \in G$ have maximal order; $G/\langle x \rangle$ is a direct sum of cyclics by induction, $G/\langle x \rangle = \langle \bar{x}_1 \rangle \oplus \cdots \oplus \langle \bar{x}_n \rangle$, with $|\bar{x}_i| = p^{r_i}$

and $1 \le r_1 \le r_2 \le \cdots \le r_n$. Choose representatives x_i of $\bar{x}_i$ such that $|x_i| = |\bar{x}_i|$. Show that $G = \langle x_1 \rangle \oplus \cdots \oplus \langle x_n \rangle \oplus \langle x \rangle$ is the desired decomposition.]

5. If G is a finitely generated abelian group such that G/G_t has rank n, and H is a subgroup of G such that H/H_t has rank m, then $m \le n$ and $(G/H)/(G/H)_t$ has rank $n - m$.

6. Let $k, m \in \mathbf{N}^*$. If $(k,m) = 1$, then $kZ_m = Z_m$ and $Z_m[k] = 0$. If $k \mid m$, say $m = kd$, then $kZ_m \cong Z_d$ and $Z_m[k] \cong Z_k$.

7. A (sub)group in which every element has order a power of a fixed prime p is called a p-(sub)group (*note:* $|0| = 1 = p^0$). Let G be an abelian torsion group.
   (a) $G(p)$ is the unique maximum p-subgroup of G (that is, every p-subgroup of G is contained in $G(p)$).
   (b) $G = \sum G(p)$, where the sum is over all primes p such that $G(p) \neq 0$. [*Hint:* If $|u| = p_1^{n_1} \cdots p_t^{n_t}$, let $m_i = |u|/p_i^{n_i}$. There exist $c_i \in \mathbf{Z}$ such that $c_1m_1 + \cdots + c_tm_t = 1$, whence $u = c_1m_1u + \cdots + c_tm_tu$; but $c_im_iu \in G(p_i)$.]
   (c) If H is another abelian torsion group, then $G \cong H$ if and only if $G(p) \cong H(p)$ for all primes p.

8. A finite abelian p-group (Exercise 7) is generated by its elements of maximal order.

9. How many subgroups of order p^2 does the abelian group $Z_{p^3} \oplus Z_{p^2}$ have?

10. (a) Let G be a finite abelian p-group (Exercise 7). Show that for each $n \ge 0$, $p^{n+1}G \cap G[p]$ is a subgroup of $p^nG \cap G[p]$.
    (b) Show that $(p^nG \cap G[p])/(p^{n+1}G \cap G[p])$ is a direct sum of copies of Z_p; let k be the number of copies.
    (c) Write G as a direct sum of cyclics; show that the number k of part (b) is the number of summands of order p^{n+1}.

11. Let G, H, and K be finitely generated abelian groups.
    (a) If $G \oplus G \cong H \oplus H$, then $G \cong H$.
    (b) If $G \oplus H \cong G \oplus K$, then $H \cong K$.
    (c) If G_1 is a free abelian group of rank $\aleph_0$, then $G_1 \oplus \mathbf{Z} \oplus \mathbf{Z} \cong G_1 \oplus \mathbf{Z}$, but $\mathbf{Z} \oplus \mathbf{Z} \not\cong \mathbf{Z}$.
    Note: there exists an infinitely generated denumerable torsion-free abelian group G such that $G \cong G \oplus G \oplus G$, but $G \not\cong G \oplus G$, whence (a) fails to hold with $H = G \oplus G$. See A.L.S. Corner [60]. Also see Exercises 3.11, 3.12, and IV.3.12.

12. (a) What are the elementary divisors of the group $Z_2 \oplus Z_9 \oplus Z_{35}$; what are its invariant factors? Do the same for $Z_{26} \oplus Z_{42} \oplus Z_{49} \oplus Z_{200} \oplus Z_{1000}$.
    (b) Determine up to isomorphism all abelian groups of order 64; do the same for order 96.
    (c) Determine all abelian groups of order n for $n \le 20$.

13. Show that the invariant factors of $Z_m \oplus Z_n$ are (m,n) and $[m,n]$ (the greatest common divisor and the least common multiple) if $(m,n) > 1$ and mn if $(m,n) = 1$.

14. If H is a subgroup of a finite abelian group G, then G has a subgroup that is isomorphic to G/H.

15. Every finite subgroup of $\mathbf{Q}/\mathbf{Z}$ is cyclic [see Exercises I.3.7 and 7].

3. THE KRULL-SCHMIDT THEOREM

The groups **Z** and Z_{p^n} (p prime) are indecomposable, in the sense that neither is a direct sum of two of its proper subgroups (Exercise I.8.1). Consequently, Theorems 2.2 and 2.6(iii) may be rephrased as: every finitely generated abelian group is the direct sum of a finite number of indecomposable groups and these indecomposable summands are uniquely determined up to isomorphism. We shall now extend this result to a large class of (not necessarily abelian) groups.[1]

For the remainder of this chapter we return to the use of *multiplicative notation* for an arbitrary group.

Definition 3.1. *A group* G *is* **indecomposable** *if* G $\neq \langle e \rangle$ *and* G *is not the (internal) direct product of two of its proper subgroups.*

Thus G is indecomposable if and only if $G \neq \langle e \rangle$ and $G \cong H \times K$ implies $H = \langle e \rangle$ or $K = \langle e \rangle$ (Exercise 1).

EXAMPLES. Every simple group (for example, A_n, $n \neq 4$) is indecomposable. However indecomposable groups need not be simple: **Z**, Z_{p^n} (p prime) and S_n are indecomposable but not simple (Exercises 2 and I.8.1).

Definition 3.2. *A group* G *is said to satisfy the* **ascending chain condition** (ACC) *on [normal] subgroups if for every chain* $G_1 < G_2 < \cdots$ *of [normal] subgroups of* G *there is an integer* n *such that* $G_i = G_n$ *for all* $i \geq n$. G *is said to satisfy the* **descending chain condition** (DCC) *on [normal] subgroups if for every chain* $G_1 > G_2 > \cdots$ *of [normal] subgroups of* G *there is an integer* n *such that* $G_i = G_n$ *for all* $i \geq n$.

EXAMPLES. Every finite group satisfies both chain conditions. **Z** satisfies the ascending but not the descending chain condition (Exercise 5) and $Z(p^\infty)$ satisfies the descending but not the ascending chain condition (Exercise 13).

Theorem 3.3. *If a group* G *satisfies either the ascending or descending chain condition on normal subgroups, then* G *is the direct product of a finite number of indecomposable subgroups.*

SKETCH OF PROOF. Suppose G is not a finite direct product of indecomposable subgroups. Let S be the set of all normal subgroups H of G such that H is a direct factor of G (that is, $G = H \times T_H$ for some subgroup T_H of G) and H is not a finite direct product of indecomposable subgroups. Clearly $G \in S$. If $H \in S$, then H is not indecomposable, whence there must exist *proper* subgroups K_H and J_H of H such that $H = K_H \times J_H$ ($= J_H \times K_H$). Furthermore, one of these groups, say K_H, must lie in S (in particular, K_H is normal in G by Exercise I.8.12). Let $f: S \to S$ be the map

[1]The results of this section are not needed in the sequel.

defined by $f(H) = K_H$. By the Recursion Theorem 6.2 of the Introduction (with $f_n = f$ for all n) there exists a function $\varphi : \mathbf{N} \to S$ such that

$$\varphi(0) = G \quad \text{and} \quad \varphi(n+1) = f(\varphi(n)) = K_{\varphi(n)}\ (n \geq 0).$$

If we denote $\varphi(n)$ by G_n, then we have a sequence of subgroups $G_0, G_1, G_2, \ldots,$ of G (all of which are in S) such that

$$G = G_0;\ G_1 = K_{G_0};\ G_2 = K_{G_1};\ \ldots;\ G_{n+1} = K_{G_n};\ \ldots$$

By construction each G_i is normal in G and

$$G \underset{\neq}{>} G_1 \underset{\neq}{>} G_2 \underset{\neq}{>} G_3 \underset{\neq}{>} \cdots.$$

If G satisfies the descending chain condition on normal subgroups, this is a contradiction. Furthermore a routine inductive argument shows that for each $n \geq 1$, $G = G_n \times J_{G_{n-1}} \times J_{G_{n-2}} \times \cdots \times J_{G_0}$ with each J_{G_i} a proper subgroup of G. Consequently, there is a properly ascending chain of normal subgroups:

$$J_{G_0} \underset{\neq}{<} J_{G_1} \times J_{G_0} \underset{\neq}{<} J_{G_2} \times J_{G_1} \times J_{G_0} \underset{\neq}{<} \cdots.$$

If G satisfies the ascending chain condition on normal subgroups, this is a contradiction. ■

In order to determine conditions under which the decomposition of Theorem 3.3 is unique, several definitions and lemmas are needed. An endomorphism f of a group G is called a **normal endomorphism** if $af(b)a^{-1} = f(aba^{-1})$ for all $a,b \in G$.

Lemma 3.4. *Let* G *be a group that satisfies the ascending [resp. descending] chain condition on normal subgroups and* f *a [normal] endomorphism of* G. *Then* f *is an automorphism if and only if* f *is an epimorphism [resp. monomorphism].*

PROOF. Suppose G satisfies the ACC and f is an epimorphism. The ascending chain of normal subgroups $\langle e \rangle < \text{Ker } f < \text{Ker } f^2 < \cdots$ (where $f^k = ff\cdots f$) must become constant, say $\text{Ker } f^n = \text{Ker } f^{n+1}$. Since f is an epimorphism, so is f^n. If $a \in G$ and $f(a) = e$, then $a = f^n(b)$ for some $b \in G$ and $e = f(a) = f^{n+1}(b)$. Consequently $b \in \text{Ker } f^{n+1} = \text{Ker } f^n$, which implies that $a = f^n(b) = e$. Therefore, f is a monomorphism and hence an automorphism.

Suppose G satisfies the DCC and f is a monomorphism. For each $k \geq 1$, $\text{Im } f^k$ is normal in G since f is a normal endomorphism. Consequently, the descending chain $G > \text{Im } f > \text{Im } f^2 > \cdots$ must become constant, say $\text{Im } f^n = \text{Im } f^{n+1}$. Thus for any $a \in G$, $f^n(a) = f^{n+1}(b)$ for some $b \in G$. Since f is a monomorphism, so is f^n and hence $f^n(a) = f^{n+1}(b) = f^n(f(b))$ implies $a = f(b)$. Therefore f is an epimorphism, and hence an automorphism. ■

Lemma 3.5. (*Fitting*) *If* G *is a group that satisfies both the ascending and descending chain conditions on normal subgroups and* f *is a normal endomorphism of* G, *then for some* $n \geq 1$, $G = \textit{Ker}\ f^n \times \textit{Im}\ f^n$.

PROOF. Since f is a normal endomorphism each $\text{Im } f^k$ ($k \geq 1$) is normal in G. Hence we have two chains of normal subgroups:

$$G > \text{Im } f > \text{Im } f^2 > \cdots \quad \text{and} \quad \langle e \rangle < \text{Ker } f < \text{Ker } f^2 < \cdots.$$

By hypothesis there is an n such that $\text{Im } f^k = \text{Im } f^n$ and $\text{Ker } f^k = \text{Ker } f^n$ for all $k \geq n$. Suppose $a \in \text{Ker } f^n \cap \text{Im } f^n$. Then $a = f^n(b)$ for some $b \in G$ and $f^{2n}(b) = f^n(f^n(b)) = f^n(a) = e$. Consequently, $b \in \text{Ker } f^{2n} = \text{Ker } f^n$ so that $a = f^n(b) = e$. Therefore, $\text{Ker } f^n \cap \text{Im } f^n = \langle e \rangle$. For any $c \in G$, $f^n(c) \in \text{Im } f^n = \text{Im } f^{2n}$, whence $f^n(c) = f^{2n}(d)$ for some $d \in G$. Thus $f^n(cf^n(d^{-1})) = f^n(c)f^{2n}(d^{-1}) = f^n(c)f^{2n}(d)^{-1} = f^n(c)f^n(c)^{-1} = e$ and hence $cf^n(d^{-1}) \in \text{Ker } f^n$. Since $c = (cf^n(d^{-1}))f^n(d)$, we conclude that $G = (\text{Ker } f^n)(\text{Im } f^n)$. Therefore $G = \text{Ker } f^n \times \text{Im } f^n$ by Definition I.8.8. ■

An endomorphism f of a group G is said to be **nilpotent** if there exists a positive integer n such that $f^n(g) = e$ for all $g \in G$.

Corollary 3.6. *If G is an indecomposable group that satisfies both the ascending and descending chain conditions on normal subgroups and f is a normal endomorphism of G, then either f is nilpotent or f is an automorphism.*

PROOF. For some $n \geq 1$, $G = \text{Ker } f^n \times \text{Im } f^n$ by Fitting's Lemma. Since G is indecomposable either $\text{Ker } f^n = \langle e \rangle$ or $\text{Im } f^n = \langle e \rangle$. The latter implies that f is nilpotent. If $\text{Ker } f^n = \langle e \rangle$, then $\text{Ker } f = \langle e \rangle$ and f is a monomorphism. Therefore, f is an automorphism by Lemma 3.4. ■

If G is a group and f, g are functions from G to G, then $f + g$ denotes the function $G \to G$ given by $a \mapsto f(a)g(a)$. Verify that the set of all functions from G to G is a group under $+$ (with identity the map $0_G:G \to G$ given by $a \mapsto e$ for all $a \in G$). When f and g are *endomorphisms* of G, $f + g$ need *not* be an endomorphism (Exercise 7). So the subset of endomorphisms is not in general a subgroup.

Corollary 3.7. *Let G ($\neq \langle e \rangle$) be an indecomposable group that satisfies both the ascending and descending chain conditions on normal subgroups. If $f_1, \ldots, f_n$ are normal nilpotent endomorphisms of G such that every $f_{i_1} + \cdots + f_{i_r}$ ($1 \leq i_1 < i_2 < \cdots < i_r \leq n$) is an endomorphism, then $f_1 + f_2 + \cdots + f_n$ is nilpotent.*

SKETCH OF PROOF. Since each $f_{i_1} + \cdots + f_{i_r}$ is an endomorphism that is normal (Exercise 8(c)), the proof will follow by induction once the case $n = 2$ is established. If $f_1 + f_2$ is not nilpotent, it is an automorphism by Corollary 3.6. Verify that the inverse g of $f_1 + f_2$ is a normal automorphism. If $g_1 = f_1g$ and $g_2 = f_2g$, then $1_G = g_1 + g_2$ and for all $x \in G$, $x^{-1} = (g_1 + g_2)(x^{-1}) = g_1(x^{-1})g_2(x^{-1})$. Hence $x = [g_1(x^{-1})g_2(x^{-1})]^{-1} = g_2(x)g_1(x) = (g_2 + g_1)(x)$ and $1_G = g_2 + g_1$. Therefore, $g_1 + g_2 = g_2 + g_1$ and $g_1(g_1 + g_2) = g_11_G = 1_Gg_1 = (g_1 + g_2)g_1$, which implies that $g_1g_2 = g_2g_1$. A separate inductive argument now shows that for each $m \geq 1$,

$$(g_1 + g_2)^m = \sum_{i=0}^{m} c_i g_1^i g_2^{m-i} \quad (c_i \in \mathbf{Z}),$$

where the c_i are the binomial coefficients (see Theorem III.1.6) and c_ih means $h + h + \cdots + h$ (c_i summands). Since each f_i is nilpotent, $g_i = f_ig$ has a nontrivial kernel, whence g_i is nilpotent by Corollary 3.6. Therefore for large enough m and all $a \in G$, $(g_1 + g_2)^m(a) = \sum_{i=0}^{m} c_ig_1{}^ig_2^{m-i}(a) = \prod_{i=0}^{m} e^{c_i} = e$. But this contradicts the facts that $g_1 + g_2 = 1_G$ and $G \neq \langle e \rangle$. ∎

The next theorem will make use of the following facts. If a group G is the internal direct product of its subgroups $G_1, \ldots, G_s$ then by the proof of Theorem I.8.6 there is an isomorphism $\varphi : G_1 \times \cdots \times G_s \cong G$ given by $(g_1, \ldots, g_s) \mapsto g_1g_2 \cdots g_s$. Consequently, every element of G may be written uniquely as a product $g_1g_2 \cdots g_s$ ($g_i \in G_i$). For each i the map $\pi_i : G \to G_i$ given by $g_1g_2 \cdots g_s \mapsto g_i$ is a well-defined epimorphism; (it is the composition of φ^{-1} with the canonical projection $G_1 \times \cdots \times G_s \to G_i$.) We shall refer to the maps π_i as the canonical epimorphisms associated with the internal direct product $G = G_1 \times \cdots \times G_s$.

Theorem 3.8. (*Krull-Schmidt*) *Let* G *be a group that satisfies both the ascending and descending chain conditions on normal subgroups. If* $G = G_1 \times G_2 \times \cdots \times G_s$ *and* $G = H_1 \times H_2 \times \cdots \times H_t$ *with each* G_i, H_j *indecomposable, then* $s = t$ *and after reindexing* $G_i \cong H_i$ *for every* i *and for each* $r < t$.

$$G = G_1 \times \cdots \times G_r \times H_{r+1} \times \cdots \times H_t.$$

REMARKS. G has at least one such decomposition by Theorem 3.3. The uniqueness statement here is stronger than simply saying that the indecomposable factors are determined up to isomorphism.

SKETCH OF PROOF OF 3.8. Let $P(0)$ be the statement $G = H_1 \times \cdots \times H_t$. For $1 \leq r \leq \min(s,t)$ let $P(r)$ be the statement: there is a reindexing of $H_1, \ldots, H_t$ such that $G_i \cong H_i$ for $i = 1,2, \ldots, r$ and $G = G_1 \times \cdots \times G_r \times H_{r+1} \times \cdots \times H_t$ (or $G = G_1 \times \cdots \times G_t$ if $r = t$). We shall show inductively that $P(r)$ is true for all r such that $0 \leq r \leq \min(s,t)$. $P(0)$ is true by hypothesis, and so we assume that $P(r-1)$ is true: after some reindexing $G_i \cong H_i$ for $i = 1, \ldots, r-1$ and $G = G_1 \times \cdots \times G_{r-1} \times H_r \times \cdots \times H_t$. Let $\pi_1, \ldots, \pi_s$ [resp. $\pi_1', \ldots, \pi_t'$] be the canonical epimorphisms associated with the internal direct product

$$G = G_1 \times \cdots \times G_s \text{ [resp. } G = G_1 \times \cdots \times G_{r-1} \times H_r \times \cdots \times H_t]$$

as in the paragraph preceding the statement of the Theorem. Let λ_i [resp. λ_i'] be the inclusion maps sending the ith factor into G. For each i let $\varphi_i = \lambda_i\pi_i : G \to G$ and let $\psi_i = \lambda_i'\pi_i' : G \to G$. Verify that the following identities hold:

$$\begin{array}{lll} \varphi_i \mid G_i = 1_{G_i}; & \varphi_i\varphi_i = \varphi_i; & \varphi_i\varphi_j = 0_G \ (i \neq j)^2; \\ \psi_1 + \cdots + \psi_t = 1_G; & \psi_i\psi_i = \psi_i; & \psi_i\psi_j = 0_G \ (i \neq j); \\ \operatorname{Im} \varphi_i = G_i; & \operatorname{Im} \psi_i = G_i \ (i < r); & \operatorname{Im} \psi_i = H_i \ (i \geq r). \end{array}$$

It follows that $\varphi_r\psi_i = 0_G$ for all $i < r$ (since $\psi_i(x) \in G_i$ so that $\varphi_r\psi_i(x) = \varphi_r1_{G_i}\psi_i(x) = \varphi_r\varphi_i\psi_i(x) = e$).

[2]See the paragraph preceding Corollary 3.7.

The preceding identities show that $\varphi_r = \varphi_r 1_G = \varphi_r(\psi_1 + \cdots + \psi_t) = \varphi_r\psi_r + \cdots + \varphi_r\psi_t$. Every "sum" of distinct $\varphi_r\psi_i$ is a normal endomorphism (Exercises 8, 9). Since $\varphi_r \mid G_r = 1_{G_r}$ is a (normal) automorphism of G_r and G_r satisfies both chain conditions on normal subgroups (Exercise 6), Corollaries 3.6 and 3.7 imply that $\varphi_r\psi_j \mid G_r$ is an automorphism of $G_r \neq \langle e \rangle$ for some j ($r \leq j \leq t$). Therefore, for every $n \geq 1$ $(\varphi_r\psi_j)^{n+1}$ is also an automorphism of G. Consequently, since $G_r \neq \langle e \rangle$ and $(\varphi_r\psi_j)^{n+1} = \varphi_r(\psi_j\varphi_r)^n\psi_j$ for all $n \geq 1$, the normal endomorphism $\psi_j\varphi_r \mid H_j : H_j \to H_j$ cannot be nilpotent. Since H_j satisfies both chain conditions (Exercise 6), $\psi_j\varphi_r \mid H_j$ must be an automorphism of H_j by Corollary 3.7. Therefore $\psi_j \mid G_r : G_r \to H_j$ is an isomorphism and so is $\varphi_r \mid H_j : H_j \to G_r$. Reindex the H_k so that we may assume $j = r$ and $G_r \cong H_r$. We have proved the first half of statement $P(r)$.

Since $G = G_1 \times \cdots \times G_{r-1} \times H_r \times \cdots \times H_t$ by the induction hypothesis the subgroup $G_1G_2 \cdots G_{r-1}H_{r+1} \cdots H_t$ is the internal direct product $G_1 \times \cdots G_{r-1} \times H_{r+1} \times \cdots \times H_t$. Observe that for $j < r$, $\psi_r(G_j) = \psi_r\psi_j(G) = \langle e \rangle$ and for $j > r$, $\psi_r(H_j) = \psi_r\psi_j(G) = \langle e \rangle$, whence $\psi_r(G_1 \cdots G_{r-1}H_{r+1} \cdots H_t) = \langle e \rangle$. Since $\psi_r \mid G_r$ is an isomorphism, we must have $G_r \cap (G_1 \cdots G_{r-1}H_{r+1} \cdots H_t) = \langle e \rangle$. It follows that the group $G^* = G_1 \cdots G_{r-1}G_rH_{r+1} \cdots H_t$ is the internal direct product

$$G^* = G_1 \times \cdots \times G_r \times H_{r+1} \times \cdots \times H_t.$$

Define a map $\theta : G \to G$ as follows. Every element $g \in G$ may be written $g = g_1 \cdots g_{r-1}h_r \cdots h_t$ with $g_i \in G_i$ and $h_j \in H_j$. Let $\theta(g) = g_1 \cdots g_{r-1}\varphi_r(h_r)h_{r+1} \cdots h_t$. Clearly $\operatorname{Im}\theta = G^*$. θ is a monomorphism (see Theorem I.8.10) that is easily seen to be normal. Therefore θ is an automorphism by Lemma 3.4 so that $G = \operatorname{Im}\theta = G^* = G_1 \times \cdots G_r \times H_{r+1} \times \cdots \times H_t$. This proves the second part of $P(r)$ and completes the inductive argument. Therefore, after reindexing $G_i \cong H_i$ for $0 \leq i \leq \min(s,t)$. If $\min(s,t) = s$, then $G_1 \times \cdots \times G_s = G = G_1 \times \cdots \times G_s \times H_{s+1} \times \cdots \times H_t$, and if $\min(s,t) = t$, then $G_1 \times \cdots \times G_s = G = G_1 \times \cdots \times G_t$. Since $G_i \neq \langle e \rangle$, $H_j \neq \langle e \rangle$ for all i,j, we must have $s = t$ in either case. ■

EXERCISES

1. A group G is indecomposable if and only if $G \neq \langle e \rangle$ and $G \cong H \times K$ implies $H = \langle e \rangle$ or $K = \langle e \rangle$.

2. S_n is indecomposable for all $n \geq 2$. [*Hint:* If $n \geq 5$ Theorems I.6.8 and I.6.10 and Exercise I.8.7 may be helpful.]

3. The additive group **Q** is indecomposable.

4. A nontrivial homomorphic image of an indecomposable group need not be indecomposable.

5. (a) **Z** satisfies the ACC but not the DCC on subgroups.
   (b) Every finitely generated abelian group satisfies the ACC on subgroups.

6. Let H,K be normal subgroups of a group G such that $G = H \times K$.
   (a) If N is a normal subgroup of H, then N is normal in G (compare Exercise I.5.10).
   (b) If G satisfies the ACC or DCC on normal subgroups, then so do H and K.

7. If f and g are endomorphisms of a group G, then $f + g$ need not be an endomorphism. [*Hint:* Let $a = (123)$, $b = (132) \varepsilon S_3$ and define $f(x) = axa^{-1}$, $g(x) = bxb^{-1}$.]

8. Let f and g be normal endomorphisms of a group G.
   (a) fg is a normal endomorphism.
   (b) $H \lhd G$ implies $f(H) \lhd G$.
   (c) If $f + g$ is an endomorphism, then it is normal.

9. Let $G = G_1 \times \cdots \times G_n$. For each i let $\lambda_i : G_i \to G$ be the inclusion map and $\pi_i : G \to G_i$ the canonical projection (see page 59). Let $\varphi_i = \lambda_i\pi_i$. Then the "sum" $\varphi_{i_1} + \cdots + \varphi_{i_k}$ of any k ($1 \leq k \leq n$) *distinct* φ_i is a normal endomorphism of G.

10. Use the Krull-Schmidt Theorem to prove Theorems 2.2 and 2.6 (iii) for *finite* abelian groups.

11. If G and H are groups such that $G \times G \cong H \times H$ and G satisfies both the ACC and DCC on normal subgroups, then $G \cong H$ [see Exercise 2.11].

12. If G,H,K and J are groups such that $G \cong H \times K$ and $G \cong H \times J$ and G satisfies both the ACC and DCC on normal subgroups, then $K \cong J$ [see Exercise 2.11]

13. For each prime p the group $Z(p^\infty)$ satisfies the descending but not the ascending chain condition on subgroups [see Exercise I.3.7].

4. THE ACTION OF A GROUP ON A SET

The techniques developed in this section will be used in the following sections to develop structure theorems for (nonabelian finite) groups.

Definition 4.1. *An* **action** *of a group* G *on a set* S *is a function* $G \times S \to S$ (*usually denoted by* $(g,x) \mapsto gx$) *such that for all* $x \varepsilon S$ *and* $g_1,g_2 \varepsilon G$:

$$ex = x \quad \textit{and} \quad (g_1g_2)x = g_1(g_2x).$$

When such an action is given, we say that G **acts on the set** S.

Since there may be many different actions of a group G on a given set S, the notation gx is ambiguous. In context, however, this will not cause any difficulty.

EXAMPLE. An action of the symmetric group S_n on the set $I_n = \{1,2.\ldots,n\}$ is given by $(\sigma,x) \to \sigma(x)$.

EXAMPLES. Let G be a group and H a subgroup. An action of the group H on the *set* G is given by $(h,x) \mapsto hx$, where hx is the product in G. The action of $h \varepsilon H$ on G is called a (left) **translation.** If K is another subgroup of G and S is the set of all left cosets of K in G, then H acts on S by translation: $(h,xK) \mapsto hxK$.

EXAMPLES. Let H be a subgroup of a group G. An action of H on the set G is given by $(h,x) \mapsto hxh^{-1}$; to avoid confusion with the product in G, this action of $h \varepsilon H$ is always denoted hxh^{-1} and *not* hx. This action of $h \varepsilon H$ on G is called **conjugation** by

h and the element hxh^{-1} is said to be a **conjugate** of x. If K is any subgroup of G and $h \varepsilon H$, then hKh^{-1} is a subgroup of G isomorphic to K (Exercise I.5.6). Hence H acts on the set S of all subgroups of G by conjugation: $(h,K) \mapsto hKh^{-1}$. The group hKh^{-1} is said to be **conjugate** to K.

Theorem 4.2. *Let* G *be a group that acts on a set* S.

(i) *The relation on* S *defined by*

$$x \sim x' \Leftrightarrow gx = x' \quad \textit{for some} \quad g \varepsilon G$$

is an equivalence relation.

(ii) *For each* $x \varepsilon S$, $G_x = \{g \varepsilon G \mid gx = x\}$ *is a subgroup of* G.

PROOF. Exercise. ■

The equivalence classes of the equivalence relation of Theorem 4.2(i) are called the **orbits**[3] of G on S; the orbit of $x \varepsilon S$ is denoted $\bar{x}$. The subgroup G_x is called variously the **subgroup fixing** x, the **isotropy group** of x or the **stabilizer** of x.

EXAMPLES. If a group G acts on itself by conjugation, then the orbit $\{gxg^{-1} \mid g \varepsilon G\}$ of $x \varepsilon G$ is called the **conjugacy class** of x. If a subgroup H acts on G by conjugation the isotropy group $H_x = \{h \varepsilon H \mid hxh^{-1} = x\} = \{h \varepsilon H \mid hx = xh\}$ is called the **centralizer of x in H** and is denoted $C_H(x)$. If $H = G$, $C_G(x)$ is simply called the **centralizer of x.** If H acts by conjugation on the set S of all subgroups of G, then the subgroup of H fixing $K \varepsilon S$, namely $\{h \varepsilon H \mid hKh^{-1} = K\}$, is called the **normalizer of K in H** and denoted $N_H(K)$. The group $N_G(K)$ is simply called the **normalizer of K.** Clearly every subgroup K is normal in $N_G(K)$; K is normal in G if and only if $N_G(K) = G$.

Theorem 4.3. *If a group* G *acts on a set* S, *then the cardinal number of the orbit of* $x \varepsilon S$ *is the index* $[G : G_x]$.

PROOF. Let $g,h \varepsilon G$. Since

$$gx = hx \Leftrightarrow g^{-1}hx = x \Leftrightarrow g^{-1}h \varepsilon G_x \Leftrightarrow hG_x = gG_x,$$

it follows that the map given by $gG_x \mapsto gx$ is a well-defined bijection of the set of cosets of G_x in G onto the orbit $\bar{x} = \{gx \mid g \varepsilon G\}$. Hence $[G : G_x] = |\bar{x}|$. ■

Corollary 4.4. *Let* G *be a finite group and* K *a subgroup of* G.

(i) *The number of elements in the conjugacy class of* $x \varepsilon G$ *is* $[G : C_G(x)]$, *which divides* $|G|$;

(ii) *if* $\bar{x}_1, \ldots, \bar{x}_n$ $(x_i \varepsilon G)$ *are the distinct conjugacy classes of* G, *then*

[3] This agrees with our previous use of the term orbit in the proof of Theorem I.6.3, where the special case of a cyclic subgroup $\langle\sigma\rangle$ of S_n acting on the set I_n was considered.

$$|G| = \sum_{i=1}^{n} [G : C_G(x_i)];$$

(iii) *the number of subgroups of* G *conjugate to* K *is* $[G : N_G(K)]$, *which divides* $|G|$.

PROOF. (i) and (iii) follow immediately from the preceding Theorem and Lagrange's Theorem I.4.6. Since conjugacy is an equivalence relation on G (Theorem 4.2), G is the disjoint union of the conjugacy classes $\bar{x}_1, \dots, \bar{x}_n$, whence (ii) follows from (i). ■

The equation $|G| = \sum_{i=1}^{n} [G : C_G(x_i)]$ as in Corollary 4.4 (ii) is called the **class equation** of the finite group G.

Theorem 4.5. *If a group* G *acts on a set* S, *then this action induces a homomorphism* $G \to A(S)$, *where* $A(S)$ *is the group of all permutations of* S.

PROOF. If $g \in G$, define $\tau_g : S \to S$ by $x \mapsto gx$. Since $x = g(g^{-1}x)$ for all $x \in S$, τ_g is surjective. Similarly $gx = gy$ $(x,y \in S)$ implies $x = g^{-1}(gx) = g^{-1}(gy) = y$, whence τ_g is injective and therefore a bijection (permutation of S). Since $\tau_{gg'} = \tau_g\tau_{g'}$: $S \to S$ for all $g,g' \in G$, the map $G \to A(S)$ given by $g \mapsto \tau_g$ is a homomorphism. ■

Corollary 4.6. (*Cayley*) *If* G *is a group, then there is a monomorphism* $G \to A(G)$. *Hence every group is isomorphic to a group of permutations. In particular every finite group is isomorphic to a subgroup of* S_n *with* $n = |G|$.

PROOF. Let G act on itself by left translation and apply Theorem 4.5 to obtain a homomorphism $\tau : G \to A(G)$. If $\tau(g) = \tau_g = 1_G$, then $gx = \tau_g(x) = x$ for all $x \in G$. In particular $ge = e$, whence $g = e$ and τ is a monomorphism. To prove the last statement note if $|G| = n$, then $A(G) \cong S_n$. ■

Recall that if G is a group, then the set Aut G of all automorphisms of G is a group with composition of functions as binary operation (Exercise I.2.15).

Corollary 4.7. *Let* G *be a group.*

(i) *For each* $g \in G$, *conjugation by* g *induces an automorphism of* G.

(ii) *There is a homomorphism* $G \to Aut\ G$ *whose kernel is* $C(G) = \{g \in G \mid gx = xg$ *for all* $x \in G\}$.

PROOF. (1) If G acts on itself by conjugation, then for each $g \in G$, the map $\tau_g : G \to G$ given by $\tau_g(x) = gxg^{-1}$ is a bijection by the proof of Theorem 4.5. It is easy to see that τ_g is also a homomorphism and hence an automorphism. (ii) Let G act on itself by conjugation. By (i) the image of the homomorphism $\tau : G \to A(G)$ of Theorem 4.5 is contained in Aut G. Clearly

$$g \in \text{Ker}\ \tau \Leftrightarrow \tau_g = 1_G \Leftrightarrow gxg^{-1} = \tau_g(x) = x \quad \text{for all} \quad x \in G.$$

But $gxg^{-1} = x$ if and only if $gx = xg$, whence Ker $\tau = C(G)$. ■

The automorphism τ_g of Corollary 4.7(i) is called the **inner automorphism** induced by g. The normal subgroup $C(G) = \text{Ker}\,\tau$ is called the **center** of G. An element $g \in G$ is in $C(G)$ if and only if the conjugacy class of g consists of g alone. Thus if G is finite and $x \in C(G)$, then $[G : C_G(x)] = 1$ (Corollary 4.4). Consequently, the class equation of G (Corollary 4.4(ii)) may be written

$$|G| = |C(G)| + \sum_{i=1}^{m} [G : C_G(x_i)],$$

where $\bar{x}_1, \ldots, \bar{x}_m$ ($x_i \in G - C(G)$) are distinct conjugacy classes of G and each $[G : C_G(x_i)] > 1$.

Proposition 4.8. *Let* H *be a subgroup of a group* G *and let* G *act on the set* S *of all left cosets of* H *in* G *by left translation. Then the kernel of the induced homomorphism* G $\to$ A(S) *is contained in* H.

PROOF. The induced homomorphism $G \to A(S)$ is given by $g \mapsto \tau_g$, where $\tau_g : S \to S$ and $\tau_g(xH) = gxH$. If g is in the kernel, then $\tau_g = 1_S$ and $gxH = xH$ for all $x \in G$; in particular for $x = e$, $geH = eH = H$, which implies $g \in H$. ■

Corollary 4.9. *If* H *is a subgroup of index* n *in a group* G *and no nontrivial normal subgroup of* G *is contained in* H, *then* G *is isomorphic to a subgroup of* S_n.

PROOF. Apply Proposition 4.8 to H; the kernel of $G \to A(S)$ is a normal subgroup of G contained in H and must therefore be $\langle e \rangle$ by hypothesis. Hence, $G \to A(S)$ is a monomorphism. Therefore G is isomorphic to a subgroup of the group of all permutations of the n left cosets of H, and this latter group is clearly isomorphic to S_n. ■

Corollary 4.10. *If* H *is a subgroup of a finite group* G *of index* p, *where* p *is the smallest prime dividing the order of* G, *then* H *is normal in* G.

PROOF. Let S be the set of all left cosets of H in G. Then $A(S) \cong S_p$ since $[G : H] = p$. If K is the kernel of the homomorphism $G \to A(S)$ of Proposition 4.8, then K is normal in G and contained in H. Furthermore G/K is isomorphic to a subgroup of S_p. Hence $|G/K|$ divides $|S_p| = p!$ But every divisor of $|G/K| = [G : K]$ must divide $|G| = |K|\,[G : K]$. Since no number smaller than p (except 1) can divide $|G|$, we must have $|G/K| = p$ or 1. However $|G/K| = [G : K] = [G : H][H : K] = p[H : K] \geq p$. Therefore $|G/K| = p$ and $[H : K] = 1$, whence $H = K$. But K is normal in G. ■

EXERCISES

1. Let G be a group and A a normal abelian subgroup. Show that G/A operates on A by conjugation and obtain a homomorphism $G/A \to \text{Aut}\,A$.
2. If H,K are subgroups of G such that $H \lhd K$, show that $K < N_G(H)$.

3. If a group G contains an element a having exactly two conjugates, then G has a proper normal subgroup $N \neq \langle e \rangle$.

4. Let H be a subgroup of G. The centralizer of H is the set $C_G(H) = \{g \in G \mid hg = gh$ for all $h \in H\}$. Show that $C_G(H)$ is a subgroup of $N_G(H)$.

5. If H is a subgroup of G, the factor group $N_G(H)/C_G(H)$ (see Exercise 4) is isomorphic to a subgroup of Aut H.

6. Let G be a group acting on a set S containing at least two elements. Assume that G is transitive; that is, given any $x,y \in S$, there exists $g \in G$ such that $gx = y$. Prove
   (a) for $x \in S$, the orbit $\bar{x}$ of x is S;
   (b) all the stabilizers G_x (for $x \in S$) are conjugate;
   (c) if G has the property: $\{g \in G \mid gx = x$ for all $x \in S\} = \langle e \rangle$ (which is the case if $G < S_n$ for some n and $S = \{1,2, \ldots, n\}$) and if $N \lhd G$ and $N < G_x$ for some $x \in S$, then $N = \langle e \rangle$;
   (d) for $x \in S$, $|S| = [G : G_x]$; hence $|S|$ divides $|G|$.

7. Let G be a group and let In G be the set of all inner automorphisms of G. Show that In G is a normal subgroup of Aut G.

8. Exhibit an automorphism of Z_6 that is *not* an inner automorphism.

9. If $G/C(G)$ is cyclic, then G is abelian.

10. Show that the center of S_4 is $\langle e \rangle$; conclude that S_4 is isomorphic to the group of all inner automorphisms of S_4.

11. Let G be a group containing an element a not of order 1 or 2. Show that G has a nonidentity automorphism. [*Hint:* Exercise I.2.2 and Corollary 4.7.]

12. Any finite group is isomorphic to a subgroup of A_n for some n.

13. If a group G contains a subgroup ($\neq G$) of finite index, it contains a normal subgroup ($\neq G$) of finite index.

14. If $|G| = pn$, with $p > n$, p prime, and H is a subgroup of order p, then H is normal in G.

15. If a normal subgroup N of order p (p prime) is contained in a group G of order p^n, then N is in the center of G.

5. THE SYLOW THEOREMS

Nonabelian finite groups are vastly more complicated than finite abelian groups, which were completely classified (up to isomorphism) in Section 2. The Sylow Theorems are a basic first step in understanding the structure of an arbitrary finite group.

Our motivation is the question: if a positive integer m divides the order of a group G, does G have a subgroup of order m? This is the converse of Lagrange's Theorem I.4.6. It is true for abelian groups (Corollary 2.4) but may be false for arbitrary groups (Exercise I.6.8). We first consider the special case when m is prime (Theorem 5.2), and then proceed to the first Sylow Theorem which states that the answer to our question is affirmative whenever m is a power of a prime. This leads naturally to a discussion of subgroups of maximal prime power order (second and third Sylow Theorems).

Lemma 5.1. *If a group* H *of order* p^n (p *prime*) *acts on a finite set* S *and if* $S_0 = \{x \varepsilon S \mid hx = x$ *for all* $h \varepsilon H\}$, *then* $|S| \equiv |S_0|$ (*mod* p).

REMARK. This lemma (and the notation S_0) will be used frequently in the sequel.[4]

PROOF OF 5.1. An orbit $\bar{x}$ contains exactly one element if and only if $x \varepsilon S_0$. Hence S can be written as a disjoint union $S = S_0 \cup \bar{x}_1 \cup \bar{x}_2 \cup \cdots \cup \bar{x}_n$, with $|\bar{x}_i| > 1$ for all i. Hence $|S| = |S_0| + |\bar{x}_1| + |\bar{x}_2| + \cdots + |\bar{x}_n|$. Now $p \mid |\bar{x}_i|$ for each i since $|\bar{x}_i| > 1$ and $|\bar{x}_i| = [H : H_{x_i}]$ divides $|H| = p^n$. Therefore $|S| \equiv |S_0| \pmod{p}$. ■

Theorem 5.2. (*Cauchy*) *If* G *is a finite group whose order is divisible by a prime* p, *then* G *contains an element of order* p.

PROOF. (J. H. McKay) Let S be the set of p-tuples of group elements $\{(a_1,a_2,\ldots,a_p) \mid a_i \varepsilon G$ and $a_1a_2\cdots a_p = e\}$. Since a_p is uniquely determined as $(a_1a_2\cdots a_{p-1})^{-1}$, it follows that $|S| = n^{p-1}$, where $|G| = n$. Since $p \mid n$, $|S| \equiv 0 \pmod{p}$. Let the group Z_p act on S by cyclic permutation; that is, for $k \varepsilon Z_p$, $k(a_1,a_2,\ldots,a_p) = (a_{k+1},a_{k+2},\ldots,a_p,a_1,\ldots,a_k)$. Verify that $(a_{k+1},a_{k+2},\ldots,a_k) \varepsilon S$ (use the fact that in a group $ab = e$ implies $ba = (a^{-1}a)(ba) = a^{-1}(ab)a = e$). Verify that for $0,k,k' \varepsilon Z_p$ and $x \varepsilon S$, $0x = x$ and $(k + k')x = k(k'x)$ (additive notation for a group action on a set!). Therefore the action of Z_p on S is well defined.

Now $(a_1,\ldots,a_p) \varepsilon S_0$ if and only if $a_1 = a_2 = \cdots = a_p$; clearly $(e,e,\ldots,e) \varepsilon S_0$ and hence $|S_0| \neq 0$. By Lemma 5.1, $0 \equiv |S| \equiv |S_0| \pmod{p}$. Since $|S_0| \neq 0$ there must be at least p elements in S_0; that is, there is $a \neq e$ such that $(a,a,\ldots,a) \varepsilon S_0$ and hence $a^p = e$. Since p is prime, $|a| = p$. ■

A group in which every element has order a power (≥ 0) of some fixed prime p is called a **p-group.** If H is a subgroup of a group G and H is a p-group, H is said to be a **p-subgroup** of G. In particular $\langle e \rangle$ is a p-subgroup of G for every prime p since $|\langle e \rangle| = 1 = p^0$.

Corollary 5.3. *A finite group* G *is a* p-*group if and only if* |G| *is a power of* p.

PROOF. If G is a p-group and q a prime which divides $|G|$, then G contains an element of order q by Cauchy's Theorem. Since every element of G has order a power of p, $q = p$. Hence $|G|$ is a power of p. The converse is an immediate consequence of Lagrange's Theorem I.4.6. ■

[4] I am indebted to R. J. Nunke for suggesting this line of proof.

Corollary 5.4. *The center* C(G) *of a nontrivial finite* p*-group* G *contains more than one element.*

PROOF. Consider the class equation of G (see page 91):

$$|G| = |C(G)| + \sum [G : C_G(x_i)].$$

Since each $[G : C_G(x_i)] > 1$ and divides $|G| = p^n$ $(n \geq 1)$, p divides each $[G : C_G(x_i)]$ and $|G|$ and therefore divides $|C(G)|$. Since $|C(G)| \geq 1$, $C(G)$ has at least p elements. ■

Lemma 5.5. *If* H *is a* p*-subgroup of a finite group* G, *then* $[N_G(H) : H] \equiv [G : H]$ (*mod* p).

PROOF. Let S be the set of left cosets of H in G and let H act on S by (left) translation. Then $|S| = [G : H]$. Also,

$$xH \in S_0 \Leftrightarrow hxH = xH \quad \text{for all} \quad h \in H$$
$$\Leftrightarrow x^{-1}hxH = H \quad \text{for all} \quad h \in H \Leftrightarrow x^{-1}hx \in H \quad \text{for all} \quad h \in H$$
$$\Leftrightarrow x^{-1}Hx = H \Leftrightarrow xHx^{-1} = H \Leftrightarrow x \in N_G(H).$$

Therefore $|S_0|$ is the number of cosets xH with $x \in N_G(H)$; that is, $|S_0| = [N_G(H) : H]$. By Lemma 5.1 $[N_G(H) : H] = |S_0| \equiv |S| = [G : H] \pmod p$. ■

Corollary 5.6. *If* H *is* p*-subgroup of a finite group* G *such that* p *divides* [G : H], *then* $N_G(H) \neq H$.

PROOF. $0 \equiv [G : H] \equiv [N_G(H) : H] \pmod p$. Since $[N_G(H) : H] \geq 1$ in any case, we must have $[N_G(H) : H] > 1$. Therefore $N_G(H) \neq H$. ■

Theorem 5.7. (*First Sylow Theorem*) *Let* G *be a group of order* $p^n m$, *with* $n \geq 1$, p *prime, and* $(p,m) = 1$. *Then* G *contains a subgroup of order* p^i *for each* $1 \leq i \leq n$ *and every subgroup of* G *of order* p^i $(i < n)$ *is normal in some subgroup of order* p^{i+1}.

PROOF. Since $p \mid |G|$, G contains an element a, and therefore, a subgroup $\langle a \rangle$ of order p by Cauchy's Theorem. Proceeding by induction assume H is a subgroup of G of order p^i $(1 \leq i < n)$. Then $p \mid [G : H]$ and by Lemma 5.5 and Corollary 5.6 H is normal in $N_G(H)$, $H \neq N_G(H)$ and $1 < |N_G(H)/H| = [N_G(H) : H] \equiv [G : H] \equiv 0 \pmod p$. Hence $p \mid |N_G(H)/H|$ and $N_G(H)/H$ contains a subgroup of order p as above. By Corollary I.5.12 this group is of the form H_1/H where H_1 is a subgroup of $N_G(H)$ containing H. Since H is normal in $N_G(H)$, H is necessarily normal in H_1. Finally $|H_1| = |H||H_1/H| = p^i p = p^{i+1}$. ■

A subgroup P of a group G is said to be a **Sylow p-subgroup** (p prime) if P is a maximal p-subgroup of G (that is, $P < H < G$ with H a p-group implies $P = H$). Sylow p-subgroups always exist, though they may be trivial, and every p-subgroup is contained in a Sylow p-subgroup (Zorn's Lemma is needed to show this for infinite groups). Theorem 5.7 shows that a finite group G has a nontrivial Sylow p-subgroup for every prime p that divides $|G|$. Furthermore, we have

Corollary 5.8. *Let* G *be a group of order* $p^n m$ *with* p *prime,* $n \geq 1$ *and* $(m,p) = 1$. *Let* H *be a* p*-subgroup of* G.

(i) H *is a Sylow* p*-subgroup of* G *if and only if* $|H| = p^n$.
(ii) *Every conjugate of a Sylow* p*-subgroup is a Sylow* p*-subgroup.*
(iii) *If there is only one Sylow* p*-subgroup* P, *then* P *is normal in* G.

SKETCH OF PROOF. (i) Corollaries I.4.6 and 5.3 and Theorem 5.7. (ii) Exercise I.5.6 and (i). (iii) follows from (ii). ■

As a converse to Corollary 5.8 (ii) we have

Theorem 5.9. (*Second Sylow Theorem*) *If* H *is a* p*-subgroup of a finite group* G, *and* P *is any Sylow* p*-subgroup of* G, *then there exists* $x \in G$ *such that* $H < xPx^{-1}$. *In particular, any two Sylow* p*-subgroups of* G *are conjugate.*

PROOF. Let S be the set of left cosets of P in G and let H act on S by (left) translation. $|S_0| \equiv |S| = [G:P] \pmod p$ by Lemma 5.1. But $p \nmid [G:P]$; therefore $|S_0| \neq 0$ and there exists $xP \in S_0$.

$$xP \in S_0 \Leftrightarrow hxP = xP \quad \text{for all} \quad h \in H$$
$$\Leftrightarrow x^{-1}hxP = P \quad \text{for all} \quad h \in H \Leftrightarrow x^{-1}Hx < P \Leftrightarrow H < xPx^{-1}.$$

If H is a Sylow p-subgroup $|H| = |P| = |xPx^{-1}|$ and hence $H = xPx^{-1}$. ■

Theorem 5.10. (*Third Sylow Theorem*) *If* G *is a finite group and* p *a prime, then the number of Sylow* p*-subgroups of* G *divides* $|G|$ *and is of the form* $kp + 1$ *for some* $k \geq 0$.

PROOF. By the second Sylow Theorem the number of Sylow p-subgroups is the number of conjugates of any one of them, say P. But this number is $[G : N_G(P)]$, a divisor of $|G|$, by Corollary 4.4. Let S be the set of all Sylow p-subgroups of G and let P act on S by conjugation. Then $Q \in S_0$ if and only if $xQx^{-1} = Q$ for all $x \in P$. The latter condition holds if and only if $P < N_G(Q)$. Both P and Q are Sylow p-subgroups of G and hence of $N_G(Q)$ and are therefore conjugate in $N_G(Q)$. But since Q is normal in $N_G(Q)$, this can only occur if $Q = P$. Therefore, $S_0 = \{P\}$ and by Lemma 5.1, $|S| \equiv |S_0| = 1 \pmod p$. Hence $|S| = kp + 1$. ■

Theorem 5.11. *If* P *is a Sylow* p*-subgroup of a finite group* G, *then* $N_G(N_G(P)) = N_G(P)$.

PROOF. Every conjugate of P is a Sylow p-subgroup of G and of any subgroup of G that contains it. Since P is normal in $N = N_G(P)$, P is the only Sylow p-subgroup of N by Theorem 5.9. Therefore,

$$x \in N_G(N) \Rightarrow xNx^{-1} = N \Rightarrow xPx^{-1} < N \Rightarrow xPx^{-1} = P \Rightarrow x \in N.$$

Hence $N_G(N_G(P)) < N$; the other inclusion is obvious. ■

EXERCISES

1. If $N \lhd G$ and N, G/N are both p-groups, then G is a p-group.

2. If G is a finite p-group, $H \lhd G$ and $H \neq \langle e \rangle$, then $H \cap C(G) \neq \langle e \rangle$.

3. Let $|G| = p^n$. For each k, $0 \leq k \leq n$, G has a *normal* subgroup of order p^k.

4. If G is an infinite p-group (p prime), then either G has a subgroup of order p^n for each $n \geq 1$ or there exists $m \in \mathbf{N}^*$ such that every finite subgroup of G has order $\leq p^m$.

5. If P is a normal Sylow p-subgroup of a finite group G and $f: G \to G$ is an endomorphism, then $f(P) < P$.

6. If H is a normal subgroup of order p^k of a finite group G, then H is contained in every Sylow p-subgroup of G.

7. Find the Sylow 2-subgroups and Sylow 3-subgroups of S_3, S_4, S_5.

8. If every Sylow p-subgroup of a finite group G is normal for every prime p, then G is the direct product of its Sylow subgroups.

9. If $|G| = p^n q$, with $p > q$ primes, then G contains a unique normal subgroup of index q.

10. Every group of order 12, 28, 56, and 200 must contain a normal Sylow subgroup, and hence is not simple.

11. How many elements of order 7 are there in a simple group of order 168?

12. Show that every automorphism of S_4 is an inner automorphism, and hence $S_4 \cong \operatorname{Aut} S_4$. [*Hint:* see Exercise 4.10. Every automorphism of S_4 induces a permutation of the set $\{P_1,P_2,P_3,P_4\}$ of Sylow 3-subgroups of S_4. If $f \in \operatorname{Aut} S_4$ has $f(P_i) = P_i$ for all i, then $f = 1_{S_4}$.]

13. Every group G of order p^2 (p prime) is abelian [*Hint:* Exercise 4.9 and Corollary 5.4].

6. CLASSIFICATION OF FINITE GROUPS

We shall classify up to isomorphism all groups of order pq (p,q primes) and all groups of small order ($n \leq 15$). Admittedly, these are not very far reaching results; but even the effort involved in doing this much will indicate the difficulty in determining the structure of an arbitrary (finite) group. The results of this section are not needed in the sequel.

Proposition 6.1. *Let* p *and* q *be primes such that* $p > q$. *If* $q \nmid p - 1$, *then every group of order* pq *is isomorphic to the cyclic group* Z_{pq}. *If* $q \mid p - 1$, *then there are (up* *to isomorphism) exactly two distinct groups of order* pq: *the cyclic group* Z_{pq} *and a nonabelian group* K *generated by elements* c *and* d *such that*

$$|c| = p; \qquad |d| = q; \qquad dc = c^s d,$$

where $s \not\equiv 1$ (*mod* p) *and* $s^q \equiv 1$ (*mod* p).

SKETCH OF PROOF. A nonabelian group K of order pq as described in the proposition does exist (Exercise 2). Given G of order pq, G contains elements a,b with $|a| = p$, $|b| = q$ by Cauchy's Theorem 5.2. Furthermore, $S = \langle a \rangle$ is normal in G (by Corollary 4.10 or by counting Sylow p-subgroups, as below). The coset bS has order q in the group G/S. Since $|G/S| = q$, G/S is cyclic with generator bS, $G/S = \langle bS \rangle$. Therefore every element of G can be written in the form $b^i a^j$ and $G = \langle a,b \rangle$.

The number of Sylow q-subgroups is $kq + 1$ and divides pq. Hence it is 1 or p. If it is 1 (as it must be if $q \nmid p - 1$), then $\langle b \rangle$ is also normal in G. Lagrange's Theorem I.4.6 shows that $\langle a \rangle \cap \langle b \rangle = \langle e \rangle$. Thus by Theorems I.3.2, I.8.6, I.8.10 and Exercise I.8.5, $G = \langle a \rangle \times \langle b \rangle \cong Z_p \oplus Z_q \cong Z_{pq}$. If the number is p, (which can only occur if $p \mid q - 1$), then $bab^{-1} = a^r$ (since $\langle a \rangle \lhd G$) and $r \not\equiv 1$ (mod p) (otherwise G would be abelian by Theorem I.3.4(v) and hence have a unique Sylow q-subgroup). Since $bab^{-1} = a^r$, it follows by induction that $b^j a b^{-j} = a^{r^j}$. In particular for $j = q$, $a = a^{r^q}$, which implies $r^q \equiv 1$ (mod p) by Theorem I.3.4 (v).

In order to complete the proof we must show that if $q \mid p - 1$ and G is the nonabelian group described in the preceding paragraph, then G is isomorphic to K. We shall need some results from number theory. The congruence $x^q \equiv 1$ (mod p) has exactly q distinct solutions modulo p (see J. E. Shockley [51; Corollary 6.1, p. 67]). If r is a solution and k is the least positive integer such that $r^k \equiv 1$ (mod p), then $k \mid q$ (see J.E. Shockley [51; Theorem 8, p. 70]). In our case $r \not\equiv 1$ (mod p), whence $k = q$. It follows that $1, r, r^2, \ldots, r^{q-1}$ are all the distinct solutions modulo p of $x^q \equiv 1$ (mod p). Consequently, $s \equiv r^t$ (mod p) for some t ($1 \leq t \leq q - 1$). If $b_1 = b^t \in G$, then $|b_1| = q$. Our work above (with b_1 in place of b) shows that $G = \langle a,b_1 \rangle$; that every element of G can be written $b_1{}^i a^j$; that $|a| = p$; and that $b_1 a b_1{}^{-1} = b^t a b^{-t} = a^{r^t} = a^s$ (Theorem I.3.4(v)). Therefore, $b_1 a = a^s b_1$. Verify that the map $G \rightarrow K$ given by $a \mapsto c$ and $b_1 \mapsto d$ is an isomorphism. ■

Corollary 6.2. *If* p *is an odd prime, then every group of order* 2p *is isomorphic either to the cyclic group* Z_{2p} *or the dihedral group* D_p.

PROOF. Apply Proposition 6.1 with $q = 2$. If G is not cyclic, the conditions on s imply $s \equiv -1$ (mod p). Hence $G = \langle c,d \rangle$, $|d| = 2$, $|c| = p$, and $dc = c^{-1}d$ by Theorem I.3.4(v). Therefore, $G \cong D_p$ by Theorem I.6.13. ■

Proposition 6.3. *There are (up to isomorphism) exactly two distinct nonabelian groups of order* 8: *the quaternion group* Q_8 *and the dihedral group* D_4.

REMARK. The quaternion group Q_8 is described in Exercise I.2.3.

SKETCH OF PROOF OF 6.3. Verify that $D_4 \not\cong Q_8$ (Exercise 10). If a group G of order 8 is nonabelian, then it cannot contain an element of order 8 or have every nonidentity element of order 2 (Exercise I.1.13). Hence G contains an element a of order 4. The group $\langle a \rangle$ of index 2 is normal. Choose $b \notin \langle a \rangle$. Then $b^2 \in \langle a \rangle$ since $|G/\langle a \rangle| = 2$. Show that the only possibilities are $b^2 = a^2$ or $b^2 = e$. Since $\langle a \rangle$ is normal in G, $bab^{-1} \in \langle a \rangle$; the only possibility is $bab^{-1} = a^3 = a^{-1}$. It follows that every element of G can be written $b^i a^j$. Hence $G = \langle a,b \rangle$. In one case we have $|a| = 4$, $b^2 = a^2$, $ba = a^{-1}b$, and $G \cong Q_8$ by Exercise I.4.14.; in the other case, $|a| = 4$, $|b| = 2$, $ba = a^{-1}b$ and $G \cong D_4$ by Theorem I.6.13. ■

Proposition 6.4. *There are (up to isomorphism) exactly three distinct nonabelian groups of order* 12: *the dihedral group* D_6, *the alternating group* A_4, *and a group* T *generated by elements* a,b *such that* $|a| = 6$, $b^2 = a^3$, *and* $ba = a^{-1}b$.

SKETCH OF PROOF. Verify that there is a group T of order 12 as stated (Exercise 5) and that no two of D_6,A_4,T are isomorphic (Exercise 6). If G is a nonabelian group of order 12, let P be a Sylow 3-subgroup of G. Then $|P| = 3$ and $[G : P] = 4$. By Proposition 4.8 there is a homomorphism $f : G \to S_4$ whose kernel K is contained in P, whence $K = P$ or $\langle e \rangle$. If $K = \langle e \rangle$, f is a monomorphism and G is isomorphic to a subgroup of order 12 of S_4, which must be A_4 by Theorem I.6.8. Otherwise $K = P$ and P is normal in G. In this case P is the unique Sylow 3-subgroup. Hence G contains only two elements of order 3. If c is one of these, then $[G : C_G(c)] = 1$ or 2 since $[G : C_G(c)]$ is the number of conjugates of c and every conjugate of c has order 3. Hence $C_G(c)$ is a group of order 12 or 6. In either case there is $d \in C_G(c)$ of order 2 by Cauchy's Theorem. Verify that $|cd| = 6$.

Let $a = cd$; then $\langle a \rangle$ is normal in G and $|G/\langle a \rangle| = 2$. Hence there is an element $b \in G$ such that $b \notin \langle a \rangle$, $b \neq e$, $b^2 \in \langle a \rangle$, and $bab^{-1} \in \langle a \rangle$. Since G is nonabelian and $|a| = 6$, $bab^{-1} = a^5 = a^{-1}$ is the only possibility; that is, $ba = a^{-1}b$. There are six possibilities for $b^2 \in \langle a \rangle$. $b^2 = a^2$ or $b^2 = a^4$ lead to contradictions; $b^2 = a$ or $b^2 = a^5$ imply $|b| = 12$ and G abelian. Therefore, the only possibilities are

(i) $|a| = 6$; $b^2 = e$; $ba = a^{-1}b$, whence $G \cong D_6$ by Theorem I.6.13;

(ii) $|a| = 6$; $b^2 = a^3$; $ba = a^{-1}b$, whence $G \cong T$ by Exercise 5(b). ■

The table below lists (up to isomorphism) all distinct groups of small order. There are 14 distinct groups of order 16 and 51 of order 32; see M. Hall and J.K. Senior [16]. There is no known formula giving the number of distinct groups of order n, for every n.

Order	*Distinct Groups*	*Reference*
1	$\langle e \rangle$	. . .
2	Z_2	Exercise I.4.3
3	Z_3	Exercise I.4.3
4	$Z_2 \oplus Z_2$, Z_4	Exercise I.4.5
5	Z_5	Exercise I.4.3
6	Z_6, D_3	Corollary 6.2
7	Z_7	Exercise I.4.3

Order	*Distinct Groups*	*Reference*
8	$Z_2 \oplus Z_2 \oplus Z_2$, $Z_2 \oplus Z_4$, Z_8, Q_8, D_4	Theorem 2.1 and Proposition 6.3
9	$Z_3 \oplus Z_3$, Z_9	Exercise 5.13 and Theorem 2.1
10	Z_{10}, D_5	Corollary 6.2
11	Z_{11}	Exercise I.4.3
12	$Z_2 \oplus Z_6$, Z_{12}, A_4, D_6, T	Theorem 2.1 and Proposition 6.4
13	Z_{13}	Exercise I.4.3
14	Z_{14}, D_7	Corollary 6.2
15	Z_{15}	Proposition 6.1

EXERCISES

1. Let G and H be groups and $\theta : H \to \operatorname{Aut} G$ a homomorphism. Let $G \times_\theta H$ be the set $G \times H$ with the following binary operation: $(g,h)(g',h') = (g[\theta(h)(g')],hh')$. Show that $G \times_\theta H$ is a group with identity element (e,e) and $(g,h)^{-1} = (\theta(h^{-1})(g^{-1}),h^{-1})$. $G \times_\theta H$ is called the **semidirect product** of G and H.

2. Let $C_p = \langle a \rangle$ and $C_q = \langle b \rangle$ be (multiplicative) cyclic groups of prime orders p and q respectively such that $p > q$ and $q \mid p - 1$. Let s be an integer such that $s \not\equiv 1$ (mod p) and $s^q \equiv 1$ (mod p), which implies $s \not\equiv 0$ (mod p). Elementary number theory shows that such an s exists (see J.E. Shockley [51; Corollary 6.1, p. 67]).
   (a) The map $\alpha : C_p \to C_p$ given by $a^i \mapsto a^{si}$ is an automorphism.
   (b) The map $\theta: C_q \to \operatorname{Aut} C_p$ given by $\theta(b^i) = \alpha^i$ (α as in part (a)) is a homomorphism ($\alpha^0 = 1_{C_p}$).
   (c) If we write a for (a,e) and b for (e,b), then the group $C_p \times_\theta C_q$ (see Exercise 1) is a group of order pq, generated by a and b subject to the relations: $|a| = p$, $|b| = q$, $ba = a^s b$, where $s \not\equiv 1$ (mod p), and $s^q \equiv 1$ (mod p). The group $C_p \times_\theta C_q$ is called the **metacyclic group.**

3. Consider the set $G = \{\pm 1, \pm i, \pm j, \pm k\}$ with multiplication given by $i^2 = j^2 = k^2 = -1$; $ij = k$; $jk = i$, $ki = j$; $ji = -k$, $kj = -i$, $ik = -j$, and the usual rules for multiplying by ± 1. Show that G is a group isomorphic to the quaternion group Q_8.

4. What is the center of the quaternion group Q_8? Show that $Q_8/C(Q_8)$ is abelian.

5. (a) Show that there is a nonabelian subgroup T of $S_3 \times Z_4$ of order 12 generated by elements a,b such that $|a| = 6$, $a^3 = b^2$, $ba = a^{-1}b$.
   (b) Any group of order 12 with generators a,b such that $|a| = 6$, $a^3 = b^2$, $ba = a^{-1}b$ is isomorphic to T.

6. No two of D_6, A_4, and T are isomorphic, where T is the group of order 12 described in Proposition 6.4 and Exercise 5.

7. If G is a nonabelian group of order p^3 (p prime), then the center of G is the subgroup generated by all elements of the form $aba^{-1}b^{-1}$ ($a,b \in G$).

8. Let p be an *odd* prime. Prove that there are, at most, two nonabelian groups of order p^3. [One has generators a,b satisfying $|a| = p^2$; $|b| = p$; $b^{-1}ab = a^{1+p}$;

the other has generators a,b,c satisfying $|a| = |b| = |c| = p$; $c = a^{-1}b^{-1}ab$; $ca = ac$; $cb = bc$.]

9. Classify up to isomorphism all groups of order 18. Do the same for orders 20 and 30.

10. Show that D_4 is not isomorphic to Q_8. [*Hint:* Count elements of order 2.]

7. NILPOTENT AND SOLVABLE GROUPS

Consider the following conditions on a finite group G.

(i) G *is the direct product of its Sylow subgroups.*
(ii) *If* m *divides* $|G|$, *then* G *has a subgroup of order* m.
(iii) *If* $|G| = mn$ *with* $(m,n) = 1$, *then* G *has a subgroup of order* m.

Conditions (ii) and (iii) may be considered as modifications of the First Sylow Theorem. It is not difficult to show that (i) $\Rightarrow$ (ii) and obviously (ii) $\Rightarrow$ (iii). The fact that every finite abelian group satisfies (i) is an easy corollary of Theorem 2.2. Every p-group satisfies (i) trivially. On the other hand, A_4 satisfies (iii) but not (ii), and S_3 satisfies (ii) but not (i) (Exercise 1). Given the rather striking results achieved thus far with finite abelian and p-groups, the classes of groups satisfying (i), (ii), and (iii) respectively would appear to be excellent candidates for investigation. We shall restrict our attention to those groups that satisfy (i) or (iii).

We shall first define nilpotent and solvable groups in terms of certain "normal series" of subgroups. In the case of finite groups, nilpotent groups are characterized by condition (i) (Proposition 7.5) and solvable ones by condition (iii) (Proposition 7.14). This approach will also demonstrate that there is a connection between nilpotent and solvable groups and commutativity. Other characterizations of nilpotent and solvable groups are given in Section 8.

Our treatment of solvable groups is purely group theoretical. Historically, however, solvable groups first occurred in connection with the problem of determining the roots of a polynomial with coefficients in a field (see Section V.9).

Let G be a group. The center $C(G)$ of G is a normal subgroup (Corollary 4.7). Let $C_2(G)$ be the inverse image of $C(G/C(G))$ under the canonical projection $G \to G/C(G)$. Then by (the proof of) Theorem I.5.11 $C_2(G)$ is normal in G and contains $C(G)$. Continue this process by defining inductively: $C_1(G) = C(G)$ and $C_i(G)$ is the inverse image of $C(G/C_{i-1}(G))$ under the canonical projection $G \to G/C_{i-1}(G)$. Thus we obtain a sequence of normal subgroups of G, called the **ascending central series** of G: $\langle e \rangle < C_1(G) < C_2(G) < \cdots$.

Definition 7.1. *A group* G *is* **nilpotent** *if* $C_n(G) = G$ *for some* n.

Every abelian group G is nilpotent since $G = C(G) = C_1(G)$.

Theorem 7.2. *Every finite* p*-group is nilpotent.*

PROOF. G and all its nontrivial quotients are p-groups, and therefore, have nontrivial centers by Corollary 5.4. This implies that if $G \neq C_i(G)$, then $C_i(G)$ is strictly contained in $C_{i+1}(G)$. Since G is finite, $C_n(G)$ must be G for some n. ■

Theorem 7.3. *The direct product of a finite number of nilpotent groups is nilpotent.*

PROOF. Suppose for convenience that $G = H \times K$, the proof for more than two factors being similar. Assume inductively that $C_i(G) = C_i(H) \times C_i(K)$ (the case $i = 1$ is obvious). Let π_H be the canonical epimorphism $H \to H/C_i(H)$ and similarly for π_K. Verify that the canonical epimorphism $\varphi : G \to G/C_i(G)$ is the composition

$$G = H \times K \xrightarrow{\pi} H/C_i(H) \times K/C_i(K) \xrightarrow{\psi} \frac{H \times K}{C_i(H) \times C_i(K)} = \frac{H \times K}{C_i(H \times K)} = G/C_i(G),$$

where $\pi = \pi_H \times \pi_K$ (Theorem I.8.10), and ψ is the isomorphism of Corollary I.8.11. Consequently,

$$\begin{aligned} C_{i+1}(G) &= \varphi^{-1}[C(G/C_i(G))] = \pi^{-1}\psi^{-1}[C(G/C_i(G))] \\ &= \pi^{-1}[C(H/C_i(H) \times K/C_i(K))] \\ &= \pi^{-1}[C(H/C_i(H)) \times C(K/C_i(K))] \\ &= \pi_H{}^{-1}[C(H/C_i(H))] \times \pi_K{}^{-1}[C(K/C_i(K))] \\ &= C_{i+1}(H) \times C_{i+1}(K). \end{aligned}$$

Thus the inductive step is proved and $C_i(G) = C_i(H) \times C_i(K)$ for all i. Since H,K are nilpotent, there exists $n \in \mathbf{N}^*$ such that $C_n(H) = H$ and $C_n(K) = K$, whence $C_n(G) = H \times K = G$. Therefore, G is nilpotent. ■

Lemma 7.4. *If* H *is a proper subgroup of a nilpotent group* G, *then* H *is a proper subgroup of its normalizer* $N_G(H)$.

PROOF. Let $C_0(G) = \langle e \rangle$ and let n be the largest index such that $C_n(G) < H$; (there is such an n since G is nilpotent and H a proper subgroup). Choose $a \in C_{n+1}(G)$ with $a \notin H$. Then for every $h \in H$, $C_nah = (C_na)(C_nh) = (C_nh)(C_na) = C_nha$ in $G/C_n(G)$ since C_na is in the center by the definition of $C_{n+1}(G)$. Thus $ah = h'ha$, where $h' \in C_n(G) < H$. Hence $aha^{-1} \in H$ and $a \in N_G(H)$. Since $a \notin H$, H is a proper subgroup of $N_G(H)$. ■

Proposition 7.5. *A finite group is nilpotent if and only if it is the direct product of its Sylow subgroups.*

PROOF. If G is the direct product of its Sylow p-subgroups, then G is nilpotent by Theorems 7.2 and 7.3. If G is nilpotent and P is a Sylow p-subgroup of G for some prime p, then either $P = G$ (and we are done) or P is a proper subgroup of G. In the latter case P is a proper subgroup of $N_G(P)$ by Lemma 7.4. Since $N_G(P)$ is its own normalizer by Theorem 5.11, we must have $N_G(P) = G$ by Lemma 7.4. Thus P is normal in G, and hence the unique Sylow p-subgroup of G by Theorem 5.9. Let

$|G| = p_1^{n_1} \cdots p_k^{n_k}$ (p_i distinct primes, $n_i > 0$) and let $P_1, P_2, \ldots, P_k$ be the corresponding (proper normal) Sylow subgroups of G. Since $|P_i| = p_i^{n_i}$ for each i, $P_i \cap P_j = \langle e \rangle$ for $i \neq j$. By Theorem I.5.3 $xy = yx$ for every $x \in P_i$, $y \in P_j$ $(i \neq j)$. It follows that for each i, $P_1 P_2 \cdots P_{i-1} P_{i+1} \cdots P_k$ is a subgroup in which every element has order dividing $p_1^{n_1} \cdots p_{i-1}^{n_{i-1}} p_{i+1}^{n_{i+1}} \cdots p_k^{n_k}$. Consequently, $P_i \cap (P_1 \cdots P_{i-1} P_{i+1} \cdots P_k) = \langle e \rangle$ and $P_1 P_2 \cdots P_k = P_1 \times \cdots \times P_k$. Since $|G| = p_1^{n_1} \cdots p_k^{n_k} = |P_1 \times \cdots \times P_k| = |P_1 \cdots P_k|$ we must have $G = P_1 P_2 \cdots P_k = P_1 \times \cdots \times P_k$. ■

Corollary 7.6. *If* G *is a finite nilpotent group and* m *divides* |G|, *then* G *has a subgroup of order* m.

PROOF. Exercise. ■

Definition 7.7. *Let* G *be a group. The subgroup of* G *generated by the set* $\{aba^{-1}b^{-1} \mid a,b \in G\}$ *is called the* **commutator subgroup** *of* G *and denoted* G′.

The elements $aba^{-1}b^{-1}$ $(a,b \in G)$ are called **commutators.** The commutators only generate G', so that G' may well contain elements that are not commutators. G is abelian if and only if $G' = \langle e \rangle$. In a sense, G' provides a measure of how much G differs from an abelian group.

Theorem 7.8. *If* G *is a group, then* G′ *is a normal subgroup of* G *and* G/G′ *is abelian. If* N *is a normal subgroup of* G, *then* G/N *is abelian if and only if* N *contains* G′.

PROOF. Let $f : G \to G$ be any automorphism. Then

$$f(aba^{-1}b^{-1}) = f(a)f(b)f(a)^{-1}f(b)^{-1} \in G'.$$

It follows that $f(G') < G'$. In particular, if f is the automorphism given by conjugation by $a \in G$, then $aG'a^{-1} = f(G') < G'$, whence G' is normal in G by Theorem I.5.1. Since $(ab)(ba)^{-1} = aba^{-1}b^{-1} \in G'$, $abG' = baG'$ and hence G/G' is abelian. If G/N is abelian, then $abN = baN$ for all $a,b \in G$, whence $ab(ba)^{-1} = aba^{-1}b^{-1} \in N$. Therefore, N contains all commutators and $G' < N$. The converse is easy. ■

Let G be a group and let $G^{(1)}$ be G'. Then for $i \geq 1$, define $G^{(i)}$ by $G^{(i)} = (G^{(i-1)})'$. $G^{(i)}$ is called ith **derived subgroup** of G. This gives a sequence of subgroups of G, each normal in the preceding one: $G > G^{(1)} > G^{(2)} > \cdots$. Actually each $G^{(i)}$ is a normal subgroup of G (Exercise 13).

Definition 7.9. *A group* G *is said to be* **solvable** *if* $G^{(n)} = \langle e \rangle$ *for some* n.

Every abelian group is trivially solvable. More generally, we have

Proposition 7.10. *Every nilpotent group is solvable.*

PROOF. Since by the definition of $C_i(G)$ $C_i(G)/C_{i-1}(G) = C(G/C_{i-1}(G))$ is abelian, $C_i(G)' < C_{i-1}(G)$ for all $i > 1$ and $C_1(G)' = C(G)' = \langle e \rangle$. For some n, $G = C_n(G)$. Therefore, $C(G/C_{n-1}(G)) = C_n(G)/C_{n-1}(G) = G/C_{n-1}(G)$ is abelian and hence $G^{(1)} = G' < C_{n-1}(G)$. Therefore, $G^{(2)} = G^{(1)\prime} < C_{n-1}(G)' < C_{n-2}(G)$; similarly $G^{(3)} < C_{n-2}(G)' < C_{n-3}(G); \ldots, G^{(n-1)} < C_2(G)' < C_1(G); G^{(n)} < C_1(G)' = \langle e \rangle$. Hence G is solvable. ∎

Theorem 7.11. (i) *Every subgroup and every homomorphic image of a solvable group is solvable.*

(ii) *If* N *is a normal subgroup of a group* G *such that* N *and* G/N *are solvable, then* G *is solvable.*

SKETCH OF PROOF. (i) If $f : G \to H$ is a homomorphism [epimorphism], verify that $f(G^{(i)}) < H^{(i)}[f(G^{(i)}) = H^{(i)}]$ for all i. Suppose f is an epimorphism, and G is solvable. Then for some n, $\langle e \rangle = f(e) = f(G^{(n)}) = H^{(n)}$, whence H is solvable. The proof for a subgroup is similar.

(ii) Let $f : G \to G/N$ be the canonical epimorphism. Since G/N is solvable, for some n $f(G^{(n)}) = (G/N)^{(n)} = \langle e \rangle$. Hence $G^{(n)} < \text{Ker } f = N$. Since $G^{(n)}$ is solvable by (i), there exists $k \in \mathbf{N}^*$ such that $G^{(n+k)} = (G^{(n)})^{(k)} = \langle e \rangle$. Therefore, G is solvable. ∎

Corollary 7.12. *If* $n \geq 5$, *then the symmetric group* S_n *is not solvable.*

PROOF. If S_n were solvable, then A_n would be solvable. Since A_n is nonabelian, $A_n' \neq (1)$. Since A_n' is normal in A_n (Theorem 7.8) and A_n is simple (Theorem I.6.10), we must have $A_n' = A_n$. Therefore $A_n^{(i)} = A_n \neq (1)$ for all $i \geq 1$, whence A_n is not solvable. ∎

NOTE. The remainder of this section is not needed in the sequel.

In order to prove a generalization of the Sylow theorems for finite solvable groups (as mentioned in the first paragraph of this section) we need some definitions and a lemma. A subgroup H of a group G is said to be **characteristic** [resp. **fully invariant**] if $f(H) < H$ for every automorphism [resp. endomorphism] $f : G \to G$. Clearly every fully invariant subgroup is characteristic and every characteristic subgroup is normal (since conjugation is an automorphism). A **minimal normal subgroup** of a group G is a nontrivial normal subgroup that contains no proper subgroup which is normal in G.

Lemma 7.13. *Let* N *be a normal subgroup of a finite group* G *and* H *any subgroup of* G.

(i) *If* H *is a characteristic subgroup of* N, *then* H *is normal in* G.

(ii) *Every normal Sylow* p*-subgroup of* G *is fully invariant.*

(iii) *If* G *is solvable and* N *is a minimal normal subgroup, then* N *is an abelian* p*-group for some prime* p.

PROOF. (i) Since $aNa^{-1} = N$ for all $a \in G$, conjugation by a is an automorphism of N. Since H is characteristic in N, $aHa^{-1} < H$ for all $a \in G$. Hence H is normal in G by Theorem I.5.1.

(ii) is an exercise. (iii) It is easy to see that N' is fully invariant in N, whence N' is normal in G by (i). Since N is a minimal normal subgroup, either $N' = \langle e \rangle$ or $N' = N$. Since N is solvable (Theorem 7.11), $N' \neq N$. Hence $N' = \langle e \rangle$ and N is a nontrivial abelian group. Let P be a nontrivial Sylow p-subgroup of N for some prime p. Since N is abelian, P is normal in N and hence fully invariant in N by (ii). Consequently P is normal in G by (i). Since N is minimal and P nontrivial we must have $P = N$. ■

Proposition 7.14. *(P. Hall) Let* G *be a finite solvable group of order* mn, *with* (m,n) = 1. *Then*

(i) G *contains a subgroup of order* m;

(ii) *any two subgroups of* G *of order* m *are conjugate;*

(iii) *any subgroup of* G *of order* k, *where* k | m, *is contained in a subgroup of order* m.

REMARKS. If m is a prime power, this theorem merely restates several results contained in the Sylow theorems. P. Hall has also proved the converse of (i): if G is a finite group such that whenever $|G| = mn$ with $(m,n) = 1$, G has a subgroup of order m, then G is solvable. The proof is beyond the scope of this book (see M. Hall [15; p. 143]).

PROOF OF 7.14. The proof proceeds by induction on $|G|$, the orders ≤ 5 being trivial. There are two cases.

CASE 1. There is a proper normal subgroup H of G whose order is not divisible by n.

(i) $|H| = m_1n_1$, where $m_1 \mid m$, $n_1 \mid n$, and $n_1 < n$. G/H is a solvable group of order $(m/m_1)(n/n_1) < mn$, with $(m/m_1,n/n_1) = 1$. Therefore by induction G/H contains a subgroup A/H of order (m/m_1) (where A is a subgroup of G — see Corollary I.5.12). Then $|A| = |H|[A : H] = (m_1n_1)(m/m_1) = mn_1 < mn$. A is solvable (Theorem 7.11) and by induction contains a subgroup of order m.

(ii) Suppose B,C are subgroups of G of order m. Since H is normal in G, HB is a subgroup (Theorem I.5.3), whose order k necessarily divides $|G| = mn$. Since $k = |HB| = |H||B|/|H \cap B| = m_1n_1m/|H \cap B|$, we have $k|H \cap B| = m_1n_1m$, whence $k \mid m_1n_1m$. Since $(m_1,n) = 1$, there are integers x,y such that $m_1x + ny = 1$, and hence $mn_1m_1x + mn_1ny = mn_1$. Consequently $k \mid mn_1$. By Lagrange's Theorem I.4.6 $m = |B|$ and $m_1n_1 = |H|$ divide k. Thus $(m,n) = 1$ implies $mn_1 \mid k$. Therefore $k = mn_1$; similarly $|HC| = mn_1$. Thus HB/H and HC/H are subgroups of G/H of order m/m_1. By induction they are conjugate: for some $\bar{x} \in G/H$ (where $\bar{x}$ is the coset of $x \in G$), $\bar{x}(HB/H)\bar{x}^{-1} = HC/H$. It follows that $xHBx^{-1} = HC$. Consequently xBx^{-1} and C are subgroups of HC of order m and are therefore conjugate in HC by induction. Hence B and C are conjugate in G.

(iii) If a subgroup K of G has order k dividing m, then $HK/H \cong K/H \cap K$ has order dividing k. Since HK/H is a subgroup of G/H, its order also divides $|G/H| = (m/m_1)(n/n_1)$. $(k,n) = 1$ implies that the order of HK/H divides m/m_1. By induction there is a subgroup A/H of G/H of order m/m_1 which contains HK/H (where $A < G$ as above). Clearly K is a subgroup of A. Since $|A| = |H||A/H| = m_1n_1(m/m_1) = mn_1 < mn$, K is contained in a subgroup of A (and hence of G) of order m by induction.

CASE 2. Every proper normal subgroup of G has order divisible by n. If H is a minimal normal subgroup (such groups exist since G is finite), then $|H| = p^r$ for some prime p by Lemma 7.13 (iii). Since $(m,n) = 1$ and $n \mid |H|$, it follows that $n = p^r$ and hence that H is a Sylow p-subgroup of G. Since H is normal in G, H is the unique Sylow p-subgroup of G. This argument shows that H is the only minimal normal subgroup of G (otherwise $n = p^r$ and $n = q^s$ for distinct primes p,q). In particular, *every nontrivial normal subgroup of* G *contains* H.

(i) Let K be a normal subgroup of G such that K/H is a minimal normal subgroup of G/H (Corollary I.5.12). By Lemma 7.13 (iii) $|K/H| = q^s$ (q prime, $q \neq p$), so that $|K| = p^rq^s$. Let S be a Sylow q-subgroup of K and let M be the normalizer of S in G. We shall show that $|M| = m$. Since H is normal in K, HS is a subgroup of K. Clearly $H \cap S = \langle e \rangle$ so that $|HS| = |H||S|/|H \cap S| = p^rq^s = |K|$, whence $K = HS$.

Since K is normal in G and $S < K$, every conjugate of S in G lies in K. Since S is a Sylow subgroup of K, all these subgroups are already conjugate in K. Let $N = N_K(S)$; then the number c of conjugates of S in G is $[G:M] = [K:N]$ by Corollary 4.4. Since $S < N < K$, $K > HN > HS = K$, so that $K = HN$ and $c = [G:M] = [K:N] = [HN:N] = [H:H \cap N]$ (Corollary I.5.9). We shall show that $H \cap N = \langle e \rangle$, which implies $c = |H| = p^r$ and hence $|M| = |G|/[G:M] = mp^r/p^r = m$. We do this by showing first that $H \cap N < C(K)$ and second that $C(K) = \langle e \rangle$.

Let $x \in H \cap N$ and $k \in K$. Since $K = HS$, $k = hs$ ($h \in H$, $s \in S$). Since H is abelian (Lemma 7.13 (iii)) and $x \in H$, we need only show $xs = sx$ in order to have $xk = kx$ and $x \in C(K)$. Now $(xsx^{-1})s^{-1} \in S$ since $x \in N = N_K(S)$. But $x(sx^{-1}s^{-1}) \in H$ since $x \in H$ and H is normal in G. Thus $xsx^{-1}s^{-1} \in H \cap S = \langle e \rangle$, which implies $xs = sx$.

It is easy to see that $C(K)$ is a characteristic subgroup of K. Since K is normal in G, $C(K)$ is normal in G by Lemma 7.13 (i). If $C(K) \neq \langle e \rangle$, then $C(K)$ necessarily concontains H. This together with $K = HS$ implies that S is normal in K. By Lemma 7.13 (ii) and (i) S is fully invariant in K and hence normal in G (since $K \lhd G$). This implies $H < S$ which is a contradiction. Hence $C(K) = \langle e \rangle$.

(ii) Let M be as in (i) and suppose B is a subgroup of G of order m. Now $|BK|$ is divisible by $|B| = m$ and $|K| = p^rq^s$. Since $(m,p) = 1$, $|BK|$ is divisible by $p^rm = nm = |G|$. Hence $G = BK$. Consequently $G/K = BK/K \cong B/B \cap K$ (Corollary I.5.9), which implies that $|B \cap K| = |B|/|G/K| = q^s$. By the Second Sylow Theorem $B \cap K$ is conjugate to S in K. Furthermore $B \cap K$ is normal in B (since $K \lhd G$) and hence B is contained in $N_G(B \cap K)$. Verify that conjugate subgroups have conjugate normalizers. Hence $N_G(B \cap K)$ and $N_G(S) = M$ are conjugate in G. Thus $|N_G(B \cap K)| = |M| = m$. But $|B| = m$; therefore $B < N_G(B \cap K)$ implies $B = N_G(B \cap K)$. Hence B and M are conjugate.

(iii) Let $D < G$, where $|D| = k$ and $k \mid m$. Let M (of order m) and H (of order p^r, with $(p,m) = 1$) be as in (i). Then $D \cap H = \langle e \rangle$ and $|DH| = |D||H|/|D \cap H| = kp^r$. We also have $|G| = mp^r$, $M \cap H = \langle e \rangle$ and $MH = G$ (since $|MH| = |M||H|/|M \cap H| = mp^r = |G|$). Hence $M(DH) = G$ and therefore $|M \cap DH| = |M||DH|/|MDH| = m(kp^r)/mp^r = k$. Let $M^* = M \cap DH$; then M^* and D are conjugate (by (ii) applied to the group DH). For some $a \in G$, $aM^*a^{-1} = D$. Since $M^* < M$, D is contained in aMa^{-1}, a conjugate of M, and thus a subgroup of order m. ■

We close this section by mentioning a longstanding conjecture of Burnside: every finite group of odd order is solvable. This remarkable result was first proved by W. Feit and J. Thompson [61] in 1963.

EXERCISES

1. (a) A_4 is not the direct product of its Sylow subgroups, but A_4 does have the property: $mn = 12$ and $(m,n) = 1$ imply there is a subgroup of order m.
(b) S_3 has subgroups of orders 1, 2, 3, and 6 but is not the direct product of its Sylow subgroups.

2. Let G be a group and $a,b \in G$. Denote the commutator $aba^{-1}b^{-1} \in G$ by $[a,b]$. Show that for any $a,b,c, \in G$, $[ab,c] = a[b,c]a^{-1}[a,c]$.

3. If H and K are subgroups of a group G, let (H,K) be the subgroup of G generated by the elements $\{hkh^{-1}k^{-1} \mid h \in H, k \in K\}$. Show that
   (a) (H,K) is normal in $H \vee K$.
   (b) If $(H,G') = \langle e \rangle$, then $(H',G) = \langle e \rangle$.
   (c) $H \lhd G$ if and only if $(H,G) < H$.
   (d) Let $K \lhd G$ and $K < H$; then $H/K < C(G/K)$ if and only if $(H,G) < K$.

4. Define a chain of subgroups $\gamma_i(G)$ of a group G as follows: $\gamma_1(G) = G$, $\gamma_2(G) = (G,G)$, $\gamma_i(G) = (\gamma_{i-1}(G),G)$ (see Exercise 3). Show that G is nilpotent if and only if $\gamma_m(G) = \langle e \rangle$ for some m.

5. Every subgroup and every quotient group of a nilpotent group is nilpotent. [*Hint:* Theorem 7.5 or Exercise 4.].

6. (Wielandt) Prove that a finite group G is nilpotent if and only if every maximal proper subgroup of G is normal. Conclude that every maximal proper subgroup has prime index. [*Hint:* if P is a Sylow p-subgroup of G, show that any subgroup containing $N_G(P)$ is its own normalizer; see Theorem 5.11.]

7. If N is a nontrivial normal subgroup of a nilpotent group G, then $N \cap C(G) \neq \langle e \rangle$.

8. If D_n is the dihedral group with generators a of order n and b of order 2, then
   (a) $a^2 \in D_n'$.
   (b) If n is odd, $D_n' \cong Z_n$.
   (c) If n is even, $D_n' \cong Z_m$, where $2m = n$.
   (d) D_n is nilpotent if and only if n is a power of 2.

9. Show that the commutator subgroup of S_4 is A_4. What is the commutator group of A_4?

10. S_n is solvable for $n \leq 4$, but S_3 and S_4 are not nilpotent.

11. A nontrivial finite solvable group G contains a normal abelian subgroup $H \neq \langle e \rangle$. If G is not solvable then G contains a normal subgroup H such that $H' = H$.

12. There is no group G such that $G' = S_4$. [*Hint:* Exercises 9 and 5.12 may be helpful.]

13. If G is a group, then the ith derived subgroup $G^{(i)}$ is a fully invariant subgroup, whence $G^{(i)}$ is normal.

14. If $N \lhd G$ and $N \cap G' = \langle e \rangle$, then $N < C(G)$.

15. If H is a maximal proper subgroup of a finite solvable group G, then $[G : H]$ is a prime power.

16. For any group G, $C(G)$ is characteristic, but not necessarily fully invariant.

17. If G is an abelian p-group, then the subgroup $G[p]$ (see Lemma 2.5) is fully invariant in G.

18. If G is a finite nilpotent group, then every minimal normal subgroup of G is contained in $C(G)$ and has prime order.

8. NORMAL AND SUBNORMAL SERIES

The usefulness of the ascending central series and the series of derived subgroups of a group suggests that other such series of subgroups should be investigated. We do this next and obtain still other characterizations of nilpotent and solvable groups, as well as the famous theorem of Jordan-Hölder.

Definition 8.1. *A* **subnormal series** *of a group* G *is a chain of subgroups* $G = G_0 > G_1 > \cdots > G_n$ *such that* G_{i+1} *is normal in* G_i *for* $0 \leq i < n$. *The* **factors** *of the series are the quotient groups* G_i/G_{i+1}. *The* **length** *of the series is the number of strict inclusions (or alternatively, the number of nonidentity factors). A subnormal series such that* G_i *is normal in* G *for all* i *is said to be* **normal**.[5]

A subnormal series need not be normal (Exercise I.5.10).

EXAMPLES. The derived series $G > G^{(1)} > \cdots > G^{(n)}$ is a normal series for any group G (see Exercise 7.13). If G is nilpotent, the ascending central series $C_1(G) < \cdots < C_n(G) = G$ is a normal series for G.

Definition 8.2. *Let* $G = G_0 > G_1 > \cdots > G_n$ *be a subnormal series. A one-step refinement of this series is any series of the form* $G = G_0 > \cdots > G_i > N > G_{i+1} > \cdots$

[5]Some authors use the terms "normal" where we use "subnormal."

$> G_n$ or $G = G_0 > \cdots > G_n > N$, where N is a normal subgroup of G_i and (if $i < n$) G_{i+1} is normal in N. A **refinement** *of a subnormal series S is any subnormal series obtained from S by a finite sequence of one-step refinements. A refinement of S is said to be* **proper** *if its length is larger than the length of S.*

Definition 8.3. *A subnormal series $G = G_0 > G_1 > \cdots > G_n = \langle e\rangle$ is a* **composition series** *if each factor G_i/G_{i+1} is simple. A subnormal series $G = G_0 > G_1 > \cdots > G_n = \langle e\rangle$ is a* **solvable series** *if each factor is abelian.*

The following fact is used frequently when dealing with composition series: if N is a normal subgroup of a group G, then every normal subgroup of G/N is of the form H/N where H is a normal subgroup of G which contains N (Corollary I.5.12). Therefore, when $G \neq N$, G/N is simple if and only if N is a maximal in the set of all normal subgroups M of G with $M \neq G$ (such a subgroup N is called a **maximal normal subgroup** of G).

Theorem 8.4. (i) *Every finite group G has a composition series.*

(ii) *Every refinement of a solvable series is a solvable series.*

(iii) *A subnormal series is a composition series if and only if it has no proper refinements.*

PROOF. (i) Let G_1 be a maximal normal subgroup of G; then G/G_1 is simple by Corollary I.5.12. Let G_2 be a maximal normal subgroup of G_1, and so on. Since G is finite, this process must end with $G_n = \langle e\rangle$. Thus $G > G_1 > \cdots > G_n = \langle e\rangle$ is a composition series.

(ii) If G_i/G_{i+1} is abelian and $G_{i+1} \lhd H \lhd G_i$, then H/G_{i+1} is abelian since it is a subgroup of G_i/G_{i+1} and G_i/H is abelian since it is isomorphic to the quotient $(G_i/G_{i+1})/(H/G_{i+1})$ by the Third Isomorphism Theorem I.5.10. The conclusion now follows immediately.

(iii) If $G_{i+1} \underset{\neq}{\lhd} H \underset{\neq}{\lhd} G_i$ are groups, then H/G_{i+1} is a proper normal subgroup of G_i/G_{i+1} and every proper normal subgroup of G_i/G_{i+1} has this form by Corollary I.5.12. The conclusion now follows from the observation that a subnormal series $G = G_0 > G_1 > \cdots > G_n = \langle e\rangle$ has a proper refinement if and only if there is a subgroup H such that for some i, $G_{i+1} \underset{\neq}{\lhd} H \underset{\neq}{\lhd} G_i$. ■

Theorem 8.5. *A group G is solvable if and only if it has a solvable series.*

PROOF. If G is solvable, then the derived series $G > G^{(1)} > G^{(2)} > \cdots > G^{(n)} = \langle e\rangle$ is a solvable series by Theorem 7.8. If $G = G_0 > G_1 > \cdots > G_n = \langle e\rangle$ is a solvable series for G, then G/G_1 abelian implies that $G_1 > G^{(1)}$ by Theorem 7.8; G_1/G_2 abelian implies $G_2 > G_1' > G^{(2)}$. Continue by induction and conclude that $G_i > G^{(i)}$ for all i; in particular $\langle e\rangle = G_n > G^{(n)}$ and G is solvable. ■

EXAMPLES. The dihedral group D_n is solvable since $D_n > \langle a\rangle > \langle e\rangle$ is a solvable series, where a is the generator of order n (so that $D_n/\langle a\rangle \cong Z_2$). Similarly if

$|G| = pq$ ($p > q$ primes), then G contains an element a of order p and $\langle a \rangle$ is normal in G (Corollary 4.10). Thus $G > \langle a \rangle > \langle e \rangle$ is a solvable series and G is solvable. More generally we have

Proposition 8.6. *A finite group G is solvable if and only if G has a composition series whose factors are cyclic of prime order.*

PROOF. A (composition) series with cyclic factors is a solvable series. Conversely, assume $G = G_0 > G_1 > \cdots > G_n = \langle e \rangle$ is a solvable series for G. If $G_0 \neq G_1$, let H_1 be a maximal normal subgroup of $G = G_0$ which contains G_1. If $H_1 \neq G_1$, let H_2 be a maximal normal subgroup of H_1 which contains G_1, and so on. Since G is finite, this gives a series $G > H_1 > H_2 > \cdots > H_k > G_1$ with each subgroup a maximal normal subgroup of the preceding, whence each factor is simple. Doing this for each pair (G_i, G_{i+1}) gives a solvable refinement $G = N_0 > N_1 > \cdots > N_r = \langle e \rangle$ of the original series by Theorem 8.4 (ii). Each factor of this series is abelian and simple and hence cyclic of prime order (Exercise I.4.3). Therefore, $G > N_1 > \cdots > N_r = \langle e \rangle$ is a composition series. ■

A given group may have many subnormal or solvable series. Likewise it may have several different composition series (Exercise 1). However we shall now show that any two composition series of a group are equivalent in the following sense.

Definition 8.7. *Two subnormal series S and T of a group G are* **equivalent** *if there is a one-to-one correspondence between the nontrivial factors of S and the nontrivial factors of T such that corresponding factors are isomorphic groups.*

Two subnormal series need not have the same number of terms in order to be equivalent, but they must have the same length (that is, the same number of nontrivial factors). Clearly, equivalence of subnormal series is an equivalence relation.

Lemma 8.8. *If S is a composition series of a group G, then any refinement of S is equivalent to S.*

PROOF. Let S be denoted $G = G_0 > G_1 > \cdots > G_n = \langle e \rangle$. By Theorem 8.4 (iii) S has no proper refinements. This implies that the only possible refinements of S are obtained by inserting additional copies of each G_i. Consequently any refinement of S has exactly the same nontrivial factors as S and is therefore equivalent to S. ■

The next lemma is quite technical. Its value will be immediately apparent in the proof of Theorem 8.10.

Lemma 8.9. *(Zassenhaus) Let A^*, A, B^*, B be subgroups of a group G such that A^* is normal in A and B^* is normal in B.*

(i) $A^*(A \cap B^*)$ *is a normal subgroup of* $A^*(A \cap B)$;
(ii) $B^*(A^* \cap B)$ *is a normal subgroup of* $B^*(A \cap B)$;
(iii) $A^*(A \cap B)/A^*(A \cap B^*) \cong B^*(A \cap B)/B^*(A^* \cap B)$.

PROOF. Since B^* is normal in B, $A \cap B^* = (A \cap B) \cap B^*$ is a normal subgroup of $A \cap B$ (Theorem I.5.3 (i)); similarly $A^* \cap B$ is normal in $A \cap B$. Consequently $D = (A^* \cap B)(A \cap B^*)$ is a normal subgroup of $A \cap B$ (Theorem I.5.3 iii) and Exercise I.5.13). Theorem I.5.3 (iii) also implies that $A^*(A \cap B)$ and $B^*(A \cap B)$ are subgroups of A and B respectively. We shall define an epimorphism $f : A^*(A \cap B) \to (A \cap B)/D$ with kernel $A^*(A \cap B^*)$. This will imply that $A^*(A \cap B^*)$ is normal in $A^*(A \cap B)$ (Theorem I.5.5) and that $A^*(A \cap B)/A^*(A \cap B^*) \cong (A \cap B)/D$ (Corollary I.5.7).

Define $f : A^*(A \cap B) \to (A \cap B)/D$ as follows. If $a \in A^*$, $c \in A \cap B$, let $f(ac) = Dc$. Then f is well defined since $ac = a_1c_1$ $(a,a_1 \in A^*; c,c_1 \in A \cap B)$ implies $c_1c^{-1} = a_1^{-1}a \in (A \cap B) \cap A^* = A^* \cap B < D$, whence $Dc_1 = Dc$. f is clearly surjective. f is an epimorphism since $f[(a_1c_1)(a_2c_2)] = f(a_1a_3c_1c_2) = Dc_1c_2 = Dc_1Dc_2 = f(a_1c_1)f(a_2c_2)$, where $a_i \in A^*$, $c_j \in A \cap B$, and $c_1a_2 = a_3c_1$ since A^* is normal in A. Finally $ac \in \operatorname{Ker} f$ if and only if $c \in D$, that is, if and only if $c = a_1c_1$, with $a_1 \in A^* \cap B$ and $c_1 \in A \cap B^*$. Hence $ac \in \operatorname{Ker} f$ if and only if $ac = (aa_1)c_1 \in A^*(A \cap B^*)$. Therefore, $\operatorname{Ker} f = A^*(A \cap B^*)$,

A symmetric argument shows that $B^*(A^* \cap B)$ is normal in $B^*(A \cap B)$ and $B^*(A \cap B)/B^*(A^* \cap B) \cong (A \cap B)/D$, whence (iii) follows immediately. ■

Theorem 8.10. *(Schreier) Any two subnormal [resp. normal] series of a group* G *have subnormal [resp. normal] refinements that are equivalent.*

PROOF. Let $G = G_0 > G_1 > \cdots > G_n$ and $G = H_0 > H_1 > \cdots > H_m$ be subnormal [resp. normal] series. Let $G_{n+1} = \langle e \rangle = H_{m+1}$ and for each $0 \leq i \leq n$ consider the groups

$$G_i = G_{i+1}(G_i \cap H_0) > G_{i+1}(G_i \cap H_1) > \cdots > G_{i+1}(G_i \cap H_j) > G_{i+1}(G_i \cap H_{j+1}) > \cdots > G_{i+1}(G_i \cap H_m) > G_{i+1}(G_i \cap H_{m+1}) = G_{i+1}.$$

For each $0 \leq j \leq m$, the Zassenhaus Lemma (applied to G_{i+1},G_i,H_{j+1}, and H_j) shows that $G_{i+1}(G_i \cap H_{j+1})$ is normal in $G_{i+1}(G_i \cap H_j)$. [If the original series were both normal, then each $G_{i+1}(G_i \cap H_j)$ is normal in G by Theorem I.5.3 (iii) and Exercises I.5.2 and I.5.13.] Inserting these groups between each G_i and G_{i+1}, and denoting $G_{i+1}(G_i \cap H_j)$ by $G(i,j)$ thus gives a subnormal [resp. normal] refinement of the series $G_0 > G_1 > \cdots > G_n$:

$$G = G(0,0) > G(0,1) > \cdots > G(0,m) > G(1,0) > G(1,1) > G(1,2) > \cdots > G(1,m) > G(2,0) > \cdots > G(n-1,m) > G(n,0) > \cdots > G(n,m),$$

where $G(i,0) = G_i$. Note that this refinement has $(n+1)(m+1)$ (not necessarily distinct) terms. A symmetric argument shows that there is a refinement of $G = H_0 > H_1 > \cdots > H_m$ (where $H(i,j) = H_{j+1}(G_i \cap H_j)$ and $H(0,j) = H_j$):

$$G = H(0,0) > H(1,0) > \cdots > H(n,0) > H(0,1) > H(1,1) > H(2,1) > \cdots > H(n,1) > H(0,2) > \cdots > H(n,m-1) > H(0,m) > \cdots > H(n,m).$$

This refinement also has $(n+1)(m+1)$ terms. For each pair (i,j) $(0 \leq i \leq n, 0 \leq j \leq m)$ there is by the Zassenhaus Lemma 8.9 (applied to G_{i+1}, G_i, H_{j+1}, and H_j) an isomorphism:

$$\frac{G(i,j)}{G(i,j+1)} = \frac{G_{i+1}(G_i \cap H_j)}{G_{i+1}(G_i \cap H_{j+1})} \cong \frac{H_{j+1}(G_i \cap H_j)}{H_{j+1}(G_{i+1} \cap H_j)} = \frac{H(i,j)}{H(i+1,j)}.$$

This provides the desired one-to-one correspondence of the factors and shows that the refinements are equivalent. ■

Theorem 8.11. (*Jordan-Hölder*) *Any two composition series of a group* G *are equivalent. Therefore every group having a composition series determines a unique list of simple groups.*

REMARK. The theorem does *not* state the existence of a composition series for a given group.

PROOF OF 8.11. Since composition series are subnormal series, any two composition series have equivalent refinements by the Theorem 8.10. But every refinement of a composition series S is equivalent to S by Lemma 8.8. It follows that any two composition series are equivalent. ■

The Jordan-Hölder Theorem indicates that some knowledge of simple groups might be useful. A major achievement in recent years has been the complete classification of all finite simple groups. This remarkable result is based on the work of a large number of group theorists. For an introduction to the problem and an outline of the method of proof, see *Finite Simple Groups* by Daniel Gorenstein (Plenum Publishing Corp., 1982). Nonabelian simple groups of small order are quite rare. It can be proved that there are (up to isomorphism) only two nonabelian simple groups of order less than 200, namely A_5 and a subgroup of S_7 of order 168 (see Exercises 13-20).

EXERCISES

1. (a) Find a normal series of D_4 consisting of 4 subgroups.
   (b) Find all composition series of the group D_4.
   (c) Do part (b) for the group A_4.
   (d) Do part (b) for the group $S_3 \times Z_2$.
   (e) Find all composition factors of S_4 and D_6.
2. If $G = G_0 > G_1 > \cdots > G_n$ is a subnormal series of a finite group G, then $|G| = \left(\prod_{i=0}^{n-1} |G_i/G_{i+1}|\right)|G_n|$.
3. If N is a simple normal subgroup of a group G and G/N has a composition series, then G has a composition series.
4. A composition series of a group is a subnormal series of maximal (finite) length.

5. An abelian group has a composition series if and only if it is finite.

6. If $H \lhd G$, where G has a composition series, then G has a composition series one of whose terms is H.

7. A solvable group with a composition series is finite.

8. If H and K are solvable subgroups of G with $H \lhd G$, then HK is a solvable subgroup of G.

9. Any group of order p^2q (p,q primes) is solvable.

10. A group G is nilpotent if and only if there is a normal series $G = G_0 > G_1 > \cdots > G_n = \langle e \rangle$ such that $G_i/G_{i+1} < C(G/G_{i+1})$ for every i.

11. (a) Show that the analogue of Theorem 7.11 is false for nilpotent groups [Consider S_3].
(b) If $H < C(G)$ and G/H is nilpotent, then G is nilpotent.

12. Prove the Fundamental Theorem of Arithmetic, Introduction, Theorem 6.7, by applying the Jordan-Hölder Theorem to the group Z_n.

13. Any simple group G of order 60 is isomorphic to A_5. [*Hint:* use Corollary 4.9; if $H < G$, then $[G:H] \geq 5$ (since $|S_n| < 60$ for $n \leq 4$); if $[G:H] = 5$ then $G \cong A_5$ by Theorem I.6.8. The assumption that there is no subgroup of index 5 leads to a contradiction.]

14. There are no nonabelian simple groups of order < 60.

15. Let G be the subgroup of S_7 generated by (1234567) and (26)(34). Show that $|G| = 168$.

Exercises 16–20 outline a proof of the fact that the group G of Exercise 15 is simple. We consider G as acting on the set $S = \{1,2,3,4,5,6,7\}$ as in the first example after Definition 4.1 and make use of Exercise 4.6.

16. The group G is transitive (see Exercise 4.6).

17. For each $x \in S$, G_x is a maximal (proper) subgroup of G. The proof of this fact proceeds in several steps:
(a) A *block* of G is a subset T of S such that for each $g \in G$ either $gT \cap T = \emptyset$ or $gT = T$, where $gT = \{gx \mid x \in T\}$. Show that if T is a block, then $|T|$ divides 7. [*Hint:* let $H = \{g \in G \mid gT = T\}$ and show that for $x \in T$, $G_x < H$ and $[H:G_x] = |T|$. Hence $|T|$ divides $[G:G_x] = [G:H][H:G_x]$. But $[G:G_x] = 7$ by Exercise 4.6(a) and Theorem 4.3.]
(b) If G_x is not maximal, then there is a block T of G such that $|T| \nmid 7$, contradicting part (a). [*Hint:* If $G_x \underset{\neq}{<} H < G$, show that H is not transitive on S (since $1 \leq [H:G_x] < |S|$, which contradicts Exercise 4.6.(d)). Let $T = \{hx \mid h \in H\}$. Since H is not transitive, $|T| < |S| = 7$ and since $H \neq G_x$, $|T| > 1$. Show that T is a block.]

18. If $(1) \neq N \lhd G$, then 7 divides $|N|$. [*Hint:* Exercise 4.6 (c) $\Rightarrow G_x \underset{\neq}{<} NG_x$ for all $x \in S \Rightarrow NG_x = G$ for all $x \in S$ by Exercise 17 $\Rightarrow N$ is transitive on $S \Rightarrow 7$ divides $|N|$ by Exercise 4.6 (d).]

19. The group G contains a subgroup P of order 7 such that the smallest normal subgroup of G containing P is G itself.

20. If $(1) \neq N \lhd G$, then $N = G$; hence G is simple. [Use Exercise I.5.19 and Exercise 18 to show $P < N$; apply Exercise 19.]

CHAPTER III

RINGS

Another fundamental concept in the study of algebra is that of a ring. The problem of classifying all rings (in a given class) up to isomorphism is far more complicated than the corresponding problem for groups. It will be partially dealt with in Chapter IX. The present chapter is concerned, for the most part, with presenting those facts in the theory of rings that are most frequently used in several areas of algebra. The first two sections deal with rings, homomorphisms and ideals. Much (but not all) of this material is simply a straightforward generalization to rings of concepts which have proven useful in group theory. Sections 3 and 4 are concerned with commutative rings that resemble the ring of integers in various ways. Divisibility, factorization, Euclidean rings, principal ideal domains, and unique factorization are studied in Section 3. In Section 4 the familiar construction of the field of rational numbers from the ring of integers is generalized and rings of quotients of an arbitrary commutative ring are considered in some detail. In the last two sections the ring of polynomials in n indeterminates over a ring R is studied. In particular, the concepts of Section 3 are studied in the context of polynomial rings (Section 6).

The approximate interdependence of the sections of this chapter is as follows:

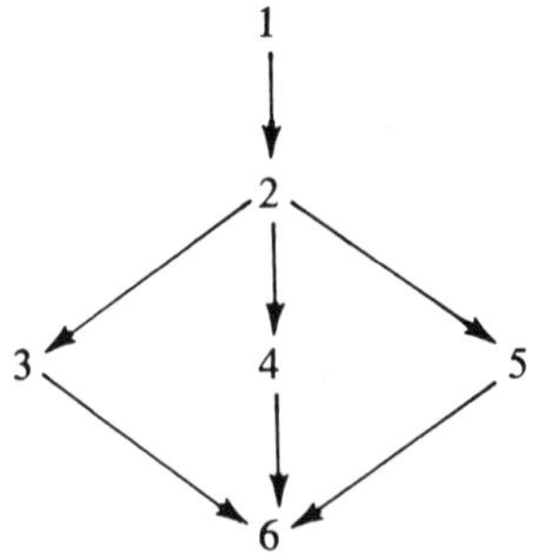

Section 6 requires only certain parts of Sections 4 and 5.

1. RINGS AND HOMOMORPHISMS

The basic concepts in the theory of rings are defined and numerous examples given. Several frequently used calculational facts are presented. The only difficulty with this material is the large quantity of terminology that must be absorbed in a short period of time.

Definition 1.1. *A* **ring** *is a nonempty set* R *together with two binary operations (usually denoted as addition* (+) *and multiplication) such that:*

(i) (R,+) *is an abelian group;*
(ii) (ab)c = a(bc) *for all* a,b,c ε R *(associative multiplication);*
(iii) a(b + c) = ab + ac *and* (a + b)c = ac + bc *(left and right distributive laws).*

If in addition:

(iv) ab = ba *for all* a,b ε R,

then R *is said to be a* **commutative ring.** *If* R *contains an element* 1_R *such that*

(v) $1_R a = a 1_R = a$ *for all* a ε R,

then R *is said to be a* **ring with identity.**

REMARK. The symbol 1_R is also used to denote the identity map $R \to R$. In context this usage will not be ambiguous.

The additive identity element of a ring is called the zero element and denoted 0. If R is a ring, $a \in R$ and $n \in \mathbf{Z}$, then na has its usual meaning for additive groups (Definition I.1.8); for example, $na = a + a + \cdots + a$ (n summands) when $n > 0$. Before giving examples of rings we record

Theorem 1.2. *Let* R *be a ring. Then*

(i) 0a = a0 = 0 *for all* a ε R;
(ii) (−a)b = a(−b) = −(ab) *for all* a,b ε R;
(iii) (−a)(−b) = ab *for all* a,b ε R;
(iv) (na)b = a(nb) = n(ab) *for all* n ε **Z** *and all* a,b ε R;
(v)
$$\left(\sum_{i=1}^{n} a_i\right)\left(\sum_{j=1}^{m} b_j\right) = \sum_{i=1}^{n}\sum_{j=1}^{m} a_ib_j \quad \textit{for all} \quad a_i,b_j \in R.$$

SKETCH OF PROOF. (i) $0a = (0 + 0)a = 0a + 0a$, whence $0a = 0$. (ii) $ab + (-a)b = (a + (-a))b = 0b = 0$, whence $(-a)b = -(ab)$ by Theorem I.1.2(iii). (ii) implies (iii). (v) is proved by induction and includes (iv) as a special case. ■

The next two definitions introduce some more terminology; after which some examples will be given.

Definition 1.3. *A nonzero element* a *in a ring* R *is said to be a* **left** [*resp. right*] **zero divisor** *if there exists a nonzero* b ε R *such that* ab = 0 [*resp.* ba = 0]. *A* **zero divisor** *is an element of* R *which is both a left and a right zero divisor.*

It is easy to verify that a ring R has no zero divisors if and only if the right and left cancellation laws hold in R; that is, for all $a,b,c \in R$ with $a \neq 0$,

$$ab = ac \quad \text{or} \quad ba = ca \qquad \Rightarrow \qquad b = c.$$

Definition 1.4. *An element* a *in a ring* R *with identity is said to be* **left** [*resp. right*] **invertible** *if there exists* c ε R [*resp.* b ε R] *such that* ca = 1_R [*resp.* ab = 1_R]. *The element* c [*resp.* b] *is called a* **left** [*resp. right*] **inverse** *of* a. *An element* a ε R *that is both left and right invertible is said to be* **invertible** *or to be a* **unit.**

REMARKS. (i) The left and right inverses of a unit a in a ring R with identity necessarily coincide (since $ab = 1_R = ca$ implies $b = 1_Rb = (ca)b = c(ab) = c1_R = c$). (ii) The set of units in a ring R with identity forms a group under multiplication.

Definition 1.5. *A commutative ring* R *with identity* $1_R \neq 0$ *and no zero divisors is called an* **integral domain.** *A ring* D *with identity* $1_D \neq 0$ *in which every nonzero element is a unit is called a* **division ring.** *A* **field** *is a commutative division ring.*

REMARKS. (i) Every integral domain and every division ring has at least two elements (namely 0 and 1_R). (ii) A ring R with identity is a division ring if and only if the nonzero elements of R form a group under multiplication (see Remark (ii) after Definition 1.4). (iii) Every field F is an integral domain since $ab = 0$ and $a \neq 0$ imply that $b = 1_Fb = (a^{-1}a)b = a^{-1}(ab) = a^{-1}0 = 0$.

EXAMPLES. The ring **Z** of integers is an integral domain. The set E of even integers is a commutative ring without identity. Each of **Q** (rationals), **R** (real numbers), and **C** (complex numbers) is a field under the usual operations of addition and multiplication. The $n \times n$ matrices over **Q** (or **R** or **C**) form a noncommutative ring with identity. The units in this ring are precisely the nonsingular matrices.

EXAMPLE. For each positive integer n the set Z_n of integers modulo n is a ring. See the example after Theorem I.1.5 for details. If n is not prime, say $n = kr$ with $k > 1$, $r > 1$, then $\bar{k} \neq \bar{0}$, $\bar{r} \neq 0$ and $\bar{k}\bar{r} = \overline{kr} = \bar{n} = \bar{0}$ in Z_n, whence $\bar{k}$ and $\bar{r}$ are zero divisors. If p is prime, then Z_p is a field by Exercise I.1.7.

EXAMPLE. Let A be an abelian group and let End A be the set of endomorphisms $f: A \to A$. Define addition in End A by $(f+g)(a) = f(a) + g(a)$. Verify that $f + g \in$ End A. Since A is abelian, this makes End A an abelian group. Let multiplication in End A be given by composition of functions. Then End A is a (possibly noncommutative) ring with identity $1_A: A \to A$.

EXAMPLE. Let G be a (multiplicative) group and R a ring. Let $R(G)$ be the additive abelian group $\sum_{g \in G} R$ (one copy of R for each $g \in G$). It will be convenient to adopt a new notation for the elements of $R(G)$. An element $x = \{r_g\}_{g \in G}$ of $R(G)$ has only finitely many nonzero coordinates, say $r_{g_1}, \ldots, r_{g_n}$ $(g_i \in G)$. Denote x by the **formal sum** $r_{g_1}g_1 + r_{g_2}g_2 + \cdots + r_{g_n}g_n$ or $\sum_{i=1}^{n} r_{g_i}g_i$. We also allow the possibility that some of the r_{g_i} are zero or that some g_i are repeated, so that an element of $R(G)$ may be written in formally different ways (for example, $r_1g_1 + 0g_2 = r_1g_1$ or $r_1g_1 + s_1g_1 = (r_1 + s_1)\,g_1$). In this notation, addition in the group $R(G)$ is given by:

$$\sum_{i=1}^{n} r_{g_i}g_i + \sum_{i=1}^{n} s_{g_i}g_i = \sum_{i=1}^{n} (r_{g_i} + s_{g_i})g_i;$$

(by inserting zero coefficients if necessary we can always assume that two formal sums involve exactly the same indices $g_1, \ldots, g_n$). Define multiplication in $R(G)$ by

$$\left(\sum_{i=1}^{n} r_i g_i\right)\left(\sum_{j=1}^{m} s_j h_j\right) = \sum_{i=1}^{n}\sum_{j=1}^{m} (r_i s_j)(g_i h_j);$$

this makes sense since there is a product defined in both R (r_is_j) and $G(g_ih_j)$ and thus the expression on the right is a formal sum as desired. With these operations $R(G)$ is a ring, called the **group ring** of G over R. $R(G)$ is commutative if and only if both R and G are commutative. If R has an identity 1_R, and e is the identity element of G, then 1_Re is the identity element of $R(G)$.

EXAMPLE. Let $\mathbf{R}$ be the field of real numbers and S the set of symbols $1,i,j,k$. Let K be the additive abelian group $\mathbf{R} \oplus \mathbf{R} \oplus \mathbf{R} \oplus \mathbf{R}$ and write the elements of K as formal sums $(a_0,a_1,a_2,a_3) = a_01 + a_1i + a_2j + a_3k$. Then $a_01 + a_1i + a_2j + a_3k = b_01 + b_1i + b_2j + b_3k$ if and only if $a_i = b_i$ for every i. We adopt the conventions that $a_01 \in K$ is identified with $a_0 \in \mathbf{R}$ and that terms with zero coefficients may be omitted (for example, $4 + 2j = 4\cdot 1 + 0i + 2j + 0k$ and $i = 0 + 1i + 0j + 0k$). Then addition in K is given by

$$\begin{aligned}&(a_0 + a_1i + a_2j + a_3k) + (b_0 + b_1i + b_2j + b_3k)\\ &= (a_0 + b_0) + (a_1 + b_1)i + (a_2 + b_2)j + (a_3 + b_3)k.\end{aligned}$$

Define multiplication in K by

$$\begin{aligned}&(a_0 + a_1i + a_2j + a_3k)(b_0 + b_1i + b_2j + b_3k)\\ &= (a_0b_0 - a_1b_1 - a_2b_2 - a_3b_3) + (a_0b_1 + a_1b_0 + a_2b_3 - a_3b_2)i\\ &\quad + (a_0b_2 + a_2b_0 + a_3b_1 - a_1b_3)j + (a_0b_3 + a_3b_0 + a_1b_2 - a_2b_1)k.\end{aligned}$$

This product formula is obtained by multiplying the formal sums term by term subject to the following relations: (i) associativity; (ii) $ri = ir$; $rj = jr$, $rk = kr$ (for all $r \in \mathbf{R}$); (iii) $i^2 = j^2 = k^2 = ijk = -1$; $ij = -ji = k$; $jk = -kj = i$; $ki = -ik = j$. Under this product K is a noncommutative division ring in which the multiplicative inverse of $a_0 + a_1i + a_2j + a_3k$ is $(a_0/d) - (a_1/d)i - (a_2/d)j - (a_3/d)k$, where $d = a_0^2 + a_1^2 + a_2^2 + a_3^2$. K is called the division ring of **real quaternions.** The quaternions may also be interpreted as a certain subring of the ring of all 2×2 matrices over the field $\mathbf{C}$ of complex numbers (Exercise 8).

Definition 1.1 shows that under multiplication the elements of a ring R form a semigroup (a monoid if R has an identity). Consequently Definition I.1.8 is applicable and exponentiation is defined in R. We have for each $a \in R$ and $n \in \mathbf{N}^*$, $a^n = a\cdots a$ (n factors) and $a^0 = 1_R$ if R has an identity. By Theorem I.1.9

$$a^ma^n = a^{m+n} \quad\text{and}\quad (a^m)^n = a^{mn}$$

Subtraction in a ring R is defined in the usual way: $a - b = a + (-b)$. Clearly $a(b - c) = ab - ac$ and $(a - b)c = ac - bc$ for all $a,b,c \in R$.

The next theorem is frequently useful in computations. Recall that if k and n are integers with $0 \leq k \leq n$, then the **binomial coefficient** $\binom{n}{k}$ is the number $n!/(n-k)!k!$, where $0! = 1$ and $n! = n(n-1)(n-2)\cdots 2\cdot 1$ for $n \geq 1$. $\binom{n}{k}$ is actually an integer (Exercise 10).

Theorem 1.6. *(Binomial Theorem). Let* R *be a ring with identity,* n *a positive integer, and* $a,b,a_1,a_2,\ldots,a_s \in R$.

(i) *If* $ab = ba$, *then* $(a+b)^n = \sum_{k=0}^{n}\binom{n}{k}a^kb^{n-k}$;

(ii) *If* $a_ia_j = a_ja_i$ *for all* i *and* j, *then*

$$(a_1 + a_2 + \cdots + a_s)^n = \sum \frac{n!}{(i_1!)\cdots(i_s!)} a_1^{i_1}a_2^{i_2}\cdots a_s^{i_s},$$

where the sum is over all s*-tuples* $(i_1,i_2,\ldots,i_s)$ *such that* $i_1 + i_2 + \cdots + i_s = n$.

SKETCH OF PROOF. (i) Use induction on n and the fact that $\binom{n}{k} + \binom{n}{k+1} = \binom{n+1}{k+1}$ for $k < n$ (Exercise 10(c)); the distributive law and the commutativity of a and b are essential. (ii) Use induction on s. The case $s = 2$ is just part (i) since $(a_1 + a_2)^n = \sum_{k=0}^{n}\binom{n}{k}a_1^ka_2^{n-k} = \sum_{k+j=n}\frac{n!}{k!j!}a_1^ka_2^j$. If the theorem is true for s, note that

$$(a_1 + \cdots + a_s + a_{s+1})^n = ((a_1 + \cdots + a_s) + a_{s+1})^n = \sum_{k=0}^{n}\binom{n}{k}(a_1 + \cdots + a_s)^ka_{s+1}^{n-k}$$

$= \sum_{k+j=n}\frac{n!}{k!j!}(a_1 + \cdots + a_s)^ka_{s+1}^j$ by part (i). Apply the induction hypothesis and compute. ■

Definition 1.7. *Let* R *and* S *be rings. A function* $f : R \to S$ *is a* **homomorphism of rings** *provided that for all* $a,b \in R$:

$$f(a+b) = f(a) + f(b) \quad \textit{and} \quad f(ab) = f(a)f(b).$$

REMARK. It is easy to see that the class of all rings together with all ring homomorphisms forms a (concrete) category.

When the context is clear then we shall frequently write "homomorphism" in place of "homomorphism of rings." A homomorphism of rings is, in particular, a homomorphism of the underlying additive groups. Consequently the same terminology is used: a **monomorphism** [resp. **epimorphism, isomorphism**] **of rings** is a homomorphism of rings which is an injective [resp. surjective, bijective] map. A monomorphism of rings $R \to S$ is sometimes called an **embedding** of R in S. An isomorphism $R \to R$ is called an **automorphism** of R.

The **kernel** of a homomorphism of rings $f: R \to S$ is its kernel as a map of additive groups; that is, $\operatorname{Ker} f = \{r \varepsilon R \mid f(r) = 0\}$. Similarly the image of f, denoted $\operatorname{Im} f$, is $\{s \varepsilon S \mid s = f(r) \text{ for some } r \varepsilon R\}$. If R and S both have identities 1_R and 1_S, we do *not* require that a homomorphism of rings map 1_R to 1_S (see Exercises 15, 16).

EXAMPLES. The canonical map $\mathbf{Z} \to Z_m$ given by $k \mapsto \bar{k}$ is an epimorphism of rings. The map $Z_3 \to Z_6$ given by $\bar{k} \mapsto \overline{4k}$ is a well-defined monomorphism of rings.

EXAMPLE. Let G and H be multiplicative groups and $f: G \to H$ a homomorphism of groups. Let R be a ring and define a map on the group rings $\bar{f}: R(G) \to R(H)$ by:

$$\bar{f}\left(\sum_{i=1}^{n} r_i g_i\right) = \sum_{i=1}^{n} r_i f(g_i).$$

Then $\bar{f}$ is a homomorphism of rings.

Definition 1.8. *Let* R *be a ring. If there is a least positive integer* n *such that* $na = 0$ *for all* $a \varepsilon R$, *then* R *is said to have* **characteristic n.** *If no such* n *exists* R *is said to have* **characteristic zero.** (*Notation: char* $R = n$).

Theorem 1.9. *Let* R *be a ring with identity* 1_R *and characteristic* $n > 0$.

(i) *If* $\varphi: \mathbf{Z} \to R$ *is the map given by* $m \mapsto m1_R$, *then* φ *is a homomorphism of rings with kernel* $\langle n \rangle = \{kn \mid k \varepsilon \mathbf{Z}\}$.

(ii) n *is the least positive integer such that* $n1_R = 0$.

(iii) *If* R *has no zero divisors* (*in particular if* R *is an integral domain*), *then* n *is prime.*

SKETCH OF PROOF. (ii) If k is the least positive integer such that $k1_R = 0$, then for all $a \varepsilon R$: $ka = k(1_R a) = (k1_R)a = 0 \cdot a = 0$ by Theorem 1.2. (iii) If $n = kr$ with $1 < k < n$, $1 < r < n$, then $0 = n1_R = (kr)1_R 1_R = (k1_R)(r1_R)$ implies that $k1_R = 0$ or $r1_R = 0$, which contradicts (ii). ∎

Theorem 1.10. *Every ring* R *may be embedded in a ring* S *with identity. The ring* S (*which is not unique*) *may be chosen to be either of characteristic zero or of the same characteristic as* R.

SKETCH OF PROOF. Let S be the additive abelian group $R \oplus \mathbf{Z}$ and define multiplication in S by

$$(r_1,k_1)(r_2,k_2) = (r_1r_2 + k_2r_1 + k_1r_2, k_1k_2), \; (r_i \varepsilon R;\ k_i \varepsilon \mathbf{Z}).$$

Verify that S is a ring with identity $(0,1)$ and characteristic zero and that the map $R \to S$ given by $r \mapsto (r,0)$ is a ring monomorphism (embedding). If char $R = n > 0$, use a similar proof with $S = R \oplus Z_n$ and multiplication defined by

$$(r_1,\bar{k}_1)(r_2,\bar{k}_2) = (r_1r_2 + k_2r_1 + k_1r_2, \bar{k}_1\bar{k}_2),$$

where $r_i \in R$ and $\bar{k}_i \in Z_n$ is the image of $k_i \in \mathbf{Z}$ under the canonical map. Then char $S = n$. ■

EXERCISES

1. (a) Let G be an (additive) abelian group. Define an operation of multiplication in G by $ab = 0$ (for all $a,b \in G$). Then G is a ring.
   (b) Let S be the set of all subsets of some fixed set U. For $A,B \in S$, define $A + B = (A - B) \cup (B - A)$ and $AB = A \cap B$. Then S is a ring. Is S commutative? Does it have an identity?

2. Let $\{R_i \mid i \in I\}$ be a family of rings with identity. Make the direct sum of abelian groups $\sum_{i \in I} R_i$ into a ring by defining multiplication coordinatewise. Does $\sum_{i \in I} R_i$ have an identity?

3. A ring R such that $a^2 = a$ for all $a \in R$ is called a **Boolean ring.** Prove that every Boolean ring R is commutative and $a + a = 0$ for all $a \in R$. [For an example of a Boolean ring, see Exercise 1(b).]

4. Let R be a ring and S a nonempty set. Then the group $M(S,R)$ (Exercise I.1.2) is a ring with multiplication defined as follows: the product of $f,g \in M(S,R)$ is the function $S \to R$ given by $s \mapsto f(s)g(s)$.

5. If A is the abelian group $\mathbf{Z} \oplus \mathbf{Z}$, then End A is a noncommutative ring (see page 116).

6. A finite ring with more than one element and no zero divisors is a division ring. (Special case: a finite integral domain is a field.)

7. Let R be a ring with more than one element such that for each nonzero $a \in R$ there is a unique $b \in R$ such that $aba = a$. Prove:
   (a) R has no zero divisors.
   (b) $bab = b$.
   (c) R has an identity.
   (d) R is a division ring.

8. Let R be the set of all 2×2 matrices over the complex field $\mathbf{C}$ of the form

$$\begin{pmatrix} z & w \\ -\bar{w} & \bar{z} \end{pmatrix},$$

   where $\bar{z},\bar{w}$ are the complex conjugates of z and w respectively (that is, $c = a + b\sqrt{-1} \Leftrightarrow \bar{c} = a - b\sqrt{-1}$). Then R is a division ring that is isomorphic to the division ring K of real quaternions. [*Hint:* Define an isomorphism $K \to R$ by letting the images of $1,i,j,k \in K$ be respectively the matrices

$$\begin{pmatrix} 1 & 0 \\ 0 & 1 \end{pmatrix}, \begin{pmatrix} \sqrt{-1} & 0 \\ 0 & -\sqrt{-1} \end{pmatrix}, \begin{pmatrix} 0 & 1 \\ -1 & 0 \end{pmatrix}, \begin{pmatrix} 0 & \sqrt{-1} \\ \sqrt{-1} & 0 \end{pmatrix}.$$

9. (a) The subset $G = \{1,-1,i,-i,j,-j,k,-k\}$ of the division ring K of real quaternions forms a group under multiplication.
(b) G is isomorphic to the quaternion group (Exercises I.4.14 and I.2.3).
(c) What is the difference between the ring K and the group ring $\mathbf{R}(G)$ ($\mathbf{R}$ the field of real numbers)?

10. Let k,n be integers such that $0 \leq k \leq n$ and $\binom{n}{k}$ the binomial coefficient $n!/(n-k)!k!$, where $0! = 1$ and for $n > 0$, $n! = n(n-1)(n-2)\cdots 2\cdot 1$.
(a) $\binom{n}{k} = \binom{n}{n-k}$
(b) $\binom{n}{k} < \binom{n}{k+1}$ for $k+1 \leq n/2$.
(c) $\binom{n}{k} + \binom{n}{k+1} = \binom{n+1}{k+1}$ for $k < n$.
(d) $\binom{n}{k}$ is an integer.
(e) if p is prime and $1 \leq k \leq p^n - 1$, then $\binom{p^n}{k}$ is divisible by p.
[*Hints:* (b) observe that $\binom{n}{k+1} = \binom{n}{k}\dfrac{n-k}{k+1}$; (d) note that $\binom{m}{0} = \binom{m}{m} = 1$ and use induction on n in part (c).]

11. (The Freshman's Dream[1]). Let R be a commutative ring with identity of prime characteristic p. If $a,b \in R$, then $(a \pm b)^{p^n} = a^{p^n} \pm b^{p^n}$ for all integers $n \geq 0$ [see Theorem 1.6 and Exercise 10; note that $b = -b$ if $p = 2$].

12. An element of a ring is **nilpotent** if $a^n = 0$ for some n. Prove that in a commutative ring $a + b$ is nilpotent if a and b are. Show that this result may be false if R is not commutative.

13. In a ring R the following conditions are equivalent.
(a) R has no nonzero nilpotent elements (see Exercise 12).
(b) If $a \in R$ and $a^2 = 0$, then $a = 0$.

14. Let R be a commutative ring with identity and prime characteristic p. The map $R \to R$ given by $r \mapsto r^p$ is a homomorphism of rings called the Frobenius homomorphism [see Exercise 11].

15. (a) Give an example of a nonzero homomorphism $f : R \to S$ of rings with identity such that $f(1_R) \neq 1_S$.
(b) If $f : R \to S$ is an epimorphism of rings with identity, then $f(1_R) = 1_S$.
(c) If $f : R \to S$ is a homomorphism of rings with identity and u is a unit in R such that $f(u)$ is a unit in S, then $f(1_R) = 1_S$ and $f(u^{-1}) = f(u)^{-1}$. [Note: there are easy examples which show that $f(u)$ need not be a unit in S even though u is a unit in R.]

16. Let $f : R \to S$ be a homomorphism of rings such that $f(r) \neq 0$ for some nonzero $r \in R$. If R has an identity and S has no zero divisors, then S is a ring with identity $f(1_R)$.

[1]Terminology due to V. O. McBrien.

17. (a) If R is a ring, then so is R^{op}, where R^{op} is defined as follows. The underlying set of R^{op} is precisely R and addition in R^{op} coincides with addition in R. Multiplication in R^{op}, denoted $\circ$, is defined by $a \circ b = ba$, where ba is the product in R. R^{op} is called the **opposite ring** of R.
(b) R has an identity if and only if R^{op} does.
(c) R is a division ring if and only if R^{op} is.
(d) $(R^{op})^{op} = R$.
(e) If S is a ring, then $R \cong S$ if and only if $R^{op} \cong S^{op}$.

18. Let $\mathbf{Q}$ be the field of rational numbers and R any ring. If $f,g : \mathbf{Q} \to R$ are homomorphisms of rings such that $f \mid \mathbf{Z} = g \mid \mathbf{Z}$, then $f = g$. [*Hint:* show that for $n \in \mathbf{Z}$ ($n \neq 0$), $f(1/n)g(n) = g(1)$, whence $f(1/n) = g(1/n)$.]

2. IDEALS

Just as normal subgroups played a crucial role in the theory of groups, so ideals play an analogous role in the study of rings. The basic properties of ideals are developed, including a characterization of principal ideals (Theorem 2.5) and the various isomorphism theorems (2.9–2.13; these correspond to the isomorphism theorems for groups). Prime and maximal ideals are characterized in several ways. Direct products in the category of rings are discussed and the Chinese Remainder Theorem is proved.

Definition 2.1. *Let* R *be a ring and* S *a nonempty subset of* R *that is closed under the operations of addition and multiplication in* R. *If* S *is itself a ring under these operations then* S *is called a* **subring** *of* R. *A subring* I *of a ring* R *is a* **left ideal** *provided*

$$r \in R \quad \text{and} \quad x \in I \quad \Rightarrow \quad rx \in I;$$

I *is a* **right ideal** *provided*

$$r \in R \quad \text{and} \quad x \in I \quad \Rightarrow \quad xr \in I;$$

I *is an* **ideal** *if it is both a left and right ideal.*

Whenever a statement is made about left ideals it is to be understood that the analogous statement holds for right ideals.

EXAMPLE. If R is any ring, then the **center** of R is the set $C = \{c \in R \mid cr = rc$ for all $r \in R\}$. C is easily seen to be a subring of R, but may not be an ideal (Exercise 6).

EXAMPLE. If $f : R \to S$ is a homomorphism of rings, then Ker f is an ideal in R (Theorem 2.8 below) and Im f is a subring of S. Im f need not be an ideal in S.

EXAMPLE. For each integer n the cyclic subgroup $\langle n \rangle = \{kn \mid k \in \mathbf{Z}\}$ is an ideal in $\mathbf{Z}$.

EXAMPLE. In the ring R of $n \times n$ matrices over a division ring D, let I_k be the set of all matrices that have nonzero entries only in column k. Then I_k is a left ideal, but not a right ideal. If J_k consist of those matrices with nonzero entries only in row k, then J_k is a right ideal but not a left ideal.

EXAMPLE. Two ideals of a ring R are R itself and the **trivial ideal** (denoted 0), which consists only of the zero element.

REMARKS. A [left] ideal I of R such that $I \neq 0$ and $I \neq R$ is called a **proper** [left] **ideal.** Observe that if R has an identity 1_R and I is a [left] ideal of R, then $I = R$ if and only if $1_R \in I$. Consequently, a nonzero [left] ideal I of R is proper if and only if I contains no units of R; (for if $u \in R$ is a unit and $u \in I$, then $1_R = u^{-1}u \in I$). In particular, a division ring D has no proper left (or right) ideals since every nonzero element of D is a unit. For the converse, see Exercise 7. The ring of $n \times n$ matrices over a division ring has proper left and right ideals (see above), but no proper (two-sided) ideals (Exercise 9).

Theorem 2.2. *A nonempty subset* I *of a ring* R *is a left* [*resp. right*] *ideal if and only if for all* a,b $\in$ I *and* r $\in$ R:

(i) $a,b \in I \quad \Rightarrow \quad a - b \in I$; *and*
(ii) $a \in I, r \in R \quad \Rightarrow \quad ra \in I$ [*resp.* $ar \in I$].

PROOF. Exercise; see Theorem I.2.5. ■

Corollary 2.3. *Let* $\{A_i \mid i \in I\}$ *be a family of* [*left*] *ideals in a ring* R. *Then* $\bigcap_{i \in I} A_i$ *is also a* [*left*] *ideal.*

PROOF. Exercise. ■

Definition 2.4. *Let* X *be a subset of a ring* R. *Let* $\{A_i \mid i \in I\}$ *be the family of all* [*left*] *ideals in* R *which contain* X. *Then* $\bigcap_{i \in I} A_i$ *is called the* [*left*] **ideal generated by X.** *This ideal is denoted* (X).

The elements of X are called **generators** of the ideal (X). If $X = \{x_1, \ldots, x_n\}$, then the ideal (X) is denoted by $(x_1,x_2, \ldots, x_n)$ and said to be **finitely generated.** An ideal (x) generated by a single element is called a **principal ideal. A principal ideal ring** is a ring in which every ideal is principal. A principal ideal ring which is an integral domain is called a **principal ideal domain.**[2]

Theorem 2.5. *Let* R *be a ring* $a \in R$ *and* $X \subset R$.

(i) *The principal ideal* (a) *consists of all elements of the form* $ra + as + na + \sum_{i=1}^{m} r_i a s_i$ $(r,s,r_i,s_i \in R;\ m \in \mathbf{N}^*;$ *and* $n \in \mathbf{Z})$.

[2]The term "principal ideal ring" is sometimes used in the literature to denote what we have called a principal ideal domain.

(ii) *If* R *has an identity, then* (a) $= \left\{\sum_{i=1}^{n} r_i a s_i \mid r_i, s_i \in R; n \in \mathbf{N}^*\right\}$.

(iii) *If* a *is in the center of* R, *then* (a) $= \{ra + na \mid r \in R, n \in \mathbf{Z}\}$.

(iv) $Ra = \{ra \mid r \in R\}$ [*resp.* $aR = \{ar \mid r \in R\}$] *is a left* [*resp. right*] *ideal in* R (*which may not contain* a). *If* R *has an identity, then* $a \in Ra$ *and* $a \in aR$.

(v) *If* R *has an identity and* a *is in the center of* R, *then* $Ra = (a) = aR$.

(vi) *If* R *has an identity and* X *is in the center of* R, *then the ideal* (X) *consists of all finite sums* $r_1a_1 + \cdots + r_na_n$ ($n \in \mathbf{N}^*$; $r_i \in R$; $a_i \in X$).

REMARK. The hypothesis of (iii) is always satisfied in a commutative ring.

SKETCH OF PROOF OF 2.5. (i) Show that the set

$$I = \left\{ra + as + na + \sum_{i=1}^{m} r_i a s_i \mid r, s, r_i, s_i \in R; n \in \mathbf{Z};\ m \in \mathbf{N}^*\right\}$$

is an ideal containing a and contained in every ideal containing a. Then $I = (a)$. (ii) follows from the facts that $ra = ra1_R$, $as = 1_R as$, and $na = n(1_R a) = (n1_R)a$, with $n1_R \in R$. ■

Let $A_1, A_2, \ldots, A_n$ be nonempty subsets of a ring R. Denote by $A_1 + A_2 + \cdots + A_n$ the set $\{a_1 + a_2 + \cdots + a_n \mid a_i \in A_i \text{ for } i = 1,2,\ldots,n\}$. If A and B are nonempty subsets of R let AB denote the set of all finite sums $\{a_1b_1 + \cdots + a_nb_n \mid n \in \mathbf{N}^*;\ a_i \in A;\ b_i \in B\}$. If A consists of a single element a, we write aB for AB. Similarly if $B = \{b\}$, we write Ab for AB. Observe that if B [resp. A] is closed under addition, then $aB = \{ab \mid b \in B\}$ [resp. $Ab = \{ab \mid a \in A\}$]. More generally let $A_1A_2\cdots A_n$ denote the set of all finite sums of elements of the form $a_1a_2\cdots a_n$ ($a_i \in A_i$ for $i = 1,2,\ldots,n$). In the special case when all A_i ($1 \leq i \leq n$) are the same set A we denote $A_1A_2\cdots A_n = AA\cdots A$ by A^n.

Theorem 2.6. *Let* $A, A_1, A_2, \ldots, A_n$, B *and* C *be* [*left*] *ideals in a ring* R.

(i) $A_1 + A_2 + \cdots + A_n$ *and* $A_1A_2\cdots A_n$ *are* [*left*] *ideals;*

(ii) $(A + B) + C = A + (B + C)$;

(iii) $(AB)C = ABC = A(BC)$;

(iv) $B(A_1 + A_2 + \cdots + A_n) = BA_1 + BA_2 + \cdots BA_n$; *and* $(A_1 + A_2 + \cdots + A_n)C = A_1C + A_2C + \cdots + A_nC$.

SKETCH OF PROOF. Use Theorem 2.2 for (i). (iii) is a bit complicated but straightforward argument using the definitions. Use induction to prove (iv) by first showing that $A(B + C) = AB + AC$ and $(A + B)C = AC + BC$. ■

Ideals play approximately the same role in the theory of rings as normal subgroups do in the theory of groups. For instance, let R be a ring and I an ideal of R. Since the additive group of R is abelian, I is a normal subgroup. Consequently, by Theorem I.5.4 there is a well-defined quotient group R/I in which addition is given by:

$$(a + I) + (b + I) = (a + b) + I.$$

R/I can in fact be made into a ring.

Theorem 2.7. *Let* R *be a ring and* I *an ideal of* R. *Then the additive quotient group* R/I *is a ring with multiplication given by*

$$(a + I)(b + I) = ab + I.$$

If R *is commutative or has an identity, then the same is true of* R/I.

SKETCH OF PROOF OF 2.7. Once we have shown that multiplication in R/I is well defined, the proof that R/I is a ring is routine. (For example, if R has identity 1_R, then $1_R + I$ is the identity in R/I.) Suppose $a + I = a' + I$ and $b + I = b' + I$. We must show that $ab + I = a'b' + I$. Since $a' \varepsilon a' + I = a + I$, $a' = a + i$ for some $i \varepsilon I$. Similarly $b' = b + j$ with $j \varepsilon I$. Consequently $a'b' = (a + i)(b + j) = ab + ib + aj + ij$. Since I is an ideal,

$$a'b' - ab = ib + aj + ij \varepsilon I.$$

Therefore $a'b' + I = ab + I$ by Corollary I.4.3, whence multiplication in R/I is well defined. ■

As one might suspect from the analogy with groups, ideals and homomorphisms of rings are closely related.

Theorem 2.8. *If* $f : R \to S$ *is a homomorphism of rings, then the kernel of* f *is an ideal in* R. *Conversely if* I *is an ideal in* R, *then the map* $\pi : R \to R/I$ *given by* $r \mapsto r + I$ *is an epimorphism of rings with kernel* I.

The map π is called the **canonical epimorphism** (or projection).

PROOF OF 2.8. Ker f is an additive subgroup of R. If $x \varepsilon$ Ker f and $r \varepsilon R$, then $f(rx) = f(r)f(x) = f(r)0 = 0$, whence $rx \varepsilon$ Ker f. Similarly, $xr \varepsilon$ Ker f. Therefore, Ker f is an ideal. By Theorem I.5.5 the map π is an epimorphism of groups with kernel I. Since $\pi(ab) = ab + I = (a + I)(b + I) = \pi(a)\pi(b)$ for all $a,b \varepsilon R$, π is also an epimorphism of rings. ■

In view of the preceding results it is not surprising that the various isomorphism theorems for groups (Theorems I.5.6–I.5.12) carry over to rings with *normal subgroups* and *groups* replaced by *ideals* and *rings* respectively. In each case the desired isomorphism is known to exist for additive abelian groups. If the groups involved are, in fact, rings and the normal subgroups ideals, then one need only verify that the known isomorphism of groups is also a homomorphism and hence an isomorphism of rings. *Caution:* in the proofs of the isomorphism theorems for groups all groups and cosets are written multiplicatively, whereas the additive group of a ring and the cosets of an ideal are written additively.

Theorem 2.9. *If* $f : R \to S$ *is a homomorphism of rings and* I *is an ideal of* R *which is contained in the kernel of* f, *then there is a unique homomorphism of rings* $\bar{f} : R/I \to S$ *such that* $\bar{f}(a + I) = f(a)$ *for all* $a \varepsilon R$. *Im* $\bar{f}$ = *Im* f *and Ker* $\bar{f}$ = (*Ker* f)/I. $\bar{f}$ *is an isomorphism if and only if* f *is an epimorphism and* I = *Ker* f.

PROOF. Exercise; see Theorem I.5.6. ■

Corollary 2.10. (*First Isomorphism Theorem*) *If* $f : R \to S$ *is a homomorphism of rings, then* f *induces an isomorphism of rings* $R/\mathit{Ker}\ f \cong \mathit{Im}\ f$.

PROOF. Exercise; see Corollary I.5.7. ■

Corollary 2.11. *If* $f: R \to S$ *is a homomorphism of rings,* I *is an ideal in* R *and* J *is an ideal in* S *such that* $f(I) \subset J$, *then* f *induces a homomorphism of rings* $\bar{f} : R/I \to S/J$, *given by* $a + I \mapsto f(a) + J$. $\bar{f}$ *is an isomorphism if and only if* $\mathit{Im}\ f + J = S$ *and* $f^{-1}(J) \subset I$. *In particular, if* f *is an epimorphism such that* $f(I) = J$ *and* $\mathit{Ker}\ f \subset I$, *then* $\bar{f}$ *is an isomorphism.*

PROOF. Exercise; see Corollary I.5.8. ■

Theorem 2.12. *Let* I *and* J *be ideals in a ring* R.

(i) (*Second Isomorphism Theorem*) *There is an isomorphisms of rings* $I/(I \cap J) \cong (I + J)/J$;

(ii) (*Third Isomorphism Theorem*) *if* $I \subset J$, *then* J/I *is an ideal in* R/I *and there is an isomorphism of rings* $(R/I)/(J/I) \cong R/J$.

PROOF. Exercise; see Corollaries I.5.9 and I.5.10. ■

Theorem 2.13. *If* I *is an ideal in a ring* R, *then there is a one-to-one correspondence between the set of all ideals of* R *which contain* I *and the set of all ideals of* R/I, *given by* $J \mapsto J/I$. *Hence every ideal in* R/I *is of the form* J/I, *where* J *is an ideal of* R *which contains* I.

PROOF. Exercise; see Theorem I.5.11, Corollary I.5.12 and Exercise 13. ■

Next we shall characterize in several ways two kinds of ideals (prime and maximal), which are frequently of interest.

Definition 2.14. *An ideal* P *in a ring* R *is said to be* **prime** *if* $P \neq R$ *and for any ideals* A,B *in* R

$$AB \subset P \quad \Rightarrow \quad A \subset P \quad \text{or} \quad B \subset P.$$

The definition of prime ideal excludes the ideal *R* for both historical and technical reasons. Here is a very useful characterization of prime ideals; other characterizations are given in Exercise 14.

Theorem 2.15. *If* P *is an ideal in a ring* R *such that* $P \neq R$ *and for all* $a,b \in R$

$$ab \in P \quad \Rightarrow \quad a \in P \quad or \quad b \in P, \qquad \textbf{(1)}$$

then P *is prime. Conversely if* P *is prime and* R *is commutative, then* P *satisfies condition* (1).

REMARK. Commutativity is necessary for the converse (Exercise 9 (b)).

PROOF OF 2.15. If A and B are ideals such that $AB \subset P$ and $A \not\subset P$, then there exists an element $a \in A - P$. For every $b \in B$, $ab \in AB \subset P$, whence $a \in P$ or $b \in P$. Since $a \notin P$, we must have $b \in P$ for all $b \in B$; that is, $B \subset P$. Therefore, P is prime. Conversely, if P is any ideal and $ab \in P$, then the principal ideal (ab) is contained in P by Definition 2.4. If R is commutative, then Theorem 2.5 implies that $(a)(b) \subset (ab)$, whence $(a)(b) \subset P$. If P is prime, then either $(a) \subset P$ or $(b) \subset P$, whence $a \in P$ or $b \in P$. ■

EXAMPLES. The zero ideal in any integral domain is prime since $ab = 0$ if and only if $a = 0$ or $b = 0$. If p is a prime integer, then the principal ideal (p) in **Z** is prime since

$$ab \in (p) \quad \Rightarrow \quad p \mid ab \quad \Rightarrow \quad p \mid a \quad \text{or} \quad p \mid b \quad \Rightarrow \quad a \in (p) \quad \text{or} \quad b \in (p).$$

Theorem 2.16. *In a commutative ring* R *with identity* $1_R \neq 0$ *an ideal* P *is prime if and only if the quotient ring* R/P *is an integral domain.*

PROOF. R/P is a commutative ring with identity $1_R + P$ and zero element $0 + P = P$ by Theorem 2.7. If P is prime, then $1_R + P \neq P$ since $P \neq R$. Furthermore, R/P has no zero divisors since

$$(a + P)(b + P) = P \quad \Rightarrow \quad ab + P = P \quad \Rightarrow \quad ab \in P \quad \Rightarrow \quad a \in P \quad \text{or}$$
$$b \in P \quad \Rightarrow \quad a + P = P \quad \text{or} \quad b + P = P.$$

Therefore, R/P is an integral domain. Conversely, if R/P is an integral domain, then $1_R + P \neq 0 + P$, whence $1_R \notin P$. Therefore, $P \neq R$. Since R/P has no zero divisors,

$$ab \in P \quad \Rightarrow \quad ab + P = P \quad \Rightarrow \quad (a + P)(b + P) = P \quad \Rightarrow \quad a + P = P \quad \text{or}$$
$$b + P = P \quad \Rightarrow \quad a \in P \quad \text{or} \quad b \in P.$$

Therefore, P is prime by Theorem 2.15. ■

Definition 2.17. *An ideal [resp. left ideal]* M *in a ring* R *is said to be* **maximal** *if* $M \neq R$ *and for every ideal [resp. left ideal]* N *such that* $M \subset N \subset R$, *either* $N = M$ *or* $N = R$.

EXAMPLE. The ideal (3) is maximal in **Z**; but the ideal (4) is not since $(4) \underset{\neq}{\subset} (2) \underset{\neq}{\subset} \mathbf{Z}$.

REMARK. If R is a ring and $\mathcal{S}$ is the set of all ideals I of R such that $I \neq R$, then $\mathcal{S}$ is partially ordered by set-theoretic inclusion. M is a maximal ideal (Definition 2.17) if and only if M is a maximal element in the partially ordered set $\mathcal{S}$ in the sense of Introduction, Section 7. More generally one sometimes speaks of an ideal I that is maximal *with respect to a given property*, meaning that under the partial ordering of set theoretic inclusion, I is maximal in the set of all ideals of R which have the given property. In this case I need not be maximal in the sense of Definition 2.17.

Theorem 2.18. *In a nonzero ring* R *with identity maximal* [*left*] *ideals always exist. In fact every* [*left*] *ideal in* R (*except* R *itself*) *is contained in a maximal* [*left*] *ideal.*

PROOF. Since 0 is an ideal and $0 \neq R$, it suffices to prove the second statement. The proof is a straightforward application of Zorn's Lemma. If A is a [left] ideal in R such that $A \neq R$, let $\mathcal{S}$ be the set of all [left] ideals B in R such that $A \subset B \neq R$. $\mathcal{S}$ is nonempty since $A \in \mathcal{S}$. Partially order $\mathcal{S}$ by set theoretic inclusion (that is, $B_1 \leq B_2 \Leftrightarrow B_1 \subset B_2$). In order to apply Zorn's Lemma we must show that every chain $\mathcal{C} = \{C_i \mid i \in I\}$ of [left] ideals in $\mathcal{S}$ has an upper bound in $\mathcal{S}$. Let $C = \bigcup_{i \in I} C_i$. We claim that C is a [left] ideal. If $a,b \in C$, then for some $i,j \in I$, $a \in C_i$ and $b \in C_j$. Since $\mathcal{C}$ is a chain, either $C_i \subset C_j$ or $C_j \subset C_i$; say the latter. Hence $a,b \in C_i$. Since C_i is a left ideal, $a - b \in C_i$ and $ra \in C_i$ for all $r \in R$ (if C_i is an ideal $ar \in C_i$ as well). Therefore, $a,b \in C$ imply $a - b$ and ra are in $C_i \subset C$. Consequently, C is a [left] ideal by Theorem 2.2. Since $A \subset C_i$ for every i, $A \subset \bigcup C_i = C$. Since each C_i is in $\mathcal{S}$, $C_i \neq R$ for all $i \in I$. Consequently, $1_R \notin C_i$ for every i (otherwise $C_i = R$), whence $1_R \notin \bigcup C_i = C$. Therefore, $C \neq R$ and hence, $C \in \mathcal{S}$. Clearly C is an upper bound of the chain $\mathcal{C}$. Thus the hypotheses of Zorn's Lemma are satisfied and hence $\mathcal{S}$ contains a maximal element. But a maximal element of $\mathcal{S}$ is obviously a maximal [left] ideal in R that contains A. ■

Theorem 2.19. *If* R *is a commutative ring such that* $R^2 = R$ (*in particular if* R *has an identity*), *then every maximal ideal* M *in* R *is prime.*

REMARK. The converse of Theorem 2.19 is false. For example, 0 is a prime ideal in **Z**, but not a maximal ideal. See also Exercise 9.

PROOF OF 2.19. Suppose $ab \in M$ but $a \notin M$ and $b \notin M$. Then each of the ideals $M + (a)$ and $M + (b)$ properly contains M. By maximality $M + (a) = R = M + (b)$. Since R is commutative and $ab \in M$, Theorem 2.5 implies that $(a)(b) \subset (ab) \subset M$. Therefore, $R = R^2 = (M + (a))(M + (b)) \subset M^2 + (a)M + M(b) + (a)(b) \subset M$. This contradicts the fact that $M \neq R$ (since M is maximal). Therefore, $a \in M$ or $b \in M$, whence M is prime by Theorem 2.15. ■

Maximal ideals, like prime ideals, may be characterized in terms of their quotient rings.

Theorem 2.20. *Let* M *be an ideal in a ring* R *with identity* $1_R \neq 0$.

(i) *If* M *is maximal and* R *is commutative, then the quotient ring* R/M *is a field.*
(ii) *If the quotient ring* R/M *is a division ring, then* M *is maximal.*

REMARKS. (i) is false if R does not have an identity (Exercise 19). If M is maximal and R is not commutative, then R/M need not be a division ring (Exercise 9).

PROOF OF 2.20. (i) If M is maximal, then M is prime (Theorem 2.19), whence R/M is an integral domain by Theorem 2.16. Thus we need only show that if $a + M \neq M$, then $a + M$ has a multiplicative inverse in R/M. Now $a + M \neq M$ implies that $a \notin M$, whence M is properly contained in the ideal $M + (a)$. Since M is maximal, we must have $M + (a) = R$. Therefore, since R is commutative, $1_R = m + ra$ for some $m \in M$ and $r \in R$, by Theorem 2.5(v). Thus $1_R - ra = m \in M$, whence

$$1_R + M = ra + M = (r + M)(a + M).$$

Thus $r + M$ is a multiplicative inverse of $a + M$ in R/M, whence R/M is a field.

(ii) If R/M is a division ring, then $1_R + M \neq 0 + M$, whence $1_R \notin M$ and $M \neq R$. If N is an ideal such that $M \underset{\neq}{\subset} N$, let $a \in N - M$. Then $a + M$ has a multiplicative inverse in R/M, say $(a + M)(b + M) = 1_R + M$. Consequently, $ab + M = 1_R + M$ and $ab - 1_R = c \in M$. But $a \in N$ and $M \subset N$ imply that $1_R \in N$. Thus $N = R$. Therefore, M is maximal. ■

Corollary 2.21. *The following conditions on a commutative ring* R *with identity* $1_R \neq 0$ *are equivalent.*

(i) R *is a field;*
(ii) R *has no proper ideals;*
(iii) 0 *is a maximal ideal in* R;
(iv) *every nonzero homomorphism of rings* $R \to S$ *is a monomorphism.*

REMARK. The analogue of Corollary 2.21 for division rings is false (Exercise 9).

PROOF OF 2.21. This result may be proved directly (Exercise 7) or as follows. $R \cong R/0$ is a field if and only if 0 is maximal by Theorem 2.20. But clearly 0 is maximal if and only if R has no proper ideals. Finally, for every ideal $I(\neq R)$ the canonical map $\pi : R \to R/I$ is a nonzero homomorphism with kernel I (Theorem 2.8). Since π is a monomorphism if and only if $I = 0$, (iv) holds if and only if R has no proper ideals. ■

We now consider (direct) products in the category of rings. Their existence and basic properties are easily proved, using the corresponding facts for groups. Coproducts of rings, however, are decidedly more complicated. Furthermore coproducts in the category of rings are of less use than, for example, coproducts (direct sums) in the category of abelian groups.

Theorem 2.22. *Let* $\{R_i \mid i \in I\}$ *be a nonempty family of rings and* $\prod_{i \in I} R_i$ *the direct product of the additive abelian groups* R_i;

(i) $\prod_{i\in I} R_i$ *is a ring with multiplication defined by* $\{a_i\}_{i\in I}\{b_i\}_{i\in I} = \{a_ib_i\}_{i\in I}$;

(ii) *if* R_i *has an identity [resp. is commutative] for every* $i \in I$, *then* $\prod_{i\in I} R_i$ *has an identity [resp. is commutative]*;

(iii) *for each* $k \in I$ *the canonical projection* $\pi_k: \prod_{i\in I} R_i \to R_k$ *given by* $\{a_i\} \mapsto a_k$, *is an epimorphism of rings;*

(iv) *for each* $k \in I$ *the canonical injection* $\iota_k : R_k \to \prod_{i\in I} R_i$, *given by* $a_k \mapsto \{a_i\}$ *(where* $a_i = 0$ *for* $i \neq k$*), is a monomorphism of rings.*

PROOF. Exercise. ■

$\prod_{i\in I} R_i$ is called the **(external) direct product** of the family of rings $\{R_i \mid i \in I\}$. If the index set is finite, say $I = \{1, \ldots, n\}$, then we sometimes write $R_1 \times R_2 \times \cdots \times R_n$ instead of $\prod R_i$.

If $\{R_i \mid i \in I\}$ is a family of rings and for each $i \in I$, A_i is an ideal in R_i, then it is easy to see that $\prod_{i\in I} A_i$ is an ideal in $\prod_{i\in I} R_i$. If $A_i = 0$ for all $i \neq k$, then the ideal $\prod_{i\in I} A_i$ is precisely $\iota_k(A_k)$. If the index set I is finite and each R_i has an identity, then every ideal in $\prod_{i\in I} R_i$ is of the form $\prod_{i\in I} A_i$ with A_i an ideal in R_i (Exercise 22).

Theorem 2.23. *Let* $\{R_i \mid i \in I\}$ *be a nonempty family of rings,* S *a ring and* $\{\varphi_i : S \to R_i \mid i \in I\}$ *a family of homomorphisms of rings. Then there is a unique homomorphism of rings* $\varphi : S \to \prod_{i\in I} R_i$ *such that* $\pi_i\varphi = \varphi_i$ *for all* $i \in I$. *The ring* $\prod_{i\in I} R_i$ *is uniquely determined up to isomorphism by this property. In other words* $\prod_{i\in I} R_i$ *is a product in the category of rings.*

SKETCH OF PROOF. By Theorem I.8.2 there is a unique homomorphism of groups $\varphi : S \to \prod_{i\in I} R_i$ such that $\pi_i\varphi = \varphi_i$ for all $i \in I$. Verify that φ is also a ring homomorphism. Thus $\prod_{i\in I} R_i$ is a product in the category of rings (Definition I.7.2) and therefore determined up to isomorphism by Theorem I.7.3. ■

Theorem 2.24. *Let* $A_1, A_2, \ldots, A_n$ *be ideals in a ring* R *such that* (i) $A_1 + A_2 + \cdots + A_n = R$ *and* (ii) *for each* k $(1 \leq k \leq n)$, $A_k \cap (A_1 + \cdots + A_{k-1} + A_{k+1} + \cdots + A_n) = 0$. *Then there is a ring isomorphism* $R \cong A_1 \times A_2 \times \cdots \times A_n$.

PROOF. By the proof of Theorem I.8.6 the map $\varphi : A_1 \times A_2 \times \cdots \times A_n \to R$ given by $(a_1, \ldots, a_n) \mapsto a_1 + a_2 + \cdots + a_n$ is an isomorphism of additive abelian groups. We need only verify that φ is a ring homomorphism. Observe that if $i \neq j$ and $a_i \in A_i$, $a_j \in A_j$, then by (ii) $a_ia_j \in A_i \cap A_j = 0$. Consequently, for all $a_i, b_i \in A_i$:

$$(a_1 + a_2 + \cdots + a_n)(b_1 + b_2 + \cdots + b_n) = a_1b_1 + \cdots + a_nb_n,$$

whence φ is a homomorphism of rings. ■

If R is a ring and $A_1, \ldots, A_n$ are ideals in R that satisfy the hypotheses of Theorem 2.24, then R is said to be the **(internal) direct product** of the ideals A_i. As in the case of groups, there is a distinction between internal and external direct products. If a ring R is the *internal* direct product of ideals $A_1, \ldots, A_n$, then each of the A_i is actually an ideal contained in R and R is *isomorphic* to the *external* direct product $A_1 \times \cdots \times A_n$. However, the external direct product $A_1 \times \cdots \times A_n$ does *not* contain the A_i, but only isomorphic copies of them (namely the $\iota_i(A_i)$ — see Theorem 2.22). Since this distinction is unimportant in practice, the adjectives "internal" and "external" will be omitted whenever the context is clear and the following notation will be used.

NOTATION. We write $R = \prod A_i$ or $R = A_1 \times A_2 \times \cdots \times A_n$ to indicate that the ring R is the internal direct product of its ideals $A_1, \ldots, A_n$.

Other characterizations of finite direct products are given in Exercise 24.

We close this section with a result that will be needed in Chapters VIII and IX. Let A be an ideal in a ring R and $a,b \in R$. The element a is said to be **congruent** to b modulo A (denoted $a \equiv b \pmod{A}$) if $a - b \in A$. Thus

$$a \equiv b \pmod{A} \quad \Leftrightarrow \quad a - b \in A \quad \Leftrightarrow \quad a + A = b + A.$$

Since R/A is a ring by Theorem 2.7,

$$\begin{aligned} &a_1 \equiv a_2 \pmod{A} \quad \text{and} \quad b_1 \equiv b_2 \pmod{A} \quad \Rightarrow \\ &a_1 + b_1 \equiv a_2 + b_2 \pmod{A} \quad \text{and} \quad a_1b_1 \equiv a_2b_2 \pmod{A}. \end{aligned}$$

Theorem 2.25. (*Chinese Remainder Theorem*) *Let* $A_1, \ldots, A_n$ *be ideals in a ring* R *such that* $R^2 + A_i = R$ *for all* i *and* $A_i + A_j = R$ *for all* $i \neq j$. *If* $b_1, \ldots, b_n \in R$, *then there exists* $b \in R$ *such that*

$$b \equiv b_i \pmod{A_i} \qquad (i = 1,2, \ldots, n).$$

Furthermore b *is uniquely determined up to congruence modulo the ideal*

$$A_1 \cap A_2 \cap \cdots \cap A_n.$$

REMARK. If R has an identity, then $R^2 = R$, whence $R^2 + A = R$ for every ideal A of R.

SKETCH OF PROOF OF 2.25. Since $A_1 + A_2 = R$ and $A_1 + A_3 = R$,

$$\begin{aligned} R^2 &= (A_1 + A_2)(A_1 + A_3) = A_1^2 + A_1A_3 + A_2A_1 + A_2A_3 \\ &\subset A_1 + A_2A_3 \subset A_1 + (A_2 \cap A_3). \end{aligned}$$

Consequently, since $R = A_1 + R^2$,

$$R = A_1 + R^2 \subset A_1 + (A_1 + (A_2 \cap A_3)) = A_1 + (A_2 \cap A_3) \subset R.$$

Therefore, $R = A_1 + (A_2 \cap A_3)$. Assume inductively that

$$R = A_1 + (A_2 \cap A_3 \cap \cdots \cap A_{k-1}).$$

Then

$$R^2 = (A_1 + (A_2 \cap \cdots \cap A_{k-1}))(A_1 + A_k) \subset A_1 + (A_2 \cap A_3 \cap \cdots \cap A_k)$$

and hence

$$R = R^2 + A_1 \subset A_1 + (A_2 \cap \cdots \cap A_k) \subset R.$$

Therefore, $R = A_1 + (A_2 \cap \cdots \cap A_k)$ and the induction step is proved. Consequently, $R = A_1 + (A_2 \cap \cdots \cap A_n) = A_1 + (\bigcap_{i \neq 1} A_i)$. A similar argument shows that for each $k = 1,2,\ldots,n$, $R = A_k + (\bigcap_{i \neq k} A_i)$. Consequently for each k there exist elements $a_k \in A_k$ and $r_k \in \bigcap_{i \neq k} A_i$ such that $b_k = a_k + r_k$. Furthermore

$$r_k \equiv b_k \pmod{A_k} \quad \text{and} \quad r_k \equiv 0 \pmod{A_i} \quad \text{for} \quad i \neq k.$$

Let $b = r_1 + r_2 + \cdots + r_n$ and use the remarks preceding the theorem to verify that $b \equiv b_i \pmod{A_i}$ for every i. Finally if $c \in R$ is such that $c \equiv b_i \pmod{A_i}$ for every i, then $b \equiv c \pmod{A_i}$ for each i, whence $b - c \in A_i$ for all i. Therefore, $b - c \in \bigcap_{i=1}^{n} A_i$ and $b \equiv c \left(\text{mod } \bigcap_{i=1}^{n} A_i\right)$. ■

The Chinese Remainder Theorem is so named because it is a generalization of the following fact from elementary number theory, which was known to Chinese mathematicians in the first century A.D.

Corollary 2.26. *Let* $m_1,m_2,\ldots,m_n$ *be positive integers such that* $(m_i,m_j) = 1$ *for* $i \neq j$. *If* $b_1,b_2,\ldots,b_n$ *are any integers, then the system of congruences*

$$x \equiv b_1 \pmod{m_1};\ x \equiv b_2 \pmod{m_2};\ \cdots;\ x \equiv b_n \pmod{m_n}$$

has an integral solution that is uniquely determined modulo $m = m_1m_2\cdots m_n$.

SKETCH OF PROOF. Let $A_i = (m_i)$; then $\bigcap_{i=1}^{n} A_i = (m)$. Show that $(m_i,m_j) = 1$ implies $A_i + A_j = \mathbf{Z}$ and apply Theorem 2.25. ■

Corollary 2.27. *If* $A_1,\ldots,A_n$ *are ideals in a ring* R, *then there is a monomorphism of rings*

$$\theta : R/(A_1 \cap \cdots \cap A_n) \to R/A_1 \times R/A_2 \times \cdots \times R/A_n.$$

If $R^2 + A_i = R$ *for all* i *and* $A_i + A_j = R$ *for all* $i \neq j$, *then* θ *is an isomorphism of rings.*

SKETCH OF PROOF. By Theorem 2.23 the canonical epimorphisms $\pi_k : R \to R/A_k$ $(k = 1,\ldots,n)$ induce a homomorphism of rings $\theta_1 : R \to R/A_1 \times \cdots \times R/A_n$ with $\theta_1(r) = (r + A_1,\ldots,r + A_n)$. Clearly $\ker \theta_1 = A_1 \cap \cdots \cap A_n$. Therefore, θ_1 induces a monomorphism of rings $\theta : R/(A_1 \cap \cdots \cap A_n) \to R/A_1 \times \cdots \times R/A_n$ (Theorem 2.9). The map θ need not be surjective (Exercise 26). However, if the hypotheses of Theorem 2.25 are satisfied and $(b_1 + A_1,\ldots,b_n + A_n) \in R/A_1$

$\times \cdots \times R/A_n$, then there exists $b \in R$ such that $b \equiv b_i \pmod{A_i}$ for all i. Thus $\theta(b + \bigcap_i A_i) = (b + A_1, \ldots, b + A_n) = (b_1 + A_1, \ldots, b_n + A_n)$, whence θ is an epimorphism. ■

EXERCISES

1. The set of all nilpotent elements in a commutative ring forms an ideal [see Exercise 1.12].

2. Let I be an ideal in a commutative ring R and let Rad $I = \{r \in R \mid r^n \in I$ for some $n\}$. Show that Rad I is an ideal.

3. If R is a ring and $a \in R$, then $J = \{r \in R \mid ra = 0\}$ is a left ideal and $K = \{r \in R \mid ar = 0\}$ is a right ideal in R.

4. If I is a left ideal of R, then $A(I) = \{r \in R \mid rx = 0$ for every $x \in I\}$ is an ideal in R.

5. If I is an ideal in a ring R, let $[R : I] = \{r \in R \mid xr \in I$ for every $x \in R\}$. Prove that $[R : I]$ is an ideal of R which contains I.

6. (a) The center of the ring S of all 2×2 matrices over a field F consists of all matrices of the form $\begin{pmatrix} a & 0 \\ 0 & a \end{pmatrix}$.
   (b) The center of S is not an ideal in S.
   (c) What is the center of the ring of all $n \times n$ matrices over a division ring?

7. (a) A ring R with identity is a division ring if and only if R has no proper left ideals. [Proposition I.1.3 may be helpful.]
   (b) If S is a ring (possibly without identity) with no proper left ideals, then either $S^2 = 0$ or S is a division ring. [*Hint:* show that $\{a \in S \mid Sa = 0\}$ is an ideal. If $cd \neq 0$, show that $\{r \in S \mid rd = 0\} = 0$. Find $e \in S$ such that $ed = d$ and show that e is a (two-sided) identity.]

8. Let R be a ring with identity and S the ring of all $n \times n$ matrices over R. J is an ideal of S if and only if J is the ring of all $n \times n$ matrices over I for some ideal I in R. [*Hint:* Given J, let I be the set of all those elements of R that appear as the row 1–column 1 entry of some matrix in J. Use the matrices $E_{r,s}$, where $1 \leq r \leq n$, $1 \leq s \leq n$, and $E_{r,s}$ has 1_R as the row r–column s entry and 0 elsewhere. Observe that for a matrix $A = (a_{ij})$, $E_{p,r}AE_{s,q}$ is the matrix with a_{rs} in the row p-column q entry and 0 elsewhere.]

9. Let S be the ring of all $n \times n$ matrices over a division ring D.
   (a) S has no proper ideals (that is, 0 is a maximal ideal). [*Hint:* apply Exercise 8 or argue directly, using the matrices $E_{r,s}$ mentioned there.]
   (b) S has zero divisors. Consequently, (i) $S \cong S/0$ is not a division ring and (ii) 0 is a prime ideal which does not satisfy condition (1) of Theorem 2.15.

10. (a) Show that **Z** is a principal ideal ring [see Theorem I.3.1].
    (b) Every homomorphic image of a principal ideal ring is also a principal ideal ring.
    (c) Z_m is a principal ideal ring for every $m > 0$.

11. If N is the ideal of all nilpotent elements in a commutative ring R (see Exercise 1), then R/N is a ring with no nonzero nilpotent elements.

12. Let R be a ring without identity and with no zero divisors. Let S be the ring whose additive group is $R \times \mathbf{Z}$ as in the proof of Theorem 1.10. Let $A = \{(r,n) \varepsilon S \mid rx + nx = 0 \text{ for every } x \varepsilon R\}$.
   (a) A is an ideal in S.
   (b) S/A has an identity and contains a subring isomorphic to R.
   (c) S/A has no zero divisors.

13. Let $f : R \to S$ be a homomorphism of rings, I an ideal in R, and J an ideal in S.
   (a) $f^{-1}(J)$ is an ideal in R that contains Ker f.
   (b) If f is an epimorphism, then $f(I)$ is an ideal in S. If f is not surjective, $f(I)$ need not be an ideal in S.

14. If P is an ideal in a not necessarily commutative ring R, then the following conditions are equivalent.
   (a) P is a prime ideal.
   (b) If $r,s \varepsilon R$ are such that $rRs \subset P$, then $r \varepsilon P$ or $s \varepsilon P$. [*Hint:* If (a) holds and $rRs \subset P$, then $(RrR)(RsR) \subset P$, whence $RrR \subset P$ or $RsR \subset P$, say $RrR \subset P$. If $A = (r)$, then $A^3 \subset RrR \subset P$, whence $r \varepsilon A \subset P$.]
   (c) If (r) and (s) are principal ideals of R such that $(r)(s) \subset P$, then $r \varepsilon P$ or $s \varepsilon P$.
   (d) If U and V are right ideals in R such that $UV \subset P$, then $U \subset P$ or $V \subset P$.
   (e) If U and V are left ideals in R such that $UV \subset P$, then $U \subset P$ or $V \subset P$.

15. The set consisting of zero and all zero divisors in a commutative ring with identity contains at least one prime ideal.

16. Let R be a commutative ring with identity and suppose that the ideal A of R is contained in a finite union of prime ideals $P_1 \cup \cdots \cup P_n$. Show that $A \subset P_i$ for some i. [*Hint:* otherwise one may assume that $A \cap P_j \not\subset \bigcup_{i \neq j} P_i$ for all j. Let $a_j \varepsilon (A \cap P_j) - (\bigcup_{i \neq j} P_i)$. Then $a_1 + a_2a_3 \cdots a_n$ is in A but not in $P_1 \cup \cdots \cup P_n$.]

17. Let $f : R \to S$ be an epimorphism of rings with kernel K.
   (a) If P is a prime ideal in R that contains K, then $f(P)$ is a prime ideal in S [see Exercise 13].
   (b) If Q is a prime ideal in S, then $f^{-1}(Q)$ is a prime ideal in R that contains K.
   (c) There is a one-to-one correspondence between the set of all prime ideals in R that contain K and the set of all prime ideals in S, given by $P \mapsto f(P)$.
   (d) If I is an ideal in a ring R, then every prime ideal in R/I is of the form P/I, where P is a prime ideal in R that contains I.

18. An ideal $M \neq R$ in a commutative ring R with identity is maximal if and only if for every $r \varepsilon R - M$, there exists $x \varepsilon R$ such that $1_R - rx \varepsilon M$.

19. The ring E of even integers contains a maximal ideal M such that E/M is *not* a field.

20. In the ring $\mathbf{Z}$ the following conditions on a nonzero ideal I are equivalent: (i) I is prime; (ii) I is maximal; (iii) $I = (p)$ with p prime.

21. Determine all prime and maximal ideals in the ring Z_m.

22. (a) If $R_1, \ldots, R_n$ are rings with identity and I is an ideal in $R_1 \times \cdots \times R_n$, then $I = A_1 \times \cdots \times A_m$, where each A_i is an ideal in R_i. [*Hint:* Given I let $A_k = \pi_k(I)$, where $\pi_k : R_1 \times \cdots \times R_n \to R_k$ is the canonical epimorphism.]
(b) Show that the conclusion of (a) need not hold if the rings R_i do not have identities.

23. An element e in a ring R is said to be **idempotent** if $e^2 = e$. An element of the center of the ring R is said to be **central.** If e is a central idempotent in a ring R with identity, then
(a) $1_R - e$ is a central idempotent;
(b) eR and $(1_R - e)R$ are ideals in R such that $R = eR \times (1_R - e)R$.

24. Idempotent elements $e_1, \ldots, e_n$ in a ring R [see Exercise 23] are said to be **orthogonal** if $e_ie_j = 0$ for $i \neq j$. If $R, R_1, \ldots, R_n$ are rings with identity, then the following conditions are equivalent:
(a) $R \cong R_1 \times \cdots \times R_n$.
(b) R contains a set of orthogonal central idempotents [Exercise 23] $\{e_1, \ldots, e_n\}$ such that $e_1 + e_2 + \cdots + e_n = 1_R$ and $e_iR \cong R_i$ for each i.
(c) R is the internal direct product $R = A_1 \times \cdots \times A_n$ where each A_i is an ideal of R such that $A_i \cong R_i$.
[*Hint:* (a) $\Rightarrow$ (b) The elements $\bar{e}_1 = (1_{R_1},0, \ldots, 0)$, $\bar{e}_2 = (0,1_{R_2},0, \ldots, 0), \ldots, \bar{e}_n = (0, \ldots, 0,1_{R_n})$ are orthogonal central idempotents in $S = R_1 \times \cdots \times R_n$ such that $\bar{e}_1 + \cdots + \bar{e}_n = 1_S$ and $\bar{e}_iS \cong R_i$. (b) $\Rightarrow$ (c) Note that $A_k = e_kR$ is the principal ideal (e_k) in R and that e_kR is itself a ring with identity e_k.]

25. If $m \in \mathbf{Z}$ has a prime decomposition $m = p_1^{k_1} \cdots p_t^{k_t}$ ($k_i > 0$; p_i distinct primes), then there is an isomorphism of rings $Z_m \cong Z_{p_1^{k_1}} \times \cdots \times Z_{p_t^{k_t}}$. [*Hint:* Corollary 2.27.]

26. If $R = \mathbf{Z}$, $A_1 = (6)$ and $A_2 = (4)$, then the map $\theta : R/A_1 \cap A_2 \to R/A_1 \times R/A_2$ of Corollary 2.27 is not surjective.

3. FACTORIZATION IN COMMUTATIVE RINGS

In this section we extend the concepts of divisibility, greatest common divisor and prime in the ring of integers to arbitrary commutative rings and study those integral domains in which an analogue of the Fundamental Theorem of Arithmetic (Introduction, Theorem 6.7) holds. The chief result is that every principal ideal domain is such a unique factorization domain. In addition we study those commutative rings in which an analogue of the division algorithm is valid (Euclidean rings).

Definition 3.1. *A nonzero element* a *of a commutative ring* R *is said to* **divide** *an element* $b \in R$ (*notation:* $a \mid b$) *if there exists* $x \in R$ *such that* $ax = b$. *Elements* a,b *of* R *are said to be* **associates** *if* $a \mid b$ *and* $b \mid a$.

Virtually all statements about divisibility may be phrased in terms of principal ideals as we now see.

Theorem 3.2. *Let* a,b *and* u *be elements of a commutative ring* R *with identity.*

(i) a | b *if and only if* (b) ⊂ (a).
(ii) a *and* b *are associates if and only if* (a) = (b).
(iii) u *is a unit if and only if* u | r *for all* r ε R.
(iv) u *is a unit if and only if* (u) = R.
(v) *The relation* "a *is an associate of b*" *is an equivalence relation on* R.
(vi) *If* a = br *with* r ε R *a unit, then* a *and* b *are associates. If* R *is an integral domain, the converse is true.*

PROOF. Exercise; Theorem 2.5(v) may be helpful for (i) and (ii). ■

Definition 3.3. *Let* R *be a commutative ring with identity. An element* c *of* R *is* **irreducible** *provided that:*

(i) c *is a nonzero nonunit;*
(ii) c = ab ⇒ a *or* b *is a unit.*

An element p *of* R *is* **prime** *provided that:*

(i) p *is a nonzero nonunit;*
(ii) p | ab ⇒ p | a *or* p | b.

EXAMPLES. If p is an ordinary prime integer, then both p and $-p$ are irreducible and prime in **Z** in the sense of Definition 3.3. In the ring Z_6, 2 is easily seen to be a prime. However $2 \in Z_6$ is not irreducible since $2 = 2\cdot 4$ and neither 2 nor 4 are units in Z_6 (indeed they are zero divisors). For an example of an irreducible element which is not prime, see Exercise 3.

There is a close connection between prime [resp. irreducible] elements in a ring R and prime [resp. maximal] principal ideals in R.

Theorem 3.4. *Let* p *and* c *be nonzero elements in an integral domain* R.

(i) p *is prime if and only if* (p) *is nonzero prime ideal;*
(ii) c *is irreducible if and only if* (c) *is maximal in the set* S *of all proper principal ideals of R.*
(iii) *Every prime element of* R *is irreducible.*
(iv) *If* R *is a principal ideal domain, then* p *is prime if and only if* p *is irreducible.*
(v) *Every associate of an irreducible* [*resp. prime*] *element of* R *is irreducible* [*resp. prime*].
(vi) *The only divisors of an irreducible element of* R *are its associates and the units of* R.

REMARK. Several parts of Theorem 3.4 are true for any commutative ring with identity, as is seen in the following proof.

SKETCH OF PROOF OF 3.4. (i) Use Definition 3.3 and Theorem 2.15. (ii) If c is irreducible then (c) is a proper ideal of R by Theorem 3.2. If $(c) \subset (d)$, then

$c = dx$. Since c is irreducible either d is a unit (whence $(d) = R$) or x is a unit (whence $(c) = (d)$ by Theorem 3.2). Hence (c) is maximal in S. Conversely if (c) is maximal in S, then c is a (nonzero) nonunit in R by Theorem 3.2. If $c = ab$, then $(c) \subset (a)$, whence $(c) = (a)$ or $(a) = R$. If $(a) = R$, then a is a unit (Theorem 3.2). If $(c) = (a)$, then $a = cy$ and hence $c = ab = cyb$. Since R is an integral domain $1 = yb$, whence b is a unit. Therefore, c is irreducible. (iii) If $p = ab$, then $p \mid a$ or $p \mid b$; say $p \mid a$. Then $px = a$ and $p = ab = pxb$, which implies that $1 = xb$. Therefore, b is a unit. (iv) If p is irreducible, use (ii), Theorem 2.19 and (i) to show that p is prime. (v) If c is irreducible and d is an associate of c, then $c = du$ with $u \in R$ a unit (Theorem 3.2). If $d = ab$, then $c = abu$, whence a is a unit or bu is a unit. But if bu is a unit, so is b. Hence d is irreducible. (vi) If c is irreducible and $a \mid c$, then $(c) \subset (a)$, whence $(c) = (a)$ or $(a) = R$ by (ii). Therefore, a is either an associate of c or a unit by Theorem 3.2. ∎

We have now developed the analogues in an arbitrary integral domain of the concepts of divisibility and prime integers in the ring **Z**. Recall that every element in **Z** is a product of a finite number of irreducible elements (prime integers or their negatives) according to the Fundamental Theorem of Arithmetic (Introduction, Theorem 6.7). Furthermore this factorization is essentially unique (except for the order of the irreducible factors). Consequently, **Z** is an example of:

Definition 3.5. *An integral domain* R *is a* **unique factorization domain** *provided that:*

(i) *every nonzero nonunit element* a *of* R *can be written* $a = c_1c_2\cdots c_n$, *with* $c_1, \ldots, c_n$ *irreducible.*

(ii) *If* $a = c_1c_2\cdots c_n$ *and* $a = d_1d_2\cdots d_m$ (c_i,d_i *irreducible*), *then* $n = m$ *and for some permutation* σ *of* $\{1,2,\ldots,n\}$, c_i *and* $d_{\sigma(i)}$ *are associates for every* i.

REMARK. Every irreducible element in a unique factorization domain is necessarily prime by (ii). Consequently, irreducible and prime elements coincide by Theorem 3.4 (iii).

Definition 3.5 is nontrivial in the sense that there are integral domains in which every element is a finite product of irreducible elements, but this factorization is not unique (that is, Definition 3.5 (ii) fails to hold); see Exercise 4. Indeed one of the historical reasons for introducing the concept of *ideal* was to obtain some sort of unique factorization theorems (for ideals) in rings of algebraic integers in which factorization of elements was not necessarily unique; see Chapter VIII.

In view of the relationship between irreducible elements and principal ideals (Theorem 3.4) and the example of the integers, it seems plausible that every principal ideal domain is a unique factorization domain. In order to prove that this is indeed the case we need:

Lemma 3.6. *If* R *is a principal ideal ring and* $(a_1) \subset (a_2) \subset \cdots$ *is a chain of ideals in* R, *then for some positive integer* n, $(a_j) = (a_n)$ *for all* $j \geq n$.

PROOF. Let $A = \bigcup_{i\geq 1} (a_i)$. We claim that A is an ideal. If $b,c \in A$, then $b \in (a_i)$ and $c \in (a_j)$. Either $i \leq j$ or $i \geq j$; say $i \geq j$. Consequently $(a_j) \subset (a_i)$ and $b,c \in (a_i)$. Since (a_i) is an ideal $b - c \in (a_i) \subset A$. Similarly if $r \in R$ and $b \in A$, then $b \in (a_i)$, whence $rb \in (a_i) \subset A$ and $br \in (a_i) \subset A$. Therefore, A is an ideal by Theorem 2.2. By hypothesis A is principal, say $A = (a)$. Since $a \in A = \bigcup(a_i)$, $a \in (a_n)$ for some n. By Definition 2.4 $(a) \subset (a_n)$. Therefore, for every $j \geq n$, $(a) \subset (a_n) \subset (a_j) \subset A = (a)$, whence $(a_j) = (a_n)$. ■

Theorem 3.7. *Every principal ideal domain* R *is a unique factorization domain.*

REMARK. The converse of Theorem 3.7 is false. For example the polynomial ring $\mathbf{Z}[x]$ can be shown to be a unique factorization domain (Theorem 6.14 below), but $\mathbf{Z}[x]$ is not a principal ideal domain (Exercise 6.1).

SKETCH OF PROOF OF 3.7. Let S be the set of all nonzero nonunit elements of R which cannot be factored as a finite product of irreducible elements. We shall first show that S is empty, whence every nonzero nonunit element of R has at least one factorization as a finite product of irreducibles. Suppose S is not empty and $a \in S$. Then (a) is a proper ideal by Theorem 3.2(iv) and is contained in a maximal ideal (c) by Theorem 2.18. The element $c \in R$ is irreducible by Theorem 3.4(ii). Since $(a) \subset (c)$, c divides a. Therefore, it is possible to choose for each $a \in S$ an irreducible divisor c_a of a (Axiom of Choice). Since R is an integral domain, c_a uniquely determines a nonzero $x_a \in R$ such that $c_a x_a = a$. We claim that $x_a \in S$. For if x_a were a unit, then $a = c_a x_a$ would be irreducible by Theorems 3.2(vi) and 3.4(v). If x_a is a nonunit and not in S, then x_a has a factorization as a product of irreducibles, whence a also does. Since $a \in S$ this is a contradiction. Hence $x_a \in S$. Furthermore, we claim that the ideal (a) is properly contained in the ideal (x_a). Since $x_a \mid a$, $(a) \subset (x_a)$ by Theorem 3.2(i). But $(a) = (x_a)$ implies that $x_a = ay$ for some $y \in R$, whence $a = x_a c_a = ayc_a$ and $1 = yc_a$. This contradicts the fact that c_a is irreducible (and hence a nonunit). Therefore $(a) \subsetneq (x_a)$.

The preceding remarks show that the function $f : S \to S$ given by $f(a) = x_a$ is well defined. By the Recursion Theorem 6.2 of the Introduction (with $f = f_n$ for all n) there exists a function $\varphi : \mathbf{N} \to S$ such that

$$\varphi(0) = a \quad \text{and} \quad \varphi(n+1) = f(\varphi(n)) = x_{\varphi(n)} \ (n \geq 0).$$

If we denote $\varphi(n)$ by a_n, we thus have a sequence of elements of S: $a, a_1, a_2, \ldots$ such that

$$a_1 = x_a;\ a_2 = x_{a_1};\ \cdots;\ a_{n+1} = x_{a_n};\ \cdots.$$

Consequently, the preceding paragraph shows that there is an ascending chain of ideals

$$(a) \subsetneq (a_1) \subsetneq (a_2) \subsetneq (a_3) \subsetneq \cdots,$$

contradicting Lemma 3.6. Therefore, the set S must be empty, whence every nonzero nonunit element in R has a factorization as a finite product of irreducibles.

Finally if $c_1c_2\cdots c_n = a = d_1d_2\cdots d_m$ (c_i,d_i irreducible), then c_1 divides some d_i by Theorem 3.4(iv). Since c_1 is a nonunit, it must be an associate of d_i by Theorem 3.4 (vi). The proof of uniqueness is now completed by a routine inductive argument. ■

Several important integral domains that we shall meet frequently have certain properties not shared by all integral domains.

Definition 3.8. *Let* N *be the set of nonnegative integers and* R *a commutative ring.* R *is a* **Euclidean ring** *if there is a function* $\varphi : R - \{0\} \to N$ *such that:*

(i) *if* $a,b \in R$ *and* $ab \neq 0$, *then* $\varphi(a) \leq \varphi(ab)$;
(ii) *if* $a,b \in R$ *and* $b \neq 0$, *then there exist* $q,r \in R$ *such that* $a = qb + r$ *with* $r = 0$, *or* $r \neq 0$ *and* $\varphi(r) < \varphi(b)$.

A Euclidean ring which is an integral domain is called a **Euclidean domain.**

EXAMPLE. The ring **Z** of integers with $\varphi(x) = |x|$ is a Euclidean domain.

EXAMPLE. If F is a field, let $\varphi(x) = 1$ for all $x \in F$, $x \neq 0$. Then F is a Euclidean domain.

EXAMPLE. If F is a field, then the ring of polynomials in one variable $F[x]$ is a Euclidean domain with $\varphi(f) =$ degree of f; see Corollary 6.4 below.

EXAMPLE. Let $\mathbf{Z}[i]$ be the following subset of the complex numbers $\mathbf{Z}[i] = \{a + bi \mid a, b \in \mathbf{Z}\}$. $\mathbf{Z}[i]$ is an integral domain called the domain of **Gaussian integers.** Define $\varphi(a + bi) = a^2 + b^2$. Clearly $\varphi(a + bi) \neq 0$ if $a + bi \neq 0$; it is also easy to show that condition (i) of the definition is satisfied. The proof that φ satisfies condition (ii) is left to the reader (Exercise 6).

Theorem 3.9. *Every Euclidean ring* R *is a principal ideal ring with identity. Consequently every Euclidean domain is a unique factorization domain.*

REMARK. The converse of Theorem 3.9 is false since there are principal ideal domains that are not Euclidean domains (Exercise 8).

PROOF OF 3.9. If I is a nonzero ideal in R, choose $a \in I$ such that $\varphi(a)$ is the least integer in the set of nonnegative integers $\{\varphi(x) \mid x \neq 0; x \in I\}$. If $b \in I$, then $b = qa + r$ with $r = 0$ or $r \neq 0$ and $\varphi(r) < \varphi(a)$. Since $b \in I$ and $qa \in I$, r is necessarily in I. Since $\varphi(r) < \varphi(a)$ would contradict the choice of a, we must have $r = 0$, whence $b = qa$. Consequently, by Theorem 2.5 $I \subset Ra \subset (a) \subset I$. Therefore $I = Ra = (a)$ and R is a principal ideal ring.

Since R itself is an ideal, $R = Ra$ for some $a \in R$. Consequently, $a = ea = ae$ for some $e \in R$. If $b \in R = Ra$, then $b = xa$ for some $x \in R$. Therefore, $be = (xa)e = x(ae) = xa = b$, whence e is a multiplicative identity element for R. The last statement of the theorem is now an immediate consequence of Theorem 3.7. ■

We close this section with some further observations on divisibility that will be used occasionally in the sequel (Sections 5, 6 and IV.6).

Definition 3.10. *Let* X *be a nonempty subset of a commutative ring* R. *An element* $d \in R$ *is a* **greatest common divisor** *of* X *provided:*

(i) $d \mid a$ *for all* $a \in X$;
(ii) $c \mid a$ *for all* $a \in X \Rightarrow c \mid d$.

Greatest common divisors do not always exist. For example, in the ring E of even integers 2 has no divisors at all, whence 2 and 4 have no (greatest) common divisor. Even when a greatest common divisor of $a_1, \ldots, a_n$ exists, it need not be unique. However, any two greatest common divisors of X are clearly associates by (ii). Furthermore any associate of a greatest common divisor of X is easily seen to be a greatest common divisor of X. If R has an identity and $a_1, a_2, \ldots, a_n$ have 1_R as a greatest common divisor, then $a_1, a_2, \ldots a_n$ are said to be **relatively prime.**

Theorem 3.11. *Let* $a_1, \ldots, a_n$ *be elements of a commutative ring* R *with identity.*

(i) $d \in R$ *is a greatest common divisor of* $\{a_1, \ldots, a_n\}$ *such that* $d = r_1a_1 + \cdots + r_na_n$ *for some* $r_i \in R$ *if and only if* $(d) = (a_1) + (a_2) + \cdots + (a_n)$;

(ii) *if* R *is a principal ideal ring, then a greatest common divisor of* $a_1, \ldots, a_n$ *exists and every one is of the form* $r_1a_1 + \cdots + r_na_n$ $(r_i \in R)$;

(iii) *if* R *is a unique factorization domain, then there exists a greatest common divisor of* $a_1, \ldots, a_n$.

REMARK. Theorem 3.11(i) does *not* state that every greatest common divisor of $a_1, \ldots, a_n$ is expressible as a linear combination of $a_1, \ldots, a_n$. In general this is not the case (Exercise 6.15). See also Exercise 12.

SKETCH OF PROOF OF 3.11. (i) Use Definition 3.10 and Theorem 2.5. (ii) follows from (i). (iii) Each a_i has a factorization: $a_i = c_1^{m_{i1}}c_2^{m_{i2}} \cdots c_t^{m_{it}}$ with $c_1, \ldots, c_t$ distinct irreducible elements and each $m_{ij} \geq 0$. Show that $d = c_1^{k_1}c_2^{k_2} \cdots c_t^{k_t}$ is a greatest common divisor of $a_1, \ldots, a_n$, where $k_j = \min\{m_{1j}, m_{2j}, m_{3j}, \ldots, m_{nj}\}$. ■

EXERCISES

1. A nonzero ideal in a principal ideal domain is maximal if and only if it is prime.

2. An integral domain R is a unique factorization domain if and only if every nonzero prime ideal in R contains a nonzero principal ideal that is prime.

3. Let R be the subring $\{a + b\sqrt{10} \mid a,b \in \mathbf{Z}\}$ of the field of real numbers.
(a) The map $N : R \to \mathbf{Z}$ given by $a + b\sqrt{10} \mapsto (a + b\sqrt{10})(a - b\sqrt{10}) = a^2 - 10b^2$ is such that $N(uv) = N(u)N(v)$ for all $u,v \in R$ and $N(u) = 0$ if and only if $u = 0$.
(b) u is a unit in R if and only if $N(u) = \pm 1$.
(c) 2, 3, $4 + \sqrt{10}$ and $4 - \sqrt{10}$ are irreducible elements of R.
(d) 2, 3, $4 + \sqrt{10}$ and $4 - \sqrt{10}$ are not prime elements of R. [*Hint:* $3 \cdot 2 = 6 = (4 + \sqrt{10})(4 - \sqrt{10})$.]

4. Show that in the integral domain of Exercise 3 every element can be factored into a product of irreducibles, but this factorization need not be unique (in the sense of Definition 3.5 (ii)).

5. Let R be a principal ideal domain.
(a) Every proper ideal is a product $P_1P_2\cdots P_n$ of maximal ideals, which are uniquely determined up to order.
(b) An ideal P in R is said to be primary if $ab \in P$ and $a \notin P$ imply $b^n \in P$ for some n. Show that P is primary if and only if for some n, $P = (p^n)$, where $p \in R$ is prime ($=$ irreducible) or $p = 0$.
(c) If $P_1, P_2, \ldots, P_n$ are primary ideals such that $P_i = (p_i^{n_i})$ and the p_i are distinct primes, then $P_1P_2\cdots P_n = P_1 \cap P_2 \cap \cdots \cap P_n$.
(d) Every proper ideal in R can be expressed (uniquely up to order) as the intersection of a finite number of primary ideals.

6. (a) If a and n are integers, $n > 0$, then there exist integers q and r such that $a = qn + r$, where $|r| \leq n/2$.
(b) The Gaussian integers $\mathbf{Z}[i]$ form a Euclidean domain with $\varphi(a + bi) = a^2 + b^2$. [*Hint:* to show that Definition 3.8(ii) holds, first let $y = a + bi$ and assume x is a positive integer. By part (a) there are integers such that $a = q_1x + r_1$ and $b = q_2x + r_2$, with $|r_1| \leq x/2$, $|r_2| \leq x/2$. Let $q = q_1 + q_2i$ and $r = r_1 + r_2i$; then $y = qx + r$, with $r = 0$ or $\varphi(r) < \varphi(x)$. In the general case, observe that for $x = c + di \neq 0$ and $\bar{x} = c - di$, $x\bar{x} > 0$. There are $q, r_0 \in \mathbf{Z}[i]$ such that $y\bar{x} = q(x\bar{x}) + r_0$, with $r_0 = 0$ or $\varphi(r_0) < \varphi(x\bar{x})$. Let $r = y - qx$; then $y = qx + r$ and $r = 0$ or $\varphi(r) < \varphi(x)$.]

7. What are the units in the ring of Gaussian integers $\mathbf{Z}[i]$?

8. Let R be the following subring of the complex numbers: $R = \{a + b(1 + \sqrt{19}\,i)/2 \mid a,b \in \mathbf{Z}\}$. Then R is a principal ideal domain that is not a Euclidean domain.

9. Let R be a unique factorization domain and d a nonzero element of R. There are only a finite number of distinct principal ideals that contain the ideal (d). [*Hint:* $(d) \subset (k) \Rightarrow k \mid d$.]

10. If R is a unique factorization domain and $a,b \in R$ are relatively prime and $a \mid bc$, then $a \mid c$.

11. Let R be a Euclidean ring and $a \in R$. Then a is a unit in R if and only if $\varphi(a) = \varphi(1_R)$.

12. Every nonempty set of elements (possibly infinite) in a commutative principal ideal ring with identity has a greatest common divisor.

13. (Euclidean algorithm). Let R be a Euclidean domain with associated function $\varphi : R - \{0\} \to \mathbf{N}$. If $a,b \in R$ and $b \neq 0$, here is a method for finding the greatest common divisor of a and b. By repeated use of Definition 3.8(ii) we have:

$$a = q_0b + r_1, \quad \text{with} \quad r_1 = 0 \quad \text{or} \quad \varphi(r_1) < \varphi(b);$$
$$b = q_1r_1 + r_2, \quad \text{with} \quad r_2 = 0 \quad \text{or} \quad \varphi(r_2) < \varphi(r_1);$$
$$r_1 = q_2r_2 + r_3, \quad \text{with} \quad r_3 = 0 \quad \text{or} \quad \varphi(r_3) < \varphi(r_2);$$
$$\vdots$$
$$r_k = q_{k+1}r_{k+1} + r_{k+2}, \quad \text{with} \quad r_{k+2} = 0 \quad \text{or} \quad \varphi(r_{k+2}) < \varphi(r_{k+1});$$
$$\vdots$$

Let $r_0 = b$ and let n be the least integer such that $r_{n+1} = 0$ (such an n exists since the $\varphi(r_k)$ form a strictly decreasing sequence of nonnegative integers). Show that r_n is the greatest common divisor a and b.

4. RINGS OF QUOTIENTS AND LOCALIZATION

In the first part of this section the familiar construction of the field of rational numbers from the ring of integers is considerably generalized. The rings of quotients so constructed from any commutative ring are characterized by a universal mapping property (Theorem 4.5). The last part of this section, which is referred to only occasionally in the sequel, deals with the (prime) ideal structure of rings of quotients and introduces localization at a prime ideal.

Definition 4.1. *A nonempty subset* S *of a ring* R *is* **multiplicative** *provided that*

$$a,b \in S \Rightarrow ab \in S.$$

EXAMPLES. The set S of all elements in a nonzero ring with identity that are not zero divisors is multiplicative. In particular, the set of all nonzero elements in an integral domain is multiplicative. The set of units in any ring with identity is a multiplicative set. If P is a prime ideal in a commutative ring R, then both P and $S = R - P$ are multiplicative sets by Theorem 2.15.

The motivation for what follows may be seen most easily in the ring **Z** of integers and the field **Q** of rational numbers. The set S of all nonzero integers is clearly a multiplicative subset of **Z**. Intuitively the field **Q** is thought of as consisting of all fractions a/b with $a \in \mathbf{Z}$ and $b \in S$, subject to the requirement

$$a/b = c/d \Leftrightarrow ad = bc \text{ (or } ad - bc = 0).$$

More precisely, **Q** may be constructed as follows (details of the proof will be supplied later). The relation on the set $\mathbf{Z} \times S$ defined by

$$(a,b) \sim (c,d) \Leftrightarrow ad - bc = 0$$

is easily seen to be an equivalence relation. **Q** is defined to be the set of equivalence classes of $\mathbf{Z} \times S$ under this equivalence relation. The equivalence class of (a,b) is denoted a/b and addition and multiplication are defined in the usual way. One verifies that these operations are well defined and that **Q** is a field. The map $\mathbf{Z} \rightarrow \mathbf{Q}$ given by $a \mapsto a/1$ is easily seen to be a monomorphism (embedding).

We shall now extend the construction just outlined to an arbitrary multiplicative subset of any commutative ring R (possibly without identity). We shall construct a commutative ring $S^{-1}R$ with identity and a homomorphism $\varphi_S : R \rightarrow S^{-1}R$. If S is the set of all nonzero elements in an integral domain R, then $S^{-1}R$ will be a field ($S^{-1}R = \mathbf{Q}$ if $R = \mathbf{Z}$) and φ_S will be a monomorphism embedding R in $S^{-1}R$.

Theorem 4.2. *Let* S *be a multiplicative subset of a commutative ring* R. *The relation defined on the set* $R \times S$ *by*

$$(r,s) \sim (r',s') \Leftrightarrow s_1(rs' - r's) = 0 \quad \textit{for some} \quad s_1 \in S$$

is an equivalence relation. Furthermore if R *has no zero divisors and* $0 \notin S$, *then*

$$(r,s) \sim (r',s') \quad \Leftrightarrow \quad rs' - r's = 0.$$

PROOF. Exercise. ■

Let S be a multiplicative subset of a commutative ring R and $\sim$ the equivalence relation of Theorem 4.2. The equivalence class of $(r,s) \in R \times S$ will be denoted r/s. The set of all equivalence classes of $R \times S$ under $\sim$ will be denoted by $S^{-1}R$. Verify that

(i) $r/s = r'/s' \quad \Leftrightarrow \quad s_1(rs' - r's) = 0$ for some $s_1 \in S$;
(ii) $tr/ts = r/s$ for all $r \in R$ and $s,t \in S$;
(iii) If $0 \in S$, then $S^{-1}R$ consists of a single equivalence class.

Theorem 4.3. *Let* S *be a multiplicative subset of a commutative ring* R *and let* $S^{-1}R$ *be the set of equivalence classes of* $R \times S$ *under the equivalence relation of Theorem* 4.2.

(i) $S^{-1}R$ *is a commutative ring with identity, where addition and multiplication are defined by*

$$r/s + r'/s' = (rs' + r's)/ss' \quad \textit{and} \quad (r/s)(r'/s') = rr'/ss'.$$

(ii) *If* R *is a nonzero ring with no zero divisors and* $0 \notin S$, *then* $S^{-1}R$ *is an integral domain.*

(iii) *If* R *is a nonzero ring with no zero divisors and* S *is the set of all nonzero elements of* R, *then* $S^{-1}R$ *is a field.*

SKETCH OF PROOF. (i) Once we know that addition and multiplication in $S^{-1}R$ are well-defined binary operations (independent of the choice of r,s,r',s'), the rest of the proof of (i) is routine. In particular, for all $s,s' \in S$, $0/s = 0/s'$ and $0/s$ is the additive identity. The additive inverse of r/s is $-r/s$. For any $s,s' \in S$, $s/s = s'/s'$ and s/s is the multiplicative identity in $S^{-1}R$.

To show that addition is well defined, observe first that since S is multiplicative $(rs' + r's)/ss'$ is an element of $S^{-1}R$. If $r/s = r_1/s_1$ and $r'/s' = r_1'/s_1'$, we must show that $(rs' + r's)/ss' = (r_1s_1' + r_1's_1)/s_1s_1'$. By hypothesis there exist $s_2,s_3 \in S$ such that

$$s_2(rs_1 - r_1s) = 0,$$
$$s_3(r's_1' - r_1's') = 0.$$

Multiply the first equation by $s_3s's_1'$ and the second by s_2ss_1. Add the resulting equations to obtain

$$s_2s_3[(rs' + r's)s_1s_1' - (r_1s_1' + r_1's_1)ss'] = 0.$$

Therefore, $(rs' + r's)/ss' = (r_1s_1' + r_1's_1)/s_1s_1'$ (since $s_2s_3 \in S$). The proof that multiplication is independent of the choice of r,s,r',s' is similar.

(ii) If R has no zero divisors and $0 \notin S$, then $r/s = 0/s$ if and only if $r = 0$ in R. Consequently, $(r/s)(r'/s') = 0$ in $S^{-1}R$ if and only if $rr' = 0$ in R. Since $rr' = 0$ if and only if $r = 0$ or $r' = 0$, it follows that $S^{-1}R$ is an integral domain. (iii) If $r \neq 0$, then the multiplicative inverse of $r/s \in S^{-1}R$ is $s/r \in S^{-1}R$. ■

The ring $S^{-1}R$ in Theorem 4.3 is called the **ring of quotients** or **ring of fractions** or **quotient ring** of R by S. An important special case occurs when S is the set of all nonzero elements in an integral domain R. Then $S^{-1}R$ is a field (Theorem 4.3(iii)) which is called the **quotient field of the integral domain** R. Thus if $R = \mathbf{Z}$, the quotient field is precisely the field $\mathbf{Q}$ of rational numbers. More generally suppose R is any nonzero commutative ring and S is the set of all nonzero elements of R that are *not* zero divisors. If S is nonempty (as is always the case if R has an identity), then $S^{-1}R$ is called the **complete** (or full) **ring of quotients** (or fractions) of the ring R.[3] Theorem 4.3 (iii) may be rephrased: if a nonzero ring R has no zero divisors, then the complete ring of quotients of R is a field. Clearly the complete ring of quotients of an integral domain is just its quotient field.

If $\varphi : \mathbf{Z} \to \mathbf{Q}$ is the map given by $n \mapsto n/1$, then φ is clearly a monomorphism that embeds $\mathbf{Z}$ in $\mathbf{Q}$. Furthermore, for every nonzero n, $\varphi(n)$ is a unit in $\mathbf{Q}$. More generally, we have:

Theorem 4.4. *Let* S *be a multiplicative subset of a commutative ring* R.

(i) *The map* $\varphi_S : R \to S^{-1}R$ *given by* $r \mapsto rs/s$ (*for any* $s \in S$) *is a well-defined homomorphism of rings such that* $\varphi_S(s)$ *is a unit in* $S^{-1}R$ *for every* $s \in S$.

(ii) *If* $0 \notin S$ *and* S *contains no zero divisors, then* φ_S *is a monomorphism. In particular, any integral domain may be embedded in its quotient field.*

(iii) *If* R *has an identity and* S *consists of units, then* φ_S *is an isomorphism. In particular, the complete ring of quotients* (= *quotient field*) *of a field* F *is isomorphic to* F.

SKETCH OF PROOF. (i) If $s,s' \in S$, then $rs/s = rs'/s'$, whence φ_S is well defined. Verify that φ_S is a ring homomorphism and that for each $s \in S$, $s/s^2 \in S^{-1}R$ is the multiplicative inverse of $s^2/s = \varphi_S(s)$. (ii) If $\varphi_S(r) = rs/s = 0$ in $S^{-1}R$, then $rs/s = 0/s$, whence $rs^2s_1 = 0$ for some $s_1 \in S$. Since $s^2s_1 \in S$, $s^2s_1 \neq 0$. Since S has no zero divisors, we must have $r = 0$. (iii) φ_S is a monomorphism by (ii). If $r/s \in S^{-1}R$ with s a unit in R, then $r/s = \varphi_S(rs^{-1})$, whence φ_S is an epimorphism. ■

In view of Theorem 4.4 (ii) it is customary to identify an integral domain R with its image under φ_S and to consider R as a subring of its quotient field. Since $1_R \in S$ in this case, $r \in R$ is thus identified with $r/1_R \in S^{-1}R$.

The next theorem shows that rings of quotients may be completely characterized by a universal mapping property. This theorem is sometimes used as a definition of the ring of quotients.

Theorem 4.5. *Let* S *be a multiplicative subset of a commutative ring* R *and let* T *be any commutative ring with identity. If* $f : R \to T$ *is a homomorphism of rings such that* $f(s)$ *is a unit in* T *for all* $s \in S$, *then there exists a unique homomorphism of rings* $\bar{f} : S^{-1}R \to T$ *such that* $\bar{f}\varphi_S = f$. *The ring* $S^{-1}R$ *is completely determined* (*up to isomorphism*) *by this property.*

SKETCH OF PROOF. Verify that the map $\bar{f} : S^{-1}R \to T$ given by $\bar{f}(r/s) = f(r)f(s)^{-1}$ is a well-defined homomorphism of rings such that $\bar{f}\varphi_S = f$. If

[3]For the noncommutative analogue, see Definition IX.4.7.

$g : S^{-1}R \to T$ is another homomorphism such that $g\varphi_S = f$, then for every $s \in S$, $g(\varphi_S(s))$ is a unit in T. Consequently, $g(\varphi_S(s)^{-1}) = g(\varphi_S(s))^{-1}$ for every $s \in S$ by Exercise 1.15. Now for each $s \in S$, $\varphi_S(s) = s^2/s$, whence $\varphi_S(s)^{-1} = s/s^2 \in S^{-1}R$. Thus for each $r/s \in S^{-1}R$:

$$\begin{aligned} g(r/s) &= g(\varphi_S(r)\varphi_S(s)^{-1}) = g(\varphi_S(r))g(\varphi_S(s)^{-1}) = g(\varphi_S(r))g(\varphi_S(s))^{-1} \\ &= f(r)\,f(s)^{-1} = \bar{f}(r/s). \end{aligned}$$

Therefore, $\bar{f} = g$.

To prove the last statement of the theorem let $\mathcal{C}$ be the category whose objects are all (f,T), where T is a commutative ring with identity and $f : R \to T$ a homomorphism of rings such that $f(s)$ is a unit in T for every $s \in S$. Define a morphism in $\mathcal{C}$ from (f_1,T_1) to (f_2,T_2) to be a homomorphism of rings $g : T_1 \to T_2$ such that $g f_1 = f_2$. Verify that $\mathcal{C}$ is a category and that a morphism g in $\mathcal{C}$ $(f_1,T_1) \to (f_2,T_2)$ is an equivalence if and only if $g : T_1 \to T_2$ is an isomorphism of rings. The preceding paragraph shows that $(\varphi_S,S^{-1}R)$ is a universal object in the category $\mathcal{C}$, whence $S^{-1}R$ is completely determined up to isomorphism by Theorem I.7.10. ■

Corollary 4.6. *Let* R *be an integral domain considered as a subring of its quotient field* F. *If* E *is a field and* f : R → E *a monomorphism of rings, then there is a unique monomorphism of fields* $\bar{f} : F \to E$ *such that* $\bar{f} \mid R = f$. *In particular any field* E_1 *containing* R *contains an isomorphic copy* F_1 *of* F *with* $R \subset F_1 \subset E_1$.

SKETCH OF PROOF. Let S be the set of all nonzero elements of R and apply Theorem 4.5 to $f : R \to E$. Then there is a homomorphism $\bar{f} : S^{-1}R = F \to E$ such that $\bar{f}\varphi_S = f$. Verify that $\bar{f}$ is a monomorphism. Since R is identified with $\varphi_S(R)$, this means that $\bar{f} \mid R = f$. The last statement of the theorem is the special case when $f : R \to E_1$ is the inclusion map. ■

Theorems 4.7-4.11 deal with the ideal structure of rings of quotients. This material will be used only in Section VIII.6. Theorem 4.13, which does not depend on Theorems 4.7-4.11, will be referred to in the sequel.

Theorem 4.7. *Let* S *be a multiplicative subset of a commutative ring* R.

(i) *If* I *is an ideal in* R, *then* $S^{-1}I = \{a/s \mid a \in I;\ s \in S\}$ *is an ideal in* $S^{-1}R$.
(ii) *If* J *is another ideal in* R, *then*

$$S^{-1}(I + J) = S^{-1}I + S^{-1}J;$$
$$S^{-1}(IJ) = (S^{-1}I)(S^{-1}J);$$
$$S^{-1}(I \cap J) = S^{-1}I \cap S^{-1}J.$$

REMARKS. $S^{-1}I$ is called the **extension** of I in $S^{-1}R$. Note that $r/s \in S^{-1}I$ need not imply that $r \in I$ since it is possible to have $a/s = r/s$ with $a \in I$, $r \notin I$.

SKETCH OF PROOF OF 4.7. Use the facts that in $S^{-1}R$, $\sum_{i=1}^{n} (c_i/s) = \left(\sum_{i=1}^{n} c_i\right)/s$; $\sum_{j=1}^{m} (a_jb_j/s) = \sum_{j=1}^{m} (a_j/s)(b_js/s)$; and

$$\sum_{k=1}^{t}(c_k/s_k) = \left(\sum_{k=1}^{t} c_k s_1 s_2 \cdots s_{k-1} s_{k+1} \cdots s_t\right)/s_1 s_2 \cdots s_t. \quad \blacksquare$$

Theorem 4.8. *Let* S *be a multiplicative subset of a commutative ring* R *with identity and let* I *be an ideal of* R. *Then* $S^{-1}I = S^{-1}R$ *if and only if* $S \cap I \neq \emptyset$.

PROOF. If $s \in S \cap I$, then $1_{S^{-1}R} = s/s \in S^{-1}I$ and hence $S^{-1}I = S^{-1}R$. Conversely, if $S^{-1}I = S^{-1}R$, then $\varphi_S^{-1}(S^{-1}I) = R$ whence $\varphi_S(1_R) = a/s$ for some $a \in I$, $s \in S$. Since $\varphi_S(1_R) = 1_R s/s$ we have $s^2 s_1 = ass_1$ for some $s_1 \in S$. But $s^2 s_1 \in S$ and $ass_1 \in I$ imply $S \cap I \neq \emptyset$. $\blacksquare$

In order to characterize the prime ideals in a ring of quotients we need a lemma. Recall that if J is an ideal in a ring of quotients $S^{-1}R$, then $\varphi_S^{-1}(J)$ is an ideal in R (Exercise 2.13). $\varphi_S^{-1}(J)$ is sometimes called the **contraction** of J in R.

Lemma 4.9. *Let* S *be a multiplicative subset of a commutative ring* R *with identity and let* I *be an ideal in* R.

(i) $I \subset \varphi_S^{-1}(S^{-1}I)$.

(ii) *If* $I = \varphi_S^{-1}(J)$ *for some ideal* J *in* $S^{-1}R$, *then* $S^{-1}I = J$. *In other words every ideal in* $S^{-1}R$ *is of the form* $S^{-1}I$ *for some ideal* I *in* R.

(iii) *If* P *is a prime ideal in* R *and* $S \cap P = \emptyset$, *then* $S^{-1}P$ *is a prime ideal in* $S^{-1}R$ *and* $\varphi_S^{-1}(S^{-1}P) = P$.

PROOF. (i) If $a \in I$, then $as \in I$ for every $s \in S$. Consequently, $\varphi_S(a) = as/s \in S^{-1}I$, whence $a \in \varphi_S^{-1}(S^{-1}I)$. Therefore, $I \subset \varphi_S^{-1}(S^{-1}I)$. (ii) Since $I = \varphi_S^{-1}(J)$ every element of $S^{-1}I$ is of the form r/s with $\varphi_S(r) \in J$. Therefore, $r/s = (1_R/s)(rs/s) = (1_R/s)\varphi_S(r) \in J$, whence $S^{-1}I \subset J$. Conversely, if $r/s \in J$, then $\varphi_S(r) = rs/s = (r/s)(s^2/s) \in J$, whence $r \in \varphi_S^{-1}(J) = I$. Thus $r/s \in S^{-1}I$ and hence $J \subset S^{-1}I$. (iii) $S^{-1}P$ is an ideal such that $S^{-1}P \neq S^{-1}R$ by Theorem 4.8. If $(r/s)(r'/s') \in S^{-1}P$, then $rr'/ss' = a/t$ with $a \in P$, $t \in S$. Consequently, $s_1trr' = s_1ss'a \in P$ for some $s_1 \in S$. Since $s_1t \in S$ and $S \cap P = \emptyset$, Theorem 2.15 implies that $rr' \in P$, whence $r \in P$ or $r' \in P$. Thus $r/s \in S^{-1}P$ or $r'/s' \in S^{-1}P$. Therefore, $S^{-1}P$ is prime by Theorem 2.15. Finally $P \subset \varphi_S^{-1}(S^{-1}P)$ by (i). Conversely if $r \in \varphi_S^{-1}(S^{-1}P)$, then $\varphi_S(r) \in S^{-1}P$. Thus $\varphi_S(r) = rs/s = a/t$ with $a \in P$ and $s, t \in S$. Consequently, $s_1str = s_1sa \in P$ for some $s_1 \in S$. Since $s_1st \in S$ and $S \cap P = \emptyset$, $r \in P$ by Theorem 2.15. Therefore, $\varphi_S^{-1}(S^{-1}P) \subset P$. $\blacksquare$

Theorem 4.10. *Let* S *be a multiplicative subset of a commutative ring* R *with identity. Then there is a one-to-one correspondence between the set* $\mathfrak{U}$ *of prime ideals of* R *which are disjoint from* S *and the set* $\mathfrak{V}$ *of prime ideals of* $S^{-1}R$, *given by* $P \mapsto S^{-1}P$.

PROOF. By Lemma 4.9(iii) the assignment $P \mapsto S^{-1}P$ defines an injective map $\mathfrak{U} \to \mathfrak{V}$. We need only show that it is surjective as well. Let J be a prime ideal of $S^{-1}R$ and let $P = \varphi_S^{-1}(J)$. Since $S^{-1}P = J$ by Lemma 4.9(ii), it suffices to show that P is prime. If $ab \in P$, then $\varphi_S(a)\varphi_S(b) = \varphi_S(ab) \in J$ since $P = \varphi_S^{-1}(J)$. Since J is prime in $S^{-1}R$, either $\varphi_S(a) \varepsilon J$ or $\varphi_S(b) \varepsilon J$ by Theorem 2.15. Consequently, either $a \varepsilon \varphi_S^{-1}(J) = P$ or $b \varepsilon P$. Therefore, P is prime by Theorem 2.15. ■

Let R be a commutative ring with identity and P a prime ideal of R. Then $S = R - P$ is a multiplicative subset of R by Theorem 2.15. The ring of quotients $S^{-1}R$ is called the **localization of R at P** and is denoted R_P. If I is an ideal in R, then the ideal $S^{-1}I$ in R_P is denoted I_P.

Theorem 4.11. *Let* P *be a prime ideal in a commutative ring* R *with identity.*

(i) *There is a one-to-one correspondence between the set of prime ideals of* R *which are contained in* P *and the set of prime ideals of* R_P, *given by* $Q \mapsto Q_P$;
(ii) *the ideal* P_P *in* R_P *is the unique maximal ideal of* R_P.

PROOF. Since the prime ideals of R contained in P are precisely those which are disjoint from $S = R - P$, (i) is an immediate consequence of Theorem 4.10. If M is a maximal ideal of R_P, then M is prime by Theorem 2.19, whence $M = Q_P$ for some prime ideal Q of R with $Q \subset P$. But $Q \subset P$ implies $Q_P \subset P_P$. Since $P_P \neq R_P$ by Theorem 4.8, we must have $Q_P = P_P$. Therefore, P_P is the unique maximal ideal in R_P. ■

Rings with a unique maximal ideal, such as R_P in Theorem 4.11, are of some interest in their own right.

Definition 4.12. *A* **local ring** *is a commutative ring with identity which has a unique maximal ideal.*

REMARK. Since every ideal in a ring with identity is contained in some maximal ideal (Theorem 2.18), the unique maximal ideal of a local ring R must contain every ideal of R (except of course R itself).

EXAMPLE. If p is prime and $n \geq 1$, then Z_{p^n} is a local ring with unique maximal ideal (p).

Theorem 4.13. *If* R *is a commutative ring with identity then the following conditions are equivalent.*

(i) R *is a local ring;*
(ii) *all nonunits of* R *are contained in some ideal* $M \neq R$;
(iii) *the nonunits of* R *form an ideal.*

SKETCH OF PROOF. If I is an ideal of R and $a \varepsilon I$, then $(a) \subset I$ by Theorem 2.5. Consequently, $I \neq R$ if and only if I consists only of nonunits (Theorem 3.2(iv)). (ii) $\Rightarrow$ (iii) and (iii) $\Rightarrow$ (i) follow from this fact. (i) $\Rightarrow$ (ii) If $a \varepsilon R$ is a nonunit, then $(a) \neq R$. Therefore, (a) (and hence a) is contained in the unique maximal ideal of R by the remark after Definition 4.12. ■

EXERCISES

1. Determine the complete ring of quotients of the ring Z_n for each $n \geq 2$.

2. Let S be a multiplicative subset of a commutative ring R with identity and let T be a multiplicative subset of the ring $S^{-1}R$. Let $S_* = \{r \varepsilon R \mid r/s \varepsilon T \text{ for some } s \varepsilon S\}$. Then S_* is a multiplicative subset of R and there is a ring isomorphism $S_*^{-1}R \cong T^{-1}(S^{-1}R)$.

3. (a) The set E of positive even integers is a multiplicative subset of $\mathbf{Z}$ such that $E^{-1}(\mathbf{Z})$ is the field of rational numbers.
(b) State and prove condition(s) on a multiplicative subset S of $\mathbf{Z}$ which insure that $S^{-1}\mathbf{Z}$ is the field of rationals.

4. If $S = \{2,4\}$ and $R = Z_6$, then $S^{-1}R$ is isomorphic to the field Z_3. Consequently, the converse of Theorem 4.3(ii) is false.

5. Let R be an integral domain with quotient field F. If T is an integral domain such that $R \subset T \subset F$, then F is (isomorphic to) the quotient field of T.

6. Let S be a multiplicative subset of an integral domain R such that $0 \notin S$. If R is a principal ideal domain [resp. unique factorization domain], then so is $S^{-1}R$.

7. Let R_1 and R_2 be integral domains with quotient fields F_1 and F_2 respectively. If $f: R_1 \rightarrow R_2$ is an isomorphism, then f extends to an isomorphism $F_1 \cong F_2$. [*Hint:* Corollary 4.6.]

8. Let R be a commutative ring with identity, I an ideal of R and $\pi: R \rightarrow R/I$ the canonical projection.
(a) If S is a multiplicative subset of R, then $\pi S = \pi(S)$ is a multiplicative subset of R/I.
(b) The mapping $\theta: S^{-1}R \rightarrow (\pi S)^{-1}(R/I)$ given by $r/s \mapsto \pi(r)/\pi(s)$ is a well-defined function.
(c) θ is a ring epimorphism with kernel $S^{-1}I$ and hence induces a ring isomorphism $S^{-1}R/S^{-1}I \cong (\pi S)^{-1}(R/I)$.

9. Let S be a multiplicative subset of a commutative ring R with identity. If I is an ideal in R, then $S^{-1}(\text{Rad } I) = \text{Rad } (S^{-1}I)$. [See Exercise 2.2.]

10. Let R be an integral domain and for each maximal ideal M (which is also prime, of course), consider R_M as a subring of the quotient field of R. Show that $\bigcap R_M = R$, where the intersection is taken over all maximal ideals M of R.

11. Let p be a prime in $\mathbf{Z}$; then (p) is a prime ideal. What can be said about the relationship of Z_p and the localization $\mathbf{Z}_{(p)}$?

12. A commutative ring with identity is local if and only if for all $r, s \varepsilon R$, $r + s = 1_R$ implies r or s is a unit.

13. The ring R consisting of all rational numbers with denominators not divisible by some (fixed) prime p is a local ring.

14. If M is a maximal ideal in a commutative ring R with identity and n is a positive integer, then the ring R/M^n has a unique prime ideal and therefore is local.

15. In a commutative ring R with identity the following conditions are equivalent: (i) R has a unique prime ideal; (ii) every nonunit is nilpotent (see Exercise 1.12);

(iii) R has a minimal prime ideal which contains all zero divisors, and all nonunits of R are zero divisors.

16. Every nonzero homomorphic image of a local ring is local.

5. RINGS OF POLYNOMIALS AND FORMAL POWER SERIES

We begin by defining and developing notation for polynomials in one indeterminate over a ring R. Next the ring of polynomials in n indeterminates over R is defined and its basic properties are developed. The last part of the section, which is not needed in the sequel, is a brief introduction to the ring of formal power series in one indeterminate over R.

Theorem 5.1. *Let* R *be a ring and let* $R[x]$ *denote the set of all sequences of elements of* R $(a_0,a_1,\ldots)$ *such that* $a_i = 0$ *for all but a finite number of indices* i.

(i) $R[x]$ *is a ring with addition and multiplication defined by:*

$$(a_0,a_1,\ldots) + (b_0,b_1,\ldots) = (a_0 + b_0,a_1 + b_1,\ldots)$$

and

$$(a_0,a_1,\ldots)(b_0,b_1,\ldots) = (c_0,c_1,\ldots),$$

where

$$c_n = \sum_{i=0}^{n} a_{n-i}b_i = a_nb_0 + a_{n-1}b_1 + \cdots + a_1b_{n-1} + a_0b_n = \sum_{k+j=n} a_kb_j.$$

(ii) *If* R *is commutative [resp. a ring with identity or a ring with no zero divisors or an integral domain], then so is* $R[x]$.

(iii) *The map* $R \to R[x]$ *given by* $r \mapsto (r,0,0,\ldots)$ *is a monomorphism of rings.*

PROOF. Exercise. If R has an identity 1_R, then $(1_R,0,0,\ldots)$ is an identity in $R[x]$. Observe that if $(a_0,a_1,\ldots), (b_0,b_1,\ldots) \in R[x]$ and k [resp. j] is the smallest index such that $a_k \neq 0$ [resp. $b_j \neq 0$], then

$$(a_0,a_1,\ldots)(b_0,b_1,\ldots) = (0,\ldots,0,a_kb_j,a_{k+1}b_j + a_kb_{j+1},\ldots). \quad \blacksquare$$

The ring $R[x]$ of Theorem 5.1 is called the **ring of polynomials** over R. Its elements are called polynomials. The notation $R[x]$ is explained below. In view of Theorem 5.1(iii) we shall identify R with its isomorphic image in $R[x]$ and write $(r,0,0,\ldots)$ simply as r. Note that $r(a_0,a_1,\ldots) = (ra_0,ra_1,\ldots)$. We now develop a more familiar notation for polynomials.

Theorem 5.2. *Let* R *be a ring with identity and denote by* x *the element* $(0,1_R,0,0,\ldots)$ *of* $R[x]$.

(i) $x^n = (0,0,\ldots,0,1_R,0,\ldots)$, *where* 1_R *is the* $(n + 1)$st *coordinate.*

(ii) *If* $r \in R$, *then for each* $n \geq 0$, $rx^n = x^nr = (0,\ldots,0,r,0,\ldots)$, *where* r *is the* $(n + 1)$st *coordinate.*

(iii) *For every nonzero polynomial* f *in* R[x] *there exists an integer* $n \in N$ *and elements* $a_0, \ldots, a_n \in R$ *such that* $f = a_0x^0 + a_1x^1 + \cdots + a_nx^n$. *The integer* n *and elements* a_i *are unique in the sense that* $f = b_0x^0 + b_1x^1 + \cdots + b_mx^m$ ($b_i \in R$) *implies* $m \geq n$; $a_i = b_i$ *for* $i = 1,2, \ldots, n$; *and* $b_i = 0$ *for* $n < i \leq m$.

SKETCH OF PROOF. Use induction for (i) and straightforward computation for (ii). (iii) If $f = (a_0,a_1, \ldots) \in R[x]$, there must be a largest index n such that $a_n \neq 0$. Then $a_0,a_1, \ldots, a_n \in R$ are the desired elements. ■

If R has an identity, then $x^0 = 1_R$ (as in any ring with identity) and we write the polynomial $f = a_0x^0 + a_1x^1 + \cdots + a_nx^n$ as $f = a_0 + a_1x + \cdots + a_nx^n$. It will be convenient to extend the notation of Theorem 5.2 to rings without identity as follows. If R is a ring without identity, then R may be embedded in a ring S with identity by Theorem 1.10. Identify R with its image under the embedding map so that R is a subring of S. Then $R[x]$ is clearly a subring of $S[x]$. Consequently, every polynomial $f = (a_0,a_1, \ldots) \in R[x]$ may be written uniquely as $f = a_0 + a_1x^1 + \cdots + a_nx^n$, where $a_i \in R \subset S$, $a_n \neq 0$, and $x = (0,1_S,0,0, \ldots) \in S[x]$. The only important difference between this and the case when R has an identity is that in this case the element x is not in $R[x]$.

Hereafter a polynomial f over a ring R (with or without identity) will always be written in the form $f = a_0 + a_1x + a_2x^2 + \cdots + a_nx^n$ ($a_i \in R$). In this notation addition and multiplication in $R[x]$ are given by the familiar rules:

$$\sum_{i=0}^{n} a_ix^i + \sum_{i=0}^{n} b_ix^i = \sum_{i=0}^{n} (a_i + b_i)x^i$$

$$\left(\sum_{i=0}^{n} a_ix^i\right)\left(\sum_{j=0}^{m} b_jx^j\right) = \sum_{k=0}^{m+n} c_kx^k, \quad \text{where} \quad c_k = \sum_{i+j=k} a_ib_j.$$

If $f = \sum_{i=0}^{n} a_ix^i \in R[x]$, then the elements $a_i \in R$ are called the **coefficients** of f. The element a_0 is called the **constant term.** Elements of R, which all have the form $r = (r, 0, 0, \ldots) = rx^0$ are called **constant polynomials.** If $f = \sum_{i=0}^{n} a_ix^i = a_0 + a_1x + \cdots + a_nx^n = a_nx^n + \cdots + a_1x + a_0$ has $a_n \neq 0$, then a_n is called the **leading coefficient** of f. If R has an identity and leading coefficient 1_R, then f is said to be a **monic polynomial**.

Let R be a ring (with identity). For historical reasons the element $x = (0,1_R,0, \ldots)$ of $R[x]$ is called an **indeterminate.** One speaks of polynomials in the indeterminate x. If S is another ring (with identity), then the indeterminate $x \in S[x]$ is *not* the same element as $x \in R[x]$. In context this ambiguous notation will cause no confusion.

If R is any ring, it is sometimes convenient to distinguish one copy of the polynomial ring over R from another. In this situation the indeterminate in one copy is denoted by one symbol, say x, and in the other copy by a different symbol, say y. In the latter case the polynomial ring is denoted $R[y]$ and its elements have the form $a_0 + a_1y + \cdots + a_ny^n$.

We shall now define polynomials in more than one indeterminate. For convenience the discussion here is restricted to the case of a finite number of indeterminates. For the general case see Exercise 4. The definition is motivated by the fact that a polynomial in one indeterminate is by definition a particular kind of sequence, that is, a function $\mathbf{N} \to R$. For each positive integer n let $\mathbf{N}^n = \mathbf{N} \times \cdots \times \mathbf{N}$ (n factors). The elements of $\mathbf{N}^n$ are ordered n tuples of elements of $\mathbf{N}$. $\mathbf{N}^n$ is clearly an additive abelian monoid under coordinate-wise addition.

Theorem 5.3. *Let* R *be a ring and denote by* $R[x_1, \ldots, x_n]$ *the set of all functions* $f : \mathbf{N}^n \to R$ *such that* $f(u) \neq 0$ *for at most a finite number of elements* u *of* $\mathbf{N}^n$.

(i) $R[x_1, \ldots, x_n]$ *is a ring with addition and multiplication defined by*

$$(f + g)(u) = f(u) + g(u) \quad \text{and} \quad (fg)(u) = \sum_{\substack{v+w=u \\ v,w \in \mathbf{N}^n}} f(v)g(w),$$

where $f,g \in R[x_1, \ldots, x_n]$ *and* $u \in \mathbf{N}^n$.

(ii) *If* R *is commutative [resp. a ring with identity or a ring without zero divisors or an integral domain], then so is* $R[x_1, \ldots, x_n]$.

(iii) *The map* $R \to R[x_1, \ldots, x_n]$ *given by* $r \mapsto f_r$, *where* $f_r(0, \ldots, 0) = r$ *and* $f(u) = 0$ *for all other* $u \in \mathbf{N}^n$, *is a monomorphism of rings.*

PROOF. Exercise. ■

The ring $R[x_1, \ldots, x_n]$ of Theorem 5.3 is called the **ring of polynomials in n indeterminates** over R. R is identified with its isomorphic image under the map of Theorem 5.3(iii) and considered as a subring of $R[x_1, \ldots, x_n]$. If $n = 1$, then $R[x_1]$ is precisely the ring of polynomials as in Theorem 5.1. As in the case of polynomials in one indeterminate, there is a more convenient notation for elements of $R[x_1, \ldots, x_n]$.

Let n be a positive integer and for each $i = 1,2, \ldots, n$, let

$$\varepsilon_i = (0, \ldots, 0,1,0, \ldots, 0) \in \mathbf{N}^n,$$

where 1 is the ith coordinate of ε_i. If $k \in \mathbf{N}$, let $k\varepsilon_i = (0, \ldots, 0,k,0, \ldots 0)$. Then every element of $\mathbf{N}^n$ may be written in the form $k_1\varepsilon_1 + k_2\varepsilon_2 + \cdots + k_n\varepsilon_n$.

Theorem 5.4. *Let* R *be a ring with identity and* n *a positive integer. For each* $i = 1,2, \ldots, n$ *let* $x_i \in R[x_1, \ldots, x_n]$ *be defined by* $x_i(\varepsilon_i) = 1_R$ *and* $x_i(u) = 0$ *for* $u \neq \varepsilon_i$.

(i) *For each integer* $k \in \mathbf{N}$, $x_i^k(k\varepsilon_i) = 1_R$ *and* $x_i^k(u) = 0$ *for* $u \neq k\varepsilon_i$;

(ii) *for each* $(k_1, \ldots, k_n) \in \mathbf{N}^n$, $x_1^{k_1}x_2^{k_2}\cdots x_n^{k_n}(k_1\varepsilon_1 + \cdots + k_n\varepsilon_n) = 1_R$ *and* $x_1^{k_1}x_2^{k_2}\cdots x_n^{k_n}(u) = 0$ *for* $u \neq k_1\varepsilon_1 + \cdots + k_n\varepsilon_n$;

(iii) $x_i^s x_j^t = x_j^t x_i^s$ *for all* $s,t \in \mathbf{N}$ *and all* $i,j = 1,2, \ldots, n$;

(iv) $x_i^t r = r x_i^t$ *for all* $r \in R$ *and all* $t \in \mathbf{N}$;

(v) *for every polynomial* f *in* $R[x_1, \ldots, x_n]$ *there exist unique elements* $a_{k_1, \ldots, k_n} \in R$, *indexed by all* $(k_1, \ldots, k_n) \in \mathbf{N}^n$ *and nonzero for at most a finite number of* $(k_1, \ldots, k_n) \in \mathbf{N}^n$, *such that*

$$f = \sum a_{k_1, \ldots, k_n} x_1^{k_1} \cdots x_n^{k_n},$$

where the sum is over all $(k_1, \ldots, k_n) \in \mathbf{N}^n$.

SKETCH OF PROOF. (v) Let $a_{k_1, \ldots, k_n} = f(k_1, \ldots, k_n)$. ■

If R is a ring with identity, then the elements $x_1,x_2,\ldots,x_n \in R[x_1,\ldots,x_n]$ as in Theorem 5.4 are called **indeterminates.** As in the case of one indeterminate symbols different than $x_1,\ldots,x_n$ may be used to denote indeterminates whenever convenient. The elements $a_0,a_1,\ldots,a_m$ in Theorem 5.4(v) are called the **coefficients** of the polynomial f. A polynomial of the form $ax_1^{k_1}x_2^{k_2}\cdots x_n^{k_n}$ $(a \in R)$ is called a **monomial** in $x_1,x_2,\ldots,x_n$. Theorem 5.4(v) shows that every polynomial is a sum of monomials. It is customary to omit those x_i that appear with exponent zero in a monomial. For example, $a_0x_1^0x_2^0x_3^0 + a_1x_1^2x_2^0x_3 + a_2x_1x_2^3x_3$ is written $a_0 + a_1x_1^2x_3 + a_2x_1x_2^3x_3$. The notation and terminology of Theorem 5.4 is extended to polynomial ring $R[x_1,\ldots,x_n]$, where R has no identity, just as in the case of one indeterminate. The ring R is embedded in a ring S with identity and $R[x_1,\ldots,x_n]$ is considered as a subring of $S[x_1,\ldots,x_n]$. If R has no identity then the indeterminates $x_1,x_2,\ldots,x_n$ and the monomials $x_1^{k_1}x_2^{k_2}\cdots x_n^{k_n}$ $(k_i \in \mathbf{N})$ are not elements of $R[x_1,\ldots,x_n]$.

If R is any ring, then the map $R[x_1] \to R[x_1,\ldots,x_n]$ defined by $\sum_{i=0}^{m} a_ix_1^i \mapsto \sum_{i=0}^{m} a_ix_1^ix_2^0\cdots x_n^0 = \sum_{i=0}^{m} a_ix_1^i \in R[x_1,\ldots,x_n]$ is easily seen to be a monomorphism of rings. Similarly, for any subset $\{i_1,\ldots,i_k\}$ of $\{1,2,\ldots,n\}$ there is a monomorphism $R[x_{i_1},\ldots,x_{i_k}] \to R[x_1,\ldots,x_n]$. $R[x_{i_1},\ldots,x_{i_k}]$ is usually identified with its isomorphic image and considered to be a subring of $R[x_1,\ldots,x_n]$.

Let $\varphi : R \to S$ be a homomorphism of rings, $f \in R[x_1,\ldots,x_n]$ and $s_1,s_2,\ldots,s_n \in S$. By Theorem 5.4 $f = \sum_{i=0}^{m} a_ix_1^{k_{i1}}\cdots x_n^{k_{in}}$ with $a_i \in R$ and $k_{ij} \in \mathbf{N}$. Omit all x_i that appear with exponent zero. Then $\varphi f(s_1,s_2,\ldots,s_n)$ is defined to be $\sum_{i=0}^{m} \varphi(a_i)s_1^{k_{i1}}\cdots s_n^{k_{in}} \in S$; that is, $\varphi f(s_1,\ldots,s_n)$ is obtained by substituting $\varphi(a_i)$ for a_i and $s_i^{k_{ij}}$ for $x_i^{k_{ij}}$ $(k_{ij} > 0)$. Since the a_i and k_{ij} are uniquely determined (Theorem 5.4), $\varphi f(s_1,\ldots,s_n)$ is a well-defined element of S. If R is a subring of S and φ is the inclusion map, we write $f(s_1,\ldots,s_n)$ instead of $\varphi f(s_1,\ldots,s_n)$.

As is the case with most interesting algebraic constructions, the polynomial ring $R[x_1,\ldots,x_n]$ can be characterized by a universal mapping property. The following Theorem and its corollaries are true in the noncommutative case if appropriate hypotheses are added (Exercise 5). They are also true for rings of polynomials in an infinite number of indeterminates (Exercise 4).

Theorem 5.5. *Let* R *and* S *be commutative rings with identity and* $\varphi : R \to S$ *a homomorphism of rings such that* $\varphi(1_R) = 1_S$. *If* $s_1,s_2,\ldots,s_n \in S$, *then there is a unique homomorphism of rings* $\bar{\varphi} : R[x_1,\ldots,x_n] \to S$ *such that* $\bar{\varphi} \mid R = \varphi$ *and* $\bar{\varphi}(x_i) = s_i$ *for* $i = 1,2,\ldots,n$. *This property completely determines the polynomial ring* $R[x_1,\ldots,x_n]$ *up to isomorphism.*

SKETCH OF PROOF. If $f \in R[x_1,\ldots,x_n]$, then

$$f = \sum_{i=0}^{m} a_ix_1^{k_{i1}} \cdots x_n^{k_{in}} \quad (a_i \in R; k_{ij} \in \mathbf{N})$$

by Theorem 5.4. The map $\bar{\varphi}$ given by $\bar{\varphi}(f) = \varphi f(s_1,\ldots,s_n)$ is clearly a well-defined map such that $\bar{\varphi} \mid R = \varphi$ and $\bar{\varphi}(x_i) = s_i$. Use the fact that φ is a homomorphism, the rules of exponentiation and the Binomial Theorem 1.6 to verify that $\bar{\varphi}$ is a homomorphism of rings. Suppose that $\psi : R[x_1, \ldots, x_n] \to S$ is a homomorphism such that $\psi \mid R = \varphi$ and $\psi(x_i) = s_i$ for each i. Then

$$\psi(f) = \psi\left(\sum_{i=0}^{m} a_i x_1^{k_{i1}} \cdots x_n^{k_{in}}\right) = \sum_{i=0}^{m} \psi(a_i)\psi(x_1^{k_{i1}}) \cdots \psi(x_n^{k_{in}})$$
$$= \sum_{i=0}^{m} \varphi(a_i)\psi(x_1)^{k_{i1}} \cdots \psi(x_n)^{k_{in}}$$
$$= \sum_{i=0}^{m} \varphi(a_i)s_1^{k_{i1}} \cdots s_n^{k_{in}} = \varphi f(s_1, s_2, \ldots, s_n) = \bar{\varphi}(f);$$

whence $\psi = \bar{\varphi}$ and $\bar{\varphi}$ is unique. Finally in order to show that $R[x_1, \ldots, x_n]$ is completely determined by this mapping property define a category $\mathcal{C}$ whose objects are all $(n + 2)$-tuples $(\psi, K, s_1, \ldots, s_n)$ where K is a commutative ring with identity, $s_i \in K$ and $\psi : R \to K$ is a homomorphism with $\psi(1_R) = 1_K$. A morphism in $\mathcal{C}$ from $(\psi, K, s_1, \ldots, s_n)$ to $(\theta, T, t_1, \ldots, t_n)$ is a homomorphism of rings $\zeta : K \to T$ such that $\zeta(1_K) = 1_T$, $\zeta\psi = \theta$ and $\zeta(s_i) = t_i$ for $i = 1, 2, \ldots, n$. Verify that ζ is an equivalence in $\mathcal{C}$ if and only if ζ is an isomorphism of rings. If $\iota : R \to R[x_1, \ldots, x_n]$ is the inclusion map, then the first part of the proof shows that $(\iota, R[x_1, \ldots, x_n], x_1, \ldots, x_n)$ is a universal object in $\mathcal{C}$. Therefore, $R[x_1, \ldots, x_n]$ is completely determined up to isomorphism by Theorem I.7.10. ∎

Corollary 5.6. *If* $\varphi : R \to S$ *is a homomorphism of commutative rings and* $s_1, s_2, \ldots, s_n \in S$, *then the map* $R[x_1, \ldots, x_n] \to S$ *given by* $f \mapsto \varphi f(s_1, \ldots, s_n)$ *is a homomorphism of rings.*

SKETCH OF PROOF OF 5.6. The proof of Theorem 5.5 showing that the assignment $f \mapsto \varphi f(s_1, \ldots, s_n)$ defines a homomorphism is valid even when R and S do not have identities. ∎

REMARKS. The map $R[x_1, \ldots, x_n] \to S$ of Corollary 5.6 is called the **evaluation** or **substitution homomorphism.** Corollary 5.6 may be false if R and S are not commutative. This is important since Corollary 5.6 is frequently used without explicit mention. For example, the frequently seen argument that if $f = gh$ $(f, g, h \in R[x])$ and $c \in R$, then $f(c) = g(c)h(c)$, need not be valid if R is not commutative (Exercise 6).

Another consequence of Theorem 5.5 can be illustrated by the following example. Let R be a commutative ring with identity and consider the polynomial

$$f = x^2y + x^3y + x^4 + xy + y^2 + r \in R[x,y].$$

Observe that $f = y^2 + (x^2 + x^3 + x)y + (x^4 + r)$, whence $f \in R[x][y]$. Similarly, $f = x^4 + yx^3 + yx^2 + yx + (y^2 + r) \in R[y][x]$. This suggests that $R[x,y]$ is isomorphic to both $R[x][y]$ and $R[y][x]$. More generally we have:

Corollary 5.7. *Let* R *be a commutative ring with identity and* n *a positive integer. For each* k $(1 \leq k < n)$ *there are isomorphisms of rings* $R[x_1, \ldots, x_k][x_{k+1}, \ldots, x_n] \cong R[x_1, \ldots, x_n] \cong R[x_{k+1}, \ldots, x_n][x_1, \ldots, x_k]$.

PROOF. The corollary may be proved by directly constructing the isomorphisms or by using the universal mapping property of Theorem 5.5 as follows. Given a homomorphism $\varphi : R \to S$ of commutative rings with identity and elements $s_1, \ldots, s_n \in S$, there exists a homomorphism $\bar{\varphi} : R[x_1, \ldots, x_k] \to S$ such that $\bar{\varphi} \mid R = \varphi$ and $\bar{\varphi}(x_i) = s_i$ for $i = 1,2, \ldots, k$ by Theorem 5.5. Applying Theorem 5.5 with $R[x_1, \ldots, x_k]$ in place of R yields a homomorphism $\bar{\bar{\varphi}} : R[x_1, \ldots, x_k]\,[x_{k+1}, \ldots, x_n] \to S$ such that $\bar{\bar{\varphi}} \mid R[x_1, \ldots, x_k] = \bar{\varphi}$ and $\bar{\bar{\varphi}}(x_i) = s_i$ for $i = k+1, \ldots, n$. By construction $\bar{\bar{\varphi}} \mid R = \bar{\varphi} \mid R = \varphi$ and $\bar{\bar{\varphi}}(x_i) = s_i$ for $i = 1,2, \ldots, n$. Suppose that $\psi : R[x_1, \ldots, x_k][x_{k+1}, \ldots, x_n] \to S$ is a homomorphism such that $\psi \mid R = \varphi$ and $\psi(x_i) = s_i$ for $i = 1,2, \ldots, n$. Then the same argument used in the proof of uniqueness in Theorem 5.5 shows that $\psi \mid R[x_1, \ldots, x_k] = \bar{\varphi}$. Therefore, the uniqueness statement of Theorem 5.5 (applied to $R[x_1, \ldots, x_k]$) implies that $\psi = \bar{\bar{\varphi}}$. Consequently, $R[x_1, \ldots, x_k][x_{k+1}, \ldots, x_n]$ has the desired universal mapping property, whence $R[x_1, \ldots, x_k][x_{k+1}, \ldots, x_n] \cong R[x_1, \ldots, x_n]$ by Theorem 5.5. The other isomorphism is proved similarly. ■

Since $R[x_1, \ldots, x_k]$ is usually considered as a subring of $R[x_1, \ldots, x_n]$ (see page 152) it is customary to identify the various polynomial rings in Corollary 5.6 under the isomorphisms stated there and write, for example, $R[x_1, \ldots, x_k][x_{k+1}, \ldots, x_n] = R[x_1, \ldots, x_n]$.

We close this section with a brief introduction to rings of formal power series, which is not needed in the sequel.

Proposition 5.8. *Let* R *be a ring and denote by* R[[x]] *the set of all sequences of elements of* R $(a_0, a_1, \ldots)$.

(i) R[[x]] *is a ring with addition and multiplication defined by:* $(a_0,a_1,\ldots) + (b_0,b_1,\ldots) = (a_0 + b_0, a_1 + b_1, \ldots)$ *and* $(a_0,a_1,\ldots)(b_0,b_1,\ldots) = (c_0,c_1,\ldots)$, *where* $c_n = \sum_{i=0}^{n} a_i b_{n-i} = \sum_{k+j=n} a_k b_j$.

(ii) *The polynomial ring* R[x] *is a subring of* R[[x]].

(iii) *If* R *is commutative* [*resp. a ring with identity or a ring with no zero divisors or an integral domain*], *then so is* R[[x]].

PROOF. Exercise; see Theorem 5.1. ■

The ring $R[[x]]$ of Proposition 5.8 is called the **ring of formal power series** over the ring R. Its elements are called power series. If R has an identity then the polynomial $x = (0,1_R,0,\ldots) \in R[[x]]$ is called an indeterminate. It is easy to verify that $x^i r = r x^i$ for all $r \in R$ and $i \in \mathbf{N}$. If $(a_0,a_1,\ldots) \in R[[x]]$, then for each n, $(a_0,a_1,\ldots,a_n,0,0,\ldots)$ is a polynomial, whence $(a_0,\ldots,a_n,0,0,\ldots) = a_0 + a_1x + a_2x^2 + \cdots + a_nx^n$ by Theorem 5.2. Consequently, we shall adopt the following notation. The power series $(a_0,a_1,\ldots) \in R[[x]]$ is denoted by the formal sum $\sum_{i=0}^{\infty} a_i x^i$. The elements a_i are called **coefficients** and a_0 is called the **constant term.** Just as in the case of polynomials this notation is used even when R does not have an identity (in which case $x \notin R[[x]]$).

Proposition 5.9. *Let* R *be a ring with identity and* $f = \sum_{i=0}^{\infty} a_i x^i \in R[[x]]$.

(i) f *is a unit in* R[[x]] *if and only if its constant term* a_0 *is a unit in* R.
(ii) *If* a_0 *is irreducible in* R, *then* f *is irreducible in* R[[x]].

REMARK. If $f \in R[[x]]$ is actually a polynomial with irreducible [resp. unit] constant term then f need not be irreducible [resp. a unit] in the polynomial ring $R[x]$ (Exercise 8).

PROOF OF 5.9. (i) If there exists $g = \sum b_i x^i \in R[[x]]$ such that

$$fg = gf = 1_R \in R[[x]],$$

it follows immediately that $a_0 b_0 = b_0 a_0 = 1_R$, whence a_0 is a unit in R. Now suppose a_0 is a unit in R. If there were an element $g = \sum b_i x^i \in R[[x]]$ such that $fg = 1_R$, then the following equations would hold:

$$\begin{aligned} a_0 b_0 &= 1_R \\ a_0 b_1 + a_1 b_0 &= 0 \\ &\vdots \\ a_0 b_n + a_1 b_{n-1} + \cdots + a_n b_0 &= 0 \\ &\vdots \end{aligned}$$

Conversely if a solution $(b_0, b_1, b_2, \ldots)$ for this system of equations in R exists, then $g = \sum_{i=0}^{\infty} b_i x^i \in R[[x]]$ clearly has the property that $fg = 1_R$. Since a_0 is a unit (with multiplicative inverse a_0^{-1}), the first equation can be solved: $b_0 = a_0^{-1}$; similarly, $b_1 = a_0^{-1}(-a_1 b_0) = a_0^{-1}(-a_1 a_0^{-1})$. Proceeding inductively, if $b_0, \ldots, b_{n-1}$ are determined in terms of the a_i, then $a_0 b_n = -a_1 b_{n-1} - \cdots - a_n b_0$ implies that $b_n = a_0^{-1}(-a_1 b_{n-1} - \cdots - a_n b_0)$. Thus, if a_0 is a unit this system of equations can be solved and there is a g such that $fg = 1_R \in R[[x]]$. A similar argument shows that there exists $h \in R[[x]]$ such that $hf = 1_R$. But $h = h 1_R = h(fg) = (hf)g = 1_R g = g$, whence g is a two-sided inverse of f. Therefore f is a unit in $R[[x]]$. (ii) is an immediate consequence of (i). ■

Corollary 5.10. *If* R *is a division ring, then the units in* R[[x]] *are precisely those power series with nonzero constant term. The principal ideal* (x) *consists precisely of the nonunits in* R[[x]] *and is the unique maximal ideal of* R[[x]]. *Thus if* R *is a field,* R[[x]] *is a local ring.*

PROOF. The first statement follows from Proposition 5.9 (i) and the fact that every nonzero element of R is a unit. Since x is in the center of $R[[x]]$,

$$(x) = \{xf \mid f \in R[[x]]\}$$

by Theorem 2.5. Consequently, every element xf of (x) has zero constant term, whence xf is a nonunit. Conversely every nonunit $f \in R[[x]]$ is necessarily of the form $f = \sum_{i=0}^{\infty} a_i x^i$ with $a_0 = 0$. Let $g = \sum_{i=0}^{\infty} b_i x^i$ where $b_i = a_{i+1}$ for all i. Then $xg = f$, whence $f \in (x)$. Therefore, (x) is the set of nonunits. Finally, since $1_R \notin (x)$, $(x) \neq R[[x]]$. Furthermore, every ideal I of $R[[x]]$ with $I \neq R[[x]]$ necessarily consists of nonunits (Remarks, p. 123). Thus every ideal of $R[[x]]$ except $R[[x]]$ is contained in (x). Therefore, (x) is the unique maximal ideal of $R[[x]]$. ■

EXERCISES

1. (a) If $\varphi : R \to S$ is a homomorphism of rings, then the map $\bar{\varphi} : R[[x]] \to S[[x]]$ given by $\bar{\varphi}(\sum a_i x^i) = \sum \varphi(a_i)x^i$ is a homomorphism of rings such that $\bar{\varphi}(R[x]) \subset S[x]$.
(b) $\bar{\varphi}$ is a monomorphism [epimorphism] if and only if φ is. In this case $\bar{\varphi} : R[x] \to S[x]$ is also a monomorphism [epimorphism].
(c) Extend the results of (a) and (b) to the polynomial rings $R[x_1, \ldots, x_n]$, $S[x_1, \ldots, x_n]$.

2. Let $\mathrm{Mat}_n R$ be the ring of $n \times n$ matrices over a ring R. Then for each $n \geq 1$:
(a) $(\mathrm{Mat}_n R)[x] \cong \mathrm{Mat}_n R[x]$.
(b) $(\mathrm{Mat}_n R)[[x]] \cong \mathrm{Mat}_n R[[x]]$.

3. Let R be a ring and G an infinite multiplicative cyclic group with generator denoted x. Is the group ring $R(G)$ (see page 117) isomorphic to the polynomial ring in one indeterminate over R?

4. (a) Let S be a nonempty set and let $\mathbf{N}^S$ be the set of all functions $\varphi : S \to \mathbf{N}$ such that $\varphi(s) \neq 0$ for at most a finite number of elements $s \in S$. Then $\mathbf{N}^S$ is a *multiplicative* abelian monoid with product defined by
$$(\varphi\psi)(s) = \varphi(s) + \psi(s) \quad (\varphi,\psi \in \mathbf{N}^S; s \in S).$$
The identity element in $\mathbf{N}^S$ is the zero function.
(b) For each $x \in S$ and $i \in \mathbf{N}$ let $x^i \in \mathbf{N}^S$ be defined by $x^i(x) = i$ and $x^i(s) = 0$ for $s \neq x$. If $\varphi \in \mathbf{N}^S$ and $x_1, \ldots, x_n$ are the only elements of S such that $\varphi(x_i) \neq 0$, then in $\mathbf{N}^S$, $\varphi = x_1^{i_1} x_2^{i_2} \cdots x_n^{i_n}$, where $i_j = \varphi(x_j)$.
(c) If R is a ring with identity let $R[S]$ be the set of all functions $f : \mathbf{N}^S \to R$ such that $f(\varphi) \neq 0$ for at most a finite number of $\varphi \in \mathbf{N}^S$. Then $R[S]$ is a ring with identity, where addition and multiplication are defined as follows:
$$(f + g)(\varphi) = f(\varphi) + g(\varphi) \quad (f,g \in R[S]; \varphi \in \mathbf{N}^S);$$
$$(fg)(\varphi) = \sum f(\theta)g(\zeta) \quad (f,g \in R[S]; \theta,\zeta,\varphi \in \mathbf{N}^S),$$
where the sum is over all pairs (θ,ζ) such that $\theta\zeta = \varphi$. $R[S]$ is called the **ring of polynomials in S over R.**
(d) For each $\varphi = x_1^{i_1} \cdots x_n^{i_n} \in \mathbf{N}^S$ and each $r \in R$ we denote by $rx_1^{i_1} \cdots x_n^{i_n}$ the function $\mathbf{N}^S \to R$ which is r at φ and 0 elsewhere. Then every nonzero element f of $R[S]$ can be written in the form $f = \sum_{i=0}^{m} r_i x_1^{k_{i1}} x_2^{k_{i2}} \cdots x_n^{k_{in}}$ with the $r_i \in R$, $x_i \in S$ and $k_{ij} \in \mathbf{N}$ all uniquely determined.
(e) If S is finite of cardinality n, then $R[S] \cong R[x_1, \ldots, x_n]$. [*Hint:* if $\mathbf{N}^n$ is considered as an additive abelian monoid as in the text, then there is an isomorphism

of monoids $\mathbf{N}^S \cong \mathbf{N}^n$ given by $\varphi \mapsto (\varphi(s_1), \ldots, \varphi(s_n))$, where $S = \{s_1, \ldots, s_n\}$.]
(f) State and prove an analogue of Theorem 5.5 for $R[S]$.

5. Let R and S be rings with identity, $\varphi : R \to S$ a homomorphism of rings with $\varphi(1_R) = 1_S$, and $s_1, s_2, \ldots, s_n \in S$ such that $s_i s_j = s_j s_i$ for all i,j and $\varphi(r)s_i = s_i\varphi(r)$ for all $r \in R$ and all i. Then there is a unique homomorphism $\bar{\varphi} : R[x_1, \ldots, x_n] \to S$ such that $\bar{\varphi}|R = \varphi$ and $\bar{\varphi}(x_i) = s_i$. This property completely determines $R[x_1, \ldots, x_n]$ up to isomorphism.

6. (a) If R is the ring of all 2×2 matrices over $\mathbf{Z}$, then for any $A \in R$,
$$(x + A)(x - A) = x^2 - A^2 \in R[x].$$
(b) There exist $C,A \in R$ such that $(C + A)(C - A) \neq C^2 - A^2$. Therefore, Corollary 5.6 is false if the rings involved are not commutative.

7. If R is a commutative ring with identity and $f = a_n x^n + \cdots + a_0$ is a zero divisor in $R[x]$, then there exists a nonzero $b \in R$ such that $ba_n = ba_{n-1} = \cdots = ba_0 = 0$.

8. (a) The polynomial $x + 1$ is a unit in the power series ring $\mathbf{Z}[[x]]$, but is not a unit in $\mathbf{Z}[x]$.
(b) $x^2 + 3x + 2$ is irreducible in $\mathbf{Z}[[x]]$, but not in $\mathbf{Z}[x]$.

9. If F is a field, then (x) is a maximal ideal in $F[x]$, but it is not the only maximal ideal (compare Corollary 5.10).

10. (a) If F is a field then every nonzero element of $F[[x]]$ is of the form $x^k u$ with $u \in F[[x]]$ a unit.
(b) $F[[x]]$ is a principal ideal domain whose only ideals are 0, $F[[x]] = (1_F) = (x^0)$ and (x^k) for each $k \geq 1$.

11. Let $\mathcal{C}$ be the category with objects all commutative rings with identity and morphisms all ring homomorphisms $f : R \to S$ such that $f(1_R) = 1_S$. Then the polynomial ring $\mathbf{Z}[x_1, \ldots, x_n]$ is a free object on the set $\{x_1, \ldots, x_n\}$ in the category $\mathcal{C}$. [*Hint:* for any R in $\mathcal{C}$ the map $\mathbf{Z} \to R$ given by $n \mapsto n1_R$ is a ring homomorphism; use Theorem 5.5.]

6. FACTORIZATION IN POLYNOMIAL RINGS

We now consider the topics introduced in Section 3 (divisibility, irreducibility, and unique factorization) in the context of polynomial rings over a commutative ring. We begin with two basic tools: the concept of the degree of a polynomial and the division algorithm. Factors of degree one of a polynomial are then studied; finding such factors is equivalent to finding roots of the polynomial. Finally we consider irreducible factors of higher degree: Eisenstein's irreducibility criterion is proved and it is shown that the polynomial domain $D[x_1, \ldots, x_n]$ is a unique factorization domain if D is.

Let R be a ring. The **degree of a nonzero monomial** $ax_1^{k_1}x_2^{k_2}\cdots x_n^{k_n} \in R[x_1, \ldots, x_n]$ is the nonnegative integer $k_1 + k_2 + \cdots + k_n$. If f is a nonzero polynomial in $R[x_1, \ldots, x_n]$, then $f = \sum_{i=0}^{m} a_i x_1^{k_{i1}} \ldots x_n^{k_{in}}$ by Theorem 5.4. The **(total) degree of the polynomial** f is the maximum of the degrees of the monomials $a_i x_1^{k_{i1}} \ldots x_n^{k_{in}}$ such that $a_i \neq 0$ $(i = 1,2,\ldots,m)$. The (total) degree of f is denoted deg f. Clearly a nonzero polynomial f has degree zero if and only if f is a constant polynomial $f = a_0 = a_0x_1^0\cdots x_n^0$. A polynomial which is a sum of monomials, *each* of which has degree k, is said to be **homogeneous of degree k.** Recall that for each k $(1 \leq k \leq n)$, $R[x_1,\ldots,x_{k-1},x_{k+1},\ldots,x_n]$ is a subring of $R[x_1,\ldots,x_n]$ (see page 152). The **degree of f in x_k** is the degree of f considered as a polynomial in one indeterminate x_k over the ring $R[x_1,\ldots,x_{k-1},x_{k+1},\ldots,x_n]$.

EXAMPLE. The polynomial $3x_1^2x_2^2x_3^2 + 3x_1x_3^4 - 6x_2^3x_3 \in \mathbf{Z}[x]$ has degree 2 in x_1, degree 3 in x_2, degree 4 in x_3 and total degree 6.

For technical reasons it is convenient to define the degree of the zero polynomial to be $-\infty$ and to adopt the following conventions about the symbol deg $0 = -\infty$: $(-\infty) < n$ and $(-\infty) + n = -\infty = n + (-\infty)$ for every integer n; $(-\infty) + (-\infty) = -\infty$.

Theorem 6.1. *Let* R *be a ring and* f,g $\in$ R[$x_1,\ldots,x_n$].

(i) *deg*(f + g) $\leq$ *max* (*deg* f, *deg* g).
(ii) *deg*(fg) $\leq$ *deg* f + *deg* g.
(iii) *If* R *has no zero divisors, deg*(fg) = *deg* f + *deg* g.
(iv) *If* n = 1 *and the leading coefficient of* f *or* g *is not a zero divisor in* R (*in particular, if it is a unit*), *then deg*(fg) = *deg* f + *deg* g.

REMARK. The theorem is also true if deg f is taken to mean "degree of f in x_k."

SKETCH OF PROOF OF 6.1. Since we shall apply this theorem primarily when $n = 1$ we shall prove only that case. (i) is easy (ii) is trivial if $f = 0$ or $g = 0$. If $0 \neq f = \sum_{i=0}^{n} a_ix^i$ has degree n and $0 \neq g = \sum_{i=0}^{m} b_ix^i$ has degree m, then $fg = a_0b_0 + \cdots + (a_{n-1}b_m + a_nb_{m-1})x^{n+m-1} + a_nb_mx^{m+n}$ has degree at most $m + n$. Since $a_n \neq 0 \neq b_m$, fg has degree $m + n$ if one of a_n,b_m is not a zero divisor. ■

Theorem 6.2. (*The Division Algorithm*) *Let* R *be a ring with identity and* f,g $\in$ R[x] *nonzero polynomials such that the leading coefficient of* g *is a unit in* R. *Then there exist unique polynomials* q,r $\in$ R[x] *such that*

$$f = qg + r \quad \textit{and} \quad \deg r < \deg g.$$

PROOF. If deg $g >$ deg f, let $q = 0$ and $r = f$. If deg $g \leq$ deg f, then $f = \sum_{i=0}^{n} a_ix^i$, $g = \sum_{i=0}^{m} b_ix^i$, with $a_n \neq 0$, $b_m \neq 0$, $m \leq n$, and b_m a unit in R. Proceed by induction on $n =$ deg f. If $n = 0$, then $m = 0$, $f = a_0$, $g = b_0$ and b_0 is a unit. Let $q = a_0b_0^{-1}$ and $r = 0$; then deg $r <$ deg g and $qg + r = (a_0b_0^{-1})b_0 = a_0 = f$.

Assume that the existence part of the theorem is true for polynomials of degree less than $n =$ deg f. A straightforward calculation shows that the polynomial $(a_nb_m^{-1}x^{n-m})g$ has degree n and leading coefficient a_n. Hence

$$f - (a_n b_m^{-1} x^{n-m})g = (a_n x^n + \cdots + a_0) - (a_n x^n + \cdots + a_n b_m^{-1} b_0 x^{n-m})$$

is a polynomial of degree less than n. By the induction hypothesis there are polynomials q' and r such that

$$f - (a_n b_m^{-1} x^{n-m})g = q'g + r \quad \text{and} \quad \deg r < \deg g.$$

Therefore, if $q = a_n b_m^{-1} x^{n-m} + q'$, then

$$f = (a_n b_m^{-1} x^{n-m})g + q'g + r = qg + r.$$

(Uniqueness) Suppose $f = q_1 g + r_1$, and $f = q_2 g + r_2$ with $\deg r_1 < \deg g$ and $\deg r_2 < \deg g$. Then $q_1 g + r_1 = q_2 g + r_2$ implies

$$(q_1 - q_2)g = r_2 - r_1.$$

Since the leading coefficient b_m of g is a unit, Theorem 6.1 implies

$$\deg(q_1 - q_2) + \deg g = \deg(q_1 - q_2)g = \deg(r_2 - r_1).$$

Since $\deg(r_2 - r_1) \leq \max(\deg r_2, \deg r_1) < \deg g$, the above equality is true only if $\deg(q_1 - q_2) = (-\infty) = \deg(r_2 - r_1)$. In other words $q_1 - q_2 = 0$ and $r_2 - r_1 = 0$. ■

Corollary 6.3. (*Remainder Theorem*) *Let* R *be a ring with identity and*

$$f(x) = \sum_{i=0}^{n} a_i x^i \in R[x].$$

For any $c \in R$ *there exists a unique* $q(x) \in R[x]$ *such that* $f(x) = q(x)(x - c) + f(c)$.

PROOF. If $f = 0$ let $q = 0$. Suppose then that $f \neq 0$. Theorem 6.2 implies that there exist unique polynomials $q(x), r(x)$ in $R[x]$ such that $f(x) = q(x)(x - c) + r(x)$ and $\deg r(x) < \deg(x - c) = 1$. Thus $r(x) = r$ is a constant polynomial (possibly 0). If $q(x) = \sum_{j=0}^{n-1} b_j x^j$, then $f(x) = q(x)(x - c) + r = -b_0 c + \sum_{k=1}^{n-1} (-b_k c + b_{k-1})x^k + b_{n-1}x^n + r$, whence

$$f(c) = -b_0 c + \sum_{k=1}^{n-1} (-b_k c + b_{k-1})c^k + b_{n-1}c^n + r$$

$$= -\sum_{k=0}^{n-1} b_k c^{k+1} + \sum_{k=1}^{n} b_{k-1} c^k + r = 0 + r = r. \quad ■$$

Corollary 6.4. *If* F *is a field, then the polynomial ring* F[x] *is a Euclidean domain, whence* F[x] *is a principal ideal domain and a unique factorization domain. The units in* F[x] *are precisely the nonzero constant polynomials.*

SKETCH OF PROOF. $F[x]$ is an integral domain by Theorem 5.1. Define $\varphi : F[x] - \{0\} \to \mathbf{N}$ by $\varphi(f) = \deg f$. Since every nonzero element of F is a unit, Theorems 6.1(iv) and 6.2 imply that $F[x]$ is a Euclidean domain. Therefore, $F[x]$ is a principal ideal domain and a unique factorization domain (Theorem 3.9). Finally Theorem 6.1 (iv) implies that every unit f in $F[x]$ has degree zero, whence f is a nonzero constant. The converse is obvious. ■

If F is a field, then $F[x_1, \ldots, x_n]$ is not a principal ideal domain (Exercise 1), but it is a unique factorization domain (Theorem 6.14 below). Before proving this latter fact we shall discuss factors of degree one in polynomial rings.

Definition 6.5. *Let* R *be a subring of a commutative ring* S, $c_1, c_2, \ldots, c_n \in S$ *and* $f = \sum_{i=0}^{m} a_i x_1^{k_{i1}} \cdots x_n^{k_{in}} \in R[x_1, \ldots, x_n]$ *a polynomial such that* $f(c_1, c_2, \ldots c_n) = 0$. *Then* $(c_1, c_2, \ldots, c_n)$ *is said to be a* **root** *or* **zero** *of* f (*or a* **solution** *of the polynomial equation* $f(x_1, \ldots, x_n) = 0$).[4]

Theorem 6.6. *Let* R *be a commutative ring with identity and* $f \in R[x]$. *Then* $c \in R$ *is a root of* f *if and only if* $x - c$ *divides* f.

SKETCH OF PROOF. We have $f(x) = q(x)(x - c) + f(c)$ by Corollary 6.3. If $x - c \mid f(x)$, then $h(x)(x - c) = f(x) = q(x)(x - c) + f(c)$ with $h \in R[x]$, whence $(h(x) - q(x))(x - c) = f(c)$. Since R is commutative, Corollary 5.6 (with $\varphi = 1_R$) implies $f(c) = (h(c) - q(c))(c - c) = 0$. Commutativity is not required for the converse; use Corollary 6.3. ∎

Theorem 6.7. *If* D *is an integral domain contained in an integral domain* E *and* $f \in D[x]$ *has degree* n, *then* f *has at most* n *distinct roots in* E.

SKETCH OF PROOF. Let $c_1, c_2, \ldots$ be the *distinct* roots of f in E. By Theorem 6.6 $f(x) = q_1(x)(x - c_1)$, whence $0 = f(c_2) = q_1(c_2)(c_2 - c_1)$ by Corollary 5.6. Since $c_1 \neq c_2$ and E is an integral domain, $q_1(c_2) = 0$. Therefore, $x - c_2$ divides q_2 and $f(x) = q_3(x)(x - c_2)(x - c_1)$. An inductive argument now shows that whenever $c_1, \ldots, c_m$ are distinct roots of f in E, then $g_m = (x - c_1)(x - c_2) \cdots (x - c_m)$ divides f. But deg $g_m = m$ by Theorem 6.1. Therefore $m \leq n$ by Theorem 6.1 again. ∎

REMARK. Theorem 6.7 may be false without the hypothesis of commutativity. For example, $x^2 + 1$ has an infinite number of distinct roots in the division ring of real quaternions (including $\pm i$, $\pm j$ and $\pm k$).

If D is a unique factorization domain with quotient field F and $f \in D[x]$, then the roots of f in F may be found via

Proposition 6.8. *Let* D *be a unique factorization domain with quotient field* F *and let* $f = \sum_{i=0}^{n} a_i x^i \in D[x]$. *If* $u = c/d \in F$ *with* c *and* d *relatively prime, and* u *is a root of* f, *then* c *divides* a_0 *and* d *divides* a_n.

[4]Commutativity is not essential in the definition provided one distinguishes "left roots" and "right roots" (the latter occur when f is written $f = \sum x_1^{k_{i1}} \cdots x_n^{k_{in}} a_i$).

SKETCH OF PROOF. $f(u) = 0$ implies that $a_0d^n = c\left(\sum_{i=1}^{n}(-a_i)c^{i-1}d^{n-i}\right)$ and $-a_nc^n = \left(\sum_{i=0}^{n-1}c^id^{n-i-1}\right)d$. Consequently, if $(c,d) = 1_R$ then $c \mid a_0$ and $d \mid a_n$ by Exercise 3.10. ■

EXAMPLE. If $f = x^4 - 2x^3 - 7x^2 - (11/3)x - 4/3 \varepsilon \mathbf{Q}[x]$, then f has the same roots in $\mathbf{Q}$ as does $3f = 3x^4 - 6x^3 - 21x^2 - 11x - 4 \varepsilon \mathbf{Z}[x]$. By Proposition 6.8 the only possible rational roots of $3f$ are $\pm 1, \pm 2, \pm 4, \pm 1/3, \pm 2/3$ and $\pm 4/3$. Substitution shows that 4 is the only rational root.

Let D be an integral domain and $f \varepsilon D[x]$. If $c \varepsilon D$ and c is a root of f, then repeated application of Theorem 6.6 together with Theorem 6.7 shows that there is a greatest integer m $(0 \leq m \leq \deg f)$ such that

$$f(x) = (x - c)^m g(x),$$

where $g(x) \varepsilon R[x]$ and $x - c \nmid g(x)$ (that is, $g(c) \neq 0$). The integer m is called the **multiplicity of the root** c of f. If c has multiplicity 1, c is said to be a **simple root.** If c has multiplicity $m > 1$, c is called a **multiple root.** In order to determine when a polynomial has multiple roots we need:

Lemma 6.9. *Let* D *be an integral domain and* $f = \sum_{i=0}^{n} a_ix^i \varepsilon D[x]$. *Let* $f' \varepsilon D[x]$ *be the polynomial* $f' = \sum_{k=1}^{n} ka_kx^{k-1} = a_1 + 2a_2x + 3a_3x^2 + \cdots + na_nx^{n-1}$. *Then for all* $f,g \varepsilon D[x]$ *and* $c \varepsilon D$:

(i) $(cf)' = cf'$;
(ii) $(f + g)' = f' + g'$;
(iii) $(fg)' = f'g + fg'$;
(iv) $(g^n)' = ng^{n-1}g'$.

PROOF. Exercise. ■

The polynomial f' is called the **formal derivative** of f. The word "formal" emphasizes the fact that the definition of f' does not involve the concept of limits.

According to Definition 3.3 a nonzero polynomial $f \varepsilon R[x]$ is **irreducible** provided f is not a unit and in every factorization $f = gh$, either g or h is a unit in $R[x]$.

Theorem 6.10. *Let* D *be an integral domain which is a subring of an integral domain* E. *Let* $f \varepsilon D[x]$ *and* $c \varepsilon E$.

(i) c *is a multiple root of* f *if and only if* $f(c) = 0$ *and* $f'(c) = 0$.
(ii) *If* D *is a field and* f *is relatively prime to* f', *then* f *has no multiple roots in* E.
(iii) *If* D *is a field,* f *is irreducible in* D[x] *and* E *contains a root of* f, *then* f *has no multiple roots in* E *if and only if* $f' \neq 0$.

PROOF. (i) $f(x) = (x - c)^m g(x)$ where m is the multiplicity of $f (m \geq 0)$ and $g(c) \neq 0$. By Lemma 6.9 $f'(x) = m(x - c)^{m-1}g(x) + (x - c)^m g'(x)$. If c is a multiple root of f, then $m > 1$, whence $f'(c) = 0$. Conversely, if $f(c) = 0$, then $m \geq 1$ (Theorem 6.6). If $m = 1$, then $f'(x) = g(x) + (x - c)g'(x)$. Consequently, if $f'(c) = 0$, then $0 = f'(c) = g(c)$ by Corollary 5.6, which is a contradiction. Therefore, $m > 1$.

(ii) By Corollary 6.4 and Theorem 3.11 $kf + hf' = 1_D$ for some $k,h \in D[x] \subset E[x]$. If c is a multiple root of f, then by Corollary 5.6 and (i) $1_D = k(c)f(c) + h(c)f'(c) = 0$, which is a contradiction. Hence c is simple root.

(iii) If f is irreducible and $f' \neq 0$, then f and f' are relatively prime since $\deg f' < \deg f$. Therefore, f has no multiple roots in E by (ii). Conversely, suppose f has no multiple roots in E and b is a root of f in E. If $f' = 0$, then b is a multiple root by (i), which is a contradiction. Hence $f' \neq 0$. ■

This completes the discussion of linear factors of polynomials. We now consider the more general question of determining the units and irreducible elements in the polynomial ring $D[x]$, where D is an integral domain. In general this is quite difficult, but certain facts are easily established:

(i) The units in $D[x]$ are precisely the constant polynomials that are units in D [see the proof of Corollary 6.4].

(ii) If $c \in D$ and c is irreducible in D, then the constant polynomial c is irreducible in $D[x]$ [use Theorem 6.1 and (i)].

(iii) Every first degree polynomial whose leading coefficient is a unit in D is irreducible in $D[x]$. In particular, every first degree polynomial over a field is irreducible.

(iv) Suppose D is a subring of an integral domain E and $f \in D[x] \subset E[x]$. Then f may be irreducible in $E[x]$ but not in $D[x]$ and vice versa, as is seen in the following examples.

EXAMPLES. $2x + 2$ is irreducible in $\mathbf{Q}[x]$ by (iii) above. However, $2x + 2 = 2(x + 1)$ and neither 2 nor $x + 1$ is a unit in $\mathbf{Z}[x]$ by (i), whence $2x + 2$ is reducible in $\mathbf{Z}[x]$. $x^2 + 1$ is irreducible over the real field, but factors over the complex field as $(x + i)(x - i)$. Since $x + i$ and $x - i$ are not units in $\mathbf{C}[x]$ by (i), $x^2 + 1$ is reducible in $\mathbf{C}[x]$.

In order to obtain what few general results there are in this area the rest of the discussion will be restricted to polynomials over a unique factorization domain D. We shall eventually prove that $D[x_1, \ldots, x_n]$ is also a unique factorization domain. The proof requires some preliminaries, which will also provide a criterion for irreducibility in $D[x]$.

Let D be a unique factorization domain and $f = \sum_{i=0}^{n} a_i x^i$ a nonzero polynomial in $D[x]$. A greatest common divisor of the coefficients $a_0, a_1, \ldots, a_n$ is called a **content** of f and is denoted $C(f)$. Strictly speaking, the notation $C(f)$ is ambiguous since greatest common divisors are not unique. But any two contents of f are necessarily associates and any associate of a content of f is also a content of f. We shall write $b \approx c$ whenever b and c are associates in D. Now $\approx$ is an equivalence relation on D and since D is an integral domain, $b \approx c$ if and only if $b = cu$ for some unit $u \in D$ by Theorem 3.2 (vi). If $a \in D$ and $f \in D[x]$, then $C(af) \approx aC(f)$ (Exercise 4). If $f \in D[x]$ and $C(f)$ is a unit in D, then f is said to be **primitive.** Clearly for any polynomial $g \in D[x]$, $g = C(g)g_1$ with g_1 primitive.

Lemma 6.11. *(Gauss) If* D *is a unique factorization domain and* f,g ε D[x], *then* C(fg) ≈ C(f)C(g). *In particular, the product of primitive polynomials is primitive.*

PROOF. $f = C(f)f_1$ and $g = C(g)g_1$ with f_1,g_1 primitive. Consequently, $C(fg) = C(C(f)f_1C(g)g_1) \approx C(f)C(g)C(f_1g_1)$. Hence it suffices to prove that f_1g_1 is primitive (that is, $C(f_1g_1)$ is a unit). If $f_1 = \sum_{i=0}^{n} a_ix^i$ and $g_1 = \sum_{j=0}^{m} b_jx^j$, then $f_1g_1 = \sum_{k=0}^{m+n} c_kx^k$ with $c_k = \sum_{i+j=k} a_ib_j$. If f_1g_1 is not primitive, then there exists an irreducible element p in R such that $p \mid c_k$ for all k. Since $C(f_1)$ is a unit $p \nmid C(f_1)$, whence there is a least integer s such that

$$p \mid a_i \quad \text{for} \quad i < s \quad \text{and} \quad p \nmid a_s.$$

Similarly there is a least integer t such that

$$p \mid b_j \quad \text{for} \quad j < t \quad \text{and} \quad p \nmid b_t.$$

Since p divides $c_{s+t} = a_0b_{s+t} + \cdots + a_{s-1}b_{t+1} + a_sb_t + a_{s+1}b_{t-1} + \cdots + a_{s+t}b_0$, p must divide a_sb_t. Since every irreducible element in D is prime, $p \mid a_s$ or $p \mid b_t$. This is a contradiction. Therefore f_1g_1 is primitive. ∎

Lemma 6.12. *Let* D *be a unique factorization domain with quotient field* F *and let* f *and* g *be primitive polynomials in* D[x]. *Then* f *and* g *are associates in* D[x] *if and only if they are associates in* F[x].

PROOF. If f and g are associates in the integral domain $F[x]$, then $f = gu$ for some unit $u \in F[x]$ (Theorem 3.2 (vi)). By Corollary 6.4 $u \in F$, whence $u = b/c$ with $b,c \in D$ and $c \neq 0$. Therefore, $cf = bg$. Since $C(f)$ and $C(g)$ are units in D,

$$c \approx cC(f) \approx C(cf) = C(bg) \approx bC(g) \approx b.$$

Therefore, $b = cv$ for some unit $v \in D$ and $cf = bg = vcg$. Consequently, $f = vg$ (since $c \neq 0$), whence f and g are associates in $D[x]$. The converse is trivial. ∎

Lemma 6.13. *Let* D *be a unique factorization domain with quotient field* F *and* f *a primitive polynomial of positive degree in* D[x]. *Then* f *is irreducible in* D[x] *if and only if* f *is irreducible in* F[x].

SKETCH OF PROOF. Suppose f is irreducible in $D[x]$ and $f = gh$ with $g,h \in F[x]$ and $\deg g \geq 1$, $\deg h \geq 1$. Then $g = \sum_{i=0}^{n} (a_i/b_i)x^i$ and $h = \sum_{j=0}^{m} (c_j/d_j)x^j$ with $a_i,b_i,c_j,d_j \in D$ and $b_i \neq 0$, $d_j \neq 0$. Let $b = b_0b_1\cdots b_n$ and for each i let $b_i^* = b_0b_1\cdots b_{i-1}b_{i+1}\cdots b_n$. If $g_1 = \sum_{i=0}^{n} a_ib_i^*x^i \in D[x]$, then $g_1 = ag_2$ with $a = C(g_1)$, $g_2 \in D[x]$ and g_2 primitive. Verify that $g = (1_D/b)g_1 = (a/b)g_2$ and $\deg g = \deg g_2$. Similarly $h = (c/d)h_2$ with $c,d \in D$, $h_2 \in D[x]$, h_2 primitive and $\deg h = \deg h_2$. Consequently, $f = gh = (a/b)(c/d)g_2h_2$, whence $bdf = acg_2h_2$. Since f is primitive by hypothesis and g_2h_2 is primitive by Lemma 6.11,

$$bd \approx bdC(f) \approx C(bdf) = C(acg_2h_2) \approx acC(g_2h_2) \approx ac.$$

As in the proof of Lemma 6.12, bd and ac associates in D imply that f and g_2h_2 are associates in $D[x]$. Consequently, f is reducible in $D[x]$, which is a contradiction. Therefore, f is irreducible in $F[x]$.

Conversely if f is irreducible in $F[x]$ and $f = gh$ with $g,h \in D[x]$, then one of g,h (say g) is a constant by Corollary 6.4. Thus $C(f) = gC(h)$. Since f is primitive, g must be a unit in D and hence in $D[x]$. Therefore, f is irreducible in $D[x]$. ■

Theorem 6.14. *If* D *is a unique factorization domain, then so is the polynomial ring* $D[x_1, \ldots, x_n]$.

REMARK. Since a field F is trivially a unique factorization domain, $F[x_1, \ldots, x_n]$ is a unique factorization domain.

SKETCH OF PROOF OF 6.14. We shall prove only that $D[x]$ is a unique factorization domain. Since $D[x_1, \ldots, x_n] = D[x_1, \ldots, x_{n-1}][x_n]$ by Corollary 5.7, a routine inductive argument then completes the proof. If $f \in D[x]$ has positive degree, then $f = C(f)f_1$ with f_1 a primitive polynomial in $D[x]$ of positive degree. Since D is a unique factorization domain, either $C(f)$ is a unit or $C(f) = c_1c_2 \cdots c_m$ with each c_i irreducible in D and hence in $D[x]$. Let F be the quotient field of D. Since $F[x]$ is a unique factorization domain (Corollary 6.4) which contains $D[x]$, $f_1 = p_1^*p_2^* \cdots p_n^*$ with each p_i^* an irreducible polynomial in $F[x]$. The proof of Lemma 6.13 shows that for each i, $p_i^* = (a_i/b_i)p_i$ with $a_i,b_i \in D$, $b_i \neq 0$, $a_i/b_i \in F$, $p_i \in D[x]$ and p_i primitive. Clearly each p_i is irreducible in $F[x]$, whence each p_i is irreducible in $D[x]$ by Lemma 6.13. If $a = a_1a_2 \cdots a_n$ and $b = b_1b_2 \cdots b_n$, then $f_1 = (a/b)p_1p_2 \cdots p_n$. Consequently, $bf_1 = ap_1p_2 \cdots p_n$. Since f_1 and $p_1p_2 \cdots p_n$ are primitive (Lemma 6.11), it follows (as in the proof of Lemma 6.12) that a and b are associates in D. Thus $a/b = u$ with u a unit in D. Therefore, if $C(f)$ is a nonunit, $f = C(f)f_1 = c_1c_2 \cdots c_m(up_1)p_2 \cdots p_n$ with each c_i,p_i, and up_1 irreducible in $D[x]$. Similarly, if $C(f)$ is a unit, f is a product of irreducible elements in $D[x]$.

(Uniqueness) Suppose f is a nonprimitive polynomial in $D[x]$ of positive degree. Verify that any factorization of f as a product of irreducible elements may be written $f = c_1c_2 \cdots c_mp_1 \cdots p_n$ with each c_i irreducible in D, $C(f) = c_1 \cdots c_m$ and each p_i irreducible (and hence primitive) in $D[x]$ of positive degree. Suppose $f = d_1 \cdots d_rq_1 \cdots q_s$ with each d_j irreducible in D, $C(f) = d_1 \cdots d_r$ and each q_j irreducible primitive in $D[x]$ of positive degree. Then $c_1c_2 \cdots c_n$ and $d_1d_2 \cdots d_r$ are associates in D. Unique factorization in D implies that $n = r$, and (after reindexing) each c_i is an associate of d_i. Consequently, $p_1p_2 \cdots p_n$ and $q_1q_2 \cdots q_s$ are associates in $D[x]$ and hence in $F[x]$. Since each p_i [resp. q_j] is irreducible in $F[x]$ by Lemma 6.13, unique factorization in $F[x]$ (Corollary 6.4) implies that $n = s$ and (after reindexing) each p_i is an associate of q_i in $F[x]$. By Lemma 6.12 each p_i is an associate of q_i in $D[x]$. ■

Theorem 6.15. *(Eisenstein's Criterion). Let* D *be a unique factorization domain with quotient field* F. *If* $f = \sum_{i=0}^{n} a_ix^i \in D[x]$, *deg* $f \geq 1$ *and* p *is an irreducible element of* D *such that*

$$p \nmid a_n; \quad p \mid a_i \quad \textit{for} \quad i = 0,1,\ldots,n-1; \quad p^2 \nmid a_0,$$

then f *is irreducible in* F[x]. *If* f *is primitive, then* f *is irreducible in* D[x].

PROOF. $f = C(f)f_1$ with f_1 primitive in $D[x]$ and $C(f) \varepsilon D$; (in particular $f_1 = f$ if f is primitive). Since $C(f)$ is a unit in F (Corollary 6.4), it suffices to show that f_1 is irreducible in $F[x]$. By Lemma 6.13 we need only prove that f_1 is irreducible in $D[x]$. Suppose on the contrary that $f_1 = gh$ with

$$g = b_rx^r + \cdots + b_0 \varepsilon D[x], \deg g = r \geq 1; \text{ and}$$
$$h = c_sx^s + \cdots + c_0 \varepsilon D[x], \deg h = s \geq 1.$$

Now p does not divide $C(f)$ (since $p \nmid a_n$), whence the coefficients of $f_1 = \sum_{i=0}^{n} a_i^*x^i$ satisfy the same divisibility conditions with respect to p as do the coefficients of f. Since p divides $a_0^* = b_0c_0$ and every irreducible in D is prime, either $p \mid b_0$ or $p \mid c_0$, say $p \mid b_0$. Since $p^2 \nmid a_0^*$, c_0 is *not* divisible by p. Now some coefficient b_k of g is not divisible by p (otherwise p would divide every coefficient of $gh = f_1$, which would be a contradiction). Let k be the least integer such that

$$p \mid b_i \quad \text{for} \quad i < k \quad \text{and} \quad p \nmid b_k.$$

Then $1 \leq k \leq r < n$. Since $a_k^* = b_0c_k + b_1c_{k-1} + \cdots + b_{k-1}c_1 + b_kc_0$ and $p \mid a_k^*$, p must divide b_kc_0, whence p divides b_k or c_0. Since this is a contradiction, f_1 must be irreducible in $D[x]$. ■

EXAMPLE. If $f = 2x^5 - 6x^3 + 9x^2 - 15 \varepsilon \mathbf{Z}[x]$, then the Eisenstein Criterion with $p = 3$ shows that f is irreducible in both $\mathbf{Q}[x]$ and $\mathbf{Z}[x]$.

EXAMPLE. Let $f = y^3 + x^2y^2 + x^3y + x \varepsilon R[x,y]$ with R a unique factorization domain. Then x is irreducible in $R[x]$ and f considered as an element of $(R[x])[y]$ is primitive. Therefore, f is irreducible in $R[x][y] = R[x,y]$ by Theorem 6.14 and Eisenstein's Criterion (with $p = x$ and $D = R[x]$).

For another application of Eisenstein's Criterion see Exercise 10. There is a lengthy method, due to Kronecker, for finding all the irreducible factors of a polynomial over a unique factorization domain, which has only a finite number of units, such as $\mathbf{Z}$ (Exercise 13). Other examples and techniques appear in Exercises 6–9.

EXERCISES

1. (a) If D is an integral domain and c is an irreducible element in D, then $D[x]$ is not a principal ideal domain. [*Hint:* consider the ideal (x,c) generated by x and c.]
   (b) $\mathbf{Z}[x]$ is not a principal ideal domain.
   (c) If F is a field and $n \geq 2$, then $F[x_1, \ldots, x_n]$ is not a principal ideal domain. [*Hint:* show that x_1 is irreducible in $F[x_1, \ldots, x_{n-1}]$.]

2. If F is a field and $f,g \varepsilon F[x]$ with $\deg g \geq 1$, then there exist unique polynomials $f_0, f_1, \ldots, f_r \varepsilon F[x]$ such that $\deg f_i < \deg g$ for all i and

$$f = f_0 + f_1g + f_2g^2 + \cdots + f_rg^r.$$

3. Let f be a polynomial of positive degree over an integral domain D.
   (a) If char $D = 0$, then $f' \neq 0$.

(b) If char $D = p \neq 0$, then $f' = 0$ if and only if f is a polynomial in x^p (that is, $f = a_0 + a_px^p + a_{2p}x^{2p} + \cdots + a_{jp}x^{jp}$).

4. If D is a unique factorization domain, $a \in D$ and $f \in D[x]$, then $C(af)$ and $aC(f)$ are associates in D.

5. Let R be a commutative ring with identity and $f = \sum_{i=0}^{n} a_ix^i \in R[x]$. Then f is a unit in $R[x]$ if and only if a_0 is a unit in R and $a_1, \ldots, a_n$ are nilpotent elements of R (Exercise 1.12).

6. [Probably impossible with the tools at hand.] Let $p \in \mathbf{Z}$ be a prime; let F be a field and let $c \in F$. Then $x^p - c$ is irreducible in $F[x]$ if and only if $x^p - c$ has no root in F. [*Hint:* consider two cases: char $F = p$ and char $F \neq p$.]

7. If $f = \sum a_ix^i \in \mathbf{Z}[x]$ and p is prime, let $\bar{f} = \sum \bar{a}_ix^i \in Z_p[x]$, where $\bar{a}$ is the image of a under the canonical epimorphism $\mathbf{Z} \to Z_p$.
 (a) If f is monic and $\bar{f}$ is irreducible in $Z_p[x]$ for some prime p, then f is irreducible in $\mathbf{Z}[x]$.
 (b) Give an example to show that (a) may be false if f is not monic.
 (c) Extend (a) to polynomials over a unique factorization domain.

8. [Probably impossible with the tools at hand.] (a) Let $c \in F$, where F is a field of characteristic p (p prime). Then $x^p - x - c$ is irreducible in $F[x]$ if and only if $x^p - x - c$ has no root in F.
 (b) If char $F = 0$, part (a) is false.

9. Let $f = \sum_{i=0} a_ix^i \in \mathbf{Z}[x]$ have degree n. Suppose that for some k ($0 < k < n$) and some prime $p : p \nmid a_n; p \nmid a_k; p \mid a_i$ for all $0 \leq i \leq k - 1$; and $p^2 \nmid a_0$. Show that f has a factor g of degree at least k that is irreducible in $\mathbf{Z}[x]$.

10. (a) Let D be an integral domain and $c \in D$. Let $f(x) = \sum_{i=0}^{n} a_ix^i \in D[x]$ and $f(x - c) = \sum_{i=0}^{n} a_i(x - c)^i \in D[x]$. Then $f(x)$ is irreducible in $D[x]$ if and only if $f(x - c)$ is irreducible.
 (b) For each prime p, the **cyclotomic polynomial** $f = x^{p-1} + x^{p-2} + \cdots + x + 1$ is irreducible in $\mathbf{Z}[x]$. [*Hint:* observe that $f = (x^p - 1)/(x - 1)$, whence $f(x + 1) = ((x + 1)^p - 1)/x$. Use the Binomial Theorem 1.6 and Eisenstein's Criterion to show that $f(x + 1)$ is irreducible in $\mathbf{Z}[x]$.]

11. If $c_0, c_1, \ldots, c_n$ are distinct elements of an integral domain D and $d_0, \ldots, d_n$ are any elements of D, then there is at most one polynomial f of degree $n + 1$ in $D[x]$ such that $f(c_i) = d_i$ for $i = 0, 1, \ldots, n$. [For the existence of f, see Exercise 12].

12. *Lagrange's Interpolation Formula.* If F is a field, $a_0, a_1, \ldots, a_n$ are distinct elements of F and $c_0, c_1, \ldots, c_n$ are any elements of F, then

$$f(x) = \sum_{i=0}^{n} \frac{(x - a_0)\cdots(x - a_{i-1})(x - a_{i+1})\cdots(x - a_n)}{(a_i - a_0)\cdots(a_i - a_{i-1})(a_i - a_{i+1})\cdots(a_i - a_n)} c_i$$

is the unique polynomial in $F[x]$ such that $f(a_i) = c_i$ for all i [see Exercise 11].

13. Let D be a unique factorization domain with a finite number of units and quotient field F. If $f \in D[x]$ has degree n and $c_0, c_1, \ldots, c_n$ are $n + 1$ distinct ele-

ments of D, then f is completely determined by $f(c_0), f(c_1), \ldots, f(c_n)$ according to Exercise 11. Here is **Kronecker's Method** for finding all the irreducible factors of f in $D[x]$.

(a) It suffices to find only those factors g of degree at most $n/2$.

(b) If g is a factor of f, then $g(c)$ is a factor of $f(c)$ for all $c \in D$.

(c) Let m be the largest integer $\leq n/2$ and choose distinct elements $c_0, c_1, \ldots, c_m \in D$. Choose $d_0, d_1, \ldots, d_m \in D$ such that d_i is a factor of $f(c_i)$ in D for all i. Use Exercise 12 to construct a polynomial $g \in F[x]$ such that $g(c_i) = d_i$ for all i; it is unique by Exercise 11.

(d) Check to see if the polynomial g of part (c) is a factor of f in $F[x]$. If not, make a new choice of $d_0, \ldots, d_m$ and repeat part (c). (Since D is a unique factorization domain with only finitely many units there are only a finite number of possible choices for $d_0, \ldots, d_m$.) If g is a factor of f, say $f = gh$, then repeat the entire process on g and h.

(e) After a finite number of steps, all the (irreducible) factors of f in $F[x]$ will have been found. If $g \in F[x]$ is such a factor (of positive degree) then choose $r \in D$ such that $rg \in D[x]$ (for example, let r be the product of the denominators of the coefficients of g). Then $r^{-1}(rg)$ and hence rg is a factor of f. Then $rg = C(rg)g_1$ with $g_1 \in D[x]$ primitive and irreducible in $F[x]$. By Lemma 6.13, g_1 is an irreducible factor of f in $D[x]$. Proceed in this manner to obtain all the nonconstant irreducible factors of f; the constants are then easily found.

14. Let R be a commutative ring with identity and $c,b \in R$ with c a unit.

(a) Show that the assignment $x \mapsto cx + b$ induces a unique automorphism of $R[x]$ that is the identity of R. What is its inverse?

(b) If D is an integral domain, then show that every automorphism of $D[x]$ that is the identity on D is of the type described in (a).

15. If F is a field, then x and y are relatively prime in the polynomial domain $F[x,y]$, but $F[x,y] = (1_F) \underset{\neq}{\supset} (x) + (y)$ [compare Theorem 3.11 (i)].

16. Let $f = a_nx^n + \cdots + a_0$ be a polynomial over the field $\mathbf{R}$ of real numbers and let $\varphi = |a_n|x^n + \cdots + |a_0| \in \mathbf{R}[x]$.

(a) If $|u| \leqslant d$, then $|f(u)| \leqslant \varphi(d)$. [Recall that $|a + b| \leqslant |a| + |b|$ and that $|a| \leqslant a'$, $|b| \leqslant b' \Rightarrow |ab| \leqslant a'b'$.]

(b) Given $a,c \in \mathbf{R}$ with $c > 0$ there exists $M \in \mathbf{R}$ such that $|f(a + h) - f(a)| \leq M|h|$ for all $h \in \mathbf{R}$ with $|h| \leq c$. [*Hint:* use part (a).]

(c) (Intermediate Value Theorem) If $a < b$ and $f(a) < d < f(b)$, then there exists $c \in \mathbf{R}$ such that $a < c < b$ and $f(c) = d$. [*Hint:* Let c be the least upper bound of $S = \{x \mid a < x < b \text{ and } f(x) \leq d\}$. Use part (b).]

(d) Every polynomial g of odd degree in $\mathbf{R}[x]$ has a real root. [*Hint:* for suitable $a,b \in \mathbf{R}$, $g(a) < 0$ and $g(b) > 0$; use part (c).]

CHAPTER IV

MODULES

Modules over a ring are a generalization of abelian groups (which are modules over **Z**). They are basic in the further study of algebra. Section 1 is mostly devoted to carrying over to modules various concepts and results of group theory. Although the classification (up to isomorphism) of modules over an arbitrary ring is quite difficult, we do have substantially complete results for free modules over a ring (Section 2) and finitely generated modules over a principal ideal domain (Section 6). Free modules, of which vector spaces over a division ring are a special case, have widespread applications and are studied thoroughly in Section 2. Projective modules (a generalization of free modules) are considered in Section 3; this material is needed only in Section VIII.6 and Chapter IX.

With the exception of Sections 2 and 6, we shall concentrate on external structures involving modules rather than on the internal structure of modules. Of particular interest are certain categorical aspects of the theory of modules: exact sequences (Section 1) and module homomorphisms (Section 4). In addition we shall study various constructions involving modules such as the tensor product (Section 5). Algebras over a commutative ring K with identity are introduced in Section 7.

The approximate interdependence of the sections of this chapter is as follows:

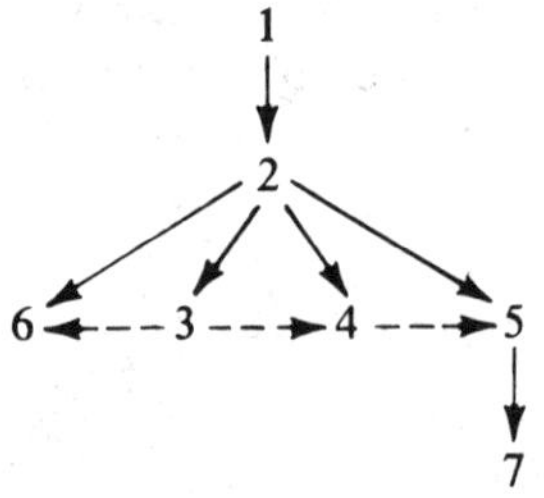

A broken arrow $A \dashrightarrow B$ indicates that an occasional result from Section A is used in Section B, but that Section B is essentially independent of Section A.

1. MODULES, HOMOMORPHISMS AND EXACT SEQUENCES

Modules over a ring are a generalization of abelian groups (which are modules over **Z**). Consequently, the first part of this section is primarily concerned with carrying over to modules various concepts and results of group theory. The remainder of the section presents the basic facts about exact sequences.

Definition 1.1. *Let* R *be a ring. A (left)* **R-module** *is an additive abelian group* A *together with a function* $R \times A \to A$ *(the image of* (r,a) *being denoted by* ra) *such that for all* $r,s \in R$ *and* $a,b \in A$:

(i) $r(a + b) = ra + rb$.
(ii) $(r + s)a = ra + sa$.
(iii) $r(sa) = (rs)a$.

If R *has an identity element* 1_R *and*

(iv) $1_R a = a$ *for all* $a \in A$,

then A *is said to be a* **unitary R-module.** *If* R *is a division ring, then a unitary* R*-module is called a (left)* **vector space.**

A (unitary) right R-module is defined similarly via a function $A \times R \to A$ denoted $(a,r) \mapsto ar$ and satisfying the obvious analogues of (i)–(iv). From now on, unless specified otherwise, "R-module" means "left R-module" and it is understood that all theorems about left R-modules also hold, *mutatis mutandis*, for right R-modules.

A given group A may have many different R-module structures (both left and right). If R is commutative, it is easy to verify that every left R-module A can be given the structure of a right R-module by defining $ar = ra$ for $r \in R$, $a \in A$ (commutativity is needed for (iii); for a generalization of this idea to arbitrary rings, see Exercise 16). Unless specified otherwise, every module A over a *commutative* ring R is assumed to be both a left and a right module with $ar = ra$ for all $r \in R$, $a \in A$.

If A is a module with additive identity element 0_A over a ring R with additive identity 0_R, then it is easy to show that for all $r \in R$, $a \in A$:

$$r0_A = 0_A \quad \text{and} \quad 0_R a = 0_A.$$

In the sequel $0_A, 0_R, 0 \in \mathbf{Z}$ and the trivial module $\{0\}$ will all be denoted 0.

It also is easy to verify that for all $r \in R$, $n \in \mathbf{Z}$ and $a \in A$:

$$(-r)a = -(ra) = r(-a) \quad \text{and} \quad n(ra) = r(na),$$

where na has its usual meaning for groups (Definition I.1.8, additive notation).

EXAMPLE. Every additive abelian group G is a unitary **Z**-module, with na ($n \in \mathbf{Z}, a \in G$) given by Definition I.1.8.

EXAMPLE. If S is a ring and R is a subring, then S is an R-module (but not vice versa!) with ra ($r \in R, a \in S$) being multiplication in S. In particular, the rings $R[x_1, \ldots, x_m]$ and $R[[x]]$ are R-modules.

EXAMPLES. If I is a *left* ideal of a ring R, then I is a left R-module with ra ($r \varepsilon R, a \varepsilon I$) being the ordinary product in R. In particular, 0 and R are R-modules. Furthermore, since I is an additive subgroup of R, R/I is an (abelian) group. R/I is an R-module with $r(r_1 + I) = rr_1 + I$. R/I need not be a ring, however, unless I is a two-sided ideal.

EXAMPLE. Let R and S be rings and $\varphi : R \to S$ a ring homomorphism. Then every S-module A can be made into an R-module by defining rx ($x \varepsilon A$) to be $\varphi(r)x$. One says that the R-module structure of A is given by **pullback along φ.**

EXAMPLE. Let A be an abelian group and End A its endomorphism ring (see p. 116). Then A is a unitary (End A)-module, with fa defined to be $f(a)$ (for $a \varepsilon A$, $f \varepsilon$ End A).

EXAMPLE. If R is a ring, every abelian group can be made into an R-module with **trivial module structure** by defining $ra = 0$ for all $r \varepsilon R$ and $a \varepsilon A$.

Definition 1.2. *Let* A *and* B *be modules over a ring* R. *A function* $f : A \to B$ *is an* **R-module homomorphism** *provided that for all* a,c ε A *and* r ε R:

$$f(a + c) = f(a) + f(c) \quad \textit{and} \quad f(ra) = rf(a).$$

If R *is a division ring, then an* R*-module homomorphism is called a* **linear transformation.**

When the context is clear R-module homomorphisms are called simply homomorphisms. Observe that an R-module homomorphism $f: A \to B$ is necessarily a homomorphism of additive abelian groups. Consequently the same terminology is used: f is an **R-module monomorphism** [resp. **epimorphism, isomorphism**] if it is injective [resp. surjective, bijective] as a map of sets. The **kernel** of f is its kernel as a homomorphism of abelian groups, namely Ker $f = \{a \varepsilon A \mid f(a) = 0\}$. Similarly the **image** of f is the set Im $f = \{b \varepsilon B \mid b = f(a) \text{ for some } a \varepsilon A\}$. Finally, Theorem I.2.3 implies:

(i) f is an R-module monomorphism if and only if Ker $f = 0$;

(ii) $f : A \to B$ is an R-module isomorphism if and only if there is an R-module homomorphism $g : B \to A$ such that $gf = 1_A$ and $fg = 1_B$.

EXAMPLES. For any modules the zero map $0 : A \to B$ given by $a \mapsto 0$ ($a \varepsilon A$) is a module homomorphism. Every homomorphism of abelian groups is a **Z**-module homomorphism. If R is a ring, the map $R[x] \to R[x]$ given by $f \mapsto xf$ (for example, $(x^2 + 1) \mapsto x(x^2 + 1)$) is an R-module homomorphism, but *not* a ring homomorphism.

REMARK. For a given ring R the class of all R-modules [resp. unitary R-modules] and R-module homomorphisms clearly forms a (concrete) category. In fact, one can define epimorphisms and monomorphisms strictly in categorical terms (objects and morphisms only — no elements); see Exercise 2.

Definition 1.3. *Let* R *be a ring,* A *an* R*-module and* B *a nonempty subset of* A. B *is a* **submodule** *of* A *provided that* B *is an additive subgroup of* A *and* $rb \in B$ *for all* $r \in R$, $b \in B$. *A submodule of a vector space over a division ring is called a* **subspace.**

Note that a submodule is itself a module. Also a submodule of a unitary module over a ring with identity is necessarily unitary.

EXAMPLES. If R is a ring and $f: A \to B$ is an R-module homomorphism, then Ker f is a submodule of A and Im f is a submodule of B. If C is any submodule of B, then $f^{-1}(C) = \{a \in A \mid f(a) \in C\}$ is a submodule of A.

EXAMPLE. Let I be a left ideal of the ring R, A an R-module and S a nonempty subset of A. Then $IS = \left\{\sum_{i=1}^{n} r_ia_i \mid r_i \in I; a_i \in S; n \in \mathbf{N}^*\right\}$ is a submodule of A (Exercise 3). Similarly if $a \in A$, then $Ia = \{ra \mid r \in I\}$ is a submodule of A.

EXAMPLE. If $\{B_i \mid i \in I\}$ is a family of submodules of a module A, then $\bigcap_{i\in I} B_i$ is easily seen to be a submodule of A.

Definition 1.4. *If* X *is a subset of a module* A *over a ring* R, *then the intersection of all submodules of* A *containing* X *is called the* **submodule generated by X** (*or* **spanned by X**).

If X is finite, and X generates the module B, B is said to be **finitely generated.** If $X = \emptyset$, then X clearly generates the zero module. If X consists of a single element, $X = \{a\}$, then the submodule generated by X is called the **cyclic (sub)module** generated by a. Finally, if $\{B_i \mid i \in I\}$ is a family of submodules of A, then the submodule generated by $X = \bigcup_{i\in I} B_i$ is called the **sum** of the modules B_i. If the index set I is finite, the sum of $B_1, \dots, B_n$ is denoted $B_1 + B_2 + \cdots + B_n$.

Theorem 1.5. *Let* R *be a ring,* A *an* R*-module,* X *a subset of* A, $\{B_i \mid i \in I\}$ *a family of submodules of* A *and* $a \in A$. *Let* $Ra = \{ra \mid r \in R\}$.

(i) Ra *is a submodule of* A *and the map* $R \to Ra$ *given by* $r \mapsto ra$ *is an* R*-module epimorphism.*

(ii) *The cyclic submodule* C *generated by* a *is* $\{ra + na \mid r \in R; n \in \mathbf{Z}\}$. *If* R *has an identity and* C *is unitary, then* $C = Ra$.

(iii) *The submodule* D *generated by* X *is*

$$\left\{\sum_{i=1}^{s} r_ia_i + \sum_{j=1}^{t} n_jb_j \mid s,t \in \mathbf{N}^*; a_i,b_j \in X ; r_i \in R; n_j \in \mathbf{Z}\right\}.$$

If R *has an identity and* A *is unitary, then*

$$D = RX = \left\{\sum_{i=1}^{s} r_ia_i \mid s \in \mathbf{N}^*; a_i \in X; r_i \in R\right\}.$$

(iv) *The sum of the family* $\{B_i \mid i \in I\}$ *consists of all finite sums* $b_{i_1} + \cdots + b_{i_n}$ *with* $b_{i_k} \in B_{i_k}$.

PROOF. Exercise; note that if R has an identity 1_R and A is unitary, then $n1_R \in R$ for all $n \in \mathbf{Z}$ and $na = (n1_R)a$ for all $a \in A$. ■

Theorem 1.6. *Let* B *be a submodule of a module* A *over a ring* R. *Then the quotient group* A/B *is an* R*-module with the action of* R *on* A/B *given by:*

$$r(a + B) = ra + B \quad \textit{for all} \quad r \in R, a \in A.$$

The map $\pi : A \to A/B$ *given by* $a \mapsto a + B$ *is an* R*-module epimorphism with kernel* B.

The map π is called the **canonical epimorphism** (or **projection**).

SKETCH OF PROOF OF 1.6. Since A is an additive abelian group, B is a normal subgroup, and A/B is a well-defined abelian group. If $a + B = a' + B$, then $a - a' \in B$. Since B is a submodule $ra - ra' = r(a - a') \in B$ for all $r \in R$. Thus $ra + B = ra' + B$ by Corollary I.4.3 and the action of R on A/B is well defined. The remainder of the proof is now easy. ■

In view of the preceding results it is not surprising that the various isomorphism theorems for groups (Theorems I.5.6–I.5.12) are valid, *mutatis mutandis*, for modules. One need only check at each stage of the proof to see that every subgroup or homomorphism is in fact a submodule or module homomorphism. For convenience we list these results here.

Theorem 1.7. *If* R *is a ring and* $f : A \to B$ *is an* R*-module homomorphism and* C *is a submodule of Ker* f, *then there is a unique* R*-module homomorphism* $\bar{f} : A/C \to B$ *such that* $\bar{f}(a + C) = f(a)$ *for all* $a \in A$; *Im* $\bar{f}$ = *Im* f *and Ker* $\bar{f}$ = *Ker* f/C. $\bar{f}$ *is an* R*-module isomorphism if and only if* f *is an* R*-module epimorphism and* C = *Ker* f. *In particular,* A/*Ker* f $\cong$ *Im* f.

PROOF. See Theorem I.5.6 and Corollary I.5.7. ■

Corollary 1.8. *If* R *is a ring and* A′ *is a submodule of the* R*-module* A *and* B′ *a submodule of the* R*-module* B *and* $f : A \to B$ *is an* R*-module homomorphism such that* $f(A') \subset B'$, *then* f *induces an* R*-module homomorphism* $\bar{f} : A/A' \to B/B'$ *given by* $a + A' \mapsto f(a) + B'$. $\bar{f}$ *is an* R*-module isomorphism if and only if Im* f + B′ = B *and* $f^{-1}(B') \subset A'$. *In particular if* f *is an epimorphism such that* f(A′) = B′ *and Ker* f ⊂ A′, *then* $\bar{f}$ *is an* R*-module isomorphism.*

PROOF. See Corollary I.5.8. ■

Theorem 1.9. *Let* B *and* C *be submodules of a module* A *over a ring* R.

(i) *There is an* R*-module isomorphism* $B/(B \cap C) \cong (B + C)/C$;

(ii) *if* $C \subset B$, *then* B/C *is a submodule of* A/C, *and there is an* R*-module isomorphism* $(A/C)/(B/C) \cong A/B$.

PROOF. See Corollaries I.5.9 and I.5.10. ■

Theorem 1.10. *If* R *is a ring and* B *is a submodule of an* R*-module* A, *then there is a one-to-one correspondence between the set of all submodules of* A *containing* B *and the set of all submodules of* A/B, *given by* $C \mapsto C/B$. *Hence every submodule of* A/B *is of the form* C/B, *where* C *is a submodule of* A *which contains* B.

PROOF. See Theorem I.5.11 and Corollary I.5.12. ■

Next we show that products and coproducts always exist in the category of R-modules.

Theorem 1.11. *Let* R *be a ring and* $\{A_i \mid i \in I\}$ *a nonempty family of* R*-modules,* $\prod_{i \in I} A_i$ *the direct product of the abelian groups* A_i, *and* $\sum_{i \in I} A_i$ *the direct sum of the abelian groups* A_i.

(i) $\prod_{i \in I} A_i$ *is an* R*-module with the action of* R *given by* $r\{a_i\} = \{ra_i\}$.

(ii) $\sum_{i \in I} A_i$ *is a submodule of* $\prod_{i \in I} A_i$.

(iii) *For each* $k \in I$, *the canonical projection* $\pi_k : \prod A_i \to A_k$ (*Theorem* I.8.1) *is an* R*-module epimorphism.*

(iv) *For each* $k \in I$, *the canonical injection* $\iota_k : A_k \to \sum A_i$ (*Theorem* I.8.4) *is an* R*-module monomorphism.*

PROOF. Exercise. ■

$\prod_{i \in I} A_i$ is called the (external) **direct product** of the family of R-modules $\{A_i \mid i \in I\}$ and $\sum_{i \in I} A_i$ is its (external) **direct sum.** If the index set is finite, say $I = \{1,2,\ldots,n\}$, then the direct product and direct sum coincide and will be written $A_1 \oplus A_2 \oplus \cdots \oplus A_n$. The maps π_k [resp. ι_k] are called the **canonical projections** [resp. **injections**].

Theorem 1.12. *If* R *is a ring,* $\{A_i \mid i \in I\}$ *a family of* R*-modules,* C *an* R*-module, and* $\{\varphi_i : C \to A_i \mid i \in I\}$ *a family of* R*-module homomorphisms, then there is a unique* R*-module homomorphism* $\varphi : C \to \prod_{i \in I} A_i$ *such that* $\pi_i\varphi = \varphi_i$ *for all* $i \in I$. $\prod_{i \in I} A_i$ *is uniquely determined up to isomorphism by this property. In other words,* $\prod_{i \in I} A_i$ *is a product in the category of* R*-modules.*

PROOF. By Theorem I.8.2 there is a unique group homomorphism $\varphi : C \to \prod A_i$ which has the desired property, given by $\varphi(c) = \{\varphi_i(c)\}_{i \in I}$. Since each φ_i is an R-

module homomorphism, $\varphi(rc) = \{\varphi_i(rc)\}_{i\varepsilon I} = \{r\varphi_i(c)\}_{i\varepsilon I} = r\{\varphi_i(c)\}_{i\varepsilon I} = r\varphi(c)$ and φ is an R-module homomorphism. Thus $\prod A_i$ is a product in the category of R-modules (Definition I.7.2) and therefore determined up to isomorphism by Theorem I.7.3. ■

Theorem 1.13. *If* R *is a ring,* $\{A_i \mid i \varepsilon I\}$ *a family of* R*-modules,* D *an* R*-module, and* $\{\psi_i : A_i \to D \mid i \varepsilon I\}$ *a family of* R*-module homomorphisms, then there is a unique* R*-module homomorphism* $\psi : \sum_{i\varepsilon I} A_i \to D$ *such that* $\psi\iota_i = \psi_i$ *for all* $i \varepsilon I$. $\sum_{i\varepsilon I} A_i$ *is uniquely determined up to isomorphism by this property. In other words,* $\sum_{i\varepsilon I} A_i$ *is a coproduct in the category of* R*-modules.*

PROOF. By Theorem I.8.5 there is a unique abelian group homomorphism $\psi : \sum A_i \to D$ with the desired property, given by $\psi(\{a_i\}) = \sum_i \psi_i(a_i)$, where the sum is taken over the finite set of indices i such that $a_i \neq 0$. It is easy to see that ψ is an R-module map. Hence $\sum A_i$ is a coproduct in the category of R-modules (Definition I.7.4), and therefore, determined up to isomorphism by Theorem I.7.5. ■

Finite direct sums occur so frequently that a further description of them will be useful. We first observe that if f and g are R-module homomorphisms from an R-module A to an R-module B, then the map $f + g : A \to B$ given by $a \mapsto f(a) + g(a)$ is also an R-module homomorphism. It is easy to verify that the set $\mathrm{Hom}_R(A,B)$ of all R-module homomorphisms $A \to B$ is an abelian group under this addition (Exercise 7). Furthermore addition of module homomorphisms is distributive with respect to composition of functions; that is,

$$h(f+g) = hf + hg \quad \text{and} \quad (f+g)k = fk + gk,$$

where $f,g : A \to B$, $h : B \to C$, $k : D \to A$.

Theorem 1.14. *Let* R *be a ring and* $A, A_1, A_2, \ldots, A_n$ R*-modules. Then* $A \cong A_1 \oplus A_2 \oplus \cdots \oplus A_n$ *if and only if for each* $i = 1,2,\ldots,n$ *there are* R*-module homomorphisms* $\pi_i : A \to A_i$ *and* $\iota_i : A_i \to A$ *such that*

(i) $\pi_i\iota_i = 1_{A_i}$ *for* $i = 1,2,\ldots,n$;
(ii) $\pi_j\iota_i = 0$ *for* $i \neq j$;
(iii) $\iota_1\pi_1 + \iota_2\pi_2 + \cdots + \iota_n\pi_n = 1_A$.

PROOF. ($\Rightarrow$) If A is the module $A_1 \oplus A_2 \oplus \cdots \oplus A_n$, then the canonical injections ι_i and projections π_i satisfy (i)–(iii) as the reader may easily verify. Likewise if $A \cong A_1 \oplus \cdots \oplus A_n$, under an isomorphism $f : A \to A_1 \oplus \cdots \oplus A_n$, then the homomorphisms $\pi_i f : A \to A_i$ and $f^{-1}\iota_i : A_i \to A$ satisfy (i)–(iii).

($\Leftarrow$) Let $\pi_i : A \to A_i$ and $\iota_i : A_i \to A$ $(i = 1,2,\ldots,n)$ satisfy (i)–(iii). Let $\pi_i' : A_1 \oplus \cdots \oplus A_n \to A_i$ and $\iota_i' : A_i \to A_1 \oplus \cdots \oplus A_n$ be the canonical projections and injections. Let $\varphi : A_1 \oplus \cdots \oplus A_n \to A$ be given by $\varphi = \iota_1\pi_1' + \iota_2\pi_2' + \cdots + \iota_n\pi_n'$ and $\psi : A \to A_1 \oplus \cdots \oplus A_n$ by $= \iota_1'\pi_1 + \iota_2'\pi_2 + \cdots + \iota_n'\pi_n$. Then

$$\varphi\psi = \left(\sum_{i=1}^n \iota_i\pi_i'\right)\left(\sum_{j=1}^n \iota_j'\pi_j\right) = \sum_{i=1}^n \sum_{j=1}^n \iota_i\pi_i'\iota_j'\pi_j = \sum_{i=1}^n \iota_i\pi_i'\iota_i'\pi_i$$

$$= \sum_{i=1}^{n} \iota_i 1_{A_i} \pi_i = \sum_{i=1}^{n} \iota_i \pi_i = 1_A.$$

Similarly $\psi\varphi = \sum_{i=1}^{n}\sum_{j=1}^{n} \iota_i'\pi_i\iota_j\pi_j' = \sum_{i=1}^{n} \iota_i'\pi_i' = 1_{A_1 \oplus \cdots \oplus A_n}$. Therefore, φ is an isomorphism by Theorem I.2.3. ■

Theorem 1.15. *Let* R *be a ring and* $\{A_i \mid i \in I\}$ *a family of submodules of an* R*-module* A *such that*

(i) A *is the sum of the family* $\{A_i \mid i \in I\}$;

(ii) *for each* $k \in I$, $A_k \cap A_k^* = 0$, *where* A_k^* *is the sum of the family* $\{A_i \mid i \neq k\}$.

Then there is an isomorphism $A \cong \sum_{i \in I} A_i$.

PROOF. Exercise; see Theorem I.8.6. ■

A module A is said to be the **(internal) direct sum** of a family of submodules $\{A_i \mid i \in I\}$ provided that A and $\{A_i\}$ satisfy the hypotheses of Theorem 1.15. As in the case of groups, there is a distinction between internal and external direct sums. If a module A is the *internal* direct sum of modules A_i, then by definition each of the A_i is actually a submodule of A and A is *isomorphic* to the *external* direct sum $\sum_{i \in I} A_i$. However the external direct sum $\sum_{i \in I} A_i$ does *not* contain the modules A_i, but only iscmorphic copies of them (namely the $\iota_i(A_i)$ — see Theorem 1.11 and Exercise I.8.10). Since this distinction is unimportant in practice, the adjectives "internal" and "external" will be omitted whenever the context is clear and the following notation will be used.

NOTATION. We write $A = \sum_{i \in I} A_i$ to indicate that the module A is the internal direct sum of the family of submodules $\{A_i \mid i \in I\}$.

Definition 1.16. *A pair of module homomorphisms,* $A \xrightarrow{f} B \xrightarrow{g} C$, *is said to be* **exact** *at* B *provided Im* f = *Ker* g. *A finite sequence of module homomorphisms,* $A_0 \xrightarrow{f_1} A_1 \xrightarrow{f_2} A_2 \xrightarrow{f_3} \cdots \xrightarrow{f_{n-1}} A_{n-1} \xrightarrow{f_n} A_n$, *is* **exact** *provided Im* f_i = *Ker* f_{i+1} *for* $i = 1,2,\ldots, n-1$. *An infinite sequence of module homomorphisms,* $\cdots \xrightarrow{f_{i-1}} A_{i-1} \xrightarrow{f_i} A_i \xrightarrow{f_{i+1}} A_{i+1} \xrightarrow{f_{i+2}} \cdots$ *is* **exact** *provided Im* f_i = *Ker* f_{i+1} *for all* $i \in \mathbf{Z}$.

When convenient we shall abuse the language slightly and refer to an exact sequence of modules rather than an exact sequence of module homomorphisms.

EXAMPLES. Note first that for any module A, there are unique module homomorphisms $0 \to A$ and $A \to 0$. If A and B are any modules then the sequences $0 \to A \xrightarrow{\iota} A \oplus B \xrightarrow{\pi} B \to 0$ and $0 \to B \xrightarrow{\iota} A \oplus B \xrightarrow{\pi} A \to 0$ are exact, where the ι's and π's are the canonical injections and projections respectively. Similarly, if C is a submodule of D, then the sequence $0 \to C \xrightarrow{i} D \xrightarrow{p} D/C \to 0$ is exact, where i is the inclusion map and p the canonical epimorphism. If $f: A \to B$ is a module homomorphism, then $A/\text{Ker } f$ [resp. $B/\text{Im } f$] is called the **coimage** of f [resp. **cokernel** of f] and denoted Coim f [resp. Coker f]. Each of the following sequences is exact: $0 \to \text{Ker } f \to A \to \text{Coim } f \to 0$, $0 \to \text{Im } f \to B \to \text{Coker } f \to 0$ and $0 \to \text{Ker } f \to A \xrightarrow{f} B \to \text{Coker } f \to 0$, where the unlabeled maps are the obvious inclusions and projections.

REMARKS. $0 \to A \xrightarrow{f} B$ is an exact sequence of module homomorphisms if and only if f is a module monomorphism. Similarly, $B \xrightarrow{g} C \to 0$ is exact if and only if g is a module epimorphism. If $A \xrightarrow{f} B \xrightarrow{g} C$ is exact, then $gf = 0$. Finally if $A \xrightarrow{f} B \xrightarrow{g} C \to 0$ is exact, then $\text{Coker } f = B/\text{Im } f = B/\text{Ker } g = \text{Coim } g \cong C$. An exact sequence of the form $0 \to A \xrightarrow{f} B \xrightarrow{g} C \to 0$ is called a **short exact sequence**; note that f is a monomorphism and g an epimorphism. The preceding remarks show that a short exact sequence is just another way of presenting a submodule ($A \cong \text{Im } f$) and its quotient module ($B/\text{Im } f = B/\text{Ker } g \cong C$).

Lemma 1.17. *(The Short Five Lemma) Let* R *be a ring and*

$$\begin{array}{ccccccccc} 0 & \to & A & \xrightarrow{f} & B & \xrightarrow{g} & C & \to & 0 \\ & & \downarrow{\scriptstyle\alpha} & & \downarrow{\scriptstyle\beta} & & \downarrow{\scriptstyle\gamma} & & \\ 0 & \to & A' & \xrightarrow{f'} & B' & \xrightarrow{g'} & C' & \to & 0 \end{array}$$

a commutative diagram of R*-modules and* R*-module homomorphisms such that each row is a short exact sequence. Then*

(i) *α,γ monomorphisms $\Rightarrow \beta$ is a monomorphism;*
(ii) *α,γ epimorphisms $\Rightarrow \beta$ is an epimorphism;*
(iii) *α,γ isomorphisms $\Rightarrow \beta$ is an isomorphism.*

PROOF. (i) Let $b \in B$ and suppose $\beta(b) = 0$; we must show that $b = 0$. By commutativity we have

$$\gamma g(b) = g'\beta(b) = g'(0) = 0.$$

This implies $g(b) = 0$, since γ is a monomorphism. By exactness of the top row at B, we have $b \in \text{Ker } g = \text{Im } f$, say $b = f(a)$, $a \in A$. By commutativity,

$$f'\alpha(a) = \beta f(a) = \beta(b) = 0.$$

By exactness of the bottom row at A', f' is a monomorphism (Theorem I.2.3(i)); hence $\alpha(a) = 0$. But α is a monomorphism; therefore $a = 0$ and hence $b = f(a) = f(0) = 0$. Thus β is a monomorphism.

(ii) Let $b' \in B'$. Then $g'(b') \in C'$; since γ is an epimorphism $g'(b') = \gamma(c)$ for some $c \in C$. By exactness of the top row at C, g is an epimorphism; hence $c = g(b)$ for some $b \in B$. By commutativity,

$$g'\beta(b) = \gamma g(b) = \gamma(c) = g'(b').$$

Thus $g'[\beta(b) - b'] = 0$ and $\beta(b) - b' \in \operatorname{Ker} g' = \operatorname{Im} f'$ by exactness, say $f'(a') = \beta(b) - b', a' \in A'$. Since α is an epimorphism, $a' = \alpha(a)$ for some $a \in A$. Consider $b - f(a) \in B$:

$$\beta[b - f(a)] = \beta(b) - \beta f(a).$$

By commutativity, $\beta f(a) = f'\alpha(a) = f'(a') = \beta(b) - b'$; hence

$$\beta[b - f(a)] = \beta(b) - \beta f(a) = \beta(b) - (\beta(b) - b') = b'$$

and β is an epimorphism.

(iii) is an immediate consequence of (i) and (ii). ∎

Two short exact sequences are said to be **isomorphic** if there is a commutative diagram of module homomorphisms

$$\begin{array}{ccccccccc} 0 & \rightarrow & A & \rightarrow & B & \rightarrow & C & \rightarrow & 0 \\ & & \downarrow{\scriptstyle f} & & \downarrow{\scriptstyle g} & & \downarrow{\scriptstyle h} & & \\ 0 & \rightarrow & A' & \rightarrow & B' & \rightarrow & C' & \rightarrow & 0 \end{array}$$

such that f,g, and h are isomorphisms. In this case, it is easy to verify that the diagram

$$\begin{array}{ccccccccc} 0 & \rightarrow & A & \rightarrow & B & \rightarrow & C & \rightarrow & 0 \\ & & \uparrow{\scriptstyle f^{-1}} & & \uparrow{\scriptstyle g^{-1}} & & \uparrow{\scriptstyle h^{-1}} & & \\ 0 & \rightarrow & A' & \rightarrow & B' & \rightarrow & C' & \rightarrow & 0 \end{array}$$

(with the same horizontal maps) is also commutative. In fact, isomorphism of short exact sequences is an equivalence relation (Exercise 14).

Theorem 1.18. *Let* R *be a ring and* $0 \rightarrow A_1 \xrightarrow{f} B \xrightarrow{g} A_2 \rightarrow 0$ *a short exact sequence of* R*-module homomorphisms. Then the following conditions are equivalent.*

(i) *There is an* R*-module homomorphism* $h : A_2 \rightarrow B$ *with* $gh = 1_{A_2}$;
(ii) *There is an* R*-module homomorphism* $k : B \rightarrow A_1$ *with* $kf = 1_{A_1}$;
(iii) *the given sequence is isomorphic (with identity maps on* A_1 *and* A_2*) to the direct sum short exact sequence* $0 \rightarrow A_1 \xrightarrow{\iota_1} A_1 \oplus A_2 \xrightarrow{\pi_2} A_2 \rightarrow 0$; *in particular* $B \cong A_1 \oplus A_2$.

A short exact sequence that satisfies the equivalent conditions of Theorem 1.18 is said to be **split** or a **split exact** sequence.

SKETCH OF PROOF OF 1.18. (i) $\Rightarrow$ (iii) By Theorem 1.13 the homomorphisms f and h induce a module homomorphism $\varphi : A_1 \oplus A_2 \rightarrow B$, given by $(a_1,a_2) \mapsto f(a_1) + h(a_2)$. Verify that the diagram

$$\begin{array}{ccccccccc}
0 & \to & A_1 & \xrightarrow{\iota_1} & A_1 \oplus A_2 & \xrightarrow{\pi_2} & A_2 & \to & 0 \\
 & & \downarrow{\scriptstyle 1_{A_1}} & & \downarrow{\scriptstyle \varphi} & & \downarrow{\scriptstyle 1_{A_2}} & & \\
0 & \to & A_1 & \xrightarrow{f} & B & \xrightarrow{g} & A_2 & \to & 0
\end{array}$$

is commutative (use the fact that $gf = 0$ and $gh = 1_{A_2}$). By the Short Five Lemma φ is an isomorphism.

(ii) $\Rightarrow$ (iii) The diagram

$$\begin{array}{ccccccccc}
0 & \to & A_1 & \xrightarrow{f} & B & \xrightarrow{g} & A_2 & \to & 0 \\
 & & \downarrow{\scriptstyle 1_{A_1}} & & \downarrow{\scriptstyle \psi} & & \downarrow{\scriptstyle 1_{A_2}} & & \\
0 & \to & A_1 & \xrightarrow{\iota_1} & A_1 \oplus A_2 & \xrightarrow{\pi_2} & A_2 & \to & 0
\end{array}$$

is commutative, where ψ is the module homomorphism given by $\psi(b) = (k(b),g(b))$ (see Theorem 1.12). Hence the short Five Lemma implies ψ is an isomorphism.

(iii) $\Rightarrow$ (i), (ii) Given a commutative diagram with exact rows and φ an isomorphism:

$$\begin{array}{ccccccccc}
0 & \to & A_1 & \underset{\pi_1}{\overset{\iota_1}{\rightleftarrows}} & A_1 \oplus A_2 & \underset{\iota_2}{\overset{\pi_2}{\rightleftarrows}} & A_2 & \to & 0 \\
 & & \downarrow{\scriptstyle 1_{A_1}} & & \downarrow{\scriptstyle \varphi} & & \downarrow{\scriptstyle 1_{A_2}} & & \\
0 & \to & A_1 & \xrightarrow{f} & B & \xrightarrow{g} & A_2 & \to & 0,
\end{array}$$

define $h : A_2 \to B$ to be $\varphi\iota_2$ and $k : B \to A_1$ to be $\pi_1\varphi^{-1}$. Use the commutativity of the diagram and the facts $\pi_i\iota_i = 1_{A_i}$, $\varphi^{-1}\varphi = 1_{A_1 \oplus A_2}$ to show that $kf = 1_{A_1}$ and $gh = 1_{A_2}$. ■

EXERCISES

Note: R is a ring.

1. If A is an abelian group and $n > 0$ an integer such that $na = 0$ for all $a \in A$, then A is a unitary Z_n-module, with the action of Z_n on A given by $\bar{k}a = ka$, where $k \in \mathbf{Z}$ and $k \mapsto \bar{k} \in Z_n$ under the canonical projection $\mathbf{Z} \to Z_n$.

2. Let $f: A \to B$ be an R-module homomorphism.

   (a) f is a monomorphism if and only if for every pair of R-module homomorphisms $g,h : D \to A$ such that $fg = fh$, we have $g = h$. [*Hint:* to prove ($\Leftarrow$), let $D = \text{Ker } f$, with g the inclusion map and h the zero map.]

   (b) f is an epimorphism if and only if for every pair of R-module homomorphisms $k,t : B \to C$ such that $kf = tf$, we have $k = t$. [*Hint:* to prove ($\Leftarrow$), let k be the canonical epimorphism $B \to B/\text{Im } f$ and t the zero map.]

3. Let I be a left ideal of a ring R and A an R-module.

   (a) If S is a nonempty subset of A, then $IS = \left\{ \sum_{i=1}^{n} r_i a_i \mid n \in \mathbf{N}^*; r_i \in I; a_i \in S \right\}$ is a submodule of A. Note that if $S = \{a\}$, then $IS = Ia = \{ra \mid r \in I\}$.

(b) If I is a two-sided ideal, then A/IA is an R/I-module with the action of R/I given by $(r + I)(a + IA) = ra + IA$.

4. If R has an identity, then every unitary cyclic R-module is isomorphic to an R-module of the form R/J, where J is a left ideal of R.

5. If R has an identity, then a nonzero unitary R-module A is **simple** if its only submodules are 0 and A.
   (a) Every simple R-module is cyclic.
   (b) If A is simple every R-module endomorphism is either the zero map or an isomorphism.

6. A finitely generated R-module need not be finitely generated as an abelian group. [*Hint:* Exercise II.1.10.]

7. (a) If A and B are R-modules, then the set $\mathrm{Hom}_R(A,B)$ of all R-module homomorphisms $A \to B$ is an abelian group with $f + g$ given on $a \varepsilon A$ by $(f + g)(a) = f(a) + g(a) \varepsilon B$. The identity element is the zero map.
   (b) $\mathrm{Hom}_R(A,A)$ is a ring with identity, where multiplication is composition of functions. $\mathrm{Hom}_R(A,A)$ is called the **endomorphism ring** of A.
   (c) A is a left $\mathrm{Hom}_R(A,A)$-module with fa defined to be

$$f(a) \quad (a \varepsilon A, f \varepsilon \mathrm{Hom}_R(A,A)).$$

8. Prove that the obvious analogues of Theorem I.8.10 and Corollary I.8.11 are valid for R-modules.

9. If $f : A \to A$ is an R-module homomorphism such that $ff = f$, then

$$A = \mathrm{Ker}\, f \oplus \mathrm{Im}\, f.$$

10. Let $A, A_1, \dots, A_n$ be R-modules. Then $A \cong A_1 \oplus \cdots \oplus A_n$ if and only if for each $i = 1,2,\dots,n$ there is an R-module homomorphism $\varphi_i : A \to A$ such that $\mathrm{Im}\, \varphi_i \cong A_i$; $\varphi_i\varphi_j = 0$ for $i \neq j$; and $\varphi_1 + \varphi_2 + \cdots + \varphi_n = 1_A$. [*Hint:* If $A \cong A_1 \oplus \cdots \oplus A_n$ let π_i, ι_i be as in Theorem 1.14 and define $\varphi_i = \iota_i\pi_i$. Conversely, given $\{\varphi_i\}$, show that $\varphi_i\varphi_i = \varphi_i$. Let $\psi_i = \varphi_i \mid \mathrm{Im}\, \varphi_i : \mathrm{Im}\, \varphi_i \to A$ and apply Theorem 1.14 with A, $\mathrm{Im}\, \varphi_i$, φ_i, and ψ_i in place of A, A_i, π_i, and ι_i.]

11. (a) If A is a module over a commutative ring R and $a \in A$, then $\mathcal{O}_a = \{r \varepsilon R \mid ra = 0\}$ is an ideal of R. If $\mathcal{O}_a \neq 0$, a is said to be a **torsion element** of A.
   (b) If R is an integral domain, then the set $T(A)$ of all torsion elements of A is a submodule of A. ($T(A)$ is called the **torsion submodule**.)
   (c) Show that (b) may be false for a commutative ring R, which is not an integral domain.

   In (d) – (f) R is an integral domain.

   (d) If $f : A \to B$ is an R-module homomorphism, then $f(T(A)) \subset T(B)$; hence the restriction f_T of f to $T(A)$ is an R-module homomorphism $T(A) \to T(B)$.
   (e) If $0 \to A \xrightarrow{f} B \xrightarrow{g} C$ is an exact sequence of R-modules, then so is $0 \to T(A) \xrightarrow{f_T} T(B) \xrightarrow{g_T} T(C)$.
   (f) If $g : B \to C$ is an R-module epimorphism, then $g_T : T(B) \to T(C)$ need not be an epimorphism. [*Hint:* consider abelian groups.]

12. (The Five Lemma). Let

$$\begin{array}{ccccccccc} A_1 & \to & A_2 & \to & A_3 & \to & A_4 & \to & A_5 \\ \downarrow{\scriptstyle \alpha_1} & & \downarrow{\scriptstyle \alpha_2} & & \downarrow{\scriptstyle \alpha_3} & & \downarrow{\scriptstyle \alpha_4} & & \downarrow{\scriptstyle \alpha_5} \\ B_1 & \to & B_2 & \to & B_3 & \to & B_4 & \to & B_5 \end{array}$$

be a commutative diagram of R-modules and R-module homomorphisms, with exact rows. Prove that:

(a) α_1 an epimorphism and α_2,α_4 monomorphisms $\Rightarrow \alpha_3$ is a monomorphism;

(b) α_5 a monomorphism and α_2,α_4 epimorphisms $\Rightarrow \alpha_3$ is an epimorphism.

13. (a) If $0 \to A \to B \xrightarrow{f} C \to 0$ and $0 \to C \xrightarrow{g} D \to E \to 0$ are short exact sequences of modules, then the sequence $0 \to A \to B \xrightarrow{gf} D \to E \to 0$ is exact.
(b) Show that every exact sequence may be obtained by splicing together suitable short exact sequences as in (a).

14. Show that isomorphism of short exact sequences is an equivalence relation.

15. If $f : A \to B$ and $g : B \to A$ are R-module homomorphisms such that $gf = 1_A$, then $B = \operatorname{Im} f \oplus \operatorname{Ker} g$.

16. Let R be a ring and R^{op} its opposite ring (Exercise III.1.17). If A is a left [resp. right] R-module, then A is a right [resp. left] R^{op}-module such that $ra = ar$ for all $a \in A$, $r \in R$, $r \in R^{op}$.

17. (a) If R has an identity and A is an R-module, then there are submodules B and C of A such that B is unitary, $RC = 0$ and $A = B \oplus C$. [*Hint:* let $B = \{1_R a \mid a \in A\}$ and $C = \{a \in A \mid 1_R a = 0\}$ and observe that for *all* $a \in A$, $a - 1_R a \in C$.]
(b) Let A_1 be another R-module, with $A_1 = B_1 \oplus C_1$ (B_1 unitary, $RC_1 = 0$). If $f : A \to A_1$ is an R-module homomorphism then $f(B) \subset B_1$ and $f(C) \subset C_1$.
(c) If the map f of part (b) is an epimorphism [resp. isomorphism], then so are $f \mid B : B \to B_1$ and $f \mid C : C \to C_1$.

18. Let R be a ring without identity. Embed R in a ring S with identity and characteristic zero as in the proof of Theorem III.1.10. Identify R with its image in S.
(a) Show that every element of S may be uniquely expressed in the form $r1_S + n1_S$ ($r \in R$, $n \in \mathbf{Z}$).
(b) If A is an R-module and $a \in A$, show that there is a unique R-module homomorphism $f: S \to A$ such that $f(1_S) = a$. [*Hint:* Let $f(r1_S + n1_S) = ra + na$.]

2. FREE MODULES AND VECTOR SPACES

In this section we study free objects in the category of modules over a ring. Such free modules, the most important examples of which are vector spaces over a division ring (Theorem 2.4), have widespread applications in many areas of mathematics. The special case of free abelian groups ($\mathbf{Z}$-modules) will serve as a model for the first part of this section. The remainder of the section consists of a discussion of the dimension (or rank) of a free module (Theorems 2.6–2.12) and an investigation of the special properties of the dimension of a vector space (Theorems and Corollaries 2.13–2.16).

A subset X of an R-module A is said to be **linearly independent** provided that for distinct $x_1, \ldots, x_n \in X$ and $r_i \in R$.

$$r_1x_1 + r_2x_2 + \cdots + r_nx_n = 0 \quad \Rightarrow \quad r_i = 0 \quad \text{for every } i.$$

A set that is not linearly independent is said to be **linearly dependent.** If A is generated as an R-module by a set Y, then we say that Y **spans** A. If R has an identity and A is unitary, Y spans A if and only if every element of A may be written as a linear combination: $r_1y_1 + r_2y_2 + \cdots + r_ny_n$ $(r_i \in R, y_i \in Y)$; see Theorem 1.5. A linearly independent subset of A that spans A is called a **basis** of A. Observe that the empty set is (vacuously) linearly independent and is a basis of the zero module (see Definition 1.4).

Theorem 2.1. *Let* R *be a ring with identity. The following conditions on a unitary* R*-module* F *are equivalent:*

(i) F *has a nonempty basis;*

(ii) F *is the internal direct sum of a family of cyclic* R*-modules, each of which is isomorphic as a left* R*-module to* R;

(iii) F *is* R*-module isomorphic to a direct sum of copies of the left* R*-module* R;

(iv) *there exists a nonempty set* X *and a function* $\iota : X \to F$ *with the following property: given any unitary* R*-module* A *and function* $f : X \to A$, *there exists a unique* R*-module homomorphism* $\bar{f} : F \to A$ *such that* $\bar{f}\iota = f$. *In other words,* F *is a free object in the category of unitary* R*-modules.*

The theorem is proved below. A unitary module F over a ring R with identity, which satisfies the equivalent conditions of Theorem 2.1, is called a **free R-module** on the set X. By Theorem 2.1 (iv), F is a free object in the category of all *unitary* left R-modules. But such an F is *not* a free object in the category of *all* left R-modules (Exercise 15). By definition the zero module is the free module on the empty set.

It is possible to define free modules in the category of all left R-modules over an arbitrary ring R (possibly without identity); see Exercise 2. Such a free module is *not* isomorphic to a direct sum of copies of R, even when R does have an identity (Exercise 2). In a few carefully noted instances below, certain results are also valid for these free modules in the category of all left R-modules. However, unless stated otherwise, the term "free module" will always mean a unitary free module in the sense of Theorem 2.1.

SKETCH OF PROOF OF 2.1. (i) $\Rightarrow$ (ii) Let X be a basis of F and $x \in X$. The map $R \to Rx$, given by $r \mapsto rx$, is an R-module epimorphism by Theorem 1.5. If $rx = 0$, then $r = 0$ by linear independence, whence the map is a monomorphism and $R \cong Rx$ as left R-modules. Verify that F is the internal direct sum of the cyclic modules Rx $(x \in X)$.

(ii) $\Rightarrow$ (iii) Theorem 1.15 and Exercise 1.8.

(iii) $\Rightarrow$ (i) Suppose $F \cong \sum R$ and the copies of R are indexed by a set X. For each $x \in X$ let θ_x be the element $\{r_i\}$ of $\sum R$, where $r_i = 0$ for $i \neq x$ and $r_x = 1_R$. Verify that $\{\theta_x \mid x \in X\}$ is a basis of $\sum R$ and use the isomorphism $F \cong \sum R$ to obtain a basis of F.

(i) $\Rightarrow$ (iv) Let X be a basis of F and $\iota : X \to F$ the inclusion map. Suppose we are given a map $f : X \to A$. If $u \in F$, then $u = \sum_{i=1}^{n} r_i x_i \ (r_i \in R, x_i \in X)$ since X spans F. If $u = \sum_{i=1}^{n} s_i x_i \ (s_i \in R)$, then $\sum_i (r_i - s_i)x_i = 0$, whence $r_i = s_i$ for every i by linear independence. Consequently, the map $\bar{f} : F \to A$ given by

$$\bar{f}(u) = \bar{f}\left(\sum_{i=1}^{n} r_i x_i\right) = \sum_{i=1}^{n} r_i f(x_i)$$

is a well-defined function such that $\bar{f}\iota = f$. Verify that $\bar{f}$ is an R-module homomorphism. Since X generates F, any R-module homomorphism $F \to A$ is uniquely determined by its action on X. Thus, if $g : F \to A$ is an R-module homomorphism such that $g\iota = f$, then for every $x \in X$, $g(x) = g(\iota(x)) = f(x) = \bar{f}(x)$, whence $g = \bar{f}$ and $\bar{f}$ is unique. Therefore, by Definition I.7.7 F is a free object on the set X in the category of unitary R-modules.

(iv) $\Rightarrow$ (iii) Given $\iota : X \to F$ construct the direct sum $\sum R$, with one copy of R for each $x \in X$. Let $Y = \{\theta_x \mid x \in X\}$ be the basis of the (unitary) R-module $\sum R$ as in the proof of (iii) $\Rightarrow$ (i). The proof of (iii) $\Rightarrow$ (i) $\Rightarrow$ (iv) shows that $\sum R$ is a free object on the set Y in the category of R-modules (with $Y \to \sum R$ the inclusion map). Since $|X| = |Y|$, the proof of Theorem I.7.8 implies that there is an R-module isomorphism $f : F \cong \sum R$ such that $f(\iota(X)) = Y$. ∎

REMARKS. (a) If F is a free R-module on a set $X(\iota : X \to F)$, then the proof of (iv) $\Rightarrow$ (iii) of Theorem 2.1 implies that $\iota(X)$ is actually a basis of F.

(b) Conversely, the proof of (i) $\Rightarrow$ (iv) of Theorem 2.1 shows that if X is a basis of a unitary module F over a ring R with identity, then F is free on X, with $\iota : X \to F$ the inclusion map.

(c) If X is any nonempty set and R is a ring with identity, then the proof of Theorem 2.1 shows how to construct a free R-module on the set X. Simply let F be the direct sum $\sum R$, with the copies of R indexed by the set X. In the notation of the proof, $\{\theta_x \mid x \in X\}$ is a basis of F so that $F = \sum_{x \in X} R\theta_x$. Since the map $\iota : X \to F$, given by $x \mapsto \theta_x$, is injective it follows easily that F is free on X in the sense of condition (iv) of Theorem 2.1. In this situation we shall usually identify X with its image under ι, writing x in place of θ_x, so that $X \subset F$. In this notation $F = \sum_{x \in X} R\theta_x$ is written as $\sum_{x \in X} Rx$ and a typical element of F has the form $r_1x_1 + \cdots + r_nx_n \ (r_i \in R; x_i \in X)$. In particular, $X = \iota(X)$ is a basis of F.

(d) The existence of free modules on a given set in the category of all modules over an arbitrary ring (possibly without identity) is proved in Exercise 2.

Corollary 2.2. *Every (unitary) module* A *over a ring* R *(with identity) is the homomorphic image of a free* R*-module* F*. If* A *is finitely generated, then* F *may be chosen to be finitely generated.*

REMARK. Corollary 2.2 and its proof are valid if the words in parentheses are deleted and "free module" is taken to mean a free module in the category of all left modules over an arbitrary ring (as defined in Exercise 2).

SKETCH OF PROOF OF 2.2. Let X be a set of generators of A and F the free R-module on the set X. Then the inclusion map $X \to A$ induces an R-module homomorphism $\bar{f}: F \to A$ such that $X \subset \operatorname{Im} \bar{f}$ (Theorem 2.1 (iv)). Since X generates A, we must have $\operatorname{Im} \bar{f} = A$. ∎

REMARK. Unlike the situation with free abelian groups, a submodule of a free module over an arbitrary ring need not be free. For instance $\{0,2,4\}$ is a submodule of Z_6, but is clearly not a free Z_6-module; compare Theorem II.1.6 and Theorem 6.1 below.

Vector spaces over a division ring D (Definition 1.1) are important, among other reasons, because *every* vector space over D is in fact a free D-module. To prove this we need

Lemma 2.3. *A maximal linearly independent subset* X *of a vector space* V *over a division ring* D *is a basis of* V.

PROOF. Let W be the subspace of V spanned by the set X. Since X is linearly independent and spans W, X is a basis of W. If $W = V$, we are done. If not, then there exists a nonzero $a \in V$ with $a \notin W$. Consider the set $X \cup \{a\}$. If $ra + r_1x_1 + \cdots + r_nx_n = 0$ $(r, r_i \in D, x_i \in X)$ and $r \neq 0$, then $a = r^{-1}(ra) = -r^{-1}r_1x_1 - \cdots - r^{-1}r_nx_n \in W$, which contradicts the choice of a. Hence $r = 0$, which implies $r_i = 0$ for all i since X is linearly independent. Consequently $X \cup \{a\}$ is a linearly independent subset of V, contradicting the maximality of X. Therefore $W = V$ and X is a basis. ∎

Theorem 2.4. *Every vector space* V *over a division ring* D *has a basis and is therefore a free* D*-module. More generally every linearly independent subset of* V *is contained in a basis of* V.

The converse of Theorem 2.4 is also true, namely, if every unitary module over a ring D with identity is free, then D is a division ring (Exercise 3.14).

SKETCH OF PROOF OF 2.4. The first statement is an immediate consequence of the second since the null set is a linearly independent subset of every vector space. Consequently, we assume that X is any linearly independent subset of V and let $\mathcal{S}$ be the set of all linearly independent subsets of V that contain X. Since $X \in \mathcal{S}$, $\mathcal{S} \neq \varnothing$. Partially order $\mathcal{S}$ by set theoretic inclusion. If $\{C_i \mid i \in I\}$ is a chain in $\mathcal{S}$ verify that the set $C = \bigcup_{i \in I} C_i$ is linearly independent and hence an element of $\mathcal{S}$. Clearly C is an upper bound for the chain $\{C_i \mid i \in I\}$. By Zorn's Lemma $\mathcal{S}$ contains a maximal element B that contains X and is necessarily a maximal linearly independent subset of V. By Lemma 2.3 B is a basis of V. ∎

Theorem 2.5. *If* V *is a vector space over a division ring* D *and* X *is a subset that spans* V, *then* X *contains a basis of* V.

SKETCH OF PROOF. Partially order the set $\mathcal{S}$ of all linearly independent subsets of X by inclusion. Zorn's Lemma implies the existence of a maximal linearly independent subset Y of X. Every element of X is a linear combination of elements of Y (otherwise, as in Lemma 2.3, we could construct a linearly independent subset of X that properly contained Y, contradicting maximality). Since X spans V, so does Y. Hence Y is a basis of V. ■

In the case of free abelian groups (**Z**-modules) we know that any two bases of a free **Z**-module have the same cardinality (Theorem II.1.2). Unfortunately, this is not true for free modules over arbitrary rings with identity (Exercise 13). We shall now show that vector spaces over a division ring and free modules over a commutative ring with identity have this property.

Theorem 2.6. *Let* R *be a ring with identity and* F *a free* R*-module with an infinite basis* X. *Then every basis of* F *has the same cardinality as* X.

PROOF. If Y is another basis of F, then we claim that Y is infinite. Suppose on the contrary that Y were finite. Since Y generates F and every element of Y is a linear combination of a finite number of elements of X, it follows that there is a finite subset $\{x_1, \ldots, x_m\}$ of X, which generates F. Since X is infinite, there exists

$$x \in X - \{x_1, \ldots, x_m\}.$$

Then for some $r_i \in R$, $x = r_1x_1 + \cdots + r_mx_m$, which contradicts the linear independence of X. Therefore, Y is infinite.

Let $K(Y)$ be the set of all finite subsets of Y. Define a map $f : X \to K(Y)$ by $x \mapsto \{y_1, \ldots, y_n\}$, where $x = r_1y_1 + \cdots + r_ny_n$ and $r_i \neq 0$ for all i. Since Y is a basis, the y_i are uniquely determined and f is a well-defined function, (which need not be injective). If Im f were finite, then $\bigcup_{S \in \operatorname{Im} f} S$ would be a finite subset of Y that would generate X and hence F. This leads to a contradiction of the linear independence of Y as in the preceding paragraph. Hence Im f is infinite.

Next we show that $f^{-1}(T)$ is a finite subset of X for every $T \in \operatorname{Im} f \subset K(Y)$. If $x \in f^{-1}(T)$, then x is contained in the submodule F_T of F generated by T; that is, $f^{-1}(T) \subset F_T$ (see Theorem 1.5). Since T is finite and each $y \in T$ is a linear combination of a finite number of elements of X, there is a finite subset S of X such that F_T is contained in the submodule F_S of F generated by S. Thus $x \in f^{-1}(T)$ implies $x \in F_S$ and x is a linear combination of elements of S (Theorem 1.5). Since $x \in X$ and $S \subset X$, this contradicts the linear independence of X unless $x \in S$. Therefore, $f^{-1}(T) \subset S$, whence $f^{-1}(T)$ is finite.

For each $T \in \operatorname{Im} f$, order the elements of $f^{-1}(T)$, say $x_1, \ldots, x_n$, and define an injective map $g_T : f^{-1}(T) \to \operatorname{Im} f \times \mathbf{N}$ by $x_k \mapsto (T,k)$. Verify that the sets $f^{-1}(T)$ ($T \in \operatorname{Im} f$) form a partition of X. It follows that the map $X \to \operatorname{Im} f \times \mathbf{N}$ defined by $x \mapsto g_T(x)$, where $x \in f^{-1}(T)$, is a well-defined injective function, whence $|X| \leq |\operatorname{Im} f \times \mathbf{N}|$. Therefore by Definition 8.3, Theorem 8.11, and Corollary 8.13 of the Introduction:

$$|X| \leq |\operatorname{Im} f \times \mathbf{N}| = |\operatorname{Im} f|\,\aleph_0 = |\operatorname{Im} f| \leq |K(Y)| = |Y|.$$

Interchanging X and Y in the preceding argument shows that $|Y| \leq |X|$. Therefore $|Y| = |X|$ by the Schroeder-Bernstein Theorem. ■

Theorem 2.7. *If V is a vector space over a division ring D, then any two bases of V have the same cardinality.*

PROOF. Let X and Y be bases of V. If either X or Y is infinite, then $|X| = |Y|$ by Theorem 2.6. Hence we assume X and Y are finite, say $X = \{x_1, \dots, x_n\}$, and $Y = \{y_1, \dots, y_m\}$. Since X and Y are bases, $0 \neq y_m = r_1x_1 + \cdots + r_nx_n$ for some $r_i \in D$. If r_k is the first nonzero r_i, then $x_k = r_k^{-1}y_m - r_k^{-1}r_{k+1}x_{k+1} - \cdots - r_k^{-1}r_nx_n$. Therefore, the set $X' = \{y_m,x_1, \dots, x_{k-1},x_{k+1}, \dots, x_n\}$ spans V (since X does). In particular

$$y_{m-1} = s_my_m + t_1x_1 + \cdots + t_{k-1}x_{k-1} + t_{k+1}x_{k+1} + \cdots + t_nx_n \ (s_m,t_i \in D).$$

Not all of the t_i are zero (otherwise $y_{m-1} - s_my_m = 0$, which contradicts the linear independence of Y). If t_j is the first nonzero t_i, then x_j is a linear combination of y_{m-1},y_m and those x_i with $i \neq j,k$. Consequently, the set $\{y_{m-1},y_m\} \cup \{x_i \mid i \neq j,k\}$ spans V (since X' does). In particular, y_{m-2} is a linear combination of y_{m-1},y_m and the x_i with $i \neq j,k$. The above process of adding a y and eliminating an x may therefore be repeated. At the end of the kth step we have a set consisting of $y_m,y_{m-1}, \dots, y_{m-k+1}$ and $n - k$ of the x_i, which spans V. If $n < m$, then at the end of n steps we would conclude that $\{y_m, \dots, y_{m-n+1}\}$ spans V. Since $m - n + 1 \geq 2$, y_1 would be a linear combination of $y_m, \dots, y_{m-n+1}$, which would contradict the linear independence of Y. Therefore, we must have $m \leq n$. A similar argument with the roles of X and Y reversed shows that $n \leq m$ and hence $m = n$. ■

Definition 2.8. *Let R be a ring with identity such that for every free R-module F, any two bases of F have the same cardinality. Then R is said to have the* **invariant dimension property** *and the cardinal number of any basis of F is called the* **dimension** *(or* **rank***) of F over R.*

Theorem 2.7 states that every division ring has the invariant dimension property. We shall follow the widespread (but not universal) practice of using "dimension" when referring to vector spaces over a division ring and "rank" when referring to free modules over other rings. The dimension of a vector space V over a division ring D will be denoted here by $\dim_D V$. The properties of $\dim_D V$ will be investigated after Corollary 2.12. Results 2.9–2.12 are not needed in the sequel, except in Sections IV.6 and VII.5.

Proposition 2.9. *Let E and F be free modules over a ring R that has the invariant dimension property. Then E $\cong$ F if and only if E and F have the same rank.*

PROOF. Exercise; see Proposition II.1.3. ■

Lemma 2.10. *Let R be a ring with identity, I ($\neq$ R) an ideal of R, F a free R-module with basis X and π : F $\to$ F/IF the canonical epimorphism. Then F/IF is a free R/I-module with basis π(X) and $|\pi(X)| = |X|$.*

Recall that $IF = \left\{\sum_{i=1}^{n} r_i a_i \mid r_i \in I,\ a_i \in F,\ n \in \mathbf{N}^*\right\}$ and that the action of R/I on F/IF is given by $(r + I)(a + IF) = ra + IF$ (Exercise 1.3).

PROOF OF 2.10. If $u + IF \in F/IF$, then $u = \sum_{j=1}^{n} r_j x_j$ with $r_j \in R$, $x_j \in X$ since $u \in F$ and X is a basis of F. Consequently, $u + IF = (\sum_j r_j x_j) + IF = \sum_j (r_j x_j + IF)$ $= \sum_j (r_j + I)(x_j + IF) = \sum_j (r_j + I)\pi(x_j)$, whence $\pi(X)$ generates F/IF as an R/I-module. On the other hand, if $\sum_{k=1}^{m} (r_k + I)\pi(x_k) = 0$ with $r_k \in R$ and $x_1, \dots, x_m$ distinct elements of X, then $0 = \sum_k (r_k + I)\pi(x_k) = \sum_k (r_k + I)(x_k + IF)$ $= \sum_k r_k x_k + IF$, whence $\sum_k r_k x_k \in IF$. Thus $\sum_k r_k x_k = \sum_j s_j u_j$ with $s_j \in I$, $u_j \in F$. Since each u_j is a linear combination of elements of X and I is an ideal, $\sum_j s_j u_j$ is a linear combination of elements of X with coefficients in I. Consequently, $\sum_{k=1}^{m} r_k x_k$ $= \sum_j s_j u_j = \sum_{t=1}^{d} c_t y_t$ with $c_t \in I$, $y_t \in X$. The linear independence of X implies that (after reindexing and inserting terms $0x_k$, $0y_t$ if necessary) $m = d$, $x_k = y_k$ and $r_k = c_k \in I$ for every k. Hence $r_k + I = 0$ in R/I for every k and $\pi(X)$ is linearly independent over R/I. Thus F/IF is a free R/I-module with basis $\pi(X)$ (Theorem 2.1). Finally if $x, x' \in X$ and $\pi(x) = \pi(x')$ in F/IF, then $(1_R + I)\pi(x) - (1_R + I)\pi(x') = 0$. If $x \neq x'$, the preceding argument implies that $1_R \in I$, which contradicts the fact that $I \neq R$. Therefore, $x = x'$ and the map $\pi : X \to \pi(X)$ is a bijection, whence $|X| = |\pi(X)|$. ■

Proposition 2.11. *Let* $f : R \to S$ *be a nonzero epimorphism of rings with identity. If* S *has the invariant dimension property, then so does* R.

PROOF. Let $I = \operatorname{Ker} f$; then $S \cong R/I$ (Corollary III.2.10). Let X and Y be bases of the free R-module F and $\pi : F \to F/IF$ the canonical epimorphism. By Lemma 2.10 F/IF is a free R/I-module (and hence a free S-module) with bases $\pi(X)$ and $\pi(Y)$ such that $|X| = |\pi(X)|$, $|Y| = |\pi(Y)|$. Since S has the invariant dimension property, $|\pi(X)| = |\pi(Y)|$. Therefore, $|X| = |Y|$ and R has the invariant dimension property. ■

Corollary 2.12. *If* R *is a ring with identity that has a homomorphic image which is a division ring, then* R *has the invariant dimension property. In particular, every commutative ring with identity has the invariant dimension property.*

PROOF. The first statement follows from Theorem 2.7 and Proposition 2.11. If R is commutative with identity, then R contains a maximal ideal M (Theorem III.2.18) and R/M is a field (Theorem III.2.20). Thus the second statement is a special case of the first. ■

We return now to vector spaces over a division ring and investigate the properties of dimension. A vector space V over a division ring D is said to be **finite dimensional** if $\dim_D V$ is finite.

Theorem 2.13. *Let* W *be a subspace of a vector space* V *over a division ring* D.

(i) $dim_D W \leq dim_D V$;
(ii) *if* $dim_D W = dim_D V$ *and* $dim_D V$ *is finite, then* $W = V$;
(iii) $dim_D V = dim_D W + dim_D(V/W)$.

SKETCH OF PROOF. (i) Let Y be a basis of W. By Theorem 2.4 there is a basis X of V containing Y. Therefore, $\dim_D W = |Y| \leq |X| = \dim_D V$. (ii) If $|Y| = |X|$ and $|X|$ is finite, then since $Y \subset X$ we must have $Y = X$, whence $W = V$. (iii) We shall show that $U = \{x + W \mid x \in X - Y\}$ is a basis of V/W. This will imply (by Definition 8.3 of the Introduction) that $\dim_D V = |X| = |Y| + |X - Y| = |Y| + |U| = \dim_D W + \dim_D(V/W)$. If $v \in V$, then $v = \sum_i r_i y_i + \sum_j s_j x_j$ $(r_i, s_i \in D;\ y_i \in Y;$ $x_j \in X - Y)$ so that $v + W = \sum_j s_j(x_j + W)$. Therefore, U spans V/W. If $\sum_j r_j(x_j + W) = 0$ $(r_j \in D;\ x_j \in X - Y)$, then $\sum_j r_j x_j \in W$, whence $\sum_j r_j x_j = \sum_k s_k y_k$ $(s_k \in D;\ y_k \in Y)$. This contradicts the linear independence of $X = Y \cup (X - Y)$ unless $r_j = 0, s_k = 0$ for all j,k. Therefore, U is linearly independent and $|U| = |X - Y|$. ■

Corollary 2.14. *If* $f: V \to V'$ *is a linear transformation of vector spaces over a division ring* D, *then there exists a basis* X *of* V *such that* $X \cap Ker\, f$ *is a basis of Ker* f *and* $\{f(x) \mid f(x) \neq 0, x \in X\}$ *is a basis of Im* f. *In particular,*

$$dim_D V = dim_D(Ker\ f) + dim_D(Im\ f).$$

SKETCH OF PROOF. To prove the first statement let $W = \operatorname{Ker} f$ and let Y,X be as in the proof of Theorem 2.13. The second statement follows from Theorem 2.13 (iii) since $V/W \cong \operatorname{Im} f$ by Theorem 1.7. ■

Corollary 2.15. *If* V *and* W *are finite dimensional subspaces of a vector space over a division ring* D, *then*

$$dim_D V + dim_D W = dim_D(V \cap W) + dim_D(V + W).$$

SKETCH OF PROOF. Let X be a basis of $V \cap W$, Y a (finite) basis of V that contains X, and Z a (finite) basis of W that contains X (Theorem 2.4). Show that $X \cup (Y - X) \cup (Z - X)$ is a basis of $V + W$, whence

$$\begin{aligned}\dim_D(V + W) = |X| + |Y - X| + |Z - X| &= \dim_D(V \cap W)\\ &+ (\dim_D V - \dim_D(V \cap W))\\ &+ (\dim_D W - \dim_D(V \cap W)).\end{aligned}$$ ■

Recall that if a division ring R is contained in a division ring S, then S is a vector space over R with rs $(s \in S, r \in R)$ the ordinary product in S. The following theorem will be needed for the study of field extensions in Chapter V.

Theorem 2.16. *Let* R,S,T *be division rings such that* R ⊂ S ⊂ T. *Then*

$$dim_R T = (dim_S T)(dim_R S).$$

Furthermore, $dim_R T$ is finite if and only if $dim_S T$ and $dim_R S$ are finite.

PROOF. Let U be a basis of T over S, and let V a basis of S over R. It suffices to show that $\{vu \mid v \in V, u \in U\}$ is a basis of T over R. For the elements vu are all distinct by the linear independence of U over S. Consequently, we may conclude that $\dim_R T = |U||V| = (\dim_S T)(\dim_R S)$. The last statement of the theorem then follows immediately since the product of two finite cardinal numbers is finite and the product of an infinite with a finite cardinal number is infinite (Introduction, Theorem 8.11).

If $u \in T$, then $u = \sum_{i=1}^{n} s_i u_i$ $(s_i \in S, u_i \in U)$ since U spans T as a vector space over S. Since S is a vector space over R each s_i may be written as $s_i = \sum_{j=1}^{m_i} r_{ij} v_j$ $(r_{ij} \in R, v_j \in V)$. Thus $u = \sum_i s_i u_i = \sum_i (\sum_j r_{ij} v_j) u_i = \sum_i \sum_j r_{ij} v_j u_i$. Therefore, $\{vu | v \in V, u \in U\}$ spans T as a vector space over R.

Suppose that $\sum_{i=1}^{n} \sum_{j=1}^{m} r_{ij}(v_j u_i) = 0$ $(r_{ij} \in R, v_j \in V, u_i \in U)$. For each i, let $s_i = \sum_{j=1}^{m} r_{ij} v_j \in S$. Then $0 = \sum_i \sum_j r_{ij}(v_j u_i) = \sum_i (\sum_j r_{ij} v_j) u_i = \sum_i s_i u_i$. The linear independence of U over S implies that for each i, $0 = s_i = \sum_j r_{ij} v_j$. The linear independence of V over R implies that $r_{ij} = 0$ for all i,j. Therefore, $\{vu \mid v \in V, u \in U\}$ is linearly independent over R and hence a basis. ■

EXERCISES

1. (a) A set of vectors $\{x_1, \dots, x_n\}$ in a vector space V over a division ring R is linearly dependent if and only if some x_k is a linear combination of the preceding x_i.
   (b) If $\{x_1,x_2,x_3\}$ is a linearly independent subset of V, then the set $\{x_1 + x_2, x_2 + x_3, x_3 + x_1\}$ is linearly independent if and only if Char $R \neq 2$. [See Definition III.1.8].

2. Let R be any ring (possibly without identity) and X a nonempty set. In this exercise an R-module F is called a **free module on** X if F is a free object on X in the category of *all* left R-modules. Thus by Definition I.7.7, F is the free module on X if there is a function $\iota : X \to F$ such that for any left R-module A and function $f : X \to A$ there is a unique R-module homomorphism $\bar{f} : F \to A$ with $\bar{f}\iota = f$.
   (a) Let $\{X_i \mid i \in I\}$ be a collection of mutually disjoint sets and for each $i \in I$, suppose F_i is a free module on X_i, with $\iota_i : X_i \to F_i$. Let $X = \bigcup_{i \in I} X_i$ and $F = \sum_{i \in I} F_i$, with $\phi_i : F_i \to F$ the canonical injection. Define $\iota : X \to F$ by $\iota(x) = \phi_i\iota_i(x)$ for $x \in X_i$; (ι is well defined since the X_i are disjoint). Prove that F is a free module on X. [*Hint:* Theorem 1.13 may be useful.]
   (b) Assume R has an identity. Let the abelian group **Z** be given the trivial R-module structure ($rm = 0$ for all $r \in R$, $m \in$ **Z**), so that $R \oplus$ **Z** is an R-module with $r(r',m) = (rr',0)$ for all $r,r' \in R$, $m \in$ **Z**. If X is any one element set, $X = \{t\}$,

let $\iota : X \to R \bigoplus \mathbf{Z}$ be given by $\iota(t) = (1_R,1)$. Prove that $R \bigoplus \mathbf{Z}$ is a free module on X. [*Hint:* given $f : X \to A$, let $A = B \bigoplus C$ as in Exercise 1.17, so that $f(t) = b + c$ $(b \in B, c \in C)$. Define $f(r,m) = rb + mc$.]

(c) If R is an arbitrary ring and X is any set, then there exists a free module on X. [*Hint.* Since X is the disjoint union of the sets $\{t\}$ with $t \in X$, it suffices by (a) to assume X has only one element. If R has an identity, use (b). If R has no identity, embed R in a ring S with identity and characteristic 0 as in the proof of Theorem III.1.10. Use Exercise 1.18 to show that S is a free R-module on X.]

3. Let R be any ring (possibly without identity) and F a free R-module on the set X, with $\iota : X \to F$, as in Exercise 2. Show that $\iota(X)$ is a set of generators of the R-module F. [*Hint:* let G be the submodule of F generated by $\iota(X)$ and use the definition of "free module" to show that there is a module homomorphism φ such that

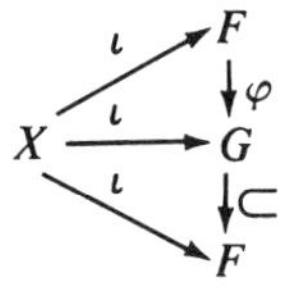

is commutative. Conclude that $\varphi = 1_F$.]

4. Let R be a principal ideal domain, A a unitary left R-module, and $p \in R$ a prime (= irreducible). Let $pA = \{pa \mid a \in A\}$ and $A[p] = \{a \in A \mid pa = 0\}$.

(a) $R/(p)$ is a field (Theorems III.2.20 and III.3.4).
(b) pA and $A[p]$ are submodules of A.
(c) A/pA is a vector space over $R/(p)$, with $(r + (p))(a + pA) = ra + pA$.
(d) $A[p]$ is a vector space over $R/(p)$, with $(r + (p))a = ra$.

5. Let V be a vector space over a division ring D and S the set of all subspaces of V, partially ordered by set theoretic inclusion.

(a) S is a complete lattice (see Introduction, Exercise 7.2; the l.u.b. of V_1,V_2 is $V_1 + V_2$ and the g.l.b. $V_1 \cap V_2$).

(b) S is a complemented lattice; that is, for each $V_1 \in S$ there exists $V_2 \in S$ such that $V = V_1 + V_2$ and $V_1 \cap V_2 = 0$, so that $V = V_1 \bigoplus V_2$.

(c) S is a modular lattice; that is, if $V_1,V_2,V_3 \in S$ and $V_3 \subset V_1$, then

$$V_1 \cap (V_2 + V_3) = (V_1 \cap V_2) + V_3.$$

6. Let $\mathbf{R}$ and $\mathbf{C}$ be the fields of real and complex numbers respectively.

(a) $\dim_{\mathbf{R}}\mathbf{C} = 2$ and $\dim_{\mathbf{R}}\mathbf{R} = 1$.
(b) There is no field K such that $\mathbf{R} \subset K \subset \mathbf{C}$.

7. If G is a nontrivial group that is not cyclic of order 2, then G has a nonidentity automorphism. [*Hint:* Exercise II.4.11 and Exercise 4(d) above.]

8. If V is a finite dimensional vector space and V^m is the vector space

$$V \bigoplus V \bigoplus \cdots \bigoplus V \quad (m \text{ summands}),$$

then for each $m \geq 1$, V^m is finite dimensional and $\dim V^m = m(\dim V)$.

9. If F_1 and F_2 are free modules over a ring with the invariant dimension property, then rank $(F_1 \oplus F_2) = \text{rank } F_1 + \text{rank } F_2$.

10. Let R be a ring with no zero divisors such that for all $r,s \in R$ there exist $a,b \in R$, not both zero, with $ar + bs = 0$.
   (a) If $R = K \oplus L$ (module direct sum), then $K = 0$ or $L = 0$.
   (b) If R has an identity, then R has the invariant dimension property.

11. Let F be a free module of infinite rank α over a ring R that has the invariant dimension property. For each cardinal β such that $0 \leq \beta \leq \alpha$, F has infinitely many proper free submodules of rank β.

12. If F is a free module over a ring with identity such that F has a basis of finite cardinality $n \geq 1$ and another basis of cardinality $n + 1$, then F has a basis of cardinality m for every $m \geq n$ $(m \in \mathbf{N}^*)$.

13. Let K be a ring with identity and F a free K-module with an infinite denumerable basis $\{e_1,e_2,\ldots\}$. Then $R = \text{Hom}_K(F,F)$ is a ring by Exercise 1.7(b). If n is any positive integer, then the free left R-module R has a basis of n elements; that is, as an R-module, $R \cong R \oplus \cdots \oplus R$ for any finite number of summands. [*Hint:* $\{1_R\}$ is a basis of one element; $\{f_1,f_2\}$ is a basis of two elements, where $f_1(e_{2n}) = e_n$, $f_1(e_{2n-1}) = 0$, $f_2(e_{2n}) = 0$ and $f_2(e_{2n-1}) = e_n$. Note that for any $g \in R$, $g = g_1f_1 + g_2f_2$, where $g_1(e_n) = g(e_{2n})$ and $g_2(e_n) = g(e_{2n-1})$.]

14. Let $f : V \to V'$ be a linear transformation of finite dimensional vector spaces V and V' such that $\dim V = \dim V'$. Then the following conditions are equivalent: (i) f is an isomorphism; (ii) f is an epimorphism; (iii) f is a monomorphism. [*Hint:* Corollary 2.14.]

15. Let R be a ring with identity. Show that R is *not* a free module on any set in the category of *all* R-modules (as defined in Exercise 2). [*Hint*. Consider a nonzero abelian group A with the trivial R-module structure ($ra = 0$ for all $r \in R$, $a \in A$). Observe that the only module homomorphism $R \to A$ is the zero map.]

3. PROJECTIVE AND INJECTIVE MODULES

Every free module is projective and arbitrary projective modules (which need not be free) have some of the same properties as free modules. Projective modules are especially useful in a categorical setting since they are defined solely in terms of modules and homomorphisms. Injectivity, which is also studied here, is the dual notion to projectivity.

Definition 3.1. *A module* P *over a ring* R *is said to be* **projective** *if given any diagram of* R*-module homomorphisms*

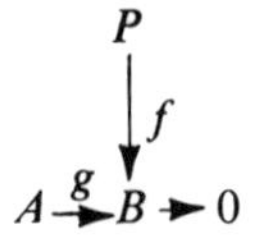

with bottom row exact (that is, g *an epimorphism), there exists an* R-*module homomorphism* h : P → A *such that the diagram*

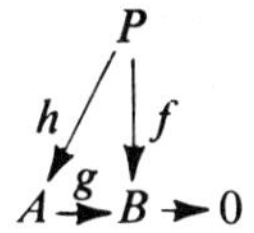

is commutative (that is, gh = f).

The theorems below will provide several examples of projective modules. We note first that if *R* has an identity and *P* is unitary, then *P* is projective if and only if for every pair of *unitary* modules *A*, *B* and diagram of *R*-module homomorphisms

$$
\begin{array}{c}
P \\
\downarrow f \\
A \xrightarrow{g} B \rightarrow 0
\end{array}
$$

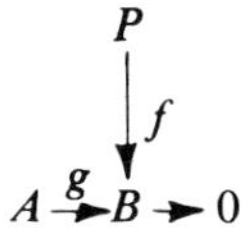

with *g* an epimorphism, there exists a homomorphism $h : P \to A$ with $gh = f$. For by Exercise 1.17, $A = A_1 \oplus A_2$ and $B = B_1 \oplus B_2$ with A_1, B_1 unitary and $RA_2 = 0 = RB_2$. Exercise 1.17 shows further that $f(P) \subset B_1$ and $g \mid A_1$ is an epimorphism $A_1 \to B_1$, so that we have a diagram of unitary modules:

$$
\begin{array}{c}
P \\
\downarrow f \\
A_1 \xrightarrow{g} B_1 \rightarrow 0
\end{array}
$$

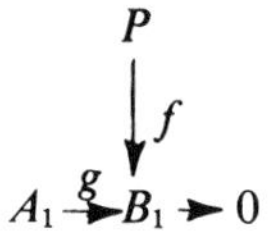

Thus the existence of $h : P \to A$ with $gh = f$ is equivalent to the existence of $h : P \to A_1$ with $gh = f$.

Theorem 3.2. *Every free module* F *over a ring* R *with identity is projective.*

REMARK. The Theorem is true if the words "with identity" are deleted and *F* is a free module in the category of all left *R*-modules (as defined in Exercise 2.2). The proof below carries over verbatim, provided Exercise 2.2 is used in place of Theorem 2.1 and the word "unitary" deleted.

PROOF OF 3.2. In view of the remarks preceding the theorem we may assume that we are given a diagram of homomorphisms of unitary *R*-modules:

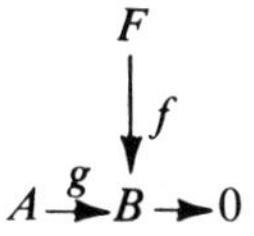

with g an epimorphism and F a free R-module on the set X ($\iota : X \to F$). For each $x \in X$, $f(\iota(x)) \in B$. Since g is an epimorphism, there exists $a_x \in A$ with $g(a_x) = f(\iota(x))$. Since F is free, the map $X \to A$ given by $x \mapsto a_x$ induces an R-module homomorphism $h : F \to A$ such that $h(\iota(x)) = a_x$ for all $x \in X$. Consequently, $gh\iota(x) = g(a_x) = f\iota(x)$ for all $x \in X$ so that $gh\iota = f\iota : X \to B$. By the uniqueness part of Theorem 2.1 (iv) we have $gh = f$. Therefore F is projective. ■

Corollary 3.3. *Every module* A *over a ring* R *is the homomorphic image of a projective* R*-module.*

PROOF. Immediate from Theorem 3.2 and Corollary 2.2. ■

Theorem 3.4. *Let* R *be a ring. The following conditions on an* R*-module* P *are equivalent.*

(i) P *is projective;*

(ii) *every short exact sequence* $0 \to A \xrightarrow{f} B \xrightarrow{g} P \to 0$ *is split exact (hence* $B \cong A \oplus P$);

(iii) *there is a free module* F *and an* R*-module* K *such that* $F \cong K \oplus P$.

REMARK. The words "free module" in condition (iii) may be interpreted in the sense of Theorem 2.1 if R has an identity and P is unitary, and in the sense of Exercise 2.2 otherwise. The proof is the same in either case.

PROOF OF 3.4. (i) $\Rightarrow$ (ii) Consider the diagram

$$\begin{array}{ccc} & & P \\ & & \downarrow 1_P \\ B & \xrightarrow{g} & P \to 0 \end{array}$$

with bottom row exact by hypothesis. Since P is projective there is an R-module homomorphism $h : P \to B$ such that $gh = 1_P$. Therefore, the short exact sequence $0 \to A \xrightarrow{f} B \underset{h}{\overset{g}{\rightleftarrows}} P \to 0$ is split exact by Theorem 1.18 and $B \cong A \oplus P$.

(ii) $\Rightarrow$ (iii) By Corollary 2.2 there is a free R-module F and an epimorphism $g : F \to P$. If $K = \operatorname{Ker} g$, then $0 \to K \xrightarrow{\subset} F \xrightarrow{g} P \to 0$ is exact. By hypothesis the sequence splits so that $F \cong K \oplus P$ by Theorem 1.18.

(iii) $\Rightarrow$ (i) Let π be the composition $F \cong K \oplus P \to P$ where the second map is the canonical projection. Similarly let ι be the composition $P \to K \oplus P \cong F$ with the first map the canonical injection. Given a diagram of R-module homomorphisms

$$\begin{array}{ccc} & & P \\ & & \downarrow f \\ A & \xrightarrow{g} & B \to 0 \end{array}$$

with exact bottom row, consider the diagram

$$\begin{array}{ccc} & & F \\ & & \pi \downarrow \uparrow \iota \\ & & P \\ & & \downarrow f \\ A & \xrightarrow{g} & B \to 0 \end{array}$$

Since F is projective by Theorem 3.2, there is an R-module homomorphism $h_1 : F \to A$ such that $gh_1 = f\pi$. Let $h = h_1\iota : P \to A$. Then $gh = gh_1\iota = (f\pi)\iota = f(\pi\iota) = f1_P = f$. Therefore, P is projective. ∎

EXAMPLE. If $R = Z_6$, then Z_3 and Z_2 are Z_6-modules (see Exercise 1.1) and there is a Z_6-module isomorphism $Z_6 \cong Z_2 \oplus Z_3$. Hence both Z_2 and Z_3 are projective Z_6-modules that are not free Z_6-modules.

Proposition 3.5. *Let* R *be a ring. A direct sum of* R*-modules* $\sum_{i \in I} P_i$ *is projective if and only if each* P_i *is projective.*

SKETCH OF PROOF. Suppose $\sum P_i$ is projective. Since the proof of (iii) ⇒ (i) in Theorem 3.4 uses only the fact that F is projective, it remains valid with $\sum_{i \in I} P_i$, $\sum_{i \neq j} P_i$ and P_j in place of F,K, and P respectively. The converse is proved by similar techniques using the diagram

$$\begin{array}{ccc} & & P_j \\ & & \iota_j \downarrow \uparrow \pi_j \\ & & \sum P_i \\ & & \downarrow f \\ A & \xrightarrow{g} & B \to 0 \end{array}$$

If each P_j is projective, then for each j there exists $h_j : P_j \to A$ such that $gh_j = f\iota_j$ By Theorem 1.13 there is a unique homomorphism $h : \sum P_i \to A$ with $h\iota_j = h_j$ for every j. Verify that $gh = f$. ∎

Recall that the dual of a concept defined in a category (that is, a concept defined in terms of objects and morphisms), is obtained by "reversing all the arrows." Pushing this idea a bit further one might say that a monomorphism is the dual of an epimorphism, since $A \to B$ is a monomorphism if and only if $0 \to A \to B$ is exact and $B \to A$ is an epimorphism if and only if $B \to A \to 0$ (arrows reversed!) is exact. This leads us to define the dual notion of projectivity as follows.

Definition 3.6. *A module* J *over a ring* R *is said to be* **injective** *if given any diagram of* R*-module homomorphisms*

$$\begin{array}{ccc} 0 \to A & \xrightarrow{g} & B \\ f\downarrow & & \\ J & & \end{array}$$

with top row exact (that is, g *a monomorphism), there exists an* R*-module homomorphism* h : B → J *such that the diagram*

$$\begin{array}{ccc} 0 \to A & \xrightarrow{g} & B \\ f\downarrow & \swarrow h & \\ J & & \end{array}$$

is commutative (that is, hg = f).

Remarks analogous to those in the paragraph following Definition 3.1 apply here to unitary injective modules over a ring with identity. It is not surprising that the duals of many (*but not all*) of the preceding propositions may be readily proved. For example since in a category products are the dual concept of coproducts (direct sums), the dual of Proposition 3.5 is

Proposition 3.7. *A direct product of* R*-modules* $\prod_{i\in I} J_i$ *is injective if and only if* J_i *is injective for every* $i \in I$.

PROOF. Exercise; see Proposition 3.5. ■

Since the concept of a free module cannot be dualized (Exercise 13), there are no analogues of Theorems 3.2 or 3.4 (iii) for injective modules. However, Corollary 3.3 can be dualized. It states, in effect, that for every module A there is a projective module P and an exact sequence $P \to A \to 0$. The dual of this statement is that for every module A there is an injective module J and an exact sequence $0 \to A \to J$; in other words, every module may be embedded in an injective module. The remainder of this section, which is not needed in the sequel, is devoted to proving this fact for unitary modules over a ring with identity. Once this has been done the dual of Theorem 3.4 (i), (ii), is easily proved (Proposition 3.13). We begin by characterizing injective R-modules in terms of left ideals (submodules) of the ring R.

Lemma 3.8. *Let* R *be a ring with identity. A unitary* R*-module* J *is injective if and only if for every left ideal* L *of* R*, any* R*-module homomorphism* L → J *may be extended to an* R*-module homomorphism* R → J.

SKETCH OF PROOF. To say that $f : L \to J$ may be extended to R means that there is a homomorphism $h : R \to J$ such that the diagram

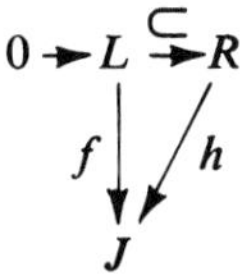

is commutative. Clearly, such an h always exists if J is injective. Conversely, suppose J has the stated extension property and suppose we are given a diagram of module homomorphisms

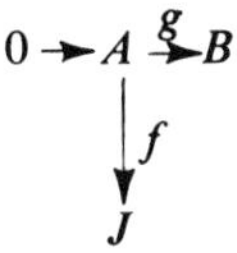

with top row exact. To show that J is injective we must find a homomorphism $h : B \to J$ with $hg = f$. Let $\mathcal{S}$ be the set of all R-module homomorphisms $h : C \to J$, where $\operatorname{Im} g \subset C \subset B$. $\mathcal{S}$ is nonempty since $fg^{-1} : \operatorname{Im} g \to J$ is an element of $\mathcal{S}$ (g is a monomorphism). Partially order $\mathcal{S}$ by extension: $h_1 \leq h_2$ if and only if $\operatorname{Dom} h_1 \subset \operatorname{Dom} h_2$ and $h_2 \mid \operatorname{Dom} h_1 = h_1$. Verify that the hypotheses of Zorn's Lemma are satisfied and conclude that $\mathcal{S}$ contains a maximal element $h : H \to J$ with $hg = f$. We shall complete the proof by showing $H = B$.

If $H \neq B$ and $b \in B - H$, then $L = \{r \in R \mid rb \in H\}$ is a left ideal of R. The map $L \to J$ given by $r \mapsto h(rb)$ is a well-defined R-module homomorphism. By hypothesis there is an R-module homomorphism $k : R \to J$ such that $k(r) = h(rb)$ for all $r \in L$. Let $c = k(1_R)$ and define a map $\bar{h} : H + Rb \to J$ by $a + rb \mapsto h(a) + rc$. We claim that $\bar{h}$ is well defined. For if $a_1 + r_1b = a_2 + r_2b \in H + Rb$, then $a_1 - a_2 = (r_2 - r_1)b \in H \cap Rb$. Hence $r_2 - r_1 \in L$ and $h(a_1) - h(a_2) = h(a_1 - a_2) = h((r_2 - r_1)b) = k(r_2 - r_1) = (r_2 - r_1)k(1_R) = (r_2 - r_1)c$. Therefore, $\bar{h}(a_1 + r_1b) = h(a_1) + r_1c = h(a_2) + r_2c = \bar{h}(a_2 + r_2b)$ and $\bar{h}$ is well defined. Verify that $\bar{h} : H + Rb \to J$ is an R-module homomorphism that is an element of the set $\mathcal{S}$. This contradicts the maximality of h since $b \notin H$ and hence $H \subsetneq H + Rb$. Therefore, $H = B$ and J is injective. ■

An abelian group D is said to be **divisible** if given any $y \in D$ and $0 \neq n \in \mathbf{Z}$, there exists $x \in D$ such that $nx = y$. For example, the additive group $\mathbf{Q}$ is divisible, but $\mathbf{Z}$ is not (Exercise 4). It is easy to prove that a direct sum of abelian groups is divisible if and only if each summand is divisible and that the homomorphic image of a divisible group is divisible (Exercise 7).

Lemma 3.9. *An abelian group* D *is divisible if and only if* D *is an injective (unitary)* **Z***-module.*

PROOF. If D is injective, $y \in D$ and $0 \neq n \in \mathbf{Z}$, let $f : \langle n \rangle \to D$ be the unique homomorphism determined by $n \mapsto y$; ($\langle n \rangle$ is a free $\mathbf{Z}$-module by Theorems I.3.2 and II.1.1). Since D is injective, there is a homomorphism $h : \mathbf{Z} \to D$ such that the diagram

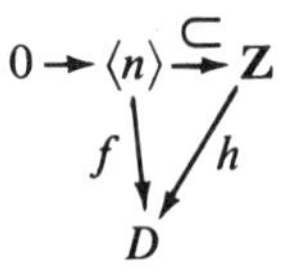

is commutative. If $x = h(1)$, then $nx = nh(1) = h(n) = f(n) = y$. Therefore, D is divisible. To prove the converse note that the only left ideals of $\mathbf{Z}$ are the cyclic groups $\langle n \rangle$, $n \in \mathbf{Z}$. If D is divisible and $f : \langle n \rangle \to D$ is a homomorphism, then there exists $x \in D$ with $nx = f(n)$. Define $h : \mathbf{Z} \to D$ by $1 \mapsto x$ and verify that h is a homomorphism that extends f. Therefore, D is injective by Lemma 3.8. ■

REMARK. A complete characterization of divisible abelian groups (injective unitary $\mathbf{Z}$-modules) is given in Exercise 11.

Lemma 3.10. *Every abelian group* A *may be embedded in a divisible abelian group.*

PROOF. By Theorem II.1.4 there is a free $\mathbf{Z}$-module F and an epimorphism $F \to A$ with kernel K so that $F/K \cong A$. Since F is a direct sum of copies of $\mathbf{Z}$ (Theorem II.1.1) and $\mathbf{Z} \subset \mathbf{Q}$, F may be embedded in a direct sum D of copies of the rationals $\mathbf{Q}$ (Theorem I.8.10). But D is a divisible group by Proposition 3.7, Lemma 3.9, and the remarks preceding it. If $f : F \to D$ is the embedding monomorphism, then f induces an isomorphism $F/K \cong f(F)/f(K)$ by Corollary I.5.8. Thus the composition $A \cong F/K \cong f(F)/f(K) \subset D/f(K)$ is a monomorphism. But $D/f(K)$ is divisible since it is the homomorphic image of a divisible group. ■

If R is a ring with identity and J is an abelian group, then $\mathrm{Hom}_{\mathbf{Z}}(R,J)$, the set of all $\mathbf{Z}$-module homomorphisms $R \to J$, is an abelian group (Exercise 1.7). Verify that $\mathrm{Hom}_{\mathbf{Z}}(R,J)$ is a unitary left R-module with the action of R defined by $(rf)(x) = f(xr)$, $(r,x \in R;\ f \in \mathrm{Hom}_{\mathbf{Z}}(R,J))$.

Lemma 3.11. *If* J *is a divisible abelian group and* R *is a ring with identity, then* $Hom_{\mathbf{Z}}$(R,J) *is an injective left* R*-module.*

SKETCH OF PROOF. By Lemma 3.8 it suffices to show that for each left ideal L of R, every R-module homomorphism $f : L \to \mathrm{Hom}_{\mathbf{Z}}(R,J)$ may be extended to an R-module homomorphism $h : R \to \mathrm{Hom}_{\mathbf{Z}}(R,J)$. The map $g : L \to J$ given by $g(a) = [f(a)](1_R)$ is a group homomorphism. Since J is an injective $\mathbf{Z}$-module by Lemma 3.9 and we have the diagram

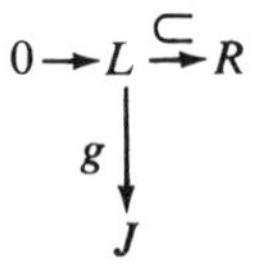

there is a group homomorphism $\bar{g} : R \to J$ such that $\bar{g} \mid L = g$. Define $h : R \to \mathrm{Hom}_{\mathbf{Z}}(R,J)$ by $r \mapsto h(r)$, where $h(r) : R \to J$ is the map given by $[h(r)](x) = \bar{g}(xr)$

($x \in R$). Verify that h is a well-defined function (that is, each $h(r)$ is a group homomorphism $R \to J$) and that h is a group homomorphism $R \to \mathrm{Hom}_{\mathbf{Z}}(R,J)$. If $s,r,x \in R$, then

$$h(sr)(x) = \bar{g}(x(sr)) = \bar{g}((xs)r) = h(r)(xs).$$

By the definition of the R-module structure of $\mathrm{Hom}_{\mathbf{Z}}(R,J)$, $h(r)(xs) = [sh(r)](x)$, whence $h(sr) = sh(r)$ and h is an R-module homomorphism. Finally suppose $r \in L$ and $x \in R$. Then $xr \in L$ and

$$h(r)(x) = \bar{g}(xr) = g(xr) = [f(xr)](1_R).$$

Since f is an R-module homomorphism and $\mathrm{Hom}_{\mathbf{Z}}(R,J)$ an R-module,

$$[f(xr)](1_R) = [xf(r)](1_R) = f(r)(1_Rx) = f(r)(x).$$

Therefore, $h(r) = f(r)$ for $r \in L$ and h is an extension of f. ■

We are now able to prove the duals of Corollary 3.3 and Theorem 3.4.

Proposition 3.12. *Every unitary module* A *over a ring* R *with identity may be embedded in an injective* R*-module.*

SKETCH OF PROOF. Since A is an abelian group, there is a divisible group J and a group monomorphism $f : A \to J$ by Lemma 3.10. The map $\bar{f} : \mathrm{Hom}_{\mathbf{Z}}(R,A) \to \mathrm{Hom}_{\mathbf{Z}}(R,J)$ given on $g \in \mathrm{Hom}_{\mathbf{Z}}(R,A)$ by $\bar{f}(g) = fg \in \mathrm{Hom}_{\mathbf{Z}}(R,J)$ is easily seen to be an R-module monomorphism. Since every R-module homomorphism is a $\mathbf{Z}$-module homomorphism, we have $\mathrm{Hom}_R(R,A) \subset \mathrm{Hom}_{\mathbf{Z}}(R,A)$. In fact, it is easy to see that $\mathrm{Hom}_R(R,A)$ is an R-submodule of $\mathrm{Hom}_{\mathbf{Z}}(R,A)$. Finally, verify that the map $A \to \mathrm{Hom}_R(R,A)$ given by $a \mapsto f_a$, where $f_a(r) = ra$, is an R-module monomorphism (in fact it is an isomorphism). Composing these maps yields an R-module monomorphism

$$A \to \mathrm{Hom}_R(R,A) \xrightarrow{\subset} \mathrm{Hom}_{\mathbf{Z}}(R,A) \xrightarrow{\bar{f}} \mathrm{Hom}_{\mathbf{Z}}(R,J).$$

Since $\mathrm{Hom}_{\mathbf{Z}}(R,J)$ is an injective R-module by Lemma 3.11, we have embedded A in an injective. ■

Proposition 3.13. *Let* R *be a ring with identity. The following conditions on a unitary* R*-module* J *are equivalent.*

(i) J *is injective;*

(ii) *every short exact sequence* $0 \to \mathrm{J} \xrightarrow{f} \mathrm{B} \xrightarrow{g} \mathrm{C} \to 0$ *is split exact (hence* $\mathrm{B} \cong \mathrm{J} \oplus \mathrm{C}$);

(iii) J *is a direct summand of any module* B *of which it is a submodule.*

SKETCH OF PROOF. (i) $\Rightarrow$ (ii) Dualize the proof of (i) $\Rightarrow$ (ii) of Theorem 3.4. (ii) $\Rightarrow$ (iii) since the sequence $0 \to J \xrightarrow{\subset} B \xrightarrow{\pi} B/J \to 0$ is split exact, there is a homomorphism $g: B/J \to B$ such that $\pi g = 1_{B/J}$. By Theorem 1.18 ((i) $\Rightarrow$ (iii)) there is an isomorphism $J \oplus B/J \cong B$ given by $(x,y) \mapsto x + g(y)$. It follows easily that B is the internal direct sum of J and $g(B/J)$. (iii) $\Rightarrow$ (i) It follows from Proposition 3.12 that J is a submodule of an injective module Q. Proposition 3.7 and (iii) imply that J is injective. ∎

EXERCISES

Note: R is a ring. If R has an identity, all R-modules are assumed to be unitary.

1. The following conditions on a ring R [with identity] are equivalent:
   (a) Every [unitary] R-module is projective.
   (b) Every short exact sequence of [unitary] R-modules is split exact.
   (c) Every [unitary] R-module is injective.

2. Let R be a ring with identity. An R-module A is injective if and only if for every left ideal L of R and R-module homomorphism $g : L \to A$, there exists $a \in A$ such that $g(r) = ra$ for every $r \in L$.

3. Every vector space over a division ring D is both a projective and an injective D-module. [See Exercise 1.]

4. (a) For each prime p, $Z(p^\infty)$ (see Exercise I.1.10) is a divisible group.
   (b) No nonzero finite abelian group is divisible.
   (c) No nonzero free abelian group is divisible.
   (d) $\mathbf{Q}$ is a divisible abelian group.

5. $\mathbf{Q}$ is not a projective $\mathbf{Z}$-module.

6. If G is an abelian group, then $G = D \oplus N$, with D divisible and N **reduced** (meaning that N has no nontrivial divisible subgroups). [*Hint:* Let D be the subgroup generated by the set theoretic union of all divisible subgroups of G.]

7. Without using Lemma 3.9 prove that:
   (a) Every homomorphic image of a divisible abelian group is divisible.
   (b) Every direct summand (Exercise I.8.12) of a divisible abelian group is divisible.
   (c) A direct sum of divisible abelian groups is divisible.

8. Every torsion-free divisible abelian group D is a direct sum of copies of the rationals $\mathbf{Q}$. [*Hint:* if $0 \neq n \in \mathbf{Z}$ and $a \in D$, then there exists a unique $b \in D$ such that $nb = a$. Denote b by $(1/n)a$. For $m, n \in \mathbf{Z}$ $(n \neq 0)$, define $(m/n)a = m(1/n)a$. Then D is a vector space over $\mathbf{Q}$. Use Theorem 2.4.]

9. (a) If D is an abelian group with torsion subgroup D_t, then D/D_t is torsion free.
   (b) If D is divisible, then so is D_t, whence $D = D_t \oplus E$, with E torsion free.

10. Let p be a prime and D a divisible abelian p-group. Then D is a direct sum of copies of $Z(p^\infty)$. [*Hint:* let X be a basis of the vector space $D[p]$ over Z_p (see Exercise 2.4). If $x \in X$, then there exists $x_1, x_2, x_3, \ldots \in D$ such that $x_1 = x$, $|x_1| = p$, $px_2 = x_1$, $px_3 = x_2, \ldots, px_{n+1} = x_n, \ldots$. If H_x is the subgroup generated by the x_i, then $H_x \cong Z(p^\infty)$ by Exercise I.3.7. Show that $D \cong \sum_{x \in X} H_x$.]

11. Every divisible abelian group is a direct sum of copies of the rationals $\mathbf{Q}$ and copies of $Z(p^\infty)$ for various primes p. [*Hint:* apply Exercise 9 to D and Exercises

7 and 8 to the torsion-free summand so obtained. The other summand D_t is a direct sum of copies of various $Z(p^\infty)$ by Exercises 7, 10 and II.2.7.]

12. Let G,H,K be divisible abelian groups.
    (a) If $G \oplus G \cong H \oplus H$, then $G \cong H$ [see Exercise 11].
    (b) If $G \oplus H \cong G \oplus K$, then $H \cong K$ [see Exercises 11 and II.2.11.].

13. If one attempted to dualize the notion of free module over a ring R (and called the object so defined "co-free") the definition would read: An R-module F is co-free on a set X if there exists a function $\iota : F \to X$ such that for any R-module A and function $f : A \to X$, there exists a unique module homomorphism $\bar{f} : A \to F$ such that $\iota\bar{f} = f$ (see Theorem 2.1(iv)). Show that for any set X with $|X| \geq 2$ no such R-module F exists. If $|X| = 1$, then 0 is the only co-free module. [*Hint:* If F exists and $|X| \geq 2$, arrive at a contradiction by considering possible images of 0 and constructing $f : R \to X$ such that $\iota\bar{f} \neq f$ for every homomorphism $\bar{f} : R \to F$.]

14. If D is a ring with identity such that every unitary D-module is free, then D is a division ring. [*Hint:* it suffices by Exercise III.2.7 and Theorem III.2.18 to show that D has no nonzero maximal left ideals. Note that every left ideal of D is a free D-module and hence a (module) direct summand of D by Theorem 3.2, Exercise 1, and Proposition 3.13.]

4. HOM AND DUALITY

We first discuss the behavior of $\text{Hom}_R(A,B)$ with respect to induced maps, exact sequences, direct sums, and direct products. The last part of the section, which is essentially independent of the first part, deals with duality.

Recall that if A and B are modules over a ring R, then $\text{Hom}_R(A,B)$ is the set of all R-module homomorphisms $f : A \to B$. If $R = \mathbf{Z}$ we shall usually write $\text{Hom}(A,B)$ in place of $\text{Hom}_{\mathbf{Z}}(A,B)$. $\text{Hom}_R(A,B)$ is an abelian group under addition and this addition is distributive with respect to composition of functions (see p. 174).

Theorem 4.1. *Let* A,B,C,D *be modules over a ring* R *and* $\varphi : C \to A$ *and* $\psi : B \to D$ R-*module homomorphisms. Then the map* $\theta : \text{Hom}_R(A,B) \to \text{Hom}_R(C,D)$ *given by* $f \mapsto \psi f \varphi$ *is a homomorphism of abelian groups.*

SKETCH OF PROOF. θ is well defined since composition of R-module homomorphisms is an R-module homomorphism. θ is a homomorphism since such composition of homomorphisms is distributive with respect to addition. ■

The map θ of Theorem 4.1 is usually denoted $\text{Hom}(\varphi,\psi)$ and called the homomorphism induced by φ and ψ. Observe that for homomorphisms $\varphi_1 : E \to C$, $\varphi_2 : C \to A$, $\psi_1 : B \to D$, $\psi_2 : D \to F$,

$$\text{Hom}(\varphi_1,\psi_2)\,\text{Hom}(\varphi_2,\psi_1) = \text{Hom}(\varphi_2\varphi_1,\psi_2\psi_1) : \text{Hom}_R(A,B) \to \text{Hom}_R(E,F).$$

There are two important special cases of the induced homomorphism. If $B = D$ and $\psi = 1_B$, then the induced map $\text{Hom}(\varphi,1_B) : \text{Hom}_R(A,B) \to \text{Hom}_R(C,B)$ is given by $f \mapsto f\varphi$ and is denoted $\bar{\varphi}$. Similarly if $A = C$ and $\varphi = 1_A$ the induced-map $\operatorname{Hom}(1_A,\psi) : \operatorname{Hom}_R(A,B) \to \operatorname{Hom}_R(A,D)$ is given by $f \mapsto \psi f$ and is denoted $\bar{\psi}$.

We now examine the behavior of Hom_R with respect to exact sequences.

Theorem 4.2. *Let* R *be a ring.* $0 \to A \xrightarrow{\varphi} B \xrightarrow{\psi} C$ *is an exact sequence of* R*-modules if and only if for every* R*-module* D

$$0 \to \mathit{Hom}_R(D,A) \xrightarrow{\bar{\varphi}} \mathit{Hom}_R(D,B) \xrightarrow{\bar{\psi}} \mathit{Hom}_R(D,C)$$

is an exact sequence of abelian groups.

PROOF. If $0 \to A \xrightarrow{\varphi} B \xrightarrow{\psi} C$ is exact we must prove: (i) $\operatorname{Ker} \bar{\varphi} = 0$ (that is, $\bar{\varphi}$ is a monomorphism); (ii) $\operatorname{Im} \bar{\varphi} \subset \operatorname{Ker} \bar{\psi}$; and (iii) $\operatorname{Ker} \bar{\psi} \subset \operatorname{Im} \bar{\varphi}$. (i) $f \in \operatorname{Ker} \bar{\varphi} \Rightarrow \varphi f = 0 \Rightarrow \varphi f(x) = 0$ for all $x \in D$. Since $0 \to A \xrightarrow{\varphi} B$ is exact, φ is a monomorphism, whence $f(x) = 0$ for all $x \in D$ and $f = 0$. Therefore, $\operatorname{Ker} \bar{\varphi} = 0$. (ii) Since $\operatorname{Im} \varphi = \operatorname{Ker} \psi$ by exactness, we have $\psi\varphi = 0$ and hence $\bar{\psi}\bar{\varphi} = \overline{\psi\varphi} = 0$. Therefore, $\operatorname{Im} \bar{\varphi} \subset \operatorname{Ker} \bar{\psi}$. (iii) $g \in \operatorname{Ker} \bar{\psi} \Rightarrow \psi g = 0 \Rightarrow \operatorname{Im} g \subset \operatorname{Ker} \psi = \operatorname{Im} \varphi$. Since φ is a monomorphism, $\varphi : A \to \operatorname{Im} \varphi$ is an isomorphism. If h is the composite $D \xrightarrow{g} \operatorname{Im} g \subset \operatorname{Im} \varphi \xrightarrow{\varphi^{-1}} A$, then $h \in \operatorname{Hom}_R(D,A)$ and $g = \varphi h = \bar{\varphi}(h)$. Therefore, $\operatorname{Ker} \bar{\psi} \subset \operatorname{Im} \bar{\varphi}$.

Conversely, assume that the Hom sequence of induced maps is exact for every D. First let $D = \operatorname{Ker} \varphi$ and let $i : D \to A$ be the inclusion map. Since $\operatorname{Ker} \bar{\varphi} = 0$ (exactness) and $\bar{\varphi}(i) = \varphi i = 0$, we must have $i = 0$, which implies that $0 = D = \operatorname{Ker} \varphi$. Therefore, $0 \to A \xrightarrow{\varphi} B$ is exact. Next let $D = A$. Since $\operatorname{Ker} \bar{\psi} = \operatorname{Im} \bar{\varphi}$ we have $0 = \bar{\psi}\bar{\varphi}(1_A) = \psi\varphi 1_A = \psi\varphi$, whence $\operatorname{Im} \varphi \subset \operatorname{Ker} \psi$. Finally let $D = \operatorname{Ker} \psi$ and let $j : D \to B$ be the inclusion map. Since $0 = \psi j = \bar{\psi}(j)$ and $\operatorname{Ker} \bar{\psi} = \operatorname{Im} \bar{\varphi}$, we have $j = \bar{\varphi}(f) = \varphi f$ for some $f : D \to A$. Therefore, for every $x \in D = \operatorname{Ker} \psi$, $x = j(x) = \varphi f(x) \in \operatorname{Im} \varphi$ and $\operatorname{Ker} \psi \subset \operatorname{Im} \varphi$. Thus $\operatorname{Ker} \psi = \operatorname{Im} \varphi$ and $0 \to A \xrightarrow{\varphi} B \xrightarrow{\psi} C$ is exact. ■

Proposition 4.3. *Let* R *be a ring.* $A \xrightarrow{\theta} B \xrightarrow{\zeta} C \to 0$ *is an exact sequence of* R*-modules if and only if for every* R*-module* D

$$0 \to \mathit{Hom}_R(C,D) \xrightarrow{\bar{\zeta}} \mathit{Hom}_R(B,D) \xrightarrow{\bar{\theta}} \mathit{Hom}_R(A,D)$$

is an exact sequence of abelian groups.

SKETCH OF PROOF. If $A \xrightarrow{\theta} B \xrightarrow{\zeta} C \to 0$ is exact, we shall show that $\operatorname{Ker} \bar{\theta} \subset \operatorname{Im} \bar{\zeta}$. If $f \in \operatorname{Ker} \bar{\theta}$, then $0 = \bar{\theta}(f) = f\theta$, whence $0 = f(\operatorname{Im} \theta) = f(\operatorname{Ker} \zeta)$. By Theorem 1.7 f induces a homomorphism $\bar{f} : B/\operatorname{Ker} \zeta \to D$ such that $\bar{f}(b + \operatorname{Ker} \zeta) = f(b)$. By Theorem 1.7 again there is an isomorphism $\varphi : B/\operatorname{Ker} \zeta \cong C$ such that $\varphi(b + \operatorname{Ker} \zeta) = \zeta(b)$. Then the map $\bar{f}\varphi^{-1} : C \to D$ is an R-module homomorphism such that $\bar{\zeta}(\bar{f}\varphi^{-1}) = f$. Hence $\operatorname{Ker} \bar{\theta} \subset \operatorname{Im} \bar{\zeta}$. The remainder of this half of the proof is analogous to that of Theorem 4.2.

Conversely if the Hom sequence is exact for every D, let $D = C/\operatorname{Im} \zeta$ and let $\pi : C \to D$ be the canonical projection. Then $\bar{\zeta}(\pi) = \pi\zeta = 0$ and $\operatorname{Ker} \bar{\zeta} = 0$ imply $\pi = 0$, whence $C = \operatorname{Im} \zeta$ and $B \xrightarrow{\zeta} C \to 0$ is exact. Similarly, show that $\operatorname{Ker} \zeta \subset \operatorname{Im} \theta$

by considering $D = B/\text{Im}\ \theta$ and the canonical epimorphism $B \to D$. Finally, if $D = C$, then $0 = \bar{\theta}\bar{\zeta}(1_C) = \zeta\theta$, whence $\text{Im}\ \theta \subset \text{Ker}\ \zeta$. Therefore, $A \xrightarrow{\theta} B \xrightarrow{\zeta} C \to 0$ is exact. ■

One sometimes summarizes the two preceding results by saying that $\text{Hom}_R(A,B)$ is **left exact.** It is *not* true in general that a short exact sequence $0 \to A \to B \to C \to 0$ induces a short exact sequence $0 \to \text{Hom}_R(D,A) \to \text{Hom}_R(D,B) \to \text{Hom}_R(D,C) \to 0$ (and similarly in the first variable; see Exercise 3). However, the next three theorems show that this result does hold in several cases.

Proposition 4.4. *The following conditions on modules over a ring* R *are equivalent.*

(i) $0 \to A \xrightarrow{\varphi} B \xrightarrow{\psi} C \to 0$ *is a split exact sequence of* R*-modules;*

(ii) $0 \to \text{Hom}_R(D,A) \xrightarrow{\bar{\varphi}} \text{Hom}_R(D,B) \xrightarrow{\bar{\psi}} \text{Hom}_R(D,C) \to 0$ *is a split exact sequence of abelian groups for every* R*-module* D;

(iii) $0 \to \text{Hom}_R(C,D) \xrightarrow{\bar{\psi}} \text{Hom}_R(B,D) \xrightarrow{\bar{\varphi}} \text{Hom}_R(A,D) \to 0$ *is a split exact sequence of abelian groups for every* R*-module* D.

SKETCH OF PROOF. (i) ⇒ (iii) By Theorem 1.18 there is a homomorphism $\alpha : B \to A$ such that $\alpha\varphi = 1_A$. Verify that the induced-homomorphism

$$\bar{\alpha} : \text{Hom}_R(A,D) \to \text{Hom}_R(B,D)$$

is such that $\bar{\varphi}\bar{\alpha} = 1_{\text{Hom}_R(A,D)}$. Consequently, $\bar{\varphi}$ is an epimorphism (Introduction, Theorem 3.1) and the Hom_R sequence is split exact by Proposition 4.3 and Theorem 1.18. (iii) ⇒ (i) If $D = A$ and $f : B \to A$ is such that $1_A = \bar{\varphi}(f) = f\varphi$, then φ is a monomorphism (Introduction, Theorem 3.1) and $0 \to A \xrightarrow{\varphi} B \xrightarrow{\psi} C \to 0$ is split exact by Proposition 4.3 and Theorem 1.18. The other implications are proved similarly. ■

Theorem 4.5. *The following conditions on a module* P *over a ring* R *are equivalent*

(i) P *is projective;*

(ii) *if* $\psi : B \to C$ *is any* R*-module epimorphism then* $\bar{\psi} : \text{Hom}_R(P,B) \to \text{Hom}_R(P,C)$ *is an epimorphism of abelian groups;*

(iii) *if* $0 \to A \xrightarrow{\varphi} B \xrightarrow{\psi} C \to 0$ *is any short exact sequence of* R*-modules, then* $0 \to \text{Hom}_R(P,A) \xrightarrow{\bar{\varphi}} \text{Hom}_R(P,B) \xrightarrow{\bar{\psi}} \text{Hom}_R(P,C) \to 0$ *is an exact sequence of abelian groups.*

SKETCH OF PROOF. (i) ⇔ (ii) The map $\bar{\psi} : \text{Hom}_R(P,B) \to \text{Hom}_R(P,C)$ (given by $g \mapsto \psi g$) is an epimorphism if and only if for every R-module homomorphism $f : P \to C$, there is an R-module homomorphism $g : P \to B$ such that the diagram

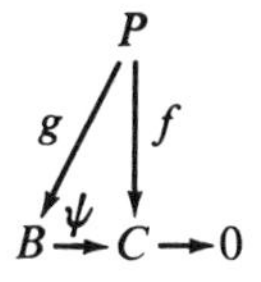

is commutative (that is, $f = \psi g = \bar{\psi}(g)$). (ii) $\Rightarrow$ (iii) Theorem 4.2. (iii) $\Rightarrow$ (ii) Given an epimorphism $\psi : B \to C$ let $A = \operatorname{Ker} \psi$ and apply (iii) to the short exact sequence $0 \to A \xrightarrow{\subset} B \xrightarrow{\psi} C \to 0$. ∎

Proposition 4.6. *The following conditions on a module* J *over a ring* R *are equivalent.*

(i) J *is injective;*

(ii) *if* $\theta: A \to B$ *is any* R*-module monomorphism, then* $\bar{\theta}: Hom_R(B,J) \to Hom_R(A,J)$ *is an epimorphism of abelian groups;*

(iii) *if* $0 \to A \xrightarrow{\theta} B \xrightarrow{\zeta} C \to 0$ *is any short exact sequence of* R*-modules, then* $0 \to Hom_R(C,J) \xrightarrow{\bar{\zeta}} Hom_R(B,J) \xrightarrow{\bar{\theta}} Hom_R(A,J) \to 0$ *is an exact sequence of abelian groups.*

PROOF. The proof is dual to that of Theorem 4.5 and is left as an exercise. ∎

Theorem 4.7. *Let* A,B, $\{A_i \mid i \in I\}$ *and* $\{B_j \mid j \in J\}$ *be modules over a ring* R. *Then there are isomorphisms of abelian groups:*

(i) $Hom_R(\sum_{i \in I} A_i, B) \cong \prod_{i \in I} Hom_R(A_i, B)$;

(ii) $Hom_R(A, \prod_{j \in J} B_j) \cong \prod_{j \in J} Hom_R(A, B_j)$.

REMARKS. If I and J are finite, then $\sum_{i \in I} A_i = \prod_{i \in I} A_i$ and $\sum_{j \in J} B_j = \prod_{j \in J} B_j$. If I and J are infinite, however, the theorem may be false if the direct product $\prod$ is replaced by the direct sum $\sum$ (see Exercise 10).

SKETCH OF PROOF OF 4.7. (i) For each $i \in I$ let $\iota_i : A_i \to \sum_{i \in I} A_i$ be the canonical injection (Theorem 1.11). Given $\{g_i\} \in \prod_{i \in I} \operatorname{Hom}_R(A_i,B)$, there is a unique R-module homomorphism $g : \sum_{i \in I} A_i \to B$ such that $g\iota_i = g_i$ for every $i \in I$ (Theorem 1.13). Verify that the map $\psi : \prod \operatorname{Hom}_R(A_i,B) \to \operatorname{Hom}_R(\sum A_i,B)$ given by $\{g_i\} \mapsto g$ is a homomorphism. Show that the map $\varphi : \operatorname{Hom}_R(\sum A_i,B) \to \prod \operatorname{Hom}_R(A_i,B)$, given by $f \mapsto \{f\iota_i\}$, is a homomorphism such that $\varphi\psi$ and $\psi\varphi$ are the respective identity maps. Thus φ is an isomorphism. (ii) is proved similarly with Theorem 1.12 in place of Theorem 1.13. ∎

In order to deal with duality and other concepts we need to consider possible module structures on the abelian group $\operatorname{Hom}_R(A,B)$. We begin with some remarks about bimodules. Let R and S be rings. An abelian group A is an R-S **bimodule** provided that A is both a left R-module and a right S-module and

$$r(as) = (ra)s \quad \text{for all} \quad a \in A, r \in R, s \in S.$$

We sometimes write ${}_RA_S$ to indicate the fact that A is an R-S bimodule. Similarly ${}_RB$ indicates a left R-module B and C_S a right S-module C.

EXAMPLES. Every ring R has associative multiplication and hence is an R-R bimodule. Every left module A over a commutative ring R is an R-R bimodule with $ra = ar$ ($a \in A$, $r \in R$).

Theorem 4.8. *Let* R *and* S *be rings and let* $_RA$, $_RB_S$, $_RC_S$, $_RD$ *be (bi)modules as indicated.*

(i) $Hom_R(A,B)$ *is a right* S*-module, with the action of* S *given by* $(fs)(a) = (f(a))s$ ($s \in S$; $a \in A$; $f \in Hom_R(A,B)$).

(ii) *If* $\varphi : A \to A'$ *is a homomorphism of left* R*-modules, then the induced map* $\bar{\varphi} : Hom_R(A',B) \to Hom_R(A,B)$ *is a homomorphism of right* S*-modules.*

(iii) $Hom_R(C,D)$ *is a left* S*-module, with the action of* S *given by* $(sg)(c) = g(cs)$ ($s \in S$; $c \in C$; $g \in Hom_R(C,D)$).

(iv) *If* $\psi : D \to D'$ *is a homomorphism of left* R*-modules, then* $\bar{\psi} : Hom_R(C,D) \to Hom_R(C,D')$ *is a homomorphism of left* S*-modules.*

SKETCH OF PROOF. (i) The verification that fs is a well-defined module homomorphism and that $\mathrm{Hom}_R(A,B)$ is actually a right S-module is tedious but straight-forward; similarly for (iii). (ii) $\bar{\varphi}$ is an abelian group homomorphism by Theorem 4.1. If $f \in \mathrm{Hom}_R(A',B)$, $a \in A$ and $s \in S$, then

$$\bar{\varphi}(fs)(a) = ((fs)\varphi)(a) = (fs)(\varphi(a)) = (f(\varphi(a)))s = (f\varphi(a))s = ((\bar{\varphi}f)(a))s.$$

Hence $\bar{\varphi}(fs) = (\bar{\varphi}f)s$ and $\bar{\varphi}$ is a right S-module homomorphism. (iv) is proved analogously. ∎

REMARK. An important special case of Theorem 4.8 occurs when R is commutative and hence every R-module C is an R-R bimodule with $rc = cr$ ($r \in R$, $c \in C$). In this case for every $r \in R$, $a \in A$, and $f \in \mathrm{Hom}_R(A,B)$ we have

$$(rf)(a) = f(ar) = f(ra) = rf(a) = (f(a))r = (fr)(a).$$

It follows that $\mathrm{Hom}_R(A,B)$ is an R-R bimodule with $rf = fr$ for all $r \in R$, $f \in \mathrm{Hom}_R(A,B)$.

Theorem 4.9. *If* A *is a unitary left module over a ring* R *with identity then there is an isomorphism of left* R*-modules* $A \cong Hom_R(R,A)$.

SKETCH OF PROOF. Since R is an R-R bimodule, the left module structure of $\mathrm{Hom}_R(R,A)$ is given by Theorem 4.8(iii). Verify that the map $\varphi : \mathrm{Hom}_R(R,A) \to A$ given by $f \mapsto f(1_R)$ is an R-module homomorphism. Define a map $\psi : A \to \mathrm{Hom}_R(R,A)$ by $a \mapsto f_a$, where $f_a(r) = ra$. Verify that ψ is a well-defined R-module homomorphism such that $\varphi\psi = 1_A$ and $\psi\varphi = 1_{\mathrm{Hom}_R(R,A)}$. ∎

Let A be a left module over a ring R. Since R is an R-R bimodule, $\mathrm{Hom}_R(A,R)$ is a right R-module by Theorem 4.8(i). $\mathrm{Hom}_R(A,R)$ is called the **dual** module of A and is denoted A^*. The elements of A^* are sometimes called **linear functionals.** Similarly if B is a right R-module, then the dual B^* of B is the *left* R-module $\mathrm{Hom}_R(B,R)$ (Exercise 4(a)).

Theorem 4.10. *Let A,B and C be left modules over a ring R.*

(i) *If* $\varphi : A \to C$ *is a homomorphism of left R-modules, then the induced map* $\bar{\varphi} : C^* = Hom_R(C,R) \to Hom_R(A,R) = A^*$ *is a homomorphism of right R-modules.*

(ii) *There is an R-module isomorphism* $(A \oplus C)^* \cong A^* \oplus C^*$.

(iii) *If R is a division ring and* $0 \to A \xrightarrow{\theta} B \xrightarrow{\zeta} C \to 0$ *is a short exact sequence of left vector spaces, then* $0 \to C^* \xrightarrow{\bar{\zeta}} B^* \xrightarrow{\bar{\theta}} A^* \to 0$ *is a short exact sequence of right vector spaces.*

PROOF. Exercise; see Theorems 2.4, 3.2, 4.1, 4.5, and 4.7. The map $\bar{\varphi}$ of (i) is called the **dual map** of φ. ■

If A is a left module over a ring R, $a \in A$, and $f \in A^* = \text{Hom}_R(A,R)$, then one frequently denotes $f(a) \in R$ by $\langle a,f \rangle$. Since f is a left R-module homomorphism,

$$\langle r_1a_1 + r_2a_2,f\rangle = r_1\langle a_1,f\rangle + r_2\langle a_2,f\rangle \qquad (r_i \in R, f \in A^*, a_i \in A). \tag{1}$$

Similarly since A^* is a right R-module with $(fr)(a) = f(a)r$, we have

$$\langle a,f_1r_1 + f_2r_2\rangle = \langle a,f_1\rangle r_1 + \langle a,f_2\rangle r_2 \qquad (r_i \in R, f_i \in A^*, a \in A). \tag{2}$$

In the proofs below we shall use the brackets notation for linear functionals as well as the **Kronecker delta notation:** for any index set I and ring R with identity the symbol δ_{ij} $(i,j \in I)$ denotes $0 \in R$ if $i \neq j$ and $1_R \in R$ if $i = j$.

Theorem 4.11. *Let F be a free left module over a ring R with identity. Let X be a basis of F and for each* $x \in X$ *let* $f_x : F \to R$ *be given by* $f_x(y) = \delta_{xy}$ $(y \in X)$. *Then*

(i) $\{f_x \mid x \in X\}$ *is a linearly independent subset of* F^* *of cardinality* $|X|$;

(ii) *if X is finite, then* F^* *is a free right R-module with basis* $\{f_x \mid x \in X\}$.

REMARKS. The homomorphisms f_x are well defined since F is free with basis X (Theorem 2.1). In part (ii), $\{ f_x \mid x \in X\}$ is called the **dual basis** to X. This theorem is clearly true for any vector space V over a division ring by Theorem 2.4. In particular, if V is finite dimensional, then Proposition 2.9 and Theorem 4.11 imply that dim V = dim V^* and $V \cong V^*$. However, if V is infinite dimensional then dim $V^* >$ dim V (Exercise 12). More generally, if F is a free module over an arbitrary ring (for example, **Z**), F^* need not be free (see Exercise 10).

PROOF OF 4.11. (i) If $f_{x_1}r_1 + f_{x_2}r_2 + \cdots + f_{x_n}r_n = 0$ $(r_i \in R;\ x_i \in X)$, then for each $j = 0,1,2, \ldots, n$,

$$0 = \langle x_j,0\rangle = \left\langle x_j, \sum_{i=1}^{n} f_{x_i}r_i \right\rangle = \sum_i \langle x_j, f_{x_i}\rangle r_i = \sum_i \delta_{ij}r_i = r_j.$$

Since $r_j = 0$ for all j, $\{ f_x \mid x \in X\}$ is linearly independent. If $x \neq y \in X$, then $f_x(x) = 1_R \neq 0 = f_y(x)$, whence $f_x \neq f_y$. Therefore, $|X| = |\{ f_x \mid x \in X\}|$.

(ii) If X is finite, say $X = \{x_1, \ldots, x_n\}$, and $f \in F^*$, let $s_i = f(x_i) = \langle x_i,f\rangle \in R$ and denote f_{x_j} by f_j. If $u \in F$, then $u = r_1x_1 + r_2x_2 + \cdots + r_nx_n \in F$ for some $r_i \in R$ and

$$\left\langle u, \sum_{j=1}^{n} f_j s_j \right\rangle = \left\langle \sum_{i=1}^{n} r_i x_i, \sum_{j} f_j s_j \right\rangle$$
$$= \sum_{i}\sum_{j} r_i \langle x_i, f_j \rangle s_j = \sum_{i}\sum_{j} r_i \delta_{ij} s_j = \sum_{i} r_i s_i$$
$$= \sum_{i} r_i \langle x_i, f \rangle = \langle \sum_{i} r_i x_i, f \rangle = \langle u, f \rangle.$$

Therefore, $f = f_1 s_1 + f_2 s_2 + \cdots + f_n s_n$ and $\{ f_i \} = \{ f_x \mid x \varepsilon X\}$ generates F^*. Hence $\{ f_x \mid x \varepsilon X\}$ is a basis and F^* is free. ■

The process of forming duals may be repeated. If A is a left R-module, then A^* is a right R-module and $A^{**} = (A^*)^* = \text{Hom}_R(\text{Hom}_R(A,R),R)$ (where the left hand Hom_R indicates all *right* R-module homomorphisms) is a left R-module (see Exercise 4(a)). A^{**} is called the **double dual** of A.

Theorem 4.12. *Let* A *be a left module over a ring* R.

(i) *There is an* R*-module homomorphism* $\theta : A \to A^{**}$.
(ii) *If* R *has an identity and* A *is free, then* θ *is a monomorphism.*
(iii) *If* R *has an identity and* A *is free with a finite basis, then* θ *is an isomorphism.*

A module A such that $\theta : A \to A^{**}$ is an isomorphism is said to be **reflexive**.

PROOF OF 4.12. (i) For each $a \varepsilon A$ let $\theta(a) : A^* \to R$ be the map defined by $[\theta(a)](f) = \langle a,f \rangle \varepsilon R$. Statement (2) after Theorem 4.10 shows that $\theta(a)$ is a homomorphism of right R-modules (that is, $\theta(a) \varepsilon A^{**}$). The map $\theta : A \to A^{**}$ given by $a \mapsto \theta(a)$ is a left R-module homomorphism by (1) after Theorem 4.10.

(ii) Let X be a basis of A. If $a \varepsilon A$, then $a = r_1 x_1 + r_2 x_2 + \cdots + r_n x_n$ ($r_i \varepsilon R$; $x_i \varepsilon X$). If $\theta(a) = 0$, then for all $f \varepsilon A^*$,

$$0 = \langle a,f \rangle = \left\langle \sum_{i=1}^{n} r_i x_i, f \right\rangle = \sum_{i} r_i \langle x_i, f \rangle.$$

In particular, for $f = f_{x_j}$ ($j = 1,2, \ldots, n$),

$$0 = \sum_{i} r_i \langle x_i, f_{x_j} \rangle = \sum_{i} r_i \delta_{ij} = r_j.$$

Therefore, $a = \sum_{i} r_i x_i = \sum_{i} 0 x_i = 0$ and θ is a monomorphism.

(iii) If X is a finite basis of A, then A^* is free on the (finite) dual basis $\{ f_x \mid x \varepsilon X\}$ by Theorem 4.11. Similarly A^{**} is free on the (finite) dual basis $\{g_x \mid x \varepsilon X\}$, where for each $x \varepsilon X$, $g_x : A^* \to R$ is the homomorphism that is uniquely determined by the condition: $g_x(f_y) = \delta_{xy}$ ($y \varepsilon X$). But $\theta(x) \varepsilon A^{**}$ is a homomorphism $A^* \to R$ such that for every $y \varepsilon X$

$$\theta(x)(f_y) = \langle x, f_y \rangle = \delta_{xy} = g_x(f_y).$$

Hence $g_x = \theta(x)$ and $\{\theta(x) \mid x \varepsilon X\}$ is a basis of A^{**}. This implies that $\text{Im}\, \theta = A^{**}$, whence θ is an epimorphism. ■

EXERCISES

Note: R is a ring.

1. (a) For any abelian group A and positive integer m, $\mathrm{Hom}(Z_m,A) \cong A[m] = \{a \in A \mid ma = 0\}$.
   (b) $\mathrm{Hom}(Z_m,Z_n) \cong Z_{(m,n)}$.
   (c) The **Z**-module Z_m has $Z_m{}^* = 0$.
   (d) For each $k \geq 1$, Z_m is a Z_{mk}-module (Exercise 1.1); as a Z_{mk}-module, $Z_m{}^* \cong Z_m$.

2. If A,B are abelian groups and m,n integers such that $mA = 0 = nB$, then every element of $\mathrm{Hom}(A,B)$ has order dividing (m,n).

3. Let $\pi : \mathbf{Z} \to Z_2$ be the canonical epimorphism. The induced map $\bar{\pi} : \mathrm{Hom}(Z_2,\mathbf{Z}) \to \mathrm{Hom}(Z_2,Z_2)$ is the zero map. Since $\mathrm{Hom}(Z_2,Z_2) \neq 0$ (Exercise 1(b)), $\bar{\pi}$ is not an epimorphism.

4. Let R,S be rings and A_R, ${}_SB_R$, ${}_SC_R$, D_R (bi)modules as indicated. Let Hom_R denote all *right* R-module homomorphisms.
   (a) $\mathrm{Hom}_R(A,B)$ is a left S-module, with the action of S given by $(sf)(a) = s(f(a))$.
   (b) If $\varphi : A \to A'$ is an homomorphism of right R-modules, then the induced map $\bar{\varphi} : \mathrm{Hom}_R(A',B) \to \mathrm{Hom}_R(A,B)$ is an homomorphism at left S-modules.
   (c) $\mathrm{Hom}_R(C,D)$ is a right S-module, with the action of S given by $(gs)(c) = g(sc)$.
   (d) If $\psi : D \to D'$ is an homomorphism of right R-modules, then $\bar{\psi} : \mathrm{Hom}_R(C,D) \to \mathrm{Hom}_R(C,D')$ is an homomorphism of right S-modules.

5. Let R be a ring with identity; then there is a *ring* isomorphism $\mathrm{Hom}_R(R,R) \cong R^{op}$ where Hom_R denotes left R-module homomorphisms (see Exercises III.1.17 and 1.7). In particular, if R is commutative, then there is a ring isomorphism $\mathrm{Hom}_R(R,R) \cong R$.

6. Let S be a nonempty subset of a vector space V over a division ring. The **annihilator** of S is the subset S^0 of V^* given by $S^0 = \{f \in V^* \mid \langle s,f \rangle = 0 \text{ for all } s \in S\}$.
   (a) $0^0 = V^*$; $V^0 = 0$; $S \neq \{0\} \Rightarrow S^0 \neq V^*$.
   (b) If W is a subspace of V, then W^0 is a subspace of V^*.
   (c) If W is a subspace of V and $\dim V$ is finite, then $\dim W^0 = \dim V - \dim W$.
   (d) Let W,V be as in (c). There is an isomorphism $W^* \cong V^*/W^0$.
   (e) Let W,V be as in (c) and identify V with V^{**} under the isomorphism θ of Theorem 4.12. Then $(W^0)^0 = W \subset V^{**}$.

7. If V is a vector space over a division ring and $f \in V^*$, let $W = \{a \in V \mid \langle a,f \rangle = 0\}$, then W is a subspace of V. If $\dim V$ is finite, what is $\dim W$?

8. If R has an identity and we denote the left R-module R by ${}_RR$ and the right R-module R by R_R, then $({}_RR)^* \cong R_R$ and $(R_R)^* \cong {}_RR$.

9. For any homomorphism $f : A \to B$ of left R-modules the diagram

$$\begin{array}{ccc} A & \xrightarrow{\theta_A} & A^{**} \\ {\scriptstyle f}\downarrow & & \downarrow{\scriptstyle f^{**}} \\ B & \xrightarrow{\theta_B} & B^{**} \end{array}$$

is commutative, where θ_A, θ_B are as in Theorem 4.12 and $\bar{f}^*$ is the map induced on $A^{**} = \mathrm{Hom}_R(\mathrm{Hom}_R(A,R),R)$ by the map $\bar{f} : \mathrm{Hom}_R(B,R) \to \mathrm{Hom}_R(A,R)$.

10. Let $F = \sum_{x \varepsilon X} \mathbf{Z}x$ be a free $\mathbf{Z}$-module with an infinite basis X. Then $\{f_x \mid x \varepsilon X\}$ (Theorem 4.11) does not form a basis of F^*. [*Hint:* by Theorems 4.7 and 4.9, $F^* \cong \prod_{x \varepsilon X} \mathbf{Z}x$; but under this isomorphism $f_y \mapsto \{\delta_{xy}x\} \varepsilon \prod_{x \varepsilon X} \mathbf{Z}x$.]
Note: $F^* = \prod \mathbf{Z}x$ is not a free $\mathbf{Z}$-module; see L. Fuchs [13; p. 168].

11. If R has an identity and P is a finitely generated projective unitary left R-module, then
   (a) P^* is a finitely generated projective right R-module.
   (b) P is reflexive.
   This proposition may be false if the words "finitely generated" are omitted; see Exercise 10.

12. Let F be a field, X an infinite set, and V the free left F-module (vector space) on the set X. Let F^X be the set of all functions. $f : X \to F$.
   (a) F^X is a (right) vector space over F (with $(f + g)(x) = f(x) + g(x)$ and $(fr)(x) = rf(x)$).
   (b) There is a vector-space isomorphism $V^* \cong F^X$.
   (c) $\dim_F F^X = |F|^{|X|}$ (see Introduction, Exercise 8.10).
   (d) $\dim_F V^* > \dim_F V$ [*Hint:* by Introduction, Exercise 8.10 and Introduction, Theorem 8.5 $\dim_F V^* = \dim_F F^X = |F|^{|X|} \geq 2^{|X|} = |P(X)| > |X| = \dim_F V$.]

5. TENSOR PRODUCTS

The tensor product $A \otimes_R B$ of modules A_R and $_RB$ over a ring R is a certain abelian group, which plays an important role in the study of multilinear algebra. It is frequently useful to view the tensor product $A \otimes_R B$ as a universal object in a certain category (Theorem 5.2). On the other hand, it is also convenient to think of $A \otimes_R B$ as a sort of dual notion to $\mathrm{Hom}_R(A,B)$. We shall do this and consider such topics as induced maps and module structures for $A \otimes_R B$ as well as the behavior of tensor products with respect to exact sequences and direct sums.

If A_R and $_RB$ are modules over a ring R, and C is an (additive) abelian group, then a **middle linear map** from $A \times B$ to C is a function $f : A \times B \to C$ such that for all $a, a_i \varepsilon A$, $b, b_i \varepsilon B$, and $r \varepsilon R$:

$$f(a_1 + a_2, b) = f(a_1,b) + f(a_2,b); \tag{3}$$

$$f(a, b_1 + b_2) = f(a,b_1) + f(a,b_2); \tag{4}$$

$$f(ar,b) = f(a,rb). \tag{5}$$

For fixed $A_R, {}_RB$ consider the category $\mathfrak{M}(A,B)$ whose objects are all middle linear maps on $A \times B$. By definition a morphism in $\mathfrak{M}(A,B)$ from the middle linear map $f : A \times B \to C$ to the middle linear map $g : A \times B \to D$ is a group homomorphism $h : C \to D$ such that the diagram

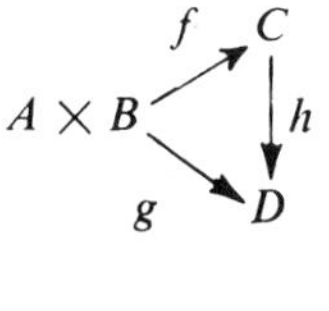

is commutative. Verify that $\mathfrak{M}(A,B)$ is a category, that 1_C is the identity morphism from f to f, and that h is an equivalence in $\mathfrak{M}(A,B)$ if and only if h is an isomorphism of groups. In Theorem 5.2 we shall construct a universal object in the category $\mathfrak{M}(A,B)$ (see Definition I.7.9). First, however, we need

Definition 5.1. *Let* A *be a right module and* B *a left module over a ring* R. *Let* F *be the free abelian group on the set* $A \times B$. *Let* K *be the subgroup of* F *generated by all elements of the following forms (for all* $a,a' \in A$; $b,b' \in B$; $r \in R$):

(i) $(a + a',b) - (a,b) - (a',b)$;
(ii) $(a,b + b') - (a,b) - (a,b')$;
(iii) $(ar,b) - (a,rb)$.

The quotient group F/K *is called the* **tensor product** *of* A *and* B; *it is denoted* $A \otimes_R B$ *(or simply* $A \otimes B$ *if* $R = \mathbf{Z}$). *The coset* $(a,b) + K$ *of the element* (a,b) *in* F *is denoted* $a \otimes b$; *the coset of* $(0,0)$ *is denoted* 0.

Since F is generated by the set $A \times B$, the quotient group $F/K = A \otimes_R B$ is generated by all elements (cosets) of the form $a \otimes b$ $(a \in A, b \in B)$. But it is *not* true that every element of $A \otimes_R B$ is of the form $a \otimes b$ (Exercise 4). For the typical element of F is a *sum* $\sum_{i=1}^{r} n_i(a_i,b_i)$ $(n_i \in \mathbf{Z}, a_i \in A, b_i \in B)$ and hence its coset in $A \otimes_R B = F/K$ is of the form $\sum_{i=1}^{r} n_i(a_i \otimes b_i)$. Furthermore, since it is possible to choose different representatives for a coset, one may have $a \otimes b = a' \otimes b'$ in $A \otimes_R B$, but $a \neq a'$ and $b \neq b'$ (Exercise 4). It is also possible to have $A \otimes_R B = 0$ even though $A \neq 0$ and $B \neq 0$ (Exercise 3).

Definition 5.1 implies that the generators $a \otimes b$ of $A \otimes_R B$ satisfy the following relations (for all $a,a_i \in A$, $b,b_i \in B$, and $r \in R$):

$$(a_1 + a_2) \otimes b = a_1 \otimes b + a_2 \otimes b; \qquad (6)$$

$$a \otimes (b_1 + b_2) = a \otimes b_1 + a \otimes b_2; \qquad (7)$$

$$ar \otimes b = a \otimes rb. \qquad (8)$$

The proof of these facts is straightforward; for example, since $(a_1 + a_2,b) - (a_1,b) - (a_2,b) \in K$, the "zero coset," we have

$$[(a_1 + a_2,b) + K] - [(a_1,b) + K] - [(a_2,b) + K] = K;$$

or in the notation $(a,b) + K = a \otimes b$,

$$(a_1 + a_2) \otimes b - a_1 \otimes b - a_2 \otimes b = 0.$$

Indeed an alternate definition of $A \otimes_R B$ is that it is the abelian group with generators all symbols $a \otimes b$ $(a \in A, b \in B)$, subject to the relations (6)–(8) above. Furthermore, since 0 is the only element of a group satisfying $x + x = x$, it is easy to see that for all $a \in A$, $b \in B$:

$$a \otimes 0 = 0 \otimes b = 0 \otimes 0 = 0.$$

Given modules A_R and ${}_RB$ over a ring R, it is easy to verify that the map $i : A \times B \to A \otimes_R B$ given by $(a,b) \mapsto a \otimes b$ is a middle linear map. The map i is called the **canonical middle linear map.** Its importance is seen in

Theorem 5.2. *Let A_R and ${}_RB$ be modules over a ring R, and let C be an abelian group. If $g : A \times B \to C$ is a middle linear map, then there exists a unique group homomorphism $\bar{g} : A \otimes_R B \to C$ such that $\bar{g}i = g$, where $i : A \times B \to A \otimes_R B$ is the canonical middle linear map. $A \otimes_R B$ is uniquely determined up to isomorphism by this property. In other words $i : A \times B \to A \otimes_R B$ is universal in the category $\mathfrak{M}(A,B)$ of all middle linear maps on $A \times B$.*

SKETCH OF PROOF. Let F be the free abelian group on the set $A \times B$, and let K be the subgroup described in Definition 5.1. Since F is free, the assignment $(a,b) \mapsto g(a,b) \in C$ determines a unique homomorphism $g_1 : F \to C$ by Theorem 2.1 (iv). Use the fact that g is middle linear to show that g_1 maps every generator of K to 0. Hence $K \subset \operatorname{Ker} g_1$. By Theorem 1.7 g_1 induces a homomorphism $\bar{g} : F/K \to C$ such that $\bar{g}[(a,b) + K] = g_1[(a,b)] = g(a,b)$. But $F/K = A \otimes_R B$ and $(a,b) + K = a \otimes b$. Therefore, $\bar{g} : A \otimes_R B \to C$ is a homomorphism such that $\bar{g}i(a,b) = \bar{g}(a \otimes b) = g(a,b)$ for all $(a,b) \in A \times B$; that is, $\bar{g}i = g$. If $h : A \otimes_R B \to C$ is any homomorphism with $hi = g$, then for any generator $a \otimes b$ of $A \otimes_R B$,

$$h(a \otimes b) = hi(a,b) = g(a,b) = \bar{g}i(a,b) = \bar{g}(a \otimes b).$$

Since h and $\bar{g}$ are homomorphisms that agree on the generators of $A \otimes_R B$, we must have $h = \bar{g}$, whence $\bar{g}$ is unique. This proves that $i : A \times B \to A \otimes_R B$ is a universal object in the category of all middle linear maps on $A \times B$, whence $A \otimes_R B$ is uniquely determined up to isomorphism (equivalence) by Theorem I.7.10. ■

Corollary 5.3. *If A_R, A_R', ${}_RB$ and ${}_RB'$ are modules over a ring R and $f : A \to A'$, $g : B \to B'$ are R-module homomorphisms, then there is a unique group homomorphism $A \otimes_R B \to A' \otimes_R B'$ such that $a \otimes b \mapsto f(a) \otimes g(b)$ for all $a \in A$, $b \in B$.*

SKETCH OF PROOF. Verify that the assignment $(a,b) \mapsto f(a) \otimes g(b)$ defines a middle linear map $h : A \times B \to C = A' \otimes_R B'$. By Theorem 5.2 there is a unique homomorphism $\bar{h} : A \otimes_R B \to A' \otimes_R B'$ such that $\bar{h}(a \otimes b) = \bar{h}i(a,b) = h(a,b) = f(a) \otimes g(b)$ for all $a \in A$, $b \in B$. ■

The unique homomorphism of Corollary 5.3 is denoted $f \otimes g : A \otimes_R B \to A' \otimes_R B'$. If $f' : A_R' \to A_R''$ and $g' : {}_RB' \to {}_RB''$ are also R-module homomorphisms, then it is easy to verify that

$$(f' \otimes g')(f \otimes g) = (f'f \otimes g'g) : A \otimes_R B \to A'' \otimes_R B''.$$

It follows readily that if f and g are R-module isomorphisms, then $f \otimes g$ is a group isomorphism with inverse $f^{-1} \otimes g^{-1}$.

Proposition 5.4. *If $A \xrightarrow{f} B \xrightarrow{g} C \to 0$ is an exact sequence of left modules over a ring R and D is a right R-module, then*

$$D \otimes_R A \xrightarrow{1_D \otimes f} D \otimes_R B \xrightarrow{1_D \otimes g} D \otimes_R C \to 0$$

is an exact sequence of abelian groups. An analogous statement holds for an exact sequence in the first variable.

PROOF. We must prove: (i) Im $(1_D \otimes g) = D \otimes_R C$; (ii) Im $(1_D \otimes f) \subset$ Ker $(1_D \otimes g)$; and (iii) Ker $(1_D \otimes g) \subset$ Im $(1_D \otimes f)$.

(i) Since g is an epimorphism by hypothesis every generator $d \otimes c$ of $D \otimes_R C$ is of the form $d \otimes g(b) = (1_D \otimes g)(d \otimes b)$ for some $b \in B$. Thus Im $(1_D \otimes g)$ contains all generators of $D \otimes_R C$, whence Im $(1_D \otimes g) = D \otimes_R C$. (ii) Since Ker $g =$ Im f we have $gf = 0$ and $(1_D \otimes g)(1_D \otimes f) = 1_D \otimes gf = 1_D \otimes 0 = 0$, whence Im $(1_D \otimes f) \subset$ Ker $(1_D \otimes g)$. (iii) Let $\pi : D \otimes_R B \to (D \otimes_R B)/\text{Im}\,(1_D \otimes f)$ be the canonical epimorphism. By (ii) and Theorem 1.7 there is a homomorphism $\alpha : (D \otimes_R B)/\text{Im}\,(1_D \otimes f) \to D \otimes_R C$ such that $\alpha(\pi(d \otimes b)) = (1_D \otimes g)(d \otimes b) = d \otimes g(b)$. We shall show that α is an isomorphism. This fact and Theorem 1.7 will imply Ker $(1_D \otimes g) =$ Im $(1_D \otimes f)$ and thus complete the proof.

We show first that the map $\beta : D \times C \to (D \otimes_R B)/\text{Im}\,(1_D \otimes f)$ given by $(d,c) \mapsto \pi(d \otimes b)$, where $g(b) = c$, is independent of the choice of b. Note that there is at least one such b since g is an epimorphism. If $g(b') = c$, then $g(b - b') = 0$ and $b - b' \in \text{Ker}\ g = \text{Im}\ f$, whence $b - b' = f(a)$ for some $a \in A$. Since $d \otimes f(a) \in$ Im $(1_D \otimes f)$ and $\pi(d \otimes f(a)) = 0$, we have

$$\begin{aligned}\pi(d \otimes b) &= \pi(d \otimes b' + f(a)) = \pi(d \otimes b' + d \otimes f(a)) \\ &= \pi(d \otimes b') + \pi(d \otimes f(a)) = \pi(d \otimes b').\end{aligned}$$

Therefore β is well defined. Verify that β is middle linear. Then by Theorem 5.2 there is a unique homomorphism $\bar{\beta} : D \otimes_R C \to (D \otimes_R B)/\text{Im}\,(1_D \otimes f)$ such that $\bar{\beta}(d \otimes c) = \bar{\beta}i(d,c) = \beta(d,c) = \pi(d \otimes b)$, where $g(b) = c$. Therefore, for any generator $d \otimes c$ of $D \otimes_R C$, $\alpha\bar{\beta}(d \otimes c) = \alpha(\pi(d \otimes b)) = d \otimes g(b) = d \otimes c$, whence $\alpha\bar{\beta}$ is the identity map. Similarly $\bar{\beta}\alpha$ is the identity so that α is an isomorphism. ■

REMARKS. If $h : A_R \to A_R'$ and $k : {}_RB \to {}_RB'$ are module epimorphisms, then Proposition 5.4 implies that $1_A \otimes k$ and $h \otimes 1_B$ are group epimorphisms. Hence $h \otimes k : A \otimes_R B \to A' \otimes_R B'$ is an epimorphism since $h \otimes k = (1_{A'} \otimes k)(h \otimes 1_B)$. However, if h and k are monomorphisms, $h \otimes 1_B$ and $1_A \otimes k$ need *not* be monomorphisms (Exercise 7).

Theorem 5.5. *Let* R *and* S *be rings and* ${}_SA_R$, ${}_RB$, C_R, ${}_RD_S$ *(bi)modules as indicated.*

(i) $A \otimes_R B$ *is a left* S*-module such that* $s(a \otimes b) = sa \otimes b$ *for all* $s \in S$, $a \in A$, $b \in B$.

(ii) *If* $f : A \to A'$ *is a homomorphism of* S-R *bimodules and* $g : B \to B'$ *is an* R*-module homomorphism, then the induced map* $f \otimes g : A \otimes_R B \to A' \otimes_R B'$ *is a homomorphism of left* S*-modules.*

(iii) $C \otimes_R D$ *is a right* S*-module such that* $(c \otimes d)s = c \otimes ds$ *for all* $c \in C$, $d \in D$, $s \in S$.

(iv) *If* $h : C \to C'$ *is an* R*-module homomorphism and* $k : D \to D'$ *a homomorphism of* R-S *bimodules, then the induced map* $h \otimes k : C \otimes_R D \to C' \otimes_R D'$ *is a homomorphism of right* S*-modules.*

SKETCH OF PROOF. (i) For each $s \in S$ the map $A \times B \to A \otimes_R B$ given by $(a,b) \mapsto sa \otimes b$ is R-middle linear, and therefore induces a unique group homomorphism $\alpha_s : A \otimes_R B \to A \otimes_R B$ such that $\alpha_s(a \otimes b) = sa \otimes b$. For each element $u = \sum_{i=1}^{n} a_i \otimes b_i \in A \otimes_R B$ define su to be the element $\alpha_s(u) = \sum_{i=1}^{n} \alpha_s(a_i \otimes b_i) = \sum_{i=1}^{n} sa_i \otimes b_i$. Since α_s is a homomorphism, this action of S is well defined (that is, independent of how u is written as a sum of generators). It is now easy to verify that $A \otimes_R B$ is a left S-module. ■

REMARK. An important special case of Theorem 5.5 occurs when R is a commutative ring and hence every R-module A is an R-R bimodule with $ra = ar$ $(r \in R, a \in A)$. In this case $A \otimes_R B$ is also an R-R bimodule with

$$r(a \otimes b) = ra \otimes b = ar \otimes b = a \otimes rb = a \otimes br = (a \otimes b)r$$

for all $r \in R$, $a \in A$, $b \in B$.

If R is a commutative ring, then the tensor product of R-modules may be characterized by a useful variation of Theorem 5.2. Let A,B,C be modules over a commutative ring R. A **bilinear map** from $A \times B$ to C is a function $f : A \times B \to C$ such that for all $a,a_i \in A$, $b,b_i \in B$, and $r \in R$:

$$f(a_1 + a_2,b) = f(a_1,b) + f(a_2,b); \tag{9}$$

$$f(a,b_1 + b_2) = f(a,b_1) + f(a,b_2); \tag{10}$$

$$f(ra,b) = rf(a,b) = f(a,rb). \tag{11}$$

Conditions (9) and (10) are simply a restatement of (3) and (4) above. For modules over a commutative ring (11) clearly implies condition (5) above, whence every bilinear map is middle linear.

EXAMPLE. If A^* is the dual of a module A over a commutative ring R, then the map $A \times A^* \to R$ given by $(a,f) \mapsto f(a) = \langle a,f \rangle$ is bilinear (see p. 204).

EXAMPLE. If A and B are modules over a commutative ring R, then so is $A \otimes_R B$ and the canonical middle linear map $i : A \times B \to A \otimes_R B$ is easily seen to be bilinear. In this context i is called the **canonical bilinear map.**

Theorem 5.6. *If* A,B,C *are modules over a commutative ring* R *and* $g : A \times B \to C$ *is a bilinear map, then there is a unique* R*-module homomorphism* $\bar{g} : A \otimes_R B \to C$ *such that* $\bar{g}i = g$*, where* $i : A \times B \to A \otimes_R B$ *is the canonical bilinear map. The module* $A \otimes_R B$ *is uniquely determined up to isomorphism by this property.*

SKETCH OF PROOF. Verify that the unique homomorphism of abelian groups $\bar{g} : A \otimes_R B \to C$ given by Theorem 5.2 is actually an R-module homomorphism. To prove the last statement let $\mathcal{B}(A,B)$ be the category of all bilinear maps on $A \times B$ (defined by replacing the groups C,D and group homomorphism $h : C \to D$ by modules and module homomorphisms in the definition of $\mathfrak{M}(A,B)$ on p. 207).

Then first part of the Theorem shows that $i : A \times B \to A \otimes_R B$ is a universal object in $\mathcal{B}(A,B)$, whence $A \otimes_R B$ is uniquely determined up to isomorphism by Theorem I.7.10. ■

Theorem 5.6 may also be used to provide an alternate definition of $A \otimes_R B$ when R is a commutative ring with identity. Let F_1 be the free *R-module* on the set $A \times B$ and K_1 the *submodule* generated by all elements of the forms:

$$\begin{aligned}&(a + a',b) - (a,b) - (a',b);\\&(a,b + b') - (a,b) - (a,b');\\&(ra,b) - r(a,b);\\&(a,rb) - r(a,b);\end{aligned}$$

where $a,a' \in A$; $b,b' \in B$; and $r \in R$; (compare Definition 5.1). We claim that there is an R-module isomorphism $A \otimes_R B \cong F_1/K_1$. The obvious analogue of the proof of Theorem 5.2 shows that the map $A \times B \to F_1/K_1$ given by $(a,b) \mapsto 1_R(a,b) + K_1$ is a universal object in the category $\mathcal{B}(A,B)$ of bilinear maps on $A \times B$. Consequently, $A \otimes_R B \cong F_1/K_1$ by Theorem 5.6.

We return now to modules over arbitrary rings.

Theorem 5.7. *If* R *is a ring with identity and* A_R, ${}_RB$ *are unitary* R*-modules, then there are* R*-module isomorphisms*

$$A \otimes_R R \cong A \quad \textit{and} \quad R \otimes_R B \cong B.$$

SKETCH OF PROOF. Since R is an R-R bimodule $R \otimes_R B$ is a left R-module by Theorem 5.5. The assignment $(r,b) \mapsto rb$ defines a middle linear map $R \times B \to B$. By Theorem 5.2 there is a group homomorphism $\alpha : R \otimes_R B \to B$ such that $\alpha(r \otimes b) = rb$. Verify that α is in fact a homomorphism of left R-modules. Then verify that the map $\beta : B \to R \otimes_R B$ given by $b \mapsto 1_R \otimes b$ is an R-module homomorphism such that $\alpha\beta = 1_B$ and $\beta\alpha = 1_{R\otimes_R B}$. Hence $\alpha : R \otimes_R B \cong B$. The isomorphism $A \otimes_R R \cong A$ is constructed similarly. ■

If R and S are rings and A_R, ${}_RB_S$, ${}_SC$ are (bi)modules, then $A \otimes_R B$ is a right S-module and $B \otimes_S C$ is a left R-module by Theorem 5.5. Consequently, both $(A \otimes_R B) \otimes {}_SC$ and $A \otimes_R (B \otimes {}_SC)$ are well-defined abelian groups.

Theorem 5.8. *If* R *and* S *are rings and* A_R, ${}_RB_S$, ${}_SC$ *are (bi)modules, then there is an isomorphism*

$$(A \otimes_R B) \otimes_S C \cong A \otimes_R (B \otimes_S C).$$

PROOF. By definition every element v of $(A \otimes_R B) \otimes_S C$ is a finite sum $\sum_{i=1}^{n} u_i \otimes c_i$ $(u_i \in A \otimes_R B,\ c_i \in C)$. Since each $u_i \in A \otimes_R B$ is a finite sum $\sum_{j=1}^{m_i} a_{ij} \otimes b_{ij}$ $(a_{ij} \in A,\ b_{ij} \in B)$, we have

$$v = \sum_i u_i \otimes c_i = \sum_i (\sum_j a_{ij} \otimes b_{ij}) \otimes c_i = \sum_i \sum_j [(a_{ij} \otimes b_{ij}) \otimes c_i].$$

Therefore, $(A \otimes_R B) \otimes_S C$ is generated by all elements of the form $(a \otimes b) \otimes c$ $(a \in A, b \in B, c \in C)$. Similarly, $A \otimes_R (B \otimes_S C)$ is generated by all $a \otimes (b \otimes c)$ with $a \in A, b \in B, c \in C$. Verify that the assignment $\left(\sum_{i=1}^{n} a_i \otimes b_i, c\right) \mapsto \sum_{i=1}^{n} [a_i \otimes (b_i \otimes c)]$ defines an S-middle linear map $(A \otimes_R B) \times C \to A \otimes_R (B \otimes_S C)$. Therefore, by Theorem 5.2 there is a homomorphism

$$\alpha : (A \otimes_R B) \otimes_S C \to A \otimes_R (B \otimes_S C)$$

with $\alpha[(a \otimes b) \otimes c] = a \otimes (b \otimes c)$ for all $a \in A, b \in B, c \in C$. Similarly there is an R-middle linear map $A \times (B \otimes_S C) \to (A \otimes_R B) \otimes_S C$ that induces a homomorphism

$$\beta : A \otimes_R (B \otimes_S C) \to (A \otimes_R B) \otimes_S C$$

such that $\beta[a \otimes (b \otimes c)] = (a \otimes b) \otimes c$ for all $a \in A, b \in B, c \in C$. For every generator $(a \otimes b) \otimes c$ of $(A \otimes_R B) \otimes_S C$, $\beta\alpha[(a \otimes b) \otimes c] = (a \otimes b) \otimes c$, whence $\beta\alpha$ is the identity map on $(A \otimes_R B) \otimes_S C$. A similar argument shows that $\beta\alpha$ is the identity on $A \otimes_R (B \otimes_S C)$. Therefore, α and β are isomorphisms. ■

In the future we shall identify $(A \otimes_R B) \otimes_S C$ and $A \otimes_R (B \otimes_S C)$ under the isomorphism of Theorem 5.8 and simply write $A \otimes_R B \otimes_S C$. It is now possible to define recursively the n-fold tensor product:

$$A^1 \otimes_{R_1} A^2 \otimes_{R_2} \cdots \otimes_{R_n} A^{n+1},$$

where $R_1, \ldots, R_n$ are rings and $A_{R_1}^1, {}_{R_1}A_{R_2}^2, \ldots, {}_{R_n}A^{n+1}$ are (bi)modules. Such iterated tensor products may also be characterized in terms of universal n-linear maps (Exercise 10).

Theorem 5.9. *Let* R *be a ring,* A *and* $\{A_i \mid i \in I\}$ *right* R*-modules,* B *and* $\{B_j \mid j \in J\}$ *left* R*-modules. Then there are group isomorphisms:*

$$(\sum_{i \in I} A_i) \otimes_R B \cong \sum_{i \in I} (A_i \otimes_R B);$$

$$A \otimes_R (\sum_{j \in J} B_j) \cong \sum_{j \in J} (A \otimes_R B_j).$$

PROOF. Let ι_k, π_k be the canonical injections and projections of $\sum_{i \in I} A_i$. By Theorem I.8.5 the family of homomorphisms $\iota_k \otimes 1_B : A_k \otimes_R B \to (\sum_{i \in I} A_i) \otimes_R B$ induce a homomorphism $\alpha : \sum_{i \in I} (A_i \otimes_R B) \to (\sum_{i \in I} A_i) \otimes_R B$ such that $\alpha[\{a_i \otimes b\}] = \sum_{i \in I_0} (\iota_i(a_i) \otimes b) = (\sum_{i \in I_0} \iota_i(a_i)) \otimes b$, where $I_0 = \{i \in I \mid a_i \otimes b \neq 0\}$. The assignment $(u,b) \mapsto \{\pi_i(u) \otimes b\}_{i \in I}$ defines a middle linear map $(\sum_{i \in I} A_i) \times B \to \sum_{i \in I} (A_i \otimes_R B)$ and thus induces a homomorphism $\beta : (\sum A_i) \otimes_R B \to \sum (A_i \otimes_R B)$ such that $\beta(u \otimes b) = \{\pi_i(u) \otimes b\}_{i \in I}$. We shall show that $\alpha\beta$ and $\beta\alpha$ are the respective identity maps, whence α is an isomorphism.

Recall that if $u \in \sum A_i$ and $I_0 = \{i \in I \mid \pi_i(u) \neq 0\}$, then $u = \sum_{i \in I_0} \iota_i\pi_i(u)$. Thus for every generator $u \otimes b$ of $(\sum A_i) \otimes_R B$ we have

$$\alpha\beta(u \otimes b) = \alpha[\{\pi_i(u) \otimes b\}] = (\sum_{i \in I_0} \iota_i\pi_i(u)) \otimes b = u \otimes b.$$

Consequently $\alpha\beta$ is the identity map.

For each $j \in I$ let $\iota_j^* : A_j \otimes_R B \to \sum_i (A_i \otimes_R B)$ be the canonical injection and verify that $\sum_i (A_i \otimes_R B)$ is generated by all elements of the form $\iota_j^*(a \otimes b) = \{\pi_i\iota_j(a) \otimes b\}_{i \in I}$ $(j \in I, a \in A_j, b \in B)$. For each such generator we have $(\pi_i\iota_j(a)) \otimes b = 0$ if $i \neq j$ and $(\pi_j\iota_j(a)) \otimes b = a \otimes b$, whence

$$\begin{aligned} \beta\alpha[\iota_j^*(a \otimes b)] &= \beta\alpha[\{\pi_i\iota_j(a) \otimes b\}] = \beta[\iota_j\pi_j\iota_j(a) \otimes b] \\ &= \beta[\iota_j(a) \otimes b] = \{\pi_i\iota_j(a) \otimes b\}_{i \in I} = \iota_j^*(a \otimes b). \end{aligned}$$

Consequently the map $\beta\alpha$ must be the identity. The second isomorphism is proved similarly. ■

Theorem 5.10. *(Adjoint Associativity) Let* R *and* S *be rings and* A_R, ${}_RB_S$, C_S *(bi)-modules. Then there is an isomorphism of abelian groups*

$$\alpha : Hom_S(A \otimes_R B, C) \cong Hom_R(A, Hom_S(B,C)),$$

defined for each $f : A \otimes_R B \to C$ *by*

$$[(\alpha f)(a)](b) = f(a \otimes b).$$

Note that $Hom_R(\_\_,\_\_)$ and $Hom_S(\_\_,\_\_)$ consist of homomorphisms of *right* modules. Recall that the *R*-module structure of $Hom_S(B,C)$ is given by: $(gr)(b) = g(rb)$ (for $r \in R$, $b \in B$, $g \in Hom_S(B,C)$; see Exercise 4.4(c)).

SKETCH OF PROOF OF 5.10. The proof is a straightforward exercise in the use of the appropriate definitions. The following items must be checked.

(i) For each $a \in A$, and $f \in Hom_S(A \otimes_R B, C)$, $(\alpha f)(a) : B \to C$ is an *S*-module homomorphism.

(ii) $(\alpha f) : A \to Hom_S(B,C)$ is an *R*-module homomorphism. Thus α is a well-defined function.

(iii) α is a group homomorphism (that is, $\alpha(f_1 + f_2) = \alpha(f_1) + \alpha(f_2)$). To show that α is an isomorphism, construct an inverse map $\beta : Hom_R(A, Hom_S(B,C)) \to Hom_S(A \otimes_R B, C)$ by defining

$$(\beta g)(a \otimes b) = [g(a)](b),$$

where $a \in A$, $b \in B$, and $g \in Hom_R(A, Hom_S(B,C))$. Verify that

(iv) βg as defined above on the generators determines a unique *S*-module homomorphism $A \otimes_R B \to C$.

(v) β is a homomorphism.

(vi) $\beta\alpha$ and $\alpha\beta$ are the respective identities. Thus α is an isomorphism. ■

We close this section with an investigation of the tensor product of free modules. Except for an occasional exercise this material will be used only in Section IX.6.

Theorem 5.11. *Let* R *be a ring with identity. If* A *is a unitary right* R*-module and* F *is a free left* R*-module with basis* Y*, then every element* u *of* $A \otimes_R F$ *may be written uniquely in the form* $u = \sum_{i=1}^{n} a_i \otimes y_i$*, where* $a_i \varepsilon A$ *and the* y_i *are distinct elements of* Y.

REMARK. Given $u = \sum_{k=1}^{t} a_k \otimes y_k$ and $v = \sum_{j=1}^{m} b_j \otimes z_j$ $(a_k,b_i \varepsilon A, y_k,z_j \varepsilon Y)$, we may, if necessary, insert terms of the form $0 \otimes y$ $(y \varepsilon Y)$ and assume that $u = \sum_{i=1}^{n} a_i \otimes y_i$ and $v = \sum_{i=1}^{n} b_i \otimes y_i$. The word "uniquely" in Theorem 5.11 means that if $\sum_{i=1}^{n} a_i \otimes y_i = \sum_{i=1}^{n} b_i \otimes y_i$, then $a_i = b_i$ for every i. In particular, if $\sum_{i=1}^{n} a_i \otimes y_i = 0 = \sum_{i=1}^{n} 0 \otimes y_i$, then $a_i = 0$ for every i.

PROOF OF 5.11. For each $y \varepsilon Y$, let A_y be a copy of A and consider the direct sum $\sum_{y\varepsilon Y} A_y$. We first construct an isomorphism $\theta : A \otimes_R F \cong \sum_{y \varepsilon Y} A_y$ as follows. Since Y is a basis, $\{y\}$ is a linearly independent set for each $y \varepsilon Y$. Consequently, the R-module epimorphism $\varphi : R \to Ry$ given by $r \mapsto ry$ (Theorem 1.5) is actually an isomorphism. Therefore, by Theorem 5.7 there is for each $y \varepsilon Y$ an isomorphism

$$A \otimes_R Ry \xrightarrow{1_A \otimes \varphi^{-1}} A \otimes_R R \cong A = A_y.$$

Thus by Theorems 5.9 and I.8.10 there is an isomorphism θ:

$$A \otimes_R F = A \otimes_R (\sum_{y\varepsilon Y} Ry) \cong \sum_{y\varepsilon Y} A \otimes_R Ry \cong \sum_{y \varepsilon Y} A_y.$$

Verify that for every $a \varepsilon A$, $z \varepsilon Y$, $\theta(a \otimes z) = \{u_y\} \varepsilon \sum A_y$, where $u_z = a$ and $u_y = 0$ for $y \neq z$; in other words, $\theta(a \otimes z) = \iota_z(a)$, with $\iota_z : A_z \to \sum A_y$ the canonical injection. Now every nonzero $v \varepsilon \sum A_y$ is a finite sum $v = \iota_{y_1}(a_1) + \cdots + \iota_{y_n}(a_n) = \theta(a_1 \otimes y_1) + \cdots + \theta(a_n \otimes y_n)$ with $y_1, \ldots, y_n$ distinct elements of Y and a_i uniquely determined nonzero elements of A. It follows that every element of $A \otimes_R F$ (which is necessarily $\theta^{-1}(v)$ for some v) may be written uniquely as $\sum_{i=1}^{n} a_i \otimes y_i$. ∎

Corollary 5.12. *If* R *is a ring with identity and* A_R *and* ${}_RB$ *are free* R*-modules with bases* X *and* Y *respectively,* then $A \otimes_R B$ *is a free (right)* R*-module with basis* $W = \{x \otimes y \mid x \varepsilon X, y \varepsilon Y\}$ *of cardinality* $|X||Y|$.

REMARKS. Since R is an R-R bimodule, so is every direct sum of copies of R. In particular, every free left R-module is also a free right R-module and vice versa. However, it is not true in general that a free (left) R-module is a free object in the category of R-R bimodules (Exercise 12).

SKETCH OF PROOF OF 5.12. By the proof of Theorem 5.11 and by Theorem 2.1 (for right R-modules) there is a group isomorphism

$$\theta : A \otimes_R B \cong \sum_{y\varepsilon Y} A_y = \sum_{y \varepsilon Y} A = \sum_{y\varepsilon Y} (\sum_{x \varepsilon X} xR).$$

Since B is an R-R bimodule by the remark preceding the proof, $A \otimes_R B$ is a right R-module by Theorem 5.5. Verify that θ is an isomorphism of right R-modules such that $\theta(W)$ is a basis of the free right R-module $\sum_Y (\sum_X xR)$. Therefore, $A \otimes_R B$ is a free right R-module with basis W. Since the elements of W are all distinct by Theorem 5.11, $|W| = |X||Y|$. ■

Corollary 5.13. *Let* S *be a ring with identity and* R *a subring of* S *that contains* 1_S. *If* F *is a free left* R*-module with basis* X, *then* $S \otimes_R F$ *is a free left* S*-module with basis* $\{1_S \otimes x \mid x \in X\}$ *of cardinality* $|X|$.

SKETCH OF PROOF. Since S is clearly an S-R bimodule, $S \otimes_R F$ is a left S-module by Theorem 5.5. The proof of Theorem 5.11 shows that there is a group isomorphism $\theta : S \otimes_R F \cong \sum_{x \in X} S_x$, with each $S_x = S$. Furthermore, if for $z \in X$, $\iota_z : S = S_z \to \sum_{x \in X} S_x$ is the canonical injection, then $\theta(1_S \otimes z) = \iota_z(1_S)$ for each $z \in X$. Verify that θ is in fact an isomorphism of left S-modules. Clearly, $\{\iota_x(1_S) \mid x \in X\}$ is a basis of cardinality $|X|$ of the free left S-module $\sum_{x \in X} S_x$, whence $S \otimes_R F$ is a free S-module with basis $\{1_S \otimes x \mid x \in X\}$ of cardinality $|X|$. ■

EXERCISES

Note: R is a ring and $\otimes = \otimes_{\mathbf{Z}}$.

1. If $R = \mathbf{Z}$, then condition (iii) of Definition 5.1 is superfluous (that is, (i) and (ii) imply (iii)).
2. Let A and B be abelian groups.
   (a) For each $m > 0$, $A \otimes Z_m \cong A/mA$.
   (b) $Z_m \otimes Z_n \cong Z_c$, where $c = (m,n)$.
   (c) Describe $A \otimes B$, when A and B are finitely generated.
3. If A is a torsion abelian group and $\mathbf{Q}$ the (additive) group of rationals, then
   (a) $A \otimes \mathbf{Q} = 0$.
   (b) $\mathbf{Q} \otimes \mathbf{Q} \cong \mathbf{Q}$.
4. Give examples to show that each of the following may actually occur for suitable rings R and modules A_R, ${}_RB$.
   (a) $A \otimes_R B \not\cong A \otimes_{\mathbf{Z}} B$.
   (b) $u \in A \otimes_R B$, but $u \neq a \otimes b$ for any $a \in A$, $b \in B$.
   (c) $a \otimes b = a_1 \otimes b_1$ but $a \neq a_1$ and $b \neq b_1$.
5. If A' is a submodule of the right R-module A and B' is a submodule of the left R-module B, then $A/A' \otimes_R B/B' \cong (A \otimes_R B)/C$, where C is the subgroup of $A \otimes_R B$ generated by all elements $a' \otimes b$ and $a \otimes b'$ with $a \in A$, $a' \in A'$, $b \in B$, $b' \in B'$.
6. Let $f : A_R \to A_R'$ and $g : {}_RB \to {}_RB'$ be R-module homomorphisms. What is the difference between the homomorphism $f \otimes g$ (as given by Corollary 5.3) and the element $f \otimes g$ of the tensor product of abelian groups

$$\mathrm{Hom}_R(A,A') \otimes \mathrm{Hom}_R(B,B')?$$

7. The usual injection $\alpha : Z_2 \to Z_4$ is a monomorphism of abelian groups. Show that $1 \otimes \alpha : Z_2 \otimes Z_2 \to Z_2 \otimes Z_4$ is the zero map (but $Z_2 \otimes Z_2 \neq 0$ and $Z_2 \otimes Z_4 \neq 0$; see Exercise 2).

8. Let $0 \to A \xrightarrow{f} B \xrightarrow{g} C \to 0$ be a short exact sequence of left R-modules and D a right R-module. Then $0 \to D \otimes_R A \xrightarrow{1_D \otimes f} D \otimes_R B \xrightarrow{1_D \otimes g} D \otimes_R C \to 0$ is a short exact sequence of abelian groups under any one of the following hypotheses:
   (a) $0 \to A \xrightarrow{f} B \xrightarrow{g} C \to 0$ is split exact.
   (b) R has an identity and D is a free right R-module.
   (c) R has an identity and D is a projective unitary right R-module.

9. (a) If I is a right ideal of a ring R with identity and B a left R-module, then there is a group isomorphism $R/I \otimes_R B \cong B/IB$, where IB is the subgroup of B generated by all elements rb with $r \varepsilon I$, $b \varepsilon B$.
   (b) If R is commutative and I,J are ideals of R, then there is an R-module isomorphism $R/I \otimes_R R/J \cong R/(I + J)$.

10. If R,S are rings, A_R, ${}_RB_S$, ${}_SC$ are (bi)modules and D an abelian group, define a *middle linear map* to be a function $f : A \times B \times C \to D$ such that

   (i) $f(a + a',b,c) = f(a,b,c) + f(a',b,c)$;
   (ii) $f(a,b + b',c) = f(a,b,c) + f(a,b',c)$;
   (iii) $f(a,b,c + c') = f(a,b,c) + f(a,b,c')$;
   (iv) $f(ar,b,c) = f(a,rb,c)$ for $r \varepsilon R$;
   (v) $f(a,bs,c) = f(a,b,sc)$ for $s \varepsilon S$.

   (a) The map $i : A \times B \times C \to (A \otimes_R B) \otimes_S C$ given by $(a,b,c) \mapsto (a \otimes b) \otimes c$ is middle linear.
   (b) The middle linear map i is *universal;* that is, given a middle linear map $g : A \times B \times C \to D$, there exists a unique group homomorphism $\bar{g} : (A \otimes_R B) \otimes_S C \to D$ such that $\bar{g}i = g$.
   (c) The map $j : A \times B \times C \to A \otimes_R (B \otimes_S C)$ given by $(a,b,c) \mapsto a \otimes (b \otimes c)$ is also a universal middle linear map.
   (d) $(A \otimes_R B) \otimes_S C \cong A \otimes_R (B \otimes_S C)$ by (b), (c), and Theorem I.7.10.
   (e) Define a middle linear function on n (bi)modules ($n \geq 4$) in the obvious way and sketch a proof of the extension of the above results to the case of n (bi)-modules (over $n - 1$ rings).
   (f) If $R = S$, R is commutative and A,B,C,D are R-modules, define a trilinear map $A \times B \times C \to D$ and extend the results of (a),(b),(c) to such maps.

11. Let A,B,C be modules over a commutative ring R.
   (a) The set $\mathfrak{L}(A,B;C)$ of all R-bilinear maps $A \times B \to C$ is an R-module with $(f + g)(a,b) = f(a,b) + g(a,b)$ and $(rf)(a,b) = rf(a,b)$.
   (b) Each one of the following R-modules is isomorphic to $\mathfrak{L}(A,B;C)$:

   (i) $\mathrm{Hom}_R(A \otimes_R B,C)$;
   (ii) $\mathrm{Hom}_R(A,\mathrm{Hom}_R(B,C))$;
   (iii) $\mathrm{Hom}_R(B,\mathrm{Hom}_R(A,C))$.

12. Assume R has an identity. Let $\mathfrak{C}$ be the category of all unitary R-R bimodules and bimodule homomorphisms (that is, group homomorphisms $f : A \to B$ such that $f(ras) = rf(a)s$ for all $r,s \varepsilon R$). Let $X = \{1_R\}$ and let $\iota : X \to R$ be the inclusion map.

(a) If R is noncommutative, then R (equipped with $\iota : X \to R$) is not a free object on the set X in the category $\mathcal{C}$.

(b) $R \otimes_{\mathbf{Z}} R$ is an R-R bimodule (Theorem 5.5). If $\iota : X \to R \otimes_{\mathbf{Z}} R$ is given by $1_R \mapsto 1_R \otimes 1_R$, then $R \otimes_{\mathbf{Z}} R$ is a free object on the set X in the category $\mathcal{C}$.

6. MODULES OVER A PRINCIPAL IDEAL DOMAIN

The chief purpose of this section, which will be used again only in Sections VII.2 and VII.4, is to determine the structure of all finitely generated modules over a principal ideal domain. Virtually all of the structure theorems for finitely generated abelian groups (Sections II.1,II.2) carry over to such modules. In fact, most of the proofs in Sections II.1 and II.2 extend immediately to modules over Euclidean domains. However, several of them must be extensively modified in order to be valid for modules over an arbitrary principal ideal domain. Consequently, we shall use a different approach in proving the structure theorems here. We shall show that just as in the case of abelian groups every finitely generated module may be decomposed in two ways as a direct sum of cyclic submodules (Theorem 6.12). Each decomposition provides a set of invariants for the given module (that is, two modules have the same invariants if and only if they are isomorphic (Corollary 6.13)). Thus each method of decomposition leads to a complete classification (up to isomorphism) of all finitely generated modules over a principal ideal domain. *Here and throughout this section "module" means "unitary module".*

We begin with free modules over a principal ideal domain R. Since R has the invariant dimension property by Corollary 2.12, the rank of a free R-module (Definition 2.8) is well defined. In particular, two free R-modules are isomorphic if and only if they have the same rank (Proposition 2.9). Furthermore we have the following generalization of Theorem II.1.6.

Theorem 6.1. *Let* F *be a free module over a principal ideal domain* R *and* G *a submodule of* F. *Then* G *is a free* R*-module and rank* G $\leq$ *rank* F.

SKETCH OF PROOF. Let $\{x_i \mid i \in I\}$ be a basis of F. Then $F = \sum_{i \in I} Rx_i$ with each Rx_i isomorphic to R (as a left R-module). Choose a well ordering $\leq$ of the set I (Introduction, Section 7). For each $i \in I$ denote the immediate successor of i by $i + 1$ (Introduction, Exercise 7.7). Let $J = I \cup \{\alpha\}$, where $\alpha \notin I$ and by definition $i < \alpha$ for all $i \in I$. Then J is well ordered and every element of I has an immediate successor in J.[1] For each $j \in J$ define F_j to be the submodule of F generated by the set $\{x_i \mid i < j\}$. Verify that the submodules F_j have the following properties:

(i) $j < k \Leftrightarrow F_j \subset F_k$;
(ii) $\bigcup_{j \in J} F_j = F$;

[1]The set J is a technical device needed to cope with the possibility that some (necessarily unique) element of I has no immediate successor in I. This occurs, for example, when I is finite.

(iii) for each $i \in I$, $F_{i+1}/F_i \cong Rx_i \cong R$. [Apply Theorem 1.7 to the canonical projection $F_{i+1} = \sum_{k<i+1} Rx_k \to Rx_i$.]

For each $j \in J$ let $G_j = G \cap F_j$ and verify that:

(iv) $j < k \Rightarrow G_j \subset G_k$;
(v) $\bigcup_{j \in J} G_j = G$;
(vi) for each $i \in I$, $G_i = G_{i+1} \cap F_i$.

Property (vi) and Theorem 1.9(i) imply that $G_{i+1}/G_i = G_{i+1}/(G_{i+1} \cap F_i) \cong (G_{i+1} + F_i)/F_i$. But $(G_{i+1} + F_i)/F_i$ is a submodule of F_{i+1}/F_i. Therefore, G_{i+1}/G_i is isomorphic to a submodule of R by (iii). But every submodule of R is necessarily an ideal of R and hence of the form $(c) = Rc$ for some $c \in R$. If $c \neq 0$, then the R-module epimorphism $R \to Rc$ of Theorem 1.5(i) is actually an isomorphism. Thus every submodule of R (and hence each G_{i+1}/G_i) is free of rank 0 or 1. By Theorems 3.2 and 3.4 the sequence $0 \to G_i \xrightarrow{\subset} G_{i+1} \to G_{i+1}/G_i \to 0$ is split exact for every $i \in I$. Theorem 1.18 and Exercise 1.15 imply that each G_{i+1} is an internal direct sum $G_{i+1} = G_i \oplus Rb_i$, where $b_i \in G_{i+1} - G_i$ and $Rb_i \cong R$ if $G_{i+1} \neq G_i$, and $b_i = 0$ if $G_{i+1} = G_i$ (that is, $G_{i+1}/G_i = 0$). Thus $b_i \in G$ is defined for each $i \in I$. Let $B = \{b_i \mid b_i \neq 0\}$. Then $|B| \leq |I| = \text{rank } F$. To complete the proof we need only show that B is a basis of G.

Suppose $u = \sum_j r_j b_j = 0$ ($j \in I$; $r_j \in R$; finite sum). Let k be the largest index (if one exists) such that $r_k \neq 0$. Then $u = \sum_{j<k} r_j b_j + r_k b_k \in G_k \oplus Rb_k = G_{k+1}$. But $u = 0$ implies that $r_k = 0$, which is a contradiction. Hence $r_j = 0$ for all j. Therefore, B is linearly independent.

Finally we must prove that B spans G. It suffices by (v) to prove that for each $k \in J$ the subset $B_k = \{b_j \in B \mid j < k\}$ of B spans G_k. We shall use transfinite induction (Introduction, Theorem 7.1). Suppose, therefore, that B_j spans G_j for all $j < k$ and let $u \in G_k$. If $k = j + 1$ for some $j \in I$, then $G_k = G_{j+1} = G_j \oplus Rb_j$ and $u = v + rb_j$ with $v \in G_j$. By the induction hypothesis v is a finite sum $v = \sum r_i b_i$ with $r_i \in R$ and $b_i \in B_j \subset B_k$. Therefore, $u = \sum r_i b_i + rb_k$, whence B_k spans G_k. Now suppose that $k \neq j + 1$ for all $j \in I$ (and this may happen; see the examples preceding Theorem 7.1 of the Introduction). Since $u \in G_k = G \cap F_k$, u is a finite sum $u = \sum r_j x_j$ with $j < k$. If t is the largest index such that $r_t \neq 0$, then $u \in F_{t+1}$ with $t + 1 < k$ by hypothesis. Therefore, $u \in G \cap F_{t+1} = G_{t+1}$ with $t + 1 < k$. By the induction hypothesis u is a linear combination of elements of B_{t+1}, which is a subset of B_k. Hence B_k spans G_k. ■

Corollary 6.2. *Let* R *be a principal ideal domain. If* A *is a finitely generated* R*-module generated by* n *elements, then every submodule of* A *may be generated by* m *elements with* m $\leq$ n.

PROOF. Exercise; see Corollary II.1.7 and Corollary 2.2. ■

Corollary 6.3. *A unitary module* A *over a principal ideal domain is free if and only if* A *is projective.*

PROOF. ($\Rightarrow$) Theorem 3.2. ($\Leftarrow$) There is a short exact sequence $0 \to K \overset{\subset}{\to} F \overset{f}{\to} A \to 0$ with F free, f an epimorphism and $K = \ker f$ by Corollary 2.2. If A is projective, then $F \cong K \oplus A$ by Theorem 3.4. Therefore, A is isomorphic to a submodule of F, whence A is free by Theorem 6.1. ■

We now develop the analogues of the order of an element in a group and of the torsion subgroup of an abelian group.

Theorem 6.4. *Let* A *be a left module over an integral domain* R *and for each* $a \in A$ *let* $\mathcal{O}_a = \{r \in R \mid ra = 0\}$.

(i) $\mathcal{O}_a$ *is an ideal of* R *for each* $a \in A$.
(ii) $A_t = \{a \in A \mid \mathcal{O}_a \neq 0\}$ *is a submodule of* A.
(iii) *For each* $a \in A$ *there is an isomorphism of left modules*

$$R/\mathcal{O}_a \cong Ra = \{ra \mid r \in R\}.$$

Let R *be a principal ideal domain and* $p \in R$ *a prime.*

(iv) *If* $p^i a = 0$ (*equivalently* $(p^i) \subset \mathcal{O}_a$), *then* $\mathcal{O}_a = (p^j)$ *with* $0 \leq j \leq i$.
(v) *If* $\mathcal{O}_a = (p^i)$, *then* $p^j a \neq 0$ *for all* j *such that* $0 \leq j < i$.

REMARK. Prime and irreducible elements coincide in a principal ideal domain by Theorem III.3.4.

SKETCH OF PROOF OF 6.4. (iii) Use Theorems 1.5(i) and 1.7. (iv) By hypothesis $\mathcal{O}_a = (r)$ for some $r \in R$. Since $p^i \in \mathcal{O}_a$, r divides p^i. Unique factorization in R (Theorem III.3.7) implies that $r = p^j u$ with $0 \leq j \leq i$ and u a unit. Hence $\mathcal{O}_a = (r) = (p^j u) = (p^j)$ by Theorem III.3.2. (v) If $p^j a = 0$ with $j < i$, then $p^j \in \mathcal{O}_a = (p^i)$, whence $p^i \mid p^j$. This contradicts unique factorization in R. ■

Let A be a module over an integral domain. The ideal $\mathcal{O}_a$ in Theorem 6.4 is called the **order ideal** of $a \in A$. The submodule A_t in Theorem 6.4 is called the **torsion submodule** of A. A is said to be a **torsion module** if $A = A_t$ and to be **torsion-free** if $A_t = 0$. Every free module is torsion-free, but not vice versa (Exercise 2).

Let A be a module over a principal ideal domain R. The order ideal of $a \in A$ is a principal ideal of R, say $\mathcal{O}_a = (r)$, and a is said to have **order** r. The element r is unique only up to multiplication by a unit (Theorem III.3.2). The cyclic submodule Ra generated by a (Theorem 1.5) is said to be **cyclic of order r**. Theorem 6.4(iii) shows that $a \in A$ has order 0 (that is, Ra is a cyclic module of order 0) if and only if $Ra \cong R$ (that is, Ra is free of rank one). Also $a \in A$ has order r, with r a unit, if and only if $a = 0$; (for $a = 1_R a = r^{-1}(ra) = r^{-1}0 = 0$).

EXAMPLE. If R is a principal ideal domain and $r \in R$, then the quotient ring $R/(r)$ is a cyclic R-module with generator $a = 1_R + (r)$. Clearly $\mathcal{O}_a = (r)$, whence a has order r and $R/(r)$ is cyclic of order r. Theorem 6.4(iii) shows that every cyclic module C over a principal ideal domain R is isomorphic to $R/(r)$, where $(r) = \mathcal{O}_a$ and a is a generator of C.

EXAMPLE. Let $R = \mathbf{Z}$ and let A be an (additive) abelian group. Suppose the group theoretic order of $a \varepsilon A$ (Definition I.3.3) is finite. Then $\mathcal{O}_a = (n)$, where $|n|$ is the group theoretic order of a. If $a \varepsilon A$ has infinite order, then $\mathcal{O}_a = (0)$. In either case $\mathbf{Z}a$ is the cyclic subgroup $\langle a \rangle$ generated by a (Theorem I.2.8). Furthermore, $\mathbf{Z}a \cong \mathbf{Z}/(n) \cong Z_n$ if $\mathcal{O}_a = (n)$, $n \neq 0$; and $\mathbf{Z}a \cong \mathbf{Z}/(0) \cong \mathbf{Z}$ if $\mathcal{O}_a = (0)$.

Theorem 6.5. *A finitely generated torsion-free module* A *over a principal ideal domain* R *is free.*

REMARK. The hypothesis that A is finitely generated is essential (Exercise II.1.10).

PROOF OF 6.5. We may assume $A \neq 0$. Let X be a finite set of nonzero generators of A. If $x \varepsilon X$, then $rx = 0$ $(r \varepsilon R)$ if and only if $r = 0$ since A is torsion-free. Consequently, there is a nonempty subset $S = \{x_1, \dots, x_k\}$ of X that is maximal with respect to the property:

$$r_1x_1 + \cdots + r_kx_k = 0 \; (r_i \varepsilon R) \quad \Rightarrow \quad r_i = 0 \quad \text{for all} \quad i.$$

The submodule F generated by S is clearly a free R-module with basis S. If $y \varepsilon X - S$, then by maximality there exist $r_y, r_1, \dots, r_k \varepsilon R$, not all zero, such that $r_yy + r_1x_1 + \cdots + r_kx_k = 0$. Then $r_yy = -\sum_{i=1}^{k} r_ix_i \varepsilon F$. Furthermore, $r_y \neq 0$ since otherwise $r_i = 0$ for every i. Since X is finite, there exists a nonzero $r \varepsilon R$ (namely $r = \prod_{y \varepsilon X - S} r_y$) such that $rX = \{rx \mid x \varepsilon X\}$ is contained in F. Therefore, $rA = \{ra \mid a \varepsilon A\} \subset F$. The map $f : A \to A$ given by $a \mapsto ra$ is easily seen to be an R-module homomorphism with image rA. Since A is torsion-free Ker $f = 0$, whence $A \cong \operatorname{Im} f = rA \subset F$. Therefore, A is free by Theorem 6.1. ∎

Determining the structure of a finitely generated module A over a principal ideal domain now proceeds in three steps. We show first that A is a direct sum of a torsion module and a free module (Theorem 6.6). Every torsion module is a direct sum of "p-primary modules" (Theorem 6.7). Finally every p-primary module is a direct sum of cyclic modules (Theorem 6.9).

Theorem 6.6. *If* A *is a finitely generated module over a principal ideal domain* R, *then* $A = A_t \oplus F$, *where* F *is a free* R*-module of finite rank and* $F \cong A/A_t$.

SKETCH OF PROOF. The quotient module A/A_t is torsion-free since for each $r \neq 0$,

$$r(a + A_t) = A_t \quad \Rightarrow \quad ra \varepsilon A_t \quad \Rightarrow \quad r_1(ra) = 0 \quad \text{for some} \quad r_1 \neq 0 \quad \Rightarrow \quad a \varepsilon A_t$$

Furthermore, A/A_t is finitely generated since A is. Therefore, A/A_t is free of finite rank by Theorem 6.5. Consequently, the exact sequence $0 \to A_t \xrightarrow{\subset} A \to A/A_t \to 0$ is split exact and $A \cong A_t \oplus A/A_t$ (Theorems 3.2 and 3.4). Under the isomorphism $A_t \oplus A/A_t \cong A$ of Theorem 3.4 the image of A_t is A_t and the image of A/A_t is a submodule F of A, which is necessarily free of finite rank. It follows that A is the internal direct sum $A = A_t \oplus F$ (see Theorem 1.15). ∎

Theorem 6.7. *Let* A *be a torsion module over a principal ideal domain* R *and for each prime* p ε R *let* A(p) = {a ε A | a *has order a power of* p}.

(i) A(p) *is a submodule of* A *for each prime* p ε R;

(ii) A = $\sum$ A(p), *where the sum is over all primes* p ε R. *If* A *is finitely generated, only finitely many of the* A(p) *are nonzero.*

PROOF. (i) Let $a,b \in A(p)$. If $\mathcal{O}_a = (p^r)$ and $\mathcal{O}_b = (p^s)$ let $k = \max(r,s)$. Then $p^k(a + b) = 0$, whence $\mathcal{O}_{a+b} = (p^i)$ with $0 \leq i \leq k$ by Theorem 6.4(iv). Therefore, $a,b \in A(p)$ imply $a + b \in A(p)$. A similar argument shows that $a \in A(p)$ and $r \in R$ imply $ra \in A(p)$. Therefore, $A(p)$ is a submodule.

(ii) Let $0 \neq a \in A$ with $\mathcal{O}_a = (r)$. By Theorem III.3.7 $r = p_1^{n_1} \cdots p_k^{n_k}$ with p_i distinct primes in R and each $n_i > 0$. For each i, let $r_i = p_1^{n_1} \cdots p_{i-1}^{n_{i-1}} p_{i+1}^{n_{i+1}} \cdots p_k^{n_k}$. Then $r_1, \ldots, r_k$ are relatively prime and there exist $s_1, \ldots, s_k \in R$ such that $s_1r_1 + \cdots + s_kr_k = 1_R$ (Theorem III.3.11). Consequently, $a = 1_Ra = s_1r_1a + \cdots + s_kr_ka$. But $p_i^{n_i}s_ir_ia = s_ira = 0$, whence $s_ir_ia \in A(p_i)$. We have proved that the submodules $A(p)$ (p prime) generate the module A.

Let $p \in R$ be prime and let A_1 be the submodule of A generated by all $A(q)$ with $q \neq p$. Suppose $a \in A(p) \cap A_1$. Then $p^ma = 0$ for some $m \geq 0$ and $a = a_1 + \cdots + a_t$ with $a_i \in A(q_i)$ for some primes $q_1, \ldots, q_t$ all distinct from p. Since $a_i \in A(q_i)$, there are integers m_i such that $q_i^{m_i}a_i = 0$, whence $(q_1^{m_1} \cdots q_t^{m_t})a = 0$. If $d = q_1^{m_1} \cdots q_t^{m_t}$, then p^m and d are relatively prime and $rp^m + sd = 1_R$ for some $r,s \in R$. Consequently, $a = 1_Ra = rp^ma + sda = 0$. Therefore, $A(p) \cap A_1 = 0$ and $A = \sum A(p)$ by Theorem 1.15. The last statement of the Theorem is a consequence of the easily verified fact that a direct sum of modules with infinitely many nonzero summands cannot be finitely generated. For each generator has only finitely many nonzero coordinates. ■

In order to determine the structure of finitely generated modules in which every element has order a power of a prime p (such as $A(p)$ in Theorem 6.7), we shall need a lemma. If A is an R-module and $r \in R$, then rA is the set $\{ra \mid a \in A\}$.

Lemma 6.8. *Let* A *be a module over a principal ideal domain* R *such that* $p^nA = 0$ *and* $p^{n-1}A \neq 0$ *for some prime* p ε R *and positive integer* n. *Let* a *be an element of* A *of order* p^n.

(i) *If* A ≠ Ra, *then there exists a nonzero* b ε A *such that* Ra ∩ Rb = 0.

(ii) *There is a submodule* C *of* A *such that* A = Ra ⊕ C.

REMARK. The following proof is quite elementary. A more elegant proof of (ii), which uses the concept of injectivity, is given in Exercise 7.

PROOF OF 6.8. (G. S. Monk) (i) If $A \neq Ra$, then there exists $c \in A - Ra$. Since $p^nc \in p^nA = 0$, there is a least *positive* integer j such that $p^jc \in Ra$, whence $p^{j-1}c \notin Ra$ and $p^jc = r_1a$ ($r_1 \in R$). Since R is a unique factorization domain $r_1 = rp^k$ for some $k \geq 0$ and $r \in R$ such that $p \nmid r$. Consequently, $0 = p^nc = p^{n-j}(p^jc) = p^{n-j}rp^ka$. Since $p \nmid r$ and $p^{n-1}a \neq 0$ (Theorem 6.4(v)), we must have $n - j + k \geq n$, whence $k \geq j \geq 1$. Therefore, $b = p^{j-1}c - rp^{k-1}a$ is a well-defined element of A.

Furthermore, $b \neq 0$ (since $p^{j-1}c \notin Ra$) and $pb = p^jc - rp^ka = p^jc - r_1a = 0$. If $Ra \cap Rb \neq 0$, then there exists $s \in R$ such that $sb \in Ra$ and $sb \neq 0$. Since $sb \neq 0$ and $pb = 0$, p does not divide s. Therefore, s and p^n are relatively prime and $sx + p^ny = 1_R$ for some $x,y \in R$ (Theorem III.3.11). Thus since $p^nA = 0$, $b = 1_Rb = sxb + p^nyb = x(sb) \in Ra$. Consequently, $p^{j-1}c = b + rp^{k-1}a \in Ra$. If $j - 1 \neq 0$, this contradicts the minimality of j, and if $j - 1 = 0$, this contradicts the fact that $c \notin Ra$. Therefore, $Ra \cap Rb = 0$.

(ii) If $A = Ra$, let $C = 0$. If $A \neq Ra$, then let $\mathcal{S}$ be the set of all submodules B of A such that $Ra \cap B = 0$. $\mathcal{S}$ is nonempty since by (i) there is a nonzero $b \in A$ such that $Ra \cap Rb = 0$. Partially order $\mathcal{S}$ by set-theoretic inclusion and verify that every chain in $\mathcal{S}$ has an upper bound in $\mathcal{S}$. By Zorn's Lemma there exists a submodule C of A that is maximal in $\mathcal{S}$. Consider the quotient module A/C. Clearly $p^n(A/C) = 0$ and $p^n(a + C) = 0$. Since $Ra \cap C = 0$ and $p^{n-1}a \neq 0$, we have $p^{n-1}(a + C) \neq C$, whence $a + C$ has order p^n in A/C and $p^{n-1}(A/C) \neq 0$. Now if A/C is *not* the cyclic R-module generated by $a + C$ (that is, $A/C \neq R(a + C)$), then by (i) there exists $d + C \in A/C$ such that $d + C \neq C$ and $R(a + C) \cap R(d + C) = C$. Since $Ra \cap C = 0$, it follows that $Ra \cap (Rd + C) = 0$. Since $d \notin C$, $Rd + C$ is in $\mathcal{S}$ and properly contains C, which contradicts the maximality of C. Therefore, A/C is the cyclic R-module generated by $a + C$ (that is, $A/C = R(a + C)$). Consequently, $A = Ra + C$, whence $A = Ra \oplus C$ by Theorem 1.15. ■

Theorem 6.9. *Let* A *be a finitely generated module over a principal ideal domain* R *such that every element of* A *has order a power of some prime* $p \in R$. *Then* A *is a direct sum of cyclic* R*-modules of orders* $p^{n_1}, \dots, p^{n_k}$ *respectively, where* $n_1 \geq n_2 \geq \cdots \geq n_k \geq 1$.

PROOF. The proof proceeds by induction on the number r of generators of A, with the case $r = 1$ being trivial. If $r > 1$, then A is generated by elements $a_1, \dots, a_r$ whose orders are respectively $p^{n_1}, p^{m_2}, p^{m_3}, \dots, p^{m_r}$. We may assume that

$$n_1 = \max\{n_1, m_2, \dots, m_r\}.$$

Then $p^{n_1}A = 0$ and $p^{n_1-1}A \neq 0$. By Lemma 6.8 there is a submodule C of A such that $A = Ra_1 \oplus C$. Let π be the canonical epimorphism $\pi : A \to C$. Since A is generated by $a_1, a_2, \dots, a_r$, C must be generated by $\pi(a_1), \pi(a_2), \dots, \pi(a_r)$. But $\pi(a_1) = 0$, whence C may be generated by $r - 1$ or fewer elements. Consequently, the induction hypothesis implies that C is a direct sum of cyclic R-modules of orders $p^{n_2}, p^{n_3}, \dots, p^{n_k}$ respectively with $n_2 \geq n_3 \geq \cdots \geq n_k \geq 1$. Thus C contains an element of order n_2. Since $p^{n_1}A = 0$, we have $p^{n_1}C = 0$, whence $n_1 \geq n_2$. Since Ra_1 is a cyclic R-module of order p^{n_1}, A is a direct sum of cyclic R-modules of orders $p^{n_1}, p^{n_2}, \dots, p^{n_k}$ respectively with $n_1 \geq n_2 \geq \cdots \geq n_k \geq 1$. ■

Theorems 6.6, 6.7, and 6.9 immediately yield a structure theorem for finitely generated modules over a principal ideal domain (see Theorem 6.12(ii) below). Just as in the case of abelian groups (Section II.2), there is a second way of decomposing a finitely generated module as a direct sum of cyclic submodules. In order to obtain this second decomposition and to prove a uniqueness theorem about each of the decompositions, we need two lemmas.

Lemma 6.10. *Let* A,B, *and* A_i ($i \in I$) *be modules over a principal ideal domain* R. *Let* $r \in R$ *and let* $p \in R$ *be prime.*

(i) $rA = \{ra \mid a \in A\}$ *and* $A[r] = \{a \in A \mid ra = 0\}$ *are submodules of* A.
(ii) $R/(p)$ *is a field and* $A[p]$ *is a vector space over* $R/(p)$.
(iii) *For each positive integer* n *there are* R-*module isomorphisms*

$$(R/(p^n))[p] \cong R/(p) \textit{ and } p^m(R/(p^n)) \cong R/(p^{n-m}) \ (0 \leq m < n).$$

(iv) *If* $A \cong \sum_{i \in I} A_i$, *then* $rA \cong \sum_{i \in I} rA_i$ *and* $A[r] \cong \sum_{i \in I} A_i[r]$.
(v) *If* $f : A \to B$ *is an* R-*module isomorphism, then* $f : A_t \cong B_t$ *and* $f : A(p) \cong B(p)$.

SKETCH OF PROOF. (ii) Exercise 2.4. (v) See Lemma II.2.5 (vii). (iii) The first example preceding Theorem 6.5 may be helpful. Verify that $(R/(p^n))[p]$ is generated as an R-module (and hence as a vector space over $R/(p)$) by the single nonzero element $p^{n-1} + (p^n)$. Therefore, $(R/(p^n))[p] \cong R/(p)$ by Theorems 2.5 and 2.1. The submodule of $R/(p^n)$ generated by $p^m + (p^n)$ is precisely $p^m(R/(p^n))$. Since $p^m + (p^n)$ has order p^{n-m}, we have $p^m(R/(p^n)) \cong R/(p^{n-m})$ by Theorem 6.4(iii). ■

Lemma 6.11. *Let* R *be a principal ideal domain. If* $r \in R$ *factors as* $r = p_1^{n_1} \cdots p_k^{n_k}$ *with* $p_1, \ldots, p_k \in R$ *distinct primes and each* $n_i > 0$, *then there is an* R-*module isomorphism*

$$R/(r) \cong R/(p_1^{n_1}) \oplus \cdots \oplus R/(p_k^{n_k}).$$

Consequently every cyclic R-*module of order* r *is a direct sum of* k *cyclic* R-*modules of orders* $p_1^{n_1}, \ldots, p_k^{n_k}$ *respectively.*

SKETCH OF PROOF. We shall prove that if $s,t \in R$ are relatively prime, then $R/(st) \cong R/(s) \oplus R/(t)$. The first part of the lemma then follows by induction on the number of distinct primes in the prime decomposition of r. The last statement of the lemma is an immediate consequence of the fact that $R/(c)$ is a cyclic R-module of order c for each $c \in R$ by Theorem 6.4. The map $\theta : R \to R$ given by $x \mapsto tx$ is an R-module monomorphism that takes the ideal (s) onto the ideal (st). By Corollary 1.8 θ induces an R-module homomorphism $R/(s) \to R/(st)$ given by $x + (s) \mapsto tx + (st)$. Similarly there is a homomorphism $R/(t) \to R/(st)$ given by $x + (t) \mapsto sx + (st)$. By the proof of Theorem 1.13 the map $\alpha : R/(s) \oplus R/(t) \to R/(st)$ given by $(x + (s), y + (t)) \mapsto [tx + sy] + (st)$ is a well-defined R-module homomorphism. Since $(s,t) = 1_R$, there exist $u,v \in R$ such that $su + tv = 1_R$ (Theorem III.3.11). If $c \in R$, then $c = suc + tvc$, whence $\alpha(vc + (s), uc + (t)) = c + (st)$. Therefore, α is an epimorphism. In order to show that α is a monomorphism we must show that

$$\alpha(x + (s), y + (t)) = 0 \quad \Rightarrow \quad x \in (s) \quad \text{and} \quad y \in (t).$$

If $\alpha(x + (s), y + (t)) = 0$, then $tx + sy = stb \in (st)$ for some $b \in R$. Hence $utx + usy = ustb$. But $y = 1_R y = (su + tv)y$, whence $utx + (y - tvy) = ustb$ and $y = ustb - utx + tvy \in (t)$. A similar argument shows that $x \in (s)$. ■

Theorem 6.12. *Let* A *be a finitely generated module over a principal ideal domain* R.

(i) A *is the direct sum of a free submodule* F *of finite rank and a finite number of cyclic torsion modules. The cyclic torsion summands (if any) are of orders* $r_1, \ldots, r_t$, *where* $r_1, \ldots, r_t$ *are (not necessarily distinct) nonzero nonunit elements of* R *such that* $r_1 \mid r_2 \mid \cdots \mid r_t$. *The rank of* F *and the list of ideals* $(r_1), \ldots, (r_t)$ *are uniquely determined by* A.

(ii) A *is the direct sum of a free submodule* E *of finite rank and a finite number of cyclic torsion modules. The cyclic torsion summands (if any) are of orders* $p_1^{s_1}, \ldots, p_k^{s_k}$, *where* $p_1, \ldots, p_k$ *are (not necessarily distinct) primes in* R *and* $s_1, \ldots, s_k$ *are (not necessarily distinct) positive integers. The rank of* E *and the list of ideals* $(p_1^{s_1}), \ldots, (p_k^{s_k})$ *are uniquely determined by* A *(except for the order of the* p_i*).*

The notation $r_1|r_2|\cdots|r_t$ means r_1 divides r_2, r_2 divides r_3, etc. The elements $r_1, \ldots, r_t$ in Theorem 6.12 are called the **invariant factors** of the module A just as in the special case of abelian groups. Similarly $p_1^{s_1}, \ldots, p_k^{s_k}$ are called the **elementary divisors** of A.

SKETCH OF PROOF OF 6.12. The existence of a direct sum decomposition of the type described in (ii) is an immediate consequence of Theorems 6.6, 6.7, and 6.9. Thus A is the direct sum of a free module and a finite family of cyclic R-modules, each of which has order a power of a prime. In the case of abelian groups these prime powers are precisely the elementary divisors of A. The method of calculating the invariant factors of an abelian group from its elementary divisors (see pp. 80-81) may be used here, *mutatis mutandis*, to prove the existence of a direct sum decomposition of A of the type described in (i). One need only make the following modifications. The role of $Z_{p^n} \cong \mathbf{Z}/(p^n)$ ($p \in \mathbf{Z}$ prime) is played by a cyclic torsion submodule of A of order p^n ($p \in R$ prime). Such a cyclic torsion module is isomorphic to $R/(p^n)$ by Theorem 6.4(iii). Lemma II.2.3 is replaced by Lemma 6.11.

The proof of the uniqueness of the direct sum decompositions in (i) and (ii) is essentially the same as the proof of the corresponding facts for abelian groups (Theorem II.2.6). The following modifications of the argument are necessary. First of all prime factorization in R is unique only up to multiplication by a unit (Definition III.3.5 and Theorem III.3.7). This causes no difficulty in $\mathbf{Z}$ since the only units are ± 1 and primes are defined to be positive. In an arbitrary principal ideal domain R, however, an element $a \in R$ may have order p and order q with p,q distinct primes. However, since $(p) = \mathcal{O}_a = (q)$, p and q are associates by Theorem III.3.2; that is, $q = pu$ with $u \in R$ a unit. Hence the uniqueness statements in (i) and (ii) deal with ideals rather than elements. Note that $a \neq 0$ implies that $\mathcal{O}_a \neq R$ and that a cyclic module Ra is free if and only if $\mathcal{O}_a = (0)$. Thus the elements r_i in (i) are nonzero nonunits. Other modifications: as above replace each finite cyclic summand $Z_n \cong \mathbf{Z}/(n)$ with $n > 1$ by a cyclic torsion module $R/(r)$ ($r \in R$ a nonzero nonunit). Replace the subgroup generated by the infinite cyclic summands $\mathbf{Z}$ by a free R-module of finite rank. Use Lemmas 6.10 and 6.11 in place of Lemmas II.2.3 and II.2.5. Instead of the counting argument on p. 79 (showing that $r = d$) use the fact that $A[p]$ is a vector space over $R/(p)$. Hence the number of summands $R/(p)$ is precisely $\dim_{R/(p)} A[p]$, which is invariant by Theorem 2.7. ■

Corollary 6.13. *Two finitely generated modules over a principal ideal domain,* A *and* B, *are isomorphic if and only if* A/A_t *and* B/B_t *have the same rank and* A *and* B *have the same invariant factors* [*resp. elementary divisors*].

PROOF. Exercise. ■

EXERCISES

Note: Unless stated otherwise, R is a principal ideal domain and all modules are unitary.

1. If R is a nonzero commutative ring with identity and every submodule of every free R-module is free, then R is a principal ideal domain. [*Hint:* Every ideal I of R is a free R-module. If $u,v \in I$ ($u \neq 0, v \neq 0$), then $uv + (-v)u = 0$, which implies that I has a basis of one element; that is, I is principal.]

2. Every free module over an arbitrary integral domain with identity is torsion-free. The converse is false (Exercise II.1.10).

3. Let A be a cyclic R-module of order $r \in R$.
   (a) If $s \in R$ is relatively prime to r, then $sA = A$ and $A[s] = 0$.
   (b) If s divides r, say $sk = r$, then $sA \cong R/(k)$ and $A[s] \cong R/(s)$.

4. If A is a cyclic R-module of order r, then (i) every submodule of A is cyclic, with order dividing r; (ii) for every ideal (s) containing (r), A has exactly one submodule, which is cyclic of order s.

5. If A is a finitely generated torsion module, then $\{r \in R \mid rA = 0\}$ is a nonzero ideal in R, say (r_1). r_1 is called the **minimal annihilator** of A. Let A be a finite abelian group with minimal annihilator $m \in \mathbf{Z}$. Show that a cyclic subgroup of A of order properly dividing m need not be a direct summand of A.

6. If A and B are cyclic modules over R of nonzero orders r and s respectively, and r is *not* relatively prime to s, then the invariant factors of $A \oplus B$ are the greatest common divisor of r,s and the least common multiple of r,s.

7. Let A and $a \in A$ satisfy the hypotheses of Lemma 6.8.
   (a) Every R-submodule of A is an $R/(p^n)$-module with $(r + (p^n))a = ra$. Conversely, every $R/(p^n)$-submodule of A is an R-submodule by pullback along $R \to R/(p^n)$.
   (b) The submodule Ra is isomorphic to $R/(p^n)$.
   (c) The only proper ideals of the ring $R/(p^n)$ are the ideals generated by $p^i + (p^n)$ $(i = 1,2,\ldots,n-1)$.
   (d) $R/(p^n)$ (and hence Ra) is an injective $R/(p^n)$-module. [*Hint:* use (c) and Lemma 3.8.]
   (e) There exists an R-submodule C of A such that $A = Ra \oplus C$. [*Hint:* Proposition 3.13.]

7. ALGEBRAS

Algebras are introduced and their basic properties developed. Tensor products are used extensively in this discussion. Algebras will be studied further in Chapter IX.

Definition 7.1. *Let* K *be a commutative ring with identity. A* **K-algebra** (*or* **algebra over K**) A *is a ring* A *such that:*

(i) (A,+) *is a unitary* (*left*) K*-module;*
(ii) k(ab) = (ka)b = a(kb) *for all* k ε K *and* a,b ε A.

A K*-algebra* A *which, as a ring, is a division ring, is called a* **division algebra**.

The classical theory of algebras deals with algebras over a field K. Such an algebra is a vector space over K and hence various results of linear algebra are applicable. An algebra over a field K that is finite dimensional as a vector space over K is called a **finite dimensional algebra** over K.

EXAMPLE. Every ring R is an additive abelian group and hence a **Z**-module. It is easy to see that R is actually a **Z**-algebra.

EXAMPLES. If K is a commutative ring with identity, then the polynomial ring $K[x_1, \dots, x_n]$ and the power series ring $K[[x]]$ are K-algebras, with the respective K-module structures given in the usual way.

EXAMPLE. If V is a vector space over a field F, then the endomorphism ring $\text{Hom}_F(V,V)$ (Exercise 1.7) is an F-algebra. The F-module structure of $\text{Hom}_F(V,V)$ is discussed in the Remark after Theorem 4.8.

EXAMPLES. Let A be a ring with identity and K a subring of the center of A such that $1_A \varepsilon K$. Then A is a K-algebra, with the K-module structure being given by multiplication in A. In particular, every commutative ring K with identity is a K-algebra.

EXAMPLE. Both the field of complex numbers **C** and the division ring of real quaternions (p. 117) are division algebras over the field **R** of real numbers.

EXAMPLE. Let G be a multiplicative group and K a commutative ring with identity. Then the group ring $K(G)$ (p. 117) is actually a K-algebra with K-module structure given by

$$k(\textstyle\sum r_i g_i) = \sum (kr_i)g_i \qquad (k,r_i \varepsilon K;\, g_i \varepsilon G).$$

$K(G)$ is called the **group algebra** of G over K.

EXAMPLE. If K is a commutative ring with identity, then the ring $\text{Mat}_n K$ of all $n \times n$ matrices over K is a K-algebra with the K-module action of K given in the usual way. More generally, if A is a K-algebra, then so is $\text{Mat}_n A$.

REMARK. Since K is commutative, every left K-module (and hence every K-algebra) A is also a right K module with $ka = ak$ for all $a \varepsilon A$, $k \varepsilon K$. This fact is implicitly assumed in Theorems 7.2 and 7.4 below, where tensor products are used.

The motivation for the next theorem, which provides another means of defining K algebras, is the fact that for any ring R the unique map $R \otimes_{\mathbf{Z}} R \to R$, defined on a generator $r \otimes s$ by $r \otimes s \mapsto rs$, is a homomorphism of additive abelian groups. Since rings are simply **Z**-algebras, this fact is a special case of

Theorem 7.2. *Let* K *be a commutative ring with identity and* A *a unitary left* K*-module. Then* A *is a* K*-algebra if and only if there exists a* K*-module homomorphism* $\pi : A \otimes_K A \to A$ *such that the diagram*

$$\begin{array}{ccc} A \otimes_K A \otimes_K A & \xrightarrow{\pi \otimes 1_A} & A \otimes_K A \\ \Big\downarrow{\scriptstyle 1_A \otimes \pi} & & \Big\downarrow{\scriptstyle \pi} \\ A \otimes_K A & \xrightarrow{\pi} & A \end{array}$$

is commutative. In this case the K*-algebra* A *has an identity if and only if there is a* K*-module homomorphism* $I : K \to A$ *such that the diagram*

$$\begin{array}{ccccc} K \otimes_K A & \xrightarrow{\zeta} & A & \xleftarrow{\theta} & A \otimes_K K \\ \Big\downarrow{\scriptstyle I \otimes 1_A} & & \Big\downarrow{\scriptstyle 1_A} & & \Big\downarrow{\scriptstyle 1_A \otimes I} \\ A \otimes_K A & \xrightarrow{\pi} & A & \xleftarrow{\pi} & A \otimes_K A \end{array}$$

is commutative, where ζ,θ *are the isomorphisms of Theorem* 5.7.

SKETCH OF PROOF. If A is a K-algebra, then the map $A \times A \to A$ given by $(a,b) \mapsto ab$ is K-bilinear, whence there is a K-module homomorphism

$$\pi : A \otimes_K A \to A$$

by Theorem 5.6. Verify that π has the required properties. If A has an identity 1_A, then the map $I : K \to A$ given by $k \mapsto k1_A$ is easily seen to be a K-module homomorphism with the required properties. Conversely, given A and the map $\pi : A \otimes_K A \to A$, define $ab = \pi(a \otimes b)$ and verify that A is a K-algebra. If $I : K \to A$ is also given, then $I(1_K)$ is an identity for A. ■

The homomorphism π of Theorem 7.2 is called the **product map** of the K-algebra A. The homomorphism I is called the **unit map.**

Definition 7.3. *Let* K *be a commutative ring with identity and* A, B K*-algebras.*

(i) *A* **subalgebra** *of* A *is a subring of* A *that is also a* K*-submodule of* A.

(ii) *A (left, right, two-sided)* **algebra ideal** *of* A *is a (left, right, two-sided) ideal of the ring* A *that is also a* K*-submodule of* A.

(iii) *A* **homomorphism** [*resp.* **isomorphism**] **of K-algebras** $f : A \to B$ *is a ring homomorphism [isomorphism] that is also a* K*-module homomorphism [isomorphism].*

REMARKS. If A is a K-algebra, an ideal of the ring A need not be an algebra ideal of A (Exercise 4). If, however, A has an identity, then for all $k \in K$ and $a \in A$

$$ka = k(1_A a) = (k1_A)a \quad \text{and} \quad ka = (ka)1_A = a(k1_A),$$

with $k1_A \in A$. Consequently, for a left [resp. right] ideal J in the ring A,

$$kJ = (k1_A)J \subset J \quad [\text{resp. } kJ = J(k1_A) \subset J].$$

Therefore, *if A has an identity, every (left, right, two-sided) ideal is also a (left, right, two-sided) algebra ideal.*

The quotient algebra of a K-algebra A by an algebra ideal I is now defined in the obvious way, as are the direct product and direct sum of a family of K-algebras.

Tensor products furnish another way to manufacture new algebras. We first observe that if A and B are K-modules, then there is a K-module isomorphism $\alpha : A \otimes_K B \to B \otimes_K A$ such that $\alpha(a \otimes b) = b \otimes a$ $(a \in A, b \in B)$; see Exercise 2.

Theorem 7.4. *Let A and B be algebras [with identity] over a commutative ring K with identity. Let π be the composition*

$$(A \otimes_K B) \otimes_K (A \otimes_K B) \xrightarrow{1_A \otimes \alpha \otimes 1_B} (A \otimes_K A) \otimes_K (B \otimes_K B) \xrightarrow{\pi_A \otimes \pi_B} A \otimes_K B,$$

where π_A, π_B are the product maps of A and B respectively. Then $A \otimes_K B$ is a K-algebra [with identity] with product map π.

PROOF. Exercise; note that for generators $a \otimes b$ and $a_1 \otimes b_1$ of $A \otimes_K B$ the product is defined to be

$$(a \otimes b)(a_1 \otimes b_1) = \pi(a \otimes b \otimes a_1 \otimes b_1) = aa_1 \otimes bb_1.$$

Thus if A and B have identities 1_A, 1_B respectively, then $1_A \otimes 1_B$ is the identity in $A \otimes_K B$. ■

The K-algebra $A \otimes_K B$ of Theorem 7.4 is called the **tensor product of the K-algebras** A and B. Tensor products of algebras are useful in studying the structure of division algebras over a field K (Section IX.6).

EXERCISES

Note: K is always a commutative ring with identity.

1. Let $\mathcal{C}$ be the category whose objects are all commutative K-algebras with identity and whose morphisms are all K-algebra homomorphisms $f: A \to B$ such that $f(1_A) = 1_B$. Then any two K-algebras A, B of $\mathcal{C}$ have a coproduct. [*Hint:* consider $A \to A \otimes_K B \leftarrow B$, where $a \mapsto a \otimes 1_B$ and $b \mapsto 1_A \otimes b$.]

2. If A and B are unitary K-modules [resp. K-algebras], then there is an isomorphism of K-modules [resp. K-algebras] $\alpha : A \otimes_K B \to B \otimes_K A$ such that $\alpha(a \otimes b) = b \otimes a$ for all $a \in A, b \in B$.

3. Let A be a ring with identity. Then A is a K-algebra with identity if and only if there is a ring homomorphism of K into the center of A such that $1_K \mapsto 1_A$.

4. Let A be a one-dimensional vector space over the rational field $\mathbf{Q}$. If we define $ab = 0$ for all $a,b \in A$, then A is a $\mathbf{Q}$-algebra. Every proper additive subgroup of A is an ideal of the ring A, but not an algebra ideal.

5. Let $\mathcal{C}$ be the category of Exercise 1. If X is the set $\{x_1, \ldots, x_n\}$, then the polynomial algebra $K[x_1, \ldots, x_n]$ is a free object on the set X in the category $\mathcal{C}$. [*Hint:* Given an algebra A in $\mathcal{C}$ and a map $g : \{x_1, \ldots, x_n\} \to A$, apply Theorem III.5.5 to the unit map $I : K \to A$ and the elements $g(x_1), \ldots, g(x_n) \in A$.]

CHAPTER V

FIELDS AND GALOIS THEORY

The first principal theme of this chapter is the structure theory of fields. We shall study a field F in terms of a specified subfield K (F is said to be an extension field of K). The basic facts about field extensions are developed in Section 1, in particular, the distinction between algebraic and transcendental extensions. For the most part we deal only with algebraic extensions in this chapter. Arbitrary field extensions are considered in Chapter VI. The structure of certain fields and field extensions is thoroughly analyzed: simple extensions (Section 1); splitting fields (normal extensions) and algebraic closures (Section 3); finite fields (Section 5); and separable algebraic extensions (Sections 3 and 6).

The Galois theory of field extensions (the other main theme of this chapter) had its historical origin in a classical problem in the theory of equations, which is discussed in detail in Sections 4 and 9. Various results of Galois theory have important applications, especially in the study of algebraic numbers (see E. Artin [48]) and algebraic geometry (see S. Lang [54]).

The key idea of Galois theory is to relate a field extension $K \subset F$ to the group of all automorphisms of F that fix K elementwise (the Galois group of the extension). A Galois field extension may be defined in terms of its Galois group (Section 2) or in terms of the internal structure of the extension (Section 3). The Fundamental Theorem of Galois theory (Section 2) states that there is a one-to-one correspondence between the intermediate fields of a (finite dimensional) Galois field extension and the subgroups of the Galois group of the extension. This theorem allows us to translate properties and problems involving fields, polynomials, and field extensions into group theoretic terms. Frequently, the corresponding problem in groups has a solution, whence the original problem in field theory can be solved. This is the case, for instance, with the classical problem in the theory of equations mentioned in the previous paragraph. We shall characterize those Galois field extensions whose Galois groups are finite cyclic (Section 7) or solvable (Section 9).

The approximate interdependence of the sections of this chapter is as follows:

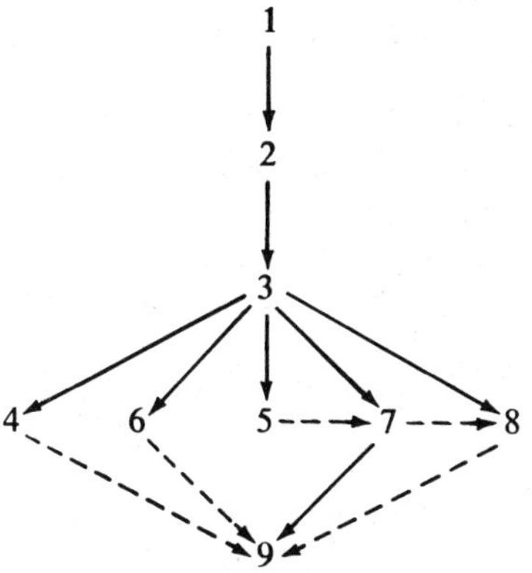

A broken arrow $A \dashrightarrow B$ indicates that an occasional result from section A is used in section B, but that section B is essentially independent of section A. *See page* xviii *for a description of a short basic course in fields and Galois theory.*

1. FIELD EXTENSIONS

The basic facts needed for the study of field extensions are presented first, followed by a discussion of simple extensions. Finally a number of essential properties of algebraic extensions are proved. In the appendix, which is not used in the sequel, several famous geometric problems of antiquity are settled, such as the trisection of an angle by ruler and compass constructions.

Definition 1.1. *A field* F *is said to be an* **extension field** *of* K *(or simply an extension of* K*) provided that* K *is a subfield of* F.

If F is an extension field of K, then it is easy to see that $1_K = 1_F$. Furthermore, F is a vector space over K (Definition IV.1.1). Throughout this chapter the dimension of the K-vector space F will be denoted by $[F : K]$ rather than $\dim_K F$ as previously. F is said to be a **finite dimensional extension** or **infinite dimensional extension** of K according as $[F : K]$ is finite or infinite.

Theorem 1.2. *Let* F *be an extension field of* E *and* E *an extension field of* K. *Then* $[F : K] = [F : E][E : K]$. *Furthermore* $[F : K]$ *is finite if and only if* $[F : E]$ *and* $[E : K]$ *are finite.*

PROOF. This is a restatement of Theorem IV.2.16. ■

In the situation $K \subset E \subset F$ of Theorem 1.2, E is said to be an **intermediate field** of K and F.

If F is a field and $X \subset F$, then the **subfield** [resp. **subring**] **generated by X** is the intersection of all subfields [resp. subrings] of F that contain X. If F is an extension field of K and $X \subset F$, then the subfield [resp. subring] generated by $K \cup X$ is called the **subfield** [resp. **subring**] **generated by X over K** and is denoted $K(X)$ [resp. $K[X]$]. Note that $K[X]$ is necessarily an integral domain.

If $X = \{u_1, \ldots, u_n\}$, then the subfield $K(X)$ [resp. subring $K[X]$] of F is denoted $K(u_1, \ldots, u_n)$ [resp. $K[u_1, \ldots, u_n]$]. The field $K(u_1, \ldots, u_n)$ is said to be a **finitely generated extension** of K (but it need not be finite dimensional over K; see Exercise 2). If $X = \{u\}$, then $K(u)$ is said to be a **simple extension** of K. A routine verification shows that neither $K(u_1, \ldots, u_n)$ nor $K[u_1, \ldots, u_n]$ depends on the order of the u_i and that $K(u_1, \ldots, u_{n-1})(u_n) = K(u_1, \ldots, u_n)$ and $K[u_1, \ldots, u_{n-1}][u_n] = K[u_1, \ldots, u_n]$ (Exercise 4). These facts will be used frequently in the sequel without explicit mention.

NOTATION. If F is a field $u,v \in F$, and $v \neq 0$, then $uv^{-1} \in F$ will sometimes be denoted by u/v.

Theorem 1.3. *If* F *is an extension field of a field* K, u, $u_i \in F$, *and* $X \subset F$, *then*

(i) *the subring* K[u] *consists of all elements of the form* f(u), *where* f *is a polynomial with coefficients in* K (*that is,* $f \in K[x]$);

(ii) *the subring* $K[u_1, \ldots, u_m]$ *consists of all elements of the form* $g(u_1,u_2, \ldots, u_m)$, *where* g *is a polynomial in* m *indeterminates with coefficients in* K (*that is,* $g \in K[x_1, \ldots, x_m]$);

(iii) *the subring* K[X] *consists of all elements of the form* $h(u_1, \ldots, u_n)$, *where each* $u_i \in X$, n *is a positive integer, and* h *is a polynomial in* n *indeterminates with coefficients in* K (*that is,* $n \in N^*$, $h \in K[x_1, \ldots, x_n]$);

(iv) *the subfield* K(u) *consists of all elements of the form* $f(u)/g(u) = f(u)g(u)^{-1}$, *where* $f,g \in K[x]$ *and* $g(u) \neq 0$;

(v) *the subfield* $K(u_1, \ldots, u_m)$ *consists of all elements of the form*

$$h(u_1, \ldots, u_m)/k(u_1, \ldots, u_m) = h(u_1, \ldots, u_m)k(u_1, \ldots, u_m)^{-1},$$

where $h,k \in K[x_1, \ldots, x_m]$ *and* $k(u_1, \ldots, u_m) \neq 0$;

(vi) *the subfield* K(X) *consists of all elements of the form*

$$f(u_1, \ldots, u_n)/g(u_1, \ldots, u_n) = f(u_1, \ldots, u_n)g(u_1, \ldots, u_n)^{-1}$$

where $n \in N^*$, $f,g \in K[x_1, \ldots, x_n]$, $u_1, \ldots, u_n \in X$ *and* $g(u_1, \ldots, u_n) \neq 0$.

(vii) *For each* $v \in K(X)$ (*resp.* K[X]) *there is a finite subset* X′ *of* X *such that* $v \in K(X')$ (*resp.* K[X′]).

SKETCH OF PROOF. (vi) Every field that contains K and X must contain the set $E = \{f(u_1, \ldots, u_n)/g(u_1, \ldots, u_n) \mid n \in N^*;\ f,g \in K[x_1, \ldots, x_n];\ u_i \in X;\ g(u_1, \ldots, u_n) \neq 0\}$, whence $K(X) \supset E$. Conversely, if $f,g \in K[x_1, \ldots, x_m]$ and $f_1,g_1 \in K[x_1, \ldots, x_n]$, then define $h,k \in K[x_1, \ldots, x_{m+n}]$ by

$$\begin{aligned} h(x_1, \ldots, x_{m+n}) &= f(x_1, \ldots, x_m)g_1(x_{m+1}, \ldots, x_{m+n}) \\ &\quad - g(x_1, \ldots, x_m)f_1(x_{m+1}, \ldots, x_{m+n}); \\ k(x_1, \ldots, x_{m+n}) &= g(x_1, \ldots, x_m)g_1(x_{m+1}, \ldots, x_{m+n}). \end{aligned}$$

Then for any $u_1, \ldots, u_m, v_1, \ldots, v_n \in X$ such that $g(u_1, \ldots, u_m) \neq 0, g_1(v_1, \ldots, v_n) \neq 0$,

$$\frac{f(u_1, \ldots, u_m)}{g(u_1, \ldots, u_m)} - \frac{f_1(v_1, \ldots, v_n)}{g_1(v_1, \ldots, v_n)} = \frac{h(u_1, \ldots, u_m,v_1, \ldots, v_n)}{k(u_1, \ldots, u_m,v_1, \ldots, v_n)} \in E.$$

Therefore, E is a group under addition (Theorem I.2.5). Similarly the nonzero elements of E form a group under multiplication, whence E is a field. Since $X \subset E$ and $K \subset E$, we have $K(X) \subset E$. Therefore, $K(X) = E$. (vii) If $u \in K(X)$, then by (vi) $u = f(u_1, \ldots, u_n)/g(u_1, \ldots, u_n) \in K(X')$, where $X' = \{u_1, \ldots, u_n\} \subset X$. ■

If L and M are subfields of a field F, the **composite** of L and M in F, denoted LM is the subfield generated by the set $L \cup M$. An immediate consequence of this definition is that $LM = L(M) = M(L)$. It is easy to show that if K is a subfield of $L \cap M$ such that $M = K(S)$ where $S \subset M$, then $LM = L(S)$ (Exercise 5). The relationships of the dimensions $[L : K]$, $[M : K]$, $[LM : K]$, etc. are considered in Exercises 20–21. The composite of any finite number of subfields $E_1, E_2, \ldots, E_n$ is defined to be the subfield generated by the set $E_1 \cup E_2 \cup \cdots \cup E_n$ and is denoted $E_1E_2 \cdots E_n$ (see Exercise 5).

The next step in the study of field extensions is to distinguish two fundamentally different situations that occur.

Definition 1.4. *Let F be an extension field of K. An element u of F is said to be* **algebraic** *over K provided that u is a root of some nonzero polynomial f ε K[x]. If u is not a root of any nonzero f ε K[x], u is said to be* **transcendental** *over K. F is called an* **algebraic extension** *of K if every element of F is algebraic over K. F is called a* **transcendental extension** *if at least one element of F is transcendental over K.*

REMARKS. If $u \in K$, then u is a root of $x - u \in K[x]$ and therefore algebraic over K. If $u \in F$ is algebraic over some subfield K' of K, then u is algebraic over K since $K'[x] \subset K[x]$. If $u \in F$ is a root of $f \in K[x]$ with leading coefficient $c \neq 0$, then u is also a root of $c^{-1}f$, which is a monic polynomial in $K[x]$. A transcendental extension may contain elements that are algebraic over K (in addition to the elements of K itself).

EXAMPLES. Let **Q**,**R** and **C** be the fields of rational, real, and complex numbers respectively. Then $i \in \mathbf{C}$ is algebraic over **Q** and hence over **R**; in fact, $\mathbf{C} = \mathbf{R}(i)$. It is a nontrivial fact that $\pi, e \in \mathbf{R}$ are transcendental over **Q**; see, for instance, I. Herstein [4].

EXAMPLE. If K is a field, then the polynomial ring $K[x_1, \ldots, x_n]$ is an integral domain (Theorem III.5.3). The quotient field of $K[x_1, \ldots, x_n]$ is denoted $K(x_1, \ldots, x_n)$. It consists of all fractions f/g, with $f,g \in K[x_1, \ldots, x_n]$ and $g \neq 0$, and the usual addition and multiplication (see Theorem III.4.3). $K(x_1, \ldots, x_n)$ is called the **field of rational functions** in $x_1, \ldots, x_n$ over K. In the field extension

$$K \subset K(x_1, \ldots, x_n)$$

each x_i is easily seen to be transcendental over K. In fact, every element of $K(x_1, \ldots, x_n)$ not in K itself is transcendental over K (Exercise 6).

In the next two theorems we shall characterize all simple field extensions up to isomorphism.

Theorem 1.5. *If F is an extension field of K and u ε F is transcendental over K, then there is an isomorphism of fields K(u) ≅ K(x) which is the identity on K.*

SKETCH OF PROOF. Since u is transcendental $f(u) \neq 0$, $g(u) \neq 0$ for all nonzero $f,g \in K[x]$. Consequently, the map $\varphi : K(x) \to F$ given by $f/g \mapsto f(u)/g(u) = f(u)g(u)^{-1}$ is a well-defined monomorphism of fields which is the identity on K. But Im $\varphi = K(u)$ by Theorem 1.3, whence $K(x) \cong K(u)$. ■

Theorem 1.6. *If* F *is an extension field of* K *and* $u \in F$ *is algebraic over* K, *then*

(i) $K(u) = K[u]$;

(ii) $K(u) \cong K[x]/(f)$, *where* $f \in K[x]$ *is an irreducible monic polynomial of degree* $n \geq 1$ *uniquely determined by the conditions that* $f(u) = 0$ *and* $g(u) = 0$ $(g \in K[x])$ *if and only if* f *divides* g;

(iii) $[K(u) : K] = n$;

(iv) $\{1_K, u, u^2, \ldots, u^{n-1}\}$ *is a basis of the vector space* K(u) *over* K;

(v) *every element of* K(u) *can be written uniquely in the form* $a_0 + a_1u + \cdots + a_{n-1}u^{n-1}$ $(a_i \in K)$.

PROOF. (i) and (ii) The map $\varphi : K[x] \to K[u]$ given by $g \mapsto g(u)$ is a nonzero ring epimorphism by Theorems III.5.5. and 1.3. Since $K[x]$ is a principal ideal domain (Corollary III.6.4), Ker $\varphi = (f)$ for some $f \in K[x]$ with $f(u) = 0$. Since u is algebraic, Ker $\varphi \neq 0$ and since $\varphi \neq 0$, Ker $\varphi \neq K[x]$. Hence $f \neq 0$ and deg $f \geq 1$. Furthermore, if c is the leading coefficient of f, then c is a unit in $K[x]$ (Corollary III.6.4), $c^{-1}f$ is monic, and $(f) = (c^{-1}f)$ (Theorem III.3.2). Consequently we may assume that f is monic. By the First Isomorphism Theorem (Corollary III.2.10),

$$K[x]/(f) = K[x]/\text{Ker } \varphi \cong \text{Im } \varphi = K[u].$$

Since $K[u]$ is an integral domain, the ideal (f) is prime in $K[x]$ by Theorem III.2.16. Theorem III.3.4 implies that f is irreducible and hence that the ideal (f) is maximal. Consequently, $K[x]/(f)$ is a field (Theorem III.2.20). Since $K(u)$ is the smallest subfield of F containing K and u and since $K(u) \supset K[u] \cong K[x]/(f)$, we must have $K(u) = K[u]$. The uniqueness of f follows from the facts that f is monic and

$$g(u) = 0 \quad \Leftrightarrow \quad g \in \text{Ker } \varphi = (f) \quad \Leftrightarrow \quad f \text{ divides } g.$$

(iv) Every element of $K(u) = K[u]$ is of the form $g(u)$ for some $g \in K[x]$ by Theorem 1.3. The division algorithm shows that $g = qf + h$ with $q,h \in K[x]$ and deg $h <$ deg f. Therefore, $g(u) = q(u)f(u) + h(u) = 0 + h(u) = h(u) = b_0 + b_1u + \cdots + b_mu^m$ with $m < n =$ deg f. Thus $\{1_K, u, \ldots, u^{n-1}\}$ spans the K-vector space $K(u)$. To see that $\{1_K, u, \ldots, u^{n-1}\}$ is linearly independent over K and hence a basis, suppose

$$a_0 + a_1u + \cdots + a_{n-1}u^{n-1} = 0 \qquad (a_i \in K).$$

Then $g = a_0 + a_1x + \cdots + a_{n-1}x^{n-1} \in K[x]$ has u as a root and has degree $\leq n - 1$. Since $f \mid g$ by (ii) and deg $f = n$, we must have $g = 0$; that is, $a_i = 0$ for all i, whence $\{1_K, u, \ldots, u^{n-1}\}$ is linearly independent. Therefore, $\{1_K, u, \ldots, u^{n-1}\}$ is a basis of $K(u)$.

(iii) is an immediate consequence of (iv). The equivalence of (iv) and (v) is a routine exercise. ■

Definition 1.7. *Let* F *be an extension field of* K *and* $u \in F$ *algebraic over* K. *The monic irreducible polynomial* f *of Theorem* 1.6 *is called* **the irreducible** (*or* **minimal** *or* **minimum) polynomial of u.** *The* **degree of u over K** *is deg* $f = [K(u) : K]$.

The following example illustrates how Theorem 1.6 and the techniques of its proof may be used for specific computations.

EXAMPLE. The polynomial $x^3 - 3x - 1$ is irreducible over **Q** (Theorem III.6.6 and Proposition III.6.8) and has real root u (Exercise III.6.16(d)). By Theorem 1.6 u has degree 3 over **Q** and $\{1,u,u^2\}$ is a basis of $\mathbf{Q}(u)$ over **Q**. The element $u^4 + 2u^3 + 3 \in \mathbf{Q}(u) = \mathbf{Q}[u]$ may be expressed as a linear combination (over **Q**) of the basis elements as follows. The division algorithm (that is, ordinary long division) in the ring $\mathbf{Q}[x]$ shows that

$$x^4 + 2x^3 + 3 = (x + 2)(x^3 - 3x - 1) + (3x^2 + 7x + 5),$$

whence

$$\begin{aligned} u^4 + 2u^3 + 3 &= (u + 2)(u^3 - 3u - 1) + (3u^2 + 7u + 5) \\ &= (u + 2)0 + (3u^2 + 7u + 5) \\ &= 3u^2 + 7u + 5. \end{aligned}$$

The multiplicative inverse of $3u^2 + 7u + 5$ in $\mathbf{Q}(u)$ may be calculated as follows. Since $x^3 - 3x - 1$ is irreducible in $\mathbf{Q}[x]$, the polynomials $x^3 - 3x - 1$ and $3x^2 + 7x + 5$ are relatively prime in $\mathbf{Q}[x]$. Consequently, by Theorem III.3.11 there exist $g(x), h(x) \in \mathbf{Q}[x]$ such that

$$(x^3 - 3x - 1)g(x) + (3x^2 + 7x + 5)h(x) = 1.$$

Therefore, since $u^3 - 3u - 1 = 0$ we have

$$(3u^2 + 7u + 5)h(u) = 1$$

so that $h(u) \in \mathbf{Q}[u]$ is the inverse of $3u^2 + 7u + 5$. The polynomials g and h may be explicitly computed via the Euclidean algorithm (Exercise III.3.13): $g(x) = -7/37x + 29/111$, and $h(x) = 7/111\ x^2 - 26/111\ x + 28/111$. Hence $h(u) = 7/111\ u^2 - 26/111\ u + 28/111$.

Suppose E is an extension field of K, F is an extension field of L, and $\sigma : K \to L$ is an isomorphism of fields. A recurrent question in the study of field extensions is: under what conditions can σ be extended to an isomorphism of E onto F. In other words, is there an isomorphism $\tau : E \to F$ such that $\tau \mid K = \sigma$? We shall answer this question now for simple extension fields and in so doing obtain criteria for two simple extensions $K(u)$ and $K(v)$ to be isomorphic (also see Exercise 16).

Recall that if $\sigma : R \to S$ is an isomorphism of rings, then the map $R[x] \to S[x]$ given by $\sum_i r_i x^i \mapsto \sum_i \sigma(r_i)x^i$ is also a ring isomorphism (Exercise III.5.1). Clearly this map extends σ. We shall denote the extended map $R[x] \to S[x]$ by σ also and the image of $f \in R[x]$ by σf.

Theorem 1.8. *Let $\sigma : K \to L$ be an isomorphism of fields, u an element of some extension field of K and v an element of some extension field of L. Assume either*

(i) *u is transcendental over K and v is transcendental over L; or*

(ii) *u is a root of an irreducible polynomial $f \in K[x]$ and v is a root of $\sigma f \in L[x]$.*

Then σ extends to an isomorphism of fields $K(u) \cong L(v)$ which maps u onto v.

SKETCH OF PROOF. (i) By the remarks preceding the theorem σ extends to an isomorphism $K[x] \cong L[x]$. Verify that this map in turn extends to an isomorphism $K(x) \rightarrow L(x)$ given by $h/g \mapsto \sigma h/\sigma g$. Therefore, by Theorem 1.5 we have $K(u) \cong K(x) \cong L(x) \cong L(v)$. The composite map extends σ and maps u onto v.

(ii) It suffices to assume that f is monic. Since $\sigma : K[x] \cong L[x]$ this implies that $\sigma f \varepsilon L[x]$ is monic irreducible. By the proof of Theorem 1.6 the maps

$$\varphi : K[x]/(f) \rightarrow K[u] = K(u) \text{ and } \psi : L[x]/(\sigma f) \rightarrow L[v] = L(v),$$

given respectively by $\varphi[g + (f)] = g(u)$ and $\psi[h + (\sigma f)] = h(v)$, are isomorphisms. The map $\theta : K[x]/(f) \rightarrow L[x]/(\sigma f)$ given by $\theta[g + (f)] = \sigma g + (\sigma f)$ is an isomorphism by Corollary III.2.11. Therefore the composite

$$K(u) \xrightarrow{\varphi^{-1}} K[x]/(f) \xrightarrow{\theta} L[x]/(\sigma f) \xrightarrow{\psi} L(v)$$

is an isomorphism of fields such that $g(u) \mapsto (\sigma g)(v)$. In particular, $\psi\theta\varphi^{-1}$ agrees with σ on K and maps u onto v (since $\sigma(1_K) = 1_L$ by Exercise III. 1.15). ∎

Corollary 1.9. *Let* E *and* F *each be extension fields of* K *and let* u ε E *and* v ε F *be algebraic over* K. *Then* u *and* v *are roots of the same irreducible polynomial* f ε K[x] *if and only if there is an isomorphism of fields* K(u) ≅ K(v) *which sends* u *onto* v *and is the identity on* K.

PROOF. (⇒) Apply Theorem 1.8 with $\sigma = 1_K$ (so that $\sigma f = f$ for all $f \varepsilon K[x]$).

(⇐) Suppose $\sigma : K(u) \cong K(v)$ with $\sigma(u) = v$ and $\sigma(k) = k$ for all $k \varepsilon K$. Let $f \varepsilon K[x]$ be the irreducible polynomial of the algebraic element u. If $f = \sum_{i=0}^{n} k_i x^i$, then $0 = f(u) = \sum_{i=0}^{n} k_i u^i$. Therefore, $0 = \sigma\left(\sum_{i=0}^{n} k_i u^i\right) = \sum_i \sigma(k_i u^i) = \sum_i \sigma(k_i)\sigma(u^i)$ $= \sum_i k_i \sigma(u)^i = \sum_{i=0}^{n} k_i v^i = f(v)$. ∎

Up to this point we have always dealt with a root of a polynomial $f \varepsilon K[x]$ in some given extension field F of K. The next theorem shows that it really is not necessary to have F given in advance.

Theorem 1.10. *If* K *is a field and* f ε K[x] *polynomial of degree* n, *then there exists a simple extension field* F = K(u) *of* K *such that:*

(i) u ε F *is a root of* f;

(ii) [K(u) : K] ≤ n, *with equality holding if and only if* f *is irreducible in* K[x];

(iii) *if* f *is irreducible in* K[x], *then* K(u) *is unique up to an isomorphism which is the identity on* K.

REMARK. In view of (iii) it is customary to speak of *the* field F obtained by **adjoining a root** of the irreducible polynomial $f \varepsilon K[x]$ to the field K.

SKETCH OF PROOF OF 1.10. We may assume that f is irreducible (if not, replace f by one of its irreducible factors). Then the ideal (f) is maximal in $K[x]$ (Theorem III.3.4 and Corollary III.6.4) and the quotient ring $F = K[x]/(f)$ is a field (Theorem III.2.20). Furthermore, the canonical projection $\pi : K[x] \to K[x]/(f) = F$, when restricted to K, is a monomorphism (since 0 is the only constant in a maximal ideal of $K[x]$). Thus F contains $\pi(K) \cong K$, and therefore may be considered as an extension field of K (providing that K is identified with $\pi(K)$ under the isomorphism). For $x \in K[x]$, let $u = \pi(x) \in$ F. Verify that $F = K(u)$ and that $f(u) = 0$ in F. Theorem 1.6 implies statement (ii) and Corollary 1.9 gives (iii). ∎

In the remainder of this section we shall develop the essential basic facts about algebraic field extensions.

Theorem 1.11. *If* F *is a finite dimensional extension field of* K, *then* F *is finitely generated and algebraic over* K.

PROOF. If $[F : K] = n$ and $u \in F$, then the set of $n + 1$ elements $\{1_K, u, u^2, \dots, u^n\}$ must be linearly dependent. Hence there are $a_i \in K$, not all zero, such that $a_0 + a_1u + a_2u^2 + \cdots + a_nu^n = 0$, which implies that u is algebraic over K. Since u was arbitrary, F is algebraic over K. If $\{v_1, \dots, v_n\}$ is a basis of F over K, then it is easy to see that $F = K(v_1, \dots, v_n)$. ∎

Theorem 1.12. *If* F *is an extension field of* K *and* X *is a subset of* F *such that* F = K(X) *and every element of* X *is algebraic over* K, *then* F *is an algebraic extension of* K. *If* X *is a finite set, then* F *is finite dimensional over* K.

PROOF. If $v \in F$, then $v \in K(u_1, \dots, u_n)$ for some $u_i \in X$ (Theorem 1.3) and there is a tower of subfields:

$$K \subset K(u_1) \subset K(u_1,u_2) \subset \cdots \subset K(u_1, \dots, u_{n-1}) \subset K(u_1, \dots, u_n).$$

Since u_i is algebraic over K, it is necessarily algebraic over $K(u_1, \dots, u_{i-1})$ for each $i \geq 2$, say of degree r_i. Since $K(u_1, \dots, u_{i-1})(u_i) = K(u_1, \dots, u_i)$ we have $[K(u_1, \dots, u_i) : K(u_1, \dots, u_{i-1})] = r_i$ by Theorem 1.6. Let r_1 be the degree of u_1 over K; then repeated application of Theorem 1.2 shows that $[K(u_1, \dots, u_n) : K] = r_1r_2 \cdots r_n$. By Theorem 1.11 $K(u_1, \dots, u_n)$ (and hence v) is algebraic over K. Since $v \in F$ was arbitrary, F is algebraic over K. If $X = \{u_1, \dots, u_n\}$ is finite, the same proof (with $F = K(u_1, \dots, u_n)$) shows that $[F : K] = r_1r_2 \cdots r_n$ is finite. ∎

Theorem 1.13. *If* F *is an algebraic extension field of* E *and* E *is an algebraic extension field of* K, *then* F *is an algebraic extension of* K.

PROOF. Let $u \in F$; since u is algebraic over E, $b_nu^n + \cdots + b_1u + b_0 = 0$ for some $b_i \in E$ $(b_n \neq 0)$. Therefore, u is algebraic over the subfield $K(b_0, \dots, b_n)$. Consequently, there is a tower of fields

$$K \subset K(b_0, \dots, b_n) \subset K(b_0, \dots, b_n)(u),$$

with $[K(b_0, \dots, b_n)(u) : K(b_0, \dots, b_n)]$ finite by Theorem 1.6 (since u is algebraic over $K(b_0, \dots, b_n)$) and $[K(b_0, \dots, b_n) : K]$ finite by Theorem 1.12 (since each $b_i \in E$ is algebraic over K). Therefore, $[K(b_0, \dots, b_n)(u) : K]$ is finite (Theorem 1.2). Hence

$u \in K(b_0, \ldots, b_n)(u)$ is algebraic over K (Theorem 1.11). Since u was arbitrary, F is algebraic over K. ■

Theorem 1.14. *Let* F *be an extension field of* K *and* E *the set of all elements of* F *which are algebraic over* K. *Then* E *is a subfield of* F *(which is, of course, algebraic over* K).

Clearly the subfield E is the unique maximal algebraic extension of K contained in F.

PROOF OF 1.14. If $u,v \in E$, then $K(u,v)$ is an algebraic extension field of K by Theorem 1.12. Therefore, since $u - v$ and uv^{-1} $(v \neq 0)$ are in $K(u,v)$, $u - v$ and $uv^{-1} \in E$. This implies that E is a field (see Theorem I.2.5). ■

APPENDIX: RULER AND COMPASS CONSTRUCTIONS

The word "ruler" is to be considered as a synonym for straightedge (as is customary in geometric discussions). We shall use field extensions to settle two famous problems of antiquity:

(A) Is it possible to trisect an arbitrary angle by ruler and compass constructions?

(B) Is it possible via ruler and compass constructions to duplicate an arbitrary cube (that is, to construct the side of a cube having twice the volume of the given cube)?

We shall assume as known all the standard ruler and compass constructions as presented in almost any plane geometry text. Example: given a straight line L and a point P not on L, the unique straight line through P and parallel L [resp. perpendicular to L] is constructible. Here and below "constructible" means "constructible by ruler and compass constructions."

Furthermore we shall adopt the viewpoint of analytic geometry as follows. Clearly we may construct with ruler and compass two perpendicular straight lines (axes). Choose a unit length. Then we can construct all points of the plane with integer coordinates (that is, locate them precisely as the intersection of suitable constructible straight lines parallel to the axes). As will be seen presently, the solution to the stated problems will result from a knowledge of what other points in the plane can be constructed via ruler and compass constructions.

If F is a subfield of the field **R** of real numbers, the **plane of F** is the subset of the plane consisting of all points (c,d) with $c \in F$, $d \in F$. If P,Q are distinct points in the plane of F, the unique line through P and Q is called a **line in F** and the circle with center P and radius the line segment PQ is called a **circle in F.** It is readily verified that every straight line in F has an equation of the form $ax + by + c = 0$ $(a,b,c \in F)$ and every circle in F an equation of the form $x^2 + y^2 + ax + by + c = 0$ $(a,b,c \in F)$ (Exercise 24).

Lemma 1.15. *Let* F *be a subfield of the field* **R** *of real numbers and let* L_1,L_2 *be nonparallel lines in* F *and* C_1,C_2 *distinct circles in* F. *Then*

(i) $L_1 \cap L_2$ *is a point in the plane of* F;

(ii) $L_1 \cap C_1 = \varnothing$ *or consists of one or two points in the plane of* $F(\sqrt{u})$ *for some* $u \in F$ $(u \geq 0)$;

(iii) $C_1 \cap C_2 = \varnothing$ *or consists of one or two points in the plane of* $F(\sqrt{u})$ *for some* $u \in F$ $(u \geq 0)$.

SKETCH OF PROOF. (i) Exercise. (iii) If the circles are $C_1 : x^2 + y^2 + a_1x + b_1y + c_1 = 0$ and $C_2 : x^2 + y^2 + a_2x + b_2y + c_2 = 0$ $(a_i,b_i,c_i \in F$ by the remarks preceding the lemma), show that $C_1 \cap C_2$ is the same as the intersection of C_1 or C_2 with the straight line $L : (a_1 - a_2)x + (b_1 - b_2)y + (c_1 - c_2) = 0$. Verify that L is a line in F; then case (iii) reduces to case (ii).

(ii) Suppose L_1 has the equation $dx + ey + f = 0$ $(d,e,f \in F)$. The case $d = 0$ is left as an exercise; if $d \neq 0$, we can assume $d = 1$ (why?), so that $x = (-ey - f)$. If $(x,y) \in L_1 \cap C_1$, then substitution gives the equation of C_1 as $0 = (-ey - f)^2 + y^2 + a_1(-ey - f) + b_1y + c_1 = Ay^2 + By + C = 0$, with $A,B,C \in F$. If $A = 0$, then $y \in F$; hence $x \in F$ and $x,y \in F(\sqrt{1}) = F$. If $A \neq 0$, we may assume $A = 1$. Then $y^2 + By + C = 0$ and completing the square yields $(y + B/2)^2 + (C - B^2/4) = 0$. This implies that either $L_1 \cap C_1 = \varnothing$ or $x,y \in F(\sqrt{u})$ with $u = -C + B^2/4 \geq 0$. ■

A real number c will be said to be **constructible** if the point $(c,0)$ can be constructed (precisely located) by a finite sequence of ruler and compass constructions that begin with points with integer coordinates. The constructibility of c (or $(c,0)$) is clearly equivalent to the constructibility (via ruler and compass) of a line segment of length $|c|$. Furthermore the point (c,d) in the plane may be constructed via ruler and compass if and only if both c and d are constructible real numbers. The integers are obviously constructible, and it is not difficult to prove the following facts (see Exercise 25):

(i) every rational number is constructible;

(ii) if $c \geq 0$ is constructible, so is $\sqrt{c}$;

(iii) if c,d are constructible, then $c \pm d$, cd, and c/d $(d \neq 0)$ are constructible, so that the constructible numbers form a subfield of the real numbers that contains the rationals.

Proposition 1.16. *If a real number* c *is constructible, then* c *is algebraic of degree a power of* 2 *over the field* **Q** *of rationals.*

PROOF. The preceding remarks show that we may as well take the plane of **Q** as given. To say that c is constructible then means that $(c,0)$ may be located (constructed) by a finite sequence of allowable ruler and compass constructions beginning with the plane of **Q**. In the course of these constructions various points of the plane will be determined as the intersections of lines and/or circles used in the construction process. For this is the only way to arrive at new points using only a ruler and compass. The first step in the process is the construction of a line or circle, either of which is completely determined by two points (center P and radius PT for the circle). Either these points are given as being in the plane of **Q** or else they may be chosen arbitrarily, in which case they may be taken to be in the plane of **Q** also. Similarly at each stage of the construction the two points that determine the line or circle used may be taken to be either points in the plane of **Q** or points constructed in previous steps. In view of Lemma 1.15 the first new point so constructed lies in the plane of an extension field $\mathbf{Q}(\sqrt{u})$ of $\mathbf{Q}$, with $u \in \mathbf{Q}$, or equivalently in the plane of an extension $\mathbf{Q}(v)$ with $v^2 \in \mathbf{Q}$. Such an extension has degree $1 = 2^0$ or 2 over $\mathbf{Q}$ (depending on whether or not $v \in \mathbf{Q}$). Similarly the next new point constructed lies in the plane of $\mathbf{Q}(v,w) = \mathbf{Q}(v)(w)$ with $w^2 \in \mathbf{Q}(v)$. It follows that a finite sequence of ruler and compass constructions gives rise to a finite tower of fields:

$$\mathbf{Q} \subset \mathbf{Q}(v_1) \subset \mathbf{Q}(v_1,v_2) \subset \cdots \subset \mathbf{Q}(v_1, \dots, v_n)$$

with $v_i^2 \in \mathbf{Q}(v_1, \dots, v_{i-1})$ and $[\mathbf{Q}(v_1, \dots, v_i) : \mathbf{Q}(v_1, \dots, v_{i-1})] = 1$ or 2 $(2 \leq i \leq n)$. The point $(c,0)$ constructed by this process then lies in the plane of $F = \mathbf{Q}(v_1, \dots, v_n)$. By Theorem 1.2, $[F : \mathbf{Q}]$ is a power of two. Therefore, c is algebraic over $\mathbf{Q}$ (Theorem 1.11). Now $\mathbf{Q} \subset \mathbf{Q}(c) \subset \mathbf{F}$ implies that $[\mathbf{Q}(c) : \mathbf{Q}]$ divides $[F : \mathbf{Q}]$ (Theorem 1.2), whence the degree $[\mathbf{Q}(c) : \mathbf{Q}]$ of c over $\mathbf{Q}$ is a power of 2. ■

Corollary 1.17. *An angle of* 60° *cannot be trisected by ruler and compass constructions.*

PROOF. If it were possible to trisect a 60° angle, we would then be able to construct a right triangle with one acute angle of 20°. It would then be possible to construct the real number (ratio) cos 20° (Exercise 25). However for any angle α, elementary trigonometry shows that

$$\cos 3\alpha = 4 \cos^3 \alpha - 3 \cos \alpha.$$

Thus if $\alpha = 20°$, then $\cos 3\alpha = \cos 60° = \frac{1}{2}$ and cos 20° is a root of the equation $\frac{1}{2} = 4x^3 - 3x$ and hence of the polynomial $8x^3 - 6x - 1$. But this polynomial is irreducible in $\mathbf{Q}[x]$ (see Theorem III. 6.6 and Proposition III.6.8). Therefore cos 20° has degree 3 over $\mathbf{Q}$ and cannot be constructible by Proposition 1.16. ■

Corollary 1.18. *It is impossible by ruler and compass constructions to duplicate a cube of side length* 1 (*that is, to construct the side of a cube of volume* 2).

PROOF. If s is the side length of a cube of volume 2, then s is a root of $x^3 - 2$, which is irreducible in $\mathbf{Q}[x]$ by Eisenstein's Criterion (Theorem III.6.15). Therefore s is not constructible by Proposition 1.16. ■

EXERCISES

Note: Unless specified otherwise F is always an extension field of the field K and $\mathbf{Q},\mathbf{R},\mathbf{C}$ denote the fields of rational, real, and complex numbers respectively.

1. (a) $[F : K] = 1$ if and only if $F = K$.
   (b) If $[F : K]$ is prime, then there are no intermediate fields between F and K.
   (c) If $u \in F$ has degree n over K, then n divides $[F : K]$.
2. Give an example of a finitely generated field extension, which is not finite dimensional. [*Hint:* think transcendental.]

3. If $u_1, \ldots, u_n \in F$ then the field $K(u_1, \ldots, u_n)$ is (isomorphic to) the quotient field of the ring $K[u_1, \ldots, u_n]$.

4. (a) For any $u_1, \ldots, u_n \in F$ and any permutation $\sigma \in S_n$, $K(u_1, \ldots, u_n) = K(u_{\sigma(1)}, \ldots, u_{\sigma(n)})$.
(b) $K(u_1, \ldots, u_{n-1})(u_n) = K(u_1, \ldots, u_n)$.
(c) State and prove the analogues of (a) and (b) for $K[u_1, \ldots, u_n]$.
(d) If each u_i is algebraic over K, then $K(u_1, \ldots, u_n) = K[u_1, \ldots, u_n]$.

5. Let L and M be subfields of F and LM their composite.
(a) If $K \subset L \cap M$ and $M = K(S)$ for some $S \subset M$, then $LM = L(S)$.
(b) When is it true that LM is the set theoretic union $L \cup M$?
(c) If $E_1, \ldots, E_n$ are subfields of F, show that

$$E_1E_2\cdots E_n = E_1(E_2(E_3(\cdots(E_{n-1}(E_n)))\cdots).$$

6. Every element of $K(x_1, \ldots, x_n)$ which is not in K is transcendental over K.

7. If v is algebraic over $K(u)$ for some $u \in F$ and v is transcendental over K, then u is algebraic over $K(v)$.

8. If $u \in F$ is algebraic of odd degree over K, then so is u^2 and $K(u) = K(u^2)$.

9. If $x^n - a \in K[x]$ is irreducible and $u \in F$ is a root of $x^n - a$ and m divides n, then prove that the degree of u^m over K is n/m. What is the irreducible polynomial for u^m over K?

10. If F is algebraic over K and D is an integral domain such that $K \subset D \subset F$, then D is a field.

11. (a) Give an example of a field extension $K \subset F$ such that $u,v \in F$ are transcendental over K, but $K(u,v) \not\cong K(x_1,x_2)$. [*Hint:* consider v over the field $K(u)$.]
(b) State and prove a generalization of Theorem 1.5 to the case of n transcendental elements $u_1, \ldots, u_n$.

12. If $d \geq 0$ is an integer that is not a square describe the field $\mathbf{Q}(\sqrt{d})$ and find a set of elements that generate the whole field.

13. (a) Consider the extension $\mathbf{Q}(u)$ of $\mathbf{Q}$ generated by a real root u of $x^3 - 6x^2 + 9x + 3$. (Why is this irreducible?) Express each of the following elements in terms of the basis $\{1,u,u^2\}$: $u^4;u^5;3u^5 - u^4 + 2;\ (u+1)^{-1};\ (u^2 - 6u + 8)^{-1}$.
(b) Do the same with respect to the basis $\{1,u,u^2,u^3,u^4\}$ of $\mathbf{Q}(u)$ where u is a real root of $x^5 + 2x + 2$ and the elements in question are: $(u^2+2)(u^3+3u);u^{-1}$; $u^4(u^4 + 3u^2 + 7u + 5);(u+2)(u^2+3)^{-1}$.

14. (a) If $F = \mathbf{Q}(\sqrt{2},\sqrt{3})$, find $[F : \mathbf{Q}]$ and a basis of F over $\mathbf{Q}$.
(b) Do the same for $F = \mathbf{Q}(i,\sqrt{3},\omega)$, where $i \in \mathbf{C}$, $i^2 = -1$, and ω is a complex (nonreal) cube root of 1.

15. In the field $K(x)$, let $u = x^3/(x+1)$. Show that $K(x)$ is a simple extension of the field $K(u)$. What is $[K(x) : K(u)]$?

16. In the field $\mathbf{C}$, $\mathbf{Q}(i)$ and $\mathbf{Q}(\sqrt{2})$ are isomorphic as vector spaces, but not as fields.

17. Find an irreducible polynomial f of degree 2 over the field Z_2. Adjoin a root u of f to Z_2 to obtain a field $Z_2(u)$ of order 4. Use the same method to construct a field of order 8.

18. A complex number is said to be an **algebraic number** if it is algebraic over **Q** and an **algebraic integer** if it is the root of a monic polynomial in $\mathbf{Z}[x]$.

(a) If u is an algebraic number, there exists an integer n such that nu is an algebraic integer.

(b) If $r \in \mathbf{Q}$ is an algebraic integer, then $r \in \mathbf{Z}$.

(c) If u is an algebraic integer and $n \in \mathbf{Z}$, then $u + n$ and nu are algebraic integers.

(d) The sum and product of two algebraic integers are algebraic integers.

19. If $u,v \in F$ are algebraic over K of degrees m and n respectively, then $[K(u,v) : K] \leq mn$. If $(m,n) = 1$, then $[K(u,v) : K] = mn$.

20. Let L and M be intermediate fields in the extension $K \subset F$.

(a) $[LM : K]$ is finite if and only if $[L : K]$ and $[M : K]$ are finite.

(b) If $[LM : K]$ is finite, then $[L : K]$ and $[M : K]$ divide $[LM : K]$ and

$$[LM : K] \leq [L : K][M : K].$$

(c) If $[L : K]$ and $[M : K]$ are finite and relatively prime, then

$$[LM : K] = [L : K][M : K].$$

(d) If L and M are algebraic over K, then so is LM.

21. (a) Let L and M be intermediate fields of the extension $K \subset F$, of finite dimension over K. Assume that $[LM : K] = [L : K][M : K]$ and prove that $L \cap M = K$.

(b) The converse of (a) holds if $[L : K]$ or $[M : K]$ is 2.

(c) Using a real and a nonreal cube root of 2 give an example where $L \cap M = K$, $[L : K] = [M : K] = 3$, but $[LM : K] < 9$.

22. F is an algebraic extension of K if and only if for every intermediate field E every monomorphism $\sigma : E \to E$ which is the identity on K is in fact an automorphism of E.

23. If $u \in F$ is algebraic over $K(X)$ for some $X \subset F$ then there exists a finite subset $X' \subset X$ such that u is algebraic over $K(X')$.

24. Let F be a subfield of **R** and P,Q points in the Euclidean plane whose coordinates lie in F.

(a) The straight line through P and Q has an equation of the form $ax + by + c = 0$, with $a,b,c \in F$.

(b) The circle with center P and radius the line segment PQ has an equation of the form $x^2 + y^2 + ax + by + c = 0$ with $a,b,c \in F$.

25. Let c,d be constructible real numbers.

(a) $c + d$ and $c - d$ are constructible.

(b) If $d \neq 0$, then c/d is constructible. [*Hint:* If $(x,0)$ is the intersection of the x axis and the straight line through $(0,1)$ that is parallel the line through $(0,d)$ and $(c,0)$, then $x = c/d$.]

(c) cd is constructible [*Hint:* use (b)].

(d) The constructible real numbers form a subfield containing **Q**.

(e) If $c \geq 0$, then $\sqrt{c}$ is constructible. [*Hint:* If y is the length of the straight line segment perpendicular to the x axis that joins $(1,0)$ with the (upper half of the) circle with center $((c + 1)/2,0)$ and radius $(c + 1)/2$ then $y = \sqrt{c}$.]

26. Let E_1 and E_2 be subfields of F and X a subset of F. If every element of E_1 is algebraic over E_2, then every element of $E_1(X)$ is algebraic over $E_2(X)$. [*Hint:* $E_1(X) \subset (E_2(X))(E_1)$; use Theorem 1.12.]

2. THE FUNDAMENTAL THEOREM

The Galois group of an arbitrary field extension is defined and the concept of a Galois extension is defined in terms of the Galois group. The remainder of the section is devoted to proving the Fundamental Theorem of Galois Theory (Theorem 2.5), which enables us to translate problems involving fields, polynomials, and extensions into group theoretical terms. An appendix at the end of the section deals with symmetric rational functions and provides examples of extensions having any given finite group as Galois group.

Let F be a field. The set Aut F of all (field) automorphisms $F \to F$ forms a group under the operation of composition of functions (Exercise 1). In general, it is not abelian. It was Galois' remarkable discovery that many questions about fields (especially about the roots of polynomials over a field) are in fact equivalent to certain group-theoretical questions in the automorphism group of the field. When these questions arise, they usually involve not only F, but also a (suitably chosen) subfield of F; in other words we deal with field extensions.

If F is an extension field of K, we have seen in Section 1 that the K-module (vector space) structure of F is of much significance. Consequently, it seems natural to consider those automorphisms of F that are also K-module maps. Clearly the set of all such automorphisms is a subgroup of Aut F.

More generally let E and F be extension fields of a field K. If $\sigma : E \to F$ is a nonzero homomorphism of fields, then $\sigma(1_E) = 1_F$ by Exercise III.1.15. If σ is also a K-module homomorphism, then for every $k \in K$

$$\sigma(k) = \sigma(k1_E) = k\sigma(1_E) = k1_F = k.$$

Conversely, if a homomorphism of fields $\sigma : E \to F$ fixes K elementwise (that is, $\sigma(k) = k$ for all $k \in K$), then σ is nonzero and for any $u \in E$,

$$\sigma(ku) = \sigma(k)\sigma(u) = k\sigma(u)$$

whence σ is a K-module homomorphism.

Definition 2.1. *Let* E *and* F *be extension fields of a field* K. *A nonzero map* $\sigma : \mathrm{E} \to \mathrm{F}$ *which is both a field and a* K*-module homomorphism is called a* **K-homomorphism.** *Similarly if a field automorphism* $\sigma \in$ *Aut* F *is a* K*-homomorphism, then* σ *is called a* **K-automorphism** *of* F. *The group of all* K*-automorphisms of* F *is called the* **Galois group** *of* F *over* K *and is denoted* $Aut_{\mathrm{K}}\mathrm{F}$.

REMARKS. K-monomorphisms and K-isomorphisms are defined in the obvious way. Here and below the identity element of $\mathrm{Aut}_K F$ and its identity subgroup will both be denoted by 1.

EXAMPLE. Let $F = K(x)$, with K any field. For each $a \in K$ with $a \neq 0$ the map $\sigma_a : F \to F$ given by $f(x)/g(x) \mapsto f(ax)/g(ax)$ is a K-automorphism of F; (this may be verified directly or via Corollaries III.2.21(iv), III.4.6, and III.5.6, and Theorem III.4.4(ii)). If K is infinite, then there are infinitely many distinct automorphisms σ_a, whence $\mathrm{Aut}_K F$ is infinite. Similarly for each $b \in K$, the map $\tau_b : F \to F$ given by $f(x)/g(x) \mapsto f(x+b)/g(x+b)$ is a K-automorphism of F. If $a \neq 1_K$ and $b \neq 0$, then $\sigma_a\tau_b \neq \tau_b\sigma_a$, whence $\mathrm{Aut}_K F$ is nonabelian. Also see Exercise 6.

Theorem 2.2. *Let* F *be an extension field of* K *and* $f \in K[x]$. *If* $u \in F$ *is a root of* f *and* $\sigma \in \mathrm{Aut}_K F$, *then* $\sigma(u) \in F$ *is also a root of* f.

PROOF. If $f = \sum_{i=1}^{n} k_i x^i$, then $f(u) = 0$ implies $0 = \sigma(f(u)) = \sigma(\sum k_i u^i) = \sum \sigma(k_i)\sigma(u^i) = \sum_i k_i \sigma(u)^i = f(\sigma(u))$. ■

One of the principal applications of Theorem 2.2 is in the situation where u is algebraic over K with irreducible polynomial $f \in K[x]$ of degree n. Then any $\sigma \in \mathrm{Aut}_K K(u)$ is completely determined by its action on u (since $\{1_K, u, u^2, \dots, u^{n-1}\}$ is a basis of $K(u)$ over K by Theorem 1.6). Since $\sigma(u)$ is a root of f by Theorem 2.2, $|\mathrm{Aut}_K K(u)| \leq m$, where m is the number of distinct roots of f in $K(u)$; ($m \leq n$ by Theorem III.6.7).

EXAMPLES. Obviously if $F = K$, then $\mathrm{Aut}_K F$ consists of the identity element alone. The converse, however, is false. For instance, if u is a real cube root of 2 (so that $\mathbf{Q} \subsetneq \mathbf{Q}(u) \subset \mathbf{R}$), then $\mathrm{Aut}_{\mathbf{Q}}\mathbf{Q}(u)$ is the identity group. For the only possible images of u are the roots of $x^3 - 2$ and the other two roots are complex. Similarly, $\mathrm{Aut}_{\mathbf{Q}}\mathbf{R}$ is the identity (Exercise 2).

EXAMPLES. $\mathbf{C} = \mathbf{R}(i)$ and $\pm i$ are the roots of $x^2 + 1$. Thus $\mathrm{Aut}_{\mathbf{R}}\mathbf{C}$ has order at most 2. It is easy to verify that complex conjugation ($a + bi \mapsto a - bi$) is a non-identity $\mathbf{R}$-automorphism of $\mathbf{C}$, so that $|\mathrm{Aut}_{\mathbf{R}}\mathbf{C}| = 2$ and hence $\mathrm{Aut}_{\mathbf{R}}\mathbf{C} \cong Z_2$. Similarly $\mathrm{Aut}_{\mathbf{Q}}\mathbf{Q}(\sqrt{3}) \cong Z_2$.

EXAMPLES. If $F = \mathbf{Q}(\sqrt{2},\sqrt{3}) = \mathbf{Q}(\sqrt{2})(\sqrt{3})$, then since $x^2 - 3$ is irreducible over $\mathbf{Q}(\sqrt{2})$ the proof of Theorem 1.2 and Theorem 1.6 show that $\{1,\sqrt{2},\sqrt{3},\sqrt{6}\}$ is a basis of F over $\mathbf{Q}$. Thus if $\sigma \in \mathrm{Aut}_{\mathbf{Q}}F$, then σ is completely determined by $\sigma(\sqrt{2})$ and $\sigma(\sqrt{3})$. By Theorem 2.2 $\sigma(\sqrt{2}) = \pm\sqrt{2}$ and $\sigma(\sqrt{3}) = \pm\sqrt{3}$ and this means that there are at most four distinct $\mathbf{Q}$-automorphisms of F. It is readily verified that each of the four possibilities is indeed a $\mathbf{Q}$-automorphism of F and that $\mathrm{Aut}_{\mathbf{Q}}F \cong Z_2 \oplus Z_2$.

It is shown in the appendix (Proposition 2.16) that for any given finite group G, there is an extension with Galois group G. It is still an open question as to whether or not every finite group is the Galois group of some extension over a *specific* field (such as $\mathbf{Q}$).

The basic idea of what is usually called Galois Theory is to set up some sort of correspondence between the intermediate fields of a field extension $K \subset F$ and the subgroups of the Galois group $\mathrm{Aut}_K F$. Although the case where F is finite dimensional over K is of the most interest, we shall keep the discussion as general as possible for as long as we can. The first step in establishing this correspondence is given by

Theorem 2.3. *Let* F *be an extension field of* K, E *an intermediate field and* H *a subgroup of* $Aut_K F$. *Then*

(i) $H' = \{v \in F \mid \sigma(v) = v$ *for all* $\sigma \in H\}$ *is an intermediate field of the extension;*
(ii) $E' = \{\sigma \in Aut_K F \mid \sigma(u) = u$ *for all* $u \in E\} = Aut_E F$ *is a subgroup of* $Aut_K F$.

PROOF. Exercise. ■

The field H' is called the **fixed field** of H in F (although this is a standard term there is no universal notation for it, but the "prime notation" will prove useful). Likewise, whenever it is convenient, we shall continue to denote the group $\mathrm{Aut}_E F$ in this context as E'. If we denote $\mathrm{Aut}_K F$ by G, it is easy to see that on the one hand,

$$F' = \mathrm{Aut}_F F = 1 \quad \text{and} \quad K' = \mathrm{Aut}_K F = G;$$

and on the other, $1' = F$ (that is, F is the fixed field of the identity subgroup). It is not necessarily true, however, that $G' = K$ (as can be seen in the first examples after Theorem 2.2, where $G = 1$ and hence $G' = F \neq K$; also see Exercise 2).

Definition 2.4. *Let* F *be an extension field of* K *such that the fixed field of the Galois group* $Aut_K F$ *is* K *itself. Then* F *is said to be a* **Galois extension** *(field) of* K *or to be* **Galois** *over* K.[1]

REMARKS. F is Galois over K if and only if for any $u \in F - K$, there exists a K-automorphism $\sigma \in \mathrm{Aut}_K F$ such that $\sigma(u) \neq u$. If F is an arbitrary extension field of K and K_0 is the fixed field of $\mathrm{Aut}_K F$ (possibly $K_0 \neq K$), then it is easy to see that F is Galois over K_0, that $K \subset K_0$, and that $\mathrm{Aut}_K F = \mathrm{Aut}_{K_0} F$.

EXAMPLES. **C** is Galois over **R** and $\mathbf{Q}(\sqrt{3})$ is Galois over **Q** (Exercise 5). If K is an infinite field, then $K(x)$ is Galois over K (Exercise 9).

Although a proof is still some distance away, it is now possible to state the Fundamental Theorem of Galois Theory, so that the reader will be able to see just where the subsequent discussion is headed. If L,M are intermediate fields of an extension with $L \subset M$, the dimension $[M : L]$ is called the **relative dimension** of L and M. Similarly, if H,J are subgroups of the Galois group with $H < J$, the index $[J : H]$ is called the **relative index** of H and J.

Theorem 2.5. *(Fundamental Theorem of Galois Theory) If* F *is a finite dimensional Galois extension of* K, *then there is a one-to-one correspondence between the set of all*

[1]A Galois extension is frequently required to be finite dimensional or at least algebraic and is defined in terms of normality and separability, which will be discussed in Section 3. In the finite dimensional case our definition is equivalent to the usual one. Our definition is essentially due to Artin, except that he calls such an extension "normal." Since this use of "normal" conflicts (in case char $F \neq 0$) with the definition of "normal" used by many other authors, we have chosen to follow Artin's basic approach, but to retain the (more or less) conventional terminology.

intermediate fields of the extension and the set of all subgroups of the Galois group $Aut_K F$ (*given by* $E \mapsto E' = Aut_E F$) *such that:*

(i) *the relative dimension of two intermediate fields is equal to the relative index of the corresponding subgroups; in particular,* $Aut_K F$ *has order* $[F : K]$;

(ii) F *is Galois over every intermediate field* E, *but* E *is Galois over* K *if and only if the corresponding subgroup* $E' = Aut_E F$ *is normal in* $G = Aut_K F$; *in this case* G/E' *is* (*isomorphic to*) *the Galois group* $Aut_K E$ *of* E *over* K.

The proof of the theorem (which begins on p. 251) requires some rather lengthy preliminaries. The rest of this section is devoted to developing these. We leave the problem of constructing Galois extension fields and the case of algebraic Galois extensions of arbitrary dimension for the next section. The reader should note that many of the propositions to be proved now apply to the general case.

As indicated in the statement of the Fundamental Theorem, the so-called Galois correspondence is given by assigning to each intermediate field E the Galois group $\mathrm{Aut}_E F$ of F over E. It will turn out that the inverse of this one-to-one correspondence is given by assigning to each subgroup H of the Galois group its fixed field in F. It will be very convenient to use the "prime notation" of Theorem 2.3, so that E' denotes $\mathrm{Aut}_E F$ and H' denotes the fixed field of H in F.

It may be helpful to visualize these priming operations schematically as follows. Let L and M be intermediate fields of the extension $K \subset F$ and let J, H be subgroups of the Galois group $G = \mathrm{Aut}_K F$.

$$
\begin{array}{ccc}
F & \longmapsto & 1 \\
\cup & & \wedge \\
M & \longmapsto & M' \\
\cup & & \wedge \\
L & \longmapsto & L' \\
\cup & & \wedge \\
K & \longmapsto & G;
\end{array}
\qquad\qquad\qquad
\begin{array}{ccc}
F & \longleftarrow & 1 \\
\cup & & \wedge \\
H' & \longleftarrow & H \\
\cup & & \wedge \\
J' & \longleftarrow & J \\
\cup & & \wedge \\
K & & G.
\end{array}
$$

Formally, the basic facts about the priming operations are given by

Lemma 2.6. *Let* F *be an extension field of* K *with intermediate fields* L *and* M. *Let* H *and* J *be subgroups of* $G = Aut_K F$. *Then:*

(i) $F' = 1$ *and* $K' = G$;
(i') $1' = F$;
(ii) $L \subset M \Rightarrow M' < L'$;
(ii') $H < J \Rightarrow J' \subset H'$;
(iii) $L \subset L''$ *and* $H < H''$ (*where* $L'' = (L')'$ *and* $H'' = (H')'$);
(iv) $L' = L'''$ *and* $H' = H'''$.

SKETCH OF PROOF. (i)–(iii) follow directly from the appropriate definitions. To prove the first part of (iv) observe that (iii) and (ii) imply $L''' < L'$ and that (iii) applied with L' in place of H implies $L' < L'''$. The other part is proved similarly. ■

REMARKS. It is quite possible for L'' to contain L properly (similarly for H'' and H). F is Galois over K (by definition) if $G' = K$. Thus since $K' = G$ in any case, F is Galois over K if and only if $K = K''$. Similarly F is Galois over an intermediate field E if and only if $E = E''$.

Let X be an intermediate field or subgroup of the Galois group. X will be called **closed** provided $X = X''$. Note that F is Galois over K if and only if K is closed.

Theorem 2.7. *If* F *is an extension field of* K, *then there is a one-to-one correspondence between the closed intermediate fields of the extension and the closed subgroups of the Galois group, given by* $E \mapsto E' = Aut_E F$.

PROOF. Exercise; the inverse of the correspondence is given by assigning to each closed subgroup H its fixed field H'. Note that by Lemma 2.6(iv) all primed objects are closed. ■

This theorem is not very helpful until we have some more specific information as to which intermediate fields and which subgroups are closed. Eventually we shall show that in an algebraic Galois extension all intermediate fields are closed and that in the finite dimensional case all subgroups of the Galois group are closed as well. We begin with some technical lemmas that give us estimates of various relative dimensions.

Lemma 2.8. *Let* F *be an extension field of* K *and* L,M *intermediate fields with* $L \subset M$. *If* $[M:L]$ *is finite, then* $[L':M'] \le [M:L]$. *In particular, if* $[F:K]$ *is finite, then* $|Aut_K F| \le [F:K]$.

PROOF. We proceed by induction on $n = [M:L]$, with the case $n = 1$ being trivial. If $n > 1$ and the theorem is true for all $i < n$, choose $u \in M$ with $u \notin L$. Since $[M:L]$ is finite, u is algebraic over L (Theorem 1.11) with irreducible polynomial $f \in L[x]$ of degree $k > 1$. By Theorems 1.6 and 1.1, $[L(u):L] = k$ and $[M:L(u)] = n/k$. Schematically we have:

$$\begin{array}{cccccc} & & M & \longmapsto & M' \\ & n/k & \cup & & \wedge \\ n & & L(u) & \longmapsto & L(u)' \\ & k & \cup & & \wedge \\ & & L & \longmapsto & L'. \end{array}$$

There are now two cases. If $k < n$, then $1 < n/k < n$ and by induction $[L':L(u)'] \le k$ and $[L(u)':M'] \le n/k$. Hence $[L':M'] = [L':L(u)'][L(u)':M'] \le k(n/k) = n = [M:L]$ and the theorem is proved. On the other hand if $k = n$, then $[M:L(u)] = 1$ and $M = L(u)$. In order to complete the proof in this case, we shall construct an injective map from the set of S of all left cosets of M' in L' to the set T of all distinct roots (in F) of the polynomial $f \varepsilon L[x]$, whence $|S| \leq |T|$. Since $|T| \leq n$ by Theorem III.6.7 and $|S| = [L' : M']$ by definition, this will show that $[L' : M'] \leq |T| \leq n = [M : L]$. The final statement of the theorem then follows immediately since $|\mathrm{Aut}_K F| = [\mathrm{Aut}_K F : 1] = [K' : F'] \leq [F : K]$.

Let $\tau M'$ be a left coset of M' in L'. If $\sigma \varepsilon M' = \mathrm{Aut}_M F$, then since $u \varepsilon M$, $\tau\sigma(u) = \tau(u)$. Thus every element of the coset $\tau M'$ has the same effect on u and maps $u \mapsto \tau(u)$. Since $\tau \varepsilon L' = \mathrm{Aut}_L F$, and u is a root of $f \varepsilon L[x]$, $\tau(u)$ is also a root of f by Theorem 2.2. This implies that the map $S \to T$ given by $\tau M' \mapsto \tau(u)$ is well defined. If $\tau(u) = \tau_0(u)$ $(\tau,\tau_0 \varepsilon L')$, then $\tau_0^{-1}\tau(u) = u$ and hence $\tau_0^{-1}\tau$ fixes u. Therefore, $\tau_0^{-1}\tau$ fixes $L(u) = M$ elementwise (see Theorem 1.6(iv)) and $\tau_0^{-1}\tau \varepsilon M'$. Consequently by Corollary I.4.3 $\tau_0 M' = \tau M'$ and the map $S \to T$ is injective. ■

Several important applications of Lemma 2.8 are treated in the appendix. We now prove an analogue of Lemma 2.8 for subgroups of the Galois group.

Lemma 2.9. *Let* F *be an extension field of* K *and let* H,J *be subgroups of the Galois group* Aut_KF *with* H < J. *If* [J : H] *is finite, then* [H′ : J′] ≤ [J : H].

PROOF. Let $[J : H] = n$ and suppose that $[H' : J'] > n$. Then there exist $u_1,u_2, \ldots, u_{n+1} \varepsilon H'$ that are linearly independent over J'. Let $\{\tau_1,\tau_2, \ldots, \tau_n\}$ be a complete set of representatives of the left cosets of H in J (that is, $J = \tau_1 H \cup \tau_2 H \cup \cdots \cup \tau_n H$ and $\tau_i^{-1}\tau_j \varepsilon H$ if and only if $i = j$) and consider the system of n homogeneous linear equations in $n + 1$ unknowns with coefficients $\tau_i(u_j)$ in the field F:

$$\begin{aligned}
\tau_1(u_1)x_1 + \tau_1(u_2)x_2 + \tau_1(u_3)x_3 + \cdots + \tau_1(u_{n+1})x_{n+1} &= 0\\
\tau_2(u_1)x_1 + \tau_2(u_2)x_2 + \tau_2(u_3)x_3 + \cdots + \tau_2(u_{n+1})x_{n+1} &= 0\\
\vdots \qquad\qquad\qquad\qquad\qquad\qquad &\ \ \vdots\\
\tau_n(u_1)x_1 + \tau_n(u_2)x_2 + \tau_n(u_3)x_3 + \cdots + \tau_n(u_{n+1})x_{n+1} &= 0
\end{aligned} \qquad (1)$$

Such a system always has a nontrivial solution (that is, one different from the zero solution $x_1 = x_2 = \cdots = x_{n+1} = 0$; see Exercise VII.2.4(d)). Among all such nontrivial solutions choose one, say $x_1 = a_1, \ldots, x_{n+1} = a_{n+1}$ with a minimal number of nonzero a_i. By reindexing if necessary we may assume that $x_1 = a_1, \ldots, x_r = a_r$, $x_{r+1} = \cdots = x_{n+1} = 0$ $(a_i \neq 0)$. Since every multiple of a solution is also a solution we may also assume $a_1 = 1_F$ (if not multiply through by a_1^{-1}).

We shall show below that the hypothesis that $u_1, \ldots, u_{n+1} \varepsilon H'$ are linearly independent over J' (that is, that $[H' : J'] > n$) implies that there exists $\sigma \varepsilon J$ such that $x_1 = \sigma a_1, x_2 = \sigma a_2, \ldots, x_r = \sigma a_r, x_{r+1} = \cdots = x_{n+1} = 0$ is a solution of the system (1) and $\sigma a_2 \neq a_2$. Since the difference of two solutions is also a solution, $x_1 = a_1 - \sigma a_1$, $x_2 = a_2 - \sigma a_2, \ldots, x_r = a_r - \sigma a_r, x_{r+1} = \cdots = x_{n+1} = 0$, is also a solution of (1). But since $a_1 - \sigma a_1 = 1_F - 1_F = 0$ and $a_2 \neq \sigma a_2$, it follows that $x_1 = 0, x_2 = a_2 - \sigma a_2, \ldots, x_r = a_r - \sigma a_r, x_{r+1} = \cdots = x_{n+1} = 0$ is a nontrivial solution of (1) $(x_2 \neq 0)$ with at most $r - 1$ nonzero entries. This contradicts the minimality of the solution $x_1 = a_1, \ldots, x_r = a_r$, $x_{r+1} = \cdots = x_{n+1} = 0$. Therefore $[H' : J'] \leq n$ as desired.

To complete the proof we must find $\sigma \in J$ with the desired properties. Now exactly one of the τ_j, say τ_1, is in H by definition; therefore $\tau_1(u_i) = u_i \in H'$ for all i. Since the a_i form a solution of (1), the first equation of the system yields:

$$u_1a_1 + u_2a_2 + \cdots + u_ra_r = 0.$$

The linear independence of the u_i over J' and the fact that the a_i are nonzero imply that some a_i, say a_2, is not in J'. Therefore there exists $\sigma \in J$ such that $\sigma a_2 \neq a_2$.

Next consider the system of equations

$$\begin{aligned}
\sigma\tau_1(u_1)x_1 + \sigma\tau_1(u_2)x_2 + \cdots + \sigma\tau_1(u_{n+1})x_{n+1} &= 0\\
\sigma\tau_2(u_1)x_1 + \sigma\tau_2(u_2)x_2 + \cdots + \sigma\tau_2(u_{n+1})x_{n+1} &= 0\\
\vdots \qquad\qquad\qquad\qquad &\ \ \vdots\\
\sigma\tau_n(u_1)x_1 + \sigma\tau_n(u_2)x_2 + \cdots + \sigma\tau_n(u_{n+1})x_{n+1} &= 0
\end{aligned} \tag{2}$$

It is obvious, since σ is an automorphism and $x_1 = a_1, \ldots, x_r = a_r, x_{r+1} = \cdots = x_{n+1} = 0$ is a solution of (1), that $x_1 = \sigma a_1, \ldots, x_r = \sigma a_r, x_{r+1} = \cdots = x_{n+1} = 0$ is a solution of (2). We claim that system (2), except for the order of the equations, is identical with system (1) (so that $x_1 = \sigma a_1, \ldots, x_r = \sigma a_r, x_{r+1} = \cdots = x_{n+1} = 0$ is a solution of (1)). To see this the reader should first verify the following two facts.

(i) For any $\sigma \in J$, $\{\sigma\tau_1, \sigma\tau_2, \ldots, \sigma\tau_n\} \subset J$ is a complete set of coset representatives of H in J;

(ii) if ζ and θ are both elements in the same coset of H in J, then (since $u_i \in H'$) $\zeta(u_i) = \theta(u_i)$ for $i = 1,2,\ldots,n+1$.

It follows from (i) that there is some reordering $i_1, \ldots, i_{n+1}$ of $1,2,\ldots,n+1$, so that for each $k = 1,2,\ldots,n+1$ $\sigma\tau_k$ and τ_{i_k} are in the same coset of H in J. By (ii) the kth equation of (2) is identical with the i_kth equation of (1). ■

Lemma 2.10. *Let* F *be an extension field of* K, L *and* M *intermediate fields with* L ⊂ M, *and* H,J *subgroups of the Galois group* Aut_KF *with* H < J.

(i) *If* L *is closed and* [M : L] *finite, then* M *is closed and* [L′ : M′] = [M : L];

(ii) *if* H *is closed and* [J : H] *finite, then* J *is closed and* [H′ : J′] = [J : H];

(iii) *if* F *is a finite dimensional Galois extension of* K, *then all intermediate fields and all subgroups of the Galois group are closed and* Aut_KF *has order* [F : K].

Note that (ii) (with $H = 1$) implies that every finite subgroup of $\mathrm{Aut}_K F$ is closed.

SKETCH OF PROOF OF 2.10. (ii) Applying successively the facts that $J \subset J''$ and $H = H''$ and Lemmas 2.8 and 2.9 yields

$$[J : H] \leq [J'' : H] = [J'' : H''] \leq [H' : J'] \leq [J : H];$$

this implies that $J = J''$ and $[H' : J'] = [J : H]$. (i) is proved similarly.

(iii) If E is an intermediate field then $[E : K]$ is finite (since $[F : K]$ is). Since F is Galois over K, K is closed and (i) implies that E is closed and $[K' : E'] = [E : K]$. In particular, if $E = F$, then $|\mathrm{Aut}_K F| = [\mathrm{Aut}_K F : 1] = [K' : F'] = [F : K]$ is finite.

Therefore, every subgroup J of $\mathrm{Aut}_K F$ is finite. Since 1 is closed (ii) implies that J is closed. ■

The first part of the Fundamental Theorem 2.5 can easily be derived from Theorem 2.7 and Lemma 2.10. In order to prove part (ii) of Theorem 2.5 we must determine which intermediate fields correspond to normal subgroups of the Galois group under the Galois correspondence. This will be done in the next lemma.

If E is an intermediate field of the extension $K \subset F$, E is said to be **stable** (relative to K and F) if every K-automorphism $\sigma \in \mathrm{Aut}_K F$ maps E into itself. If E is stable and $\sigma^{-1} \in \mathrm{Aut}_K F$ is the inverse automorphism, then σ^{-1} also maps E into itself. This implies that $\sigma \mid E$ is in fact a K-automorphism of E (that is, $\sigma \mid E \in \mathrm{Aut}_K E$) with inverse $\sigma^{-1} \mid E$. It will turn out that in the finite dimensional case E is stable if and only if E is Galois over K.

Lemma 2.11. *Let* F *be an extension field of* K.

(i) *If* E *is a stable intermediate field of the extension, then* $E' = \mathrm{Aut}_E F$ *is a normal subgroup of the Galois group* $\mathrm{Aut}_K F$;

(ii) *if* H *is a normal subgroup of* $\mathrm{Aut}_K F$, *then the fixed field* H' *of* H *is a stable intermediate field of the extension.*

PROOF. (i) If $u \in E$ and $\sigma \in \mathrm{Aut}_K F$, then $\sigma(u) \in E$ by stability and hence $\tau\sigma(u) = \sigma(u)$ for any $\tau \in E' = \mathrm{Aut}_E F$. Therefore, for any $\sigma \in \mathrm{Aut}_K F$, $\tau \in E'$ and $u \in E$, $\sigma^{-1}\tau\sigma(u) = \sigma^{-1}\sigma(u) = u$. Consequently, $\sigma^{-1}\tau\sigma \in E'$ and hence E' is normal in $\mathrm{Aut}_K F$.

(ii) If $\sigma \in \mathrm{Aut}_K F$ and $\tau \in H$, then $\sigma^{-1}\tau\sigma \in H$ by normality. Therefore, for any $u \in H'$, $\sigma^{-1}\tau\sigma(u) = u$, which implies that $\tau\sigma(u) = \sigma(u)$ for all $\tau \in H$. Thus $\sigma(u) \in H'$ for any $u \in H'$, which means that H' is stable. ■

In the next three lemmas we explore in some detail the relationships between stable intermediate fields and Galois extensions and the relationship of both to the Galois group.

Lemma 2.12. *If* F *is a Galois extension field of* K *and* E *is a stable intermediate field of the extension, then* E *is Galois over* K.

PROOF. If $u \in E - K$, then there exists $\sigma \in \mathrm{Aut}_K F$ such that $\sigma(u) \neq u$ since F is Galois over K. But $\sigma \mid E \in \mathrm{Aut}_K E$ by stability. Therefore, E is Galois over K by the Remarks after Definition 2.4. ■

Lemma 2.13. *If* F *is an extension field of* K *and* E *is an intermediate field of the extension such that* E *is algebraic and Galois over* K, *then* E *is stable* (*relative to* F *and* K).

REMARK. The hypothesis that E is algebraic is essential; see Exercise 13.

PROOF OF 2.13. If $u \in E$, let $f \in K[x]$ be the irreducible polynomial of u and let $u = u_1, u_2, \ldots, u_r$ be the distinct roots of f that lie in E. Then $r \leq n = \deg f$ by Theorem III.6.7. If $\tau \in \mathrm{Aut}_K E$, then it follows from Theorem 2.2 that τ simply permutes the u_i. This implies that the coefficients of the monic polynomial $g(x) = (x - u_1)(x - u_2)\cdots(x - u_r) \in E[x]$ are fixed by every $\tau \in \mathrm{Aut}_K E$. Since E is Galois over K, we must have $g \in K[x]$. Now $u = u_1$ is a root of g and hence $f \mid g$ (Theorem 1.6(ii)). Since g is monic and $\deg g \leq \deg f$, we must have $f = g$. Consequently, all the roots of f are distinct and lie in E. Now if $\sigma \in \mathrm{Aut}_K F$, then $\sigma(u)$ is a root of f by Theorem 2.2, whence $\sigma(u) \in E$. Therefore, E is stable relative to F and K. ■

Let E be an intermediate field of the extension $K \subset F$. A K-automorphism $\tau \in \mathrm{Aut}_K E$ is said to be **extendible** to F if there exists $\sigma \in \mathrm{Aut}_K F$ such that $\sigma \mid E = \tau$. It is easy to see that the extendible K-automorphisms form a subgroup of $\mathrm{Aut}_K E$. Recall that if E is stable, $E' = \mathrm{Aut}_E F$ is a normal subgroup of $G = \mathrm{Aut}_K F$ (Lemma 2.11). Consequently, the quotient group G/E' is defined.

Lemma 2.14. *Let* F *be an extension field of* K *and* E *a stable intermediate field of the extension. Then the quotient group* $\mathrm{Aut}_K F/\mathrm{Aut}_E F$ *is isomorphic to the group of all* K-*automorphisms of* E *that are extendible to* F.

SKETCH OF PROOF. Since E is stable, the assignment $\sigma \mapsto \sigma \mid E$ defines a group homomorphism $\mathrm{Aut}_K F \to \mathrm{Aut}_K E$ whose image is clearly the subgroup of all K-automorphisms of E that are extendible to F. Observe that the kernel is $\mathrm{Aut}_E F$ and apply the First Isomorphism Theorem I.5.7. ■

PROOF OF THEOREM 2.5. (Fundamental Theorem of Galois Theory) Theorem 2.7 shows that there is a one-to-one correspondence between closed intermediate fields of the extension and closed subgroups of the Galois group. But in this case all intermediate fields and all subgroups are closed by Lemma 2.10(iii). Statement (i) of the theorem follows immediately from Lemma 2.10(i).

(ii) F is Galois over E since E is closed (that is, $E = E''$). E is finite dimensional over K (since F is) and hence algebraic over K by Theorem 1.11. Consequently, if E is Galois over K, then E is stable by Lemma 2.13. By Lemma 2.11(i) $E' = \mathrm{Aut}_E F$ is normal in $\mathrm{Aut}_K F$. Conversely if E' is normal in $\mathrm{Aut}_K F$, then E'' is a stable intermediate field (Lemma 2.11(ii)). But $E = E''$ since all intermediate fields are closed and hence E is Galois over K by Lemma 2.12.

Suppose E is an intermediate field that is Galois over K (so that E' is normal in $\mathrm{Aut}_K F$). Since E and E' are closed and $G' = K$ (F is Galois over K), Lemma 2.10 implies that $|G/E'| = [G : E'] = [E'' : G'] = [E : K]$. By Lemma 2.14 $G/E' = \mathrm{Aut}_K F/\mathrm{Aut}_E F$ is isomorphic to a subgroup (of order $[E : K]$) of $\mathrm{Aut}_K E$. But part (i) of the theorem shows that $|\mathrm{Aut}_K E| = [E : K]$ (since E is Galois over K). This implies that $G/E' \cong \mathrm{Aut}_K E$. ■

The modern development of Galois Theory owes a great deal to Emil Artin. Although our treatment is ultimately due to Artin (via I. Kaplansky) his approach differs from the one given here in terms of emphasis. Artin's viewpoint is that the basic object is a given field F together with a (finite) group G of automorphisms of F. One then constructs the subfield K of F as the fixed field of G (the proof that the subset of F fixed elementwise by G *is* a field is a minor variation of the proof of Theorem 2.3).

Theorem 2.15. *(Artin) Let F be a field, G a group of automorphisms of F and K the fixed field of G in F. Then F is Galois over K. If G is finite, then F is a finite dimensional Galois extension of K with Galois group G.*

PROOF. In any case G is a subgroup of $\mathrm{Aut}_K F$. If $u \in F - K$, then there must be a $\sigma \in G$ such that $\sigma(u) \neq u$. Therefore, the fixed field of $\mathrm{Aut}_K F$ is K, whence F is Galois over K. If G is finite, then Lemma 2.9 (with $H = 1$, $J = G$) shows that $[F : K] = [1' : G'] \leq [G : 1] = |G|$. Consequently, F is finite dimensional over K, whence $G = G''$ by Lemma 2.10(iii). Since $G' = K$ (and hence $G'' = K'$) by hypothesis, we have $\mathrm{Aut}_K F = K' = G'' = G$. ∎

APPENDIX: SYMMETRIC RATIONAL FUNCTIONS

Let K be a field, $K[x_1, \ldots, x_n]$ the polynomial domain and $K(x_1, \ldots, x_n)$ the field of rational functions (see the example preceding Theorem 1.5). Since $K(x_1, \ldots, x_n)$ is by definition the quotient field of $K[x_1, \ldots, x_n]$, we have $K[x_1, \ldots, x_n] \subset K(x_1, \ldots, x_n)$ (under the usual identification of f with $f/1_K$). Let S_n be the symmetric group on n letters. A rational function $\varphi \in K(x_1, \ldots, x_n)$ is said to be **symmetric** in $x_1, \ldots, x_n$ over K if for every $\sigma \in S_n$,

$$\varphi(x_1, x_2, \ldots, x_n) = \varphi(x_{\sigma(1)}, x_{\sigma(2)}, \ldots, x_{\sigma(n)}).$$

Trivially every constant polynomial is a symmetric function. If $n = 4$, then the polynomials $f_1 = x_1 + x_2 + x_3 + x_4$, $f_2 = x_1x_2 + x_1x_3 + x_1x_4 + x_2x_3 + x_2x_4 + x_3x_4$, $f_3 = x_1x_2x_3 + x_1x_2x_4 + x_1x_3x_4 + x_2x_3x_4$ and $f_4 = x_1x_2x_3x_4$ are all symmetric functions. More generally the **elementary symmetric functions** in $x_1, \ldots, x_n$ over K are defined to be the polynomials:

$$f_1 = x_1 + x_2 + \cdots + x_n = \sum_{i=1}^{n} x_i;$$
$$f_2 = \sum_{1 \leq i < j \leq n} x_i x_j;$$
$$f_3 = \sum_{1 \leq i < j < k \leq n} x_i x_j x_k;$$
$$\vdots$$
$$f_k = \sum_{1 \leq i_1 < \cdots < i_k \leq n} x_{i_1} x_{i_2} \cdots x_{i_k};$$
$$\vdots$$
$$f_n = x_1 x_2 \cdots x_n.$$

The verification that the f_i are indeed symmetric follows from the fact that they are simply the coefficients of y in the polynomial $g(y) \in K[x_1, \ldots, x_n][y]$, where

$$\begin{aligned} g(y) &= (y - x_1)(y - x_2)(y - x_3) \cdots (y - x_n) \\ &= y^n - f_1 y^{n-1} + f_2 y^{n-2} - \cdots + (-1)^{n-1} f_{n-1} y + (-1)^n f_n. \end{aligned}$$

If $\sigma \in S_n$, then the assignments $x_i \mapsto x_{\sigma(i)}$ $(i = 1, 2, \ldots, n)$ and

$$f(x_1, \ldots, x_n)/g(x_1, \ldots, x_n) \mapsto f(x_{\sigma(1)}, \ldots, x_{\sigma(n)})/g(x_{\sigma(1)}, \ldots, x_{\sigma(n)})$$

define a K-automorphism of the field $K(x_1, \ldots, x_n)$ which will also be denoted σ (Exercise 16). The map $S_n \rightarrow \mathrm{Aut}_K K(x_1, \ldots, x_n)$ given by $\sigma \mapsto \sigma$ is clearly a monomorphism of groups, whence S_n may be considered to be a subgroup of the Galois group $\mathrm{Aut}_K K(x_1, \ldots, x_n)$. Clearly, the fixed field E of S_n in $K(x_1, \ldots, x_n)$ consists precisely of the symmetric functions; that is, the set of all symmetric functions is a subfield of $K(x_1, \ldots, x_n)$ containing K. Therefore, by Artin's Theorem 2.15 $K(x_1, \ldots, x_n)$ is a Galois extension of E with Galois group S_n and dimension $|S_n| = n!$.

Proposition 2.16. *If* G *is a finite group, then there exists a Galois field extension with Galois group isomorphic to* G.

PROOF. Cayley's Theorem II.4.6 states that for $n = |G|$, G is isomorphic to a subgroup of S_n (also denoted G). Let K be any field and E the subfield of symmetric rational functions in $K(x_1, \ldots, x_n)$. The discussion preceding the theorem shows that $K(x_1, \ldots, x_n)$ is a Galois extension of E with Galois group S_n. The proof of the Fundamental Theorem 2.5 shows that $K(x_1, \ldots, x_n)$ is a Galois extension of the fixed field E_1 of G such that $\mathrm{Aut}_{E_1} K(x_1, \ldots, x_n) = G$. ■

The remainder of this appendix (which will be used only in the appendix to Section 9) is devoted to proving two classical theorems about symmetric functions. Throughout this discussion n is a positive integer, K an arbitrary field, E the subfield of symmetric rational functions in $K(x_1, \ldots, x_n)$ and $f_1, \ldots, f_n \in E$ the elementary symmetric functions in $x_1, \ldots, x_n$ over K. We have a tower of fields:

$$K \subset K(f_1, \ldots, f_n) \subset E \subset K(x_1, \ldots, x_n).$$

In Theorem 2.18 we shall show that $E = K(f_1, \ldots, f_n)$.

If $u_1, \ldots, u_r \in K(x_1, \ldots, x_n)$, then every element of $K(u_1, \ldots, u_r)$ is of the form $g(u_1, \ldots, u_r)/h(u_1, \ldots, u_r)$ with $g, h \in K[x_1, \ldots, x_r]$ by Theorem 1.3. Consequently, an element of $K(u_1, \ldots, u_r)$ [resp. $K[u_1, \ldots, u_r]$] is usually called a rational function [resp. polynomial] in $u_1, \ldots, u_r$ over K. Thus the statement $E = K(f_1, \ldots, f_n)$ may be rephrased as: every rational symmetric function is in fact a rational function of the elementary symmetric functions $f_1, \ldots, f_n$ over K. In order to prove that $E = K(f_1, \ldots, f_n)$ we need

Lemma 2.17. *Let* K *be a field,* $f_1, \ldots, f_n$ *the elementary symmetric functions in* $x_1, \ldots, x_n$ *over* K *and* k *an integer with* $1 \leq k \leq n - 1$. *If* $h_1, \ldots, h_k \in K[x_1, \ldots, x_n]$ *are the elementary symmetric functions in* $x_1, \ldots, x_k$, *then each* h_j *can be written as a polynomial over* K *in* $f_1, f_2, \ldots, f_n$ *and* $x_{k+1}, x_{k+2}, \ldots, x_n$.

SKETCH OF PROOF. The theorem is true when $k = n - 1$ since in that case $h_1 = f_1 - x_n$ and $h_j = f_j - h_{j-1}x_n$ $(2 \leq j \leq n)$. Complete the proof by induction on k in reverse order: assume that the theorem is true when $k = r + 1$ and $r + 1 \leq n - 1$. Let $g_1, \ldots, g_{r+1}$ be the elementary symmetric functions in $x_1, \ldots, x_{r+1}$ and $h_1, \ldots, h_r$ the elementary symmetric functions in $x_1, \ldots, x_r$. Since $h_1 = g_1 - x_{r+1}$ and $h_j = g_j - h_{j-1}x_{r+1}$ $(2 \leq j \leq r)$, it follows that the theorem is also true for $k = r$. ■

Theorem 2.18. *If* K *is a field,* E *the subfield of all symmetric rational functions in* $K(x_1, \ldots, x_n)$ *and* $f_1, \ldots, f_n$ *the elementary symmetric functions, then* $E = K(f_1, \ldots, f_n)$.

SKETCH OF PROOF. Since $[K(x_1, \ldots, x_n) : E] = n!$ and $K(f_1, \ldots, f_n) \subset E \subset K(x_1, \ldots, x_n)$, it suffices by Theorem 1.2 to show that $[K(x_1, \ldots, x_n) : K(f_1, \ldots, f_n)] \leq n!$. Let $F = K(f_1, \ldots, f_n)$ and consider the tower of fields:

$$F \subset F(x_n) \subset F(x_{n-1},x_n) \subset \cdots \subset F(x_2, \ldots, x_n) \subset F(x_1, \ldots, x_n) = K(x_1, \ldots, x_n).$$

Since $F(x_k,x_{k+1}, \ldots, x_n) = F(x_{k+1}, \ldots, x_n)(x_k)$, it suffices by Theorems 1.2 and 1.6 to show that x_n is algebraic over F of degree $\leq n$ and for each $k < n$, x_k is algebraic of degree $\leq k$ over $F(x_{k+1}, \ldots, x_n)$. To do this, let $g_n(y) \in F[y]$ be the polynomial

$$g_n(y) = (y - x_1)(y - x_2)\cdots(y - x_n) = y^n - f_1y^{n-1} + \cdots + (-1)^nf_n.$$

Since $g_n \in F[y]$ has degree n and x_n is a root of g_n, x_n is algebraic of degree at most n over $F = K(f_1, \ldots, f_n)$ by Theorem 1.6. Now for each k $(1 \leq k < n)$ define a monic polynomial:

$$g_k(y) = g_n(y)/(y - x_{k+1})\cdots(y - x_n) = (y - x_1)(y - x_2)\cdots(y - x_k).$$

Clearly each $g_k(y)$ has degree k, x_k is a root of $g_k(y)$ and the coefficients of $g_k(y)$ are precisely the elementary symmetric functions in $x_1, \ldots, x_k$. By Lemma 2.17 each $g_k(y)$ lies in $F(x_{k+1}, \ldots, x_n)[y]$, whence x_k is algebraic of degree at most k over $F(x_{k+1}, \ldots, x_n)$. ■

We shall now prove an analogue of Theorem 2.18 for symmetric *polynomial* functions, namely: every symmetric polynomial in $x_1, \ldots, x_n$ over K is in fact a polynomial in the elementary symmetric functions $f_1, \ldots, f_n$ over K. In other words, every symmetric polynomial in $K[x_1, \ldots, x_n]$ lies in $K[f_1, \ldots, f_n]$. First we need

Lemma 2.19. *Let* K *be a field and* E *the subfield of all symmetric rational functions in* $K(x_1, \ldots, x_n)$. *Then the set* $X = \{x_1^{i_1}x_2^{i_2}\cdots x_n^{i_n} \mid 0 \leq i_k < k$ *for each* $k\}$ *is a basis of* $K(x_1, \ldots, x_n)$ *over* E.

SKETCH OF PROOF. Since $[K(x_1, \ldots, x_n) : E] = n!$ and $|X| = n!$, it suffices to show that X spans $K(x_1, \ldots, x_n)$ (see Theorem IV.2.5). Consider the tower of fields $E \subset E(x_n) \subset E(x_{n-1},x_n) \subset \cdots \subset E(x_1, \ldots, x_n) = K(x_1, \ldots, x_n)$. Since x_n is algebraic of degree $\leq n$ over E (by the proof of Theorem 2.18), the set $\{x_n{}^j \mid 0 \leq j < n\}$ spans $E(x_n)$ over E (Theorem 1.6). Since $E(x_{n-1},x_n) = E(x_n)(x_{n-1})$, and x_{n-1} is algebraic of degree $\leq n - 1$ over $E(x_n)$, the set $\{x_{n-1}^i \mid 0 \leq i < n - 1\}$ spans $E(x_{n-1},x_n)$ over $E(x_n)$. The argument in the second paragraph of the proof of Theorem IV.2.16 shows that the set $\{x_{n-1}^i x_n{}^j \mid 0 \leq i < n - 1; 0 \leq j < n\}$ spans $E(x_{n-1},x_n)$ over E. This is the first step in an inductive proof, which is completed by similar arguments. ■

Proposition 2.20. *Let* K *be a field and let* $f_1, \ldots, f_n$ *be the elementary symmetric functions in* $K(x_1, \ldots, x_n)$.

(i) *Every polynomial in* $K[x_1, \ldots, x_n]$ *can be written uniquely as a linear combination of the* $n!$ *elements* $x_1^{i_1}x_2^{i_2}\cdots x_n^{i_n}$ $(0 \leq i_k < k$ *for each* $k)$ *with coefficients in* $K[f_1, \ldots, f_n]$;

(ii) *every symmetric polynomial in* $K[x_1, \ldots, x_n]$ *lies in* $K[f_1, \ldots, f_n]$.

PROOF. Let $g_k(y)$ $(k = 1, \ldots, n)$ be as in the proof of Theorem 2.18. As noted there the coefficients of $g_k(y)$ are polynomials (over K) in $f_1, \ldots, f_n$ and $x_{k+1}, \ldots, x_n$. Since g_k is monic of degree k and $g_k(x_k) = 0$, $x_k{}^k$ can be expressed as a polynomial over K in $f_1, \ldots, f_n, x_{k+1}, \ldots, x_n$ and $x_k{}^i$ $(i \leq k - 1)$. If we proceed step by step beginning with $k = 1$ and substitute this expression for $x_k{}^k$ in a polynomial $h \in K[x_1, \ldots, x_n]$, the result is a polynomial in $f_1, \ldots, f_n, x_1, \ldots, x_n$ in which the highest exponent of any x_k is $k - 1$. In other words h is a linear combination of the $n!$ elements $x_1^{i_1}x^{i_2}\cdots x_n{}^{i_n}$ $(i_k < k$ for each $k)$ with coefficients in $K[f_1, \ldots, f_n]$. Furthermore these coefficient polynomials are uniquely determined since

$$\{x_1{}^{i_1}\cdots x_n{}^{i_n} \mid 0 \leq i_k < k \quad \text{for each } k\}$$

is linearly independent over $E = K(f_1, \ldots, f_n)$ by Lemma 2.19. This proves (i) and also implies that if a polynomial $h \in K[x_1, \ldots, x_n]$ is a linear combination of the $x_1{}^{i_1}\cdots x_n{}^{i_n}$ $(i_k < k)$ with coefficients in $K(f_1, \ldots, f_n)$, then the coefficients are in fact polynomials in $K[f_1, \ldots, f_n]$. In particular, if h is a symmetric polynomial (that is, $h \in E = K(f_1, \ldots, f_n)$), then $h = hx_1{}^0x_2{}^0\cdots x_n{}^0$ necessarily lies in $K[f_1, \ldots, f_n]$. This proves (ii). ■

EXERCISES

Note: Unless stated otherwise F is always an extension field of the field K and E is an intermediate field of the extension.

1. (a) If F is a field and $\sigma : F \to F$ a (ring) homomorphism, then $\sigma = 0$ or σ is a monomorphism. If $\sigma \neq 0$, then $\sigma(1_F) = 1_F$.
   (b) The set Aut F of all field automorphisms $F \to F$ forms a group under the operation of composition of functions.
   (c) $\text{Aut}_K F$, the set of all K-automorphisms of F is a subgroup of Aut F.
2. $\text{Aut}_{\mathbf{Q}}\mathbf{R}$ is the identity group. [*Hint:* Since every positive element of $\mathbf{R}$ is a square, it follows that an automorphism of $\mathbf{R}$ sends positives to positives and hence that it preserves the order in $\mathbf{R}$. Trap a given real number between suitable rational numbers.]
3. If $0 \leq d \in \mathbf{Q}$, then $\text{Aut}_{\mathbf{Q}}\mathbf{Q}(\sqrt{d})$ is the identity or is isomorphic to Z_2.
4. What is the Galois group of $\mathbf{Q}(\sqrt{2},\sqrt{3},\sqrt{5})$ over $\mathbf{Q}$?
5. (a) If $0 \leq d \in \mathbf{Q}$, then $\mathbf{Q}(\sqrt{d})$ is Galois over $\mathbf{Q}$.
   (b) $\mathbf{C}$ is Galois over $\mathbf{R}$.
6. Let $f/g \in K(x)$ with $f/g \notin K$ and f,g relatively prime in $K[x]$ and consider the extension of K by $K(x)$.
   (a) x is algebraic over $K(f/g)$ and $[K(x) : K(f/g)] = \max(\deg f, \deg g)$. [*Hint:* x is a root of the nonzero polynomial $\varphi(y) = (f/g)g(y) - f(y) \in K(f/g)[y]$; show that φ has degree max (deg f,deg g). Show that φ is irreducible as follows.

Since f/g is transcendental over K (why?) we may for convenience replace $K(f/g)$ by $K(z)$ (z an indeterminate) and consider $\varphi = zg(y) - f(y) \varepsilon K(z)[y]$. By Lemma III.6.13 φ is irreducible in $K(z)[y]$ provided it is irreducible in $K[z][y]$. The truth of this latter condition follows from the fact that φ is linear in z and f,g are relatively prime.]

(b) If $E \neq K$ is an intermediate field, then $[K(x) : E]$ is finite.

(c) The assignment $x \mapsto f/g$ induces a homomorphism $\sigma : K(x) \to K(x)$ such that $\varphi(x)/\psi(x) \mapsto \varphi(f/g)/\psi(f/g)$. σ is a K automorphism of $K(x)$ if and only if max (deg f,deg g) $= 1$.

(d) $\mathrm{Aut}_K K(x)$ consists of all those automorphisms induced (as in (c)) by the assignment

$$x \mapsto (ax + b)/(cx + d),$$

where $a,b,c,d \varepsilon K$ and $ad - bc \neq 0$.

7. Let G be the subset of $\mathrm{Aut}_K K(x)$ consisting of the three automorphisms induced (as in 6 (c)) by $x \mapsto x$, $x \mapsto 1_K/(1_K - x)$, $x \mapsto (x - 1_K)/x$. Then G is a subgroup of $\mathrm{Aut}_K K(x)$. Determine the fixed field of G.

8. Assume char $K = 0$ and let G be the subgroup of $\mathrm{Aut}_K K(x)$ that is generated by the automorphism induced by $x \mapsto x + 1_K$. Then G is an infinite cyclic group. Determine the fixed field E of G. What is $[K(x) : E]$?

9. (a) If K is an infinite field, then $K(x)$ is Galois over K. [*Hint:* If $K(x)$ is not Galois over K, then $K(x)$ is finite dimensional over the fixed field E of $\mathrm{Aut}_K K(x)$ by Exercise 6(b). But $\mathrm{Aut}_E K(x) = \mathrm{Aut}_K K(x)$ is infinite by Exercise 6(d), which contradicts Lemma 2.8.]
(b) If K is finite, then $K(x)$ is *not* Galois over K. [*Hint:* If $K(x)$ were Galois over K, then $\mathrm{Aut}_K K(x)$ would be infinite by Lemma 2.9. But $\mathrm{Aut}_K K(x)$ is finite by Exercise 6(d).]

10. If K is an infinite field, then the only closed subgroups of $\mathrm{Aut}_K K(x)$ are itself and its finite subgroups. [*Hint:* see Exercises 6(b) and 9.]

11. In the extension of $\mathbf{Q}$ by $\mathbf{Q}(x)$, the intermediate field $\mathbf{Q}(x^2)$ is closed, but $\mathbf{Q}(x^3)$ is not.

12. If E is an intermediate field of the extension such that E is Galois over K, F is Galois over E, and every $\sigma \varepsilon \mathrm{Aut}_K E$ is extendible to F, then F is Galois over K.

13. In the extension of an infinite field K by $K(x,y)$, the intermediate field $K(x)$ is Galois over K, but not stable (relative to $K(x,y)$ and K). [See Exercise 9; compare this result with Lemma 2.13.]

14. Let F be a finite dimensional Galois extension of K and let L and M be two intermediate fields.
(a) $\mathrm{Aut}_{LM} F = \mathrm{Aut}_L F \cap \mathrm{Aut}_M F$;
(b) $\mathrm{Aut}_{L \cap M} F = \mathrm{Aut}_L F \vee \mathrm{Aut}_M F$;
(c) What conclusion can be drawn if $\mathrm{Aut}_L F \cap \mathrm{Aut}_M F = 1$?

15. If F is a finite dimensional Galois extension of K and E is an intermediate field, then there is a unique smallest field L such that $E \subset L \subset F$ and L is Galois over K; furthermore

$$\mathrm{Aut}_L F = \bigcap_\sigma \sigma(\mathrm{Aut}_E F)\sigma^{-1},$$

where σ runs over $\mathrm{Aut}_K F$.

16. If $\sigma \in S_n$, then the map $K(x_1, \ldots, x_n) \to K(x_1, \ldots, x_n)$ given by

$$\frac{f(x_1, \ldots, x_n)}{g(x_1, \ldots, x_n)} \mapsto \frac{f(x_{\sigma(1)}, \ldots, x_{\sigma(n)})}{g(x_{\sigma(1)}, \ldots, x_{\sigma(n)})}$$

is a K-automorphism of $K(x_1, \ldots, x_n)$.

3. SPLITTING FIELDS, ALGEBRAIC CLOSURE AND NORMALITY

We turn now to the problem of identifying and/or constructing Galois extensions. Splitting fields, which constitute the principal theme of this section, will enable us to do this. We first develop the basic properties of splitting fields and algebraic closures (a special case of splitting fields). Then algebraic Galois extensions are characterized in terms that do not explicitly mention the Galois group (Theorem 3.11), and the Fundamental Theorem is extended to the infinite dimensional algebraic case (Theorem 3.12). Finally normality and other characterizations of splitting fields are discussed. The so-called fundamental theorem of algebra (every polynomial equation over the complex numbers has a solution) is proved in the appendix.

Let F be a field and $f \in F[x]$ a polynomial of *positive* degree. f is said to **split** over F (or to split in $F[x]$) if f can be written as a product of linear factors in $F[x]$; that is, $f = u_0(x - u_1)(x - u_2)\cdots(x - u_n)$ with $u_i \in F$.

Definition 3.1. *Let* K *be a field and* $f \in K[x]$ *a polynomial of positive degree. An extension field* F *of* K *is said to be a* **splitting field over K of the polynomial** f *if* f *splits in* F[x] *and* $F = K(u_1, \ldots, u_n)$ *where* $u_1, \ldots, u_n$ *are the roots of* f *in* F.

Let S *be a set of polynomials of positive degree in* K[x]. *An extension field* F *of* K *is said to be a* **splitting field over K of the set S of polynomials** *if every polynomial in* S *splits in* F[x] *and* F *is generated over* K *by the roots of all the polynomials in* S.

EXAMPLES. The only roots of $x^2 - 2$ over $\mathbf{Q}$ are $\sqrt{2}$ and $-\sqrt{2}$ and $x^2 - 2 = (x - \sqrt{2})(x + \sqrt{2})$. Therefore $\mathbf{Q}(\sqrt{2}) = \mathbf{Q}(\sqrt{2}, -\sqrt{2})$ is a splitting field of $x^2 - 2$ over $\mathbf{Q}$. Similarly $\mathbf{C}$ is a splitting field of $x^2 + 1$ over $\mathbf{R}$. However, if u is a root of an irreducible $f \in K[x]$, $K(u)$ need not be a splitting field of f. For instance if u is the real cube root of 2 (the others being complex), then $\mathbf{Q}(u) \subset \mathbf{R}$, whence $\mathbf{Q}(u)$ is not a splitting field of $x^3 - 2$ over $\mathbf{Q}$.

REMARKS. If F is a splitting field of S over K, then $F = K(X)$, where X is the set of all roots of polynomials in the subset S of $K[x]$. Theorem 1.12 immediately implies that F is algebraic over K (and finite dimensional if S, and hence X, is a finite set). Note that if S is finite, say $S = \{f_1, f_2, \ldots, f_n\}$, then a splitting field of S coincides with a splitting field of the single polynomial $f = f_1 f_2 \cdots f_n$ (Exercise 1). This fact will be used frequently in the sequel without explicit mention. Thus the splitting field of a set S of polynomials will be chiefly of interest when S either consists of a single polynomial or is infinite. It will turn out that every [finite dimensional] algebraic Galois extension is in fact a particular kind of splitting field of a [finite] set of polynomials.

The obvious question to be answered next is whether every set of polynomials has a splitting field. In the case of a single polynomial (or equivalently a finite set of polynomials), the answer is relatively easy.

Theorem 3.2. *If* K *is a field and* $f \varepsilon K[x]$ *has degree* $n \geq 1$, *then there exists a splitting field* F *of* f *with* $[F : K] \leq n!$

SKETCH OF PROOF. Use induction on $n = \deg f$. If $n = 1$ or if f splits over K, then $F = K$ is a splitting field. If $n > 1$ and f does not split over K, let $g \varepsilon K[x]$ be an irreducible factor of f of degree greater than one. By Theorem 1.10 there is a simple extension field $K(u)$ of K such that u is a root of g and $[K(u) : K] = \deg g > 1$. Then by Theorem III. 6.6, $f = (x - u)h$ with $h \varepsilon K(u)[x]$ of degree $n - 1$. By induction there exists a splitting field F of h over $K(u)$ of dimension at most $(n - 1)!$ Show that F is a splitting field of f over K (Exercise 3) of dimension $[F : K] = [F : K(u)][K(u) : K] \leq (n - 1)! (\deg g) \leq n!$ ■

Proving the existence of a splitting field of an infinite set of polynomials is considerably more difficult. We approach the proof obliquely by introducing a special case of such a splitting field (Theorem 3.4) which is of great importance in its own right.

Note: The reader who is interested only in splitting fields of a single polynomial (i.e. finite dimensional splitting fields) should skip to Theorem 3.8. Theorem 3.12 should be omitted and Theorems 3.8–3.16 read in the finite dimensional case. The proof of each of these results is either divided in two cases (finite and infinite dimensional) or is directly applicable to both cases. The only exception is the proof of (ii) ⇒ (i) in Theorem 3.14; an alternate proof is suggested in Exercise 25.

Theorem 3.3. *The following conditions on a field* F *are equivalent.*

(i) *Every nonconstant polynomial* $f \varepsilon F[x]$ *has a root in* F;

(ii) *every nonconstant polynomial* $f \varepsilon F[x]$ *splits over* F;

(iii) *every irreducible polynomial in* F[x] *has degree one;*

(iv) *there is no algebraic extension field of* F *(except* F *itself);*

(v) *there exists a subfield* K *of* F *such that* F *is algebraic over* K *and every polynomial in* K[x] *splits in* F[x].

PROOF. Exercise; see Section III. 6 and Theorems 1.6, 1.10, 1.12 and 1.13. ■

A field that satisfies the equivalent conditions of Theorem 3.3 is said to be **algebraically closed.** For example, we shall show that the field **C** of complex numbers is algebraically closed (Theorem 3.19).

Theorem 3.4. *If* F *is an extension field of* K, *then the following conditions are equivalent.*

(i) F *is algebraic over* K *and* F *is algebraically closed;*
(ii) F *is a splitting field over* K *of the set of all* [*irreducible*] *polynomials in* K[x].

PROOF. Exercise; also see Exercises 9, 10. ∎

An extension field F of a field K that satisfies the equivalent conditions of Theorem 3.4 is called an **algebraic closure** of K. For example, $\mathbf{C} = \mathbf{R}(i)$ is an algebraic closure of $\mathbf{R}$. Clearly, if F is an algebraic closure of K and S is any set of polynomials in $K[x]$, then the subfield E of F generated by K and all roots of polynomials in S is a splitting field of S over K by Theorems 3.3 and 3.4. Thus the existence of arbitrary splitting fields over a field K is equivalent to the existence of an algebraic closure of K.

The chief difficulty in proving that every field K has an algebraic closure is set-theoretic rather than algebraic. The basic idea is to apply Zorn's Lemma to a suitably chosen *set* of algebraic extension fields of K.[2] To do this we need

Lemma 3.5. *If* F *is an algebraic extension field of* K, *then* $|F| \leq \aleph_0|K|$.

SKETCH OF PROOF. Let T be the set of monic polynomials of positive degree in $K[x]$. We first show that $|T| = \aleph_0|K|$. For each $n \in \mathbf{N}^*$ let T_n be the set of all polynomials in T of degree n. Then $|T_n| = |K^n|$, where $K^n = K \times K \times \cdots K$ (n factors), since every polynomial $f = x^n + a_{n-1}x^{n-1} + \cdots + a_0 \in T_n$ is completely determined by its n coefficients $a_0, a_1, \ldots, a_{n-1} \in K$. For each $n \in \mathbf{N}^*$ let $f_n : T_n \to K^n$ be a bijection. Since the sets T_n [resp. K^n] are mutually disjoint, the map $f : T = \bigcup_{n\in\mathbf{N}^*} T_n \to \bigcup_{n\in\mathbf{N}^*} K^n$, given by $f(u) = f_n(u)$ for $u \in T_n$, is a well-defined bijection. Therefore $|T| = |\bigcup_{n\in\mathbf{N}^*} K^n| = \aleph_0|K|$ by Introduction, Theorem 8.12(ii).

Next we show that $|F| \leq |T|$, which will complete the proof. For each irreducible $f \in T$, choose an ordering of the distinct roots of f in F. Define a map $F \to T \times \mathbf{N}^*$ as follows. If $a \in F$, then a is algebraic over K by hypothesis, and there exists a unique irreducible monic polynomial $f \in T$ with $f(a) = 0$ (Theorem 1.6). Assign to $a \in F$ the pair $(f,i) \in T \times \mathbf{N}^*$ where a is the ith root of f in the previously chosen ordering of the roots of f in F. Verify that this map $F \to T \times \mathbf{N}^*$ is well defined and injective. Since T is infinite, $|F| \leq |T \times \mathbf{N}^*| = |T||\mathbf{N}^*| = |T|\aleph_0 = |T|$ by Theorem 8.11 of the Introduction. ∎

Theorem 3.6. *Every field* K *has an algebraic closure. Any two algebraic closures of* K *are* K*-isomorphic.*

SKETCH OF PROOF. Choose a set S such that $\aleph_0|K| < |S|$ (this can always be done by Theorem 8.5 of the Introduction). Since $|K| \leq \aleph_0|K|$ (Introduction, Theorem 8.11) there is by Definition 8.4 of the Introduction an injective map $\theta : K \to S$. Consequently we may assume $K \subset S$ (if not, replace S by the union of $S - \operatorname{Im}\theta$ and K).

[2]As anyone familiar with the paradoxes of set theory (Introduction, Section 2) might suspect, the class of *all* algebraic extension fields of K need not be a set, and therefore, cannot be used in such an argument.

Let $\mathcal{S}$ be the class of all fields E such that E is a subset of S and E is an algebraic extension field of K. Such a *field* E is completely determined by the subset E of S and the binary operations of addition and multiplication in E. Now addition [resp. multiplication] is a function $\varphi : E \times E \to E$ [resp. $\psi : E \times E \to E$]. Hence φ [resp. ψ] may be identified with its graph, a certain subset of $E \times E \times E \subset S \times S \times S$ (see Introduction, Section 4). Consequently, there is an injective map τ from $\mathcal{S}$ into the *set* P of all subsets of the set $S \times (S \times S \times S) \times (S \times S \times S)$, given by $E \mapsto (E,\varphi,\psi)$. Now Im τ is actually a set since Im τ is a subclass of the *set* P. Since $\mathcal{S}$ is the image of Im τ under the function $\tau^{-1} : \text{Im } \tau \to \mathcal{S}$, the axioms of set theory guarantee that $\mathcal{S}$ is in fact a set.

Note that $\mathcal{S} \neq \varnothing$ since $K \in \mathcal{S}$. Partially order the set $\mathcal{S}$ by defining $E_1 \leq E_2$ if and only if E_2 is an extension field of E_1. Verify that every chain in $\mathcal{S}$ has an upper bound (the union of the fields in the chain will do). Therefore by Zorn's Lemma there exists a maximal element F of $\mathcal{S}$.

We claim that F is algebraically closed. If not, then some $f \in F[x]$ does not split over F. Thus there is a proper algebraic extension $F_0 = F(u)$ of F, where u is a root of f which does not lie in F (Theorem 1.10). Furthermore F_0 is an algebraic extension of K by Theorem 1.13. Therefore, $|F_0 - F| \leq |F_0| \leq \aleph_0|K| < |S|$ by Lemma 3.5. Since $|F| \leq |F_0| < |S|$ and $|S| = |(S - F) \cup F| = |S - F| + |F|$, we must have $|S| = |S - F|$ by Theorem 8.10 of the Introduction. Thus $|F_0 - F| < |S - F|$ and the identity map on F may be extended to an injective map of sets $\zeta : F_0 \to S$. Then $F_1 = \text{Im } \zeta$ may be made into a field by defining $\zeta(a) + \zeta(b) = \zeta(a + b)$ and $\zeta(a)\zeta(b) = \zeta(ab)$. Clearly F_1 is an extension field of F, $F_1 \subset S$ and $\zeta : F_0 \to F_1$ is an F-isomorphism of fields. Consequently, since F_0 is a proper algebraic extension of F (and hence of K), so is F_1. This means that $F_1 \in \mathcal{S}$ and $F < F_1$, which contradicts the maximality of F. Therefore, F is algebraically closed and algebraic over K and hence an algebraic closure of K. The uniqueness statement of the theorem is proved in Corollary 3.9 below. ∎

Corollary 3.7. *If* K *is a field and* S *a set of polynomials (of positive degree) in* K[x], *then there exists a splitting field of* S *over* K.

PROOF. Exercise. ∎

We turn now to the question of the uniqueness of splitting fields and algebraic closures. The answer will be an immediate consequence of the following result on the extendibility of isomorphisms (see Theorem 1.8 and the remarks preceding it).

Theorem 3.8. *Let* $\sigma : K \to L$ *be an isomorphism of fields,* $S = \{f_i\}$ *a set of polynomials (of positive degree) in* K[x], *and* $S' = \{\sigma f_i\}$ *the corresponding set of polynomials in* L[x]. *If* F *is a splitting field of* S *over* K *and* M *is a splitting field of* S′ *over* L, *then* σ *is extendible to an isomorphism* $F \cong M$.

SKETCH OF PROOF. Suppose first that S consists of a single polynomial $f \in K[x]$ and proceed by induction on $n = [F : K]$. If $n = 1$, then $F = K$ and f splits over K. This implies that σf splits over L and hence that $L = M$. Thus σ itself is the desired isomorphism $F = K \xrightarrow{\sigma} L = M$. If $n > 1$, then f must have an irreducible factor g of degree greater than 1. Let u be a root of g in F. Then verify that σg is irreducible in $L[x]$. If v is a root of σg in M, then by Theorem 1.8 σ extends to an isomorphism $\tau : K(u) \cong L(v)$ with $\tau(u) = v$. Since $[K(u) : K] = \deg g > 1$ (Theorem 1.6), we must have $[F : K(u)] < n$ (Theorem 1.2). Since F is a splitting field of f over $K(u)$ and M is a splitting field of σf over $L(v)$ (Exercise 2), the induction hypothesis implies that τ extends to an isomorphism $F \cong M$.

If S is arbitrary, let $\mathcal{S}$ consist of all triples (E,N,τ), where E is an intermediate field of F and K, N is an intermediate field of M and L, and $\tau : E \to N$ is an isomorphism that extends σ. Define $(E_1,N_1,\tau_1) \leq (E_2,N_2,\tau_2)$ if $E_1 \subset E_2$, $N_1 \subset N_2$ and $\tau_2 \mid E_1 = \tau_1$. Verify that $\mathcal{S}$ is a nonempty partially ordered set in which every chain has an upper bound in $\mathcal{S}$. By Zorn's Lemma there is a maximal element (F_0,M_0,τ_0) of $\mathcal{S}$. We claim that $F_0 = F$ and $M_0 = M$, so that $\tau_0 : F \cong M$ is the desired extension of σ. If $F_0 \neq F$, then some $f \in S$ does not split over F_0. Since all the roots of f lie in F, F contains a splitting field F_1 of f over F_0. Similarly, M contains a splitting field M_1 of $\tau_0 f = \sigma f$ over M_0. The first part of the proof shows that τ_0 can be extended to an isomorphism $\tau_1 : F_1 \cong M_1$. But this means that $(F_1,M_1,\tau_1) \in \mathcal{S}$ and $(F_0,M_0,\tau_0) < (F_1,M_1,\tau_1)$ which contradicts the maximality of (F_0,M_0,τ_0). A similar argument using τ_0^{-1} works if $M_0 \neq M$. ■

Corollary 3.9. *Let* K *be a field and* S *a set of polynomials (of positive degree) in* K[x]. *Then any two splitting fields of* S *over* K *are* K*-isomorphic. In particular, any two algebraic closures of* K *are* K*-isomorphic.*

SKETCH OF PROOF. Apply Theorem 3.8 with $\sigma = 1_K$. The last statement is then an immediate consequence of Theorem 3.4(ii). ■

In order to characterize Galois extensions in terms of splitting fields, we must first consider a phenomenon that occurs only in the case of fields of nonzero characteristic. Recall that if K is any field, f is a nonzero polynomial in $K[x]$, and c is a root of f, then $f = (x - c)^m g(x)$ where $g(c) \neq 0$ and m is a uniquely determined positive integer. The element c is a **simple** or **multiple root** of f according as $m = 1$ or $m > 1$ (see p. 161).

Definition 3.10. *Let* K *be a field and* f ε K[x] *an irreducible polynomial. The polynomial* f *is said to be* **separable** *if in some splitting field of* f *over* K *every root of* f *is a simple root.*

If F *is an extension field of* K *and* u ε F *is algebraic over* K, *then* u *is said to be* **separable** *over* K *provided its irreducible polynomial is separable. If every element of* F *is separable over* K, *then* F *is said to be a* **separable extension** *of* K.

REMARKS. (i) In view of Corollary 3.9 it is clear that a separable polynomial $f \in K[x]$ has no multiple roots in *any* splitting field of f over K. (ii) Theorem III.6.10 shows that an irreducible polynomial in $K[x]$ is separable if and only if its derivative is nonzero, whence every irreducible polynomial is separable if char $K = 0$ (Exercise III.6.3). *Hence every algebraic extension field of a field of characteristic* 0 *is separable.* (iii) Separability is defined here only for *irreducible* polynomials. (iv) According to Definition 3.10 a separable extension field of K is necessarily algebraic over K. There is a definition of separability for possibly nonalgebraic extension fields that agrees with this one in the algebraic case (Section VI.2). Throughout this chapter, however, we shall use only Definition 3.10.

EXAMPLES. $x^2 + 1 \in \mathbf{Q}[x]$ is separable since $x^2 + 1 = (x + i)(x - i)$ in $\mathbf{C}[x]$. On the other hand, the polynomial $x^2 + 1$ over Z_2 has no simple roots; in fact it is not even irreducible since $x^2 + 1 = (x + 1)^2$ in $Z_2[x]$.

Theorem 3.11. *If* F *is an extension field of* K, *then the following statements are equivalent.*

(i) F *is algebraic and Galois over* K;

(ii) F *is separable over* K *and* F *is a splitting field over* K *of a set* S *of polynomials in* K[x];

(iii) F *is a splitting field over* K *of a set* T *of separable polynomials in* K[x].

REMARKS. If F is finite dimensional over K, then statements (ii) and (iii) can be slightly sharpened. In particular (iii) may be replaced by: F is a splitting field over K of a polynomial $f \in K[x]$ whose irreducible factors are separable (Exercise 13).

PROOF OF 3.11. (i) $\Rightarrow$ (ii) and (iii). If $u \in F$ has irreducible polynomial f, then the first part of the proof of Lemma 2.13 (with $E = F$) carries over verbatim and shows that f splits in $F[x]$ into a product of distinct linear factors. Hence u is separable over K. Let $\{v_i \mid i \in I\}$ be a basis of F over K and for each $i \in I$ let $f_i \in K[x]$ be the irreducible polynomial of v_i. The preceding remarks show that each f_i is separable and splits in $F[x]$. Therefore F is a splitting field over K of $S = \{f_i \mid i \in I\}$.

(ii) $\Rightarrow$ (iii) Let $f \in S$ and let $g \in K[x]$ be a monic irreducible factor of f. Since f splits in $F[x]$, g must be the irreducible polynomial of some $u \in F$. Since F is separable over K, g is necessarily separable. It follows that F is a splitting field over K of the set T of separable polynomials consisting of all monic irreducible factors (in $K[x]$) of polynomials in S (see Exercise 4).

(iii) $\Rightarrow$ (i) F is algebraic over K since any splitting field over K is an algebraic extension. If $u \in F - K$, then $u \in K(v_1, \dots, v_n)$ with each v_i a root of some $f_i \in T$ by the definition of a splitting field and Theorem 1.3(vii). Thus $u \in E = K(u_1, \dots, u_r)$ where the u_i are all the roots of $f_1, \dots, f_n$ in F. Hence $[E : K]$ is finite by Theorem 1.12. Since each f_i splits in F, E is a splitting field over K of the finite set $\{f_1, \dots, f_n\}$, or equivalently, of $f = f_1 f_2 \cdots f_n$. Assume for now that the theorem is true in the finite dimensional case. Then E is Galois over K and hence there exists $\tau \in \mathrm{Aut}_K E$ such that $\tau(u) \neq u$. Since F is a splitting field of T over E (Exercise 2), τ extends to an automorphism $\sigma \in \mathrm{Aut}_K F$ such that $\sigma(u) = \tau(u) \neq u$ by Theorem 3.8. Therefore, u (which was an arbitrary element of $F - K$) is not in the fixed field of $\mathrm{Aut}_K F$; that is, F is Galois over K.

The argument in the preceding paragraph shows that we need only prove the theorem when $[F : K]$ is finite. In this case there exist a finite number of polynomials $g_1, \dots, g_t \in T$ such that F is a splitting field of $\{g_1, \dots, g_t\}$ over K (otherwise F would be infinite dimensional over K). Furthermore $\mathrm{Aut}_K F$ is a finite group by Lemma 2.8. If K_0 is the fixed field of $\mathrm{Aut}_K F$, then F is a Galois extension of K_0 with

$[F:K_0] = |\mathrm{Aut}_K F|$ by Artin's Theorem 2.15 and the Fundamental Theorem. Thus in order to show that F is Galois over K (that is, $K = K_0$) it suffices to show that $[F:K] = |\mathrm{Aut}_K F|$.

We proceed by induction on $n = [F:K]$, with the case $n = 1$ being trivial. If $n > 1$, then one of the g_i, say g_1, has degree $s > 1$ (otherwise all the roots of the g_i lie in K and $F = K$). Let $u \in F$ be a root of g_1; then $[K(u):K] = \deg g_1 = s$ by Theorem 1.6 and the number of distinct roots of g_1 is s since g_1 is separable. The second paragraph of the proof of Lemma 2.8 (with $L = K$, $M = K(u)$ and $f = g_1$) shows that there is an injective map from the set of all left cosets of $H = \mathrm{Aut}_{K(u)}F$ in $\mathrm{Aut}_K F$ to the set of all roots of g_1 in F, given by $\sigma H \mapsto \sigma(u)$. Therefore, $[\mathrm{Aut}_K F : H] \leq s$. Now if $v \in F$ is any other root of g_1, there is an isomorphism $\tau : K(u) \cong K(v)$ with $\tau(u) = v$ and $\tau \mid K = 1_K$ by Corollary 1.9. Since F is a splitting field of $\{g_1, \ldots, g_t\}$ over $K(u)$ and over $K(v)$ (Exercise 2), τ extends to an automorphism $\sigma \in \mathrm{Aut}_K F$ with $\sigma(u) = v$ (Theorem 3.8). Therefore, every root of g_1 is the image of some coset of H and $[\mathrm{Aut}_K F : H] = s$. Furthermore, F is a splitting field over $K(u)$ of the set of all irreducible factors h_j (in $K(u)[x]$) of the polynomials g_i (Exercise 4). Each h_j is clearly separable since it divides some g_i. Since $[F:K(u)] = n/s < n$, the induction hypothesis implies that $[F:K(u)] = |\mathrm{Aut}_{K(u)}F| = |H|$. Therefore,

$$[F:K] = [F:K(u)][K(u):K] = |H|s = |H|[\mathrm{Aut}_K F : H] = |\mathrm{Aut}_K F|$$

and the proof is complete. ■

Theorem 3.12. (*Generalized Fundamental Theorem*) *If* F *is an algebraic Galois extension field of* K, *then there is a one-to-one correspondence between the set of all intermediate fields of the extension and the set of all closed subgroups of the Galois group* $\mathrm{Aut}_K F$ *(given by* $E \mapsto E' = \mathrm{Aut}_E F$*) such that:*

(ii′) F *is Galois over every intermediate field* E, *but* E *is Galois over* K *if and only if the corresponding subgroup* E′ *is normal in* $G = \mathrm{Aut}_K F$; *in this case* G/E' *is (isomorphic to) the Galois group* $\mathrm{Aut}_K E$ *of* E *over* K.

REMARKS. Compare this Theorem, which is proved below, with Theorem 2.5. The analogue of (i) in the Fundamental Theorem is false in the infinite dimensional case (Exercise 16). If $[F:K]$ is infinite there are always subgroups of $\mathrm{Aut}_K F$ that are not closed. The proof of this fact depends on an observation of Krull[64]: when F is algebraic over K, it is possible to make $\mathrm{Aut}_K F$ into a compact topological group in such a way that a subgroup is topologically closed if and only if it is closed in the sense of Section 2 (that is, $H = H''$). It is not difficult to show that some infinite compact topological groups contains subgroups that are not topologically closed. A fuller discussion, with examples, is given in P. J. McCarthy [40; pp. 60–63]. Also see Exercise 5.11 below.

PROOF OF 3.12. In view of Theorem 2.7 we need only show that every intermediate field E is closed in order to establish the one-to-one correspondence. By Theorem 3.11 F is the splitting field over K of a set T of separable polynomials. Therefore, F is also a splitting field of T over E (Exercise 2). Hence by Theorem 3.11 again, F is Galois over E; that is, E is closed.

(ii′) Since every intermediate field E is algebraic over K, the first paragraph of the proof of Theorem 2.5(ii) carries over to show that E is Galois over K if and only if E' is normal in $\mathrm{Aut}_K F$.

If $E = E''$ is Galois over K, so that E' is normal in $G = \mathrm{Aut}_K F$, then E is a stable intermediate field by Lemma 2.11. Therefore, Lemma 2.14 implies that $G/E' = \mathrm{Aut}_K F/\mathrm{Aut}_E F$ is isomorphic to the subgroup of $\mathrm{Aut}_K E$ consisting of those automorphisms that are extendible to F. But F is a splitting field over K (Theorem 3.11) and hence over E also (Exercise 2). Therefore, every K-automorphism in $\mathrm{Aut}_K E$ extends to F by Theorem 3.8 and $G/E' \cong \mathrm{Aut}_K E$. ∎

We return now to splitting fields and characterize them in terms of a property that has already been used on several occasions.

Definition 3.13. *An algebraic extension field* F *of* K *is* **normal** *over* K (*or a* **normal extension**) *if every irreducible polynomial in* K[x] *that has a root in* F *actually splits in* F[x].

Theorem 3.14. *If* F *is an algebraic extension field of* K, *then the following statements are equivalent.*

(i) F *is normal over* K;
(ii) F *is a splitting field over* K *of some set of polynomials in* K[x];
(iii) *if* $\overline{\mathrm{K}}$ *is any algebraic closure of* K *containing* F, *then for any* K*-monomorphism of fields* $\sigma: \mathrm{F} \to \overline{\mathrm{K}}$, *Im* $\sigma = \mathrm{F}$ *so that* σ *is actually a* K*-automorphism of* F.

REMARKS. The theorem remains true if the algebraic closure $\overline{K}$ in (iii) is replaced by any normal extension of K containing F (Exercise 21). See Exercise 25 for a direct proof of (ii) ⇒ (i) in the finite dimensional case.

PROOF OF 3.14. (i) ⇒ (ii) F is a splitting field over K of $\{f_i \in K[x] \mid i \in I\}$, where $\{u_i \mid i \in I\}$ is a basis of F over K and f_i is the irreducible polynomial of u_i.

(ii) ⇒ (iii) Let F be a splitting field of $\{f_i \mid i \in I\}$ over K and $\sigma: F \to \overline{K}$ a K-monomorphism of fields. If $u \in F$ is a root of f_i, then so is $\sigma(u)$ (same proof as Theorem 2.2). By hypothesis f_i splits in F, say $f_i = c(x - u_1)\cdots(x - u_n)$ $(u_i \in F; c \in K)$. Since $\overline{K}[x]$ is a unique factorization domain (Corollary III.6.4), $\sigma(u_i)$ must be one of $u_1, \ldots, u_n$ for every i (see Theorem III.6.6). Since σ is injective, it must simply permute the u_i. But F is generated over K by all the roots of all the f_i. It follows from Theorem 1.3 that $\sigma(F) = F$ and hence that $\sigma \in \mathrm{Aut}_K F$.

(iii) ⇒ (i) Let $\overline{K}$ be an algebraic closure of F (Theorem 3.6). Then $\overline{K}$ is algebraic over K (Theorem 1.13). Therefore $\overline{K}$ is an algebraic closure of K containing F (Theorem 3.4). Let $f \in K[x]$ be irreducible with a root $u \in F$. By construction $\overline{K}$ contains all the roots of f. If $v \in \overline{K}$ is any root of f then there is a K-isomorphism of fields $\sigma: K(u) \cong K(v)$ with $\sigma(u) = v$ (Corollary 1.9), which extends to a K-automorphism of $\overline{K}$ by Theorems 3.4 and 3.8 and Exercise 2. $\sigma \mid F$ is a monomorphism $F \to \overline{K}$ and by hypothesis $\sigma(F) = F$. Therefore, $v = \sigma(u) \in F$, which implies that f splits in F. Hence F is normal over K. ∎

Corollary 3.15. *Let* F *be an algebraic extension field of* K. *Then* F *is Galois over* K *if and only if* F *is normal and separable over* K. *If char* K = 0, *then* F *is Galois over* K *if and only if* F *is normal over* K.

PROOF. Exercise; use Theorems 3.11 and 3.14. ■

Theorem 3.16. *If* E *is an algebraic extension field of* K, *then there exists an extension field* F *of* E *such that*

(i) F *is normal over* K;
(ii) *no proper subfield of* F *containing* E *is normal over* K;
(iii) *if* E *is separable over* K, *then* F *is Galois over* K;
(iv) [F : K] *is finite if and only if* [E : K] *is finite.*

The field F *is uniquely determined up to an* E-*isomorphism.*

The field F in Theorem 3.16 is sometimes called the **normal closure** of E over K.

PROOF OF 3.16. (i) Let $X = \{u_i \mid i \in I\}$ be a basis of E over K and let $f_i \in K[x]$ be the irreducible polynomial of u_i. If F is a splitting field of $S = \{f_i \mid i \in I\}$ over E, then F is also a splitting field of S over K (Exercise 3), whence F is normal over K by Theorem 3.14. (iii) If E is separable over K, then each f_i is separable. Therefore F is Galois over K by Theorem 3.11. (iv) If $[E : K]$ is finite, then so is X and hence S. This implies that $[F : K]$ is finite (by the Remarks after Definition 3.1). (ii) A subfield F_0 of F that contains E necessarily contains the root u_i of $f_i \in S$ for every i. If F_0 is normal over K (so that each f_i splits in F_0 by definition), then $F \subset F_0$ and hence $F = F_0$.

Finally let F_1 be another extension field of E with properties (i) and (ii). Since F_1 is normal over K and contains each u_i, F_1 must contain a splitting field F_2 of S over K with $E \subset F_2$. F_2 is normal over K (Theorem 3.14), whence $F_2 = F_1$ by (ii). Therefore both F and F_1 are splitting fields of S over K and hence of S over E (Exercise 2). By Theorem 3.8 the identity map on E extends to an E-isomorphism $F \cong F_1$. ■

APPENDIX: THE FUNDAMENTAL THEOREM OF ALGEBRA

The theorem referred to in the title states that the field **C** of complex numbers is algebraically closed (that is, every polynomial equation over **C** can be completely solved.) Every known proof of this fact depends at some point on results from analysis. We shall assume:

(A) every positive real number has a real positive square root;

(B) every polynomial in **R**$[x]$ of odd degree has a root in **R** (that is, every irreducible polynomial in **R**$[x]$ of degree greater than one has even degree).

Assumption (A) follows from the construction of the real numbers from the rationals and assumption (B) is a corollary of the Intermediate Value Theorem of elementary calculus; see Exercise III.6.16. We begin by proving a special case of a theorem that will be discussed below (Proposition 6.15).

Lemma 3.17. *If* F *is a finite dimensional separable extension of an infinite field* K, *then* F = K(u) *for some* u ε F.

SKETCH OF PROOF. By Theorem 3.16 there is a finite dimensional Galois extension field F_1 of K that contains F. The Fundamental Theorem 2.5 implies that $\mathrm{Aut}_K F_1$ is finite and that the extension of K by F_1 has only finitely many intermediate fields. Therefore, there can be only a finite number of intermediate fields in the extension of K by F.

Since $[F:K]$ is finite, we can choose $u \varepsilon F$ such that $[K(u):K]$ is maximal. If $K(u) \neq F$, there exists $v \varepsilon F - K(u)$. Consider all intermediate fields of the form $K(u + av)$ with $a \varepsilon K$. Since K is infinite and there are only finitely many intermediate fields, there exist $a,b \varepsilon K$ such that $a \neq b$ and $K(u + av) = K(u + bv)$. Therefore $(a - b)v = (u + av) - (u + bv) \varepsilon K(u + av)$. Since $a \neq b$, we have $v = (a - b)^{-1}(a - b)v \varepsilon K(u + av)$, whence $u = (u + av) - av \varepsilon K(u + av)$. Consequently $K \subset K(u) \underset{\neq}{\subset} K(u + av)$, whence $[K(u + av):K] > [K(u):K]$. This contradicts the choice of u. Hence $K(u) = F$. ■

Lemma 3.18. *There are no extension fields of dimension* 2 *over the field of complex numbers.*

SKETCH OF PROOF. It is easy to see that any extension field F of dimension 2 over $\mathbf{C}$ would necessarily be of the form $F = \mathbf{C}(u)$ for any $u \varepsilon F - \mathbf{C}$. By Theorem 1.6 u would be the root of an irreducible monic polynomial $f \varepsilon \mathbf{C}[x]$ of degree 2. To complete the proof we need only show that no such f can exist.

For each $a + bi \varepsilon \mathbf{C} = \mathbf{R}(i)$ the positive real numbers $|(a + \sqrt{a^2 + b^2})/2|$ and $|(-a + \sqrt{a^2 + b^2})/2|$ have real positive square roots c and d respectively by assumption (A). Verify that with a proper choice of signs $(\pm c \pm di)^2 = a + bi$. Hence every element of $\mathbf{C}$ has a square root in $\mathbf{C}$. Consequently, if $f = x^2 + sx + t \varepsilon \mathbf{C}[x]$, then f has roots $(-s \pm \sqrt{s^2 - 4t})/2$ in $\mathbf{C}$, whence f splits over $\mathbf{C}$. Thus there are no irreducible monic polynomials of degree 2 in $\mathbf{C}[x]$. ■

Theorem 3.19. *(The Fundamental Theorem of Algebra) The field of complex numbers is algebraically closed.*

PROOF. In order to show that every nonconstant $f \varepsilon \mathbf{C}[x]$ splits over $\mathbf{C}$, it suffices by Theorem 1.10 to prove that $\mathbf{C}$ has no finite dimensional extensions except itself. Since $[\mathbf{C}:\mathbf{R}] = 2$ and char $\mathbf{R} = 0$ every finite dimensional extension field E_1 of $\mathbf{C}$ is a finite dimensional separable extension of $\mathbf{R}$ (Theorem 1.2). Consequently, E_1 is contained in a finite dimensional Galois extension field F of $\mathbf{R}$ by Theorem 3.16. We need only show that $F = \mathbf{C}$ in order to conclude $E_1 = \mathbf{C}$.

The Fundamental Theorem 2.5 shows that $\mathrm{Aut}_{\mathbf{R}}F$ is a finite group. By Theorems II.5.7 and 2.5 $\mathrm{Aut}_{\mathbf{R}}F$ has a Sylow 2-subgroup H of order 2^n $(n \geq 0)$ and odd index, whose fixed field E has odd dimension, $[E:\mathbf{R}] = [\mathrm{Aut}_{\mathbf{R}}F : H]$. E is separable over $\mathbf{R}$ (since char $\mathbf{R} = 0$), whence $E = \mathbf{R}(u)$ by Lemma 3.17. Thus the irreducible polynomial of u has odd degree $[E:\mathbf{R}] = [\mathbf{R}(u):\mathbf{R}]$ (Theorem 1.6). This degree must be 1 by assumption (B). Therefore, $u \in \mathbf{R}$ and $[\mathrm{Aut}_{\mathbf{R}}F : H] = [E : \mathbf{R}] = 1$, whence $\mathrm{Aut}_{\mathbf{R}}F = H$ and $|\mathrm{Aut}_{\mathbf{R}}F| = 2^n$. Consequently, the subgroup $\mathrm{Aut}_{\mathbf{C}}F$ of $\mathrm{Aut}_{\mathbf{R}}F$ has order 2^m for some m $(0 \leq m \leq n)$.

Suppose $m > 0$. Then by the First Sylow Theorem II.5.7 $\mathrm{Aut}_{\mathbf{C}}F$ has a subgroup J of index 2; let E_0 be the fixed field of J. By the Fundamental Theorem E_0 is an extension of $\mathbf{C}$ with dimension $[\mathrm{Aut}_{\mathbf{C}}F : J] = 2$, which contradicts Lemma 3.18. Therefore, $m = 0$ and $\mathrm{Aut}_{\mathbf{C}}F = 1$. The Fundamental Theorem 2.5 implies that $[F : \mathbf{C}] = [\mathrm{Aut}_{\mathbf{C}}F : 1] = |\mathrm{Aut}_{\mathbf{C}}F| = 1$, whence $F = \mathbf{C}$. ∎

Corollary 3.20. *Every proper algebraic extension field of the field of real numbers is isomorphic to the field of complex numbers.*

PROOF. If F is an algebraic extension of $\mathbf{R}$ and $u \in F - \mathbf{R}$ has irreducible polynomial $f \in R[x]$ of degree greater than one, then f splits over $\mathbf{C}$ by Theorem 3.19. If $v \in \mathbf{C}$ is a root of f, then by Corollary 1.9 the identity map on $\mathbf{R}$ extends to an isomorphism $\mathbf{R}(u) \cong \mathbf{R}(v) \subset \mathbf{C}$. Since $[\mathbf{R}(v) : \mathbf{R}] = [\mathbf{R}(u) : \mathbf{R}] > 1$ and $[\mathbf{C} : \mathbf{R}] = 2$, we must have $[\mathbf{R}(v) : \mathbf{R}] = 2$ and $\mathbf{R}(v) = \mathbf{C}$. Therefore, F is an algebraic extension of the algebraically closed field $\mathbf{R}(u) \cong \mathbf{C}$. But an algebraically closed field has no algebraic extensions except itself (Theorem 3.3). Thus $F = \mathbf{R}(u) \cong \mathbf{C}$. ∎

EXERCISES

Note: Unless stated otherwise F is always an extension field of the field K and S is a set of polynomials (of positive degree) in $K[x]$.

1. F is a splitting field over K of a finite set $\{f_1, \ldots, f_n\}$ of polynomials in $K[x]$ if and only if F is a splitting field over K of the single polynomial $f = f_1f_2\cdots f_n$.

2. If F is a splitting field of S over K and E is an intermediate field, then F is a splitting field of S over E.

3. (a) Let E be an intermediate field of the extension $K \subset F$ and assume that $E = K(u_1, \ldots, u_r)$ where the u_i are (some of the) roots of $f \in K[x]$. Then F is a splitting field of f over K if and only if F is a splitting field of f over E.
(b) Extend part (a) to splitting fields of arbitrary sets of polynomials.

4. If F is a splitting field over K of S, then F is also a splitting field over K of the set T of all irreducible factors of polynomials in S.

5. If $f \in K[x]$ has degree n and F is a splitting field of f over K, then $[F : K]$ divides $n!$.

6. Let K be a field such that for every extension field F the maximal algebraic extension of K contained in F (see Theorem 1.14) is K itself. Then K is algebraically closed.

7. If F is algebraically closed and E consists of all elements in F that are algebraic over K, then E is an algebraic closure of K [see Theorem 1.14].
8. No finite field K is algebraically closed. [*Hint:* If $K = \{a_0, \ldots, a_n\}$ consider $a_1 + (x - a_0)(x - a_1)\cdots(x - a_n) \in K[x]$, where $a_1 \neq 0$.]

9. F is an algebraic closure of K if and only if F is algebraic over K and for every algebraic extension E of K there exists a K-monomorphism $E \to F$.

10. F is an algebraic closure of K if and only if F is algebraic over K and for every algebraic field extension E of another field K_1 and isomorphism of fields $\sigma : K_1 \to K$, σ extends to a monomorphism $E \to F$.

11. (a) If $u_1, \ldots, u_n \in F$ are separable over K, then $K(u_1, \ldots, u_n)$ is a separable extension of K.
(b) If F is generated by a (possibly infinite) set of separable elements over K, then F is a separable extension of K.

12. Let E be an intermediate field.
(a) If $u \in F$ is separable over K, then u is separable over E.
(b) If F is separable over K, then F is separable over E and E is separable over K.

13. Suppose $[F : K]$ is finite. Then the following conditions are equivalent:
(i) F is Galois over K;
(ii) F is separable over K and a splitting field of a polynomial $f \in K[x]$;
(iii) F is a splitting field over K of a polynomial $f \in K[x]$ whose irreducible factors are separable.

14. (Lagrange's Theorem on Natural Irrationalities). If L and M are intermediate fields such that L is a finite dimensional Galois extension of K, then LM is finite dimensional and Galois over M and $\mathrm{Aut}_M LM \cong \mathrm{Aut}_{L \cap M} L$.

15. Let E be an intermediate field.
(a) If F is algebraic Galois over K, then F is algebraic Galois over E. [Exercises 2.9 and 2.11 show that the "algebraic" hypothesis is necessary.]
(b) If F is Galois over E, E is Galois over K and F is a splitting field over E of a family of polynomials in $K[x]$, then F is Galois over K [see Exercise 2.12].

16. Let F be an algebraic closure of the field $\mathbf{Q}$ of rational numbers and let $E \subset F$ be a splitting field over $\mathbf{Q}$ of the set $S = \{x^2 + a \mid a \in \mathbf{Q}\}$ so that E is algebraic and Galois over $\mathbf{Q}$ (Theorem 3.11).
(a) $E = \mathbf{Q}(X)$ where $X = \{\sqrt{p} \mid p = -1$ or p is a prime integer$\}$.
(b) If $\sigma \in \mathrm{Aut}_{\mathbf{Q}}E$, then $\sigma^2 = 1_E$. Therefore, the group $\mathrm{Aut}_{\mathbf{Q}}E$ is actually a vector space over Z_2 [see Exercises I.1.13 and IV.1.1].
(c) $\mathrm{Aut}_{\mathbf{Q}}E$ is infinite and not denumerable. [*Hint:* for each subset Y of X there exists $\sigma \in \mathrm{Aut}_{\mathbf{Q}}E$ such that $\sigma(\sqrt{p}) = -\sqrt{p}$ for $\sqrt{p} \in Y$ and $\sigma(\sqrt{p}) = \sqrt{p}$ for $\sqrt{p} \in X - Y$. Therefore, $|\mathrm{Aut}_{\mathbf{Q}}E| = |P(X)| > |X|$ by Introduction, Theorem 8.5. But $|X| = \aleph_0$.]
(d) If B is a basis of $\mathrm{Aut}_{\mathbf{Q}}E$ over Z_2, then B is infinite and not denumerable.
(e) $\mathrm{Aut}_{\mathbf{Q}}E$ has an infinite nondenumerable number of subgroups of index 2. [*Hint:* If $b \in B$, then $B - \{b\}$ generates a subgroup of index 2.]
(f) The set of extension fields of $\mathbf{Q}$ contained in E of dimension 2 over $\mathbf{Q}$ is denumerable.
(g) The set of closed subgroups of index 2 in $\mathrm{Aut}_{\mathbf{Q}}E$ is denumerable.
(h) $[E : \mathbf{Q}] \leq \aleph_0$, whence $[E : \mathbf{Q}] < |\mathrm{Aut}_{\mathbf{Q}}E|$.

17. If an intermediate field E is normal over K, then E is stable (relative to F and K).

18. Let F be normal over K and E an intermediate field. Then E is normal over K if and only if E is stable [see Exercise 17]. Furthermore $\mathrm{Aut}_K F/E' \cong \mathrm{Aut}_K E$.

19. Part (ii) or (ii)′ of the Fundamental Theorem (2.5 or 3.12) is equivalent to: an intermediate field E is normal over K if and only if the corresponding subgroup E' is normal in $G = \mathrm{Aut}_K F$ in which case $G/E' \cong \mathrm{Aut}_K E$. [See Exercise 18.]

20. If F is normal over an intermediate field E and E is normal over K, then F need not be normal over K. [*Hint:* Let $\sqrt[4]{2}$ be a real fourth root of 2 and consider $\mathbf{Q}(\sqrt[4]{2}) \supset \mathbf{Q}(\sqrt{2}) \supset \mathbf{Q}$; use Exercise 23.] Compare Exercise 2.

21. Let F be algebraic over K. F is normal over K if and only if for every K-monomorphism of fields $\sigma : F \to N$, where N is any normal extension of K containing F, $\sigma(F) = F$ so that σ is a K-automorphism of F. [*Hint:* Adapt the proof of Theorem 3.14, using Theorem 3.16.]

22. If F is algebraic over K and every element of F belongs to an intermediate field that is normal over K, then F is normal over K.

23. If $[F : K] = 2$, then F is normal over K.

24. An algebraic extension F of K is normal over K if and only if for every irreducible $f \in K[x]$, f factors in $F[x]$ as a product of irreducible factors all of which have the same degree.

25. Let F be a splitting field of $f \in K[x]$. Without using Theorem 3.14 show that F is normal over K. [*Hints:* if an irreducible $g \in K[x]$ has a root $u \in F$, but does not split in F, then show that there is a K-isomorphism $\varphi : K(u) \cong K(v)$, where $v \notin F$ and v is a root of g. Show that φ extends to an isomorphism $F \cong F(v)$. This contradicts the fact that $[F : K] < [F(v) : K]$.]

4. THE GALOIS GROUP OF A POLYNOMIAL

The primary purpose of this section is to provide some applications and examples of the concepts introduced in the preceding sections. With two exceptions this material is not needed in the sequel. Definition 4.1 and Theorem 4.12, which depends only on Theorem 4.2, are used in Section 9, where we shall consider the solvability by radicals of a polynomial equation.

Definition 4.1. *Let* K *be a field. The* **Galois group of a polynomial** $f \in K[x]$ *is the group* $\mathrm{Aut}_K F$, *where* F *is a splitting field of* f *over* K.

By virtue of Corollary 3.9, the Galois group of f is independent of the choice of F. Before giving any examples we first develop some useful facts. Recall that a subgroup G of the symmetric group S_n is said to be **transitive** if given any $i \neq j\,(1 \leq i,j \leq n)$, there exists $\sigma \in G$ such that $\sigma(i) = j$.

Theorem 4.2. *Let* K *be a field and* $f \in K[x]$ *a polynomial with Galois group* G.

(i) G *is isomorphic to a subgroup of some symmetric group* S_n.

(ii) *If* f *is (irreducible) separable of degree* n, *then* n *divides* $|G|$ *and* G *is isomorphic to a transitive subgroup of* S_n;

SKETCH OF PROOF. (i) If $u_1, \ldots, u_n$ are the distinct roots of f in some splitting field F $(1 \leq n \leq \deg f)$, then Theorem 2.2 implies that every $\sigma \in \mathrm{Aut}_K F$ induces a unique permutation of $\{u_1, \ldots, u_n\}$ (but not necessarily vice versa!). Consider S_n as the group of all permutations of $\{u_1, \ldots, u_n\}$ and verify that the assignment of $\sigma \in \mathrm{Aut}_K F$ to the permutation it induces defines a monomorphism $\mathrm{Aut}_K F \rightarrow S_n$. (Note that $F = K(u_1, \ldots, u_n)$.)

(ii) F is Galois over K (Theorem 3.11) and $[K(u_1) : K] = n = \deg f$ (Theorem 1.6). Therefore, G has a subgroup of index n by the Fundamental Theorem 2.5, whence $n \mid |G|$. For any $i \neq j$ there is a K-isomorphism $\sigma : K(u_i) \cong K(u_j)$ such that $\sigma(u_i) = u_j$ (Corollary 1.9). σ extends to a K-automorphism of F by Theorem 3.8, whence G is isomorphic to a transitive subgroup of S_n. ∎

Hereafter the Galois group of polynomial f will frequently be identified with the isomorphic subgroup of S_n and considered as a group of permutations of the roots of f. Furthermore we shall deal primarily with polynomials $f \in K[x]$ all of whose roots are distinct in some splitting field. This implies that the irreducible factors of f are separable. Consequently by Theorem 3.11 (and Exercise 3.13) the splitting field F of f is Galois over K. If the Galois groups of such polynomials can always be calculated, then it is possible (in principle at least) to calculate the Galois group of an arbitrary polynomial (Exercise 1).

Corollary 4.3. *Let* K *be a field and* $f \in K[x]$ *an irreducible polynomial of degree* 2 *with Galois group* G. *If* f *is separable (as is always the case when char* $K \neq 2$), *then* $G \cong Z_2$; *otherwise* $G = 1$.

SKETCH OF PROOF. Note that $S_2 = Z_2$. Use Remark (ii) after Definition 3.10 and Theorem 4.2. ∎

Theorem 4.2 (ii) immediately yields the fact that the Galois group of a separable polynomial of degree 3 is either S_3 or A_3 (the only transitive subgroups of S_3). In order to get a somewhat sharper result, we introduce a more general consideration.

Definition 4.4. *Let* K *be a field with char* $K \neq 2$ *and* $f \in K[x]$ *a polynomial of degree* n *with* n *distinct roots* $u_1, \ldots, u_n$ *in some splitting field* F *of* f *over* K. *Let* $\Delta = \prod_{i<j}(u_i - u_j) = (u_1 - u_2)(u_1 - u_3)\cdots(u_{n-1} - u_n) \in F$; *the* **discriminant** *of* f *is the element* $D = \Delta^2$.

Note that Δ is an element of a specific splitting field F and therefore, *a priori*, $D = \Delta^2$ is also in F. However, we have

Proposition 4.5. *Let* K, f, F *and* Δ *be as in Definition* 4.4.

(i) *The discriminant* Δ^2 *of* f *actually lies in* K.
(ii) *For each* $\sigma \in Aut_K F < S_n$, σ *is an even* [*resp. odd*] *permutation if and only if* $\sigma(\Delta) = \Delta$ [*resp.* $\sigma(\Delta) = -\Delta$].

SKETCH OF PROOF. For (ii) see the proof of Theorem I.6.7. Assuming (ii) note that for every $\sigma \in \mathrm{Aut}_K F$, $\sigma(\Delta^2) = \sigma(\Delta)^2 = (\pm\Delta)^2 = \Delta^2$. Therefore, $\Delta^2 \in K$ since F is Galois over K (Theorem 3.11; Exercise 3.13). ■

Corollary 4.6. *Let* K, f, F, Δ *be as in Definition* 4.4 (*so that* F *is Galois over* K) *and consider* $G = Aut_K F$ *as a subgroup of* S_n. *In the Galois correspondence* (*Theorem* 2.5) *the subfield* $K(\Delta)$ *corresponds to the subgroup* $G \cap A_n$. *In particular,* G *consists of even permutations if and only if* $\Delta \in K$.

PROOF. Exercise. ■

Corollary 4.7. *Let* K *be a field and* $f \in K[x]$ *an* (*irreducible*) *separable polynomial of degree* 3. *The Galois group of* f *is either* S_3 *or* A_3. *If char* $K \neq 2$, *it is* A_3 *if and only if the discriminant of* f *is the square of an element of* K.

PROOF. Exercise; use Theorem 4.2 and Corollary 4.6. ■

If the base field K is a subfield of the field of real numbers, then the discriminant of a cubic polynomial $f \in K[x]$ can be used to find out how many real roots f has (Exercise 2).

Let f be as in Corollary 4.7. If the Galois group of f is $A_3 \cong Z_3$ there are, of course, no intermediate fields. If it is S_3, then there are four proper intermediate fields, $K(\Delta)$, $K(u_1)$, $K(u_2)$, and $K(u_3)$ where u_1,u_2,u_3 are the roots of f. $K(\Delta)$ corresponds to A_3 and $K(u_i)$ corresponds to the subgroup $\{(1),(jk)\}$ $(i \neq j,k)$ of S_3, which has order 2 and index 3 (Exercise 3).

Except in the case of characteristic 2, then, computing the Galois group of a separable cubic reduces to computing the discriminant and determining whether or not it is a square in K. The following result is sometimes helpful.

Proposition 4.8. *Let* K *be a field with char* $K \neq 2,3$. *If* $f(x) = x^3 + bx^2 + cx + d \in K[x]$ *has three distinct roots in some splitting field, then the polynomial* $g(x) = f(x - b/3) \in K[x]$ *has the form* $x^3 + px + q$ *and the discriminant of* f *is* $-4p^3 - 27q^2$.

SKETCH OF PROOF. Let F be a splitting field of f over K and verify that $u \in F$ is a root of f if and only if $u + b/3$ is a root of $g = f(x - b/3)$. This implies that g has the same discriminant as f. Verify that g has the form $x^3 + px + q$ $(p,q \in K)$. Let v_1,v_2,v_3 be the roots of g in F. Then $(x - v_1)(x - v_2)(x - v_3) = g(x) = x^3 + px + q$ which implies

$$v_1 + v_2 + v_3 = 0;$$
$$v_1v_2 + v_1v_3 + v_2v_3 = p;$$
$$-v_1v_2v_3 = q.$$

Since each v_i is a root of g

$$v_i^3 = -pv_i - q \qquad (i = 1,2,3).$$

The fact that the discriminant Δ^2 of g is $-4p^3 - 27q^2$ now follows from a gruesome computation involving the definition $\Delta^2 = (v_1 - v_2)^2(v_1 - v_3)^2(v_2 - v_3)^2$, the equations above and the fact that $(v_i - v_j)^2 = (v_i + v_j)^2 - 4v_iv_j$. ■

EXAMPLE. The polynomial $x^3 - 3x + 1 \in \mathbf{Q}[x]$ is irreducible by Theorem III.6.6 and Proposition III.6.8 and separable since char $\mathbf{Q} = 0$. The discriminant is $-4(-3)^3 - 27(1)^2 = 108 - 27 = 81$ which is a square in $\mathbf{Q}$. Hence the Galois group is A_3 by Corollary 4.7.

EXAMPLE. If $f(x) = x^3 + 3x^2 - x - 1 \in \mathbf{Q}[x]$, then

$$g(x) = f(x - 3/3) = f(x - 1) = x^3 - 4x + 2,$$

which is irreducible by Eisenstein's Criterion (Theorem III.6.15). By Proposition 4.8 the discriminant of f is $-4(-4)^3 - 27(2)^2 = 256 - 108 = 148$, which is not a square in $\mathbf{Q}$. Therefore the Galois group is S_3.

We turn now to polynomials of degree four (quartics) over a field K. As above, we shall deal only with those $f \in K[x]$ that have distinct roots u_1,u_2,u_3,u_4 in some splitting field F. Consequently, F is Galois over K and the Galois group of f may be considered as a group of permutations of $\{u_1,u_2,u_3,u_4\}$ and a subgroup of S_4. The subset $V = \{(1),(12)(34),(13)(24),(14)(23)\}$ is a normal subgroup of S_4 (Exercise I.6.7), which will play an important role in the discussion. Note that V is isomorphic to the four group $Z_2 \oplus Z_2$ and $V \cap G$ is a normal subgroup of $G = \mathrm{Aut}_K F < S_4$.

Lemma 4.9. *Let* K, f, F, u_i, V, *and* $G = Aut_K F < S_4$ *be as in the preceding paragraph. If* $\alpha = u_1u_2 + u_3u_4$, $\beta = u_1u_3 + u_2u_4$, $\gamma = u_1u_4 + u_2u_3 \in F$, *then under the Galois correspondence (Theorem 2.5) the subfield* $K(\alpha,\beta,\gamma)$ *corresponds to the normal subgroup* $V \cap G$. *Hence* $K(\alpha,\beta,\gamma)$ *is Galois over* K *and* $Aut_K K(\alpha,\beta,\gamma) \cong G/(G \cap V)$.

SKETCH OF PROOF. Clearly every element in $G \cap V$ fixes α,β,γ and hence $K(\alpha,\beta,\gamma)$. In order to complete the proof it suffices, in view of the Fundamental Theorem, to show that every element of G not in V moves at least one of α,β,γ. For instance if $\sigma = (12) \in G$ and $\sigma(\beta) = \beta$, then $u_2u_3 + u_1u_4 = u_1u_3 + u_2u_4$ and hence $u_2(u_3 - u_4) = u_1(u_3 - u_4)$. Consequently, $u_1 = u_2$ or $u_3 = u_4$, either of which is a contradiction. Therefore $\sigma(\beta) \neq \beta$. The other possibilities are handled similarly. [*Hint:* Rather than check all 20 possibilities show that it suffices to consider only one representative from each coset of V in S_4]. ■

Let K, f, F, u_i and α,β,γ be as in Lemma 4.9. The elements α,β,γ play a crucial role in determining the Galois groups of arbitrary quartics. The polynomial $(x - \alpha)(x - \beta)(x - \gamma) \in K(\alpha,\beta,\gamma)[x]$ is called the **resolvant cubic** of f. The resolvant cubic is actually a polynomial over K:

Lemma 4.10. *If* K *is a field and* $f = x^4 + bx^3 + cx^2 + dx + e \in K[x]$, *then the resolvant cubic of* f *is the polynomial* $x^3 - cx^2 + (bd - 4e)x - b^2e + 4ce - d^2 \in K[x]$.

SKETCH OF PROOF. Let f have roots $u_1, \ldots, u_4$ in some splitting field F. Then use the fact that $f = (x - u_1)(x - u_2)(x - u_3)(x - u_4)$ to express b,c,d,e in terms of the u_i. Expand the resolvant cubic $(x - \alpha)(x - \beta)(x - \gamma)$ and make appropriate substitutions, using the definition of α,β,γ (Lemma 4.9) and the expressions for b,c,d,e obtained above. ■

We are now in a position to compute the Galois group of any (irreducible) separable quartic $f \in K[x]$. Since its Galois group G is a transitive subgroup of S_4 whose order is divisible by 4 (Theorem 4.2), G must have order 24, 12, 8 or 4. Verify that the only transitive subgroups of orders 24, 12, and 4 are S_4, A_4, $V\ (\cong Z_2 \oplus Z_2)$ and the various cyclic subgroups of order 4 generated by 4-cycles; see Exercise I.4.5 and Theorem I.6.8. One transitive subgroup of S_4 of order 8 is the dihedral group D_4 generated by (1234) and (24) (page 50). Since D_4 is not normal in S_4, and since every subgroup of order 8 is a Sylow 2-subgroup, it follows from the second and third Sylow Theorems that S_4 has precisely three subgroups of order 8, each isomorphic to D_4.

Proposition 4.11. *Let* K *be a field and* $f \in K[x]$ *an (irreducible) separable quartic with Galois group* G *(considered as a subgroup of* S_4*). Let* α,β,γ *be the roots of the resolvant cubic of* f *and let* $m = [K(\alpha,\beta,\gamma) : K]$. *Then:*

(i) $m = 6 \Leftrightarrow G = S_4$;

(ii) $m = 3 \Leftrightarrow G = A_4$;

(iii) $m = 1 \Leftrightarrow G = V$;

(iv) $m = 2 \Leftrightarrow G \cong D_4$ *or* $G \cong Z_4$; *in this case* $G \cong D_4$ *if* f *is irreducible over* $K(\alpha,\beta,\gamma)$ *and* $G \cong Z_4$ *otherwise.*

SKETCH OF PROOF. Since $K(\alpha,\beta,\gamma)$ is a splitting field over K of a cubic, the only possibilities for m are 1,2,3, and 6. In view of this and the discussion preceding the theorem, it suffices to prove only the implications $\Leftarrow$ in each case. We use the fact that $m = [K(\alpha,\beta,\gamma) : K] = |G/G \cap V|$ by Lemma 4.9.

If $G = A_4$, then $G \cap V = V$ and $m = |G/V| = |G|/|V| = 3$. Similarly, if $G = S_4$, then $m = 6$. If $G = V$, then $G \cap V = G$ and $m = |G/G| = 1$. If $G \cong D_4$, then $G \cap V = V$ since V is contained in every Sylow 2-subgroup of S_4 and $m = |G/V| = |G|/|V| = 2$. If G is cyclic of order 4, then G is generated by a 4-cycle whose square must be in V so that $|G \cap V| = 2$ and $m = |G/G \cap V| = |G|/|G \cap V| = 2$.

Since f is either irreducible or reducible and $D_4 \not\cong Z_4$, it suffices to prove the converse of the last statement. Let u_1,u_2,u_3,u_4 be the roots of f in some splitting field F and suppose $G \cong D_4$, so that $G \cap V = V$. Since V is a transitive subgroup and $G \cap V = \mathrm{Aut}_{K(\alpha,\beta,\gamma)}F$ (Lemma 4.9), there exists for each pair $i \neq j$ $(1 \leq i,j \leq 4)$ a $\sigma \in G \cap V$ which induces an isomorphism $K(\alpha,\beta,\gamma)(u_i) \cong K(\alpha,\beta,\gamma)(u_j)$ such that $\sigma(u_i) = u_j$ and $\sigma \mid K(\alpha,\beta,\gamma)$ is the identity. Consequently for each $i \neq j$, u_i and u_j are roots of the same irreducible polynomial over $K(\alpha,\beta,\gamma)$ by Corollary 1.9. It follows that f is irreducible over $K(\alpha,\beta,\gamma)$. On the other hand if $G \cong Z_4$, then $G \cap V =$

$\text{Aut}_{K(\alpha,\beta,\gamma)}F$ has order 2 and is not transitive. Hence for some $i \neq j$ there is no $\sigma \in G \cap V$ such that $\sigma(u_i) = u_j$. But since F is a splitting field over $K(\alpha,\beta,\gamma)(u_i)$ and $K(\alpha,\beta,\gamma)(u_j)$, if there were an isomorphism $K(\alpha,\beta,\gamma)(u_i) \cong K(\alpha,\beta,\gamma)(u_j)$, which was the identity on $K(\alpha,\beta,\gamma)$ and sent u_i to u_j, it would be the restriction of some $\sigma \in \text{Aut}_{K(\alpha,\beta,\gamma)}F = G \cap V$ by Theorem 3.8. Therefore, no such isomorphism exists, whence u_i and u_j cannot be roots of the same irreducible polynomial over $K(\alpha,\beta,\gamma)$ by Corollary 1.9. Consequently, f must be reducible over $K(\alpha,\beta,\gamma)$. ■

EXAMPLE. The polynomial $f = x^4 + 4x^2 + 2 \in \mathbf{Q}[x]$ is irreducible by Eisenstein's Criterion (Theorem III.6.15); f is separable since char $\mathbf{Q} = 0$. Using Lemma 4.10 the resolvant cubic is found to be $x^3 - 4x^2 - 8x + 32 = (x - 4)(x^2 - 8)$ so that $\alpha = 4$, $\beta = \sqrt{8}$, $\gamma = -\sqrt{8}$ and $\mathbf{Q}(\alpha,\beta,\gamma) = \mathbf{Q}(\sqrt{8}) = \mathbf{Q}(2\sqrt{2}) = \mathbf{Q}(\sqrt{2})$ is of dimension 2 over $\mathbf{Q}$. Hence the Galois group is (isomorphic to) D_4 or Z_4. A substitution $z = x^2$ reduces f to $z^2 + 4z + 2$ whose roots are easily seen to be $z = -2 \pm \sqrt{2}$; thus the roots of f are $x = \pm\sqrt{z} = \pm\sqrt{-2 \pm \sqrt{2}}$. Hence

$$f = \left(x - \sqrt{-2 + \sqrt{2}}\right)\left(x + \sqrt{-2 + \sqrt{2}}\right)\left(x - \sqrt{-2 - \sqrt{2}}\right)\left(x + \sqrt{-2 - \sqrt{2}}\right)$$
$$= (x^2 - (-2 + \sqrt{2}))\left(x^2 - (-2 - \sqrt{2})\right) \in \mathbf{Q}(\sqrt{2})[x].$$

Therefore, f is reducible over $\mathbf{Q}(\sqrt{2})$ and hence the Galois group is cyclic of order 4 by Proposition 4.11 (iv).

EXAMPLE. To find the Galois group of $f = x^4 - 10x^2 + 4 \in \mathbf{Q}[x]$ we first verify that f is irreducible (and hence separable as well). Now f has no roots in $\mathbf{Q}$, and thus no linear or cubic factors, by Theorem III.6.6 and Proposition III.6.8. To check for quadratic factors it suffices by Lemma III.6.13 to show that f has no quadratic factors in $\mathbf{Z}[x]$. It is easy to verify that there are no integers a,b,c,d such that $f = (x^2 + ax + b)(x^2 + cx + d)$. Thus f is irreducible in $\mathbf{Q}[x]$. The resolvant cubic of f is $x^3 + 10x^2 - 16x - 160 = (x + 10)(x + 4)(x - 4)$, all of whose roots are in $\mathbf{Q}$. Therefore, $m = [\mathbf{Q}(\alpha,\beta,\gamma) : \mathbf{Q}] = 1$ and the Galois group of f is V ($\cong Z_2 \oplus Z_2$) by Proposition 4.11.

EXAMPLE. The polynomial $x^4 - 2 \in \mathbf{Q}[x]$ is irreducible (and separable) by Eisenstein's Criterion. The resolvant cubic is $x^3 + 8x = x(x + 2\sqrt{2}i)(x - 2\sqrt{2}i)$ and $\mathbf{Q}(\alpha,\beta,\gamma) = \mathbf{Q}(\sqrt{2}i)$ has dimension 2 over $\mathbf{Q}$. Verify that $x^4 - 2$ is irreducible over $\mathbf{Q}(\sqrt{2}i)$ (since $\sqrt{2}, \sqrt[4]{2} \notin \mathbf{Q}(\sqrt{2}i)$). Therefore the Galois group is isomorphic to the dihedral group D_4 by Proposition 4.11.

EXAMPLE. Consider $f = x^4 - 5x^2 + 6 \in \mathbf{Q}[x]$. Observe that f is reducible over $\mathbf{Q}$, namely $f = (x^2 - 2)(x^2 - 3)$. Thus Proposition 4.11 is not applicable here. Clearly $F = \mathbf{Q}(\sqrt{2},\sqrt{3})$ is a splitting field of f over $\mathbf{Q}$ and since $x^2 - 3$ is irreducible over $\mathbf{Q}(\sqrt{2})$, $[F : \mathbf{Q}] = [F : \mathbf{Q}(\sqrt{2})]\,[\mathbf{Q}(\sqrt{2}) : \mathbf{Q}] = 2 \cdot 2 = 4$. Therefore $\text{Aut}_{\mathbf{Q}}F$, the Galois group of f, has order 4 by the Fundamental Theorem. It follows from the proof of Theorem 4.2 and Corollary 4.3 that $\text{Aut}_{\mathbf{Q}}\mathbf{Q}(\sqrt{2})$ consists of two elements: the identity map 1 and a map σ with $\sigma(\sqrt{2}) = -\sqrt{2}$. By Corollary 1.9, 1 and σ each extend to a $\mathbf{Q}$-automorphism of F in two different ways (depending on whether $\sqrt{3} \mapsto \sqrt{3}$ or $\sqrt{3} \mapsto -\sqrt{3}$). This gives four distinct elements of $\text{Aut}_{\mathbf{Q}}F$ (determined by the four possible combinations: $\sqrt{2} \mapsto \pm\sqrt{2}$ and $\sqrt{3} \mapsto \pm\sqrt{3}$). Since $|\text{Aut}_{\mathbf{Q}}F| = 4$ and each of these automorphisms has order 2 the Galois group of f must be isomorphic to the four group $Z_2 \oplus Z_2$ by Exercise I.4.5.

Determining the intermediate fields and corresponding subgroups of the Galois group of a separable quartic is more complicated than doing the same for a separable cubic. Among other things one may have $K(u_i) = K(u_j)$ even though $u_i \neq u_j$ (see the last example above). There is no easily stated proposition to cover the quartic case and each situation must be attacked on an *ad hoc* basis.

EXAMPLE. Let $F \subset \mathbf{C}$ be a splitting field over $\mathbf{Q}$ of $f = x^4 - 2 \in \mathbf{Q}[x]$. If u is the positive real fourth root of 2, then the roots of f are $u, -u, ui, -ui$. In order to consider the Galois group $G = \mathrm{Aut}_{\mathbf{Q}}F$ of f as a subgroup of S_4, we must choose an ordering of the roots, say $u_1 = u$, $u_2 = -u$, $u_3 = ui$, $u_4 = -ui$. We know from the third example after Proposition 4.11 that G is one of the three subgroups of order 8 in S_4, each of which is isomorphic to the dihedral group D_4. Observe that complex conjugation is an $\mathbf{R}$-automorphism of $\mathbf{C}$ which clearly sends $u \mapsto u$, $-u \mapsto -u$, $ui \mapsto -ui$ and $-ui \mapsto ui$. Thus it induces a $\mathbf{Q}$-automorphism τ of $F = \mathbf{Q}(u,ui)$. As an element of S_4, $\tau = (34)$. Now every subgroup of order 8 in S_4 is conjugate to D_4 (Second Sylow Theorem) and an easy calculation shows that the only one containing (34) is the subgroup D generated by $\sigma = (1324)$ and $\tau = (34)$. It is easy to see that $F = \mathbf{Q}(u,ui) = \mathbf{Q}(u,i)$, so that every $\mathbf{Q}$-automorphism of F is completely determined by its action on u and i. Thus the elements of D may be described either in terms of σ and τ or by their action on u and i. This information is summarized in the table:

	(1)	(34)	(1324)	(12)(34)	(1423)	(13)(24)	(12)	(14)(23)
		τ	σ	σ^2	σ^3	$\sigma\tau$	$\sigma^2\tau$	$\sigma^3\tau$
$u \mapsto$	u	u	ui	$-u$	$-ui$	ui	$-u$	$-ui$
$i \mapsto$	i	$-i$	i	i	i	$-i$	$-i$	$-i$

It is left to the reader to verify that the subgroup lattice of D and the lattice of intermediate fields are as given below, with fields and subgroups in the same relative position corresponding to one another in the Galois correspondence.

Subgroup lattice ($H \to K$ means $H < K$):

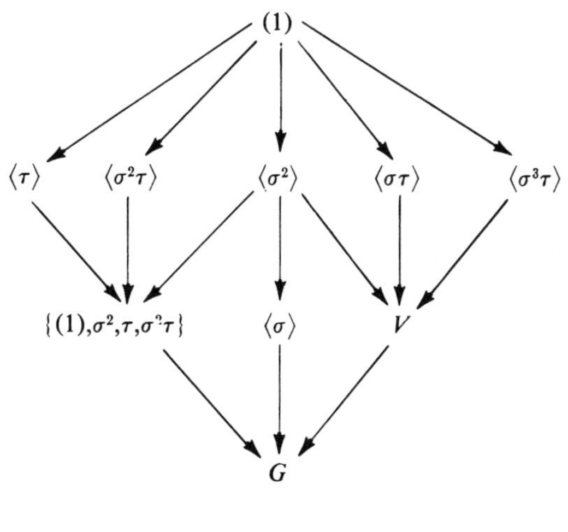

Intermediate field lattice ($M \to N$ means $M \subset N$):

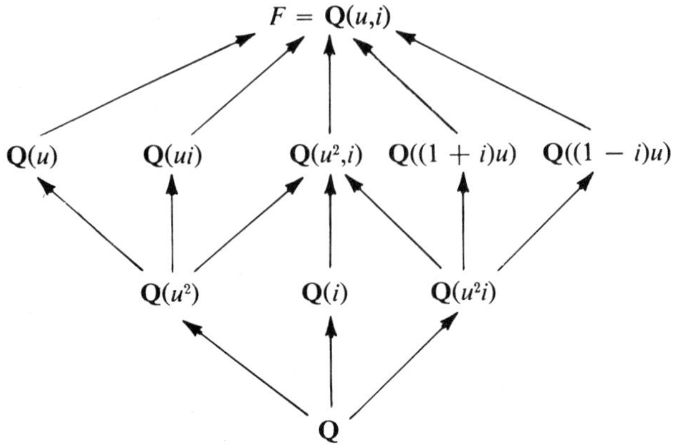

Specific techniques for computing Galois groups of polynomials of degree greater than 4 over arbitrary fields are rather scarce. We shall be content with a very special case.

Theorem 4.12. *If* p *is prime and* f *is an irreducible polynomial of degree* p *over the field of rational numbers which has precisely two nonreal roots in the field of complex numbers, then the Galois group of* f *is (isomorphic to)* S_p.

SKETCH OF PROOF. Let G be the Galois group of f considered as a subgroup of S_p. Since $p \mid |G|$ (Theorem 4.2), G contains an element σ of order p by Cauchy's Theorem II.5.2. σ is a p-cycle by Corollary I.6.4. Now complex conjugation $(a + bi \mapsto a - bi)$ is an $\mathbf{R}$-automorphism of $\mathbf{C}$ that moves every nonreal element. Therefore, by Theorem 2.2 it interchanges the two nonreal roots of f and fixes all the others. This implies that G contains a transposition $\tau = (ab)$. Since σ can be written $\sigma = (aj_2 \cdots j_p)$, some power of σ is of the form $\sigma^k = (abi_3 \cdots i_p) \in G$. By changing notation, if necessary, we may assume $\tau = (12)$ and $\sigma^k = (123 \cdots p)$. But these two elements generate S_p by Exercise I.6.4. Therefore $G = S_p$. ■

EXAMPLE. An inspection of the graph of $f = x^5 - 4x + 2 \in \mathbf{Q}[x]$ shows that it has only three real roots. The polynomial f is irreducible by Eisenstein's Criterion (Theorem III.6.15) and its Galois group is S_5 by Theorem 4.12.

It is still an open question as to whether or not there exists for every finite group G a Galois extension field of $\mathbf{Q}$ with Galois group G. If $G = S_n$, however, the answer is affirmative (Exercise 14).

EXERCISES

Note: Unless stated otherwise K is a field, $f \in K[x]$ and F is a splitting field of f over K.

1. Suppose $f \in K[x]$ splits in F as $f = (x - u_1)^{n_1} \cdots (x - u_k)^{n_k}$ (u_i distinct; $n_i \geq 1$). Let $v_0, \ldots, v_k$ be the coefficients of the polynomial $g = (x - u_1)(x - u_2) \ldots (x - u_k)$ and let $E = K(v_0, \ldots v_k)$. Then

(a) F is a splitting field of g over E.
(b) F is Galois over E.
(c) $\mathrm{Aut}_E F = \mathrm{Aut}_K F$.

2. Suppose K is a subfield of $\mathbf{R}$ (so that F may be taken to be a subfield of $\mathbf{C}$) and that f is irreducible of degree 3. Let D be the discriminant of f. Then
   (a) $D > 0$ if and only if f has three real roots.
   (b) $D < 0$ if and only if f has precisely one real root.

3. Let f be a separable cubic with Galois group S_3 and roots $u_1, u_2, u_3 \in F$. Then the distinct intermediate fields of the extension of K by F are F, $K(\Delta)$, $K(u_1)$, $K(u_2)$, $K(u_3)$, K. The corresponding subgroups of the Galois group are $1, A_3, T_1, T_2, T_3$ and S_3 where $T_i = \{(1),(jk) \mid j \neq i \neq k\}$.

4. If char $K \neq 2,3$ then the discriminant of $x^3 + bx^2 + cx + d$ is $-4c^3 - 27d^2 + b^2(c^2 - 4bd) + 18bcd$.

5. If char $K \neq 2$ and $f \in K[x]$ is a cubic whose discriminant is a square in K, then f is either irreducible or factors completely in K.

6. Over any base field K, $x^3 - 3x + 1$ is either irreducible or splits over K.

7. S_4 has no transitive subgroup of order 6.

8. Let f be an (irreducible) separable quartic over K and u a root of f. There is no field properly between K and $K(u)$ if and only if the Galois group of f is either A_4 or S_4.

9. Let $x^4 + ax^2 + b \in K[x]$ (with char $K \neq 2$) be irreducible with Galois group G.
   (a) If b is a square in K, then $G = V$.
   (b) If b is not a square in K and $b(a^2 - 4b)$ is a square in K, then $G \cong Z_4$.
   (c) If neither b nor $b(a^2 - 4b)$ is a square in K, then $G \cong D_4$.

10. Determine the Galois groups of the following polynomials over the fields indicated:
    (a) $x^4 - 5$ over $\mathbf{Q}$; over $\mathbf{Q}(\sqrt{5})$; over $\mathbf{Q}(\sqrt{5}i)$.
    (b) $(x^3 - 2)(x^2 - 3)(x^2 - 5)(x^2 - 7)$ over $\mathbf{Q}$.
    (c) $x^3 - x - 1$ over $\mathbf{Q}$; over $\mathbf{Q}(\sqrt{23}i)$.
    (d) $x^3 - 10$ over $\mathbf{Q}$; over $\mathbf{Q}(\sqrt{2})$.
    (e) $x^4 + 3x^3 + 3x - 2$ over $\mathbf{Q}$.
    (f) $x^5 - 6x + 3$ over $\mathbf{Q}$.
    (g) $x^3 - 2$ over $\mathbf{Q}$.
    (h) $(x^3 - 2)(x^2 - 5)$ over $\mathbf{Q}$.
    (i) $x^4 - 4x^2 + 5$ over $\mathbf{Q}$.
    (j) $x^4 + 2x^2 + x + 3$ over $\mathbf{Q}$.

11. Determine all the subgroups of the Galois group and all of the intermediate fields of the splitting field (over $\mathbf{Q}$) of the polynomial $(x^3 - 2)(x^2 - 3) \in \mathbf{Q}[x]$.

12. Let K be a subfield of the real numbers and $f \in K[x]$ an irreducible quartic. If f has exactly two real roots, the Galois group of f is S_4 or D_4.

13. Assume that $f(x) \in K[x]$ has distinct roots $u_1, u_2, \ldots, u_n$ in the splitting field F and let $G = \mathrm{Aut}_K F < S_n$ be the Galois group of f. Let $y_1, \ldots, y_n$ be indeterminates and define:

$$g(x) = \prod_{\sigma \varepsilon S_n} (x - (u_{\sigma(1)}y_1 + u_{\sigma(2)}y_2 + \cdots + u_{\sigma(n)}y_n))$$

(a) Show that

$$g(x) = \prod_{\sigma \varepsilon S_n} (x - (u_1 y_{\sigma(1)} + u_2 y_{\sigma(2)} + \cdots + u_n y_{\sigma(n)})).$$

(b) Show that $g(x) \varepsilon K[y_1, \ldots, y_n, x]$.

(c) Suppose $g(x)$ factors as $g_1(x)g_2(x)\cdots g_r(x)$ with $g_i(x) \varepsilon K(y_1, \ldots, y_n)[x]$ monic irreducible. If $x - \sum_i u_{\sigma(i)}y_i$ is a factor of $g_1(x)$, then show that

$$g_1(x) = \prod_{\tau \varepsilon G} (x - \sum_i u_{\tau\sigma(i)}y_i).$$

Show that this implies that $\deg g_i(x) = |G|$.

(d) If $K = \mathbf{Q}$, $f \varepsilon \mathbf{Z}[x]$ is monic, and p is a prime, let $\bar{f} \varepsilon Z_p[x]$ be the polynomial obtained from f by reducing the coefficients of $f(\text{mod } p)$. Assume $\bar{f}$ has distinct roots $\bar{u}_1, \ldots, \bar{u}_n$ in some splitting field $\bar{F}$ over Z_p. Show that

$$\bar{g}(x) = \prod_{\tau \varepsilon S_n} (x - \sum_i \bar{u}_i y_{\tau(i)}) \varepsilon \bar{F}[x, y_1, \ldots, y_n].$$

If the $\bar{u}_i$ are suitably ordered, then prove that the Galois group $\bar{G}$ of $\bar{f}$ is a subgroup of the Galois group G of f.

(e) Show that $x^6 + 22x^5 - 9x^4 + 12x^3 - 37x^2 - 29x - 15 \varepsilon \mathbf{Q}[x]$ has Galois group S_6. [*Hint:* apply (d) with $p = 2, 3, 5$.]

(f) The Galois group of $x^5 - x - 1 \varepsilon \mathbf{Q}[x]$ is S_5.

14. Here is a method for constructing a polynomial $f \varepsilon \mathbf{Q}[x]$ with Galois group S_n for a given $n > 3$. It depends on the fact that there exist irreducible polynomials of every degree in $Z_p[x]$ (p prime; Corollary 5.9 below). First choose $f_1, f_2, f_3 \varepsilon \mathbf{Z}[x]$ such that:

(i) $\deg f_1 = n$ and $\bar{f}_1 \varepsilon Z_2[x]$ is irreducible (notation as in 13(d)).

(ii) $\deg f_2 = n$ and $\bar{f}_2 \varepsilon Z_3[x]$ factors in $Z_3[x]$ as gh with g an irreducible of degree $n - 1$ and h linear;

(iii) $\deg f_3 = n$ and $\bar{f}_3 \varepsilon Z_5[x]$ factors as gh or gh_1h_2 with g an irreducible quadratic in $Z_5[x]$ and h, h_1, h_2 irreducible polynomials of odd degree in $Z_5[x]$.

(a) Let $f = -15f_1 + 10f_2 + 6f_3$. Then $f \equiv f_1 \pmod 2$, $f \equiv f_2 \pmod 3$, and $f \equiv f_3 \pmod 5$.

(b) The Galois group G of f is transitive (since $\bar{f}$ is irreducible in $Z_2[x]$).

(c) G contains a cycle of the type $\zeta = (i_1 i_2 \cdots i_{n-1})$ and element $\sigma\lambda$ where σ is a transposition and λ a product of cycles of odd order. Therefore $\sigma \varepsilon G$, whence $(i_k i_n) \varepsilon G$ for some $k (1 \leq k \leq n - 1)$ by Exercise I.6.3 and transitivity.

(d) $G = S_n$ (see part (c) and Exercise I.6.4(b)).

5. FINITE FIELDS

In this section finite fields (sometimes called **Galois fields**) are characterized in terms of splitting fields and their structure completely determined. The Galois group of an extension of a finite field by a finite field is shown to be cyclic and its generator is given explicitly.

We begin with two theorems and a lemma that apply to fields which need *not* be finite. In each case, of course, we are interested primarily in the implications for finite fields.

Theorem 5.1. *Let* F *be a field and let* P *be the intersection of all subfields of* F. *Then* P *is a field with no proper subfields. If char* F $=$ p *(prime), then* P $\cong Z_p$. *If char* F $= 0$, *then* P $\cong \mathbf{Q}$, *the field of rational numbers.*

The field P is called the **prime subfield** of F.

SKETCH OF PROOF OF 5.1. Note that every subfield of F must contain 0 and 1_F. It follows readily that P is a field that has no proper subfields. Clearly P contains all elements of the form $m1_F$ $(m \in \mathbf{Z})$. To complete the proof one may either show directly that $P = \{m1_F \mid m \in \mathbf{Z}\}$ if char $F = p$ and $P = \{(m1_F)(n1_F)^{-1} \mid m,n \in \mathbf{Z}, n \neq 0\}$ if char $F = 0$ or one may argue as follows. By Theorem III.1.9 the map $\varphi: \mathbf{Z} \to P$ given by $m \mapsto m1_F$ is a ring homomorphism with kernel (n), where $n =$ char F and $n = 0$ or n is prime. If $n = p$ (prime), then $Z_p \cong \mathbf{Z}/(p) = \mathbf{Z}/\text{Ker } \varphi \cong \text{Im } \varphi \subset P$. Since Z_p is a field and P has no proper subfields, we must have $Z_p \cong \text{Im } \varphi = P$. If $n = 0$, then $\varphi: \mathbf{Z} \to P$ is a monomorphism and by Corollary III.4.6 there is a monomorphism of fields $\bar{\varphi}: \mathbf{Q} \to P$. As before, we must have $\mathbf{Q} \cong \text{Im } \bar{\varphi} = P$. ■

Corollary 5.2. *If* F *is a finite field, then char* F $=$ p $\neq 0$ *for some prime* p *and* $|\text{F}| = \text{p}^\text{n}$ *for some integer* n ≥ 1.

PROOF. Theorem III.1.9 and Theorem 5.1 imply that F has prime characteristic $p \neq 0$. Since F is a finite dimensional vector space over its prime subfield Z_p, $F \cong Z_p \oplus \cdots \oplus Z_p$ (n summands) by Theorem IV.2.4 and hence $|F| = p^n$. ■

In the sequel the prime subfield of a field F of characteristic p will always be identified with Z_p under the isomorphism of Theorem 5.1. For example, we shall write $Z_p \subset F$; in particular, 1_F coincides with $1 \in Z_p$.

Theorem 5.3. *If* F *is a field and* G *is a finite subgroup of the multiplicative group of nonzero elements of* F, *then* G *is a cyclic group. In particular, the multiplicative group of all nonzero elements of a finite field is cyclic.*

PROOF. If $G\,(\neq 1)$ is a finite abelian group, $G \cong Z_{m_1} \oplus Z_{m_2} \oplus \cdots \oplus Z_{m_k}$ where $m_1 > 1$ and $m_1 \mid m_2 \mid \cdots \mid m_k$ by Theorem II.2.1. Since $m_k(\sum Z_{m_i}) = 0$, it follows that every $u \in G$ is a root of the polynomial $x^{m_k} - 1_F \in F[x]$ (G is a *multiplicative* group). Since this polynomial has at most m_k distinct roots in F (Theorem III.6.7), we must have $k = 1$ and $G \cong Z_{m_k}$. ■

Corollary 5.4. *If* F *is a finite field, then* F *is a simple extension of its prime subfield* Z_p; *that is,* F $= \text{Z}_\text{p}(\text{u})$ *for some* u $\in$ F.

SKETCH OF PROOF. Let u be a generator of the multiplicative group of nonzero elements of F. ■

Lemma 5.5. *If* F *is a field of characteristic* p *and* $r \geq 1$ *is an integer, then the map* $\varphi : F \to F$ *given by* $u \mapsto u^{p^r}$ *is a* Z_p*-monomorphism of fields. If* F *is finite, then* φ *is a* Z_p*-automorphism of* F.

SKETCH OF PROOF. The key fact is that for characteristic p, $(u \pm v)^{p^r} = u^{p^r} \pm v^{p^r}$ for all $u,v \in F$ (Exercise III.1.11). Since $1_F \mapsto 1_F$, φ fixes each element in the prime subfield Z_p of F. ■

We can now give a useful characterization of finite fields.

Proposition 5.6. *Let* p *be a prime and* $n \geq 1$ *an integer. Then* F *is a finite field with* p^n *elements if and only if* F *is a splitting field of* $x^{p^n} - x$ *over* Z_p.

PROOF. If $|F| = p^n$, then the multiplicative group of nonzero elements of F has order $p^n - 1$ and hence every nonzero $u \in F$ satisfies $u^{p^n-1} = 1_F$. Thus every nonzero $u \in F$ is a root of $x^{p^n-1} - 1_F$ and therefore a root of $x(x^{p^n-1} - 1_F) = x^{p^n} - x \in Z_p[x]$ as well. Since $0 \in F$ is also a root of $x^{p^n} - x$, $x^{p^n} - x$ has p^n distinct roots in F (that is, it splits over F by Theorem III.6.7) and these roots are precisely the elements of F. Therefore, F is a splitting field of $x^{p^n} - x$ over Z_p.

If F is a splitting field of $f = x^{p^n} - x$ over Z_p, then since char F = char $Z_p = p$, $f' = -1$ and f is relatively prime to f'. Therefore f has p^n distinct roots in F by Theorem III.6.10(ii). If φ is the monomorphism of Lemma 5.5 (with $r = n$), it is easy to see that $u \in F$ is a root of f if and only if $\varphi(u) = u$. Use this fact to verify that the set E of all roots of f in F is a subfield of F of order p^n, which necessarily contains the prime subfield Z_p of F. Since F is a splitting field, it is generated over Z_p by the roots of f (that is, the elements of E). Therefore, $F = Z_p(E) = E$. ■

Corollary 5.7. *If* p *is a prime and* $n \geq 1$ *an integer, then there exists a field with* p^n *elements. Any two finite fields with the same number of elements are isomorphic.*

PROOF. Given p and n, a splitting field F of $x^{p^n} - x$ over Z_p exists by Theorem 3.2 and has order p^n by Proposition 5.6. Since every finite field of order p^n is a splitting field of $x^{p^n} - x$ over Z_p by Proposition 5.6, any two such are isomorphic by Corollary 3.9. ■

Corollary 5.8. *If* K *is a finite field and* $n \geq 1$ *is an integer, then there exists a simple extension field* $F = K(u)$ *of* K *such that* F *is finite and* $[F : K] = n$. *Any two* n*-dimensional extension fields of* K *are* K*-isomorphic.*

SKETCH OF PROOF. Given K of order p^r let F be a splitting field of $f = x^{p^{rn}} - x$ over K. By Proposition 5.6 every $u \varepsilon K$ satisfies $u^{p^r} = u$ and it follows inductively that $u^{p^{rn}} = u$ for all $u \varepsilon K$. Therefore, F is actually a splitting field of f over Z_p (Exercise 3.3). The proof of Proposition 5.6 shows that F consists of precisely the p^{nr} distinct roots of f. Thus $p^{nr} = |F| = |K|^{[F:K]} = (p^r)^{[F:K]}$, whence $[F:K] = n$. Corollary 5.4 implies that F is a simple extension of K. If F_1 is another extension field of K with $[F_1 : K] = n$, then $[F_1 : Z_p] = n[K : Z_p] = nr$, whence $|F_1| = p^{nr}$. By Proposition 5.6 F_1 is a splitting field of $x^{p^{nr}} - x$ over Z_p and hence over K. Consequently, F and F_1 are K-isomorphic by Corollary 3.9. ■

Corollary 5.9. *If* K *is a finite field and* $n \geq 1$ *an integer, then there exists an irreducible polynomial of degree* n *in* K[x].

PROOF. Exercise; use Corollary 5.8 and Theorem 1.6. ■

Proposition 5.10. *If* F *is a finite dimensional extension field of a finite field* K, *then* F *is finite and is Galois over* K. *The Galois group* $Aut_K F$ *is cyclic.*

SKETCH OF PROOF. Let Z_p be the prime subfield of K. Then F is finite dimensional over Z_p (Theorem 1.2), say of dimension n, which implies that $|F| = p^n$. By the proof of Proposition 5.6 and Exercise 3.2 F is a splitting field over Z_p and hence over K, of $x^{p^n} - x$, all of whose roots are distinct. Theorem 3.11 implies that F is Galois over K. The map $\varphi : F \to F$ given by $u \mapsto u^p$ is a Z_p-automorphism by Lemma 5.5. Clearly φ^n is the identity and no lower power k of φ can be the identity (for this would imply that $x^{p^k} - x$ had p^n distinct roots in F with $k < n$, contradicting Theorem III.6.7). Since $|\mathrm{Aut}_{Z_p} F| = n$ by the Fundamental Theorem, $\mathrm{Aut}_{Z_p} F$ must be the cyclic group generated by φ. Since $\mathrm{Aut}_K F$ is a subgroup of $\mathrm{Aut}_{Z_p} F$, $\mathrm{Aut}_K F$ is cyclic by Theorem I.3.5. ■

EXERCISES

Note: F always denotes an extension field of a field K.

1. If K is a finite field of characteristic p, describe the structure of the additive group of K.

2. (Fermat) If $p \varepsilon \mathbf{Z}$ is prime, then $a^p = a$ for all $a \varepsilon Z_p$ or equivalently, $c^p \equiv c$ (mod p) for all $c \varepsilon \mathbf{Z}$.

3. If $|K| = p^n$, then every element of K has a unique pth root in K.

4. If the roots of a monic polynomial $f \varepsilon K[x]$ (in some splitting field of f over K) are distinct and form a field, then char $K = p$ and $f = x^{p^n} - x$ for some $n \geq 1$.

5. (a) Construct a field with 9 elements and give its addition and multiplication tables.
   (b) Do the same for a field of 25 elements.

6. If $|K| = q$ and $(n,q) = 1$ and F is a splitting field of $x^n - 1_K$ over K, then $[F:K]$ is the least positive integer k such that $n \mid (q^k - 1)$.

7. If $|K| = q$ and $f \varepsilon K[x]$ is irreducible, then f divides $x^{q^n} - x$ if and only if $\deg f$ divides n.

8. If $|K| = p^r$ and $|F| = p^n$, then $r \mid n$ and $\mathrm{Aut}_K F$ is cyclic with generator φ given by $u \mapsto u^{p^r}$.

9. If $n \geq 3$, then $x^{2^n} + x + 1$ is reducible over Z_2.

10. Every element in a finite field may be written as the sum of two squares.

11. Let F be an algebraic closure of Z_p (p prime).
    (a) F is algebraic Galois over Z_p.
    (b) The map $\varphi : F \to F$ given by $u \mapsto u^p$ is a nonidentity Z_p-automorphism of F.
    (c) The subgroup $H = \langle \varphi \rangle$ is a proper subgroup of $\mathrm{Aut}_{Z_p} F$ whose fixed field is Z_p, which is also the fixed field of $\mathrm{Aut}_{Z_p} F$ by (a).

12. If K is finite and F is an algebraic closure of K, then $\mathrm{Aut}_K F$ is abelian. Every element of $\mathrm{Aut}_K F$ (except 1_F) has infinite order.

6. SEPARABILITY

Our study of separability will be greatly facilitated by the simultaneous consideration of a concept that is, in a sense, the complete opposite of separability. Consequently the section begins with purely inseparable extensions, which are characterized in several different ways in Theorem 6.4. These ideas are then used to prove all the important facts about separability of algebraic extensions (principally Theorem 6.7). The degree of (in)separability of an algebraic extension is discussed in detail (most of this material, however, is not needed in the sequel). Finally the Primitive Element Theorem is proved (Proposition 6.15). This result is independent of the rest of the section and may be read at any time.

Definition 6.1. *Let* F *be an extension field of* K. *An algebraic element* $u \varepsilon F$ *is* **purely inseparable** *over* K *if its irreducible polynomial* f *in* K[x] *factors in* F[x] *as* $f = (x - u)^m$. F *is a* **purely inseparable extension** *of* K *if every element of* F *is purely inseparable over* K.

Thus u is separable over K if its irreducible polynomial f of degree n has n distinct roots (in some splitting field) and purely inseparable over K if f has precisely one root. It is possible to have an element that is neither separable nor purely inseparable over K.

Theorem 6.2. *Let* F *be an extension field of* K. *Then* $u \varepsilon F$ *is both separable and purely inseparable over* K *if and only if* $u \varepsilon K$.

PROOF. The element $u \varepsilon F$ is separable and purely inseparable over K if and only if its irreducible polynomial is of the form $(x - u)^m$ and has m distinct roots in some splitting field. Clearly this occurs only when $m = 1$ so that $x - u \varepsilon K[x]$ and $u \varepsilon K$. ■

If char $K = 0$, then every algebraic element over K is separable over K. Therefore, Theorem 6.2 implies that the only elements that are purely inseparable over K are the elements of K itself. Thus purely inseparable extensions of K are trivial if char $K = 0$. Consequently, we usually restrict our attention to the case of nonzero (prime) characteristic. We shall frequently use the following fact about characteristic p without explicit mention: if char $K = p \neq 0$ and $u,v \in K$, then $(u \pm v)^{p^n} = u^{p^n} \pm v^{p^n}$ for all $n \geq 0$ (Exercise III.1.11). In order to characterize purely inseparable extensions we need:

Lemma 6.3. *Let* F *be an extension field of* K *with char* $K = p \neq 0$. *If* $u \in F$ *is algebraic over* K, *then* u^{p^n} *is separable over* K *for some* $n \geq 0$.

SKETCH OF PROOF. Use induction on the degree of u over K. If deg $u = 1$ or u is separable, the lemma is true. If f is the irreducible polynomial of a nonseparable u of degree greater than one, then $f' = 0$ (Theorem III.6.10), whence f is a polynomial in x^p (Exercise III.6.3). Therefore, u^p is algebraic of degree less than deg u over K, whence by induction $(u^p)^{p^m}$ is separable over K for some $m \geq 0$. ∎

Theorem 6.4. *If* F *is an algebraic extension field of a field* K *of characteristic* $p \neq 0$, *then the following statements are equivalent:*

(i) F *is purely inseparable over* K;
(ii) *the irreducible polynomial of any* $u \in F$ *is of the form* $x^{p^n} - a \in K[x]$;
(iii) *if* $u \in F$, *then* $u^{p^n} \in K$ *for some* $n \geq 0$;
(iv) *the only elements of* F *which are separable over* K *are the elements of* K *itself;*
(v) F *is generated over* K *by a set of purely inseparable elements.*

SKETCH OF PROOF OF 6.4. (i) ⇒ (ii) Let $(x - u)^m$ be the irreducible polynomial of $u \in F$ and let $m = np^r$ with $(n,p) = 1$. Then $(x - u)^m = (x - u)^{p^r n} = (x^{p^r} - u^{p^r})^n$ by Exercise III.1.11. Since $(x - u)^m \in K[x]$, the coefficient of $x^{p^r(n-1)}$, namely $\pm nu^{p^r}$ (Theorem III.1.6), must lie in K. Now $(p,n) = 1$ implies that $u^{p^r} \in K$ (Exercise 1). Since $(x - u)^m = (x^{p^r} - u^{p^r})^n$ is irreducible in $K[x]$, we must have $n = 1$ and $(x - u)^m = x^{p^r} - a$, where $a = u^{p^r} \in K$.

The implications (ii) ⇒ (iii) and (i) ⇒ (v) are trivial. (iii) ⇒ (i) by Exercise III.1.11; (i) ⇒ (iv) by Theorem 6.2; and (iv) ⇒ (iii) by Lemma 6.3. (v) ⇒ (iii) If u is purely inseparable over K, then the proof of (i) ⇒ (ii) shows that $u^{p^n} \in K$ for some $n \geq 0$. If $u \in F$ is arbitrary use Theorem 1.3 and Exercise III.1.11. ∎

Corollary 6.5. *If* F *is a finite dimensional purely inseparable extension field of* K *and char* $K = p \neq 0$, *then* $[F : K] = p^n$ *for some* $n \geq 0$.

PROOF. By Theorem 1.11 $F = K(u_1, \ldots, u_m)$. By hypothesis each u_i is purely inseparable over K and hence over $K(u_1, \ldots, u_{i-1})$ as well (Exercise 2). Theorems 1.6 and 6.4 (ii) imply that every step in the tower $K \subset K(u_1) \subset K(u_1,u_2) \subset \cdots \subset K(u_1, \ldots, u_m) = F$ has dimension a power of p. Therefore $[F : K] = p^n$ by Theorem 1.2. ∎

One more preliminary is needed for the principal theorem on separability.

Lemma 6.6 *If* F *is an extension field of* K, X *is a subset of* F *such that* F = K(X), *and every element of* X *is separable over* K, *then* F *is a separable extension of* K.

PROOF. If $v \varepsilon F$, then there exist $u_1, \ldots, u_n \varepsilon X$ such that $v \varepsilon K(u_1, \ldots, u_n)$ by Theorem 1.3. Let $f_i \varepsilon K[x]$ be the irreducible separable polynomial of u_i and E a splitting field of $\{f_1, \ldots, f_n\}$ over $K(u_1, \ldots, u_n)$. Then E is also a splitting field of $\{f_1, \ldots, f_n\}$ over K (Exercise 3.3). By Theorem 3.11 E is separable (in fact Galois) over K, which implies that $v \varepsilon K(u_1, \ldots, u_n) \subset E$ is separable over K. ■

Theorem 6.7. *Let* F *be an algebraic extension field of* K, S *the set of all elements of* F *which are separable over* K, *and* P *the set of all elements of* F *which are purely inseparable over* K.

(i) S *is a separable extension field of* K.
(ii) F *is purely inseparable over* S.
(iii) P *is a purely inseparable extension field of* K.
(iv) P ∩ S = K.
(v) F *is separable over* P *if and only if* F = SP.
(vi) *If* F *is normal over* K, *then* S *is Galois over* K, F *is Galois over* P *and* $Aut_K S \cong Aut_P F = Aut_K F$.

REMARKS. It is clear that S is the unique largest subfield of F separable over K and that S contains every intermediate field that is separable over K; similarly for P and purely inseparable intermediate fields. If char $K = 0$, then $S = F$ and $P = K$ (Theorem 6.2).

SKETCH OF PROOF OF 6.7. (i) If $u, v \varepsilon S$ and $v \neq 0$, then $K(u,v)$ is separable over K by Lemma 6.6, which implies that $u - v, uv^{-1} \varepsilon S$. Therefore, S is a subfield. Lemma 6.3 and Theorem 6.4 imply (ii). (iii) is a routine exercise using Exercise III.1.11 if char $K = p$ and the fact that $P = K$ if char $K = 0$. Theorem 6.2 implies (iv).

(v) If F is separable over P, then F is separable over the composite field SP (Exercise 3.12) and purely inseparable over SP ((ii) and Exercise 2). Therefore, $F = SP$ by Theorem 6.2. Conversely, if $F = SP = P(S)$, then F is separable over P by Exercise 3.12 and Lemma 6.6.

(vi) We show first that the fixed field K_0 of $\text{Aut}_K F$ is in fact P, which immediately implies that F is Galois over P and $\text{Aut}_P F = \text{Aut}_K F$. Let $u \varepsilon F$ have irreducible polynomial f over K and let $\sigma \varepsilon \text{Aut}_K F$; $\sigma(u)$ is a root of f (Theorem 2.2). If $u \varepsilon P$, then $f = (x - u)^m$ and hence $\sigma(u) = u$. Therefore, $P \subset K_0$. If $u \varepsilon K_0$ and $v \varepsilon F$ is any other root of f, then there is a K-isomorphism $\tau : K(u) \to K(v)$ such that $\tau(u) = v$ (Corollary 1.9). By Theorems 3.8 and 3.14 and Exercise 3.2 τ extends to a K-automorphism of F. Since $u \varepsilon K_0$, we have $u = \tau(u) = v$. Since f splits in $F[x]$ by normality, this argument shows that $f = (x - u)^m$ for some m. Therefore, $u \varepsilon P$ and $P \supset K_0$. Hence $P = K_0$.

Every $\sigma \varepsilon \text{Aut}_P F = \text{Aut}_K F$ must send separable elements to separable elements (Theorem 2.2). Therefore, the assignment $\sigma \mapsto \sigma \mid S$ defines a homomorphism $\theta : \text{Aut}_P F \to \text{Aut}_K S$. Since F is normal over S, θ is an epimorphism (Theorems 3.8

and 3.14 and Exercise 3.2). Since F is Galois over P, $F = SP$ by (v), which implies that θ is a monomorphism. Hence $\text{Aut}_P F \cong \text{Aut}_K S$. Finally suppose $u \in S$ is fixed by all $\sigma \in \text{Aut}_K S$. Since θ is an epimorphism, u is in the fixed field P of $\text{Aut}_P F$, whence $u \in P \cap S = K$. Therefore, S is Galois over K. ■

Corollary 6.8. *If* F *is a separable extension field of* E *and* E *is a separable extension field of* K, *then* F *is separable over* K.

PROOF. If S is as in Theorem 6.7, then $E \subset S$ and F is purely inseparable over S. But F is separable over E and hence over S (Exercise 3.12). Therefore, $F = S$ by Theorem 6.2. ■

Let F be a field of characteristic $p \neq 0$. Lemma 5.5 shows that for each $n \geq 1$, the set $F^{p^n} = \{u^{p^n} \mid u \in F\}$ is a subfield of F. By Theorem 6.4 (iii), F is purely inseparable over F^{p^n} and hence over any intermediate field as well (Exercise 2).

Corollary 6.9. *Let* F *be an algebraic extension field of* K, *with char* $K = p \neq 0$. *If* F *is separable over* K, *then* $F = KF^{p^n}$ *for each* $n \geq 1$. *If* $[F : K]$ *is finite and* $F = KF^p$, *then* F *is separable over* K. *In particular,* $u \in F$ *is separable over* K *if and only if* $K(u^p) = K(u)$.

SKETCH OF PROOF. Let S be as in Theorem 6.7. If $[F : K]$ is finite, then $F = K(u_1, \ldots, u_m) = S(u_1, \ldots, u_m)$ by Theorem 1.11. Since each u_i is purely inseparable over S (Theorem 6.7), there is an $n \geq 1$ such that $u_i^{p^n} \in S$ for every i. Since $F = S(u_1, \ldots, u_m)$, Exercise III.1.11 and Theorem 1.3 imply that $F^{p^n} \subset S$. Clearly every element of S is purely inseparable over F^{p^n}, and hence over KF^{p^n}. S is separable over K, and hence over KF^{p^n}. Therefore $S = KF^{p^n}$ by Theorem 6.2. Use the fact that char $K = p$ and Theorem 1.3 to show that for any $t \geq 1$, $F^{p^t} = [K(u_1, \ldots, u_m)]^{p^t} = K^{p^t}(u_1^{p^t}, \ldots, u_m^{p^t})$. Consequently for any $t \geq 1$ we have $KF^{p^t} = K(K^{p^t}(u_1^{p^t}, \ldots, u_m^{p^t}) = K(u_1^{p^t}, \ldots, u_m^{p^t})$. Note that this argument works for *any* generators $u_1, \ldots, u_m$ of F over K. Now if $F = KF^p$, then $K(u_1, \ldots, u_m) = F = KF^p = K(u_1^p, \ldots, u_m^p)$. An iterated argument with the generators $u_i^{p^t}$ in place of u_i $[t = 1, 2, \ldots, n]$ shows that $F = K(u_1, \ldots, u_m) = K(u_1^{p^n}, \ldots, u_m^{p^n}) = KF^{p^n} = S$, whence F is separable over K. Conversely, if F is separable over K, then F is both separable and purely inseparable over KF^{p^n} (for any $n \geq 1$). Therefore $F = KF^{p^n}$ by Theorem 6.2. ■

Next we consider separability and inseparability from a somewhat different point of view. Although Proposition 6.12 is used at one point in Section 7, all that is really essential for understanding the sequel is Definition 6.10 and the subsequent remarks.

Definition 6.10. *Let* F *be an algebraic extension field of* K *and* S *the largest subfield of* F *separable over* K *(as in Theorem 6.7). The dimension* $[S : K]$ *is called the* **separable degree** *of* F *over* K *and is denoted* $[F : K]_s$. *The dimension* $[F : S]$ *is called the* **inseparable degree** *(or* **degree of inseparability***) of* F *over* K *and is denoted* $[F : K]_i$.

REMARKS. $[F : K]_s = [F : K]$ and $[F : K]_i = 1$ if and only if F is separable over K. $[F : K]_s = 1$ and $[F : K]_i = [F : K]$ if and only if F is purely inseparable over K. In any case, $[F:K] = [F:K]_s[F:K]_i$ by Theorem 1.2. If $[F:K]$ is finite and char $K = p \neq 0$, then $[F:K]_i$ is a power of p by Corollary 6.5 and Theorem 6.7(ii). The following lemma will enable us to give an alternate description of $[F:K]_s$ and to show that for any intermediate field E, $[F:E]_s[E:K]_s = [F:K]_s$.

Lemma 6.11. *Let* F *be an extension field of* E, E *an extension field of* K *and* N *a normal extension field of* K *containing* F. *If* r *is the cardinal number of distinct* E*-monomorphisms* F $\to$ N *and* t *is the cardinal number of distinct* K*-monomorphisms* E $\to$ N, *then* rt *is the cardinal number of distinct* K*-monomorphisms* F $\to$ N.

PROOF. For convenience we assume that r, t are finite. The same proof will work in the general case with only slight modifications of notation. Let $\tau_1, \ldots, \tau_r$ be all the distinct E-monomorphisms $F \to N$ and $\sigma_1, \ldots, \sigma_t$ all the distinct K-monomorphisms $E \to N$. Each σ_i extends to a K-automorphism of N (Theorems 3.8 and 3.14 and Exercise 3.2) which will also be denoted σ_i. Each composite map $\sigma_i\tau_j$ is a K-monomorphism $F \to N$. If $\sigma_i\tau_j = \sigma_a\tau_b$, then $\sigma_a^{-1}\sigma_i\tau_j = \tau_b$ which implies that $\sigma_a^{-1}\sigma_i \mid E = 1_E$. Consequently, we have $\sigma_i = \sigma_a$ and $i = a$. Since σ_i is injective $\sigma_i\tau_j = \sigma_i\tau_b$ implies that $\tau_j = \tau_b$ and $j = b$. Therefore, the rt K-monomorphisms $\sigma_i\tau_j : F \to N\,(1 \leq i \leq t, 1 \leq j \leq r)$ are all distinct. Let $\sigma : F \to N$ be any K-monomorphism. Then $\sigma \mid E = \sigma_i$ for some i and $\sigma_i^{-1}\sigma$ is a K-monomorphism $F \to N$, which is the identity on E. Therefore, $\sigma_i^{-1}\sigma = \tau_j$ for some j, whence $\sigma = \sigma_i\tau_j$. Thus the rt distinct maps $\sigma_i\tau_j$ are all of the K-monomorphisms $F \to N$. ■

Proposition 6.12. *Let* F *be a finite dimensional extension field of* K *and* N *a normal extension field of* K *containing* F. *The number of distinct* K*-monomorphisms* F $\to$ N *is precisely* $[\mathrm{F} : \mathrm{K}]_s$, *the separable degree of* F *over* K.

SKETCH OF PROOF. Let S be the maximal subfield of F separable over K (Theorem 6.7(i)). Every K-monomorphism $S \to N$ extends to a K-automorphism of N (Theorems 3.8 and 3.14 and Exercise 3.2) and hence (by restriction) to a K-monomorphism $F \to N$. We claim that the number of distinct K-monomorphisms $F \to N$ is the same as the number of distinct K-monomorphisms $S \to N$. This is trivially true if char $K = 0$ since $F = S$ in that case. So let char $K = p \neq 0$ and suppose σ, τ are K-monomorphisms $F \to N$ such that $\sigma \mid S = \tau \mid S$. If $u \in F$, then $u^{p^n} \in S$ for some $n \geq 0$ by Theorems 6.4 and 6.7(ii). Therefore,

$$\sigma(u)^{p^n} = \sigma(u^{p^n}) = \tau(u^{p^n}) = \tau(u)^{p^n},$$

whence $\sigma(u) = \tau(u)$. Thus $\sigma \mid S = \tau \mid S$ implies $\sigma = \tau$, which proves our claim. Consequently, it suffices to assume that F is separable over K (that is, $F = S$), in which case we have $[F:K] = [F:K]_s$, $[F:E] = [F:E]_s$ and $[E:K] = [E:K]_s$ for any intermediate field E (Exercise 3.12).

Proceed now by induction on $n = [F:K] = [F:K]_s$ with the case $n = 1$ being trivial. If $n > 1$ choose $u \notin F - K$; then $[K(u):K] = r > 1$. If $r < n$ use the induction hypothesis and Lemma 6.11 (with $E = K(u)$) to prove the theorem. If $r = n$ then $F = K(u)$ and $[F:K]$ is the degree of the (separable) irreducible polynomial $f \in K[x]$ of u. Every K-monomorphism $\sigma : F \to N$ is completely determined by $v = f(u)$.

Since v is a root of f (as in Theorem 2.2) there are at most $[F : K] = \deg f$ such K-monomorphisms. Since f splits in N by normality and is separable, Corollary 1.9 shows that there are exactly $[F : K]$ distinct K-monomorphisms $F \to N$. ■

Corollary 6.13. *If* F *is an extension field of* E *and* E *is an extension field of* K, *then*

$$[F : E]_s[E : K]_s = [F : K]_s \quad \textit{and} \quad [F : E]_i[E : K]_i = [F : K]_i.$$

PROOF. Exercise; use Lemma 6.11 and Proposition 6.12. ■

Corollary 6.14. *Let* $f \in K[x]$ *be an irreducible monic polynomial over a field* K, F *a splitting field of* f *over* K *and* u_1 *a root of* f *in* F. *Then*

(i) *every root of* f *has multiplicity* $[K(u_1) : K]_i$ *so that in* F[x],

$$f = [(x - u_1)\cdots(x - u_n)]^{[K(u_1):K]_i},$$

where $u_1, \ldots, u_n$ *are all the distinct roots of* f *and* $n = [K(u_1) : K]_s$;
(ii) $u_1^{[K(u_1):K]_i}$ *is separable over* K.

SKETCH OF PROOF. Assume char $K = p \neq 0$ since the case char $K = 0$ is trivial. (i) For any $i > 1$ there is a K-isomorphism $\sigma : K(u_1) \cong K(u_i)$ with $\sigma(u_1) = u_i$ that extends to a K-isomorphism σ of F (Corollary 1.9, Theorem 3.8, and Exercise 3.2). Since $f \in K[x]$ we have by Theorem 2.2

$$(x - u_1)^{r_1}\cdots(x - u_n)^{r_n} = f = \sigma f = (x - \sigma(u_1))^{r_1}\cdots(x - \sigma(u_n))^{r_n}.$$

Since $u_1, \ldots, u_n$ are distinct and σ is injective, unique factorization in $K[x]$ implies that $(x - u_i)^{r_i} = (x - \sigma(u_1))^{r_1}$, whence $r_1 = r_i$. This shows that every root of f has multiplicity $r = r_1$ so that $f = (x - u_1)^r\cdots(x - u_n)^r$ and $[K(u_1) : K] = \deg f = nr$. Now Corollary 1.9 and Theorem 2.2 imply that there are n distinct K-monomorphisms $K(u_1) \to F$, whence $[K(u_1) : K]_s = n$ by Proposition 6.12 and Theorem 3.14. Therefore,

$$[K(u_1) : K]_i = [K(u_1) : K]/[K(u_1) : K]_s = nr/n = r.$$

(ii) Since r is a power of $p = $ char K, we have $f = (x - u_1)^r\cdots(x - u_n)^r = (x^r - u_1^r)\cdots(x^r - u_n^r)$. Thus f is a polynomial in x^r with coefficients in K, say $f = \sum_{i=0}^{n} a_i x^{ri}$. Consequently, u_1^r is a root of $g(x) = \sum_{i=0}^{n} a_i x^i = (x - u_1^r)\cdots(x - u_n^r) \in K[x]$. Since $u_1, \ldots, u_n$ are distinct, $g(x) \in K[x]$ is separable. Therefore $u_1^r = u_1^{[K(u_1):K]_i}$ is separable over K. ■

The following result is independent of the preceding material and is not needed in the sequel.

Proposition 6.15. (*Primitive Element Theorem*) *Let* F *be a finite dimensional extension field of* K.

(i) *If* F *is separable over* K, *then* F *is a simple extension of* K.

(ii) (*Artin*) *More generally,* F *is a simple extension of* K *if and only if there are only finitely many intermediate fields.*

REMARK. An element u such that $F = K(u)$ is said to be **primitive.**

SKETCH OF PROOF OF 6.15. The first paragraph of the proof of Lemma 3.17, which is valid even if the field K is finite, shows that a separable extension has only finitely many intermediate fields. Thus it suffices to prove (ii). Since (ii) clearly holds if K is finite (Corollary 5.8), we assume that K is infinite. One implication of (ii) is proved in the second paragraph of the proof of Lemma 3.17. Conversely assume $F = K(u)$ with u algebraic over K (since $[F : K]$ is finite). Let E be an intermediate field and $g \in E[x]$ the irreducible monic polynomial of u over E. If $g = x^n + a_{n-1}x^{n-1} + \cdots + a_1x + a_0$, then $[F : E] = n$. Show that $E = K(a_0, a_1, \ldots, a_{n-1})$ by verifying that $[F : K(a_0, \ldots, a_{n-1})] = n$. Thus every intermediate field E is uniquely determined by the irreducible monic polynomial g of u over E. If f is the monic irreducible polynomial of u over K, then $g \mid f$ by Theorem 1.6. Since f factors uniquely in any splitting field (Corollary III.6.4), f can have only a finite number of distinct monic divisors. Consequently, there are only a finite number of intermediate fields. ■

EXERCISES

Note: Unless stated otherwise F is always an extension field of a field K.

1. Let char $K = p \neq 0$ and let $n \geq 1$ be an integer such that $(p,n) = 1$. If $v \in F$ and $nv \in K$, then $v \in K$.

2. If $u \in F$ is purely inseparable over K, then u is purely inseparable over any intermediate field E. Hence if F is purely inseparable over K, then F is purely inseparable over E.

3. If F is purely inseparable over an intermediate field E and E is purely inseparable over K, then F is purely inseparable over K.

4. If $u \in F$ is separable over K and $v \in F$ is purely inseparable over K, then $K(u,v) = K(u + v)$. If $u \neq 0$, $v \neq 0$, then $K(u,v) = K(uv)$.

5. If char $K = p \neq 0$ and $a \in K$ but $a \notin K^p$, then $x^{p^n} - a \in K[x]$ is irreducible for every $n > 1$.

6. If $f \in K[x]$ is monic irreducible, $\deg f \geq 2$, and f has all its roots equal (in a splitting field), then char $K = p \neq 0$ and $f = x^{p^n} - a$ for some $n \geq 1$ and $a \in K$.

7. Let F, K, S, P be as in Theorem 6.7 and suppose E is an intermediate field. Then
   (a) F is purely inseparable over E if and only if $S \subset E$.
   (b) If F is separable over E, then $P \subset E$.
   (c) If $E \cap S = K$, then $E \subset P$.

8. If char $K = p \neq 0$ and $[F : K]$ is finite and not divisible by p, then F is separable over K.

9. Let char $K = p \neq 0$. Then an algebraic element $u \in F$ is separable over K if and only if $K(u) = K(u^{p^n})$ for all $n \geq 1$.

10. Let char $K = p \neq 0$ and let $f \in K[x]$ be irreducible of degree n. Let m be the largest nonnegative integer such that f is a polynomial in x^{p^m} but is not a polynomial in $x^{p^{m+1}}$. Then $n = n_0 p^m$. If u is a root of f, then $[K(u) : K]_s = n_0$ and $[K(u) : K]_i = p^m$.

11. If $f \in K[x]$ is irreducible of degree $m > 0$, and char K does not divide m, then f is separable.

12. F is purely inseparable over K if and only if F is algebraic over K and for any extension field E of F, the only K-monomorphism $F \to E$ is the inclusion map.

13. (a) The following conditions on a field K are equivalent:

    (i) every irreducible polynomial in $K[x]$ is separable;
    (ii) every algebraic closure $\overline{K}$ of K is Galois over K;
    (iii) every algebraic extension field of K is separable over K;
    (iv) either char $K = 0$ or char $K = p$ and $K = K^p$.

    A field K that satisfies (i)–(iv) is said to be **perfect.**
    (b) Every finite field is perfect.

14. If $F = K(u,v)$ with u,v algebraic over K and u separable over K, then F is a simple extension of K.

15. Let char $K = p \neq 0$ and assume $F = K(u,v)$ where $u^p \in K$, $v^p \in K$ and $[F : K] = p^2$. Then F is not a simple extension of K. Exhibit an infinite number of intermediate fields.

16. Let F be an algebraic extension of K such that every polynomial in $K[x]$ has a root in F. Then F is an algebraic closure of K. [*Hint:* Theorems 3.14 and 6.7 and Proposition 6.15 may be helpful.]

7. CYCLIC EXTENSIONS

The basic idea in Sections 7–9 is to analyze Galois field extensions whose Galois groups have a prescribed structure (for example, cyclic or solvable). In this section we shall characterize most finite dimensional Galois extensions with cyclic Galois groups (Propositions 7.7 and 7.8; Theorem 7.11). In order to do this it is first necessary to develop some information about the trace and norm.

Definition 7.1. *Let* F *be a finite dimensional extension field of* K *and* $\overline{K}$ *an algebraic closure of* K *containing* F. *Let* $\sigma_1, \dots, \tau_r$ *be all the distinct* K*-monomorphisms* $F \to \overline{K}$. *If* $u \in F$, *the* **norm** *of* u, *denoted,* $N_K^F(u)$ *is the element*

$$N_K^F(u) = (\sigma_1(u)\sigma_2(u)\cdots\sigma_r(u))^{[F:K]_i}.$$

The **trace** *of* u, *denoted* $T_K^F(u)$, *is the element*

$$T_K^F(u) = [F : K]_i(\sigma_1(u) + \sigma_2(u) + \cdots + \sigma_r(u)).$$

REMARKS. Theorem 7.3 below shows that the definition does not depend on the choice of $\overline{K}$. It can be shown that an equivalent definition is obtained if one replaces $\bar{K}$ by any normal extension of K containing F (Exercise 1). $\bar{K}$ is normal over K (Theorems 3.4 and 3.14), whence $r = [F:K]_s$ is finite by Proposition 6.12. If the context is clear $N_K{}^F$ and $T_K{}^F$ will sometimes be written simply as N and T.

Note that the trace is essentially the additive analogue of the norm. In many instances this means that a proof involving the one will translate directly into a proof of the analogous fact for the other. There are some exceptions, however. For instance if F is not separable over K, then char $K = p \neq 0$ and $[F:K]_i = p^t$ $(t \geq 1)$. Consequently, $T_K{}^F(u) = 0$ for every $u \in F$, but $N_K{}^F(u)$ may not be zero.

EXAMPLE. Let $F = \mathbf{C}$ and $K = \mathbf{R}$ and take $\bar{K} = \mathbf{C}$. It is easy to see that the only $\mathbf{R}$-monomorphisms $\mathbf{C} \to \mathbf{C}$ are the identity and complex conjugation. Consequently $N(a + bi) = [(a + bi)(a - bi)]^1 = a^2 + b^2$.

The principal applications to be given here of the norm and trace occur when F is Galois over K. In this case the Galois group is finite and there is a more convenient description of the norm and trace, which is sometimes taken as a definition.

Theorem 7.2. *If* F *is a finite dimensional Galois extension field of* K *and*

$$Aut_K F = \{\sigma_1, \ldots, \sigma_n\},$$

then for any $u \in F$,

$$N_K{}^F(u) = \sigma_1(u)\sigma_2(u)\cdots\sigma_n(u); \quad \text{and}$$
$$T_K{}^F(u) = \sigma_1(u) + \sigma_2(u) + \cdots + \sigma_n(u).$$

PROOF. Let $\bar{K}$ be an algebraic closure of K which contains F. Since F is normal over K (Corollary 3.15), the K-monomorphisms $F \to \bar{K}$ are precisely the elements of $\mathrm{Aut}_K F$ by Theorem 3.14. Since F is also separable over K (Corollary 3.15), $[F:K]_i = 1$. The conclusion of the theorem now follows directly from Definition 7.1. ■

Suppose F is Galois over K and $\mathrm{Aut}_K F = \{\sigma_1, \ldots, \sigma_n\}$. Since $\mathrm{Aut}_K F$ is a group, the elements $\sigma_i\sigma_1, \sigma_i\sigma_2, \ldots, \sigma_i\sigma_n$ (for any fixed $\sigma_i \in \mathrm{Aut}_K F$) are simply $\sigma_1, \sigma_2, \ldots, \sigma_n$ in a possibly different order. This implies that for any $u \in F$, $N_K{}^F(u)$ and $T_K{}^F(u)$ are fixed by every $\sigma_i \in \mathrm{Aut}_K F$. *Therefore,* $N_K{}^F(u)$ *and* $T_K{}^F(u)$ *must lie in* K. The next theorem shows that this is true even if F is not Galois over K. The first two parts will be used frequently; *the last two parts are not needed in the sequel.*

Theorem 7.3. *Let* F *be a finite dimensional extension field of* K. *Then for all* $u,v \in F$:

(i) $N_K{}^F(u)N_K{}^F(v) = N_K{}^F(uv)$ *and* $T_K{}^F(u) + T_K{}^F(v) = T_K{}^F(u + v)$;

(ii) *if* $u \in K$, *then* $N_K{}^F(u) = u^{[F:K]}$ *and* $T_K{}^F(u) = [F:K]u$;

(iii) $N_K{}^F(u)$ *and* $T_K{}^F(u)$ *are elements of* K. *More precisely,*

$$N_K{}^F(u) = ((-1)^n a_0)^{[F:K(u)]} \in K \text{ and } T_K{}^F(u) = -[F:K(u)]a_{n-1} \in K,$$

where $f = x^n + a_{n-1}x^{n-1} + \cdots + a_0 \in K[x]$ *is the irreducible polynomial of* u;

(iv) *if* E *is an intermediate field, then*

$$N_K{}^E(N_E{}^F(u)) = N_K{}^F(u) \text{ and } T_K{}^E(T_E{}^F(u)) = T_K{}^F(u).$$

SKETCH OF PROOF. (i) and (ii) follow directly from Definition 7.1 and the facts that $r = [F:K]_s$ and $[F:K]_s[F:K]_i = [F:K]$.

(iii) Let $E = K(u)$. An algebraic closure $\overline{K}$ of K which contains F is also an algebraic closure of E. The proof of Lemma 6.11 shows that the distinct K-monomorphisms $F \to \overline{K}$ are precisely the maps $\sigma_k\tau_j$ $(1 \leq k \leq t; 1 \leq j \leq r)$, where the σ's are all the K-automorphisms of $\overline{K}$ whose restrictions to E are distinct and the τ's are all the distinct E-monomorphisms $F \to \overline{K}$. Thus by Proposition 6.12, $t = [E:K]_s$, whence $n = [E:K] = t[E:K]_i$ (Corollary 6.13).

Use (ii) and Corollary 6.13 to show that $N_K{}^F(u) = \left(\prod_{k=1}^{t} \sigma_k(u)\right)^{[F:E][E:K]_i}$ and $T_K{}^F(u) = [F:E][E:K]_i\left(\sum_{k=1}^{t} \sigma_k(u)\right)$. Since $\sigma_i : K(u) \cong K(\sigma_i(u))$ Corollary 1.9 implies that $\sigma_1(u), \ldots, \sigma_t(u)$ are all the distinct roots of f. By Corollary 6.14

$$f = [(x - \sigma_1(u))(x - \sigma_2(u)) \cdots (x - \sigma_t(u))]^{[E:K]_i}$$
$$= \left[x^t - \left(\sum_{k=1}^{t} \sigma_k(u)\right)x^{t-1} + \cdots + \left((-1)^t \prod_{k=1}^{t} \sigma_k(u)\right)\right]^{[E:K]_i}.$$

If $[E:K]_i = 1$, then $n = t$ and the conclusion is immediate. If $[E:K]_i > 1$, then $[E:K]_i$ is a positive power of $p =$ char K. It is easy to calculate a_0 and to see that $a_{n-1} = 0 = T_K{}^F(u)$; use Exercise III.1.11.

(iv) Use the notation in the first paragraph of the proof of (iii), with E *any* intermediate field. Apply the appropriate definitions and use Corollary 6.13. ∎

In addition to the trace and norm we shall need

Definition 7.4. *Let* S *be a nonempty set of automorphisms of a field* F. S *is* **linearly independent** *provided that for any* $a_1, \ldots, a_n \in F$ *and* $\sigma_1, \ldots, \sigma_n \in S$ $(n \geq 1)$:

$$a_1\sigma_1(u) + \cdots + a_n\sigma_n(u) = 0 \textit{ for all } u \in F \quad \Rightarrow \quad a_i = 0 \textit{ for every } i.$$

Lemma 7.5. *If* S *is a set of distinct automorphisms of a field* F, *then* S *is linearly independent.*

PROOF. If S is not linearly independent then there exist nonzero $a_i \in F$ and distinct $\sigma_i \in S$ such that

$$a_1\sigma_1(u) + a_2\sigma_2(u) + \cdots + a_n\sigma_n(u) = 0 \quad \text{for all} \quad u \in F. \tag{1}$$

Among all such "dependence relations" choose one with n minimal; clearly $n > 1$. Since σ_1 and σ_2 are distinct, there exists $v \in F$ with $\sigma_1(v) \neq \sigma_2(v)$. Applying (1) to the element uv (for any $u \in F$) yields:

$$a_1\sigma_1(u)\sigma_1(v) + a_2\sigma_2(u)\sigma_2(v) + \cdots + a_n\sigma_n(u)\sigma_n(v) = 0; \tag{2}$$

and multiplying (1) by $\sigma_1(v)$ gives:

$$a_1\sigma_1(u)\sigma_1(v) + a_2\sigma_2(u)\sigma_1(v) + \cdots + a_n\sigma_n(u)\sigma_1(v) = 0. \tag{3}$$

The difference of (2) and (3) is a relation:

$$a_2[\sigma_2(v) - \sigma_1(v)]\sigma_2(u) + a_3[\sigma_3(v) - \sigma_1(v)]\sigma_3(u) + \cdots + a_n[\sigma_n(v) - \sigma_1(v)]\sigma_n(u) = 0$$

for all $u \in F$. Since $a_2 \neq 0$ and $\sigma_2(v) \neq \sigma_1(v)$ not all the coefficients are zero and this contradicts the minimality of n. ■

An extension field F of a field K is said to be **cyclic** [resp. **abelian**] if F is algebraic and Galois over K and $\mathrm{Aut}_K F$ is a cyclic [resp. abelian] group. If in this situation $\mathrm{Aut}_K F$ is a finite cyclic group of order n, then F is said to be a **cyclic extension of degree n** (and $[F : K] = n$ by the Fundamental Theorem 2.5). For example, Theorem 5.10 states that every finite dimensional extension of a finite field is a cyclic extension. The next theorem is the crucial link between cyclic extensions and the norm and trace.

Theorem 7.6. *Let* F *be a cyclic extension field of* K *of degree* n, σ *a generator of* $\mathrm{Aut}_K F$ *and* $u \in F$. *Then*

(i) $T_K^F(u) = 0$ *if and only if* $u = v - \sigma(v)$ *for some* $v \in F$;

(ii) (*Hilbert's Theorem* 90) $N_K^F(u) = 1_K$ *if and only if* $u = v\sigma(v)^{-1}$ *for some nonzero* $v \in F$.

SKETCH OF PROOF. For convenience write $\sigma(x) = \sigma x$. Since σ generates $\mathrm{Aut}_K F$, it has order n and $\sigma, \sigma^2, \sigma^3, \dots, \sigma^{n-1}, \sigma^n = 1_F = \sigma^0$ are n distinct automorphisms of F. By Theorem 7.2, $T(u) = u + \sigma u + \sigma^2 u + \cdots + \sigma^{n-1}u$ and $N(u) = u(\sigma u)(\sigma^2 u)\cdots(\sigma^{n-1}u)$.

(i) If $u = v - \sigma v$, then use the definition and the facts that

$$T(v - \sigma v) = T(v) - T(\sigma v) \quad \text{and} \quad \sigma^n(v) = v$$

to show that $T(u) = 0$. Conversely suppose $T(u) = 0$. Choose $w \in F$ such that $T(w) = 1_K$ as follows. By Lemma 7.5 (since $1_K \neq 0$) there exists $z \in F$ such that

$$0 \neq 1_F z + \sigma z + \sigma^2 z + \cdots + \sigma^{n-1} z = T(z).$$

Since $T(z) \in K$ by the remarks after Theorem 7.2, we have $\sigma[T(z)^{-1}z] = T(z)^{-1}\sigma(z)$. Consequently, if $w = T(z)^{-1}z$, then

$$\begin{aligned} T(w) &= T(z)^{-1}z + T(z)^{-1}\sigma z + \cdots + T(z)^{-1}\sigma^{n-1}z \\ &= T(z)^{-1}T(z) = 1_K. \end{aligned}$$

Now let

$$\begin{aligned} v = {} & uw + (u + \sigma u)(\sigma w) + (u + \sigma u + \sigma^2 u)(\sigma^2 w) \\ & + (u + \sigma u + \sigma^2 u + \sigma^3 u)(\sigma^3 w) + \cdots + (u + \sigma u + \cdots + \sigma^{n-2}u)(\sigma^{n-2}w). \end{aligned}$$

Use the fact that σ is an automorphism and that

$$0 = T(u) = u + \sigma u + \sigma^2 u + \cdots + \sigma^{n-1}u,$$

which implies that $u = -(\sigma u + \sigma^2 u + \cdots + \sigma^{n-1}u)$, to show that

$$\begin{aligned} v - \sigma v &= uw + u(\sigma w) + u(\sigma^2 w) + u(\sigma^3 w) + \cdots + u(\sigma^{n-2}w) \\ &\quad + u(\sigma^{n-1}w) = uT(w) = u1_K = u. \end{aligned}$$

(ii) If $u = v\sigma(v)^{-1}$, then since σ is an automorphism of order n, $\sigma^n(v^{-1}) = v^{-1}$, $\sigma(v^{-1}) = \sigma(v)^{-1}$ and for each $1 \leq i \leq n-1$, $\sigma^i(v\sigma(v)^{-1}) = \sigma^i(v)\sigma^{i+1}(v)^{-1}$. Hence:

$$N(u) = (v\sigma(v)^{-1})(\sigma v\sigma^2(v)^{-1})(\sigma^2 v\sigma^3(v)^{-1})\cdots(\sigma^{n-1}v\sigma^n(v)^{-1}) = 1_K.$$

Conversely suppose $N(u) = 1_K$, which implies $u \neq 0$. By Lemma 7.5 there exists $y \in F$ such that the element v given by

$$v = uy + (u\sigma u)\sigma y + (u\sigma u\sigma^2 u)\sigma^2 y + \cdots + (u\sigma u \cdots \sigma^{n-2}u)\sigma^{n-2}y$$
$$+ (u\sigma u \cdots \sigma^{n-1}u)\sigma^{n-1}y$$

is nonzero. Since the last summand of v is $N(u)\sigma^{n-1}y = 1_K\sigma^{n-1}y = \sigma^{n-1}y$, it is easy to verify that $u^{-1}v = \sigma v$, whence $u = v\sigma(v)^{-1}$ ($\sigma(v) \neq 0$ since $v \neq 0$ and σ is injective). ■

We now have at hand all the necessary equipment for an analysis of cyclic extensions. We begin by reducing the problem to simpler form.

Proposition 7.7. *Let* F *be a cyclic extension field of* K *of degree* n *and suppose* $n = mp^t$ *where* $0 \neq p = \text{char } K$ *and* $(m,p) = 1$. *Then there is a chain of intermediate fields* $F \supset E_0 \supset E_1 \supset \cdots \supset E_{t-1} \supset E_t = K$ *such that* F *is a cyclic extension of* E_0 *of degree* m *and for each* $0 \leq i \leq t$, E_{i-1} *is a cyclic extension of* E_i *of degree* p.

SKETCH OF PROOF. By hypothesis F is Galois over K and $\text{Aut}_K F$ is cyclic (abelian) so that every subgroup is normal. Recall that every subgroup and quotient group of a cyclic group is cyclic (Theorem I.3.5). Consequently, the Fundamental Theorem 2.5(ii) implies that for any intermediate field E, F is cyclic over E and E is cyclic over K. It follows that for any pair L,M of intermediate fields with $L \subset M$, M is a cyclic extension of L; in particular, M is algebraic Galois over L.

Let H be the unique (cyclic) subgroup of order m of $\text{Aut}_K F$ (Exercise I.3.6) and let E_0 be its fixed field (so that $H = H'' = E_0' = \text{Aut}_{E_0} F$). Then F is cyclic over E_0 of degree m and E_0 is cyclic over K of degree p^t. Since $\text{Aut}_K E_0$ is cyclic of order p^t it has a chain of subgroups

$$1 = G_0 < G_1 < G_2 < \cdots < G_{t-1} < G_t = \text{Aut}_K E_0$$

with $|G_i| = p^i$, $[G_i : G_{i-1}] = p$ and G_i/G_{i-1} cyclic of order p (see Theorem I.3.4(vii)). For each i let E_i be the fixed field of G_i (relative to E_0 and $\text{Aut}_K E_0$). The Fundamental Theorem 2.5 implies that: (i) $E_0 \supset E_1 \supset E_2 \supset \cdots \supset E_{t-1} \supset E_t = K$; (ii) $[E_{i-1} : E_i] = [G_i : G_{i-1}] = p$; and (iii) $\text{Aut}_{E_i} E_{i-1} \cong G_i/G_{i-1}$. Therefore, E_{i-1} is a cyclic extension of E_i of degree p $(0 \leq i \leq t-1)$. ■

Let F be a cyclic extension field of K of degree n. In view of Proposition 7.7 we may, at least in principle, restrict our attention to just two cases: (i) $n = \text{char } K = p \neq 0$; (ii) $\text{char } K = 0$ or $\text{char } K = p \neq 0$ and $(p,n) = 1$ (that is, $\text{char } K \nmid n$). The first of these is treated in

Proposition 7.8. *Let* K *be a field of characteristic* $p \neq 0$. F *is a cyclic extension field of* K *of degree* p *if and only if* F *is a splitting field over* K *of an irreducible polynomial of the form* $x^p - x - a \in K[x]$. *In this case* $F = K(u)$ *where* u *is any root of* $x^p - x - a$.

PROOF. ($\Rightarrow$) If σ is a generator of the cyclic group $\text{Aut}_K F$, then

$$T_K^F(1_K) = [F : K]1_K = p1_K = 0$$

by Theorem 7.3(ii), whence $1_K = v - \sigma(v)$ for some $v \in F$ by Theorem 7.6(i). If $u = -v$, then $\sigma(u) = u + 1_K \neq u$, whence $u \notin K$. Since $[F : K] = p$ there are no intermediate fields, and we must have $F = K(u)$. Note that $\sigma(u^p) = (u + 1_K)^p = u^p + 1_K{}^p = u^p + 1_K$ which implies that $\sigma(u^p - u) = (u^p + 1_K) - (u + 1_K) = u^p - u$. Since F is Galois over K and $\mathrm{Aut}_K F = \langle\sigma\rangle$, $a = u^p - u$ must be in K. Therefore, u is a root of $x^p - x - a \in K[x]$, which is necessarily the irreducible polynomial of u over K since the degree of u over K is $[K(u) : K] = [F : K] = p$.

Recall that the prime subfield Z_p of K consists of the p distinct elements $0, 1 = 1_K$, $2 = 1_K + 1_K, \ldots, p - 1 = 1_K + \cdots + 1_K$ (Theorem 5.1). The first paragraph of the proof of Theorem 5.6 shows that $i^p = i$ for all $i \in Z_p$. Since u is a root of $x^p - x - a$, we have for each $i \in Z_p$: $(u + i)^p - (u + i) - a = u^p + i^p - u - i - a = (u^p - u - a) + (i^p - i) = 0 + 0 = 0$. Thus $u + i \in K(u) = F$ is a root of $x^p - x - a$ for each $i \in Z_p$, whence F contains p distinct roots of $x^p - x - a$. Therefore, $F = K(u)$ is a splitting field over K of $x^p - x - a$. Finally if $u + i$ $(i \in Z_p \subset K)$ is any root of $x^p - x - a$, then clearly $K(u + i) = K(u) = F$.

$(\Leftarrow)$ Suppose F is a splitting field over K of $x^p - x - a \in K[x]$. We shall not assume that $x^p - x - a$ is irreducible and shall prove somewhat more than is stated in the theorem. If u is a root of $x^p - x - a$, then the preceding paragraph shows that $K(u)$ contains p distinct roots of $x^p - x - a$: $u, u + 1, \ldots, u + (p - 1) \in K(u)$. But $x^p - x - a$ has at most p roots in F and these roots generate F over K. Therefore, $F = K(u)$, the irreducible factors of $x^p - x - a$ are separable and F is Galois over K (Theorem 3.11 and Exercise 3.13). Every $\tau \in \mathrm{Aut}_K F = \mathrm{Aut}_K K(u)$ is completely determined by $\tau(u)$. Theorem 2.2 implies that $\tau(u) = u + i$ for some $i \in Z_p \subset K$. Verify that the assignment $\tau \mapsto i$ defines a monomorphism of groups $\theta : \mathrm{Aut}_K F \to Z_p$. Consequently, $\mathrm{Aut}_K F \cong \mathrm{Im}\,\theta$ is either 1 or Z_p. If $\mathrm{Aut}_K F = 1$, then $[F : K] = 1$ by the Fundamental Theorem 2.5, whence $u \in K$ and $x^p - x - a$ splits in $K[x]$. Thus if $x^p - x - a$ is irreducible over K, we must have $\mathrm{Aut}_K F \cong Z_p$. In this case, therefore, F is cyclic over K of degree p. ■

Corollary 7.9. *If* K *is a field of characteristic* p $\neq 0$ *and* $x^p - x - a \in K[x]$, *then* $x^p - x - a$ *is either irreducible or splits in* $K[x]$.

PROOF. We use the notation of Proposition 7.8. In view of the last paragraph of that proof it suffices to prove that if $\mathrm{Aut}_K F \cong \mathrm{Im}\,\theta = Z_p$, then $x^p - x - a$ is irreducible. If u and $v = u + i$ $(i \in Z_p \subset K)$ are roots of $x^p - x - a$, then there exists $\tau \in \mathrm{Aut}_K F$ such that $\tau(u) = v$ and hence $\tau : K(u) \cong K(v)$ (choose τ with $\theta(\tau) = i$). Therefore, u and v are roots of the same irreducible polynomial in $K[x]$ (Corollary 1.9). Since v was arbitrary this implies that $x^p - x - a$ is irreducible. ■

Proposition 7.8 completely describes the structure of a cyclic extension of the first type mentioned on p. 293. In order to determine the structure of a cyclic extension of degree n of the second type it will be necessary to introduce an additional assumption on the ground field K.

Let K be a field and n a positive integer. An element $\zeta \in K$ is said to be an **nth root of unity** provided $\zeta^n = 1_K$ (that is, ζ is a root of $x^n - 1_K \in K[x]$). It is easy to see that the set of all nth roots of unity in K forms a multiplicative subgroup of the multiplicative group of nonzero elements of K. This subgroup is cyclic by Theorem 5.3 and has order at most n by Theorem III.6.7. $\zeta \in K$ is said to be a **primitive nth root of unity** provided ζ is an nth root of unity and ζ has order n in the multiplicative group of nth roots of unity. In particular, a primitive nth root of unity generates the cyclic group of all nth roots of unity.

REMARKS. If char $K = p$ and $p \mid n$, then $n = p^k m$ with $(p,m) = 1$ and $m < n$. Thus $x^n - 1_K = (x^m - 1_K)^{p^k}$ (Exercise III.1.11). Consequently the nth roots of unity in K coincide with the mth roots of unity in K. Since $m < n$, there can be no primitive nth root of unity in K. Conversely, if char $K \nmid n$ (in particular, if char $K = 0$), then $nx^{n-1} \neq 0$, whence $x^n - 1_K$ is relatively prime to its derivative. Therefore $x^n - 1_K$ has n distinct roots in any splitting field F of $x^n - 1_K$ over K (Theorem III.6.10). Thus the cyclic group of nth roots of unity in F has order n and F (but not necessarily K) contains a primitive nth root of unity. Note that if K does contain a primitive nth root of unity, then K contains n distinct roots of $x^n - 1_K$, whence $F = K$.

EXAMPLES. 1_K is an nth root of unity in the field K for all $n \geq 1$. If char $K = p \neq 0$ and $n = p^k$, then 1_K is the only nth root of unity in K. The subfield $\mathbf{Q}(i)$ of $\mathbf{C}$ contains both primitive fourth roots of unity ($\pm i$) but no cube roots of unity except 1, (the others being $-1/2 \pm \sqrt{3}\, i/2$). For each $n > 0$, $e^{2\pi i/n} \in \mathbf{C}$ is a primitive nth root of unity.

In order to finish our characterization of cyclic extensions we need

Lemma 7.10. *Let* n *be a positive integer and* K *a field which contains a primitive* n*th root of unity* ζ.

(i) *If* $d \mid n$, *then* $\zeta^{n/d} = \eta$ *is a primitive* d*th root of unity in* K.

(ii) *If* $d \mid n$ *and* u *is a nonzero root of* $x^d - a \in K[x]$, *then* $x^d - a$ *has* d *distinct roots, namely* $u, \eta u, \eta^2 u, \ldots, \eta^{d-1}u$, *where* $\eta \in K$ *is a primitive* d*th root of unity. Furthermore* K(u) *is a splitting field of* $x^d - a$ *over* K *and is Galois over* K.

PROOF. (i) ζ generates a multiplicative cyclic group of order n by definition. If $d \mid n$, then $\eta = \zeta^{n/d}$ has order d by Theorem I.3.4, whence η is a primitive dth root of unity. (ii) If u is a root of $x^d - a$, then so is $\eta^i u$. The elements $\eta^0 = 1_K, \eta, \ldots, \eta^{d-1}$ are distinct (Theorem I.3.4). Consequently since $\eta \in K$, the roots $u, \eta u, \ldots, \eta^{d-1}u$ of $x^d - a$ are distinct elements of $K(u)$. Thus $K(u)$ is a splitting field of $x^d - a$ over K. The irreducible factors of $x^d - a$ are separable since all the roots are distinct, whence $K(u)$ is Galois over K by Theorem 3.11 and Exercise 3.13. ■

Theorem 7.11. *Let* n *be a positive integer and* K *a field which contains a primitive* n*th root of unity* ζ. *Then the following conditions on an extension field* F *of* K *are equivalent.*

(i) F *is cyclic of degree* d, *where* $d \mid n$;

(ii) F *is a splitting field over* K *of a polynomial of the form* $x^n - a \in K[x]$ (*in which case* $F = K(u)$, *for any root* u *of* $x^n - a$);

(iii) F *is a splitting field over* K *of an irreducible polynomial of the form* $x^d - b \in K[x]$, *where* $d \mid n$ (*in which case* $F = K(v)$, *for any root* v *of* $x^d - b$).

PROOF. (ii) $\Rightarrow$ (i) Lemma 7.10 shows that $F = K(u)$ and F is Galois over K for any root u of $x^n - a$. If $\sigma \in \mathrm{Aut}_K F = \mathrm{Aut}_K K(u)$, then σ is completely determined by $\sigma(u)$, which is a root of $x^n - a$ by Theorem 2.2. Therefore, $\sigma(u) = \zeta^i u$ for some $i\,(0 \leq i \leq n-1)$ by Lemma 7.10. Verify that the assignment $\sigma \mapsto \zeta^i$ defines a monomorphism from $\mathrm{Aut}_K F$ to the multiplicative cyclic group (of order n) of nth roots of unity in K. Consequently, $\mathrm{Aut}_K F$ is a cyclic group whose order d divides n (Theorem I.3.5 and Corollary I.4.6). Hence F is cyclic of degree d over K.

(i) $\Rightarrow$ (iii) By hypothesis $\mathrm{Aut}_K F$ is cyclic of order $d = [F:K]$ with generator σ. Let $\eta = \zeta^{n/d} \in K$ be a primitive dth root of unity. Since $N_K{}^F(\eta) = \eta^{[F:K]} = \eta^d = 1_K$, Theorem 7.6(ii) implies that $\eta = w\sigma(w)^{-1}$ for some $w \in F$. If $v = w^{-1}$, then $\sigma(v) = \eta v$ and $\sigma(v^d) = (\eta v)^d = \eta^d v^d = v^d$. Since F is Galois over K, $v^d = b$ must lie in K so that v is a root of $x^d - b \in K[x]$. By Lemma 7.10 $K(v) \subset F$ and $K(v)$ is a splitting field over K of $x^d - b$ (whose distinct roots are $v, \eta v, \dots, \eta^{d-1}v$). Furthermore for each $i\,(0 \leq i \leq d-1)$, $\sigma^i(v) = \eta^i v$ so that $\sigma^i : K(v) \cong K(\eta^i v)$. By Corollary 1.9 v and $\eta^i v$ are roots of the same irreducible polynomial over K. Consequently, $x^d - b$ is irreducible in $K[x]$. Therefore, $[K(v):K] = d = [F:K]$, whence $F = K(v)$.

(iii) $\Rightarrow$ (ii) If $v \in F$ is a root of $x^d - b \in K[x]$, then $F = K(v)$ by Lemma 7.10. Now $(\zeta v)^n = \zeta^n v^n = 1_K v^{d(n/d)} = b^{n/d} \in K$ so that ζv is a root of $x^n - a \in K[x]$, where $a = b^{n/d}$. By Lemma 7.10 again $K(\zeta v)$ is a splitting field of $x^n - a$ over K. But $\zeta \in K$ implies that $F = K(v) = K(\zeta v)$. ■

It is clear that the primitive nth roots of unity play an important role in the proof of the preceding results. Characterization of the splitting fields of polynomials of the form $x^n - a \in K[x]$ is considerably more difficult when K does not contain a primitive nth root of unity. The case when $a = 1_K$ is considered in Section 8.

EXERCISES

1. If $\overline{K}$ is replaced by any normal extension N of K containing F in Definition 7.1, then this new definition of norm and trace is equivalent to the original one. In particular, the new definition does not depend on the choice of N. See Exercise 3.21.

2. Let F be a finite dimensional extension of a finite field K. The norm $N_K{}^F$ and the trace $T_K{}^F$ (considered as maps $F \to K$) are surjective.

3. Let $\overline{\mathbf{Q}}$ be a (fixed) algebraic closure of $\mathbf{Q}$ and $v \in \overline{\mathbf{Q}}$, $v \notin \mathbf{Q}$. Let E be a subfield of $\overline{\mathbf{Q}}$ maximal with respect to the condition $v \notin E$. Prove that every finite dimensional extension of E is cyclic.

4. Let K be a field, $\overline{K}$ an algebraic closure of K and $\sigma \in \mathrm{Aut}_K \overline{K}$. Let
$$F = \{u \in \overline{K} \mid \sigma(u) = u\}.$$
Then F is a field and every finite dimensional extension of F is cyclic.

5. If F is a cyclic extension of K of degree p^n (p prime) and L is an intermediate field such that $F = L(u)$ and L is cyclic over K of degree p^{n-1}, then $F = K(u)$.

6. If char $K = p \neq 0$, let $K_p = \{u^p - u \mid u \in K\}$.
   (a) A cyclic extension field F of K of degree p exists if and only if $K \neq K_p$.

(b) If there exists a cyclic extension of degree p of K, then there exists a cyclic extension of degree p^n for every $n \geq 1$. [*Hint:* Use induction; if E is cyclic over K of degree p^{n-1} with $\mathrm{Aut}_K E$ generated by σ, show that there exist $u,v \in E$ such that $T_K^E(v) = 1_K$ and $\sigma(u) - u = v^p - v$. Then $x^p - x - u \in E[x]$ is irreducible and if w is a root, then $K(w)$ is cyclic of degree p^n over K.]

7. If n is an odd integer such that K contains a primitive nth root of unity and char $K \neq 2$, then K also contains a primitive $2n$th root of unity.

8. If F is a finite dimensional extension of $\mathbf{Q}$, then F contains only a finite number of roots of unity.

9. Which roots of unity are contained in the following fields: $\mathbf{Q}(i)$, $\mathbf{Q}(\sqrt{2})$, $\mathbf{Q}(\sqrt{3})$, $\mathbf{Q}(\sqrt{5})$, $\mathbf{Q}(\sqrt{-2})$, $\mathbf{Q}(\sqrt{-3})$?

10. (a) Let p be a prime and assume either (i) char $K = p$ or (ii) char $K \neq p$ and K contains a primitive pth root of unity. Then $x^p - a \in K[x]$ is either irreducible or splits in $K[x]$.
(b) If char $K = p \neq 0$, then for any root u of $x^p - a \in K[x]$, $K(u) \neq K(u^p)$ if and only if $[K(u) : K] = p$.

8. CYCLOTOMIC EXTENSIONS

Except for Theorem 8.1 this section is not needed in the sequel. We shall examine splitting fields of the polynomial $x^n - 1_K$, with special attention to the case $K = \mathbf{Q}$. These splitting fields turn out to be abelian extensions whose Galois groups are well known.

A splitting field F over a field K of $x^n - 1_K \in K[x]$ (where $n \geq 1$) is called a **cyclotomic extension of order n.** If char $K = p \neq 0$ and $n = mp^t$ with $(p,m) = 1$, then $x^n - 1_K = (x^m - 1)^{p^t}$ (Exercise III.1.11) so that a cyclotomic extension of order n coincides with one of order m. Thus we shall usually assume that char K does not divide n (that is, char $K = 0$ or is relatively prime to n).

The dimension of a cyclotomic extension field of order n is related to the **Euler function** φ of elementary number theory, which assigns to each positive integer n the number $\varphi(n)$ of integers i such that $1 \leq i \leq n$ and $(i,n) = 1$. For example, $\varphi(6) = 2$ and $\varphi(p) = p - 1$ for every prime p. Let $\bar{i}$ be the image of $i \in \mathbf{Z}$ under the canonical projection $\mathbf{Z} \to Z_n$. It is easily verified that $(i,n) = 1$ if and only if $\bar{i}$ is a unit in the ring Z_n (Exercise 1). Therefore the multiplicative group of units in Z_n has order $\varphi(n)$; for the structure of this group see Exercise 4.

Theorem 8.1. *Let* n *be a positive integer,* K *a field such that char* K *does not divide* n *and* F *a cyclotomic extension of* K *of order* n.

(i) F $=$ K(ζ), *where* $\zeta \in$ F *is a primitive* n*th root of unity.*
(ii) F *is an abelian extension of dimension* d, *where* d $\mid \varphi$(n) (φ *the Euler function*); *if* n *is prime* F *is actually a cyclic extension.*
(iii) Aut_KF *is isomorphic to a subgroup of order* d *of the multiplicative group of units of* Z_n.

REMARKS. Recall that an abelian extension is an algebraic Galois extension whose Galois group is abelian. The dimension of F over K may be strictly less than $\varphi(n)$. For example, if ζ is a primitive 5th root of unity in **C**, then $\mathbf{R} \subset \mathbf{R}(\zeta) \subset \mathbf{C}$, whence, $[\mathbf{R}(\zeta) : \mathbf{R}] = 2 < 4 = \varphi(5)$. If $K = \mathbf{Q}$, then the structure of the group $\mathrm{Aut}_{\mathbf{Q}}F$ is completely determined in Exercise 7.

SKETCH OF PROOF OF 8.1. (i) The remarks preceding Lemma 7.10 show that F contains a primitive nth root of unity ζ. By definition $1_K, \zeta, \ldots, \zeta^{n-1} \in K(\zeta)$ are the n distinct roots of $x^n - 1_K$, whence $F = K(\zeta)$. (ii) and (iii). Since the irreducible factors of $x^n - 1_K$ are clearly separable, Theorem 3.11 and Exercise 3.13 imply that F is Galois over K. If $\sigma \in \mathrm{Aut}_K F$, then σ is completely determined by $\sigma(\zeta)$. For some $i\,(1 \leq i \leq n-1)$, $\sigma(\zeta) = \zeta^i$ by Theorem 2.2. Similarly $\sigma^{-1}(\zeta) = \zeta^j$ so that $\zeta = \sigma^{-1}\sigma(\zeta) = \zeta^{ij}$. By Theorem I.3.4(v), $ij \equiv 1 \pmod{n}$ and hence $\bar{\imath} \in Z_n$ is a unit (where $i \mapsto \bar{\imath}$ under the canonical projection $\mathbf{Z} \to Z_n$). Verify that the assignment $\sigma \mapsto \bar{\imath}$ defines a monomorphism f from $\mathrm{Aut}_K F$ to the (abelian) multiplicative group of units of the ring Z_n (which has order $\varphi(n)$ by Exercise 1). Therefore, $\mathrm{Aut}_K F \cong \mathrm{Im}\, f$ is abelian with order d dividing $\varphi(n)$. Thus $[F : K] = d$ by the Fundamental Theorem 2.5. If n is prime, then Z_n is a field and $\mathrm{Aut}_K F \cong \mathrm{Im}\, f$ is cyclic by Theorem 5.3. ∎

Let n be a positive integer, K a field such that char K does not divide n, and F a cyclotomic extension of order n of K. The **nth cyclotomic polynomial** over K is the monic polynomial $g_n(x) = (x - \zeta_1)(x - \zeta_2)\cdots(x - \zeta_r)$ where $\zeta_1, \ldots, \zeta_r$ are all the distinct primitive nth roots of unity in F.

EXAMPLES. $g_1(x) = x - 1_K$ and $g_2(x) = (x - (-1_K)) = x + 1_K$. If $K = \mathbf{Q}$, then $g_3(x) = (x - (-1/2 + \sqrt{3}i/2))(x - (-1/2 - \sqrt{3}i/2)) = x^2 + x + 1$ and $g_4(x) = (x - i)(x + i) = x^2 + 1$. These examples suggest several properties of the cyclotomic polynomials.

Proposition 8.2. *Let* n *be a positive integer,* K *a field such that char* K *does not divide* n *and* $g_n(x)$ *the* n*th cyclotomic polynomial over* K.

(i) $x^n - 1_K = \prod_{d|n} g_d(x)$.

(ii) *The coefficients of* $g_n(x)$ *lie in the prime subfield* P *of* K. *If char* K = 0 *and* P *is identified with the field* **Q** *of rationals, then the coefficients are actually integers.*

(iii) *Deg* $g_n(x) = \varphi(n)$, *where* φ *is the Euler function.*

PROOF. (i) Let F be a cyclotomic extension of K of order n and $\zeta \in F$ a primitive nth root of unity. Lemma 7.10 (applied to F) shows that the cyclic group $G = \langle \zeta \rangle$ of all nth roots of unity contains all dth roots of unity for every divisor d of n. Clearly $\eta \in G$ is a primitive dth root of unity (where $d \mid n$) if and only if $|\eta| = d$. Therefore for each divisor d of n, $g_d(x) = \prod_{\substack{\eta \in G \\ |\eta| = d}} (x - \eta)$ and

$$x^n - 1_K = \prod_{\eta \in G} (x - \eta) = \prod_{\substack{d \\ d|n}} \Big(\prod_{\substack{\eta \in G \\ |\eta| = d}} (x - \eta) \Big) = \prod_{\substack{d \\ d|n}} g_d(x).$$

(ii) We prove the first statement by induction on n. Clearly $g_1(x) = x - 1_K \varepsilon P[x]$. Assume that (ii) is true for all $k < n$ and let $f(x) = \prod_{\substack{d \\ d|n \\ d<n}} g_d(x)$. Then $f \varepsilon P[x]$ by the induction hypothesis and in $F[x]$, $x^n - 1_K = f(x)g_n(x)$ by (i). On the other hand $x^n - 1_K \varepsilon P[x]$ and f is monic. Consequently, the division algorithm in $P[x]$ implies that $x^n - 1_K = fh + r$ for some $h, r \varepsilon P[x] \subset F[x]$. Therefore by the uniqueness of quotient and remainder (of the division algorithm applied in $F[x]$) we must have $r = 0$ and $g_n(x) = h \varepsilon P[x]$. This completes the induction. If char $K = 0$ and $P = \mathbf{Q}$, then a similar inductive argument using the division algorithm in $\mathbf{Z}[x]$ and $\mathbf{Q}[x]$ (instead of $P[x]$, $F[x]$) shows that $g_n(x) \varepsilon \mathbf{Z}[x]$.

(iii) deg g_n is clearly the number of primitive nth roots of unity. Let ζ be such a primitive root so that every other (primitive) root is a power of ζ. Then ζ^i $(1 \leq i \leq n)$ is a primitive nth root of unity (that is, a generator of G) if and only if $(i,n) = 1$ by Theorem I.3.6. But the number of such i is by definition precisely $\varphi(n)$. ■

REMARKS. Part (i) of the theorem gives a recursive method for determining $g_n(x)$ since

$$g_n(x) = \frac{x^n - 1_K}{\prod_{\substack{d \\ d|n \\ d<n}} g_d(x)}.$$

For example if p is prime, then $g_p(x) = (x^p - 1_K)/g_1(x) = (x^p - 1_K)/(x - 1_K) = x^{p-1} + x^{p-2} + \cdots + x^2 + x + 1_K$. Using the example preceding Theorem 8.2 we have for $K = \mathbf{Q}$:

$$\begin{aligned} g_6(x) &= (x^6 - 1)/g_1(x)g_2(x)g_3(x) \\ &= (x^6 - 1)/(x - 1)(x + 1)(x^2 + x + 1) \\ &= x^2 - x + 1; \end{aligned}$$

similarly

$$\begin{aligned} g_{12}(x) &= (x^{12} - 1)/(x - 1)(x + 1)(x^2 + x + 1)(x^2 + 1)(x^2 - x + 1) \\ &= x^4 - x^2 + 1. \end{aligned}$$

When the base field is the field $\mathbf{Q}$, we can strengthen the previous results somewhat.

Proposition 8.3. *Let* F *be a cyclotomic extension of order* n *of the field* $\mathbf{Q}$ *of rational numbers and* $g_n(x)$ *the* n*th cyclotomic polynomial over* $\mathbf{Q}$. *Then*

(i) $g_n(x)$ *is irreducible in* $\mathbf{Q}[x]$.

(ii) $[F : \mathbf{Q}] = \varphi(n)$, *where* φ *is the Euler function.*

(iii) $Aut_{\mathbf{Q}}F$ *is isomorphic to the multiplicative group of units in the ring* Z_n.

SKETCH OF PROOF. (i) It suffices by Lemma III.6.13 to show that the monic polynomial $g_n(x)$ is irreducible in $\mathbf{Z}[x]$. Let h be an irreducible factor of g_n in $\mathbf{Z}[x]$ with deg $h \geqslant 1$. Then $g_n(x) = f(x)h(x)$ with $f,h \varepsilon \mathbf{Z}[x]$ monic. Let ζ be a root of h and p any prime integer such that $(p,n) = 1$.

We shall show first that ζ^p is also a root of h. Since ζ is a root of $g_n(x)$, ζ is a primitive nth root of unity. The proof of Proposition 8.2(iii) implies that ζ^p is also a primitive nth root of unity and therefore a root of either f or h. Suppose ζ^p is not a root of h. Then ζ^p is a root of $f(x) = \sum_{i=0}^{r} a_i x^i$ and hence ζ is a root of $f(x^p) = \sum_{i=0}^{r} a_i x^{ip}$. Since h is irreducible in $\mathbf{Q}[x]$ (Lemma III.6.13) and has ζ as a root, h must divide $f(x^p)$ by Theorem 1.6, say $f(x^p) = h(x)k(x)$ with $k \in \mathbf{Q}[x]$. By the division algorithm in $\mathbf{Z}[x]$, $f(x^p) = h(x)k_1(x) + r_1(x)$ with $k_1, r_1 \in \mathbf{Z}[x]$. The uniqueness statement of the division algorithm in $\mathbf{Q}[x]$ shows that $k(x) = k_1(x) \in \mathbf{Z}[x]$. Recall that the canonical projection $\mathbf{Z} \to Z_p$ (denoted on elements by $b \mapsto \bar{b}$) induces a ring epimorphism $\mathbf{Z}[x] \to Z_p[x]$ defined by $g = \sum_{i=0}^{t} b_i x^i \mapsto \bar{g} = \sum_{i=0}^{t} \bar{b}_i x^i$ (Exercise III.5.1). Consequently, in $Z_p[x]$, $\bar{f}(x^p) = \bar{h}(x)\bar{k}(x)$. But in $Z_p[x]$, $\bar{f}(x^p) = \bar{f}(x)^p$ (since char $Z_p = p$). Therefore,

$$\bar{f}(x)^p = \bar{h}(x)\bar{k}(x) \in Z_p[x].$$

Consequently, some irreducible factor of $\bar{h}(x)$ of positive degree must divide $\bar{f}(x)^p$ and hence $\bar{f}(x)$ in $Z_p[x]$. On the other hand, since $g_n(x)$ is a factor of $x^n - 1$, we have $x^n - 1 = g_n(x)r(x) = f(x)h(x)r(x)$ for some $r(x) \in \mathbf{Z}[x]$. Thus in $Z_p[x]$

$$x^n - \bar{1} = \overline{x^n - 1} = \bar{f}(x)\bar{h}(x)\bar{r}(x).$$

Since $\bar{f}$ and $\bar{h}$ have a common factor, $x^n - \bar{1} \in Z_p[x]$ must have a multiple root. This contradicts the fact that the roots of $x^n - \bar{1}$ are all distinct since $(p,n) = 1$ (see the Remarks preceding Lemma 7.10). Therefore ζ^p is a root of $h(x)$.

If $r \in \mathbf{Z}$ is such that $1 \leq r \leq n$ and $(r,n) = 1$, then $r = p_1^{k_1} \cdots p_s^{k_s}$ where $k_i > 0$ and each p_i is a prime such that $(p_i,n) = 1$. Repeated application of the fact that ζ^p is a root of h whenever ζ is, shows that ζ^r is a root of $h(x)$. But the ζ^r ($1 \leq r \leq n$ and $(r,n) = 1$) are precisely all of the primitive nth roots of unity by the proof of Proposition 8.2(iii). Thus $h(x)$ is divisible by $\prod_{\substack{1 \leq r \leq n \\ (r,n)=1}} (x - \zeta^r) = g_n(x)$, whence $g_n(x) = h(x)$. Therefore, $g_n(x)$ is irreducible.

(ii) Lemma 7.10 shows that $F = \mathbf{Q}(\zeta)$, whence

$$[F : \mathbf{Q}] = [\mathbf{Q}(\zeta) : \mathbf{Q}] = \deg g_n = \varphi(n)$$

by Proposition 8.2 and (i). (iii) is a consequence of (ii), Theorem 8.1, and Exercise 1. ■

REMARK. A nontrivial theorem of Kronecker states that every abelian extension of $\mathbf{Q}$ is contained in a cyclotomic extension.

EXERCISES

1. If $i \in \mathbf{Z}$, let $\bar{i}$ denote the image of i in Z_n under the canonical projection $\mathbf{Z} \to Z_n$. Prove that $\bar{i}$ is a unit in the ring Z_n if and only if $(i,n) = 1$. Therefore the multiplicative group of units in Z_n has order $\varphi(n)$.

2. Establish the following properties of the Euler function φ.
   (a) If p is prime and $n > 0$, then $\varphi(p^n) = p^n(1 - 1/p) = p^{n-1}(p - 1)$.
   (b) If $(m,n) = 1$, then $\varphi(mn) = \varphi(m)\varphi(n)$.

(c) If $n = p_1^{k_1}\cdots p_r^{k_r}$ (p_i distinct primes; $k_i > 0$), then $\varphi(n) = n(1 - 1/p_1)(1 - 1/p_2)\cdots(1 - 1/p_r)$.

(d) $\sum_{d|n} \varphi(d) = n$.

(e) $\varphi(n) = \sum_{d|n} d\mu(n/d)$, where μ is the Moebius function defined by

$$\mu(n) = \begin{cases} 1 & \text{if } n = 1 \\ (-1)^t & \text{if } n \text{ is a product of } t \text{ distinct primes} \\ 0 & \text{if } p^2 \text{ divides } n \text{ for some prime } p. \end{cases}$$

3. Let φ be the Euler function.
   (a) $\varphi(n)$ is even for $n > 2$.
   (b) Find all $n > 0$ such that $\varphi(n) = 2$.
   (c) Find all pairs (n,p) (where $n,p > 0$, and p is prime) such that $\varphi(n) = n/p$. [See Exercise 2.]

4. (a) If p is an odd prime and $n > 0$, then the multiplicative group of units in the ring Z_{p^n} is cyclic of order $p^{n-1}(p - 1)$.
   (b) Part (a) is also true if $p = 2$ and $1 \leq n \leq 2$.
   (c) If $n \geq 3$, then the multiplicative group of units in Z_{2^n} is isomorphic to $Z_2 \oplus Z_{2^{n-2}}$.

5. If $f(x) = \sum_{i=0}^{t} a_i x^i$, let $f(x^s)$ be the polynomial $\sum_{i=0}^{t} a_i x^{is}$. Establish the following properties of the cyclotomic polynomials $g_n(x)$ over $\mathbf{Q}$.
   (a) If p is prime and $k \geq 1$, then $g_{p^k}(x) = g_p(x^{p^{k-1}})$.
   (b) If $n = p_1^{r_1}\cdots p_k^{r_k}$ (p_i distinct primes; $r_i > 0$), then

$$g_n(x) = g_{p_1\cdots p_k}(x^{p_1^{r_1-1}\cdots p_k^{r_k-1}}).$$

   (c) If n is odd, then $g_{2n}(x) = g_n(-x)$.
   (d) If p is a prime and $p \nmid n$, then $g_{pn}(x) = g_n(x^p)/g_n(x)$.
   (e) $g_n(x) = \prod_{d|n} (x^{n/d} - 1)^{\mu(d)}$, where μ is the Moebius function of Exercise 2 (e).
   (f) $g_n(1) = p$ if $n = p^k$ $(k > 0)$, 0 if $n = 1$, and 1 otherwise.

6. Calculate the nth cyclotomic polynomials over $\mathbf{Q}$ for all positive n with $n \leq 20$.

7. Let F_n be a cyclotomic extension of $\mathbf{Q}$ of order n. Determine the structure of $\mathrm{Aut}_{\mathbf{Q}}F_n$ for every n. [*Hint:* if $U_n{}^*$ denotes the multiplicative group of units in Z_n, then show that $U_n{}^* = \prod_{i=1}^{r} U_{p_i{}^{n_i}}{}^*$ where n has prime decomposition $n = p_1^{n_1}\cdots p_r^{n_r}$. Apply Exercise 4.]

8. Let F_n be a cyclotomic extension of $\mathbf{Q}$ of order n.
   (a) Determine $\mathrm{Aut}_{\mathbf{Q}}F_5$ and all intermediate fields.
   (b) Do the same for F_8.
   (c) Do the same for F_7; if ζ is a primitive 7th root of unity what is the irreducible polynomial over $\mathbf{Q}$ of $\zeta + \zeta^{-1}$?

9. If $n > 2$ and ζ is a primitive nth root of unity over $\mathbf{Q}$, then $[\mathbf{Q}(\zeta + \zeta^{-1}) : \mathbf{Q}] = \varphi(n)/2$.

10. (Wedderburn) A finite division ring D is a field. Here is an outline of the proof (in which E^* denotes the multiplicative group of nonzero elements of a division ring E).

(a) The center K of D is a field and D is a vector space over K, whence $|D| = q^n$ where $q = |K| \geq 2$.

(b) If $0 \neq a \varepsilon D$, then $N(a) = \{d \varepsilon D \mid da = ad\}$ is a subdivision ring of D containing K. Furthermore, $|N(a)| = q^r$ where $r \mid n$.

(c) If $0 \neq a \varepsilon D - K$, then $N(a)^*$ is the centralizer of a in the group D^* and $[D^* : N(a)^*] = (q^n - 1)/(q^r - 1)$ for some r such that $1 \leq r < n$ and $r \mid n$.

(d) $q^n - 1 = q - 1 + \sum_r (q^n - 1)/(q^r - 1)$, where the last sum taken over a finite number of integers r such that $1 \leq r < n$ and $r \mid n$. [*Hint:* use the class equation of D^*; see pp. 90–91.]

(e) For each primitive nth root of unity $\zeta \varepsilon \mathbf{C}$, $|q - \zeta| > q - 1$, where $|a + bi| = \sqrt{a^2 + b^2}$ for $a + bi \varepsilon \mathbf{C}$. Consequently, $|g_n(q)| > q - 1$, where g_n is the nth cyclotomic polynomial over $\mathbf{Q}$.

(f) The equation in (d) is impossible unless $n = 1$, whence $K = D$. [*Hint:* Use Proposition 8.2 to show that for each positive divisor r of n with $r \neq n$, $f_r(x) = (x^n - 1)/(x^r - 1)$ is in $\mathbf{Z}[x]$ and $f_r(x) = g_n(x)h_r(x)$ for some $h_r(x) \varepsilon \mathbf{Z}[x]$. Consequently, for each such r $g_n(q)$ divides $f_r(q)$ in $\mathbf{Z}$, whence $g_n(q) \mid (q - 1)$ by (d). This contradicts (e).]

9. RADICAL EXTENSIONS

Galois theory had its historical origin in a classical problem in the theory of equations, which may be intuitively but reasonably accurately stated as follows. Given a field K, does there exist an explicit "formula" (involving only field operations and the extraction of nth roots) which gives all the solutions of an *arbitrary* polynomial equation $f(x) = 0$ ($f \varepsilon K[x]$)? If the degree of f is at most four, the answer is affirmative (for example, the familiar "quadratic formula" when $\deg f = 2$ and char $K \neq 2$; see also Exercise 5). We shall show, however, that the answer is negative in general (Proposition 9.8). In doing so we shall characterize certain field extensions whose Galois groups are solvable (Theorem 9.4 and Proposition 9.6).

The first task is to formulate a precise statement of the classical problem in field-theoretic terms. Throughout the discussion we shall work in a fixed algebraic closure of the given base field K. Intuitively the existence of a "formula" for solving a *specific* polynomial equation $f(x) = 0$ means that there is a finite sequence of steps, each step being a field operation (addition, multiplication, inverses) or the extraction of an nth root, which yields all solutions of the given equation. Performing a field operation leaves the base field unchanged, but the extraction of an nth root of an element c in a field E amounts to constructing an extension field $E(u)$ with $u^n \varepsilon E$ (that is, $u = \sqrt[n]{c}$). Thus the existence of a "formula" for solving $f(x) = 0$ would in effect imply the existence of a finite tower of fields

$$K = E_0 \subset E_1 \subset \cdots \subset E_n$$

such that E_n contains a splitting field of f over K and for each $i \geq 1$, $E_i = E_{i-1}(u_i)$ with some positive power of u_i lying in E_{i-1}.

Conversely suppose that there exists such a tower of fields and that E_n contains a splitting field of f (that is, E_n contains all solutions of $f(x) = 0$). Then

$$E_n = K(u_1, \ldots, u_n)$$

and each solution is of the form

$$f(u_1, \ldots, u_n)/g(u_1, \ldots, u_n) \qquad (f,g \in K[x_1, \ldots, x_n])$$

by Theorem 1.3. Thus each solution is expressible in terms of a finite number of elements of K, a finite number of field operations and $u_1, \ldots, u_n$ (which are obtained by extracting roots). But this amounts to saying that there is a "formula" for the solutions of the particular given equations. These considerations motivate the next two definitions.

Definition 9.1. *An extension field* F *of a field* K *is a* **radical extension** *of* K *if* $F = K(u_1, \ldots, u_n)$, *some power of* u_1 *lies in* K *and for each* $i \geq 2$, *some power of* u_i *lies in* $K(u_1, \ldots, u_{i-1})$.

REMARKS. If $u_i^m \in K(u_1, \ldots, u_{i-1})$ then u_i is a root of

$$x^m - u_i^m \in K(u_1, \ldots, u_{i-1})[x].$$

Hence $K(u_1, \ldots, u_i)$ is finite dimensional algebraic over $K(u_1, \ldots, u_{i-1})$ by Theorem 1.12. Therefore every radical extension F of K is finite dimensional algebraic over K by Theorems 1.2 and 1.11.

Definition 9.2. *Let* K *be a field and* $f \in K[x]$. *The equation* $f(x) = 0$ *is* **solvable by radicals** *if there exists a radical extension* F *of* K *and a splitting field* E *of* f *over* K *such that* $F \supset E \supset K$.

Definition 9.2 is the first step in the formulation of the classical problem of finding a "formula" for the solutions of $f(x) = 0$ that is valid for *every* polynomial $f \in K[x]$ of a given degree r (such as the quadratic formula for $r = 2$). For whatever the precise definition of such a "formula" might be, it is clear from the discussion preceding Definition 9.1 that the existence of such a "formula" should imply that every polynomial equation of degree r is solvable by radicals.

Thus in order to demonstrate the nonexistence of such a formula, it suffices to prove that a specific polynomial equation is not solvable by radicals. We shall now develop the necessary information in order to do this (Corollary 9.5) and shall leave the precise formulation of the classical problem for the appendix.

Lemma 9.3. *If* F *is a radical extension field of* K *and* N *is a normal closure of* F *over* K (*Theorem* 3.16), *then* N *is a radical extension of* K.

SKETCH OF PROOF. The proof consists of combining two facts. (i) If F is any finite dimensional extension of K (not necessarily radical) and N is the normal closure of F over K, then N is the composite field $E_1E_2 \cdots E_r$, where each E_i is a subfield of N which is K-isomorphic to F. (ii) If $E_1, \ldots, E_r$ are each radical extensions of

K (as is the case here since F is radical), then the composite field $E_1E_2\cdots E_r$ is a radical extension of K. These statements are justified as follows.

(i) Let $\{w_1, \ldots, w_n\}$ be a basis of F over K and let f_i be the irreducible polynomial of w_i over K. The proof of Theorem 3.16 shows that N is a splitting field of $\{f_1, \ldots, f_n\}$ over K. Let v be any root of f_j in N. Then there is a K-isomorphism $\sigma : K(w_j) \cong K(v)$ such that $\sigma(w_j) = v$ by Theorem 1.8. By Theorem 3.8 σ extends to a K-automorphism τ of N. Clearly $\tau(F)$ is a subfield of N which is K-isomorphic to F and contains $\tau(w_j) = \sigma(w_j) = v$. In this way we can find for every root v of every f_j a subfield E of N such that $v \in E$ and E is K-isomorphic to F. If $E_1, \ldots, E_r$ are the subfields so obtained, then $E_1E_2\cdots E_r$ is a subfield of N which contains all the roots of $f_1, f_2, \ldots, f_n$, whence $E_1E_2\cdots E_r = N$.

(ii) Suppose $r = 2$, $E_1 = K(u_1, \ldots, u_k)$ and $E_2 = K(v_1, \ldots, v_m)$ as in Definition 9.1. Then $E_1E_2 = K(u_1, \ldots, u_k, v_1, \ldots, v_m)$ is clearly a radical extension of K. The general case is similar. ■

Theorem 9.4. *If* F *is a radical extension field of* K *and* E *is an intermediate field, then* Aut_KE *is a solvable group.*

PROOF. If K_0 is the fixed field of E relative to the group $\mathrm{Aut}_K E$, then E is Galois over K_0, $\mathrm{Aut}_{K_0} E = \mathrm{Aut}_K E$ and F is a radical extension of K_0 (Exercise 1). Thus we may assume to begin with that E is algebraic Galois over K. Let N be a normal closure of F over K (Theorem 3.16). Then N is a radical extension of K by Lemma 9.3 and E is a stable intermediate field by Lemma 2.13. Consequently, restriction $(\sigma \mapsto \sigma \mid E)$ induces a homomorphism $\theta : \mathrm{Aut}_K N \to \mathrm{Aut}_K E$. Since N is a splitting field over K (and hence over E) every $\sigma \in \mathrm{Aut}_K E$ extends to a K-automorphism of N by Theorem 3.8. Therefore θ is an epimorphism. Since the homomorphic image of a solvable group is solvable (Theorem II.7.11), it suffices to prove that $\mathrm{Aut}_K N$ is solvable. If K_1 is the fixed field of N relative to $\mathrm{Aut}_K N$, then N is a radical Galois extension of K_1 (Exercise 1) and $\mathrm{Aut}_{K_1} N = \mathrm{Aut}_K N$. Therefore, we may return to our original notation and with no loss of generality assume that $F = E$ and F is a Galois radical extension of K.

If $F = K(u_1, \ldots, u_n)$ with $u_1^{m_1} \in K$ and $u_i^{m_i} \in K(u_1, \ldots, u_{i-1})$ for $i \geq 2$, then we may assume that char K does not divide m_i. This is obvious if char $K = 0$. If char $K = p \neq 0$ and $m_i = rp^t$ with $(r,p) = 1$, then $u_i^{rp^t} \in K(u_1, \ldots, u_{i-1})$ so that u_i^r is purely inseparable over $K(u_1, \ldots, u_{i-1})$. But F is Galois and thus separable over K (Theorem 3.11), whence F is separable over $K(u_1, \ldots, u_{i-1})$ (Exercise 3.12). Therefore $u_i^r \in K(u_1, \ldots, u_{i-1})$ by Theorem 6.2, and we may assume $m_i = r$.

If $m = m_1m_2\cdots m_n$, then by the previous paragraph char K ($=$ char F) does not divide m. Consider the cyclotomic extension $F(\zeta)$ of F, where ζ is a primitive mth root of unity (Theorem 8.1). The situation is this:

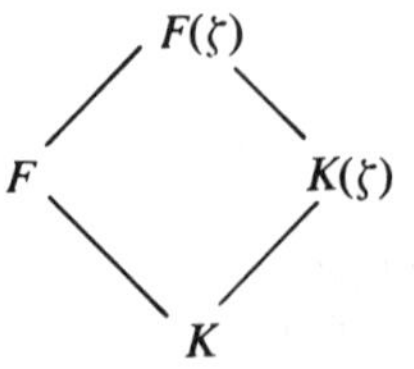

where $F(\zeta)$ is Galois over F (Theorem 8.1) and hence over K as well (Exercise 3.15(b)). The Fundamental Theorem 2.5 shows that $\text{Aut}_K F \cong \text{Aut}_K F(\zeta)/\text{Aut}_F F(\zeta)$. Consequently, it suffices by Theorem II.7.11 to prove that $\text{Aut}_K F(\zeta)$ is solvable. Observe that $K(\zeta)$ is an abelian Galois extension of K (Theorem 8.1), whence $\text{Aut}_K K(\zeta) \cong \text{Aut}_K F(\zeta)/\text{Aut}_{K(\zeta)} F(\zeta)$ by the Fundamental Theorem 2.5. If we knew that $\text{Aut}_{K(\zeta)} F(\zeta)$ were solvable, then Theorem II.7.11 would imply that $\text{Aut}_K F(\zeta)$ is solvable (since $\text{Aut}_K K(\zeta)$ is abelian, hence solvable). Thus we need only prove that $\text{Aut}_{K(\zeta)} F(\zeta)$ is solvable.

By assumption, $F(\zeta)$ is Galois over K and hence over any intermediate field. Let $E_0 = K(\zeta)$ and

$$E_i = K(\zeta, u_1, \ldots, u_i) \qquad (i = 1, 2, \ldots, n)$$

so that $E_n = K(\zeta, u_1, \ldots, u_n) = F(\zeta)$. Let $H_i = \text{Aut}_{E_i} F(\zeta)$, the corresponding subgroup of $\text{Aut}_{K(\zeta)} F(\zeta)$ under the Galois correspondence. Schematically we have:

$$\begin{array}{ccc}
F(\zeta) = E_n & \longmapsto & H_n = 1 \\
\vdots & & \vdots \\
E_i & \longmapsto & H_i = \text{Aut}_{E_i} F(\zeta) \\
\cup & & \\
E_{i-1} & \longmapsto & H_{i-1} = \text{Aut}_{E_{i-1}} F(\zeta) \\
\vdots & & \vdots \\
K(\zeta) = E_0 & \longmapsto & H_0 = \text{Aut}_{K(\zeta)} F(\zeta)
\end{array}$$

By Lemma 7.10(i) $K(\zeta)$ contains a primitive m_ith root of unity for each i $(i = 1, 2, \ldots, n)$. Since $u_i^{m_i} \in E_{i-1}$ and $E_i = E_{i-1}(u_i)$, each E_i is a cyclic extension of E_{i-1} by Lemma 7.10 (ii) (with $d = m_i$) and Theorem 7.11(ii) (with $n = m_i$). In particular, E_i is Galois over E_{i-1}. The Fundamental Theorem 2.5 implies that for each $i = 1, 2, \ldots, n$ $H_i \lhd H_{i-1}$ and $H_{i-1}/H_i \cong \text{Aut}_{E_{i-1}} E_i$, whence H_{i-1}/H_i is cyclic abelian. Consequently,

$$1 = H_n < H_{n-1} < \cdots < H_0 = \text{Aut}_{K(\zeta)} F(\zeta)$$

is a solvable series (Definition II.8.3). Therefore, $\text{Aut}_{K(\zeta)} F(\zeta)$ is solvable by Theorem II.8.5. ■

Corollary 9.5. *Let* K *be a field and* f $\in$ K[x]. *If the equation* f(x) = 0 *is solvable by radicals, then the Galois group of* f *is a solvable group.*

PROOF. Immediate from Theorem 9.4 and Definition 9.2. ■

EXAMPLE. The polynomial $f = x^5 - 4x + 2 \in \mathbf{Q}[x]$ has Galois group S_5 (see the example following Theorem 4.12), which is not a solvable group (Corollary II.7.12). Therefore, $x^5 - 4x + 2 = 0$ is not solvable by radicals and there can be no "formula" (involving only field operations and extraction of roots) for its solutions.

Observe that the base field plays an important role here. The polynomial $x^5 - 4x + 2 = 0$ is not solvable by radicals over $\mathbf{Q}$, but it is solvable by radicals over the field $\mathbf{R}$ of real numbers. In fact, every polynomial equation over $\mathbf{R}$ is solvable by radicals since all the solutions lie in the algebraic closure $\mathbf{C} = \mathbf{R}(i)$ which is a radical extension of $\mathbf{R}$.

We close this section by proving a partial converse to Theorem 9.4. There is no difficulty if K has characteristic zero. But if char K is positive, it will be necessary to place some restrictions on it (or alternatively to redefine "radical extension" — see Exercise 2).

Proposition 9.6. *Let* E *be a finite dimensional Galois extension field of* K *with solvable Galois group* $Aut_K E$. *Assume that char* K *does not divide* [E : K]. *Then there exists a radical extension* F *of* K *such that* F ⊃ E ⊃ K.

REMARK. The requirement that E be Galois over K is essential (Exercise 3).

SKETCH OF PROOF OF 9.6. Since $\text{Aut}_K E$ is a finite solvable group, it has a normal subgroup H of prime index p by Proposition II.8.6. Since E is Galois over K, $|\text{Aut}_K E| = [E : K]$ (Theorem 2.5), so that char $K \nmid p$. Let $N = E(\zeta)$ be a cyclotomic extension of E, where ζ is a primitive pth root of unity (Theorem 8.1). Let $M = K(\zeta)$; then we have

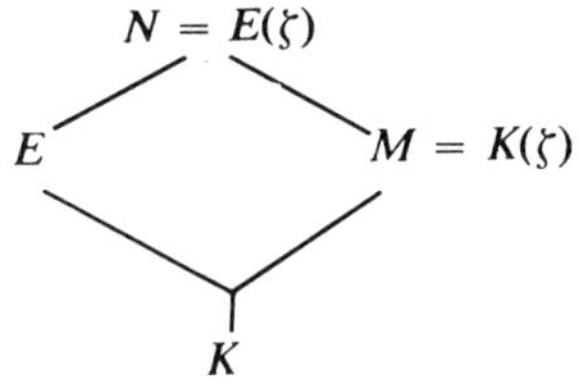

N is finite dimensional Galois over E (Theorem 8.1) and hence over K as well (Exercise 3.15(b)). Now M is clearly a radical extension of K. Consequently, it will suffice (by Exercise 4) to show that there is a radical extension of M that contains N.

First observe that E is a stable intermediate field of N and K (Lemma 2.13). Thus restriction ($\sigma \mapsto \sigma \mid E$) induces a homomorphism $\theta : \text{Aut}_M N \to \text{Aut}_K E$. If $\sigma \in \text{Aut}_M N$, then $\sigma(\zeta) = \zeta$. Hence if $\sigma \in \text{Ker } \theta$, we have $\sigma = 1_N$. Therefore θ is a monomorphism.

We now prove the theorem by induction on $n = [E : K]$. The case $n = 1$ is trivial. Assume the theorem is true for all extensions of dimension $k < n$ and consider the two possibilities:

(i) $\text{Aut}_M N$ is isomorphic under θ to a proper subgroup of $\text{Aut}_K E$;
(ii) $\theta : \text{Aut}_M N \cong \text{Aut}_K E$.

In either case $\text{Aut}_M N$ is solvable (Theorem II.7.11) and N is a finite dimensional Galois extension of K and hence of M. In case (i) $[N : M] = |\text{Aut}_M N| < |\text{Aut}_K E| = [E : K] = n$, whence the inductive hypothesis implies that there is a radical extension of M that contains N. As remarked in the first paragraph, this proves the theorem in case (i). In case (ii), let $J = \theta^{-1}(H)$. Since H is normal of index p in $\text{Aut}_K E$, J is normal of index p in $\text{Aut}_M N$. Furthermore J is solvable by Theorem II.7.11. If P is the fixed field of J (relative to $\text{Aut}_M N$), then we have:

$$\begin{array}{ccc} N & \longleftrightarrow & 1 \\ \cup & & \triangle \\ P & \longleftrightarrow & J = \mathrm{Aut}_P N \\ \cup & & \triangle \\ M & \longleftrightarrow & \mathrm{Aut}_M N \end{array}$$

The Fundamental Theorem 2.5 implies that P is Galois over M and that $\mathrm{Aut}_M P \cong \mathrm{Aut}_M N/J$. But $[\mathrm{Aut}_M N : J] = p$ by construction, whence $\mathrm{Aut}_M P \cong Z_p$. Therefore, P is a cyclic extension of M and $P = M(u)$, where u is a root of some (irreducible) $x^p - a \in M[x]$ (Theorem 7.11). Thus P is a radical extension of M and $[N : P] < [N : M] = [F : K] = n$. Since $\mathrm{Aut}_P N = J$ is solvable and N is Galois over P (Theorem 2.5), the induction hypothesis implies that there is a radical extension F of P that contains N. F is a clearly radical extension of M (Exercise 4). This completes the proof of case (ii). ■

Corollary 9.7. *Let* K *be a field and* f $\in$ K[x] *a polynomial of degree* n > 0, *where char* K *does not divide* n! (*which is always true when char* K = 0). *Then the equation* f(x) = 0 *is solvable by radicals if and only if the Galois group of* f *is solvable.*

SKETCH OF PROOF. ($\Leftarrow$) Let E be a splitting field of f over K. In view of Proposition 9.6 we need only show that E is Galois over K and char $K \nmid [E : K]$. Since char $K \nmid n!$ the irreducible factors of f are separable by Theorem III.6.10 and Exercise III.6.3, whence E is Galois over K (Theorem 3.11 and Exercise 3.13). Since every prime that divides $[E : K]$ necessarily divides $n!$ (Theorem 3.2), we conclude that char $K \nmid [E : K]$. ■

APPENDIX: THE GENERAL EQUATION OF DEGREE n

The motivation for our discussion can best be seen by examining polynomial equations of degree 2 over a field K with char $K \neq 2$. Here and below there will be no loss of generality in restricting consideration to monic polynomials. If t_1 and t_2 are indeterminates, then the equation

$$x^2 - t_1x + t_2 = 0$$

over the field $K(t_1,t_2)$ of rational functions in t_1,t_2 is called the **general quadratic equation** over K. Any (monic) quadratic equation over K may be obtained from the general quadratic equation by substituting appropriate elements of K for t_1 and t_2. It is easy to verify that the solutions of the general quadratic equation (in some algebraic closure of $K(t_1,t_2)$) are given by:

$$x = \frac{t_1 \pm \sqrt{t_1^2 - 4t_2}}{2},$$

where $n = n1_K$ for $n \in \mathbf{Z}$. This is the well known quadratic formula. It shows that the solutions of the general quadratic equation lie in the radical extension field $K(t_1,t_2)(u)$ with $u^2 = t_1^2 - 4t_2$. In order to find the solutions of $x^2 - bx + c = 0$ ($b,c \in K$) one need only substitute b,c for t_1,t_2. Clearly the solutions lie in the radical extension $K(u)$ with $u^2 = b^2 - 4c \in K$. We now generalize these ideas to polynomial equations of arbitrary degree.

Let K be a field and n a positive integer. Consider the field $K(t_1, \ldots, t_n)$ of rational functions over K in the indeterminates $t_1, \ldots, t_n$. The polynomial

$$p_n(x) = x^n - t_1x^{n-1} + t_2x^{n-2} + \cdots + (-1)^{n-1}t_{n-1}x + (-1)^n t_n \in K(t_1, \ldots, t_n)[x]$$

is called the **general polynomial of degree n** over K and the equation $p_n(x) = 0$ is called the **general equation of degree n** over K.[3] Note that any (monic) polynomial of degree n in $K[x]$, say $f(x) = x^n + a_1x^{n-1} + \cdots + a_{n-1}x + a_n$ may be obtained from the general polynomial $p_n(x)$ by substituting $(-1)^i a_i$ for t_i.

The preceding discussion makes the following definition quite natural. We say that there is a **formula** for the solutions of the general equation of degree n provided that this equation is solvable by radicals over the field $K(t_1, \ldots, t_n)$. If $p_n(x) = 0$ is solvable by radicals, then the solutions of any (monic) polynomial equation of degree n over K may be found by appropriate substitutions in the solutions of $p_n(x) = 0$. Having precisely formulated it, we can now settle the classical problem with which this section was introduced.

Proposition 9.8. *(Abel) Let* K *be a field and* n *a positive integer. The general equation of degree* n *is solvable by radicals only if* $n \leq 4$.

REMARKS. The words "only if" in Proposition 9.8 may be replaced by "if and only if" when char $K = 0$. If radical extensions are defined as in Exercise 2, then "only if" may be replaced by "if and only if" for every characteristic. The fact that the general equation of degree n is not solvable by radicals for $n \geq 5$ does *not* exclude the possibility that a particular polynomial equation over K of degree $n \geq 5$ is solvable by radicals.

SKETCH OF PROOF OF 9.8. Let the notation be as above and let $u_1, \ldots, u_n$ be the roots of $p_n(x)$ in some splitting field $F = K(t_1, \ldots, t_n)(u_1, \ldots, u_n)$. Since $p_n(x) = (x - u_1)(x - u_2)\cdots(x - u_n)$ in F, a direct calculation shows that

$$t_1 = \sum_{i=1}^{n} u_i;\ t_2 = \sum_{1 \leq i < j \leq n} u_iu_j;\ \ldots,\ t_n = u_1u_2\cdots u_n;$$

that is, $t_i = f_i(u_1, \ldots, u_n)$ where $f_1, \ldots, f_n$ are the elementary symmetric functions in n indeterminates (see the appendix to Section 2). It follows that $F = K(u_1, \ldots, u_n)$. Now consider a new set of indeterminates $\{x_1, \ldots, x_n\}$ and the field $K(x_1, \ldots, x_n)$. Let E be the subfield of all symmetric rational functions in $K(x_1, \ldots, x_n)$. The basic idea of the proof is to construct an isomorphism of fields $F \cong K(x_1, \ldots, x_n)$ such that $K(t_1, \ldots, t_n)$ is mapped onto E. Then the Galois group $\mathrm{Aut}_{K(t_1,\ldots,t_n)}F$, of $p_n(x)$ will be isomorphic to $\mathrm{Aut}_E K(x_1, \ldots, x_n)$. But $\mathrm{Aut}_E K(x_1, \ldots, x_n)$ is isomorphic to

[3]The signs $(-1)^i$ are inserted for convenience in order to simplify certain calculations

S_n (see p. 253). S_n is solvable if and only if $n \leq 4$ (Corollary II.7.12 and Exercise II.7.10). Therefore, if $p_n(x) = 0$ is solvable by radicals then $n \leq 4$ by Corollary 9.5. [Conversely if $n \leq 4$ and char $K = 0$, then $p_n(x) = 0$ is solvable by radicals by Corollary 9.7.]

In order to construct the isomorphism $F \cong K(x_1, \ldots, x_n)$ we first observe that the subfield E of $K(x_1, \ldots, x_n)$ is precisely $K(f_1, \ldots, f_n)$ by Theorem 2.18, where $f_1, \ldots, f_n$ are the elementary symmetric functions. Next we establish a ring isomorphism $K[t_1, \ldots, t_n] \cong K[f_1, \ldots, f_n]$ as follows. By Theorem III.5.5 the assignment $g(t_1, \ldots, t_n) \mapsto g(f_1, \ldots, f_n)$ (in particular $t_i \mapsto f_i$) defines an epimorphism of rings $\theta : K[t_1, \ldots, t_n] \rightarrow K[f_1, \ldots, f_n]$. Suppose $g(t_1, \ldots, t_n) \mapsto 0$, so that $g(f_1, \ldots, f_n) = 0$ in $K[f_1, \ldots, f_n] \subset K(x_1, \ldots, x_n)$. By definition

$$f_k = f_k(x_1, \ldots, x_n) = \sum_{1 \leq i_1 < \ldots < i_k \leq n} x_{i_1} x_{i_2} \cdots x_{i_k}$$

and hence $0 = g(f_1, \ldots, f_n) = g(f_1(x_1, \ldots, x_n), \ldots, f_n(x_1, \ldots, x_n))$. Since $g(f_1, \ldots, f_n)$ is a polynomial in the indeterminates $x_1, \ldots, x_n$ over K and $F = K(u_1, \ldots, u_n)$ is a field containing K, substitution of u_i for x_i yields:

$$0 = g(f_1(u_1, \ldots, u_n), \ldots, f_n(u_1, \ldots, u_n)) = g(t_1, \ldots, t_n);$$

thus θ is a monomorphism and hence an isomorphism. Furthermore θ extends to an isomorphism of quotient fields $\theta : K(t_1, \ldots, t_n) \cong K(f_1, \ldots, f_n) = E$ (Exercise III.4.7). Now $F = K(u_1, \ldots, u_n)$ is a splitting field over $K(t_1, \ldots, t_n)$ of $p_n(x)$ and under the obvious map on polynomials induced by $\theta, p_n(x) \mapsto \bar{p}_n(x) = x^n - f_1 x^{n-1} + f_2 x^{n-2} - \cdots + (-1)^n f_n = (x - x_1)(x - x_2) \cdots (x - x_n)$ (see p. 252). Clearly $K(x_1, \ldots, x_n)$ is a splitting field of $\bar{p}_n(x)$ over $K(f_1, \ldots, f_n) = E$. Therefore by Theorem 3.8 the isomorphism θ extends to an isomorphism $F \cong K(x_1, \ldots, x_n)$ which by construction maps $K(t_1, \ldots, t_n)$ onto E as desired. ■

EXERCISES

1. If F is a radical extension field of K and E is an intermediate field, then F is a radical extension of E.

2. Suppose that "radical extension" is defined as follows: F is a radical extension of K if there is a finite tower of fields $K = E_0 \subset E_1 \subset \cdots \subset E_n = F$ such that for each $1 \leq i \leq n$, $E_i = E_{i-1}(u_i)$ and one of the following is true: (i) $u_i^{m_i} \in E_{i-1}$ for some $m_i > 0$; (ii) char $K = p$ and $u^p - u \in E_{i-1}$. State and prove the analogues of Theorem 9.4. Proposition 9.6, Corollary 9.7, and Proposition 9.8.

3. Let K be a field, $f \in K[x]$ an irreducible polynomial of degree $n \geq 5$ and F a splitting field of f over K. Assume that $\mathrm{Aut}_K F \cong S_n$. (See the example following Theorem 4.12). Let u be a root of f in F. Then

   (a) $K(u)$ is not Galois over K; $[K(u) : K] = n$ and $\mathrm{Aut}_K K(u) = 1$ (and hence is solvable).

   (b) Every normal closure over K that contains u also contains an isomorphic copy of F.

   (c) There is no radical extension field E of K such that $E \supset K(u) \supset K$.

4. If F is a radical extension field of E and E is a radical extension field of K, then F is a radical extension of K.

5. (Cardan) Let K be a field with char $K \neq 2,3$ and consider the cubic equation $x^3 + a_1x^2 + a_2x + a_3 = 0$ $(a_i \in K)$. Let $p = a_2 - \dfrac{a_1^2}{3}$ and $q = \dfrac{2a_1^3}{27} - \dfrac{a_1a_2}{3} + a_3$. Let $P = \sqrt[3]{-q/2 + \sqrt{p^3/27 + q^2/4}}$ and $Q = \sqrt[3]{-q/2 - \sqrt{p^3/27 + q^2/4}}$ (with cube roots chosen properly). Then the solutions of the given equation are $P + Q - a_1/3$; $\omega P + \omega^2 Q - a_1/3$; and $\omega^2 P + \omega Q - a_1/3$ where ω is a primitive cube root of unity.

CHAPTER VI

THE STRUCTURE OF FIELDS

In this chapter we shall analyze arbitrary extension fields of a given field. Since algebraic extensions were studied in some detail in Chapter V, the emphasis here will be on transcendental extensions. As the first step in this analysis, we shall show that every field extension $K \subset F$ is in fact a two-step extension $K \subset E \subset F$, with F algebraic over E and E purely transcendental over K (Section 1). The basic concept used here is that of a transcendence base, whose cardinality (called the transcendence degree) turns out to be an invariant of the extension of K by F (Section 1). The notion of separability is extended to (possibly) nonalgebraic extensions in Section 2 and separable extensions are characterized in several ways.

1. TRANSCENDENCE BASES

The first part of this section is concerned with the concept of algebraic independence, which generalizes the idea of linear independence. A transcendence base of a field F over a subfield K is the analogue (with respect to algebraic independence) of a vector space basis of F over K (with respect to linear independence). The cardinality of a transcendence base of F over K (the transcendence degree) is shown to be an invariant and its properties are studied. In this section we shall frequently use the notation u/v for uv^{-1}, where u,v are elements of a field and $v \neq 0$. Throughout this section K denotes a field.

Definition 1.1. *Let F be an extension field of K and S a subset of F. S is* **algebraically dependent** *over K if for some positive integer n there exists a nonzero polynomial $f \in K[x_1, \dots, x_n]$ such that $f(s_1, \dots, s_n) = 0$ for some distinct $s_1, \dots, s_n \in S$. S is* **algebraically independent** *over K if S is not algebraically dependent over K.*

REMARKS. The phrase "over K" is frequently omitted when the context is clear. A subset S of F is algebraically independent over K if for all $n > 0$, $f \varepsilon K[x_1, \ldots, x_n]$ and distinct $s_1, \ldots, s_n \varepsilon S$,

$$f(s_1, \ldots, s_n) = 0 \quad \Rightarrow \quad f = 0.$$

Every subset of an algebraically independent set is algebraically independent. In particular, the null set is algebraically independent. Every subset of K is clearly algebraically dependent. The set $\{u\}$ is algebraically dependent over K if and only if u is algebraic over K. Clearly every element of an algebraically independent set is necessarily transcendental over K. Hence if F is algebraic over K, the null set is the only algebraically independent subset of F.

Algebraic (in)dependence may be viewed as an extension of the concept of linear (in)dependence. For a set S is linearly dependent over K provided that for some positive integer n there is a nonzero polynomial f *of degree one* in $K[x_1, \ldots, x_n]$ such that $f(s_1, \ldots, s_n) = 0$ for some distinct $s_i \varepsilon S$. Consequently, every algebraically *in*dependent set is also linearly independent, but *not* vice versa; (see the Example after Definition 1.4 below).

EXAMPLE. Let K be a field. In the field of rational functions $K(x_1, \ldots, x_n)$ the set of indeterminates $\{x_1, \ldots, x_n\}$ is algebraically independent over K. More generally, we have:

Theorem 1.2. *Let* F *be an extension field of* K *and* $\{s_1, \ldots, s_n\}$ *a subset of* F *which is algebraically independent over* K. *Then there is* K*-isomorphism* $K(s_1, \ldots, s_n) \cong K(x_1, \ldots, x_n)$.

SKETCH OF PROOF. The assignment $x_i \mapsto s_i$ defines a K-epimorphism of rings $\theta : K[x_1, \ldots, x_n] \to K[s_1, \ldots, s_n]$ by Theorems III.5.5 and V.1.3. The algebraic independence of $\{s_1, \ldots, s_n\}$ implies that θ is a monomorphism. By Corollary III.4.6 θ extends to a K-monomorphism of fields (also denoted θ) $K(x_1, \ldots, x_n) \to K(s_1, \ldots, s_n)$ such that $\theta(f/g) = f(s_1, \ldots, s_n)/g(s_1, \ldots, s_n) = f(s_1, \ldots, s_n)g(s_1, \ldots, s_n)^{-1}$. θ is an epimorphism by Theorem V.1.3(v). ■

Corollary 1.3. *For* $i = 1,2$ *let* F_i *be an extension field of* K_i *and* $S_i \subset F_i$ *with* S_i *algebraically independent over* K_i. *If* $\varphi : S_1 \to S_2$ *is an injective map of sets and* $\sigma : K_1 \to K_2$ *a monomorphism of fields, then* σ *extends to a monomorphism of fields* $\bar{\sigma} : K_1(S_1) \to K_2(S_2)$ *such that* $\bar{\sigma}(s) = \varphi(s)$ *for every* $s \varepsilon S_1$. *Furthermore if* φ *is bijective and* σ *an isomorphism, then* $\bar{\sigma}$ *is an isomorphism.*

REMARK. In particular, the corollary implies that every permutation of an algebraically independent set S over a field K extends to a K-automorphism of $K(S)$; (just let $K_1 = K = K_2$ and $\sigma = 1_K$).

SKETCH OF PROOF OF 1.3. For each $n \geq 1$ σ induces a monomorphism of rings $K_1[x_1, \ldots, x_n] \to K_2[x_1, \ldots, x_n]$ (also denoted σ; see p. 235). Every element of

$K_1(S_1)$ is of the form $f(s_1,\ldots,s_n)/g(s_1,\ldots,s_n)$ $(s_i \in S_1)$ by Theorem V.1.3. For convenience we write φs for $\varphi(s)$ and define $\bar{\sigma}: K_1(S_1) \to K_2(S_2)$ by

$$f(s_1,\ldots,s_n)/g(s_1,\ldots,s_n) \mapsto \sigma f(\varphi s_1,\ldots,\varphi s_n)/\sigma g(\varphi s_1,\ldots,\varphi s_n) \in K(S_2).$$

For any finite subset $\{s_1,\ldots,s_r\}$ of S_1 the restriction of $\bar{\sigma}$ to $K_1(s_1,\ldots,s_r)$ is the composition

$$K_1(s_1,\ldots,s_r) \xrightarrow{\theta_1^{-1}} K_1(x_1,\ldots,x_r) \xrightarrow{\hat{\sigma}} K_2(x_1,\ldots,x_r) \xrightarrow{\theta_2} K_2(\varphi s_1,\ldots,\varphi s_r),$$

where the θ_i are the K_i-isomorphims of Theorem 1.2 and $\hat{\sigma}$ is the unique monomorphism of quotient fields induced by $\sigma: K_1[x_1,\ldots,x_r] \to K_2[x_1,\ldots,x_r]$ and given by $\hat{\sigma}(f/g) = (\sigma f)/(\sigma g)$ (Corollary III.4.6). It follows that $\bar{\sigma}$ is a well-defined monomorphism of fields. By construction $\bar{\sigma}$ extends σ and agrees with φ on S_1. If σ is an isomorphism then so is each $\hat{\sigma}$, whence each $\theta_2\hat{\sigma}\theta_1^{-1}$ is an isomorphism. If φ is bijective as well then it follows that $\bar{\sigma}$ is an isomorphism. ■

Definition 1.4. *Let* F *be an extension field of* K. *A* **transcendence base** (*or* **basis**) *of* F *over* K *is a subset* S *of* F *which is algebraically independent over* K *and is maximal (with respect to set-theoretic inclusion) in the set of all algebraically independent subsets of* F.

The fact that transcendence bases always exist follows immediately from a Zorn's Lemma argument (Exercise 2). If we recall the analogy between algebraic and linear independence, then a transcendence base is the analogue of a vector-space basis (since such a basis is precisely a maximal linearly independent subset by Lemma IV.2.3). Note, however, that a transcendence base is *not* a vector-space basis, although as a linearly independent set it is contained in a basis (Theorem IV.2.4).

EXAMPLE. If $f/g = f(x)/g(x) \in K(x)$ with $f,g \neq 0$, then the nonzero polynomial $h(y_1,y_2) = g(y_1)y_2 - f(y_1) \in K[y_1,y_2]$ is such that $h(x,f/g) = g(x)[f(x)/g(x)] - f(x) = 0$. Thus $\{x,f/g\}$ is algebraically dependent in $K(x)$. This argument shows that $\{x\}$ is a transcendence base of $K(x)$ over K. The set $\{x\}$ is not a basis since $\{1_K,x,x^2,x^3,\ldots\}$ is linearly independent in $K(x)$.

In order to obtain a useful characterization of transcendence bases we need

Theorem 1.5. *Let* F *be an extension field of* K, S *a subset of* F *algebraically independent over* K, *and* $u \in F - K(S)$. *Then* $S \cup \{u\}$ *is algebraically independent over* K *if and only if* u *is transcendental over* K(S).

PROOF. ($\Leftarrow$) If there exist distinct $s_1,\ldots,s_{n-1} \in S$ and an $f \in K[x_1,\ldots,x_n]$ such that $f(s_1,\ldots,s_{n-1},u) = 0$, then u is a root of $f(s_1,\ldots,s_{n-1},x_n) \in K(S)[x_n]$. Now $f \in K[x_1,\ldots,x_n] = K[x_1,\ldots,x_{n-1}][x_n]$, whence $f = h_rx_n^r + h_{r-1}x_n^{r-1} + \cdots + h_1x_n + h_0$ with each $h_i \in K[x_1,\ldots,x_{n-1}]$. Since u is transcendental over $K(S)$, we have $f(s_1,\ldots,s_{n-1},x_n) = 0$. Consequently, $h_i(s_1,\ldots,s_{n-1}) = 0$ for every i. The algebraic independence of S implies that $h_i = 0$ for every i, whence $f = 0$. Therefore $S \cup \{u\}$ is algebraically independent.

($\Rightarrow$) Suppose $f(u) = 0$ where $f = \sum_{i=0}^{n} a_i x^i \in K(S)[x]$. By Theorem V.1.3 there is a finite subset $\{s_1, \ldots, s_r\}$ of S such that $a_i \in K(s_1, \ldots, s_r)$ for every i, whence $a_i = f_i(s_1, \ldots, s_r)/g_i(s_1, \ldots, s_r)$ for some $f_i, g_i \in K[x_1, \ldots, x_r]$. Let $g = g_1 g_2 \cdots g_n \in K[x_1, \ldots, x_r]$ and for each i let $\bar{f}_i = f_i g_1 \cdots g_{i-1} g_{i+1} \cdots g_n \in K[x_1, \ldots, x_r]$. Then $a_i = \bar{f}_i(s_1, \ldots, s_r)/g(s_1, \ldots, s_r)$ and

$$\begin{aligned} f(x) &= \sum a_i x^i = \sum \bar{f}_i(s_1, \ldots, s_r)/g(s_1, \ldots, s_r) x^i \\ &= g(s_1, \ldots, s_r)^{-1}(\sum \bar{f}_i(s_1, \ldots, s_r) x^i). \end{aligned}$$

(All we have done is to factor out a "common denominator" for the coefficients of f.) Let $h(x_1, \ldots, x_r, x) = \sum \bar{f}_i(x_1, \ldots, x_r) x^i \in K[x_1, \ldots, x_r, x]$. Since $f(u) = 0$ and $g(s_1, \ldots, s_r)^{-1} \neq 0$, we must have $h(s_1, \ldots, s_r, u) = 0$. The algebraic independence of $S \cup \{u\}$ implies that $h = 0$, whence $\bar{f}_i = 0$ for every i. Thus each $a_i = 0$ and $f = 0$. Therefore u is transcendental over $K(S)$. ■

Corollary 1.6. *Let F be an extension field of K and S a subset of F that is algebraically independent over K. Then S is a transcendence base of F over K if and only if F is algebraic over K(S).*

PROOF. Exercise. ■

REMARKS. A field F is called a **purely transcendental** extension of a field K if $F = K(S)$, where $S \subset F$ and S is algebraically independent over K. In this case S is necessarily a transcendence base of F over K by Corollary 1.6. If F is an arbitrary extension field of K, let S be a transcendence base of F over K and let $E = K(S)$. Corollary 1.6 shows that F is algebraic over E and E is purely transcendental over K. Finally Corollary 1.6 and the remarks after Definition 1.1 show that F is an algebraic extension of K if and only if the null set is a transcendence base of F over K. In this case the null set is clearly the unique transcendence base of F over K.

Corollary 1.7. *If F is an extension field of K and F is algebraic over K(X) for some subset X of F (in particular, if F = K(X)), then X contains a transcendence base of F over K.*

PROOF. Let S be a maximal algebraically independent subset of X (S exists by a routine Zorn's Lemma argument). Then every $u \in X - S$ is algebraic over $K(S)$ by Theorem 1.5, whence $K(X)$ is algebraic over $K(S)$ by Theorem V.1.12. Consequently, F is algebraic over $K(S)$ by Theorem V.1.13. Therefore, S is a transcendence base of F over K by Corollary 1.6. ■

As one might suspect from the analogy with linear independence and bases, any two transcendence bases have the same cardinality. As in the case of vector spaces, we break the proof into two parts.

Theorem 1.8. *Let F be an extension field of K. If S is a finite transcendence base of F over K, then every transcendence base of F over K has the same number of elements as S.*

SKETCH OF PROOF. Let $S = \{s_1, \ldots, s_n\}$ and let T be any transcendence base. We claim that some $t_1 \varepsilon T$ is transcendental over $K(s_2, \ldots, s_n)$. Otherwise every element of T is algebraic over $K(s_2, \ldots, s_n)$, whence $K(s_2, \ldots, s_n)(T)$ is algebraic over $K(s_2, \ldots, s_n)$ by Theorem V.1.12. Since F is algebraic over $K(T)$ by Corollary 1.6, F is necessarily algebraic over $K(T)(s_2, \ldots, s_n) = K(s_2, \ldots, s_n)(T)$. Therefore, F is algebraic over $K(s_2, \ldots, s_n)$ by Theorem V.1.13. In particular, s_1 is algebraic over $K(s_2, \ldots, s_n)$, which is a contradiction (Theorem 1.5). Hence some $t_1 \varepsilon T$ is transcendental over $K(s_2, \ldots, s_n)$. Consequently, $\{t_1, s_2, \ldots, s_n\}$ is algebraically independent by Theorem 1.5.

Now if s_1 were transcendental over $K(t_1, s_2, \ldots, s_n)$, then $\{t_1, s_1, s_2, \ldots, s_n\}$ would be algebraically independent by Theorem 1.5. This is obviously impossible since S is a transcendence base. Therefore, s_1 is algebraic over $K(t_1, s_2, \ldots, s_n)$. Consequently, $K(S)(t_1) = K(t_1, s_2, \ldots, s_n)(s_1)$ is algebraic over $K(t_1, s_2, \ldots, s_n)$ (Theorem V.1.12), whence F is algebraic over $K(t_1, s_2, \ldots, s_n)$ (Theorem V.1.13 and Corollary 1.6). Therefore, $\{t_1, s_2, \ldots, s_n\}$ is a transcendence base of F over K by Corollary 1.6.

A similar argument shows that some $t_2 \varepsilon T$ is transcendental over $K(t_1, s_3, \ldots, s_n)$, whence $\{t_2, t_1, s_3, \ldots, s_n\}$ is a transcendence base. Proceeding inductively (inserting a t_i and omitting an s_i at each stage) we eventually obtain $t_1, t_2, \ldots, t_n \varepsilon T$ such that $\{t_1, \ldots, t_n\}$ is a transcendence base of F over K. Clearly, we must have $T = \{t_1, \ldots, t_n\}$ and hence $|S| = |T|$. ■

Theorem 1.9. *Let* F *be an extension field of* K. *If* S *is an infinite transcendence base of* F *over* K, *then every transcendence base of* F *over* K *has the same cardinality as* S.

PROOF. If T is another transcendence base, then T is infinite by Theorem 1.8. If $s \varepsilon S$, then s is algebraic over $K(T)$ by Corollary 1.6. The coefficients of the irreducible polynomial f of s over $K(T)$ all lie in $K(T_s)$ for some finite subset T_s of T (Theorem V.1.3). Consequently, $f \varepsilon K(T_s)[x]$ and s is algebraic over $K(T_s)$. Choose such a finite subset T_s of T for each $s \varepsilon S$.

We shall show that $\bigcup_{s \varepsilon S} T_s$ is a transcendence base of F over K. Since $\bigcup_s T_s \subset T$, this will imply that $\bigcup_s T_s = T$. As a subset of T the set $\bigcup_s T_s$ is algebraically independent. Furthermore every element of S is algebraic over $K(\bigcup_s T_s)$. Consequently, $K(\bigcup_s T_s)(S)$ is algebraic over $K(\bigcup_s T_s)$ by Theorem V.1.12. Since $K(S) \subset K(\bigcup_s T_s)(S)$, every element of $K(S)$ is algebraic over $K(\bigcup_s T_s)$. Since F is algebraic over $K(S)$ by Corollary 1.6, F is also algebraic over $K(\bigcup_s T_s)$ (see Theorem V.1.13). Therefore, by Corollary 1.6 again $\bigcup_s T_s$ is a transcendence base, whence $\bigcup_s T_s = T$.

Finally we shall show that $|T| \leq |S|$. The sets T_s need not be mutually disjoint and we remedy this as follows. Well order the set S (Introduction, Section 7) and denote its first element by 1. Let $T_1' = T_1$ and for each $1 < s \varepsilon S$, define $T_s' = T_s - \bigcup_{i<s} T_i$. Clearly each T_s' is finite. Verify that $\bigcup_s T_s = \bigcup_s T_s'$ and that the T_s' are mutually disjoint. For each $s \in S$, choose a fixed ordering of the elements of T_s' : $t_1, t_2,$

$\ldots, t_{k_s}$. The assignment $t_i \mapsto (s,i)$ defines an injective map $\bigcup_s T_s' \to S \times \mathbf{N}^*$. Therefore by Definitions 8.3 and 8.4 and Theorem 8.11 of the Introduction we have:

$$|T| = |\bigcup_s T_s| = |\bigcup_s T_s'| \leq |S \times \mathbf{N}^*| = |S||\mathbf{N}^*| = |S|\aleph_0 = |S|.$$

Reversing the roles of S and T in the preceding argument shows that $|S| \leq |T|$, whence $|S| = |T|$ by the Schroeder-Bernstein Theorem 8.6 of the Introduction. ■

Definition 1.10. *Let* F *be an extension field of* K. *The* **transcendence degree** *of* F *over* K *(denoted* tr.d.F/K*) is the cardinal number* |S|, *where* S *is any transcendence base of* F *over* K.

The two preceding theorems show that tr.d.F/K is independent of the choice of S. In the analogy between algebraic and linear independence tr.d.F/K is the analogue of the vector space dimension $[F:K]$. The remarks and examples after Definition 1.4 show that tr.d.$F/K \leq [F:K]$ and that tr.d.$F/K = 0$ if and only if F is algebraic over K.

Theorem 1.11. *If* F *is an extension field of* E *and* E *an extension field of* K, *then*

$$tr.d.\mathrm{F/K} = (tr.d.\mathrm{F/E}) + (tr.d.\mathrm{E/K}).$$

PROOF. Let S be a transcendence base of E over K and T a transcendence base of F over E. Since $S \subset E$, S is algebraically dependent over E, whence $S \cap T = \varnothing$. It suffices to show that $S \cup T$ is a transcendence base of F over K, since in that case Definition 1.10 and Definition 8.3 of the Introduction imply

$$\text{tr.d.}F/K = |S \cup T| = |T| + |S| = (\text{tr.d.}F/E) + (\text{tr.d.}E/K).$$

First of all every element of E is algebraic over $K(S)$ (Corollary 1.6) and hence over $K(S \cup T)$. Thus $K(S \cup T)(E)$ is algebraic over $K(S \cup T)$ by Theorem V.1.12. Since

$$K(S \cup T) = K(S)(T) \subset E(T) \subset K(S \cup T)(E),$$

$E(T)$ is algebraic over $K(S \cup T)$. But F is algebraic over $E(T)$ (Corollary 1.6) and therefore algebraic over $K(S \cup T)$ by Theorem V.1.13. Consequently, it suffices by Corollary 1.6 to show that $S \cup T$ is algebraically independent over K.

Let f be a polynomial over K in $n + m$ variables (denoted for convenience $x_1, \ldots, x_n, y_1, \ldots, y_m$) such that $f(s_1, \ldots, s_n, t_1, \ldots, t_m) = 0$ for some distinct $s_1, \ldots, s_n \in S$, $t_1, \ldots, t_m \in T$. Let $g = g(y_1, \ldots, y_m) = f(s_1, \ldots, s_n, y_1, \ldots, y_m) \in K(S)[y_1, \ldots, y_m] \subset E[y_1, \ldots, y_m]$. Since $g(t_1, \ldots, t_m) = 0$, the algebraic independence of T over E implies that $g = 0$. Now $f = f(x_1, \ldots, x_n, y_1, \ldots, y_m)$ $= \sum_{i=1}^{r} h_i(x_1, \ldots, x_n)k_i(y_1, \ldots, y_m)$ with $h_i \in K[x_1, \ldots, x_n]$, $k_i \in K[y_1, \ldots, y_m]$. Hence $0 = g(y_1, \ldots, y_m) = f(s_1, \ldots, s_n, y_1, \ldots, y_m)$ implies that $h_i(s_1, \ldots, s_n) = 0$ for every i. The algebraic independence of S over K implies that $h_i = 0$ for all i, whence $f(x_1, \ldots, x_n, y_1, \ldots, y_m) = 0$. Therefore $S \cup T$ is algebraically independent over K. ■

If K_1 and K_2 are fields with algebraic closures, F_1,F_2 respectively, then Theorem V.3.8 implies that every isomorphism $K_1 \cong K_2$ extends to an isomorphism $F_1 \cong F_2$. Under suitable hypotheses this result can now be extended to the case where the fields F_i are algebraically closed, but not necessarily algebraic over K_i.

Theorem 1.12. *Let* F_1 *[resp.* F_2*] be an algebraically closed field extension of a field* K_1 *[resp.* K_2*]. If* $tr.d.F_1/K_1 = tr.d.F_2/K_2$*, then every isomorphism of fields* $K_1 \cong K_2$ *extends to an isomorphism* $F_1 \cong F_2$.

PROOF. Let S_i be a transcendence base of F_i over K_i. Since $|S_1| = |S_2|$, $\sigma : K_1 \cong K_2$ extends to an isomorphism $\bar{\sigma} : K_1(S_1) \cong K_2(S_2)$ by Corollary 1.3. F_i is algebraically closed and algebraic over $K_i(S_i)$ (Corollary 1.6) and hence an algebraic closure of $K_i(S_i)$. Therefore $\bar{\sigma}$ extends to an isomorphism $F_1 \cong F_2$ by Theorems V.3.4 and V.3.8. ■

EXERCISES

Note: F is always an extension field of a field K.

1. (Exchange property) Let S be a subset of F. If $u \in F$ is algebraic over $K(S)$ and u is not algebraic over $K(S - \{v\})$, where $v \in S$, then v is algebraic over $K((S - \{v\}) \cup \{u\})$.

2. (a) Use Zorn's Lemma to show that every field extension possesses a transcendence base.
(b) Every algebraically independent subset of F is contained in a transcendence base.

3. $\{x_1, \dots, x_n\}$ is a transcendence base of $K(x_1, \dots, x_n)$.

4. If E_1,E_2 are intermediate fields, then
(i) $\text{tr.d.}E_1E_2/K \geq \text{tr.d.}E_i/K$ for $i = 1,2$;
(ii) $\text{tr.d.}E_1E_2/K \leq (\text{tr.d.}E_1/K) + (\text{tr.d.}E_2/K)$.

5. If $F = K(u_1, \dots, u_n)$ is a finitely generated extension of K and E is an intermediate field, then E is a finitely generated extension of K. [*Note:* the algebraic case is trivial by Theorems V.1.11 and V.1.12.]

6. (a) If S is a transcendence base of the field **C** of complex numbers over the field **Q** of rationals, then S is infinite. [*Hint:* Show that if S is finite, then
$$|\mathbf{Q}(S)| = |\mathbf{Q}(x_1, \dots, x_n)| = |\mathbf{Q}[x_1, \dots, x_n]| = |\mathbf{Q}| < |\mathbf{C}|$$
(see Exercises 8.3 and 8.9 of the Introduction and Theorem 1.2). But Lemma V.3.5 implies $|\mathbf{Q}(S)| = |\mathbf{C}|$.]
(b) There are infinitely many distinct automorphisms of the field **C**.
(c) $\text{tr.d.}\mathbf{C}/\mathbf{Q} = |\mathbf{C}|$.

7. If F is algebraically closed and E an intermediate field such that $\text{tr.d.}E/K$ is finite, then any K-monomorphism $E \to F$ extends to a K-automorphism of F.

8. If F is algebraically closed and $\text{tr.d.}F/K$ is finite, then every K-monomorphism $F \to F$ is in fact an automorphism.

2. LINEAR DISJOINTNESS AND SEPARABILITY

The chief purpose of this section is to extend the concept of separability to (possibly) nonalgebraic field extensions. This more general concept of separability will agree with our previous definition in the case of algebraic extensions (Theorem 2.8). We first introduce the idea of linear disjointness and develop its basic properties (Theorems 2.2–2.7). Separability is defined in terms of linear disjointness and characterized in several ways (Theorem 2.10). Other properties of separability are developed in the corollaries of Theorem 2.10.

In the following discussion all fields are assumed to be subfields of some (fixed) algebraically closed field C.

Definition 2.1. *Let* C *be an algebraically closed field with subfields* K,E,F *such that* $K \subset E \cap F$. E *and* F *are* **linearly disjoint** *over* K *if every subset of* E *which is linearly independent over* K *is also linearly independent over* F.

REMARKS. An alternate definition in terms of tensor products is given in Exercise 1. Note that a subset X of E is linearly independent over a subfield of C if and only if every finite subset of X is. Consequently, when proving linear disjointness, we need only deal with finite linearly independent sets.

EXAMPLE. If $K \subset E$ then E and K are trivially linearly disjoint over K. This fact will be used in several proofs. Other less trivial examples appear in the theorems and exercises below.

The wording of Definition 2.1 suggests that the definition of linear disjointness is in fact symmetric in E and F. We now prove this fact.

Theorem 2.2. *Let* C *be an algebraically closed field with subfields* K,E,F *such that* $K \subset E \cap F$. *Then* E *and* F *are linearly disjoint over* K *if and only if* F *and* E *are linearly disjoint over* K.

PROOF. It suffices to assume E and F linearly disjoint and show that F and E are linearly disjoint. Suppose $X \subset F$ is linearly independent over K, but not over E so that $r_1u_1 + \cdots + r_nu_n = 0$ for some $u_i \in X$ and $r_i \in E$ not all zero. Choose a subset of $\{r_1, \dots, r_n\}$ which is maximal with respect to linear independence over K; reindex if necessary so that this set is $\{r_1, r_2, \dots, r_t\}$ $(t \geq 1)$. Then for each $j > t$, $r_j = \sum_{i=1}^{t} a_{ij}r_i$ with $a_{ij} \in K$ (Exercise IV.2.1). After a harmless change of index we have:

$$0 = \sum_{j=1}^{n} r_ju_j = \sum_{j=1}^{t} r_ju_j + \sum_{j=t+1}^{n} \left(\sum_{i=1}^{t} a_{ij}r_i\right)u_j$$
$$= \sum_{k=1}^{t} \left(u_k + \sum_{j=t+1}^{n} a_{kj}u_j\right)r_k.$$

Since E and F are linearly disjoint, $\{r_1, \dots, r_t\}$ is linearly independent over F which implies that $u_k + \sum_{j=t+1}^{n} a_{kj}u_j = 0$ for every $k \leq t$. This contradicts the linear independence of X over K. Therefore X is linearly independent over E. ■

The following lemma and theorem provide some useful criteria for two fields to be linearly disjoint.

Lemma 2.3. *Let* C *be an algebraically closed field with subfields* K,E,F *such that* $K \subset E \cap F$. *Let* R *be a subring of* E *such that* $K(R) = E$ *and* $K \subset R$ *(which implies that* R *is a vector space over* K*). Then the following conditions are equivalent:*

(i) E *and* F *are linearly disjoint over* K;

(ii) *every subset of* R *that is linearly independent over* K *is also linearly independent over* F;

(iii) *there exists a basis of* R *over* K *which is linearly independent over* F.

REMARK. The lemma is true with somewhat weaker hypotheses (Exercise 2) but this is all that we shall need.

PROOF OF 2.3. (i) $\Rightarrow$ (ii) and (i) $\Rightarrow$ (iii) are trivial. (ii) $\Rightarrow$ (i) Let $X = \{u_1, \ldots, u_n\}$ be a finite subset of E which is linearly independent over K. We must show that X is linearly independent over F. Since $u_i \in E = K(R)$ each u_i is of the form $u_i = c_i d_i^{-1} = c_i/d_i$, where $c_i = f_i(r_1, \ldots, r_{t_i})$, $0 \neq d_i = g_i(r_1, \ldots, r_{t_i})$ with $r_j \in R$ and $f_i, g_i \in K[x_1, \ldots, x_{t_i}]$ (Theorem V.1.3). Let $d = d_1 d_2 \cdots d_n$ and for each i let $v_i = c_i d_1 \cdots d_{i-1} d_{i+1} \cdots d_n \in R$. Then $u_i = v_i d^{-1}$ and the subset $X' = \{v_1, \ldots, v_n\}$ of R is linearly independent over a subfield of C if and only if X is. By hypothesis X and hence X' is linearly independent over K. Consequently, (ii) implies that X' is linearly independent over F, whence X is linearly independent over F.

(iii) $\Rightarrow$ (ii) Let U be a basis of R over K which is linearly independent over F. We must show that every finite subset X of R that is linearly independent over K is also linearly independent over F. Since X is finite, there is a finite subset U_1 of U such that X is contained in the K-subspace V of R spanned by U_1; (note that U_1 is a basis of V over K). Let V_1 be the vector space spanned by U_1 over F. U and hence U_1 is linearly independent over F by (iii). Therefore U_1 is a basis of V_1 over F and $\dim_K V = \dim_F V_1$. Now X is contained in some finite basis W of V over K (Theorem IV.2.4). Since W certainly spans V_1 as a vector space over F, W contains a basis W_1 of V_1 over F. Thus $|W_1| \leq |W| = \dim_K V = \dim_F V_1 = |W_1|$, whence $W = W_1$. Therefore, the subset X of W is necessarily linearly independent over F. ■

Theorem 2.4. *Let* C *be an algebraically closed field with subfields* K,E,L,F *such that* $K \subset E$ *and* $K \subset L \subset F$. *Then* E *and* F *are linearly disjoint over* K *if and only if* (i) E *and* L *are linearly disjoint over* K *and* (ii) EL *and* F *are linearly disjoint over* L.

PROOF. The situation looks like this:

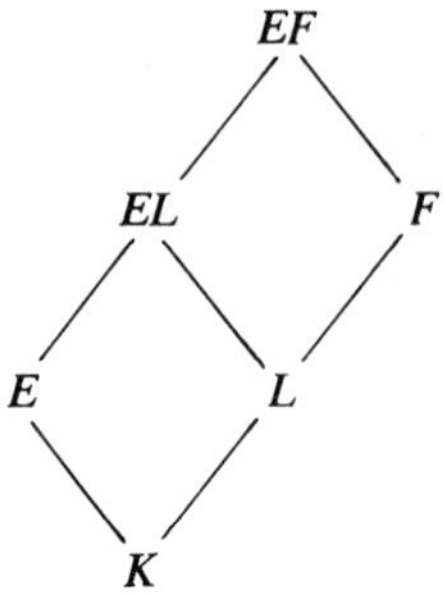

($\Leftarrow$) If a subset X of E is linearly independent over K, then X is linearly independent over L by (i). Therefore (since $X \subset E \subset EL$), X is linearly independent over F by (ii).

($\Rightarrow$) If E and F are linearly disjoint over K, then E and L are automatically linearly disjoint over K. To prove (ii) observe that $EL = L(R)$, where R is the subring $L[E]$ of C generated by L and E. By Theorem V.1.3 every element of R is of the form $f(e_1, \dots, e_n)$ ($e_i \in E; f \in L[x_1, \dots, x_n]$). Therefore, any basis U of E over K spans R considered as a vector space over L. Since E and L are linearly disjoint over K, U is linearly independent over L. Hence U is a basis of R over L. But U is linearly independent over F by the linear disjointness of E and F. Therefore, EL and F are linearly disjoint over L by Lemma 2.3. ■

Next we explore linear disjointness with respect to certain extension fields of K that will play an important part in the definition of separability.

Definition 2.5. *Let* K *be a field of characteristic* $p \neq 0$ *and let* C *be an algebraically closed field containing* K. *For each integer* $n \geq 0$

$$K^{1/p^n} = \{u \in C \mid u^{p^n} \in K\}.$$
$$K^{1/p^\infty} = \bigcup_{n \geq 0} K^{1/p^n} = \{u \in C \mid u^{p^n} \in K \ \textit{for some} \ n \geq 0\}.$$

REMARKS. Since $(u \pm v)^{p^n} = u^{p^n} \pm v^{p^n}$ in a field of characteristic p (Exercise III.1.11) each K^{1/p^n} is actually a field. Since $K = K^{1/p^0} \subset K^{1/p^n} \subset K^{1/p^m} \subset K^{1/p^\infty}$ for all n,m such that $0 \leq n \leq m$, it follows readily that K^{1/p^∞} is also a field. The fact that C is algebraically closed implies that K^{1/p^n} is a splitting field over K of the set of polynomials $\{x^{p^n} - k \mid k \in K\}$ (Exercise 5). In particular, every $k \in K$ is of the form v^{p^n} for some $v \in K^{1/p^n}$. Since K^{1/p^n} is a splitting field over K, it is essentially independent of C (that is, another choice C' would yield an isomorphic copy of K^{1/p^n} by Theorem V.3.8).

Lemma 2.6. *If* F *is an extension field of* K *of characteristic* $p \neq 0$ *and* C *is an algebraically closed field containing* F, *then for any* $n \geq 0$ *a subset* X *of* F *is linearly independent over* K^{1/p^n} *if and only if* $X^{p^n} = \{u^{p^n} \mid u \in X\}$ *is linearly independent over* K. *Furthermore* X *is linearly independent over* K^{1/p^∞} *if and only if* X *is linearly independent over* K^{1/p^n} *for all* $n \geq 0$.

SKETCH OF PROOF. Every $a \in K$ is of the form $a = v^{p^n}$ for some $v \in K^{1/p^n}$ (Exercise 5). For the first statement note that $\sum_i a_i u_i^{p^n} = 0$ $(a_i \in K;\ u_i \in X) \Leftrightarrow$ $\sum_i v_i^{p^n} u_i^{p^n} = 0$ $(v_i \in K^{1/p^n}$ and $v_i^{p^n} = a_i) \Leftrightarrow (\sum_i v_i u_i)^{p^n} = 0 \Leftrightarrow \sum_i v_i u_i = 0$. For the second statement observe that if $\sum_{i=1}^{t} w_i u_i = 0$ $(w_i \in K^{1/p^\infty};\ u_i \in X)$, then for a large enough n, $w_1, \ldots, w_t \in K^{1/p^n}$. ■

Theorem 2.7. *Let* F *be a field contained in an algebraically closed field* C. *If* F *is a purely transcendental extension of a field* K *of characteristic* $p \neq 0$, *then* F *and* K^{1/p^n} *are linearly disjoint over* K *for all* $n \geq 0$ *and* F *and* K^{1/p^∞} *are linearly disjoint over* K.

PROOF. Let $F = K(S)$ with S a transcendence base of F over K. If $S = \varnothing$, then $F = K$ and every linearly independent subset of F over K consists of exactly one nonzero element of K. Such a nonzero singleton is clearly linearly independent over any subfield of C whence the theorem is true if $S = \varnothing$. If S is not empty let M be the set of monomials over S (that is, the set of all finite products of elements of S). Then M is linearly independent over K since S is algebraically independent over K. By Theorem V.1.3 M spans the subring $K[S]$ (considered as a vector space over K). Therefore, M is a basis of $K[S]$ over K. The algebraic independence of S implies that for every $n \geq 0$, $M^{p^n} = \{m^{p^n} \mid m \in M\}$ is linearly independent over K. By Lemma 2.6 M is linearly independent over K^{1/p^n} for every n and hence over K^{1/p^∞}. Therefore, for each $0 \leq n \leq \infty$, F and K^{1/p^n} are linearly disjoint over K by Lemma 2.3 (with $K[S]$, F, K^{1/p^n} in place of R, E, F respectively). ■

The next theorem shows the connection between linear disjointness and separable algebraic extensions and will motivate a definition of separability in the case of arbitrary (possibly nonalgebraic) extensions.

Theorem 2.8. *Let* F *be an algebraic extension field of a field* K *of characteristic* $p \neq 0$ *and* C *an algebraically closed field containing* F. *Then* F *is separable over* K *if and only if* F *and* $K^{1/p}$ *are linearly disjoint over* K.

PROOF. We shall prove here only that separability implies that F and $K^{1/p}$ are linearly disjoint. The other half of the proof will be an easy consequence of a result below (see the Remarks after Theorem 2.10). Let $X = \{u_1, \ldots, u_n\}$ be a finite subset of F which is linearly independent over K. We must show that X is linearly independent over $K^{1/p}$. The subfield $E = K(u_1, \ldots, u_n)$ is finite dimensional over K (Theorem V.1.12) and has a basis $\{u_1, \ldots, u_n, u_{n+1}, \ldots, u_r\}$ which contains X (Theorem, IV.2.4). If $v \in E$ and k is a positive integer, then $v^k = \sum_{i=1}^{r} a_i u_i$ $(a_i \in K)$ and hence $v^{kp} = (\sum_i a_i u_i)^p = \sum a_i^p u_i^p$. Since v is separable over K, $K(v)$ is both separable algebraic and purely inseparable over $K(v^p)$ (Theorem V.6.4 and Lemma V.6.6), whence $K(v) = K(v^p) = K[v^p]$ (Theorems V.1.6 and V.6.2). Thus v is a linear combination of the v^{kp} and hence of the u_i^p. Therefore E is spanned by $\{u_1^p, \ldots, u_r^p\}$. Since $[E : K] = r$, $\{u_1^p, \ldots, u_r^p\}$ must be a basis by Theorems IV.2.5 and IV.2.7.

Thus $\{u_1^p, \ldots, u_r^p\}$ and hence X^p is linearly independent over K. By Lemma 2.6 X is linearly independent over $K^{1/p}$, whence F and $K^{1/p}$ are linearly disjoint over K. ∎

Definition 2.9. *Let* F *be an extension field of* K. *A transcendence base* S *of* F *over* K *is called a* **separating transcendence base** *of* F *over* K *if* F *is separable algebraic over* K(S). *If* F *has a separating transcendence base over* K, *then* F *is said to be* **separably generated** *over* K.

REMARKS. Recall that F is algebraic over $K(S)$ (Corollary 1.6). If F is separably generated over K, it is not true that every transcendence base of F over K is necessarily a separating transcendence base (Exercise 8).

EXAMPLES. If F is separable algebraic over K, then the null set is a separating transcendence base. Every purely transcendental extension is trivially separably generated since $F = K(S)$.

In order to make the principal theorem meaningful in the case of characteristic zero we define (for any field K of characteristic 0) $K^{1/0} = K^{1/0^n} = K^{1/0^\infty} = K$.

Theorem 2.10. *If* F *is an extension field of a field* K *of characteristic* $p \geq 0$ *and* C *is an algebraically closed field containing* F, *then the following conditions are equivalent.*

(i) F *and* $K^{1/p}$ *are linearly disjoint over* K;

(ii) F *and* K^{1/p^n} *are linearly disjoint over* K *for some* $n \geq 1$;

(iii) F *and* K^{1/p^∞} *are linearly disjoint over* K;

(iv) *every finitely generated intermediate field* E *is separably generated over* K;

(v) K_0 *and* F *are linearly disjoint over* K, *where* K_0 *is the fixed field (relative to* C and K) *of* $\mathrm{Aut}_K C$.

REMARKS. The theorem is proved below. The implication (i) ⇒ (iv) provides a proof of the second half of Theorem 2.8 as follows. For every $u \in F$, $K(u)$ is a finitely generated intermediate field and thus separably generated over K. But F (and hence $K(u)$) is assumed algebraic over K and the only transcendence base of an algebraic extension is the null set. Therefore $K(u)$ is separable algebraic over $K(\varnothing) = K$. Hence every $u \in F$ is separable algebraic over K.

SKETCH OF PROOF OF 2.10. Except in proving (iii) ⇔ (v) we shall assume that char $K = p \neq 0$ since the case when char $K = 0$ is trivial otherwise. (iii) ⇒ (ii) ⇒ (i) is immediate since $K^{1/p} \subset K^{1/p^n} \subset K^{1/p^\infty}$ for every $n \geq 1$.

(i) ⇒ (iv) Let $E = K(s_1, \ldots, s_n)$ and tr.d.$E/K = r$. By Corollary 1.7 $r \leq n$ and some subset of $\{s_1, \ldots, s_n\}$ is a transcendence basis of E over K, say $\{s_1, \ldots, s_r\}$. If $r = n$, then $\{s_1, \ldots, s_n\}$ is trivially a separating transcendence base, whence (iv) holds. If $r < n$, then s_{r+1} is algebraic over $K(s_1, \ldots, s_r)$ (Corollary 1.6) and therefore the root of an irreducible monic polynomial $f(x) = \sum_{i=1}^{m} a_i x^i \in K(s_1, \ldots, s_r)[x]$. A "least common denominator argument" such as that used in the proof of Theorem

1.5 shows that $f(x) = d^{-1}\left(\sum_{i=1}^{m} v_i x^i\right)$ with $0 \neq d \in K[s_1, \ldots, s_r]$, $v_i = h_i(s_1, \ldots, s_r)$ and $h_i \in K[x_1, \ldots, x_r]$. Thus $f_1 = \sum_{i=0}^{m} h_i(x_1, \ldots, x_r)x_{r+1}^i \in K[x_1, \ldots, x_{r+1}]$ and $f_1(s_1, \ldots, s_r, s_{r+1}) = 0$. It follows that there exists a polynomial $g \in K[x_1, \ldots, x_{r+1}]$ of least positive degree such that $g(s_1, \ldots, s_{r+1}) = 0$. Clearly g is irreducible in $K[x_1, \ldots, x_{r+1}]$. Recall that x_i is said to occur in $g(x_1, \ldots, x_n)$ if some nonzero term of g contains a factor x_i^m with $m \geq 1$.

We claim that some x_i occurs in g with an exponent that is not divisible by p. Otherwise $g = c_0 + c_1m_1(x_1, \ldots, x_{r+1})^p + \cdots + c_km_k(x_1, \ldots, x_{r+1})^p$, where $c_j \in K$, the c_j are not all zero, and each $m_j(x_1, \ldots, x_{r+1})$ is a monomial in $x_1, \ldots, x_{r+1}$. Let $m_0(x_1, \ldots, x_{r+1}) = 1_K$ and for each $j \geq 0$ choose $d_j \in K^{1/p}$ such that $d_j^p = c_j$. Then $g = \left(\sum_{j=0}^{k} d_jm_j(x_1, \ldots, x_{r+1})\right)^p$ and $g(s_1, \ldots, s_{r+1}) = 0$ imply that

$$\sum_{j=0}^{k} d_jm_j(s_1, \ldots, s_{r+1}) = 0,$$

whence the subset $\{m_j(s_1, \ldots, s_{r+1}) \mid j \geq 0\}$ of F is linearly dependent over $K^{1/p}$. But $\{m_j(s_1, \ldots, s_{r+1}) \mid j \geq 0\}$ is necessarily linearly *independent* over K (otherwise there would exist a $g_1 \in K[x_1, \ldots, x_{r+1}]$ with $\deg g_1 < \deg g$ and $g_1(s_1, \ldots, s_{r+1}) = 0$). This fact contradicts the linear disjointness of F and $K^{1/p}$. Therefore some x_i, say x_1, occurs in g with an exponent that is not divisible by p.

The polynomial $g(x,s_2, \ldots, s_{r+1}) \in K(s_2, \ldots, s_{r+1})[x]$ is necessarily nonzero. Otherwise, since x_1 occurs in $g(x_1, \ldots, x_{r+1})$ by the previous paragraph, we could obtain a polynomial $g_2 \in K[x_1, \ldots, x_{r+1}]$ such that $0 < \deg g_2 < \deg g$ and $g_2(s_1,s_2, \ldots, s_{r+1}) = 0$. Such a g_2 would contradict the choice of g. Therefore, $g(x,s_2, \ldots, s_{r+1}) \neq 0$. Since $g(s_1,s_2, \ldots, s_{r+1}) = 0$, s_1 is algebraic over $K(s_2, \ldots, s_{r+1})$. But $s_2, \ldots, s_{r+1}$ are obviously algebraic over $K(s_2, \ldots, s_{r+1})$ and E is algebraic over $K(s_1, \ldots, s_{r+1})$. By Theorems V.1.12 and V.1.13 E is algebraic over $K(s_2, \ldots, s_{r+1})$. Since tr.d.$E/K = r$, $\{s_2, \ldots, s_{r+1}\}$ is a transcendence base of E over K (Corollary 1.7).

The proof of Theorem 1.2 shows that the assignment $x_i \mapsto s_i$ determines a K-isomorphism $\phi : K[x_2, \ldots, x_{r+1}] \cong K[s_2, \ldots, s_{r+1}]$. Clearly ϕ extends to a K-isomorphism $K[x_1,x_2, \ldots, x_{r+1}] = K[x_2, \ldots, x_{r+1}][x_1] \cong K[s_2, \ldots, s_{r+1}][x]$ such that $x_1 \mapsto x$ and $g(x_1, \ldots, x_{r+1}) \mapsto g(x,s_2, \ldots, s_{r+1})$. Since ϕ is an isomorphism, $g(x,s_2, \ldots, s_{r+1})$ must be irreducible in $K[s_2, \ldots, s_{r+1}][x]$. Consequently $g(x,s_2, \ldots, s_{r+1})$ is primitive in $K[s_2, \ldots, s_{r+1}][x]$ and hence irreducible in $K(s_2, \ldots, s_{r+1})[x]$ by Lemma III.6.13 and Theorem III.6.14. Since ϕ is an isomorphism x must occur in $g(x,s_2, \ldots, s_{r+1})$ with an exponent not divisible by p. Thus the derivative of $g(x,s_2, \ldots, s_{r+1})$ is nonzero (Exercise III.6.3), whence $g(x,s_2, \ldots, s_{r+1})$ is separable by Theorem III.6.10. Therefore s_1 is separable algebraic over $K(s_2, \ldots, s_{r+1})$ and hence over $K(s_2, \ldots, s_n)$. In particular, $E = K(s_1, \ldots, s_n)$ is separable algebraic over $K(s_2, \ldots, s_n)$ by Lemma V.6.6. Thus if $\{s_2, \ldots, s_n\}$ is a transcendence base of E over K, then E is separably generated over K. If not, then $\{s_2, \ldots, s_n\}$ contains a transcendence base (Corollary 1.7), which we may assume (after reindexing if necessary) to be $\{s_2, \ldots, s_{r+1}\}$. A repetition of the preceding argument (with s_{i+1} in place of s_i for $i = 1,2, \ldots, r + 1$ and possibly more reindexing) shows that s_2 (and hence $K(s_2, \ldots, s_n)$) is separable algebraic over $K(s_3, \ldots, s_n)$. Hence E is separable algebraic over $K(s_3, \ldots, s_n)$ by Corollary V.6.8. Continuing this process we must eventually find $s_1, \ldots, s_t$ such that E is separable algebraic over $K(s_{t+1}, \ldots, s_n)$

and $\{s_{t+1}, \ldots, s_n\}$ is a transcendence base of E over K. Therefore E is separably generated over K.

(iv) $\Rightarrow$ (iii) Let W be a finite subset of F that is linearly independent over K. We must show that W is linearly independent over K^{1/p^∞}. Let $E = K(W)$. We need only show that E and K^{1/p^∞} are linearly disjoint over K, since this fact immediately implies that W is linearly independent over K^{1/p^∞}. Since W is finite, E has a separating transcendence base S over K by (iv). We shall prove the linear disjointness of E and K^{1/p^∞} by applying Theorem 2.4 to the extensions $K \subset K^{1/p^\infty}$ and $K \subset K(S) \subset E$ as follows. $K(S)$ and K^{1/p^∞} are linearly disjoint over K by Theorem 2.7. Let X be a subset of E that is linearly independent over $K(S)$. Since E is separable algebraic over $K(S)$, X is linearly independent over $K(S)^{1/p}$ by the half of Theorem 2.8 already proved. Therefore X^p is linearly independent over $K(S)$ by Lemma 2.6. The last three sentences form the heart of an inductive argument which shows that X^{p^m} is linearly independent over $K(S)$ for all $m \geq 0$; (note that $(X^{p^r})^p = X^{p^{r+1}}$). Hence X is linearly independent over $K(S)^{1/p^m}$ for all $m \geq 0$ by Lemma 2.6 again. Therefore X is linearly independent over $K(S)^{1/p^\infty}$ and hence over its subfield $K^{1/p^\infty}K(S)$. We have proved that E and $K^{1/p^\infty}K(S)$ are linearly disjoint over $K(S)$. Consequently E and K^{1/p^∞} are linearly disjoint over K by Theorem 2.4.

(iii) $\Leftrightarrow$ (v). It suffices to prove that $K_0 = K^{1/p^\infty}$. Let $u \in K_0$. If u is transcendental over K, then there exists $v \in C$ with $v \neq u$ and v transcendental over K (for example, take $v = u^2$). The composition $K(u) \cong K(x) \cong K(v)$ (where the isomorphisms are given by Theorem V.1.5) is a K-isomorphism σ such that $\sigma(u) = v$. We thus have $1 = \text{tr.d.}K(x)/K = \text{tr.d.}K(u)/K = \text{tr.d.}K(v)/K$. Theorem 1.11 (and Introduction, Lemma 8.9 if $\text{tr.d.}C/K(u)$ is infinite) implies that $\text{tr.d.}C/K(u) = \text{tr.d.}C/K(v)$. Therefore σ extends to a K-automorphism of C by Theorem 1.12. But $\sigma(u) = v \neq u$, which contradicts the fact that $u \in K_0$. Therefore, u must be algebraic over K with irreducible polynomial $f \in K[x]$. If $v \in C$ is another root of f, then there is a K-isomorphism $\tau : K(u) \cong K(v)$ such that $\tau(u) = v$ (Corollary V.1.9). An argument similar to the one in the transcendental case shows that τ extends to a K-automorphism of C. Since $u \in K_0$ we must have $u = \tau(u) = v$, whence f has only one root in C. Thus u is purely inseparable over K. If char $K = 0$, then f (which is necessarily separable) must have degree 1. Hence $u \in K = K^{1/0^\infty}$. If char $K = p \neq 0$, then $u^{p^n} \in K$ for some $n \geq 0$ by Theorem V.6.4. Thus $u \in K^{1/p^n} \subset K^{1/p^\infty}$. We have proved that $K_0 \subset K^{1/p^\infty}$. Conversely suppose that char $K = p \neq 0$, $u \in K^{1/p^n} \subset K^{1/p^\infty}$ and $\sigma \in \text{Aut}_K C$. Then $\sigma(u)^{p^n} = \sigma(u^{p^n}) = u^{p^n}$, whence $0 = \sigma(u)^{p^n} - u^{p^n} = (\sigma(u) - u)^{p^n}$ and $\sigma(u) = u$. Therefore, $K^{1/p^\infty} \subset K_0$. ■

Definition 2.11. *An extension field* F *of a field* K *is said to be* **separable** *over* K (*or a* **separable extension** *of* K) *if* F *satisfies the equivalent conditions of Theorem* 2.10.

REMARKS. Theorem 2.8 shows that this definition is compatible with our previous use of the term "separable" in the case of algebraic extensions (Definition V.3.10). Since the first condition of Theorem 2.10 is trivially satisfied when char $K = 0$, every extension field of characteristic 0 is separable.

The basic properties of separability are developed in the following corollaries of Theorem 2.10.

Corollary 2.12. *(Mac Lane's Criterion) If* F *is an extension field of a field* K *and* F *is separably generated over* K, *then* F *is separable over* K. *Conversely, if* F *is separable and finitely generated over* K, *say* $F = K(u_1, \ldots, u_n)$, *then* F *is separably generated over* K. *In fact some subset of* $\{u_1, \ldots, u_n\}$ *is a separating transcendence base of* F *over* K.

SKETCH OF PROOF. The proof of (iv) $\Rightarrow$ (iii) $\Rightarrow$ (i) in Theorem 2.10 is valid here with $F = E$ since it uses only the fact that E is separably generated. The last two statements are consequences of the proof of (i) $\Rightarrow$ (iv) in Theorem 2.10. ■

Corollary 2.13. *Let* F *be an extension field of* K *and* E *an intermediate field.*

(i) *If* F *is separable over* K, *then* E *is separable over* K;
(ii) *if* F *is separable over* E *and* E *is separable over* K, *then* F *is separable over* K;
(iii) *if* F *is separable over* K *and* E *is algebraic over* K, *then* F *is separable over* E.

REMARK. (iii) may be false if E is not algebraic over K (see Exercise 8).

SKETCH OF PROOF OF 2.13. (ii) Use Theorems 2.4 and 2.10. (iii) If char $K = p \neq 0$, let X be a subset of F which is linearly independent over E. Extend X to a basis U of F over E and let V be a basis of E over K. The proof of Theorem IV.2.16 shows that $UV = \{uv \mid u \in U, v \in V\}$ is a basis of F over K, whence UV is linearly independent over $K^{1/p}$ by separability. Lemma 2.6 implies that $(UV)^p = \{u^pv^p \mid u \in U, v \in V\}$ is linearly independent over K. We claim furthermore that V^p is a basis of E over K. For E is separable over K by (i). Consequently, the linear disjointness of E and $K^{1/p}$ shows that V is linearly independent over $K^{1/p}$, whence V^p is linearly independent over K by Lemma 2.6. Since $E = KE^p$ by Corollary V.6.9, V^p necessarily spans E over K. Therefore, V^p is a basis of E over K. To complete the proof we must show that X is linearly independent over $E^{1/p}$. If $\sum_i a_iu_i = 0$ $(a_i \in E^{1/p}; u_i \in X \subset U)$, then $\sum_i a_i^pu_i^p = 0$. Since each $a_i^p \in E$ is of the form $\sum_j c_{ij}v_j^p$ $(c_{ij} \in K; v_j \in V)$ we have $0 = \sum_i (\sum_j c_{ij}v_j^p)u_i^p = \sum_{i,j} c_{ij}u_i^pv_j^p$. The linear independence of $(UV)^p$ implies that $c_{ij} = 0$ for all i,j and hence that $a_i = 0$ for all i. Therefore, X is linearly independent over $E^{1/p}$. ■

EXERCISES

Note. E and F are always extension fields of a field K, and C is an algebraically closed field containing E and F.

1. The subring $E[F]$ generated by E and F is a vector space over K in the obvious way. The tensor product $E \otimes_K F$ is also a K-vector space (see Theorem IV.5.5 and Corollary IV.5.12). E and F are linearly disjoint over K if and only if the K-linear transformation $E \otimes_K F \to E[F]$ (given on generators of $E \otimes_K F$ by $a \otimes b \mapsto ab$) is an isomorphism.
2. Assume E and F are the quotient fields of integral domains R and S respectively. Then C is an R-module and an S-module in the obvious way.

(a) E and F are linearly disjoint over K if and only if every subset of R that is linearly independent over K is also linearly independent over S.

(b) Assume further that R is a vector space over K. Then E and F are linearly disjoint over K if and only if every basis of R over K is linearly independent over F.

(c) Assume that both R and S are vector spaces over K. Then E and F are linearly disjoint over K if and only if for every basis X of R over K and basis Y of S over K, the set $\{uv \mid u \in X; v \in Y\}$ is linearly independent over K.

3. Use Exercise 1 to prove Theorem 2.2.

4. Use Exercise 1 and the associativity of the tensor product to prove Theorem 2.4.

5. If char $K = p \neq 0$, then
   (a) K^{1/p^n} is a field for every $n \geq 0$. See Exercise III.1.11.
   (b) K^{1/p^∞} is a field.
   (c) K^{1/p^n} is a splitting field over K of $\{x^{p^n} - k \mid k \in K\}$.

6. If $\{u_1, \ldots, u_n\}$ is algebraically independent over F, then F and $K(u_1, \ldots, u_n)$ are linearly disjoint over K.

7. If E is a purely transcendental extension of K and F is algebraic over K, then E and F are linearly disjoint over K.

8. Let $K = Z_p$, $F = Z_p(x)$, and $E = Z_p(x^p)$.
   (a) F is separably generated and separable over K.
   (b) $E \neq F$.
   (c) F is algebraic and purely inseparable over E.
   (d) $\{x^p\}$ is a transcendence base of F over K which is not a separating transcendence base.

9. Let char $K = p \neq 0$ and let u be transcendental over K. Suppose F is generated over K by $\{u, v_1, v_2, \ldots\}$, where v_i is a root of $x^{p^i} - u \in K(u)[x]$ for $i = 1, 2, \ldots$. Then F is separable over K, but F is not separably generated over K.

10. (a) K is a perfect field if and only if every field extension of K is separable (see Exercise V.6.13).
    (b) (Mac Lane) Assume K is a perfect field, F is not perfect and tr.d.$F/K = 1$. Then F is separably generated over K.

11. F is purely inseparable over K if and only if the only K-monomorphism $F \to C$ is the inclusion map.

12. E and F are **free** over K if every subset X of E that is algebraically independent over K is also algebraically independent over F.

(a) The definition is symmetric (that is, E and F are free over K if and only if F and E are free over K).

(b) If E and F are linearly disjoint over K, then E and F are free over K. Show by example that the converse is false.

(c) If E is separable over K and E and F are free over K, then EF is separable over F.

(d) If E and F are free over K and both separable over K, then EF is separable over K.

CHAPTER VII

LINEAR ALGEBRA

Linear algebra is an essential tool in many branches of mathematics and has wide applications. A large part of the subject consists of the study of homomorphisms of (finitely generated) free modules (in particular, linear transformations of finite dimensional vector spaces). There is a crucial relationship between such homomorphisms and matrices (Section 1). The investigation of the connection between two matrices that represent the same homomorphism (relative to different bases) leads to the concepts of equivalence and similarity of matrices (Sections 2 and 4). Certain important invariants of matrices under similarity are considered in Section 5. Determinants of matrices (Section 3) are quite useful at several points in the discussion.

Since there is much interest in the applications of linear algebra, a great deal of material of a calculational nature is included in this chapter. For many readers the inclusion of such material will be well worth the burden of additional length. However, the chapter is so arranged that the reader who wishes only to cover the important basic facts of the theory may do so in a relatively short time. He need only omit those results labeled as propositions and observe the comments in the text as to which material is needed in the sequel. The approximate interdependence of the sections of this chapter is as follows:

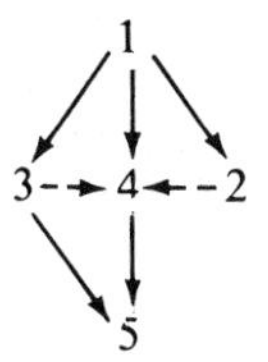

As usual a broken arrow $A \dashrightarrow B$ indicates that an occasional result of Section A is used in Section B, but that Section B is essentially independent of Section A.

1. MATRICES AND MAPS

The basic properties of matrices are briefly reviewed. Then the all important relationship between matrices and homomorphisms of free modules is explored. Except in Theorem 1.1 *all rings are assumed to have identity*, but no other restrictions are imposed. Except for the discussion of duality at the end of the section all of this material is needed in the remainder of the chapter.

Let R be a ring. An array of elements of the form

$$\begin{pmatrix} a_{11} & a_{12} & a_{13} & \cdots & a_{1m} \\ a_{21} & a_{22} & a_{23} & \cdots & a_{2m} \\ \cdot & & & & \\ \cdot & & & & \\ \cdot & & & & \\ a_{n1} & a_{n2} & a_{n3} & \cdots & a_{nm} \end{pmatrix},$$

with $a_{ij} \varepsilon R$, n rows (horizontal), and m columns (vertical), is called an **n × m matrix** over R. An $n \times n$ matrix is called a **square matrix.** For brevity of notation an arbitrary matrix is usually denoted by a capital letter, A,B,C or by (a_{ij}), which indicates that the i-jth entry (row i, column j) is the element $a_{ij} \varepsilon R$. Two $n \times m$ matrices (a_{ij}) and (b_{ij}) are **equal** if and only if $a_{ij} = b_{ij}$ in R for all i,j. The elements $a_{11},a_{22},a_{33},\ldots$ are said to form the **main diagonal** of the matrix (a_{ij}). An $n \times n$ matrix with $a_{ij} = 0$ for all $i \neq j$ is called a **diagonal matrix.** If R has an identity element, the **identity matrix** I_n is the $n \times n$ diagonal matrix with 1_R in each entry on the main diagonal; that is, $I_n = (\delta_{ij})$ where δ is the Kronecker delta. The $n \times m$ matrices with all entries 0 are called **zero matrices.** The set of all $n \times n$ matrices over R is denoted $\mathbf{Mat_nR}$. The **transpose** of an $n \times m$ matrix $A = (a_{ij})$ is the $m \times n$ matrix $A^t = (b_{ij})$ (note size!) such that $b_{ij} = a_{ji}$ for all i,j.

If $A = (a_{ij})$ and $B = (b_{ij})$ are $n \times m$ matrices, then the **sum** $A + B$ is defined to be the $n \times m$ matrix (c_{ij}), where $c_{ij} = a_{ij} + b_{ij}$. If $A = (a_{ij})$ is an $m \times n$ matrix and $B = (b_{ij})$ is an $n \times p$ matrix then the **product** AB is defined to be the $m \times p$ matrix (c_{ij}) where $c_{ij} = \sum_{k=1}^{n} a_{ik}b_{kj}$. Multiplication is not commutative in general. If $A = (a_{ij})$ is an $n \times m$ matrix and $r \varepsilon R$, rA is the $n \times m$ matrix (ra_{ij}) and Ar is the $n \times m$ matrix $(a_{ij}r)$; rI_n is called a **scalar matrix.**

If the matrix product AB is defined, then so is the product of transpose matrices B^tA^t. If R is commutative, then $(AB)^t = B^tA^t$. This conclusion may be false if R is noncommutative (Exercise 1).

Theorem 1.1. *If* R *is a ring, then the set of all* n × m *matrices over* R *forms an* R-R *bimodule under addition, with the* n × m *zero matrix as the additive identity. Multiplication of matrices, when defined, is associative and distributive over addition. For each* n > 0, *Mat*$_n$R *is a ring. If* R *has an identity, so does Mat*$_n$R *(namely the identity matrix* I$_n$*).*

PROOF. Exercise. ■

One of the important uses of matrices is in describing homomorphisms of free modules.

Theorem 1.2. *Let* R *be a ring with identity. Let* E *be a free left* R*-module with a finite basis of* n *elements and* F *a free left* R*-module with a finite basis of* m *elements. Let* M *be the left* R*-module of all* n $\times$ m *matrices over* R. *Then there is an isomorphism of abelian groups:*

$$Hom_{\mathrm{R}}(\mathrm{E,F}) \cong \mathrm{M}.$$

If R *is commutative this is an isomorphism of left* R*-modules.*

PROOF. Let $\{u_1, \ldots, u_n\}$ be a basis of E, $\{v_1, \ldots, v_m\}$ a basis of F and $f \in \mathrm{Hom}_R(E,F)$. There are elements r_{ij} of R such that

$$\begin{aligned} f(u_1) &= r_{11}v_1 + r_{12}v_2 + \cdots + r_{1m}v_m; \\ f(u_2) &= r_{21}v_1 + r_{22}v_2 + \cdots + r_{2m}v_m; \\ &\ \ \vdots \\ f(u_n) &= r_{n1}v_1 + r_{n2}v_2 + \cdots + r_{nm}v_m. \end{aligned}$$

The r_{ij} are uniquely determined since $\{v_1, \ldots, v_m\}$ is a basis of F. Define a map $\beta : \mathrm{Hom}_R(E,F) \to M$ by $f \mapsto A$, where A is the $n \times m$ matrix (r_{ij}). It is easy to verify that β is an additive homomorphism. If $\beta(f) = 0$, then $f(u_i) = 0$ for every basis element u_i, whence $f = 0$. Thus β is a monomorphism. Given a matrix $(r_{ij}) \in M$, define $f: E \to F$ by $f(u_i) = r_{i1}v_1 + r_{i2}v_2 + \cdots + r_{im}v_m$ $(i = 1,2, \ldots, n)$. Since E is free, this uniquely determines f as an element of $\mathrm{Hom}_R(E,F)$ by Theorem IV.2.1. By construction $\beta(f) = (r_{ij})$. Therefore β is surjective and hence an isomorphism. If R is commutative, then $\mathrm{Hom}_R(E,F)$ is a left R-module with $(rf)(x) = r(f(x))$ by the Remark after Theorem IV.4.8. It is easy to verify that β is an R-module isomorphism. ■

Let R,E,F and β be as in Theorem 1.2. The **matrix of a homomorphism** $f \in \mathrm{Hom}_R(E,F)$ **relative to the ordered bases** $U = \{u_1, \ldots, u_n\}$ of E and $V = \{v_1, \ldots, v_m\}$ of F is the $n \times m$ matrix $(r_{ij}) = \beta(f)$ as in the proof of Theorem 1.2. Thus the ith row of the matrix of f consists of the coefficients of $f(u_i) \in F$ relative to the ordered basis $\{v_1, \ldots, v_m\}$. In the special case when $E = F$ and $U = V$ we refer to the **matrix of the endomorphism** f **relative to the ordered basis** U.

REMARK. Let E,F,f,U,V be as in the previous paragraph. The image under f of an arbitrary element of E may be conveniently calculated from the matrix $A = (r_{ij})$ of f as follows. If $u = x_1u_1 + x_2u_2 + \cdots + x_nu_n \in E$ $(x_i \in R)$, then

$$\begin{aligned} f(u) &= f\left(\sum_{i=1}^{n} x_iu_i\right) = \sum_{i=1}^{n} x_i f(u_i) = \sum_{i=1}^{n} x_i\left(\sum_{j=1}^{m} r_{ij}v_j\right) \\ &= \sum_{j=1}^{m}\left(\sum_{i=1}^{n} x_ir_{ij}\right)v_j = \sum_{j=1}^{m} y_jv_j, \end{aligned}$$

where $y_j = \sum_{i=1}^{n} x_ir_{ij}$. Thus if X is the $1 \times n$ matrix $(x_1\, x_2 \cdots x_n)$ and Y is the $1 \times m$ matrix $(y_1\, y_2 \cdots y_m)$, then Y is precisely the matrix product XA. X and Y are sometimes called **row vectors.**

Theorem 1.3. *Let* R *be a ring with identity and let* E,F,G, *be free left* R*-modules with finite ordered bases* $U = \{u_1, \ldots, u_n\}$, $V = \{v_1, \ldots, v_m\}$, $W = \{w_1, \ldots, w_p\}$ *respectively. If* $f \in Hom_R(E,F)$ *has* $n \times m$ *matrix* A *(relative to bases* U *and* V*) and* $g \in Hom_R(F,G)$ *has* $m \times p$ *matrix* B *(relative to bases* V *and* W*), then* $gf \in Hom_R(E,G)$ *has* $n \times p$ *matrix* AB *(relative to bases* U *and* W*).*

PROOF. If $A = (r_{ij})$ and $B = (s_{kj})$, then for each $i = 1,2, \ldots, n$

$$gf(u_i) = g\left(\sum_{k=1}^{m} r_{ik}v_k\right) = \sum_{k=1}^{m} r_{ik}g(v_k) = \sum_{k=1}^{m} r_{ik}\left(\sum_{j=1}^{p} s_{kj}w_j\right)$$
$$= \sum_{j=1}^{p}\left(\sum_{k=1}^{m} r_{ik}s_{kj}\right)w_j.$$

Therefore the matrix of gf relative to U and W has i-jth entry $\sum_{k=1}^{m} r_{ik}s_{kj}$. But this is precisely the i-jth entry of the matrix AB. ■

Let R be a ring with identity and E a free left R-module with a finite basis U of n elements. Then $\mathrm{Hom}_R(E,E)$ is a ring with identity, where the product of maps f and g is simply the composite function $fg : E \to E$ (Exercise IV.1.7). We wish to note for future reference the connection between the ring $\mathrm{Hom}_R(E,E)$ and the matrix ring $\mathrm{Mat}_n R$. If S and T are any rings, then a function $\theta : S \to T$ is said to be an **anti-isomorphism** if θ is an isomorphism of *additive groups* such that $\theta(s_1s_2) = \theta(s_2)\theta(s_1)$ for all $s_i \in S$. The map $\mathrm{Hom}_R(E,E) \to \mathrm{Mat}_n R$ which assigns to each $f \in \mathrm{Hom}_R(E,E)$ its matrix (relative to U) is an anti-isomorphism of rings by Theorems 1.2 and 1.3. It would be convenient if $\mathrm{Hom}_R(E,E)$ were actually isomorphic to some matrix ring. In order to show that this is indeed the case, we need a new concept.

If R is a ring, then the **opposite ring** of R, denoted R^{op}, is the ring that has the same set of elements as R, the same addition as R, and multiplication $\circ$ given by

$$a \circ b = ba,$$

where ba is the product in R; (see Exercise III.1.17). The map given by $r \mapsto r$ is clearly an anti-isomorphism $R \to R^{op}$. If $A = (a_{ij})$ and $B = (b_{ij})$ are $n \times n$ matrices over R, then A and B may also be considered to be matrices over R^{op}. Note that in $\mathrm{Mat}_n R$, $AB = (c_{ij})$ where $c_{ij} = \sum_{k=1}^{n} a_{ik}b_{kj}$; but in $\mathrm{Mat}_n R^{op}$, $AB = (d_{ij})$, where

$$d_{ij} = \sum_{k=1}^{n} a_{ik} \circ b_{kj} = \sum_{k=1}^{n} b_{kj}a_{ik}.$$

Theorem 1.4. *Let* R *be a ring with identity and* E *a free left* R*-module with a finite basis of* n *elements. Then there is an isomorphism of rings:*

$$Hom_R(E,E) \cong Mat_n(R^{op}).$$

In particular, this isomorphism exists for every n*-dimensional vector space* E *over a division ring* R*, in which case* R^{op} *is also a division ring.*

REMARK. The conclusion of Theorem 1.4 takes a somewhat nicer form when R is commutative, since in that case $R = R^{op}$.

SKETCH OF PROOF OF 1.4. Let $\phi : \mathrm{Hom}_R(E,E) \to \mathrm{Mat}_n R$ be the anti-isomorphism that assigns to each map f its matrix relative to the given basis. Verify that the map $\psi : \mathrm{Mat}_n R \to \mathrm{Mat}_n R^{op}$ given by $\psi(A) = A^t$ is an anti-isomorphism of rings. Then the composite map $\psi\phi : \mathrm{Hom}_R(E,E) \to \mathrm{Mat}_n R^{op}$ is an *isomorphism* of rings. The last statement of the theorem is a consequence of Theorem IV.2.4 and Exercise III.1.17. ■

Let R be a ring with identity and $A \varepsilon \mathrm{Mat}_n R$. A is said to be **invertible** or **nonsingular** if there exists $B \varepsilon \mathrm{Mat}_n R$ such that $AB = I_n = BA$. The **inverse matrix** B, if it exists, is easily seen to be unique; it is usually denoted A^{-1}. Clearly $B = A^{-1}$ is invertible and $(A^{-1})^{-1} = A$. The product AC of two invertible matrices is invertible with $(AC)^{-1} = C^{-1}A^{-1}$. If A is an invertible matrix over a commutative ring, then so is its transpose and $(A^t)^{-1} = (A^{-1})^t$ (Exercise 1).

The matrix of a homomorphism of free R-modules clearly depends on the choice of (ordered) bases in both the domain and range. Consequently, it will be helpful to know the relationship between matrices that represent the same map relative to different pairs of ordered bases.

Lemma 1.5. *Let R be a ring with identity and E,F free left R-modules with ordered bases U,V respectively such that* $|U| = n = |V|$. *Let* $A \varepsilon Mat_n R$. *Then A is invertible if and only if A is the matrix of an isomorphism* $f: E \to F$ *relative to U and V. In this case* A^{-1} *is the matrix of* f^{-1} *relative to V and U.*

SKETCH OF PROOF. An R-module homomorphism $f: E \to F$ is an isomorphism if and only if there exists an R-module homomorphism $f^{-1} : F \to E$ such that $f^{-1}f = 1_E$ and $ff^{-1} = 1_F$ (see Theorem I.2.3). Suppose f is an isomorphism with matrix A relative to U and V. Let B be the matrix of f^{-1} relative to V and U. Schematically we have

$$\begin{array}{llllll} \text{map:} & & f & & f^{-1} & \\ \text{module:} & E & \longrightarrow & F & \longrightarrow & E \\ \text{basis:} & U & & V & & U \\ \text{matrix:} & & A & & B & \end{array}$$

By Theorem 1.3 AB is the matrix of $f^{-1}f = 1_E$ relative to U. But I_n is clearly the matrix of 1_E relative to U. Hence $AB = I_n$ by the proof of Theorem 1.2. Similarly $BA = I_n$, whence A is invertible and $B = A^{-1}$. The converse implication is left as an exercise. ■

Theorem 1.6. *Let R be a ring with identity. Let E and F be free left R-modules with finite ordered bases U and V respectively such that* $|U| = n$, $|V| = m$. *Let* $f \varepsilon Hom_R(E,F)$ *have* $n \times m$ *matrix A relative to U and V. Then f has* $n \times m$ *matrix B relative to another pair of ordered bases of E and F if and only if* $B = PAQ$ *for some invertible matrices P and Q.*

PROOF. ($\Rightarrow$) If B is the $n \times m$ matrix of f relative to the bases U' of E and V' of F, then $|U'| = n$ and $|V'| = m$. Let P be the $n \times n$ matrix of the identity map 1_E relative to the ordered bases U' and U. P is invertible by Lemma 1.5. Similarly let Q be the $m \times m$ invertible matrix of 1_F relative to V and V' (note order). Schematically we have:

map:		1_E		f		1_F	
module:	E	$\longrightarrow$	E	$\longrightarrow$	F	$\longrightarrow$	F
basis:	U'		U		V		V'
matrix:		P		A		Q	

By Theorem 1.3 the matrix of $f = 1_F f 1_E$ relative to U' and V' is precisely PAQ. Therefore $B = PAQ$ by the proof of Theorem 1.2.

($\Leftarrow$) We are given U,V,f,A as above and $B = PAQ$ with P,Q invertible. Let $g : E \rightarrow E$ be the isomorphism with matrix P relative to U and $h : F \rightarrow F$ the isomorphism with matrix Q^{-1} relative to V (Lemma 1.5). If $U = \{u_1, \ldots, u_n\}$, then $g(U) = \{g(u_1), \ldots, g(u_n)\}$ is also an ordered basis of E and P is the matrix of 1_E relative to the ordered bases $g(U)$ and U. Similarly Q^{-1} is the matrix of 1_F relative to the ordered bases $h(V)$ and V, whence $Q = (Q^{-1})^{-1}$ is the matrix of 1_F relative to V and $h(V)$ (Lemma 1.5). Schematically we have

map:		1_E		f		1_F	
module:	E	$\longrightarrow$	E	$\longrightarrow$	F	$\longrightarrow$	F
basis:	$g(U)$		U		V		$h(V)$
matrix:		P		A		Q	

By Theorem 1.3 the matrix of $f = 1_F f 1_E$ relative to the ordered bases $g(U)$ and $h(V)$ is $PAQ = B$. ■

Corollary 1.7 *Let* R *be a ring with identity and* E *a free left* R*-module with an ordered basis* U *of finite cardinality* n. *Let* A *be the* $n \times n$ *matrix of* $f \in Hom_R(E,E)$ *relative to* U. *Then* f *has* $n \times n$ *matrix* B *relative to another ordered basis of* E *if and only if* $B = PAP^{-1}$ *for some invertible matrix* P.

SKETCH OF PROOF. If $E = F, U = V$, and $U' = V'$ in the proof of Theorem 1.6, then $Q = P^{-1}$ by Lemma 1.5. ■

The preceding results motivate:

Definition 1.8. *Let* R *be a ring with identity. Two matrices* $A,B \in Mat_n R$ *are said to be* **similar** *if there exists an invertible matrix* P *such that* $B = PAP^{-1}$. *Two* $n \times m$ *matrices* C,D *are said to be* **equivalent** *if there exists invertible matrices* P *and* Q *such that* $D = PCQ$.

Theorem 1.6 and Corollary 1.7 may now be reworded in terms of equivalence and similarity. Equivalence and similarity are each equivalence relations (Exercise 7) and will be studied in more detail in Sections 2 and 4.

We close this section with a discussion of right modules and duality.

If R is commutative, then the preceding results are equally valid for *right* R-modules. There are important cases, however, in which R is not commutative (for example, vector spaces over a division ring). In order to prove the analogue of Theorem 1.3 for right modules in the noncommutative case it is necessary to define the matrix of a homomorphism somewhat differently.

Let R be a ring with identity and let E and F be free *right* R-modules with finite ordered bases $U = \{u_1, \ldots, u_n\}$ and $V = \{v_1, \ldots, v_m\}$ respectively. The **matrix of the homomorphism** $f \in \operatorname{Hom}_R(E,F)$ relative to U and V is defined to be the $m \times n$ matrix (note size):

$$\begin{pmatrix} s_{11} & s_{12} & \cdots & s_{1n} \\ s_{21} & s_{22} & \cdots & s_{2n} \\ \cdot & & & \\ \cdot & & & \\ \cdot & & & \\ s_{m1} & s_{m2} & \cdots & s_{mn} \end{pmatrix},$$

where the $s_{ij} \in R$ are uniquely determined by the equations:

$$\begin{aligned} f(u_1) &= v_1 s_{11} + v_2 s_{21} + v_3 s_{31} + \cdots + v_m s_{m1} \\ &\cdot \\ &\cdot \\ &\cdot \\ f(u_n) &= v_1 s_{1n} + v_2 s_{2n} + v_3 s_{3n} + \cdots + v_m s_{mn}. \end{aligned}$$

Thus the coefficients of $f(u_j)$ with respect to the ordered basis V form the jth *column* of the $m \times n$ matrix (s_{ij}) of f (compare the proof of Theorem 1.2).

The action of f may be described in terms of matrices as follows. Let $u = u_1x + u_2x_2 + \cdots + u_nx_n$ ($x_i \in R$) be any element of E and let X be the $n \times 1$ matrix (or **column vector**) $\begin{pmatrix} x_1 \\ \cdot \\ \cdot \\ \cdot \\ x_n \end{pmatrix}$. Let A be the matrix of f relative to the bases U and V. Then $f(u) = v_1y_1 + v_2y_2 + \cdots + v_my_m$, where $y_i \in R$ and $\begin{pmatrix} y_1 \\ \cdot \\ \cdot \\ \cdot \\ y_m \end{pmatrix}$ is the $m \times 1$ matrix (column vector) AX.

The analogues of results 1.2–1.5 above are now easily proved, in particular,

Theorem 1.9. *Let* R *be a ring with identity and* E,F *free right* R*-modules with finite bases* U *and* V *of cardinality* n *and* m *respectively. Let* N *be the right* R*-module of all* m $\times$ n *matrices over* R.

(i) *There is an isomorphism of abelian groups* $\operatorname{Hom}_R(E,F) \cong N$, *which is an isomorphism of right* R*-modules if* R *is commutative;*

(ii) *let* G *be a free right* R*-module with a finite basis* W *of cardinality* p. *If* $f \in Hom_R(E,F)$ *has* $m \times n$ *matrix* A *(relative to* U *and* V*) and* $g \in Hom_R(F,G)$ *has* $p \times m$ *matrix* B *(relative to* V *and* W*), then* $gf \in Hom_R(E,G)$ *has* $p \times n$ *matrix* BA *(relative to* U *and* W*);*

(iii) *there is an isomorphism of rings* $Hom_R(E,E) \cong Mat_n R$.

PROOF. Exercise; see Theorems 1.2–1.4. Note that for right modules (iii) is actually an isomorphism rather than an anti-isomorphism. ■

Proposition 1.10. *Let* R *be a ring with identity and* $f : E \to F$ *a homomorphism of finitely generated free left* R*-modules. If* A *is the matrix of* f *relative to (ordered) bases* U *and* V*, then* A *is also the matrix of the dual homomorphism* $\bar{f} : F^* \to E^*$ *of free right* R*-modules relative to the dual bases* V^* *and* U^*.

REMARK. Dual maps and dual bases are defined in Theorems IV.4.10 and IV.4.11. If R is commutative (for example, a field) it is customary to consider the dual M^* of a left R-module M as a *left* R-module (with $rm^* = m^*r$ for $r \in R$, $m^* \in M^*$ as usual). In this case the matrix of the dual map $\bar{f}$ is the transpose A^t (Exercise 8).

PROOF OF 1.10. Recall that the dual basis $V^* = \{v_1^*, \ldots, v_m^*\}$ of $F^* = \mathrm{Hom}_R(F,R)$ is determined by:

$$v_i^*(v_j) = \delta_{ij} \quad \text{(Kronecker delta; } 1 \leq i,j \leq m),$$

and similarly for the dual basis $U^* = \{u_1^*, \ldots, u_n^*\}$ of E^* (Theorem IV.4.11). According to the definition of the matrix of a map of right R-modules we must show that for each $j = 1,2, \ldots, m$, $\bar{f}(v_j^*) = \sum_{i=1}^{n} u_i^* r_{ij}$, where $A = (r_{ij})$ is the $n \times m$ matrix of $f : E \to F$ relative to U and V. Since both sides of the preceding equation are maps $E \to R$, it suffices to check their action on each $u_k \in U$. By Theorem IV.4.10 we have:

$$\bar{f}(v_j^*)(u_k) = v_j^*(f(u_k)) = v_j^*\left(\sum_{t=1}^{m} r_{kt}v_t\right) = \sum_{t=1}^{m} r_{kt}v_j^*(v_t) = r_{kj}.$$

On the other hand,

$$\left(\sum_{i=1}^{n} u_i^* r_{ij}\right)(u_k) = \sum_{i=1}^{n} u_i^*(u_k) r_{ij} = r_{kj}. \quad ■$$

EXERCISES

Note: All matrices are assumed to have entries in a ring R with identity.

1. Let R be commutative.

 (a) If the matrix product AB is defined, then so is the product B^tA^t and $(AB)^t = B^tA^t$.

 (b) If A is invertible, then so is A^t and $(A^t)^{-1} = (A^{-1})^t$.

 (c) If R is not commutative, then (a) and (b) may be false.

2. A matrix $(a_{ij}) \varepsilon \operatorname{Mat}_n R$ is said to be

$$\textbf{(upper) triangular} \Leftrightarrow a_{ij} = 0 \quad \text{for} \quad j < i;$$
$$\textbf{strictly triangular} \Leftrightarrow a_{ij} = 0 \quad \text{for} \quad j \leq i.$$

Prove that the set of all diagonal matrices is a subring of $\operatorname{Mat}_n R$ which is (ring) isomorphic to $R \times \cdots \times R(n$ factors). Show that the set T of all triangular matrices is a subring of $\operatorname{Mat}_n R$ and the set I of all strictly triangular matrices is an ideal in T. Identify the quotient ring T/I.

3. (a) The center of the ring $\operatorname{Mat}_n R$ consists of all matrices of the form rI_n, where r is in the center of R. [*Hint:* every matrix in the center of $\operatorname{Mat}_n R$ must commute with each of the matrices $B_{r,s}$, where $B_{r,s}$ has 1_R in position (r,s) and 0 elsewhere.]
(b) The center of $\operatorname{Mat}_n R$ is isomorphic to the center of R.

4. The set of all $m \times n$ matrices over R is a free R-module with a basis of mn elements.

5. A matrix $A \varepsilon \operatorname{Mat}_n R$ is **symmetric** if $A = A^t$ and **skew-symmetric** if $A = -A^t$.
(a) If A and B are [skew] symmetric, then $A + B$ is [skew] symmetric.
(b) Let R be commutative. If A,B are symmetric, then AB is symmetric if and only if $AB = BA$. Also show that for any matrix $B \varepsilon \operatorname{Mat}_n R$, BB^t and $B + B^t$ are symmetric and $B - B^t$ is skew-symmetric.

6. If R is a division ring and $A,B \varepsilon \operatorname{Mat}_n R$ are such that $BA = I_n$, then $AB = I_n$ and $B = A^{-1}$. [*Hint:* use linear transformations.]

7. Similarity of matrices is an equivalence relation on $\operatorname{Mat}_n R$. Equivalence of matrices is an equivalence relation on the set of all $m \times n$ matrices over R.

8. Let E,F be finite dimensional (left) vector spaces over a field and consider the dual spaces to be *left* vector spaces in the usual way. If A is the matrix of a linear transformation $f : E \to F$, then A^t is the matrix of the dual map $\bar{f} : F^* \to E^*$.

2. RANK AND EQUIVALENCE

The main purpose of this section is to find necessary and sufficient conditions for matrices over a division ring or a principal ideal domain to be equivalent. One such condition involves the concept of rank. In addition, useful sets of canonical forms for such matrices are presented (Theorem 2.6 and Proposition 2.11). Finally, practical techniques are developed for finding these canonical forms and for calculating the inverse of an invertible matrix over a division ring. Applications to finitely generated abelian groups are considered in an appendix, which is not needed in the sequel.

Definition 2.1. *Let* $f : E \to F$ *be a linear transformation of (left) vector spaces over a division ring* D. *The* **rank** *of* f *is the dimension of* $\operatorname{Im} f$ *and the* **nullity** *of* f *is the dimension of* $\operatorname{Ker} f$.

REMARK. If $f : E \to F$ is as in Definition 2.1, then by Corollary IV.2.14, $(\text{rank } f) + (\text{nullity } f) = \dim_D E$.

If R is a ring with identity and n a positive integer, then R^n will denote the free R-module $R \oplus \cdots \oplus R$ (n summands). The **standard (ordered) basis** of R^n consists of the elements $\varepsilon_1 = (1_R,0,\ldots,0)$, $\varepsilon_2 = (0,1_R,0,\ldots,0),\ldots,\varepsilon_n = (0,\ldots,0,1_R)$.

Definition 2.2. *The* **row space** *[resp.* **column space***] of an* $n \times m$ *matrix* A *over a ring* R *with identity is the submodule of the free left [resp. right] module* R^m *[resp.* R^n*] generated by the rows [resp. columns] of* A *considered as elements of* R^m *[resp.* R^n*]. If* R *is a division ring, then the* **row rank** *[resp.* **column rank***] of* A *is the dimension of the row [resp. column] space of* A.

Theorem 2.3. *Let* $f: E \rightarrow F$ *be a linear transformation of finite dimensional left [resp. right] vector spaces over a division ring* D. *If* A *is the matrix of* f *relative to some pair of ordered bases, then the rank of* f *is equal to the row [resp. column] rank of* A.

REMARK. "Row rank" is replaced by "column rank" in the case of right vector spaces because of the definition of the matrix of a map of right vector spaces (p. 333).

PROOF OF 2.3. Let A be the $n \times m$ [resp. $m \times n$] matrix of f relative to ordered bases $U = \{u_1,\ldots,u_n\}$ of E and $V = \{v_1,\ldots,v_m\}$ of F. Then under the usual isomorphism $F \cong D^m$ given by $\sum_i r_iv_i \mapsto (r_1,\ldots,r_m)$ the elements $f(u_1),\ldots,f(u_n)$ are mapped onto the rows [resp. columns] of A (considered as vectors in D^m). Since $\operatorname{Im} f$ is spanned by $f(u_1),\ldots,f(u_n)$, $\operatorname{Im} f$ is isomorphic to the row [resp. column] space of A, whence the rank of f is equal to the row [resp. column] rank of A. ∎

We now digress briefly to prove that the row and column rank of a matrix over a division ring are in fact equal. This fact, which is proved in Corollary 2.5, is not essential for understanding the sequel since "row rank" is all that is actually used hereafter.

Proposition 2.4. *Any linear transformation* $f: E \rightarrow F$ *of finite dimensional left vector spaces over a division ring* D *has the same rank as its dual map* $\bar{f}: F^* \rightarrow E^*$.

The dual map is defined in Theorem IV.4.10.

PROOF OF 2.4. Let rank $f = r$. By Corollary IV.2.14 there is a basis $X = \{u_1,\ldots,u_n\}$ such that $\{u_{r+1},\ldots,u_n\}$ is a basis of $\operatorname{Ker} f$ and $Y_1 = \{f(u_1),\ldots,f(u_r)\}$ is a basis of $\operatorname{Im} f$. Extend Y_1 to a basis $Y = \{t_1 = f(u_1),\ldots,t_r = f(u_r),t_{r+1},\ldots,t_m\}$ of F. Consider the dual bases X^* of E^* and Y^* of F^* (Theorem IV.4.11). Verify that for each $i = 1,2,\ldots,m$,

$$\bar{f}(t_i^*)(u_j) = t_i^*(f(u_j)) = \begin{cases} t_i^*(t_j) = \delta_{ij} & \text{if } j = 1,2,\ldots,r; \\ t_i^*(0) = 0 & \text{if } j = r+1,r+2,\ldots,n. \end{cases}$$

where δ_{ij} is the Kronecker delta. Consequently for each $j = 1,2,\ldots,n$,

$$\bar{f}(t_i^*)(u_j) = \begin{cases} \delta_{ij} = u_i^*(u_j) & \text{if } i = 1,2,\ldots,r \\ 0 & \text{if } i = r+1,r+2,\ldots,m. \end{cases}$$

Therefore, $\bar{f}(t_i^*) = u_i^*$ for $i = 1,2,\ldots,r$ and $\bar{f}(t_i^*) = 0$ for $i = r+1,\ldots,m$. Im $\bar{f}$ is spanned by $\bar{f}(Y^*)$ and hence by $\{u_1^*,\ldots,u_r^*\}$. Since $\{u_1^*,\ldots,u_r^*\}$ is a subset of X^*, it is linearly independent in E^*. Therefore $\{u_1^*,\ldots,u_r^*\}$ is a basis of Im $\bar{f}$, whence rank $\bar{f} = r =$ rank f. ■

Corollary 2.5. *If* A *is an* n × m *matrix over a division ring* D, *then row rank* A = *column rank* A.

PROOF. Let $f : D^n \to D^m$ be a linear transformation of left vector spaces with matrix A relative to the standard bases. Then the dual map $\bar{f}$ of right vector spaces also has matrix A (Proposition 1.10). By Theorem 2.3 and Proposition 2.4 row rank A = rank f = rank $\bar{f}$ = column rank A. ■

REMARK. Corollary 2.5 immediately implies that row rank A = row rank A^t for any matrix A over a field.

In view of Corollary 2.5 we shall hereafter omit the adjectives "row" and "column" and refer simply to the **rank of a matrix** over a division ring.

In Theorem 2.6 below equivalent matrices over a division ring D will be characterized in terms of rank and in terms of the following matrices. If m,n are positive integers, then $E_0^{n,m}$ is defined to be the $n \times m$ zero matrix. For each r ($1 \le r \le \min(n,m)$), $E_r^{n,m}$ is defined to be the $n \times m$ matrix whose first r rows are the standard basis vectors $\varepsilon_1,\ldots,\varepsilon_r$ of D^m and whose remaining rows are zero:

$$E_r^{n,m} = \begin{pmatrix} 1_R & 0 & 0 & \cdots & \cdots & \cdots & 0 \\ 0 & 1_R & 0 & \cdots & \cdots & \cdots & 0 \\ \vdots & & & & & & \vdots \\ 0 & \cdots & 0 & 1_R & 0 & \cdots & 0 \\ 0 & \cdots & 0 & 0 & 0 & \cdots & 0 \\ \vdots & & & & & & \vdots \\ 0 & \cdots & \cdots & \cdots & \cdots & \cdots & 0 \end{pmatrix} = \begin{pmatrix} I_r & 0 \\ 0 & 0 \end{pmatrix}.$$

Clearly rank $E_r^{n,m} = r$. Furthermore if $E_r^{n,m}$ is the matrix of an R-module homomorphism $f : E \to F$ of free R-modules, relative to bases $\{u_1,\ldots,u_n\}$ of E and $\{v_1,\ldots,v_m\}$ of F, then

$$f(u_i) = \begin{cases} v_i & \text{if } i = 1,2,\ldots,r; \\ 0 & \text{if } i = r+1, r+2,\ldots,n. \end{cases}$$

An immediate consequence of Theorem 1.6 and Theorem 2.6 below is that every linear transformation of finite dimensional vector spaces has this convenient form for some pair of bases (Exercise 6).

A set of **canonical forms** for an equivalence relation R on a set X is a subset C of X that consists of exactly one element from each equivalence class of R. In other words, for every $x \in X$ there is a unique $c \in C$ such that x is equivalent to c under R. We now show that the matrices $E_r^{n,m}$ form a set of canonical forms for the relation of equivalence on the set of all $n \times m$ matrices over a division ring.

Theorem 2.6. *Let* M *be the set of all* n × m *matrices over a division ring* D *and let* A,B ε M.

(i) A *is equivalent to* $E_r^{n,m}$ *if and only if rank* A = r.
(ii) A *is equivalent to* B *if and only if rank* A = *rank* B.
(iii) *The matrices* $E_r^{n,m}$ (r = 1,2, . . . , *min* (n,m)) *constitute a set of canonical forms for the relation of equivalence on* M.

SKETCH OF PROOF. (i) A is the matrix of some linear transformation $f : D^n \to D^m$ relative to some pair of bases by Theorem 1.2. If rank $A = r$, then Corollary IV.2.14 implies that there exist bases $U = \{u_1, \ldots, u_n\}$ of D^n and $V = \{v_1, \ldots, v_m\}$ of D^m such that $f(u_i) = v_i$ for $i = 1,2, \ldots, r$ and $f(u_i) = 0$ for $i = r + 1, \ldots, n$. Clearly the matrix of f relative to U and V is $E_r^{n,m}$. Therefore A is equivalent to $E_r^{n,m}$ by Theorem 1.6. Conversely suppose A is equivalent to $E_r^{n,m}$. By Theorem 1.6 there is a linear transformation $g : D^n \to D^m$ such that A is the matrix of g relative to one pair of bases and $E_r^{n,m}$ is the matrix of g relative to another pair of bases. Consequently, rank A = rank g = rank $E_r^{n,m} = r$ by Theorem 2.3. (ii) and (iii) are consequences of (i). ■

The following definition, theorem, and corollaries have a number of useful consequences, including practical methods for constructing:
(i) canonical forms under equivalence for matrices over a principal ideal domain (Proposition 2.11);
(ii) the canonical forms $E_r^{n,m}$ under equivalence for matrices over a division ring;
(iii) the inverse of an invertible matrix over a division ring (Proposition 2.12).
Proposition 2.11 is used only in the proof of Proposition 4.9 below. The remainder of the material is independent of Proposition 2.11 and is not needed in the sequel.

We shall frequently consider the rows [resp. columns] of a given $n \times m$ matrix over a ring R as being elements of R^m [resp. R^n]. We shall speak of adding a scalar multiple of one row [resp. column] to another; for example,

$$r(a_1,a_2, \ldots, a_m) + (b_1, \ldots, b_m) = (ra_1 + b_1, \ldots, ra_m + b_m).$$

Definition 2.7. *Let* A *be a matrix over a ring* R *with identity. Each of the following is called an* **elementary row operation** *on* A:

(i) *interchange two rows of* A;
(ii) *left multiply a row of* A *by a unit* c ε R;
(iii) *for* r ε R *and* i ≠ j, *add* r *times row* j *to row* i.

Elementary column operations *on* A *are defined analogously (with left multiplication in* (ii), (iii) *replaced by right multiplication). An* n × n **elementary (transformation) matrix** *is a matrix that is obtained by performing exactly one elementary row (or column) operation on the identity matrix* $\mathbf{I}_n$.

Theorem 2.8. *Let* A *be an* n × m *matrix over a ring* R *with identity and let* E_n [*resp.* E_m] *be the elementary matrix obtained by performing an elementary row* [*resp. column*] *operation* T *on* I_n [*resp.* I_m]. *Then* E_nA [*resp.* AE_m] *is the matrix obtained by performing the operation* T *on* A.

PROOF. Exercise. ■

Corollary 2.9. *Every* $n \times n$ *elementary matrix* E *over a ring* R *with identity is invertible and its inverse is an elementary matrix.*

SKETCH OF PROOF. Verify that I_n may be obtained from E by performing a single elementary row operation T. If F is the elementary matrix obtained by performing T on I_n, then $FE = I_n$ by Theorem 2.8. Verify directly that $EF = I_n$. ■

Corollary 2.10. *If* B *is the matrix obtained from an* $n \times m$ *matrix* A *over a ring* R *with identity by performing a finite sequence of elementary row and column operations, then* B *is equivalent to* A.

PROOF. Since each row [column] operation used to obtain B from A is given by left [right] multiplication by an appropriate elementary matrix (Theorem 2.8), we have $B = (E_p \cdots E_1)A(F_1 \cdots F_q) = PAQ$ with each E_i F_j an elementary matrix and $P = E_p \cdots E_1$, $Q = F_1 \cdots F_q$. P and Q are products of invertible matrices (Corollary 2.9) and hence invertible. ■

We now consider canonical forms under equivalence of matrices over a principal ideal domain R. The rank of a free module over R is a well-defined invariant by Corollary IV.2.12. Since every submodule of a free R-module is free (Theorem IV.6.1), we may define the **rank of a homomorphism** $f : E \to F$ of free R-modules to be the rank of $\operatorname{Im} f$. Similarly the **row rank of a matrix** A over R is defined to be the rank of the row space A (see Definition 2.2). The proof of Theorem 2.3 is easily seen to be valid here, whence the rank of a map f of finitely generated free R-modules is the row rank of any matrix of f relative to some pair of bases. Consequently, if A is equivalent to a matrix B, then row rank A = row rank B. For A and B are matrices of the same homomorphism $f : R^n \to R^m$ relative to different pairs of bases by Theorem 1.6, whence row rank A = rank f = row rank B. Here is the analogue of Theorem 2.6 for matrices over a principal ideal domain.

Proposition 2.11. *If* A *is an* $n \times m$ *matrix of rank* $r > 0$ *over a principal ideal domain* R, *then* A *is equivalent to a matrix of the form* $\begin{pmatrix} L_r & 0 \\ 0 & 0 \end{pmatrix}$, *where* L *is an* $r \times r$ *diagonal matrix with nonzero diagonal entries* $d_1, \ldots, d_r$ *such that* $d_1 \mid d_2 \mid \cdots \mid d_r$. *The ideals* $(d_1), \ldots, (d_r)$ *in* R *are uniquely determined by the equivalence class of* A.

REMARKS. The proposition provides sets of canonical forms for the relation of equivalence on the set of $n \times m$ matrices over a principal ideal domain (Exercise 5). If R is actually a Euclidean domain, then the following proof together with Exercise 7 and Theorem 2.8 shows that the matrix $\begin{pmatrix} L_r & 0 \\ 0 & 0 \end{pmatrix}$ may be obtained from A by a finite sequence of *elementary* row and column operations.

SKETCH OF PROOF OF 2.11. (i) Recall that $a,b \in R$ are *associates* if $a \mid b$ and $b \mid a$. By Theorem III.3.2 a and b are associates if and only if $a = bu$ with u a unit. We say that $c \in R$ is a *proper divisor* of $a \in R$ if $c \mid a$ and c is not an associate of a (that is, $a \nmid c$). By a slight abuse of language we say that two proper divisors c_1 and c_2

of an element a are *distinct* if c_1 and c_2 are not associates. Now R is a unique factorization domain by Theorem III.3.7. If $a = p_1^{n_1}p_2^{n_2}\cdots p_t^{n_t}$ (p_i distinct irreducibles and each $n_i > 0$), then every divisor of a is an associate of an element of the form $p_1^{k_1}p_2^{k_2}\cdots p_t^{k_t}$ with $0 \leq k_i \leq n_i$ for each i. Consequently a nonzero element of R has only finitely many distinct proper divisors.

(ii) If a and b are nonzero elements of R, let c be their greatest common divisor. By Definition III.3.10 and Theorem III.3.11 there exist $r,s \in R$ such that $ar + bs = c$, $ca_1 = a$ and $cb_1 = b$, whence $a_1r + b_1s = 1_R$ and $ba_1 - ab_1 = 0$. Consequently the $m \times m$ matrix

$$T = \begin{pmatrix} r & -b_1 & \\ s & a_1 & 0 \\ & 0 & I_{m-2} \end{pmatrix}$$

is invertible with inverse

$$T^{-1} = \begin{pmatrix} a_1 & b_1 & \\ -s & r & 0 \\ & 0 & I_{m-2} \end{pmatrix}.$$

If the first row of A is $(a,b,a_{13},\ldots,a_{1m})$, then A is equivalent to $AT = I_nAT$, whose first row is $(c,0,a_{13},\ldots,a_{1m})$. If the first column[1] of A is $(a,d,a_{31},\ldots,a_{n1})^t$, then an analogous procedure yields an invertible matrix S such that A is equivalent to SA and SA has first column $(e,0,a_{31},\ldots,a_{n1})^t$, where e is the greatest common divisor of a and d. A matrix such as S or T is called a **secondary matrix**.

(iii) Since $A \neq 0$ a suitable sequence of row and column interchanges and multiplications on the right by secondary matrices changes A into a matrix A_1 which has first row $(a_1,0,0,\ldots,0)$ with $a_1 \neq 0$. A is equivalent to A_1 by (ii) and Corollary 2.10.

(iv) If a_1 divides all entries in the first column of A_1, then a finite sequence of elementary row operations produces a matrix B of the form

$$B = \begin{pmatrix} a_1 & 0 & \cdots & 0 \\ 0 & b_{22} & \cdots & b_{2m} \\ \cdot & & & \\ \cdot & & & \\ \cdot & & & \\ 0 & b_{n2} & \cdots & b_{nm} \end{pmatrix},$$

which is equivalent to A_1, and hence to A, by Corollary 2.10.

(v) If a_1 does not divide some first column entry b of A_1, then a sequence of row and column interchanges and multiplications on the left by secondary matrices changes A_1 into a matrix A_2 which has first *column* $(a_2,0,0,\ldots,0)^t$ with a_2 a common divisor of a_1 and b (see (ii)). Note that A_2 may well have many nonzero entries in the

[1]For typographical convenience we shall frequently write an $n \times 1$ column vector as the transpose of a $1 \times n$ row vector; for example, $\begin{pmatrix} a_1 \\ a_2 \end{pmatrix} = (a_1a_2)^t$.

first *row*. However since $a_2 \mid a_1$, $a_2 \mid b$ and $a_1 \nmid b$, a_2 is a proper divisor of a_1 by (i). A_2 is equivalent to A_1, and hence to A, by (ii) and Corollary 2.10.

(vi) If a_2 divides every entry in the first *row* of A_2 then a sequence of elementary column operations produces a matrix equivalent to A and of the same general form as B above.

(vii) If a_2 fails to divide some entry k in the first row of A_2, then repeat (iii) and obtain a matrix A_3 which is equivalent to A and has first row $(a_3,0,0,\dots,0)$ with a_3 a common divisor of a_2 and k. A_3 may have nonzero entries in its first column. But since $a_3 \mid a_2$, $a_3 \mid k$ and $a_2 \nmid k$, a_3 is a proper divisor of a_2 by (i). Furthermore, a_2 and a_3 are distinct proper divisors of a_1 by (v). A_3 is equivalent to A_2, and hence to A, by (ii) and Corollary 2.10.

(viii) Since a_1 has only finitely many distinct proper divisors, a finite number of repetitions of (iii)–(vii) must yield a matrix C which is equivalent to A and has the form

$$C = \begin{pmatrix} s_1 & 0 & \cdots & 0 \\ 0 & c_{22} & \cdots & c_{2m} \\ \cdot & \cdot & & \cdot \\ \cdot & \cdot & & \cdot \\ \cdot & \cdot & & \cdot \\ 0 & c_{n2} & \cdots & c_{nm} \end{pmatrix}$$

with $s_1 \neq 0$.

(ix) If s_1 does not divide some c_{ij}, add row i to row 1 and repeat (iii)–(viii). The result is a matrix D that is equivalent to A, has the same general form as the matrix C above, and has for its (1,1) entry an element s_2 which is a common divisor of s_1 and c_{ij} and a proper divisor of s_1.

(x) If s_2 does not divide every entry in D, then a repetition of (ix) yields a matrix that is equivalent to A, has the same general form as C and has (1,1) entry s_3 such that s_3 is a proper divisor of s_2, whence s_2 and s_3 are distinct proper divisors of s_1. Since s_1 has only finitely many distinct proper divisors, a finite number of repetitions of this process produces a matrix that is equivalent to A, has the same general form as C, and has a (1,1) entry which divides all other entries of the matrix.

(xi) Use induction and (x) to show that A is equivalent to a diagonal matrix $F = \begin{pmatrix} L_r & 0 \\ 0 & 0 \end{pmatrix}$ as in the statement of the theorem. Since the rank of F is obviously r, the rank of A is r by Theorem 2.6.

(xii) (uniqueness) Let A and F be as in (xi), with $d_1,\dots,d_r$, the diagonal elements of L_r. Suppose M is a matrix equivalent to A (so that rank $M = r$) and N is a matrix equivalent to M of the form $\begin{pmatrix} L_r' & 0 \\ 0 & 0 \end{pmatrix}$ where L_r' is an $r \times r$ diagonal matrix with nonzero diagonal entries k_i such that $k_1 \mid k_2 \mid \cdots \mid k_r$. By Theorem 1.2 F is the matrix of a homomorphism $f : R^m \to R^n$ relative to bases $\{u_1,\dots,u_n\}$ of R^n and $\{v_1,\dots,v_m\}$ of R^m. Consequently, $f(u_i) = d_iv_i$ for $i = 1,2,\dots,r$ and $f(u_i) = 0$ for $i = r+1,\dots,n$, whence $\text{Im } f = Rd_1v_1 \oplus \cdots \oplus Rd_rv_r$. By the analogue for modules of Corollary I.8.11, $R^m/\text{Im } f \cong Rv_1/Rd_1v_1 \oplus \cdots \oplus Rv_r/Rd_rv_r \oplus Rv_{r+1} \oplus \cdots \oplus Rv_n \cong R/(d_1) \oplus \cdots \oplus R/(d_r) \oplus R \oplus \cdots \oplus R$ (m summands; $d_1|d_2|\cdots|d_r$). Since F is equivalent to N by hypothesis, Theorem 1.6 implies that N is the matrix of f relative to a different pair of bases. A repetition of the preceding argument then shows that $R^m/\text{Im } f \cong R/(k_1) \oplus \cdots \oplus R/(k_r) \oplus R \oplus \cdots \oplus R$

(m summands; $k_1|k_2|\cdots|k_r$). The structure Theorem IV.6.12 for modules over a principal ideal domain implies that $(d_i) = (k_i)$ for $i = 1,2,\ldots,r$. ■

A simplified version of the techniques used in the proof of Proposition 2.11 may be used to obtain the canonical form $E_r^{n,m}$ of an $n \times m$ matrix A over a division ring D. If $A = 0 = E_0^{n,m}$, there is nothing to prove. If a_{ij} is a nonzero entry in A, then interchanging rows i and 1 and columns j and 1 moves a_{ij} to position (1,1). Multiplying row 1 by a_{ij}^{-1} yields a matrix with first row of the form $(1_R,c_2,\ldots,c_m)$. Subtract suitable multiples of row 1 [resp. column 1] from each subsequent row [resp. column] and obtain a matrix of the form:

$$\begin{pmatrix} 1_R & 0 & \cdots & 0 \\ 0 & c_{22} & \cdots & c_{2m} \\ \cdot & & & \\ \cdot & & & \\ \cdot & & & \\ 0 & c_{n2} & \cdots & c_{nm} \end{pmatrix}.$$

If every $c_{ij} = 0$, we are done. If some $c_{ij} \neq 0$, then we may repeat the above procedure on the $(n-1) \times (m-1)$ submatrix (c_{ij}). Since row [column] operations on rows $2,\ldots,n$ [columns $2,\ldots,m$] do not affect the first row or column, we obtain a matrix

$$\begin{pmatrix} 1_R & 0 & 0 & \cdots & 0 \\ 0 & 1_R & 0 & \cdots & 0 \\ 0 & 0 & d_{33} & \cdots & d_{3m} \\ \cdot & & & & \\ \cdot & & & & \\ \cdot & & & & \\ 0 & 0 & d_{n3} & \cdots & d_{nm} \end{pmatrix}.$$

Continuing this process eventually yields the matrix $E_r^{n,m}$ for some r. By Corollary 2.10 A is equivalent to $E_r^{n,m}$, whence $r = \text{rank } A$ and $E_r^{n,m}$ is the canonical form of A under equivalence by Theorem 2.6.

A modified version of the preceding technique gives a constructive method for finding the inverse of an invertible matrix, as is seen in the proof of:

Proposition 2.12. *The following conditions on an* $n \times n$ *matrix* A *over a division ring* D *are equivalent:*

(i) *rank* $A = n$;
(ii) A *is equivalent to the identity matrix* I_n;
(iii) A *is invertible;*
(iv) A *is the product of elementary transformation matrices.*

SKETCH OF PROOF. (i) $\Leftrightarrow$ (ii) by Theorem 2.6 since $E_n^{n,n} = I_n$. (i) $\Rightarrow$ (iii) The rows of any matrix of rank n are necessarily linearly independent (see Theorem IV.2.5 and Definition 2.2.) Consequently, the first row of $A = (a_{ij})$ is not the zero vector and $a_{1j} \neq 0$ for some j. Interchange columns j and 1 and multiply the new column 1 by a_{1j}^{-1}. Subtracting suitable multiples of column 1 from each succeeding column yields a matrix

$$B = \begin{pmatrix} 1_D & 0 & \cdots & 0 \\ b_{21} & b_{22} & \cdots & b_{2n} \\ \cdot & & & \\ \cdot & & & \\ \cdot & & & \\ b_{n1} & b_{n2} & \cdots & b_{nn} \end{pmatrix}.$$

B is equivalent to A by Corollary 2.10. Assume inductively that there is a sequence of elementary column operations that changes A to a (necessarily equivalent) matrix

$$C = \begin{pmatrix} & I_{k-1} & & 0 & \\ c_{k1} & \cdots & c_{kk} & \cdots & c_{kn} \\ \cdot & & & & \cdot \\ \cdot & & & & \cdot \\ \cdot & & & & \cdot \\ c_{n1} & \cdots & c_{nk} & \cdots & c_{nn} \end{pmatrix}.$$

For some $j \geq k$, $c_{kj} \neq 0$ since otherwise row k would be a linear combination of rows $1,2,\ldots,k-1$. This would contradict the fact that rank C = rank $A = n$ by Theorem 2.6. Interchange columns j and k, multiply the new column k by c_{kj}^{-1} and subtract a suitable multiple of column k from each of columns $1,2,\ldots,k-1$, $k+1,\ldots,n$. The result is a matrix D that is equivalent to A (Corollary 2.10):

$$D = \begin{pmatrix} & I_k & & 0 & \\ d_{k+1\,1} & \cdots & d_{k+1\,k+1} & \cdots & d_{k+1\,n} \\ \cdot & & & & \cdot \\ \cdot & & & & \cdot \\ \cdot & & & & \cdot \\ d_{n1} & \cdots & d_{n\,k+1} & \cdots & d_{nn} \end{pmatrix}.$$

This completes the induction and shows that when $k = n$, A is changed to I_n by a finite sequence of elementary *column* operations. Therefore by Theorem 2.8 $A(F_1F_2\cdots F_t) = I_n$ with each F_i an elementary matrix. The matrix $F_1F_2\cdots F_t$ is a two-sided inverse of A by Exercise 1.7, whence A is invertible. Corollary 2.9 and the fact that $A = F_t^{-1}\cdots F_2^{-1}F_1^{-1}$ show that (i) $\Rightarrow$ (iv). (iii) $\Rightarrow$ (i) by Lemma 1.5 and Theorem 2.3. (iv) $\Rightarrow$ (iii) by Corollary 2.9. ∎

REMARK. The proof of (i) $\Rightarrow$ (iii) shows that $A^{-1} = F_1F_2\cdots F_t$ is the matrix obtained by performing on I_n the same sequence of elementary column operations used to change A to I_n. As a rule this is a more convenient way of computing inverses than the use of determinants (Section 3).

APPENDIX: ABELIAN GROUPS DEFINED BY GENERATORS AND RELATIONS

An abelian group G is said to be the **abelian group defined by the generators** $a_1,\ldots,a_m$ ($a_i \in G$) **and the relations**

$$\begin{aligned} r_{11}a_1 + r_{12}a_2 + \cdots + r_{1m}a_m &= 0, \\ r_{21}a_1 + r_{22}a_2 + \cdots + r_{2m}a_m &= 0, \\ &\;\;\vdots \\ r_{n1}a_1 + r_{n2}a_2 + \cdots + r_{nm}a_m &= 0, \end{aligned}$$

($r_{ij} \in \mathbf{Z}$) provided that $G \cong F/K$, where F is the free abelian group on the set $\{a_1, \ldots, a_m\}$ and K is the subgroup of F generated by $b_1 = r_{11}a_1 + \cdots + r_{1m}a_m$, $b_2 = r_{21}a_1 + \cdots + r_{2m}a_m, \ldots, b_n = r_{n1}a_1 + \cdots + r_{nm}a_m$. Note that the same symbol a_i denotes both an element of the group G and a basis element of the free abelian group F (see Theorem II.1.1). This definition is consistent with the concept of generators and relations discussed in Section I.9 (see Exercise 10).

The basic problem is to determine the structure of the abelian group G defined by a given finite set of generators and relations. Since G is finitely generated, G is necessarily a direct sum of cyclic groups (Theorem II.2.1). We shall now determine the orders of these cyclic summands.

Let G be the group defined by generators $a_1, \ldots, a_m$ and relations $\sum_j r_{ij}a_j = 0$ as above. We shall denote this situation by the $n \times m$ matrix $A = (r_{ij})$. The rows of A represent the generators $b_1, \ldots, b_n$ of the subgroup K relative to the ordered basis $\{a_1, \ldots, a_m\}$ of F. We claim that elementary row and column operations performed on A have the following effect.

(i) If $B = (s_{ij})$ is obtained from A by an elementary row operation, then the elements $c_1 = s_{11}a_1 + \cdots + s_{1m}a_m, \ldots, c_n = s_{n1}a_1 + \cdots + s_{nm}a_m$ of F (that is, the rows of B) generate the subgroup K. (Exercise 11 (a)).

(ii) If $B = (s_{ij})$ is obtained from A by an elementary column operation, then there is an easily determined basis $\{a_1', \ldots, a_m'\}$ of F such that $b_i = s_{i1}a_1' + s_{i2}a_2' + \cdots + s_{im}a_m'$ for every i (Exercise 11 (b), (c)).

If $K \neq 0$, then by Proposition 2.11 and Exercise 7, A may be changed via a finite sequence of elementary row and column operations, to a diagonal matrix

$$\begin{pmatrix} d_1 & & & & & & \\ & \ddots & & & 0 & & \\ & & d_r & & & & \\ & & & & 0 & & \\ & 0 & & & 0 & & \\ & & & & & \ddots & \\ & & & & & & \end{pmatrix}$$

such that $d_i \neq 0$ for all i and $d_1 \mid d_2 \mid d_3 \mid \cdots \mid d_r$. In other words a finite sequence of elementary operations yields a basis $\{u_1, \ldots, u_m\}$ of F such that $\{d_1u_1, d_2u_2, \ldots, d_ru_r\}$ generates K. Consequently by Corollary I.8.11

$$\begin{aligned} G \cong F/K &\cong (\mathbf{Z}u_1 \oplus \cdots \oplus \mathbf{Z}u_m)/(\mathbf{Z}d_1u_1 \oplus \cdots \oplus \mathbf{Z}d_ru_r \oplus 0 \oplus \cdots \oplus 0) \\ &\cong \mathbf{Z}/d_1\mathbf{Z} \oplus \cdots \oplus \mathbf{Z}/d_r\mathbf{Z} \oplus \mathbf{Z}/0 \oplus \cdots \oplus \mathbf{Z}/0 \\ &\cong Z_{d_1} \oplus \cdots \oplus Z_{d_r} \oplus \mathbf{Z} \oplus \cdots \oplus \mathbf{Z}, \end{aligned}$$

where the rank of $(\mathbf{Z} \oplus \cdots \oplus \mathbf{Z})$ is $m - r$ and $d_1 \mid d_2 \mid \cdots \mid d_r$ (see Theorem II.2.6).

EXAMPLE. Determine the structure of the abelian group G defined by generators a,b,c and relations $3a + 9b + 9c = 0$ and $9a - 3b + 9c = 0$. Let F be the free abelian group $\mathbf{Z}a + \mathbf{Z}b + \mathbf{Z}c$ and K the subgroup generated by $b_1 = 3a + 9b + 9c$ and $b_2 = 9a - 3b + 9c$. Then G is isomorphic to F/K and we have the matrix

$$A = \begin{pmatrix} 3 & 9 & 9 \\ 9 & -3 & 9 \end{pmatrix}.$$

We indicate below the various stages in the diagonalization of the matrix A by elementary operations; (sometimes several operations are performed in a single step). At each stage we indicate the basis of F and the generators of K represented by the given matrix; (this can be tricky; see Exercise 11).

Matrix	Ordered basis of F	Generators of K, expressed as linear combinations of this basis
$\begin{pmatrix} 3 & 9 & 9 \\ 9 & -3 & 9 \end{pmatrix}$	$a; b; c$	$b_1 = 3a + 9b + 9c$ $b_2 = 9a - 3b + 9c$
$\begin{pmatrix} 3 & 0 & 9 \\ 9 & -30 & 9 \end{pmatrix}$	$a + 3b; b; c$	$b_1 = 3(a + 3b) + 9c$ $b_2 = 9(a + 3b) - 30b + 9c$
$\begin{pmatrix} 3 & 0 & 0 \\ 9 & -30 & -18 \end{pmatrix}$	$a + 3b + 3c; b; c$	$b_1 = 3(a + 3b + 3c)$ $b_2 = 9(a + 3b + 3c) - 30b - 18c$
$\begin{pmatrix} 3 & 0 & 0 \\ 0 & -30 & -18 \end{pmatrix}$	$a + 3b + 3c; b; c$	$b_1 = 3(a + 3b + 3c)$ $b_2 - 3b_1 = -30b - 18c$
$\begin{pmatrix} 3 & 0 & 0 \\ 0 & 18 & 30 \end{pmatrix}$	$a + 3b + 3c; c; b$	$b_1 = 3(a + 3b + 3c)$ $-(b_2 - 3b_1) = 18c + 30b$
$\begin{pmatrix} 3 & 0 & 0 \\ 0 & 18 & 12 \end{pmatrix}$	$a + 3b + 3c; c + b; b$	$b_1 = 3(a + 3b + 3c)$ $-b_2 + 3b_1 = 18(c + b) + 12b$
$\begin{pmatrix} 3 & 0 & 0 \\ 0 & 6 & 12 \end{pmatrix}$	$a + 3b + 3c; c + b;$ $b + (c + b)$	$b_1 = 3(a + 3b + 3c)$ $-b_2 + 3b_1 = 6(c + b) + 12(2b + c)$
$\begin{pmatrix} 3 & 0 & 0 \\ 0 & 6 & 0 \end{pmatrix}$	$a + 3b + 3c; 5b + 3c;$ $2b + c$	$b_1 = 3(a + 3b + 3c)$ $-b_2 + 3b_1 = 6(5b + 3c)$

Therefore $G \cong F/K \cong \mathbf{Z}/3\mathbf{Z} \oplus \mathbf{Z}/6\mathbf{Z} \oplus \mathbf{Z}/0\mathbf{Z} \cong Z_3 \oplus Z_6 \oplus \mathbf{Z}$. If $\bar{v} \in G$ is the image of $v + K \in F/K$ under the isomorphism $F/K \cong G$, then G is the internal direct sum of a cyclic subgroup of order three with generator $\overline{a + 3b + 3c}$, a cyclic subgroup of order six with generator $\overline{5b + 3c}$, and an infinite cyclic subgroup with generator $\overline{2b + c}$.

EXERCISES

1. Let $f, g : E \to E$, $h : E \to F$, $k : F \to G$ be linear transformations of left vector spaces over a division ring D with $\dim_D E = n$, $\dim_D F = m$, $\dim_D G = p$.
   (a) Rank $(f + g) \leq$ rank $f +$ rank g.
   (b) Rank $(kh) \leq \min \{\text{rank } h, \text{rank } k\}$.
   (c) Nullity $kh \leq$ nullity $h +$ nullity k.
   (d) Rank $f +$ rank $g - n \leq$ rank $fg \leq \min \{\text{rank } f, \text{rank } g\}$.

(e) Max $\{\text{nullity } g, \text{nullity } h\} \leq \text{nullity } hg$.
(f) If $m \neq n$, then (e) is false for h and k.

2. An $n \times m$ matrix A over a division ring D has an $m \times n$ left inverse B (that is, $BA = I_m$) if and only if rank $A = m$. A has an $m \times n$ right inverse C (with $AC = I_n$) if and only if rank $A = n$.

3. If $(c_{i1}, c_{i2} \cdots c_{im})$ is a nonzero row of a matrix (c_{ij}), then its **leading entry** is c_{it} where t is the first integer such that $c_{it} \neq 0$. A matrix $C = (c_{ij})$ over a division ring D is said to be in **reduced row echelon form** provided: (i) for some $r \geq 0$ the first r rows of C are nonzero (row vectors) and all other rows are zero; (ii) the leading entry of each nonzero row is 1_D; (iii) if $c_{ij} = 1_D$ is the leading entry of row i, then $c_{kj} = 0$ for all $k \neq i$; (iv) if $c_{1j_1}, c_{2j_2}, \ldots, c_{rj_r}$ are the leading entries of rows $1,2,\ldots,r$, then $j_1 < j_2 < \cdots < j_r$.
(a) If C is in reduced row echelon form, then rank C is the number of nonzero rows.
(b) If A is any matrix over D, then A may be changed to a matrix in reduced row echelon form by a finite sequence of elementary *row* operations.

4. (a) The system of n linear equations in m unknowns x_i over a field K

$$\begin{aligned} a_{11}x_1 + a_{12}x_2 + \cdots + a_{1m}x_m &= b_1 \\ &\vdots \\ a_{n1}x_1 + a_{n2}x_2 + \cdots + a_{nm}x_m &= b_n \end{aligned}$$

has a (simultaneous) solution if and only if the matrix equation $AX = B$ has a solution X, where A is the $n \times m$ matrix (a_{ij}), X is the $m \times 1$ column vector $(x_1 x_2 \cdots x_m)^t$ and B is the $n \times 1$ column vector $(b_1 b_2 \cdots b_n)^t$.
(b) If A_1,B_1 are matrices obtained from A,B respectively by performing the same sequence of elementary *row* operations on both A_1 and B_1 then X is a solution of $AX = B$ if and only if X is a solution of $A_1X = B_1$.
(c) Let C be the $n \times (m + 1)$ matrix

$$\begin{pmatrix} a_{11} & \cdots & a_{1m} & b_1 \\ \vdots & & & \\ a_{n1} & \cdots & a_{nm} & b_n \end{pmatrix}.$$

Then $AX = B$ has solution if and only if rank A = rank C. In this case the solution is unique if and only if rank $A = m$. [*Hint:* use (b) and Exercise 3.]
(d) The system $AX = B$ is **homogeneous** if B is the zero column vector. A homogeneous system $AX = B$ has a nontrivial solution (that is, not all $x_i = 0$) if and only if rank $A < m$ (in particular, if $n < m$).

5. Let R be a principal ideal domain. For each positive integer r and sequence of nonzero ideals $I_1 \supset I_2 \supset \cdots \supset I_r$ choose a sequence $d_1, \ldots, d_r \in R$ such that $(d_i) = I_j$ and $d_1 \mid d_2 \mid \cdots \mid d_r$. For a given pair of positive integers (n,m), let S be the set of all $n \times m$ matrices of the form $\begin{pmatrix} L_r & 0 \\ 0 & 0 \end{pmatrix}$, where $r = 1,2,\ldots, \min(n,m)$ and L_r is an $r \times r$ diagonal matrix with main diagonal one of the chosen se-

quences $d_1, \dots, d_r$. Show that S is a set of canonical forms under equivalence for the set of all $n \times m$ matrices over R.

6. (a) If $f : E \to F$ is a linear transformation of finite dimensional vector spaces over a division ring, then there exist bases $\{u_1, \dots, u_n\}$ of E and $\{v_1, \dots, v_m\}$ of F and an integer r ($r \leq \min(m,n)$) such that $f(u_i) = v_i$ for $i = 1,2,\dots,r$ and $f(u_i) = 0$ for $i = r+1,\dots,n$.
(b) State and prove a similar result for free modules of finite rank over a principal ideal domain [see Proposition 2.11].

7. Let R be a Euclidean domain with "degree function" $\phi : R - \{0\} \to \mathbf{N}$ (Definition III.3.8). (For example, let $R = \mathbf{Z}$).
(a) If $A = \begin{pmatrix} a & b \\ c & d \end{pmatrix}$ is a 2×2 matrix over R then A can be changed to a diagonal matrix D by a finite sequence of elementary row and column operations. [*Hint:* If $a \neq 0$, $b \neq 0$, then $b = aq + r$ with $r = 0$, or $r \neq 0$ and $\phi(r) < \phi(a)$. Performing suitable elementary column operations yields:
$$\begin{pmatrix} a & b \\ c & d \end{pmatrix} \to \begin{pmatrix} a & b-aq \\ c & d-cq \end{pmatrix} = \begin{pmatrix} a & r \\ c & * \end{pmatrix} \to \begin{pmatrix} r & a \\ * & c \end{pmatrix}.$$
Since $\phi(r) < \phi(a)$, repetitions of this argument change A to $B = \begin{pmatrix} s & 0 \\ u & * \end{pmatrix}$ with $\phi(s) < \phi(a)$ if $s \neq 0$. If $u \neq 0$, a similar argument with rows changes B to $C = \begin{pmatrix} t & w \\ 0 & * \end{pmatrix}$ with $\phi(t) < \phi(s) < \phi(a)$ if $t \neq 0$; (and possibly $w \neq 0$). Since the degrees of the (1, 1) entries are strictly decreasing, a repetition of these arguments must yield a diagonal matrix $D = \begin{pmatrix} d_1 & 0 \\ 0 & d_2 \end{pmatrix}$ after a finite number of steps.]
(b) If A is invertible, then A is a product of elementary matrices. [*Hint:* By (a) and the proof of Corollary 2.10 $D = PAQ$ with P,Q invertible, whence D is invertible and d_1,d_2 are units in R. Thus $A = P^{-1}\begin{pmatrix} d_1 & 0 \\ 0 & 1_R \end{pmatrix}\begin{pmatrix} 1_R & 0 \\ 0 & d_2 \end{pmatrix}Q^{-1}$; use Corollary 2.9.]
(c) Every $n \times m$ secondary matrix (see the proof of Proposition 2.11) over a Euclidean domain is a product of elementary matrices.

8. (a) An invertible matrix over a principal ideal domain is a product of elementary and secondary matrices.
(b) An invertible matrix over a Euclidean domain is a product of elementary matrices [see Exercise 7].

9. Let $n_1, n_2, \dots, n_t, n$ be positive integers such that $n_1 + n_2 + \cdots + n_t = n$ and for each i let M_i be an $n_i \times n_i$ matrix. Let M be the $n \times n$ matrix
$$\begin{pmatrix} M_1 & & & & 0 \\ & M_2 & & & \\ & & \cdot & & \\ & & & \cdot & \\ 0 & & & & M_t \end{pmatrix},$$

where the main diagonal of each M_i lies on the main diagonal of M. For each permutation σ of $\{1,2,\ldots,t\}$, M is similar to the matrix

$$\sigma M = \begin{pmatrix} M_{\sigma 1} & & & & \\ & M_{\sigma 2} & & 0 & \\ & & \cdot & & \\ & 0 & & \cdot & \\ & & & & \cdot \\ & & & & M_{\sigma t} \end{pmatrix}.$$

[*Hint:* If $t = 3$, $\sigma = (13)$, and $P = \begin{pmatrix} 0 & & I_{n3} \\ & I_{n2} & \\ I_{n1} & & 0 \end{pmatrix}$, then $P^{-1} = \begin{pmatrix} 0 & & I_{n1} \\ & I_{n2} & \\ I_{n3} & & 0 \end{pmatrix}$ and $PMP^{-1} = \sigma M$. In the general case adapt the proof of results 2.8–2.10.]

10. Given the set $\{a_1,\ldots,a_n\}$ and the words $w_1,w_2,\ldots,w_r$ (on the a_i), let F^* be the free (nonabelian multiplicative) group on the set $\{a_1,\ldots,a_n\}$ and let M be the normal subgroup generated by the words $w_1,w_2,\ldots,w_r$ (see Section I.9). Let N be the normal subgroup generated by all words of the form $a_ia_ja_i^{-1}a_j^{-1}$.
   (a) F^*/M is the group defined by generators $\{a_1,\ldots,a_n\}$ and relations $\{w_1 = w_2 = \cdots = w_r = e\}$ (Definition I.9.4).
   (b) F^*/N is the free abelian group on $\{a_1,\ldots,a_n\}$ (see Exercise II.1.12).
   (c) $F^*/(M \vee N)$ is (in multiplicative notation) the abelian group defined by generators $\{a_1,\ldots,a_n\}$ and relations $\{w_1 = w_2 = \cdots = w_r = e\}$ (see p. 343).
   (d) There are group epimorphisms $F^* \to F^*/N \to F^*/(M \vee N)$.

11. Let F be a free abelian group with basis $\{a_1,\ldots,a_m\}$. Let K be the subgroup of F generated by $b_1 = r_{11}a_1 + \cdots + r_{1m}a_m, \ldots, b_n = r_{n1}a_1 + \cdots + r_{nm}a_m$ ($r_{ij} \in \mathbf{Z}$).
   (a) For each i, both $\{b_1,\ldots,b_{i-1},-b_i,b_{i+1},\ldots,b_n\}$ and $\{b_1,\ldots,b_{i-1},b_i + rb_j, b_{i+1},\ldots,b_n\}$ ($r \in \mathbf{Z}$; $i \neq j$) generate K. [See Lemma II.1.5.]
   (b) For each i $\{a_1,\ldots,a_{i-1},-a_i,a_{i+1},\ldots,a_n\}$ is a basis of F relative to which $b_j = r_{j1}a_1 + \cdots + r_{j,i-1}a_{i-1} - r_{ji}(-a_i) + r_{j,i+1}a_{i+1} + \cdots + r_{jm}a_m$.
   (c) For each i and $j \neq i$ $\{a_1,\ldots,a_{j-1},a_j - ra_i,a_{j+1},\ldots,a_m\}$ ($r \in \mathbf{Z}$) is a basis of F relative to which $b_k = r_{k1}a_1 + \cdots + r_{k,i-1}a_{i-1} + (r_{ki} + rr_{kj})a_i + r_{k,i+1}a_{i+1} + \cdots + r_{k,j-1}a_{j-1} + r_{kj}(a_j - ra_i) + r_{k,j+1}a_{j+1} + \cdots + r_{km}a_m$.

12. Determine the structure of the abelian group G defined by generators $\{a,b\}$ and relations $2a + 4b = 0$ and $3b = 0$. Do the same for the group with generators $\{a,b,c,d\}$ and relations $2a + 3b = 4a = 5c + 11d = 0$ and for the group with generators $\{a,b,c,d,e\}$ and relations $\{a - 7b + 14d - 21e = 0; 5a - 7b - 2c + 10d - 15e = 0; 3a - 3b - 2c + 6d - 9e = 0; a - b + 2d - 3e = 0\}$.

3. DETERMINANTS

The determinant function $\mathrm{Mat}_n R \to R$ is defined as a particular kind of R-multilinear function and its elementary properties are developed (Theorem 3.5). The remainder of the section is devoted to techniques for calculating determinants and the connection between determinants and invertibility. With minor exceptions this material is not needed in the sequel. Throughout this section *all rings are commutative with identity and all modules are unitary.*

If B is an R-module and $n \geq 1$ an integer, B^n will denote the R-module $B \oplus B \oplus \cdots \oplus B$ (n summands). Of course, the underlying set of the module B^n is just the cartesian product $B \times \cdots \times B$.

Definition 3.1. *Let* $B_1, \ldots, B_n$ *and* C *be modules over a commutative ring* R *with identity. A function* $f: B_1 \times \cdots \times B_n \to C$ *is said to be* **R-multilinear** *if for each* $i = 1,2, \ldots, n$ *and all* $r,s \in R$, $b_j \in B_j$ *and* $b,b' \in B_i$:

$$f(b_1, \ldots, b_{i-1}, rb + sb', b_{i+1}, \ldots, b_n) = rf(b_1, \ldots, b_{i-1}, b, b_{i+1}, \ldots, b_n) + sf(b_1, \ldots, b_{i-1}, b', b_{i+1}, \ldots, b_n).$$

If $C = R$, *then* f *is called an* **n-linear** *or* **R-multilinear form.** *If* $C = R$ *and* $B_1 = B_2 = \cdots = B_n = B$, *then* f *is called an* **R-multilinear form on B.**

The 2-linear functions are usually called **bilinear** (see Theorem IV.5.6). Let B and C be R-modules and $f: B^n \to C$ an R-multilinear function. Then f is said to be **symmetric** if

$$f(b_{\sigma 1}, \ldots, b_{\sigma n}) = f(b_1, \ldots, b_n) \quad \text{for every permutation } \sigma \in S_n,$$

and **skew-symmetric** if

$$f(b_{\sigma 1}, \ldots, b_{\sigma n}) = (\operatorname{sgn} \sigma) f(b_1, \ldots, b_n) \quad \text{for every} \quad \sigma \in S_n.$$

f is said to be **alternating** if

$$f(b_1, \ldots, b_n) = 0 \quad \text{whenever} \quad b_i = b_j \quad \text{for some} \quad i \neq j.$$

EXAMPLE. Let B be the free R-module $R \oplus R$ and let $d: B \times B \to R$ be defined by $((a_{11},a_{12}),(a_{21},a_{22})) \mapsto a_{11}a_{22} - a_{12}a_{21}$. Then d is a skew-symmetric alternating bilinear form on B. If one thinks of the elements of B as rows of 2×2 matrices over R, then d is simply the ordinary determinant function.

Theorem 3.2. *If* B *and* C *are modules over a commutative ring* R *with identity, then every alternating* R*-multilinear function* $f: B^n \to C$ *is skew-symmetric.*

SKETCH OF PROOF. In the special case when $n = 2$ and $\sigma = (1\ 2)$, we have:

$$0 = f(b_1 + b_2, b_1 + b_2) = f(b_1,b_1) + f(b_1,b_2) + f(b_2,b_1) + f(b_2,b_2) = 0 + f(b_1,b_2) + f(b_2,b_1) + 0,$$

whence $f(b_2,b_1) = -f(b_1,b_2) = (\operatorname{sgn} \sigma) f(b_1,b_2)$. In the general case, show that it suffices to assume σ is a transposition. Then the proof is an easy generalization of the case $n = 2$. ■

Our chief interest is in alternating n-linear forms on the free R-module R^n. Such a form is a function from $(R^n)^n = R^n \oplus \cdots \oplus R^n$ (n summands) to R.

Theorem 3.3. *If* R *is a commutative ring with identity and* r ε R, *then there exists a unique alternating* R*-multilinear form* $f : (R^n)^n \to R$ *such that* $f(\varepsilon_1,\varepsilon_2,\ldots,\varepsilon_n) = r$, *where* $\{\varepsilon_1,\ldots,\varepsilon_n\}$ *is the standard basis of* R^n.

REMARK. The standard basis is defined after Definition 2.1. The following facts may be helpful in understanding the proof. Since the elements of R^n may be identified with $1 \times n$ row vectors, it is clear that there is an R-module isomorphism $(R^n)^n \cong \mathrm{Mat}_n R$ given by $(X_1,X_2,\ldots,X_n) \mapsto A$, where A is the matrix with rows $X_1,X_2,\ldots,X_n$. If $\{\varepsilon_1,\ldots,\varepsilon_n\}$ is the standard basis of R^n, then $(\varepsilon_1,\varepsilon_2,\ldots,\varepsilon_n) \mapsto I_n$ under this isomorphism. Thus the multilinear form f of Theorem 3.3 may be thought of as a function whose n arguments are the rows of $n \times n$ matrices.

PROOF OF 3.3. (Uniqueness) If such an alternating n-linear form f exists and if $(X_1,\ldots,X_n) \varepsilon (R^n)^n$, then for each i there exist $a_{ij} \varepsilon R$ such that $X_i = (a_{i1},a_{i2},\ldots,a_{in}) = \sum_{j=1}^{n} a_{ij}\varepsilon_j$. (In other words, under the isomorphism $(R^n)^n \cong \mathrm{Mat}_n R$, $(X_1,\ldots,X_n) \mapsto (a_{ij})$.) Therefore by multilinearity,

$$\begin{aligned} f(X_1,\ldots,X_n) &= f(\sum_{j_1} a_{1j_1}\varepsilon_{j_1}, \sum_{j_2} a_{2j_2}\varepsilon_{j_2},\ldots,\sum_{j_n} a_{nj_n}\varepsilon_{j_n}) \\ &= \sum_{j_1}\sum_{j_2}\cdots\sum_{j_n} a_{1j_1}a_{2j_2}\cdots a_{nj_n} f(\varepsilon_{j_1},\varepsilon_{j_2},\ldots,\varepsilon_{j_n}). \end{aligned}$$

Since f is alternating the only possible nonzero terms in the final sum are those where $j_1,j_2,\ldots,j_n$ are all distinct; that is, $\{j_1,\ldots,j_n\}$ is simply the set $\{1,2,\ldots,n\}$ in some order, so that for some permutation $\sigma \varepsilon S_n$, $(j_1,\ldots,j_n) = (\sigma 1,\ldots,\sigma n)$. Consequently by Theorem 3.2,

$$\begin{aligned} f(X_1,\ldots,X_n) &= \sum_{\sigma\varepsilon S_n} a_{1\sigma 1}a_{2\sigma 2}\cdots a_{n\sigma n} f(\varepsilon_{\sigma 1},\varepsilon_{\sigma 2},\ldots,\varepsilon_{\sigma n}) \\ &= \sum_{\sigma\varepsilon S_n} (\mathrm{sgn}\,\sigma)a_{1\sigma 1}\cdots a_{n\sigma n} f(\varepsilon_1,\varepsilon_2,\ldots,\varepsilon_n). \end{aligned}$$

Since $f(\varepsilon_1,\ldots,\varepsilon_n) = r$, we have

$$f(X_1,\ldots,X_n) = \sum_{\sigma\varepsilon S_n} (\mathrm{sgn}\,\sigma) r a_{1\sigma 1}a_{2\sigma 2}\cdots a_{n\sigma n}. \tag{1}$$

Equation (1) shows that $f(X_1,\ldots,X_n)$ is uniquely determined by $X_1,\ldots,X_n$ and r.

(Existence) It suffices to *define* a function $f : (R^n)^n \to R$ by formula (1) (where $X_i = (a_{i1},\ldots,a_{in})$) and verify that f is an alternating n-linear form with $f(\varepsilon_1,\ldots,\varepsilon_n) = r$. Since for each fixed k every summand of $\sum_{\sigma\varepsilon S_n} (\mathrm{sgn}\,\sigma) r a_{1\sigma 1}\cdots a_{n\sigma n}$ contains exactly one factor a_{ij} with $i = k$, it follows easily that f is R-multilinear. Since $\varepsilon_i = \sum_{j=1}^{n} \delta_{ij}\varepsilon_j$ (Kronecker delta), $f(\varepsilon_1,\ldots,\varepsilon_n) = r$. Finally we must show that $f(X_1,\ldots,X_n) = 0$ if $X_i = X_j$ and $i \neq j$. Assume for convenience of notation that $i = 1, j = 2$. If $\rho = (12)$, then the map $A_n \to S_n$ given by $\sigma \mapsto \sigma\rho$ is an injective function whose image is the set of all odd permutations (since σ even implies $\sigma\rho$ odd and $|A_n| = |S_n|/2$). Thus S_n is a union of mutually disjoint pairs $\{\sigma,\sigma\rho\}$ with $\sigma \varepsilon A_n$. If σ is even, then the summand of $f(X_1,X_1,X_3,\ldots,X_n)$ corresponding to σ is

$$+\, r a_{1\sigma 1}a_{2\sigma 2}a_{3\sigma 3}\cdots a_{n\sigma n}.$$

Since $X_1 = X_2$, $a_{1\sigma 1} = a_{2\sigma 1}$, and $a_{2\sigma 2} = a_{1\sigma 2}$, whence the summand corresponding to the odd permutation $\sigma\rho$ is:

$$\begin{aligned} -ra_{1\sigma\rho 1}a_{2\sigma\rho 2}a_{3\sigma\rho 3}\cdots a_{n\sigma\rho n} &= -ra_{1\sigma 2}a_{2\sigma 1}a_{3\sigma 3}\cdots a_{n\sigma n} \\ &= -ra_{1\sigma 1}a_{2\sigma 2}a_{3\sigma 3}\cdots a_{n\sigma n}. \end{aligned}$$

Thus the summands of $f(X_1,X_1,X_3,\ldots,X_n)$ cancel pairwise and

$$f(X_1,X_1,X_3,\ldots,X_n) = 0.$$

Therefore f is alternating. ■

We can now use Theorem 3.3 and the Remark following it to define determinants. In particular, we shall frequently identify $\text{Mat}_n R$ and $(R^n)^n$ under the isomorphism (given in the Remark), which maps $(\varepsilon_1,\ldots,\varepsilon_n) \mapsto I_n$. Consequently, a **multilinear form on** $\mathbf{Mat}_n\mathbf{R}$ is an R-multilinear form on $(R^n)^n$ whose arguments are the rows of $n \times n$ matrices considered as elements of R^n.

Definition 3.4. *Let* R *be a commutative ring with identity. The unique alternating* R*-multilinear form* $d : \text{Mat}_n R \to R$ *such that* $d(I_n) = 1_R$ *is called the* **determinant function** *on* $\text{Mat}_n R$. *The* **determinant** *of a matrix* $A \in \text{Mat}_n R$ *is the element* $d(A) \in R$ *and is denoted* $|A|$.

Theorem 3.5. *Let* R *be a commutative ring with identity and* $A,B \in \text{Mat}_n R$.

(i) *Every alternating* R*-multilinear form* f *on* $\text{Mat}_n R$ *is a unique scalar multiple of the determinant function* d.

(ii) *If* $A = (a_{ij})$, *then* $|A| = \sum_{\sigma\in S_n} (\text{sgn}\,\sigma)a_{1\sigma 1}a_{2\sigma 2}\cdots a_{n\sigma n}$.

(iii) $|AB| = |A||B|$.

(iv) *If* A *is invertible in* $\text{Mat}_n R$, *then* $|A|$ *is a unit in* R.

(v) *If* A *and* B *are similar, then* $|A| = |B|$.

(vi) $|A^t| = |A|$.

(vii) *If* $A = (a_{ij})$ *is triangular, then* $|A| = a_{11}a_{22}\cdots a_{nn}$.

(viii) *If* B *is obtained by interchanging two rows [columns] of* A, *then* $|B| = -|A|$. *If* B *is obtained by multiplying one row [column] of* A *by* $r \in R$, *then* $|B| = r|A|$. *If* B *is obtained by adding a scalar multiple of row* i *[column* i*] to row* j *[column* j*]* ($i \neq j$), *then* $|B| = |A|$.

SKETCH OF PROOF. (i) Let $f(I_n) = r \in R$. Let d be the determinant function. Verify that the function $rd : \text{Mat}_n R \to R$ given by $A \mapsto r|A| = rd(A)$ is also an alternating R-multilinear form on $\text{Mat}_n R$ such that $rd(I_n) = r$, whence $f = rd$ by the uniqueness statement of Theorem 3.3. The uniqueness of r follows immediately.

(ii) is simply a restatement of equation (1) in the proof of Theorem 3.3. (iii) Let B be fixed and denote the *columns* of B by $Y_1,Y_2,\ldots,Y_n$. If C is any $n \times m$ matrix with *rows* $X_1,\ldots,X_n$, then the (i,j) entry of CB is precisely the element (1×1 matrix) X_iY_j. Thus the ith row of CB is $(X_iY_1,X_iY_2,\ldots,X_iY_n)$. Use this fact to verify that the map $\text{Mat}_n R \to R$ given by $C \mapsto |CB|$ is an alternating R-multilinear form f on $\text{Mat}_n R$. By (i) $f = rd$ for some $r \in R$. Consequently, $|CB| = f(C) = rd(C) = r|C|$. In particular, $|B| = |I_nB| = r|I_n| = r$, whence $|AB| = r|A| = |A||B|$.

(iv) $AA^{-1} = I_n$ implies $|A||A^{-1}| = |AA^{-1}| = |I_n| = 1$ by (iii). Hence $|A|$ is a unit in R with $|A|^{-1} = |A^{-1}|$. (v) Similarly, $B = PAP^{-1}$ implies $|B| = |P||A||P|^{-1} = |A|$ since R is commutative.

(vi) Let $A = (a_{ij})$. If $i_1, \ldots, i_n$ are the integers $1,2,\ldots,n$ in some order, then since R is commutative any product $a_{i_11}a_{i_22}\cdots a_{i_nn}$ may be written as $a_{1j_1}a_{2j_2}\cdots a_{nj_n}$. If σ is the permutation such that $\sigma(k) = i_k$, then σ^{-1} is the permutation such that $\sigma^{-1}(k) = j_k$. Furthermore, it is easy to see that for any $\sigma \varepsilon S_n$, sgn σ = sgn σ^{-1}. Let $A^t = (b_{ij})$; then since S_n is a group,

$$\begin{aligned} |A^t| &= \sum_{\sigma\varepsilon S_n} (\text{sgn } \sigma) b_{1\sigma1}\cdots b_{n\sigma n} = \sum_{\sigma\varepsilon S_n} (\text{sgn } \sigma) a_{\sigma11}\cdots a_{\sigma nn} \\ &= \sum_{\sigma^{-1}\varepsilon S_n} (\text{sgn } \sigma^{-1}) a_{1\sigma^{-1}1}\cdots a_{n\sigma^{-1}n} = |A|. \end{aligned}$$

(vii) By hypothesis either $a_{ij} = 0$ for all $j < i$ or $a_{ij} = 0$ for all $j > i$. In either case show that if $\sigma \varepsilon S_n$ and $\sigma \neq (1)$, then $a_{1\sigma1}\cdots a_{n\sigma n} = 0$, whence

$$|A| = \sum_{\sigma\varepsilon S_n} (\text{sgn } \sigma) a_{1\sigma1}\cdots a_{n\sigma n} = a_{11}a_{22}\cdots a_{nn}.$$

(viii) Let $X_1, \ldots, X_i, \ldots, X_j, \ldots, X_n$ be the rows of A. If B has rows $X_1, \ldots X_j, \ldots, X_i, \ldots, X_n$, then since d is skew-symmetric by Theorem 3.2,

$$\begin{aligned} |B| &= d(X_1, \ldots, X_j, \ldots, X_i, \ldots, X_n) \\ &= -d(X_1, \ldots, X_i, \ldots, X_j, \ldots, X_n) = -|A|. \end{aligned}$$

Similarly if B has rows $X_1, \ldots, X_i, \ldots, rX_i + X_j, \ldots, X_n$ then since d is multilinear and alternating

$$\begin{aligned} |B| &= d(X_1, \ldots, X_i, \ldots, rX_i + X_j, \ldots, X_n) \\ &= rd(X_1, \ldots, X_i, \ldots, X_i, \ldots, X_n) + d(X_1, \ldots, X_i, \ldots, X_j, \ldots, X_n) \\ &= r0 + |A| = |A|. \end{aligned}$$

The other statement is proved similarly; use (v) for the corresponding statements about columns. ■

If R is a field, then the last part of Theorem 3.5 provides a method of calculating $|A|$. Use elementary row and column operations to change A into a diagonal matrix $B = (b_{ij})$, keeping track at each stage (via (viii)) of what happens to $|A|$. By (viii), $|B| = r|A|$ for some $0 \neq r \varepsilon R$. Hence $r|A| = b_{11}b_{22}\cdots b_{nn}$ by (vii) and

$$|A| = r^{-1}b_{11}\cdots b_{nn}.$$

More generally the determinant of an $n \times n$ matrix A over any commutative ring with identity may be calculated as follows. For each pair (i,j) let A_{ij} be the $(n-1) \times (n-1)$ matrix obtained by deleting row i and column j from A. Then $|A_{ij}| \varepsilon R$ is called the **minor** of $A = (a_{ij})$ at position (i,j) and $(-1)^{i+j}|A_{ij}| \varepsilon R$ is called the **cofactor** of a_{ij}.

Proposition 3.6. *If* A *is an* n × n *matrix over a commutative ring* R *with identity, then for each* i = 1,2, . . . , n,

$$|A| = \sum_{j=1}^{n} (-1)^{i+j} a_{ij}|A_{ij}|$$

and for each j = 1,2, . . . , n,

$$|A| = \sum_{i=1}^{n} (-1)^{i+j} a_{ij} |A_{ij}|.$$

The first [second] formula for $|A|$ is called the **expansion of |A| along row i [column j]**.

PROOF OF 3.6. We let j be fixed and prove the second statement. By Theorem 3.3 and Definition 3.4 it suffices to show that the map $\phi : \text{Mat}_n R \to R$ given by $A = (a_{ij}) \mapsto \sum_{i=1}^{n} (-1)^{i+j} a_{ij} |A_{ij}|$ is an alternating R-multilinear form such that $\phi(I_n) = 1_R$. Let $X_1, \ldots, X_n$ be the rows of A. If $X_k = X_t$ with $1 \leq k < t \leq n$, then $|A_{ij}| = 0$ for $i \neq k,t$ since it is the determinant of a matrix with two identical rows. Since A_{kj} may be obtained from A_{tj} by interchanging row t successively with rows $t-1, \ldots, k+1$, $|A_{kj}| = (-1)^{t-k-1}|A_{tj}|$ by Theorem 3.5. Thus $\phi(A) = (-1)^{k+j}|A_{kj}| + (-1)^{t+j}|A_{tj}| = (-1)^{k+j+t-k-1}|A_{tj}| + (-1)^{t+j}|A_{tj}| = 0$. Hence ϕ is alternating. If for some k, $X_k = rY_k + sW_k$, let $B = (b_{ij})$ and $C = (c_{ij})$ be the matrices with rows $X_1, \ldots, X_{n-1}, Y_k, X_{k+1}, \ldots, X_n$, and $X_1, \ldots, X_{k-1}, W_k, X_{k+1}, \ldots, X_n$ respectively. To prove that ϕ is R-multilinear we need only show that $\phi(A) = r\phi(B) + s\phi(C)$. If $i = k$, then $|A_{kj}| = |B_{kj}| = |C_{kj}|$, whence $a_{kj}|A_{kj}| = (rb_{kj} + sc_{kj})|A_{kj}| = rb_{kj}|B_{kj}| + sc_{kj}|C_{kj}|$. If $i \neq k$, then since each $|A_{ij}|$ is a multilinear function of the rows of A_{ij} and $a_{ij} = b_{ij} = c_{ij}$ for $i \neq k$, we have $a_{ij}|A_{ij}| = a_{ij}(r|B_{ij}| + s|C_{ij}|) = rb_{ij}|B_{ij}| + sc_{ij}|C_{ij}|$. It follows that $\phi(A) = r\phi(B) + s\phi(C)$; hence ϕ is R-multilinear. Obviously $\phi(I_n) = 1_R$. Therefore, ϕ is the determinant function. The first statement of the theorem follows readily through the use of transposes. ■

Proposition 3.7. *If* A = (a$_{ij}$) *is an* n × n *matrix over a commutative ring* R *with identity and* $A^a = (b_{ij})$ *is the* n × n *matrix with* $b_{ij} = (-1)^{i+j}|A_{ji}|$, *then* $AA^a = |A|I_n = A^aA$. *Furthermore* A *is invertible in* Mat_nR *if and only if* |A| *is a unit in* R, *in which case* $A^{-1} = |A|^{-1}A^a$.

The matrix A^a is called the **classical adjoint** of A. Note that if R is a field, then $|A|$ is a unit if and only if $|A| \neq 0$.

PROOF OF 3.7. The (i,j) entry of AA^a is $c_{ij} = \sum_{k=1}^{n} (-1)^{j+k} a_{ik}|A_{jk}|$. If $i = j$, then $c_{ii} = |A|$ by Proposition 3.6. If $i \neq j$ (say $i < j$) and A has rows $X_1, \ldots, X_n$, let $B = (b_{ij})$ be the matrix with rows $X_1, \ldots, X_i, \ldots, X_{j-1}, X_i, X_{j+1}, \ldots, X_n$. Then $b_{ik} = a_{ik} = b_{jk}$ and $|A_{jk}| = |B_{jk}|$ for all k; in particular, $|B| = 0$ since the determinant is an alternating form. Hence

$$c_{ij} = \sum_{k=1}^{n} (-1)^{j+k} a_{ik}|A_{jk}| = \sum_{k=1}^{n} (-1)^{j+k} b_{jk}|B_{jk}| = |B| = 0.$$

Therefore, $c_{ij} = \delta_{ij}|A|$ (Kronecker delta) and $AA^a = |A|I_n$. In particular, the last statement holds with A^t in place of A : $A^t(A^t)^a = |A^t|I_n$. Since $(A^a)^t = (A^t)^a$, we have $|A|I_n = |A^t|I_n = A^t(A^t)^a = A^t(A^a)^t = (A^aA)^t$, whence $A^aA = (|A|I_n)^t = |A|I_n$. Thus if $|A|$ is a unit in R, $|A|^{-1}A^a \in \text{Mat}_n R$ and clearly $(|A|^{-1}A^a)A = I_n = A(|A|^{-1}A^a)$. Hence A is invertible with (necessarily unique) inverse $A^{-1} = |A|^{-1}A^a$. Conversely if A is invertible, then $|A|$ is a unit by Theorem 3.5. ■

Corollary 3.8. (*Cramer's Rule*) *Let* $A = (a_{ij})$ *be the matrix of coefficients of the system of* n *linear equations in* n *unknowns*

$$\begin{array}{ccc} a_{11}x_1 + a_{12}x_2 + \cdots + a_{1n}x_n & = & b_1 \\ \cdot & & \cdot \\ \cdot & & \cdot \\ \cdot & & \cdot \\ a_{n1}x_1 + a_{n2}x_2 + \cdots + a_{nn}x_n & = & b_n \end{array}$$

over a field K. *If* $|A| \neq 0$, *then the system has a unique solution which is given by:*

$$x_j = |A|^{-1}\left(\sum_{i=1}^{n} (-1)^{i+j} b_i |A_{ij}|\right) \qquad j = 1,2,\ldots,n.$$

PROOF. Clearly the given system has a solution if and only if the matrix equation $AX = B$ has a solution, where X and B are the column vectors $X = (x_1 \cdots x_n)^t$, $B = (b_1 \cdots b_n)^t$. Since $|A| \neq 0$, A is invertible by Proposition 3.7, whence $X = A^{-1}B$ is a solution. It is the unique solution since $AY = B$ implies $Y = A^{-1}B$. To obtain the formula for x_j simply compute, using the equation

$$X = A^{-1}B = (|A|^{-1}A^a)B = |A|^{-1}(A^aB). \quad \blacksquare$$

EXERCISES

Note: Unless stated otherwise all matrices have entries in a commutative ring R with identity.

1. If $r + r \neq 0$ for all nonzero $r \in R$, then prove that an n-linear form $B^n \to R$ is alternating if and only if it is skew-symmetric. What if char $R = 2$?

2. (a) If $m > n$, then every alternating R-multilinear form on $(R^n)^m$ is zero.
   (b) If $m < n$, then there is a nonzero alternating R-multilinear form on $(R^n)^m$.

3. Use Exercise 2 to prove directly that if there is an R-module isomorphism $R^m \cong R^n$, then $m = n$.

4. If $A \in \mathrm{Mat}_n R$, then $|A^a| = |A|^{n-1}$ and $(A^a)^a = |A|^{n-2}A$.

5. If R is a field and $A,B \in \mathrm{Mat}_n R$ are invertible then the matrix $A + rB$ is invertible for all but a finite number of $r \in R$.

6. Let A be an $n \times n$ matrix over a field. Without using Proposition 3.7 prove that A is invertible if and only if $|A| \neq 0$. [*Hint:* Theorems 2.6 and 3.5 (viii) and Proposition 2.12.]

7. Let F be a free R-module with basis $U = \{u_1, \ldots, u_n\}$. If $\phi : F \to F$ is an R-module endomorphism with matrix A relative to U, then the **determinant of the endomorphism** ϕ is defined to be $|A| \in R$ and is denoted $|\phi|$.
   (a) $|\phi|$ is independent of the choice of U.
   (b) $|\phi|$ is the unique element of R such that $f(\phi(b_1),\phi(b_2),\ldots,\phi(b_n)) = |\phi| f(b_1,\ldots,b_n)$ for every alternating R-multilinear form on F^n and all $b_i \in F$.

8. Suppose that $(b_1, \ldots, b_n)$ is a solution of the system of homogeneous linear equations

$$a_{11}x_1 + \cdots + a_{1n}x_n = 0$$
$$\vdots$$
$$a_{n1}x_1 + \cdots + a_{nn}x_n = 0$$

and that $A = (a_{ij})$ is the $n \times n$ matrix of coefficients. Then $|A|b_i = 0$ for every i. [*Hint:* If B_i is the $n \times n$ diagonal matrix with diagonal entries $1_R, \ldots, 1_R, b_i, 1_R, \ldots, 1_R$, then $|AB_i| = |A|b_i$. To show that $|AB_i| = 0$ add b_j times column j of AB_i to column i for every $j \neq i$. The resulting matrix has determinant $|AB_i|$ and (k,i) entry $a_{k1}b_1 + a_{k2}b_2 + \cdots + a_{kn}b_n = 0$ for $k = 1,2,\ldots,n$.]

4. DECOMPOSITION OF A SINGLE LINEAR TRANSFORMATION AND SIMILARITY

The structure of a finite dimensional vector space E over a field K relative to a linear transformation $E \to E$ is investigated. The linear transformation induces a decomposition of E as a direct sum of certain subspaces and associates with each such decomposition of E a set of polynomial invariants in $K[x]$ (Theorem 4.2). These sets of polynomial invariants enable one to choose various bases of E relative to each of which the matrix of the given linear transformation is of a certain type (Theorem 4.6). This leads to several different sets of canonical forms for the relation of similarity in $\mathrm{Mat}_n K$ (Corollary 4.7).

Note. The results of this section depend heavily on the structure theorems for finitely generated modules over a principal ideal domain (Section IV.6).

Let K be a field and $\phi: E \to E$ a linear transformation of an n-dimensional K-vector space E. We first recall some facts about the structure of $\mathrm{Hom}_K(E,E)$ and $\mathrm{Mat}_n K$. $\mathrm{Hom}_K(E,E)$ is not only a ring with identity (Exercise IV.1.7), but also a vector space over K with $(k\psi)(u) = k\psi(u)$ $(k \in K, u \in E, \psi \in \mathrm{Hom}_K(E,E))$; see the Remark after Theorem IV.4.8). Therefore if $f = \sum k_i x^i$ is a polynomial in $K[x]$, then $f(\phi) = \sum k_i \phi^i$ is a well-defined element of $\mathrm{Hom}_K(E,E)$ (where $\phi^0 = 1_E$ as usual). Similarly the ring $\mathrm{Mat}_n K$ is also a vector space over K. If $A \in \mathrm{Mat}_n K$, then $f(A) = \sum k_i A^i$ is a well-defined $n \times n$ matrix over K (with $A^0 = I_n$).

Theorem 4.1. *Let* E *be an* n*-dimensional vector space over a field* K, $\phi: \mathrm{E} \to \mathrm{E}$ *a linear transformation and* A *an* $\mathrm{n} \times \mathrm{n}$ *matrix over* K.

(i) *There exists a unique monic polynomial of positive degree,* $\mathrm{q}_\phi \in \mathrm{K[x]}$, *such that* $\mathrm{q}_\phi(\phi) = 0$ *and* $\mathrm{q}_\phi \mid \mathrm{f}$ *for all* $\mathrm{f} \in \mathrm{K[x]}$ *such that* $\mathrm{f}(\phi) = 0$.

(ii) *There exists a unique monic polynomial of positive degree,* $\mathrm{q_A} \in \mathrm{K[x]}$, *such that* $\mathrm{q_A(A)} = 0$ *and* $\mathrm{q_A} \mid \mathrm{f}$ *for all* $\mathrm{f} \in \mathrm{K[x]}$ *such that* $\mathrm{f(A)} = 0$.

(iii) *If* A *is the matrix of* ϕ *relative to some basis of* E, *then* $\mathrm{q_A} = \mathrm{q}_\phi$.

PROOF. (i) By Theorem III.5.5 there is a unique (nonzero) ring homomorphism $\zeta = \zeta_\phi : K[x] \to \mathrm{Hom}_K(E,E)$ such that $x \mapsto \phi$ and $k \mapsto k1_E$ for all $k \in K$. Consequently, if $f \in K[x]$, then $\zeta(f) = f(\phi)$. ζ is easily seen to be a linear transformation of K-vector spaces. Since $\dim_K E$ is finite, $\mathrm{Hom}_K(E,E)$ is finite dimensional over K by Theorems IV.2.1, IV. 2.4, IV.4.7, and IV.4.9. Thus Im ζ is necessarily finite dimensional over K. Since $K[x]$ is infinite dimensional over K, we must have $\operatorname{Ker} \zeta \neq 0$ by Corollary IV.2.14. Since $K[x]$ is a principal ideal domain whose units are precisely the nonzero elements of K (Corollary III.6.4), $\operatorname{Ker} \zeta = (q)$ for some monic $q \in K[x]$. Since ζ is not the zero map, $(q) \neq K[x]$, whence $\deg q \geq 1$. If $\operatorname{Ker} \zeta = (q_1)$ with $q_1 \in K[x]$ monic, then $q \mid q_1$ and $q_1 \mid q$ by Theorem III.3.2, whence $q = q_1$ since both are monic. Therefore $q_\phi = q$ has the stated properties.

(ii) The proof is the same as (i) with A in place of ϕ and $\operatorname{Mat}_n K$ in place of $\operatorname{Hom}_K(E,E)$. $q_A \in K[x]$ is the unique monic polynomial such that $(q_A) = \operatorname{Ker} \zeta_A$, where $\zeta_A : K[x] \to \operatorname{Mat}_n K$ is the unique ring homomorphism given by $f \mapsto f(A)$.

(iii) Let A be the matrix of ϕ relative to a basis U of E and let $\theta : \operatorname{Hom}_K(E,E) \cong \operatorname{Mat}_n R$ be the isomorphism of Theorem 1.2, so that $\theta(\phi) = A$. Then the diagram

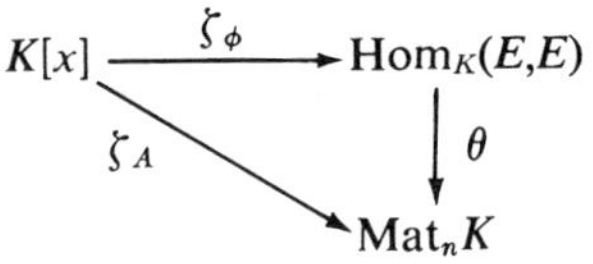

is commutative by Theorem III.5.5 since $\theta\zeta_\phi(x) = \theta(\phi) = A = \zeta_A(x)$ and $\theta\zeta_\phi(k) = \theta(k1_E) = kI_n = \zeta_A(k)$ for all $k \in K$. Since θ is an isomorphism, $(q_\phi) = \operatorname{Ker} \zeta_\phi = \operatorname{Ker} \theta\zeta_\phi = \operatorname{Ker} \zeta_A = (q_A)$. Therefore, $q_\phi \mid q_A$ and $q_A \mid q_\phi$, whence $q_\phi = q_A$ since both are monic. ∎

If K, E, and ϕ are as in Theorem 4.1, then the polynomial q_ϕ [resp. q_A] is called the **minimal polynomial** of the linear transformation ϕ [matrix A]. In general, q_ϕ is *not* irreducible. Corollary 1.7 and Theorem 4.1(iii) immediately imply that similar matrices have the same minimal polynomial.

Let K, E, and ϕ be as above. Then ϕ induces a (left) $K[x]$-module structure on E as follows. If $f \in K[x]$ and $u \in E$, then $f(\phi) \in \operatorname{Hom}_K(E,E)$ and fu is defined by $fu = f(\phi)(u)$. A K-subspace F of E is said to be **invariant** under ϕ (or **ϕ-invariant**) if $\phi(F) \subset F$. Clearly F is a ϕ-invariant K-subspace if and only if F is a $K[x]$-submodule of E. In particular, for any $v \in E$ the subspace $E(\phi,v)$ spanned by the set $\{\phi^i(v) \mid i \geq 0\}$ is ϕ-invariant. It is easy to see that $E(\phi,v)$ is precisely the cyclic $K[x]$-submodule $K[x]v$ generated by v. $E(\phi,v)$ is said to be a **ϕ-cyclic (sub)space.**

Theorem 4.2. *Let $\phi : E \to E$ be a linear transformation of an n-dimensional vector space E over a field K.*

(i) *There exist monic polynomials of positive degree $q_1, q_2, \dots, q_t \in K[x]$ and ϕ-cyclic subspaces $E_1, \dots, E_t$ of E such that $E = E_1 \oplus E_2 \oplus \cdots \oplus E_t$ and $q_1 \mid q_2 \mid \cdots \mid q_t$. Furthermore q_i is the minimal polynomial of $\phi \mid E_i : E_i \to E_i$. The sequence $(q_1, \dots, q_t)$ is uniquely determined by E and ϕ and q_t is the minimal polynomial of ϕ.*

(ii) *There exist monic irreducible polynomials $p_1, \dots, p_s \in K[x]$ and ϕ-cyclic subspaces $E_{11}, \dots, E_{1k_1}, E_{21}, \dots, E_{2k_2}, E_{31}, \dots, E_{sk_s}$ of E such that $E = \sum_{i=1}^{s}\sum_{j=1}^{k_i} E_{ij}$ and for each i there is a nonincreasing sequence of integers $m_{i1} \geq m_{i2} \geq \cdots \geq m_{ik_i} \geq 0$ such that $p_i^{m_{ij}}$ is the minimal polynomial of $\phi \mid E_{ij} : E_{ij} \to E_{ij}$. The family of poly-*

nomials $\{p_i^{m_{ij}} \mid 1 \leq i \leq s; 1 \leq j \leq k_i\}$ *is uniquely determined by* E *and* ϕ *and* $p_1^{m_{11}}p_2^{m_{21}}\cdots p_s^{m_{s1}}$ *is the minimal polynomial of* ϕ.

The polynomials $q_1, \ldots, q_t$ in part (i) of the theorem are called the **invariant factors of the linear transformation** ϕ. The prime power polynomials $p_i^{m_{ij}}$ in part (ii) are called the **elementary divisors** of ϕ.

SKETCH OF PROOF OF 4.2. (i) As indicated above E is a left module over the principal ideal domain $K[x]$ with $fu = f(\phi)(u)$ ($f \varepsilon K[x]$, $u \varepsilon E$). Since E is finite dimensional over K and $K \subset K[x]$, E is necessarily a finitely generated nonzero $K[x]$-module. If q_ϕ is the minimal polynomial of ϕ, then $q_\phi \neq 0$ and $q_\phi E = 0$, whence E is a torsion $K[x]$-module. By Theorem IV.6.12(i) E is the internal direct sum $E = E_1 \oplus \cdots \oplus E_t$, where each E_i is a nonzero cyclic $K[x]$-module of order q_i ($q_i \varepsilon K[x]$) and $q_1 \mid q_2 \mid \cdots \mid q_t$. By the remarks preceding the theorem each E_i is a ϕ-cyclic subspace. Since E_i has order q_i, there is a $K[x]$-module isomorphism $E_i \cong K[x]/(q_i)$ by Theorem IV.6.4 and the example following it. Since $E_i \neq 0$ and every nonzero ideal in $K[x]$ has a unique monic generator (Theorem III.3.2 and Corollary III.6.4), we may assume that each q_i is monic of positive degree. The uniqueness statement of Theorem IV.6.12(i) and the fact that $q_1 \mid q_2 \mid \cdots \mid q_t$ imply that $q_1, \ldots, q_t$ are uniquely determined by the $K[x]$-module E (that is, by E and ϕ). Use the $K[x]$-module structure of E_i and the fact that E_i is cyclic of order q_i to verify that the minimal polynomial of $\phi \mid E_i$ is q_i. Finally $q_tE = q_t(\phi)E_1 \oplus \cdots \oplus q_t(\phi)E_t = 0$, whence $(q_t) \subset (q_\phi)$. Since $q_\phi E = 0$, we have $q_\phi E_t = 0$, whence $(q_\phi) \subset (q_t)$. Consequently, $q_t = q_\phi$ since both are monic and $(q_t) = (q_\phi)$. The second part of the theorem is proved similarly by decomposing E as a direct sum of cyclic $K[x]$-submodules of prime power orders (Theorem IV.6.12(ii)). ■

REMARK. If $\phi = 0$, then the proof of Theorem 4.2 shows that the minimal polynomial of ϕ is x and its invariant factors [resp. elementary divisors] are $q_1 = x$, $q_2 = x, \ldots, q_n = x$. (Exercise 2).

The proof of Theorem 4.2 shows that the invariant factors and elementary divisors of a linear transformation $\phi : E \to E$ are simply the invariant factors and elementary divisors of the $K[x]$-module E. Consequently, one can obtain the elementary divisors from the invariant factors and vice versa just as in the proof of Theorem IV.6.12 (see also pp. 80–81). A technique for calculating the invariant factors of a specific linear transformation is discussed in Proposition 4.9 below.

EXAMPLE. Let $K = \mathbf{Q}$ and $\dim_K E = 15$ and suppose the invariant factors of ϕ are $q_1 = x^4 - x^2 - 2$, $q_2 = x^5 - x^3 - 2x$ and $q_3 = x^6 - x^4 - 2x^2$. Then $q_1 = (x^2 - 2)(x^2 + 1)$, $q_2 = xq_1$ and $q_3 = xq_2$, whence the elementary divisors of ϕ are: $x^2 - 2, x^2 + 1, x, x^2 - 2, x^2 + 1, x^2, x^2 - 2, x^2 + 1$. See the proof of Theorem IV.6.12 and also p. 80. Conversely if the elementary divisors of a linear transformation ψ are $x - 1, x - 1, x - 2, x - 3, (x - 2)^2, x^2 + 1, x^2 + 1, x^2 + 1$, and $(x - 1)^3$, then the invariant factors are $q_1 = (x - 1)(x^2 + 1)$, $q_2 = (x - 1)(x - 2)(x^2 + 1)$ and $q_3 = (x - 3)(x - 2)^2(x^2 + 1)(x - 1)^3$.

In view of Theorem 4.2 the next step in our analysis should be an investigation of ϕ-cyclic spaces.

Theorem 4.3. *Let* $\phi : E \to E$ *be a linear transformation of a finite dimensional vector space* E *over a field* K. *Then* E *is a* ϕ*-cyclic space and* ϕ *has minimal polynomial* $q = x^r + a_{r-1}x^{r-1} + \cdots + a_0 \in K[x]$ *if and only if* $\dim_K E = r$ *and* E *has an ordered basis* V *relative to which the matrix of* ϕ *is*

$$A = \begin{pmatrix} 0 & 1_K & 0 & 0 & 0 & \cdots & 0 & 0 \\ 0 & 0 & 1_K & 0 & 0 & \cdots & 0 & 0 \\ 0 & 0 & 0 & 1_K & 0 & \cdots & 0 & 0 \\ \cdot & & & & & & & \cdot \\ \cdot & & & & & & & \cdot \\ \cdot & & & & & & & \cdot \\ 0 & 0 & 0 & 0 & 0 & \cdots & 0 & 1_K \\ -a_0 & -a_1 & -a_2 & -a_3 & -a_4 & \cdots & -a_{r-2} & -a_{r-1} \end{pmatrix}.$$

In this case $V = \{v,\phi(v),\phi^2(v), \ldots, \phi^{r-1}(v)\}$ *for some* $v \in E$.

The matrix A is called the **companion matrix** of the monic polynomial $q \in K[x]$.[2] Note that if $q = x + a_0$, then $A = (-a_0)$.

PROOF OF 4.3. ($\Rightarrow$) If E is ϕ-cyclic, then the remarks preceding Theorem 4.2 show that for some $v \in E$, E is the cyclic $K[x]$-module $K[x]v$, with the $K[x]$-module structure induced by ϕ. If $k_0v + k_1\phi(v) + \cdots + k_{r-1}\phi^{r-1}(v) = 0$ $(k_i \in K)$, then $f = k_0 + k_1x + \cdots + k_{r-1}x^{r-1}$ is a polynomial such that $f(\phi)(v) = 0$, whence $f(\phi) = 0$ on $E = K[x]v$. Since $\deg f \leq r - 1 < \deg q$ and $q \mid f$ by Theorem 4.1(i), we must have $k_i = 0$ for all i. Therefore, $\{v,\phi(v), \ldots, \phi^{r-1}(v)\}$ is linearly independent. If $fv = f(\phi)(v)$ ($f \in K[x]$) is an arbitrary element of $E = K[x]v$, then by the division algorithm $f = qh + s$, where $s = \sum_{i=1}^{t} k_ix^i$ has degree t with $t < \deg q$. Consequently, $f(\phi) = q(\phi)h(\phi) + s(\phi) = 0 + s(\phi) = s(\phi)$ and $fv = f(\phi)(v) = s(\phi)(v) = k_0 + k_1\phi(v) + \cdots + k_t\phi^t(v)$ with $t \leq s - 1$. Therefore,

$$\{v,\phi(v), \ldots, \phi^{r-1}(v)\}$$

spans E and hence is a basis. Since $q(\phi) = 0$ we have $\phi(\phi^{r-1}(v)) = \phi^r(v) = -a_0v - a_1\phi(v) - \cdots - a_{r-1}\phi^{r-1}(v)$. It follows immediately that the matrix of ϕ relative to $\{v,\phi(v), \ldots, \phi^{r-1}(v)\}$ is the companion matrix of q.

($\Leftarrow$) If A is the matrix of ϕ relative to the basis $\{v = v_1, v_2, \ldots, v_r\}$, then a simple computation shows that $v_i = \phi^{i-1}(v)$ for $i = 2, \ldots, r$ and that $\phi^r(v) = \phi(v_r) = -a_0v - a_1\phi(v) - \cdots - a_{r-1}\phi^{r-1}(v)$. Consequently, E is the ϕ-cyclic space generated by v and $E = K[x]v$. Since $q(\phi)(v) = 0$, $q(\phi) = 0$ on E. Since

$$\{v,\phi(v), \ldots, \phi^{r-1}(v)\}$$

is linearly independent there can be no nonzero $f \in K[x]$ of degree less than r such that $f(\phi) = 0$. A routine division algorithm argument now implies that q is the minimal polynomial of ϕ. ■

[2] If E is considered as a *right* K-vector space and matrices of maps are constructed accordingly (as on p. 333) then the companion matrix of q must be defined to be A^t in order to make the theorem true.

Corollary 4.4. *Let* $\psi : E \to E$ *be a linear transformation of a finite dimensional vector space* E *over a field* K. *Then* E *is a* ψ*-cyclic space and* ψ *has minimal polynomial* $q = (x - b)^r$ $(b \in K)$ *if and only if* $\dim_K E = r$ *and* E *has an ordered basis relative to which the matrix of* ψ *is*

$$B = \begin{pmatrix} b & 1_K & 0 & 0 & \cdots & 0 & 0 \\ 0 & b & 1_K & 0 & \cdots & 0 & 0 \\ 0 & 0 & b & 1_K & \cdots & 0 & 0 \\ \cdot & & & & & & \cdot \\ \cdot & & & & & & \cdot \\ \cdot & & & & & & \cdot \\ 0 & 0 & 0 & 0 & \cdots & b & 1_K \\ 0 & 0 & 0 & 0 & \cdots & 0 & b \end{pmatrix}.$$

The $r \times r$ matrix B is called the **elementary Jordan matrix** associated with $(x - b)^r \in K[x]$. Note that for $r = 1$, $B = (b)$.

SKETCH OF PROOF OF 4.4. Let $\phi = \psi - b1_E \in \mathrm{Hom}_K(E,E)$. Then $q = (x - b)^r$ is the minimal polynomial of ψ if and only if x^r is the minimal polynomial of ϕ (for example, $\phi^r = (\psi - b1_E)^r = q(\psi) = 0$). E has two $K[x]$-module structures induced by ϕ and ψ respectively. For every $f \in K[x]$ and $v \in E$, $f(x)v$ in the ϕ-structure is the same element as $f(x - b)v$ in the ψ-structure. Therefore, E is ϕ-cyclic if and only if E is ψ-cyclic. Since $\psi = \phi + b1_E$, Theorem 1.2 shows that the matrix of ϕ relative to a given (ordered) basis of E is the companion matrix A of x^r if and only if the matrix of ψ relative to the same basis is the elementary Jordan matrix $B = A + bI_n$ associated with $(x - b)^r$. To complete the proof simply apply Theorem 4.3 to ϕ and translate the result into statements about ψ, using the facts just developed. ■

In order to use the preceding results to obtain a set of canonical forms for the relation of similarity on $\mathrm{Mat}_n K$ we need

Lemma 4.5. *Let* $\phi : E \to E$ *be a linear transformation of an* n*-dimensional vector space* E *over a field* K. *For each* $i = 1, \ldots, t$ *let* M_i *be an* $n_i \times n_i$ *matrix over* K, *with* $n_1 + n_2 + \cdots + n_t = n$. *Then* $E = E_1 \oplus E_2 \oplus \cdots \oplus E_t$, *where each* E_i *is a* ϕ*-invariant subspace of* E *and for each* i, M_i *is the matrix of* $\phi \mid E_i$ *relative to some ordered basis of* E_i, *if and only if the matrix of* ϕ *relative to some ordered basis of* E *is*

$$M = \begin{pmatrix} M_1 & & & & 0 \\ & M_2 & & & \\ & & \cdot & & \\ & & & \cdot & \\ 0 & & & & M_t \end{pmatrix}$$

where the main diagonal of each M_i *lies on the main diagonal of* M.

A matrix of the form M as in Lemma 4.5 is said to be the **direct sum of the matrices** $M_1, \ldots, M_t$ (in this order).

SKETCH OF PROOF OF 4.5. ($\Rightarrow$) For each i let V_i be an ordered basis of E_i such that the matrix of $\phi \mid E_i$ relative to V_i is M_i. Since $E = E_1 \oplus \cdots \oplus E_t$, it follows easily that $V = \bigcup_{i=1}^{t} V_i$ is a basis of E. Verify that M is the matrix of ϕ relative to V (where V is ordered in the obvious way). ($\Leftarrow$) Conversely suppose $U = \{u_1, \ldots, u_n\}$ is a basis of E and M the matrix of ϕ relative to U. Let E_1 be the subspace of E with basis $\{u_1, \ldots, u_{n_1}\}$ and for $i > 1$ let E_i be the subspace of E with basis $\{u_{r+1}, \ldots, u_{r+n_i}\}$ where $r = n_1 + n_2 + \cdots + n_{i-1}$. Then $E = E_1 \oplus E_2 \oplus \cdots \oplus E_t$, each E_i is ϕ-invariant and M_i is the matrix of $\phi \mid E_i$ relative to $\{u_{r+1}, \ldots, u_{r+n_i}\}$. ■

Theorem 4.6. *Let $\phi : E \to E$ be a linear transformation of an n-dimensional vector space E over a field K.*

(i) *E has a basis relative to which the matrix of ϕ is the direct sum of the companion matrices of the invariant factors $q_1, \ldots, q_t \in K[x]$ of ϕ.*

(ii) *E has a basis relative to which the matrix of ϕ is the direct sum of the companion matrices of the elementary divisors $p_1^{m_{11}}, \ldots, p_s^{m_{sk_s}} \in K[x]$ of ϕ.*

(iii) *If the minimal polynomial q of ϕ factors as $q = (x - b_1)^{r_1}(x - b_2)^{r_2} \cdots (x - b_d)^{r_d}$ ($b_i \in K$), which is always the case if K is algebraically closed, then every elementary divisor of ϕ is of the form $(x - b_i)^j$ ($j \leq r_i$) and E has a basis relative to which the matrix of ϕ is the direct sum of the elementary Jordan matrices associated with the elementary divisors of ϕ.*

The proof, which is an immediate consequence of results 4.2–4.5 (and unique factorization in $K[x]$ for (iii)), is left to the reader. The next corollary immediately yields two (or three if K is algebraically closed) sets of canonical forms for the relation of similarity on $\mathrm{Mat}_n K$.

Corollary 4.7. *Let A be an $n \times n$ matrix over a field K.*

(i) *A is similar to a matrix D such that D is the direct sum of the companion matrices of a unique family of polynomials $q_1, \ldots, q_t \in K[x]$ such that $q_1 \mid q_2 \mid \cdots \mid q_t$. The matrix D is uniquely determined.*

(ii) *A is similar to a matrix M such that M is the direct sum of the companion matrices of a unique family of prime power polynomials $p_1^{m_{11}}, \ldots, p_s^{m_{sk_s}} \in K[x]$, where each p_i is prime (irreducible) in $K[x]$. M is uniquely determined except for the order of the companion matrices of the $p_i^{m_{ij}}$ along its main diagonal.*

(iii) *If K is algebraically closed, then A is similar to a matrix J such that J is a direct sum of the elementary Jordan matrices associated with a unique family of polynomials of the form $(x - b)^m$ ($b \in K$). J is uniquely determined except for the order of the elementary Jordan matrices along its main diagonal.*

The proof is given below. The matrix D in part (i), is said to be in **rational canonical form** or to be the rational canonical form of the matrix A. Similarly, the matrix M

in part (ii) is said to be in **primary rational canonical form** and the matrix J in (iii) is said to be in **Jordan canonical form.**[3] The word "rational" refers to the fact that the similarity of matrices occurs in the given field K and not in an extension field of K (see Exercise 7). The uniquely determined polynomials $q_1, \ldots, q_t$ in part (i) are called the **invariant factors of the matrix** A. Similarly, the unique prime power polynomials $p_i^{m_{ij}}$ in part (ii) are called the **elementary divisors of the matrix** A.

SKETCH OF PROOF OF 4.7. (ii) Let $\phi : K^n \to K^n$ be the linear transformation with matrix A relative to the standard basis (Theorem 1.2). Corollary 1.7 and Theorem 4.6 show that A is similar to the matrix D that is the direct sum in some order of the companion matrices of the elementary divisors $p_i^{m_{ij}}$ of ϕ. If A is also similar to D_1, where D_1 is the direct sum of the companion matrices of a family of prime power polynomials $f_1, \ldots, f_b \in K[x]$, then D_1 is the matrix of ϕ relative to some basis of K^n (Corollary 1.7). By Theorem 4.3 and Lemma 4.5 $K^n = E_1 \oplus E_2 \oplus \cdots \oplus E_b$, where each E_i is a ϕ-cyclic subspace and f_i is the minimal polynomial of $\phi \mid E_i$. The uniqueness statement of Theorem 4.2 implies that the polynomials f_i are precisely the elementary divisors $p_i^{m_{ij}}$ of ϕ, whence D differs from D_1 only in the order of the companion matrices of the $p_i^{m_{ij}}$ along the main diagonal. The proof of (i) and (iii) is similar, except that in (i) a stronger uniqueness statement is possible since the invariant factors (unlike the elementary divisors) may be uniquely ordered by divisibility. ∎

Corollary 4.8. *Let $\phi : E \to E$ be a linear transformation of an n-dimensional vector space E over a field K.*

(i) *If ϕ has matrix $A \in Mat_n K$ relative to some basis, then the invariant factors [resp. elementary divisors] of ϕ are the invariant factors [elementary divisors] of A.*

(ii) *Two matrices in $Mat_n K$ are similar if and only if they have the same invariant factors [resp. elementary divisors].*

PROOF. Exercise. ∎

REMARK. If k is an element of a field K, then the matrix kI_n is a direct sum of the 1×1 companion matrices of the irreducible polynomials $x - k, \ldots, x - k$. Therefore, $x - k, \ldots, x - k$ are the elementary divisors of kI_n by Corollary 4.7. Consequently, if $k_1 \neq k_2$, then $k_1 I_n$ and $k_2 I_n$ are not similar by Corollary 4.8. Thus if K is infinite there are infinitely many distinct equivalence classes under similarity in $\text{Mat}_n K$. On the other hand, there are only $n + 1$ distinct equivalence classes under equivalence in $\text{Mat}_n K$ by Theorem 2.6.

EXAMPLE. Let E be a finite dimensional real vector space and $\phi : E \to E$ a linear transformation with invariant factors $q_1 = x^4 - 4x^3 + 5x^2 - 4x + 4 = (x-2)^2(x^2+1) \in \mathbf{R}[x]$ and $q_2 = x^7 + 6x^6 + 14x^5 - 20x^4 + 25x^3 - 22x^2 + 12x - 8 = (x-2)^3(x^2+1)^2 \in \mathbf{R}[x]$. By Theorem 4.6(i) $\dim_R E = 11$ and the minimal polynomial of ϕ is q_2. The remarks after Theorem 4.2 show that the elementary divisors

[3]Warning: rational and Jordan canonical forms are defined somewhat differently by some authors.

of ϕ in $\mathbf{R}[x]$ are $(x-2)^3 = x^3 - 6x^2 + 12x - 8$, $(x-2)^2 = x^2 - 4x + 4$, $(x^2+1)^2 = x^4 + 2x^2 + 1$, and $x^2 + 1$. By Theorem 4.6 E has two bases relative to which the respective matrices of ϕ are

$$D = \left(\begin{array}{ccccccccccc}
0 & 1 & 0 & 0 & & & & & & & \\
0 & 0 & 1 & 0 & & & & 0 & & & \\
0 & 0 & 0 & 1 & & & & & & & \\
-4 & 4 & -5 & 4 & & & & & & & \\
 & & & & 0 & 1 & 0 & 0 & 0 & 0 & 0 \\
 & & & & 0 & 0 & 1 & 0 & 0 & 0 & 0 \\
 & & & & 0 & 0 & 0 & 1 & 0 & 0 & 0 \\
 & & 0 & & 0 & 0 & 0 & 0 & 1 & 0 & 0 \\
 & & & & 0 & 0 & 0 & 0 & 0 & 1 & 0 \\
 & & & & 0 & 0 & 0 & 0 & 0 & 0 & 1 \\
 & & & & 8 & -12 & 22 & -25 & 20 & -14 & -6
\end{array}\right),$$

$$M = \left(\begin{array}{ccccccccccc}
0 & 1 & 0 & & & & & & & & \\
0 & 0 & 1 & & & & & 0 & & & \\
8 & -12 & 6 & & & & & & & & \\
 & & & 0 & 1 & & & & & & \\
 & & & -4 & 4 & & & & & & \\
 & & & & & 0 & 1 & 0 & 0 & & \\
 & & & & & 0 & 0 & 1 & 0 & & \\
 & 0 & & & & 0 & 0 & 0 & 1 & & \\
 & & & & & -1 & 0 & -2 & 0 & & \\
 & & & & & & & & & 0 & 1 \\
 & & & & & & & & & -1 & 0
\end{array}\right).$$

The matrix D is in rational canonical form and M is in primary rational canonical form. If E is actually a complex vector space and $\psi : E \to E$ is a linear transformation with the same invariant factors $q_1 = (x-2)^2(x^2+1) \in \mathbf{C}[x]$ and $q_2 = (x-2)^3(x^2+1)^2 \in \mathbf{C}[x]$, then since $x^2 + 1 = (x+i)(x-i)$ in $\mathbf{C}[x]$, the elementary divisors of ψ in $\mathbf{C}[x]$ are $(x-2)^3$, $(x-2)^2$, $(x+i)^2$, $(x+i)$, $(x-i)^2$, and $(x-i)$. Therefore, relative to some basis of E, ψ has the following matrix in Jordan canonical form

$$J = \left(\begin{array}{ccccccccccc}
2 & 1 & 0 & & & & & & & & \\
0 & 2 & 1 & & & & & & & & \\
0 & 0 & 2 & & & & & 0 & & & \\
 & & & 2 & 1 & & & & & & \\
 & & & 0 & 2 & & & & & & \\
 & & & & & -i & 1 & & & & \\
 & & & & & 0 & -i & & & & \\
 & 0 & & & & & & -i & & & \\
 & & & & & & & & i & 1 & \\
 & & & & & & & & 0 & i & \\
 & & & & & & & & & & i
\end{array}\right).$$

REMARK. The invariant factors in $K[x]$ of a matrix $A \in \operatorname{Mat}_n K$ are the same as the invariant factors of A in $F[x]$, where F is an extension field of K (Exercise 6). As the previous example illustrates, however, the elementary divisors of A over K may differ from the elementary divisors of A over F.

We close this section by presenting a method of calculating the invariant factors of a given matrix A, and hence by Corollary 4.8 of any linear transformation that has matrix A relative to some basis. This method is a consequence of

Proposition 4.9. *Let* A *be an* n $\times$ n *matrix over a field* K. *Then the matrix of polynomials* $xI_n - A \in Mat_n K[x]$ *is equivalent (over* K[x]) *to a diagonal matrix* D *with nonzero diagonal entries* $f_1, \ldots, f_n \in K[x]$ *such that each* f_i *is monic and* $f_1 \mid f_2 \mid \cdots \mid f_n$. *Those polynomials* f_i *which are not constants are the invariant factors of* A.

REMARK. If K is a field, then $K[x]$ is a Euclidean domain (Corollary III.6.4). Consequently, the following proof together with the Remarks after Proposition 2.11 show that the matrix D may be obtained from $xI_n - A$ by a finite sequence of elementary row and column operations. Thus Proposition 4.9 actually provides a constructive method for finding invariant factors. An example is given after the proof.

SKETCH OF PROOF OF 4.9. Let $\phi : K^n \to K^n$ be the K-linear transformation with matrix $A = (a_{ij})$ relative to the standard basis $\{\varepsilon_i\}$ of K^n. As usual K^n is a $K[x]$-module with structure induced by ϕ. Let F be a free $K[x]$-module with basis $U = \{u_1, \ldots, u_n\}$ and let $\pi : F \to K^n$ be the unique $K[x]$-module homomorphism such that $\pi(u_i) = \varepsilon_i$ for $i = 1,2, \ldots, n$ (Theorem IV.2.1). Let $\psi : F \to F$ be the unique $K[x]$-module homomorphism such that $\psi(u_i) = xu_i - \sum_{j=1}^{n} a_{ij}u_j$. Then the matrix of ψ relative to the basis U is $xI_n - A$.

We claim that the sequence of $K[x]$-modules $F \xrightarrow{\psi} F \xrightarrow{\pi} K^n \to 0$ is exact. Clearly π is a $K[x]$-module epimorphism. Since A is the matrix of ϕ and the $K[x]$-module structure of K^n is induced by ϕ,

$$\pi(xu_i) = x\pi(u_i) = x\varepsilon_i = \phi(\varepsilon_i) = \sum_{j=1}^{n} a_{ij}\varepsilon_j.$$

Consequently, for each i

$$\begin{aligned}\pi\psi(u_i) &= \pi\left(xu_i - \sum_{j=1}^{n} a_{ij}u_j\right) = \pi(xu_i) - \sum_j a_{ij}\pi(u_j)\\ &= \sum_j a_{ij}\varepsilon_j - \sum_j a_{ij}\varepsilon_j = 0,\end{aligned}$$

whence $\operatorname{Im}\psi \subset \operatorname{Ker}\pi$. To show that $\operatorname{Ker}\pi \subset \operatorname{Im}\psi$ it suffices to prove that every element w of F is of the form $w = \psi(v) + \sum_{j=1}^{n} k_j u_j$ ($v \in F$, $k_j \in K$). For in this case if $w \in \operatorname{Ker}\pi$, then

$$0 = \pi(w) = \pi\psi(v) + \pi(\sum_j k_j u_j) = 0 + \sum_j k_j\varepsilon_j.$$

Since $\{\varepsilon_j\}$ is a basis of K^n, $k_j = 0$ for all j. Consequently, $w = \psi(v)$ and hence $\operatorname{Ker}\pi \subset \operatorname{Im}\psi$. Since every element of F is a sum of terms of the form fu_i with $f \in K[x]$, we need only show that for each i and t, there exist $v_{it} \in F$ and $k_j \in K$ such that $x^t u_i = \psi(v_{it}) + \sum_{j=1}^{n} k_j u_j$. For each i and $t = 1$, we have $xu_i = \psi(u_i) + \sum_j a_{ij}u_j$

$(a_{ij} \in K)$. Proceeding inductively assume that for each j there exist $v_{j,t-1} \in F$ and $k_{jr} \in K$ such that $x^{t-1}u_j = \psi(v_{j,t-1}) + \sum_{r=1}^{n} k_{jr}u_r$. Then for each i

$$\begin{aligned} x^t u_i &= x^{t-1}(xu_i) = x^{t-1}(\psi(u_i) + \sum_j a_{ij}u_j) = \psi(x^{t-1}u_i) + \sum_j a_{ij}x^{t-1}u_j \\ &= \psi(x^{t-1}u_i) + \sum_j a_{ij}(\psi(v_{j,t-1}) + \sum_r k_{jr}u_r) \\ &= \psi(x^{t-1}u_i + \sum_j a_{ij}v_{j,t-1}) + \sum_r (\sum_j a_{ij}k_{jr})u_r. \end{aligned}$$

Thus $x^t u_i = \psi(v_{it}) + \sum_r c_r u_r$ with $v_{it} = x^{t-1}u_i + \sum_j a_{ij}v_{j,t-1} \in F$ and $c_r = \sum_j a_{ij}k_{jr} \in K$ and the induction is complete. Therefore $F \xrightarrow{\psi} F \xrightarrow{\pi} K^n \to 0$ is exact and hence $K^n \cong F/\operatorname{Ker} \pi = F/\operatorname{Im} \psi$.

Since $K[x]$ is a principal ideal domain, Proposition 2.11 shows that $xI_n - A$ is equivalent to a diagonal matrix $D = \begin{pmatrix} L_r & 0 \\ 0 & 0 \end{pmatrix}$, where r is the rank of $xI_n - A$ and L_r is an $r \times r$ diagonal matrix with nonzero diagonal entries $f_1, \dots, f_r \in K[x]$ such that $f_1 \mid f_2 \mid \cdots \mid f_r$. We may assume each f_i is monic (if necessary, perform suitable elementary row operations on D). Clearly the determinant $|xI_n - A|$ in $K[x]$ is a monic polynomial of degree n. In particular, $|xI_n - A| \neq 0$. By Definition 1.8 and Theorem 3.5(iii), (iv), $|D|$ is a unit multiple of $|xI_n - A|$, whence $|D| \neq 0$. Consequently, all the diagonal entries of D are nonzero. Thus $L_r = D$ and $r = n$. Since D is equivalent to $xI_n - A$, D is the matrix of ψ relative to some pair of ordered bases $V = \{v_1, \dots, v_n\}$ and $W = \{w_1, \dots, w_n\}$ of F (Theorem 1.6). This means that $\psi(v_i) = f_i w_i$ for each i and $\operatorname{Im} \psi = K[x]f_1w_1 \oplus \cdots \oplus K[x]f_nw_n$. Consequently,

$$\begin{aligned} K^n \cong F/\operatorname{Ker} \pi = F/\operatorname{Im} \psi &= \frac{K[x]w_1 \oplus \cdots \oplus K[x]w_n}{K[x]f_1w_1 \oplus \cdots \oplus K[x]f_nw_n} \\ &\cong K[x]w_1/K[x]f_1w_1 \oplus \cdots \oplus K[x]w_n/K[x]f_nw_n \\ &\cong K[x]/(f_1) \oplus \cdots \oplus K[x]/(f_n), \end{aligned}$$

where each f_i is monic and $f_1 \mid f_2 \mid \cdots \mid f_n$. For some t $(0 \leq t \leq n)$, $f_1 = f_2 = \cdots = f_t = 1_K$ and $f_{t+1}, \dots, f_n$ are nonconstant. Thus for $i \leq t$, $K[x]/(f_i) = K[x]/(1_K) = 0$ and for $i > t$, $K[x]/(f_i)$ is a cyclic $K[x]$-module of order f_i. Therefore, K^n is the internal direct sum of nonzero torsion cyclic $K[x]$-submodules (ϕ-cyclic subspaces) $E_{t+1}, \dots, E_n$ of orders $f_{t+1}, \dots, f_n$ respectively such that $f_{t+1} \mid f_{t+2} \mid \cdots \mid f_n$. Since the $K[x]$-module structure of K^n is induced by ϕ, $0 = f_iE_i = f_i(\phi)E_i$. It follows readily that f_i is the minimal polynomial of $\phi \mid E_i$. Therefore, $f_{t+1}, \dots, f_n$ are the invariant factors of ϕ (and hence of A) by Theorem 4.2. ∎

EXAMPLE. If $\phi : \mathbf{Q}^3 \to \mathbf{Q}^3$ is a linear transformation and relative to some basis the matrix of ϕ is $A = \begin{pmatrix} 0 & 4 & 2 \\ -1 & -4 & -1 \\ 0 & 0 & -2 \end{pmatrix}$, then $xI_3 - A = \begin{pmatrix} x & -4 & -2 \\ 1 & x+4 & 1 \\ 0 & 0 & x+2 \end{pmatrix}$.

Performing suitable elementary row and column operations yields:

$$\begin{pmatrix} x & -4 & -2 \\ 1 & x+4 & 1 \\ 0 & 0 & x+2 \end{pmatrix} \to \begin{pmatrix} 1 & x+4 & 1 \\ x & -4 & -2 \\ 0 & 0 & x+2 \end{pmatrix} \to$$

$$\begin{pmatrix} 1 & x+4 & 1 \\ 0 & -4-x(x+4) & -2-x \\ 0 & 0 & x+2 \end{pmatrix} \rightarrow \begin{pmatrix} 1 & 0 & 0 \\ 0 & -(x+2)^2 & -(x+2) \\ 0 & 0 & x+2 \end{pmatrix} \rightarrow$$

$$\begin{pmatrix} 1 & 0 & 0 \\ 0 & -(x+2)^2 & 0 \\ 0 & 0 & x+2 \end{pmatrix} \rightarrow \begin{pmatrix} 1 & 0 & 0 \\ 0 & x+2 & 0 \\ 0 & 0 & (x+2)^2 \end{pmatrix}.$$

Therefore by Corollary 4.8 and Proposition 4.9 the invariant factors of A and ϕ are $x + 2$ and $(x + 2)^2$ and their minimal polynomial is $(x + 2)^2$.

EXERCISES

Note: Unless stated otherwise, E is an n-dimensional vector space over a field K.

1. If A and B are $n \times n$ matrices over K with minimum polynomials q_1 and q_2 respectively, then the minimal polynomial of the direct sum of A and B (a $2n \times 2n$ matrix) is the least common multiple of q_1 and q_2.

2. The 0 linear transformation $E \rightarrow E$ has invariant factors [resp. elementary divisors] $q_1 = x, q_2 = x, \ldots, q_n = x$.

3. (a) Let a,b,c be distinct elements of K and let $D \in \text{Mat}_6K$ be the diagonal matrix with main diagonal a,a,a,b,b,c. Then the invariant factors of D are $q_1 = x - a$, $q_2 = (x - a)(x - b)$ and $q_3 = (x - a)(x - b)(x - c)$.
(b) Describe the invariant factors of any diagonal matrix in Mat_nK.

4. If q is the minimal polynomial of a linear transformation $\phi : E \rightarrow E$, with $\dim_K E = n$, then $\deg q \leq n$.

5. The minimal polynomial of the companion matrix of a monic polynomial $f \in K[x]$ is precisely f.

6. Let F be an extension field of K. The invariant factors in $K[x]$ of a matrix $A \in \text{Mat}_nK$ are the same as the invariant factors in $F[x]$ of A considered as a matrix over F. [*Hint:* A K-basis of K^n is an F-basis of F^n. Use linear transformations.]

7. Let F be an extension field of K. $A,B \in \text{Mat}_nK \subset \text{Mat}_nF$ are similar over F if and only if they are similar over K [see Exercise 6].

8. $A \in \text{Mat}_nK$ is similar to a diagonal matrix if and only if the elementary divisors of A are all linear.

9. If $A \in \text{Mat}_nK$ is nilpotent (that is, $A^r = 0$ for some $r > 0$), then A is similar to a matrix all of whose entries are zero except for certain entries 1_K on the diagonal next *above* the main diagonal.

10. Find all possible [primary] rational canonical forms for a matrix $A \in \text{Mat}_n\mathbf{Q}$ such that (i) A is 6×6 with minimal polynomial $(x - 2)^2(x + 3)$; (ii) A is 7×7 with minimal polynomial $(x^2 + 1)(x - 7)$. Find all possible Jordan canonical forms of A considered as a matrix over $\mathbf{C}$.

11. If A is the companion matrix of a monic polynomial $f \varepsilon K[x]$, with $\deg f = n$, show explicitly that $A - xI_n$ is similar to a diagonal matrix with main diagonal $1_K, 1_K, \ldots, 1_K, f$.

12. $A \varepsilon \mathrm{Mat}_n K$ is *idempotent* provided $A^2 = A$. Show that two idempotent matrices in $\mathrm{Mat}_n K$ are similar if and only if they are equivalent.

13. An $n \times n$ matrix A is similar to its transpose A^t.

5. THE CHARACTERISTIC POLYNOMIAL, EIGENVECTORS AND EIGENVALUES

In this section we investigate some more invariants of a linear transformation of a finite dimensional vector space over a field. Since several of these results are valid more generally we shall deal whenever possible with free modules of finite rank over a commutative ring with identity.

If A is an $n \times n$ matrix over a commutative ring K with identity, then $xI_n - A$ is an $n \times n$ matrix over $K[x]$, whence the determinant $|xI_n - A|$ is an element of $K[x]$. The **characteristic polynomial of the matrix** A is the polynomial $p_A = |xI_n - A| \varepsilon K[x]$. Clearly, p_A is a monic polynomial of degree n. If $B \varepsilon \mathrm{Mat}_n K$ is similar to A, say $B = PAP^{-1}$, then since xI_n is in the center of the ring $\mathrm{Mat}_n K[x]$,

$$\begin{aligned} p_B = |xI_n - B| &= |xI_n - PAP^{-1}| = |P(xI_n - A)P^{-1}| \\ &= |P||xI_n - A||P|^{-1} = |xI_n - A| = p_A; \end{aligned}$$

that is, *similar matrices have the same characteristic polynomial.*

Let $\phi : E \to E$ be an endomorphism of a free K-module E of finite rank n (see Definition IV.2.8 and Corollary IV.2.12). The **characteristic polynomial of the endomorphism** ϕ (denoted p_ϕ) is defined to be p_A, where A is any matrix of ϕ relative to some ordered basis. Since any two matrices representing ϕ are similar by Corollary 1.7, p_ϕ is independent of the choice of A.

Lemma 5.1. (i) *If* $A_1, A_2, \ldots, A_r$ *are square matrices (of various sizes) over a commutative ring* K *with identity and* $p_i \varepsilon K[x]$ *is the characteristic polynomial of* A_i, *then* $p_1 p_2 \cdots p_r \varepsilon K[x]$ *is the characteristic polynomial of the matrix direct sum of* $A_1, A_2, \ldots, A_r$.

(ii) *The companion matrix* C *of a monic polynomial* $f \varepsilon K[x]$ *has characteristic polynomial* f.

SKETCH OF PROOF. (i) If $A \varepsilon \mathrm{Mat}_n K$ and $B \varepsilon \mathrm{Mat}_m K$, then

$$\begin{pmatrix} A & 0 \\ 0 & B \end{pmatrix} = \begin{pmatrix} A & 0 \\ 0 & I_m \end{pmatrix}\begin{pmatrix} I_n & 0 \\ 0 & B \end{pmatrix}, \quad \text{whence} \quad \begin{vmatrix} A & 0 \\ 0 & B \end{vmatrix} = \begin{vmatrix} A & 0 \\ 0 & I_m \end{vmatrix}\begin{vmatrix} I_n & 0 \\ 0 & B \end{vmatrix} = |A||B|.$$

An inductive argument now shows that the determinant of a direct sum of matrices $B_1, \ldots, B_k$ is $|B_1||B_2| \cdots |B_k|$. (ii) To show that f is the characteristic polynomial of C, expand $|xI_n - C|$ along the last row. ■

Theorem 5.2. *Let $\phi : E \to E$ be a linear transformation of an n-dimensional vector space over a field K with characteristic polynomial $p_\phi \in K[x]$, minimal polynomial $q_\phi \in K[x]$, and invariant factors $q_1, \ldots, q_t \in K[x]$.*

(i) *The characteristic polynomial is the product of the invariant factors; that is,* $p_\phi = q_1q_2\cdots q_t = q_1q_2\cdots q_{t-1}q_\phi$.

(ii) *(Cayley-Hamilton) ϕ is root of its characteristic polynomial; that is,* $p_\phi(\phi) = 0$.

(iii) *An irreducible polynomial in* K[x] *divides p_ϕ if and only if it divides q_ϕ. Conclusions* (i)–(iii) *are valid,* mutatis mutandis, *for any matrix* $A \in Mat_nK$.

PROOF. By Theorem 4.6 ϕ has a basis relative to which ϕ has the matrix D that is the direct sum of the companion matrices of $q_1, \ldots, q_t$. Therefore, $p_\phi = p_D = q_1q_2\cdots q_t$ by Lemma 5.1. Furthermore, $q_\phi = q_t$ by Theorem 4.2, whence $p_\phi(\phi) = 0$ since $q_\phi(\phi) = 0$. (iii) is an immediate consequence of (i) and the fact that $q_1 \mid q_2 \mid \cdots \mid q_t$. The analogous statements about $A \in \mathrm{Mat}_nK$ are proved similarly using Corollaries 4.7 and 4.8. ■

REMARK. The Cayley-Hamilton Theorem (Theorem 5.2(ii)) is valid over any commutative ring with identity (Exercise 2).

Definition 5.3. *Let $\phi : E \to E$ be a linear transformation of a vector space E over a field K. A nonzero vector $u \in E$ is an* **eigenvector** *(or* **characteristic vector** *or* **proper vector)** *of ϕ if $\phi(u) = ku$ for some $k \in K$. An element $k \in K$ is an* **eigenvalue** *(or* **proper value** *or* **characteristic value)** *of ϕ if $\phi(u) = ku$ for some nonzero $u \in E$.*

It is quite possible for two distinct (even linearly independent) eigenvectors to have the same eigenvalue. On the other hand, a set of eigenvectors whose corresponding eigenvalues are all distinct is necessarily linearly independent (Exercise 8).

Theorem 5.4. *Let $\phi : E \to E$ be a linear transformation of a finite dimensional vector space E over a field K. Then the eigenvalues of ϕ are the roots in K of the characteristic polynomial p_ϕ of ϕ.*

REMARK. The characteristic polynomial $p_\phi \in K[x]$ need not have any roots in K, in which case ϕ has no eigenvalues or eigenvectors.

SKETCH OF PROOF OF 5.4. Let A be the matrix of ϕ relative to some ordered basis. If $k \in K$, then $kI_n - A$ is the matrix of $k1_E - \phi$ relative to the same basis. If $\phi(u) = ku$ for some nonzero $u \in E$, then $(k1_E - \phi)(u) = 0$, whence $k1_E - \phi$ is not a monomorphism. Therefore, $kI_n - A$ is not invertible (Lemma 1.5) and hence $|kI_n - A| = 0$ by Proposition 3.7 or Exercise 3.6. Thus k is a root of $p_\phi = |xI_n - A|$. Conversely, if k is a root of p_ϕ, then $|kI_n - A| = 0$. Consequently, $k1_E - \phi$ is not an isomorphism by Lemma 1.5 and Proposition 3.7 (or Exercise 3.6). Since E is finite dimensional, $k1_E - \phi$ is not a monomorphism (Exercise IV.2.14). Therefore, there is a nonzero $u \in E$ such that $(k1_E - \phi)(u) = 0$, whence $\phi(u) = ku$ and k is an eigenvalue of ϕ. ■

If $k \varepsilon K$ is an eigenvalue of an endomorphism ϕ of a K-vector space E, then it is easy to see that $C(\phi,k) = \{v \varepsilon E \mid \phi(v) = kv\}$ is a nonzero subspace of E; $C(\phi,k)$ is called the **eigenspace** or **characteristic space** of k.

Theorem 5.5. *Let $\phi : E \to E$ be a linear transformation of a finite dimensional vector space E over a field K. Then ϕ has a diagonal matrix D relative to some ordered basis of E if and only if the eigenvectors of ϕ span E. In this case the diagonal entries of D are the eigenvalues of ϕ and each eigenvalue $k \varepsilon K$ appears on the diagonal $\dim_K C(\phi,k)$ times.*

PROOF. By Theorem IV.2.5 the eigenvectors of ϕ span E if and only if E has a basis consisting of eigenvectors. Clearly $U = \{u_1, \ldots, u_n\}$ is a basis of eigenvectors with corresponding eigenvalue $k_1, \ldots, k_n \varepsilon K$ if and only if the matrix of ϕ relative to U is the diagonal matrix D with main diagonal $k_1,k_2, \ldots, k_n$. In this case suppose that $v = \sum_{i=1}^{n} r_i u_i$ is an eigenvector of ϕ with $\phi(v) = kv$. Since U is linearly independent and $\sum_{i=1}^{n} kr_i u_i = kv = \phi(v) = \sum_{i=1}^{n} r_i\phi(u_i) = \sum_{i=1}^{n} r_i k_i u_i$, we have $kr_i = r_i k_i$ for all i. Thus for each i such that $r_i \neq 0$, $k = k_i$; (since $v \neq 0$, at least one $r_i \neq 0$). Therefore, $k_1, \ldots, k_n$ are the only eigenvalues of ϕ. Furthermore, if k is an eigenvalue of ϕ that appears t times on the diagonal of D and $u_{i_1}, \ldots, u_{i_t}$ are those elements of U with eigenvalue k, then this argument shows that $\{u_{i_1}, \ldots, u_{i_t}\}$ spans $C(\phi,k)$. Since $\{u_{i_1}, \ldots, u_{i_t}\}$ is linearly independent it is a basis of $C(\phi,k)$. Therefore, $\dim_K C(\phi,k) = t$. ■

The **eigenvalues and eigenvectors of an n × n matrix** A over a field K are defined to be respectively the eigenvalues and eigenvectors of the unique linear transformation $\phi : K^n \to K^n$ that has matrix A relative to the standard basis. Theorem 5.4 shows that the eigenvalues of A are the eigenvalues of any endomorphism of an n-dimensional vector space over K which has matrix A relative to some basis.

We close this section with a brief discussion of another invariant of a matrix under similarity.

Proposition 5.6. *Let K be a commutative ring with identity. Let ϕ be an endomorphism of a free K-module of rank n and let $A = (a_{ij}) \varepsilon \mathrm{Mat}_n K$ be the matrix of ϕ relative to some ordered basis. If the characteristic polynomial of ϕ and A is $p_\phi = p_A = x^n + c_{n-1}x^{n-1} + \cdots + c_1x + c_0 \varepsilon K[x]$, then*

$$(-1)^n c_0 = |A| \quad \textit{and} \quad -c_{n-1} = a_{11} + a_{22} + \cdots + a_{nn}.$$

PROOF. $c_0 = p_\phi(0) = |0I_n - A| = |-A| = (-1)^n|A|$ by Theorem 3.5(viii). Expand $p_\phi = |xI_n - A|$ along the first row. One term of this expansion is $(x - a_{11})(x - a_{22})\cdots(x - a_{nn}) = x^n - (a_{11} + a_{22} + \cdots + a_{nn})x^{n-1} + b_{n-2}x^{n-2} + \cdots + b_0$ for some $b_i \varepsilon K$. No other term of this expansion contains any terms with a factor of x^{n-1}, whence $-c_{n-1} = a_{11} + \cdots + a_{nn}$. ■

Let K be a commutative ring with identity. The **trace of an n × n matrix** $A = (a_{ij})$ over K is $a_{11} + a_{22} + \cdots + a_{nn} \varepsilon K$ and is denoted $\operatorname{Tr}A$. The **trace of an endomorphism** ϕ of a free K-module of rank n (denoted $\operatorname{Tr}\phi$) is $\operatorname{Tr}A$, where A is the matrix of ϕ relative to some ordered basis. Since $p_\phi = p_A$ is independent of the choice of the matrix A, so is $\operatorname{Tr}\phi$ by Proposition 5.6. Similar matrices have the same trace by Corollary 1.7 (or by an easy direct argument using (iii) below). It is easy to see that for any $A,B \varepsilon \operatorname{Mat}_n K$ and $k \varepsilon K$:

(i) $\operatorname{Tr}(A + B) = \operatorname{Tr}A + \operatorname{Tr}B$;
(ii) $\operatorname{Tr}(kA) = k\operatorname{Tr}A$;
(iii) $\operatorname{Tr}(AB) = \operatorname{Tr}(BA)$.

The connection between the trace as defined here and the trace function of Galois Theory (Definition V.7.1) is explored in Exercise 9.

EXERCISES

Note: Unless stated otherwise K is a commutative ring with identity.

1. Prove directly that a matrix over K and its transpose have the same characteristic polynomial.

2. (Cayley-Hamilton) If ϕ is an endomorphism of a free K-module E of finite rank, then $p_\phi(\phi) = 0$. [*Hint:* if A is the matrix of ϕ and $B = xI_n - A$, then $B^aB = |B|I_n = p_\phi I_n$ in $\operatorname{Mat}_n K[x]$. If E is a $K[x]$-module with structure induced by ϕ and ψ is the $K[x]$-module endomorphism $E \to E$ with matrix B, then $\psi(u) = xu - \phi(u) = \phi(u) - \phi(u) = 0$ for all $u \varepsilon E$.]

3. If A is an $n \times m$ matrix over K and B an $m \times n$ matrix over K, then $x^m p_{AB} = x^n p_{BA}$. Furthermore, if $m = n$, then $p_{AB} = p_{BA}$. [*Hint:* let C,D be the $(m + n) \times (m + n)$ matrices over $K[x]$: $C = \begin{pmatrix} xI_n & A \\ B & I_m \end{pmatrix}$ and $D = \begin{pmatrix} I_n & 0 \\ -B & xI_m \end{pmatrix}$ and observe that $|CD| = |DC|$.]

4. (a) Exhibit three 3×3 matrices over $\mathbf{Q}$ no two of which are similar such that -2 is the only eigenvalue of each of the matrices.
(b) Exhibit a 4×4 matrix whose eigenvalues over $\mathbf{R}$ are ± 1 and whose eigenvalues over $\mathbf{C}$ are ± 1 and $\pm i$.

5. Let K be a field and $A \varepsilon \operatorname{Mat}_n K$.
(a) 0 is an eigenvalue of A if and only if A is not invertible.
(b) If $k_1, \ldots, k_r \varepsilon K$ are the (not necessarily distinct) eigenvalues of A and $f \varepsilon K[x]$, then $f(A) \varepsilon \operatorname{Mat}_n K$ has eigenvalues $f(k_1), \ldots, f(k_r)$.

6. If ϕ and ψ are endomorphisms of a finite dimensional vector space over an algebraically closed field K such that $\phi\psi = \psi\phi$, then ϕ and ψ have a common eigenvector.

7. (a) Let ϕ and ψ be endomorphisms of a finite dimensional vector space E such that $\phi\psi = \psi\phi$. If E has a basis of eigenvectors of ϕ and a basis of eigenvectors of ψ, then E has a basis consisting of vectors that are eigenvectors for both ϕ and ψ.
(b) Interpret (a) as a statement about matrices that are similar to a diagonal matrix.

8. Let $\phi : E \to E$ be a linear transformation of a vector space E over a field K. If U is a set of eigenvectors of ϕ whose corresponding eigenvalues are all distinct, then U is linearly independent. [*Hint:* If U were linearly dependent, there would be a relation $r_1u_1 + \cdots + r_tu_t = 0$ ($u_i \varepsilon U$; $0 \neq r_i \varepsilon K$) with t minimal. Apply the transformation $k_1 1_E - \phi$, where $\varphi(u_1) = k_1u_1$, and reach a contradiction.]

9. Let F be an extension field of a field K and $u \varepsilon F$. Let $\phi : F \to F$ be the endomorphism of the vector space F given by $v \mapsto uv$.
   (a) Then $\mathrm{Tr}\phi$ is the trace of u, $T_K{}^F(u)$, as in Definition V.7.1. [*Hint:* first try the case when $F = K(u)$].
   (b) The determinant of ϕ is the norm of u, $N_K{}^F(u)$.

10. Let K be a field and $A \varepsilon \mathrm{Mat}_n K$.
   (a) If A is nilpotent (that is, $A^m = 0$ for some m), then $\mathrm{Tr}A^r = 0$ for all $r \geq 1$. [*Hint:* the minimal polynomial of A^r has the form x^t and A^r is similar to a matrix in rational or Jordan canonical form.]
   (b) If char $K = 0$ and $\mathrm{Tr}\, A^r = 0$ for all $r \geq 1$, then A is nilpotent.

CHAPTER VIII

COMMUTATIVE RINGS AND MODULES

For the most part this chapter is a brief introduction to what is frequently called commutative algebra. We begin with chain conditions (Section 1) and prime ideals (Section 2), both of which play a central role in the study of commutative rings. Actually no commutativity restrictions are made in Section 1 since this material is also essential in the study of arbitrary rings (Chapter IX).

The theory of commutative rings follows a familiar pattern: we attempt to obtain a structure theory for those rings that possess, at least in some generalized form, properties that have proven useful in various well-known rings. Thus primary decomposition of ideals (the analogue of factorization of elements in an integral domain) is considered in Sections 2 and 3. We then study rings that share certain desirable properties with the ring of integers, such as Dedekind domains (Section 6) and Noetherian rings (Section 4). The analysis of Dedekind domains requires some knowledge about ring extensions (Section 5). This information is also used in proving the Hilbert Nullstellensatz (Section 7), a famous classical result dealing with ideals of the polynomial ring $K[x_1, \dots, x_n]$.

Except in Section 1, all rings are commutative. The approximate interdependence of the sections of this chapter (subject to the remarks below) is as follows:

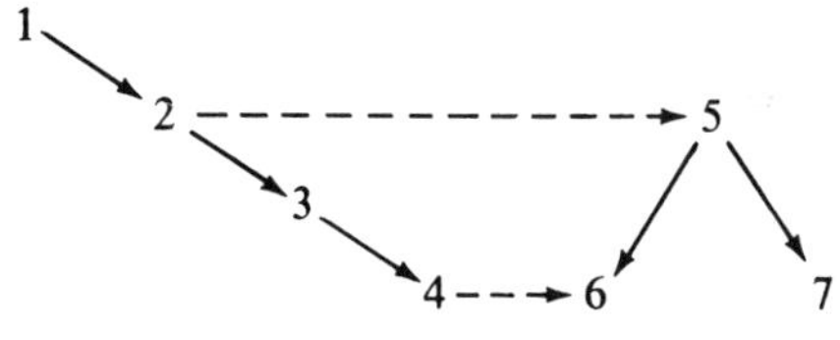

A broken arrow $A \dashrightarrow B$ indicates that an occasional result from Section A is used in Section B, but that Section B is essentially independent of Section A. Section 1 is not needed for Section 5 but is needed for Section 4. Only one important result in Section 4 depends on Sections 2 and 3. This dependence can be eliminated by using an alternate proof, which is indicated in the exercises.

1. CHAIN CONDITIONS

In this section we summarize the basic facts about the ascending and descending chain conditions for modules and rings that will be needed in the remainder of this chapter and in Chapter IX. *Rings are* not *assumed to be commutative, nor to have identity elements.*

Definition 1.1. *A module* A *is said to satisfy the* **ascending chain condition** (ACC) **on submodules** (*or to be* **Noetherian**) *if for every chain* $A_1 \subset A_2 \subset A_3 \subset \cdots$ *of submodules of A, there is an integer* n *such that* $A_i = A_n$ *for all* $i \geq n$.

A module B *is said to satisfy the* **descending chain condition (DCC) on submodules** (*or to be* **Artinian**) *if for every chain* $B_1 \supset B_2 \supset B_3 \supset \cdots$ *of submodules of* B, *there is an integer* m *such that* $B_i = B_m$ *for all* $i \geq m$.

EXAMPLE. The **Z**-module (abelian group) **Z** satisfies the ascending but not the descending chain condition on submodules (Exercise II.3.5). The **Z**-module $Z(p^\infty)$ satisfies the descending but not the ascending chain condition (Exercise II.3.13).

If a ring R is considered as a left [resp. right] module over itself, then it is easy to see that the submodules of R are precisely the left [resp. right] ideals of R. Consequently, in this case it is customary to speak of chain conditions on left or right ideals rather than submodules.

Definition 1.2. *A ring* R *is* **left** [*resp.* **right**] **Noetherian** *if* R *satisfies the ascending chain condition on left* [*resp. right*] *ideals.* R *is said to be* **Noetherian** *if* R *is both left and right Noetherian.*

A ring R *is* **left** [*resp.* **right**] **Artinian** *if* R *satisfies the descending chain condition on left* [*resp. right*] *ideals.* R *is said to be* **Artinian** *if* R *is both left and right Artinian.*

In other words, a ring R is (left or right) Noetherian if it is a (left or right) Noetherian R-module, and similarly for Artinian. Consequently, all subsequent definitions and results about modules that satisfy the ascending or descending chain condition on submodules apply, *mutatis mutandis*, to (left or right) Noetherian or Artinian rings.

EXAMPLES. A division ring D is both Noetherian and Artinian since the only left or right ideals are D and 0, (Exercise III.2.7). Every commutative principal ideal ring is Noetherian (Lemma III.3.6); special cases include **Z**, Z_n, and $F[x]$ with F a field.

EXAMPLE. The ring $\mathrm{Mat}_n D$ of all $n \times n$ matrices over a division ring is both Noetherian and Artinian (Corollary 1.12 below).

REMARKS. A right Noetherian [Artinian] ring need not be left Noetherian [Artinian] (Exercise 1). Exercise II.3.5 shows that a Noetherian ring need not be Artinian. However every left [right] Artinian ring with identity is left [right] Noetherian (Exercise IX.3.13 below).

A maximal element in a partially ordered set $(C,\leq)$ was defined in Section 7 of the Introduction. A **minimal element** is defined similarly: $b \varepsilon C$ is minimal if for every *$c \varepsilon C$ which is comparable to b, $b \leq c$.* Note that it is not necessarily true that $b \leq c$ for *all* $c \varepsilon C$. Furthermore, C may contain many minimal elements or none at all.

Definition 1.3. *A module* A *is said to satisfy the* **maximum condition** [*resp.* **minimum condition**] **on submodules** *if every nonempty set of submodules of* A *contains a maximal* [*resp. minimal*] *element (with respect to set theoretic inclusion).*

Theorem 1.4. *A module* A *satisfies the ascending* [*resp. descending*] *chain condition on submodules if and only if* A *satisfies the maximal* [*resp. minimal*] *condition on submodules.*

PROOF. Suppose A satisfies the minimal condition on submodules and $A_1 \supset A_2 \supset \cdots$ is a chain of submodules. Then the set $\{A_i \mid i \geq 1\}$ has a minimal element, say A_n. Consequently, for $i \geq n$ we have $A_n \supset A_i$ by hypothesis and $A_n \subset A_i$ by minimality, whence $A_i = A_n$ for each $i \geq n$. Therefore, A satisfies the descending chain condition.

Conversely suppose A satisfies the descending chain condition, and S is a nonempty set of submodules of A. Then there exists $B_0 \varepsilon S$. If S has no minimal element, then for each submodule B in S there exists at least one submodule B' in S such that $B \underset{\neq}{\supset} B'$. For each B in S, choose one such B' (Axiom of Choice). This choice then defines a function $f : S \to S$ by $B \mapsto B'$. By the Recursion Theorem 6.2 of the Introduction (with $f = f_n$ for all n) there is a function $\varphi : \mathbf{N} \to S$ such that

$$\varphi(0) = B_0 \quad \text{and} \quad \varphi(n+1) = f(\varphi(n)) = \varphi(n)'.$$

Thus if $B_n \varepsilon S$ denotes $\varphi(n)$, then there is a sequence $B_0, B_1, \ldots$ such that $B_0 \underset{\neq}{\supset} B_1 \underset{\neq}{\supset} B_2 \underset{\neq}{\supset} \cdots$. This contradicts the descending chain condition. Therefore, S must have a minimal element, whence A satisfies the minimum condition.

The proof for the ascending chain and maximum conditions is analogous. ■

Theorem 1.5. *Let* $0 \to A \xrightarrow{f} B \xrightarrow{g} C \to 0$ *be a short exact sequence of modules. Then* B *satisfies the ascending* [*resp. descending*] *chain condition on submodules if and only if* A *and* C *satisfy it.*

SKETCH OF PROOF. If B satisfies the ascending chain condition, then so does its submodule $f(A)$. By exactness A is isomorphic to $f(A)$, whence A satisfies the ascending chain condition. If $C_1 \subset C_2 \subset \cdots$ is a chain of submodules of C, then $g^{-1}(C_1) \subset g^{-1}(C_2) \subset \cdots$ is a chain of submodules of B. Therefore, there is an n such that $g^{-1}(C_i) = g^{-1}(C_n)$ for all $i \geq n$. Since g is an epimorphism by exactness, it follows that $C_i = C_n$ for all $i \geq n$. Therefore, C satisfies the ascending chain condition.

Suppose A and C satisfy the ascending chain condition and $B_1 \subset B_2 \subset \cdots$ is a chain of submodules of B. For each i let

$$A_i = f^{-1}(f(A) \cap B_i) \quad \text{and} \quad C_i = g(B_i).$$

Let $f_i = f \mid A_i$ and $g_i = g \mid B_i$. Verify that for each i the following sequence is exact:

$$0 \to A_i \xrightarrow{f_i} B_i \xrightarrow{g_i} C_i \to 0.$$

Verify that $A_1 \subset A_2 \subset \cdots$ and $C_1 \subset C_2 \subset \cdots$. By hypothesis there exists an integer n such that $A_i = A_n$ and $C_i = C_n$ for all $i \geq n$. For each $i \geq n$ there is a commutative diagram with exact rows:

$$\begin{array}{ccccccccc} 0 & \to & A_n & \xrightarrow{f_n} & B_n & \xrightarrow{g_n} & C_n & \to & 0 \\ & & \downarrow{\scriptstyle \alpha} & & \downarrow{\scriptstyle \beta_i} & & \downarrow{\scriptstyle \gamma} & & \\ 0 & \to & A_i & \xrightarrow{f_i} & B_i & \xrightarrow{g_i} & C_i & \to & 0, \end{array}$$

where α and γ are the respective identity maps and β_i is the inclusion map. The Short Five Lemma IV.1.17 implies that β_i is the identity map, whence B satisfies the ascending chain condition. The proof for descending chain condition is analogous. ∎

Corollary 1.6. *If* A *is a submodule of a module* B, *then* B *satisfies the ascending [resp. descending] chain condition if and only if* A *and* B/A *satisfiy it.*

PROOF. Apply Theorem 1.5 to the sequence $0 \to A \xrightarrow{\subset} B \to B/A \to 0$. ∎

Corollary 1.7. *If* $A_1, \ldots, A_n$ *are modules, then the direct sum* $A_1 \oplus A_2 \oplus \cdots \oplus A_n$ *satisfies the ascending [resp. descending] chain condition on submodules if and only if each* A_i *satisfies it.*

SKETCH OF PROOF. Use induction on n. If $n = 2$, apply Theorem 1.5 to the sequence $0 \to A_1 \xrightarrow{\iota_1} A_1 \oplus A_2 \xrightarrow{\pi_2} A_2 \to 0$. ∎

Theorem 1.8. *If* R *is a left Noetherian [resp. Artinian] ring with identity, then every finitely generated unitary left* R*-module* A *satisfies the ascending [resp. descending] chain condition on submodules.*

An analogous statement is true with "left" replaced by "right."

PROOF OF 1.8. If A is finitely generated, then by Corollary IV.2.2 there is a free R-module F with a finite basis and an epimorphism $\pi : F \to A$. Since F is a direct sum of a finite number of copies of R by Theorem IV.2.1, F is left Noetherian [resp. Artinian] by Corollary 1.7. Therefore $A \cong F/\text{Ker } \pi$ is Noetherian [resp. Artinian] by Corollary 1.6. ∎

Here is a characterization of the ascending chain condition that has no analogue for the descending chain condition.

Theorem 1.9. *A module A satisfies the ascending chain condition on submodules if and only if every submodule of A is finitely generated. In particular, a commutative ring R is Noetherian if and only if every ideal of R is finitely generated.*

PROOF. ($\Rightarrow$) If B is a submodule of A, let S be the set of all finitely generated submodules of B. Since S is nonempty ($0 \in S$), S contains a maximal element C by Theorem 1.4. C is finitely generated by $c_1,c_2,\ldots,c_n$. For each $b \in B$ let D_b be the submodule of B generated by $b,c_1,c_2,\ldots,c_n$. Then $D_b \in S$ and $C \subset D_b$. Since C is maximal, $D_b = C$ for every $b \in B$, whence $b \in D_b = C$ for every $b \in B$ and $B \subset C$. Since $C \subset B$ by construction, $B = C$ and thus B is finitely generated.

($\Leftarrow$) Given a chain of submodules $A_1 \subset A_2 \subset A_3 \subset \cdots$, then it is easy to verify that $\bigcup_{i\geq 1} A_i$ is also a submodule of A and therefore finitely generated, say by $a_1,\ldots,a_k$. Since each a_i is an element of some A_j, there is an index n such that $a_i \in A_n$ for $i = 1,2,\ldots,k$. Consequently, $\bigcup A_i \subset A_n$, whence $A_i = A_n$ for $i \geq n$. ■

We close this section by carrying over to modules the principal results of Section II.8 on subnormal series for groups. This material is introduced in order to prove Corollary 1.12, which will be useful in Chapter IX. We begin with a host of definitions, most of which are identical to those given for groups in Section II.8.

A **normal series** for a module A is a chain of submodules: $A = A_0 \supset A_1 \supset A_2 \supset \cdots \supset A_n$. The **factors** of the series are the quotient modules

$$A_i/A_{i+1} \quad (i = 0,1,\ldots,n-1).$$

The **length** of the series is the number of proper inclusions (= number of nontrivial factors). A **refinement** of the normal series $A_0 \supset A_1 \supset \cdots \supset A_n$ is a normal series obtained by inserting a finite number of additional submodules between the given ones. A **proper refinement** is one which has length larger than the original series. Two normal series are **equivalent** if there is a one-to-one correspondence between the nontrivial factors such that corresponding factors are isomorphic modules. Thus equivalent series necessarily have the same length. A **composition series** for A is a normal series $A = A_0 \supset A_1 \supset A_2 \supset \cdots \supset A_n = 0$ such that each factor A_k/A_{k+1} $(k = 0,1,\ldots,n-1)$ is a nonzero module with no proper submodules.[1]

The various results in Section II.8 carry over readily to modules. For example, *a composition series has no proper refinements* and therefore is equivalent to any of its refinements (see Theorems IV.1.10 and II.8.4 and Lemma II.8.8). Theorems of Schreier, Zassenhaus, and Jordan-Hölder are valid for modules:

Theorem 1.10. *Any two normal series of a module A have refinements that are equivalent. Any two composition series of A are equivalent.*

PROOF. See the corresponding results for groups (Lemma II.8.9 and Theorems II.8.10 and II.8.11). ■

[1]If R has an identity, then a nonzero unitary module with no proper submodules is said to be **simple.** In this case a composition series is a normal series $A = A_0 \supset \cdots \supset A_n = 0$ with simple factors. If R has no identity simplicity is defined somewhat differently; see Definition IX.1.1 and the subsequent Remarks.

Theorem 1.11. *A nonzero module A has a composition series if and only if A satisfies both the ascending and descending chain conditions on submodules.*

PROOF. ($\Rightarrow$) Suppose A has a composition series S of length n. If either chain condition fails to hold, one can find submodules

$$A = A_0 \underset{\neq}{\supset} A_1 \underset{\neq}{\supset} A_2 \underset{\neq}{\supset} \cdots \underset{\neq}{\supset} A_n \underset{\neq}{\supset} A_{n+1},$$

which form a normal series T of length $n + 1$. By Theorem 1.10 S and T have refinements that are equivalent. This is a contradiction since equivalent series have equal length. For every refinement of the composition series S has the same length n as S, but every refinement of T necessarily has length at least $n + 1$. Therefore, A satisfies both chain conditions.

($\Leftarrow$) If B is a nonzero submodule of A, let $S(B)$ be the set of all submodules C of B such that $C \neq B$. Thus if B has no proper submodules, $S(B) = \{0\}$. Also define $S(0) = \{0\}$. For each B there is a maximal element B' of $S(B)$ by Theorem 1.4. Let S be the set of all submodules of A and define a map $f : S \to S$ by $f(B) = B'$; (the Axiom of Choice is needed for the simultaneous selection of the B'). By the Recursion Theorem 6.2 of the Introduction (with $f = f_n$ for all n) there is a function $\varphi : \mathbf{N} \to S$ such that

$$\varphi(0) = A \quad \text{and} \quad \varphi(n + 1) = f(\varphi(n)) = \varphi(n)'.$$

If A_i denotes $\varphi(i)$, then $A \supset A_1 \supset A_2 \supset \cdots$ is a descending chain by construction, whence for some n, $A_i = A_n$ for all $i \geq n$. Since $A_{n+1} = A_n' = f(A_n)$, the definition of f shows that $A_{n+1} = A_n$ only if $A_n = 0 = A_{n+1}$. Let m be the smallest integer such that $A_m = 0$. Then $m \leq n$ and $A_k \neq 0$ for all $k < m$. Furthermore for each $k < m$, A_{k+1} is a maximal submodule of A_k such that $A_k \underset{\neq}{\supset} A_{k+1}$. Consequently, each A_k/A_{k+1} is nonzero and has no proper submodules by Theorem IV.1.10. Therefore, $A \supset A_1 \supset \cdots \supset A_m = 0$ is a composition series for A. ■

Corollary 1.12. *If D is a division ring, then the ring* $Mat_n D$ *of all* $n \times n$ *matrices over D is both Artinian and Noetherian.*

SKETCH OF PROOF. In view of Definition 1.2 and Theorem 1.11 it suffices to show that $R = \text{Mat}_n D$ has a composition series of left R-modules and a composition of right R-modules. For each i let $e_i \in R$ be the matrix with 1_D in position (i,i) and 0 elsewhere. Verify that $Re_i = \{Ae_i \mid A \in R\}$ is a left ideal (submodule) of R consisting of all matrices in R with column j zero for all $j \neq i$. Show that Re_i is a minimal nonzero left ideal (that is, has no proper submodules). One way to do this is via elementary transformation matrices (Definition VII.2.7 and Theorem VII.2.8). Let $M_0 = 0$ and for $i \geq 1$ let $M_i = R(e_1 + e_2 + \cdots + e_i)$. Verify that each M_i is a left ideal of R and that $M_i/M_{i-1} \cong Re_i$, whence $R = M_n \supset M_{n-1} \supset \cdots \supset M_1 \supset M_0 = 0$ is a composition series of left R-modules. A similar argument with the right ideals $e_iR = \{e_iA \mid A \in R\}$ shows that R has a composition series of right R-modules. ■

EXERCISES

1. (a) The ring of all 2×2 matrices $\begin{pmatrix} a & b \\ 0 & c \end{pmatrix}$ such that a is an integer and b,c are rational is right Noetherian but not left Noetherian.
   (b) The ring of all 2×2 matrices $\begin{pmatrix} d & r \\ 0 & s \end{pmatrix}$ such that d is rational and r,s are real is right Artinian but not left Artinian.
2. If I is a nonzero ideal in a principal ideal domain R, then the ring R/I is both Noetherian and Artinian.
3. Let S be a multiplicative subset of a commutative Noetherian ring R with identity. Then the ring $S^{-1}R$ is Noetherian.
4. Let R be a commutative ring with identity. If an ideal I of R is not finitely generated, then there is an infinite properly ascending chain of ideals $J_1 \underset{\neq}{\subset} J_2 \underset{\neq}{\subset} \cdots$ such that $J_k \subset I$ for all k. The union of the J_k need not be I.
5. Every homomorphic image of a left Noetherian [resp. Artinian] ring is left Noetherian [resp. Artinian].
6. A ring R is left Noetherian [resp. Artinian] if and only if $\text{Mat}_n R$ is left Noetherian [resp. Artinian] for every $n \geq 1$ [nontrivial].
7. An Artinian integral domain is a field. [*Hint:* to find an inverse for $a \neq 0$, consider $(a) \supset (a^2) \supset (a^3) \supset \cdots$.]

2. PRIME AND PRIMARY IDEALS

Our main purpose is to study the ideal structure of certain commutative rings. The basic properties of prime ideals are developed. The radical of an ideal is introduced and primary ideals are defined. Finally primary decomposition of ideals is discussed. Except for Theorem 2.2, *all rings are commutative*.

We begin with some background material that will serve both as a motivation and as a source of familiar examples of the concepts to be introduced. The motivation for much of this section arises from the study of principal ideal domains. In particular such a domain D is a unique factorization domain (Theorem III.3.7).

The unique factorization property of D can be stated in terms of ideals: every proper ideal of D is a product of maximal (hence prime) ideals, which are determined uniquely up to order (Exercise III.3.5). Every nonzero prime ideal of D is of the form (p) with p prime ($=$ irreducible) by Theorem III.3.4 and $(p)^n = (p^n)$. Consequently, every proper ideal (a) of D can be written uniquely (up to order) in the form

$$(a) = (p_1^{n_1})(p_2^{n_2})\cdots(p_r^{n_r}) = (p_1^{n_1}) \cap (p_2^{n_2}) \cap \cdots \cap (p_r^{n_r}),$$

where each $n_i > 0$ and the p_i are distinct primes (Exercise III.3.5). Now an ideal $Q = (p^n)$ (p prime) has the property: $ab \in Q$ and $a \notin Q$ imply $b^k \in Q$ for some k (Exercise III.3.5). Such an ideal is called primary. The preceding discussion shows that every ideal in a principal ideal domain is the intersection of a finite number of primary ideals in a unique way. Furthermore there is an obvious connection between these primary ideals and the prime ideals of D; in fact every primary ideal $(p^n) = (p)^n$ is a power of a prime ideal.

In the approach just outlined the viewpoint has switched from consideration of unique factorization of elements as products of primes in D to a consideration of the "primary decomposition" of ideals in the principal ideal domain D. We shall now investigate the "primary decomposition" of ideals in more general commutative rings (where, for instance, ideals need not be principal and primary ideals may not be powers of prime ideals). We begin with some facts about prime ideals.

Theorem 2.1. *An ideal* P $(\neq R)$ *in a commutative ring* R *is prime if and only if* $R - P$ *is a multiplicative set.*

PROOF. This is simply a restatement of Theorem III.2.15; see Definition III.4.1. ■

REMARK. The set of all prime ideals in a ring R is called the **spectrum** of R.

Theorem 2.2. *If* S *is a multiplicative subset of a ring* R *which is disjoint from an ideal* I *of* R, *then there exists an ideal* P *which is maximal in the set of all ideals of* R *disjoint from* S *and containing* I. *Furthermore any such ideal* P *is prime.*

The theorem is frequently used in the case $I = 0$.

SKETCH OF PROOF OF 2.2. The set $\mathcal{S}$ of all ideals of R that are disjoint from S and contain I is nonempty since $I \in \mathcal{S}$. Since $S \neq \emptyset$ (Definition III.4.1) every ideal in $\mathcal{S}$ is properly contained in R. $\mathcal{S}$ is partially ordered by inclusion. By Zorn's Lemma there is an ideal P which is maximal in $\mathcal{S}$. Let A,B be ideals of R such that $AB \subset P$. If $A \not\subset P$ and $B \not\subset P$, then each of the ideals $P + A$ and $P + B$ properly contains P and hence must meet S. Consequently, for some $p_i \in P$, $a \in A$, $b \in B$

$$p_1 + a = s_1 \in S \quad \text{and} \quad p_2 + b = s_2 \in S.$$

Thus $s_1s_2 = p_1p_2 + p_1b + ap_2 + ab \in P + AB \subset P$. This is a contradiction since $s_1s_2 \in S$ and $S \cap P = \emptyset$. Therefore $A \subset P$ or $B \subset P$, whence P is prime. ■

Theorem 2.3. *Let* K *be a subring of a commutative ring* R. *If* $P_1, \ldots, P_n$ *are prime ideals of* R *such that* $K \subset P_1 \cup P_2 \cup \cdots \cup P_n$, *then* $K \subset P_i$ *for some* i.

REMARK. In the case $n \leq 2$, the following proof does not use the hypothesis that each P_i is prime; the hypothesis is needed for $n > 2$.

PROOF OF 2.3. Assume $K \not\subset P_i$ for every i. It then suffices to assume that $n > 1$ and n is minimal; that is, for each i, $K \not\subset \bigcup_{j \neq i} P_j$. For each i there exists $a_i \in K - \bigcup_{j \neq i} P_j$. Since $K \subset \bigcup_i P_i$, each $a_i \in P_i$. The element $a_1 + a_2a_3 \cdots a_n$ lies in K

and hence in $\bigcup_i P_i$. Therefore $a_1 + a_2a_3\cdots a_n = b_j$ with $b_j \in P_j$. If $j > 1$, then $a_1 \in P_j$, which is a contradiction. If $j = 1$, then $a_2a_3\cdots a_n \in P_1$, whence $a_i \in P_1$ for some $i > 1$ by Theorem III.2.15. This also is a contradiction. ■

Proposition 2.4. *If* R *is a commutative ring with identity and* P *is an ideal which is maximal in the set of all ideals of* R *which are not finitely generated, then* P *is prime.*

PROOF. Suppose $ab \in P$ but $a \notin P$ and $b \notin P$. Then $P + (a)$ and $P + (b)$ are ideals properly containing P and therefore finitely generated (by maximality). Consequently $P + (a) = (p_1 + r_1a, \ldots, p_n + r_na)$ and $P + (b) = (p_1' + r_1'b, \ldots, p_m' + r_m'b)$ for some $p_i,p_i' \in P$ and $r_i,r_i' \in R$ (see Theorems III.2.5 and III.2.6). If $J = \{r \in R \mid ra \in P\}$, then J is an ideal. Since $ab \in P$, $(p_i' + r_i'b)a = p_i'a + r_i'ab \in P$ for all i, whence $P \subsetneq P + (b) \subset J$. By maximality, J is finitely generated, say $J = (j_1, \ldots, j_k)$. If $x \in P$, then $x \in P + (a)$ and hence for some $s_i \in R$, $x = \sum_{i=1}^{n} s_i(p_i + r_ia) = \sum_{i=1}^{n} s_ip_i + \sum_{i=1}^{n} s_ir_ia$. Consequently, $(\sum_i s_ir_i)a = x - \sum_i s_ip_i \in P$, whence $\sum_i s_ir_i \in J$. Thus for some $t_i \in R$, $\sum_{i=1}^{n} s_ir_i = \sum_{i=1}^{k} t_ij_i$ and $x = \sum_{i=1}^{n} s_ip_i + \sum_{i=1}^{k} t_ij_ia$. Therefore, P is generated by $p_1, \ldots, p_n, j_1a, \ldots, j_ka$, which is a contradiction. Thus $a \in P$ or $b \in P$ and P is prime by Theorem III.2.15. ■

Definition 2.5. *Let* I *be an ideal in a commutative ring* R. *The* **radical** (*or* **nilradical**) *of* I, *denoted Rad* I, *is the ideal* $\bigcap$ P, *where the intersection is taken over all prime ideals* P *which contain* I. *If the set of prime ideals containing* I *is empty, then Rad* I *is defined to be* R.

REMARKS. If R has an identity, every ideal I ($\neq R$) is contained in a maximal ideal M by Theorem III.2.18. Since $M \neq R$ and M is necessarily prime by Theorem III.2.19, Rad $I \neq R$. Despite the inconsistency of terminology, the radical of the zero ideal is sometimes called the **nilradical** or **prime radical of the ring** R.

EXAMPLES. In any integral domain the zero ideal is prime; hence Rad $0 = 0$. In the ring **Z**, Rad $(12) = (2) \cap (3) = (6)$ and Rad $(4) = (2) =$ Rad (32).

Theorem 2.6. *If* I *is an ideal in a commutative ring* R, *then Rad* I $= \{r \in R \mid r^n \in I$ *for some* $n > 0\}$.

PROOF. If Rad $I = R$, then $\{r \in R \mid r^n \in I\} \subset$ Rad I. Assume Rad $I \neq R$. If $r^n \in I$ and P is any prime ideal containing I, then $r^n \in P$ whence $r \in P$ by Theorem III.2.15. Thus $\{r \in R \mid r^n \in I\} \subset$ Rad I.

Conversely, if $t \in R$ and $t^n \notin I$ for all $n > 0$, then $S = \{t^n + x \mid n \in \mathbf{N}^*; x \in I\}$ is a multiplicative set such that $S \cap I = \varnothing$. By Theorem 2.2 there is a prime ideal P disjoint from S that contains I. By construction, $t \notin P$ and hence $t \notin$ Rad I. Thus $t \notin \{r \in R \mid r^n \in I\}$ implies $t \notin$ Rad I, whence Rad $I \subset \{r \in R \mid r^n \in I\}$. ■

Theorem 2.7. *If* I, $I_1, I_2, \ldots, I_n$ *are ideals in a commutative ring* R, *then:*

(i) $Rad(Rad\,I) = Rad\,I$;

(ii) $Rad\,(I_1I_2\cdots I_n) = Rad\left(\bigcap_{j=1}^{n} I_j\right) = \bigcap_{j=1}^{n} Rad\,I_j$;

(iii) $Rad\,(I^m) = Rad\,I$.

SKETCH OF PROOF. In each case we prove one of the two required containments. (i) If $r \in$ Rad (Rad I), then $r^n \in$ Rad I and hence $r^{nm} = (r^n)^m \in I$ for some $n, m > 0$. Therefore, $r \in$ Rad I and Rad(Rad I) $\subset$ Rad I. (ii) If $r \in \bigcap_j$ Rad I_j, then there are $m_1, m_2, \ldots, m_n > 0$ such that $r^{m_j} \in I_j$ for each j. If $m = m_1 + m_2 + \cdots + m_n$, then $r^m = r^{m_1}r^{m_2}\cdots r^{m_n} \in I_1I_2\cdots I_n$, whence $\bigcap_j$ Rad $I_j \subset$ Rad $(I_1\cdots I_n)$. Finally since $I_1\cdots I_n \subset \bigcap_j I_j$, we have Rad$(I_1\cdots I_n) \subset$ Rad$(\bigcap_j I_j)$. (iii) is a special case of (ii). ■

Definition 2.8. *An ideal* Q ($\neq$ R) *in a commutative ring* R *is* **primary** *if for any* a,b $\in$ R:

$$ab \in Q \quad \textit{and} \quad a \notin Q \quad \Rightarrow \quad b^n \in Q \quad \text{for some} \quad n > 0.$$

EXAMPLE. Every prime ideal is clearly primary. If p is a prime integer and $n \geq 2$ a positive integer, then $(p)^n = (p^n)$ is a primary ideal in **Z** which is not prime (Exercise 17). In general, a power P^n of a prime ideal P need not be primary.

EXAMPLE. If F is a field, the ideal (x,y) is maximal in $F[x,y]$ (Exercise 12) and therefore prime (Theorem III.2.19). Furthermore $(x,y)^2 = (x^2,xy,y^2) \subsetneq (x^2,y) \subsetneq (x,y)$. The ideal (x^2,y) is primary and (x,y) is the only (proper) prime ideal containing (x^2,y) (Exercise 12). Hence the primary ideal (x^2,y) is not a power of any prime ideal in $F[x,y]$.

In the rest of this section all rings have identity.

Theorem 2.9. *If* Q *is a primary ideal in a commutative ring* R, *then* Rad Q *is a prime ideal.*

PROOF. Suppose $ab \in$ Rad Q and $a \notin$ Rad Q. Then $a^nb^n = (ab)^n \in Q$ for some n. Since $a \notin$ Rad Q, $a^n \notin Q$. Since Q is a primary, there is an integer $m > 0$ such that $(b^n)^m \in Q$, whence $b \in$ Rad Q. Therefore, Rad Q is prime by Theorem III.2.15. ■

In view of Theorem 2.9 we shall adopt the following terminology. If Q is a primary ideal in a commutative ring R, then the radical P of Q is called the **associated prime ideal** of Q. One says that Q is a **primary ideal belonging to the prime P** or that Q is **primary for P** or that Q is **P-primary.** For a given primary ideal Q, the associated prime ideal Rad Q is clearly unique. However, a given prime ideal P may be the associated prime of several different primary ideals.

EXAMPLE. If p is a prime in **Z**, then each of the primary ideals $(p^2), (p^3), \ldots$ belongs to the prime ideal (p). In the ring $\mathbf{Z}[x,y]$ the ideals (x^2,y), (x^2,y^2), (x^2,y^3). etc. are all primary ideals belonging to the prime ideal (x,y) (Exercise 13).

Theorem 2.10. *Let* Q *and* P *be ideals in a commutative ring* R. *Then* Q *is primary for* P *if and only if:*

(i) $Q \subset P \subset Rad\, Q$; *and*
(ii) *if* $ab \in Q$ *and* $a \notin Q$, *then* $b \in P$.

SKETCH OF PROOF. Suppose (i) and (ii) hold. If $ab \in Q$ with $a \notin Q$, then $b \in P \subset \text{Rad } Q$, whence $b^n \in Q$ for some $n > 0$. Therefore Q is primary. To show that Q is primary for P we need only show $P = \text{Rad } Q$. By (i), $P \subset \text{Rad } Q$. If $b \in \text{Rad } Q$, let n be the least integer such that $b^n \in Q$. If $n = 1$, $b \in Q \subset P$. If $n > 1$, then $b^{n-1}b = b^n \in Q$, with $b^{n-1} \notin Q$ by the minimality of n. By (ii), $b \in P$. Thus $b \in \text{Rad } Q$ implies $b \in P$, whence $\text{Rad } Q \subset P$. The converse implication is easy. ■

Theorem 2.11. *If* $Q_1, Q_2, \ldots, Q_n$ *are primary ideals in a commutative ring* R, *all of which are primary for the prime ideal* P, *then* $\bigcap_{i=1}^{n} Q_i$ *is also a primary ideal belonging to* P.

PROOF. Let $Q = \bigcap_{i=1}^{n} Q_i$. Then by Theorem 2.7(ii), $\text{Rad } Q = \bigcap_{i=1}^{n} \text{Rad } Q_i = \bigcap_{i=1}^{n} P = P$; in particular, $Q \subset P \subset \text{Rad } Q$. If $ab \in Q$ and $a \notin Q$, then $ab \in Q_i$ and $a \notin Q_i$ for some i. Since Q_i is P-primary, $b \in P$ by Theorem 2.10(ii). Consequently, Q itself is P-primary by Theorem 2.10. ■

Definition 2.12. *An ideal* I *in a commutative ring* R *has a* **primary decomposition** *if* $I = Q_1 \cap Q_2 \cap \cdots \cap Q_n$ *with each* Q_i *primary. If no* Q_i *contains* $Q_1 \cap \cdots \cap Q_{i-1} \cap Q_{i+1} \cap \cdots \cap Q_n$ *and the radicals of the* Q_i *are all distinct, then the primary decomposition is said to be* **reduced** (*or* **irredundant**).

Theorem 2.13. *Let* I *be an ideal in a commutative ring* R. *If* I *has a primary decomposition, then* I *has a reduced primary decomposition.*

PROOF. If $I = Q_1 \cap \cdots \cap Q_n$ (Q_i primary) and some Q_i contains $Q_1 \cap \cdots \cap Q_{i-1} \cap Q_{i+1} \cap \cdots \cap Q_n$, then $I = Q_1 \cap \cdots \cap Q_{i-1} \cap Q_{i+1} \cap \cdots \cap Q_n$ is also a primary decomposition. By thus eliminating the superfluous Q_i (and reindexing) we have $I = Q_1 \cap \cdots \cap Q_k$ with no Q_i containing the intersection of the other Q_j. Let $P_1, \ldots, P_r$ be the distinct prime ideals in the set $\{\text{Rad } Q_1, \ldots, \text{Rad } Q_k\}$. Let Q_i' $(1 \leq i \leq r)$ be the intersection of all the Q's that belong to the prime P_i. By Theorem 2.11 each Q_i' is primary for P_i. Clearly no Q_i' contains the intersection of all the other Q_j'. Therefore, $I = \bigcap_{i=1}^{k} Q_i = \bigcap_{i=1}^{r} Q_i'$, whence I has a reduced primary decomposition. ■

At this point there are two obvious questions to ask. Which ideals have a reduced primary decomposition? Is a reduced primary decomposition unique in any way? Both questions will be answered in a more general setting in the next section (Theorems 3.5 and 3.6).

EXERCISES

Note: R is always a commutative ring.

1. Let R be a commutative Artinian ring with identity.
   (a) Every prime ideal of R is maximal [*Hint:* Theorems III.2.16 and III.2.20 and Exercises 1.5 and 1.7].
   (b) R has only a finite number of distinct prime ideals.

2. If R has an identity and $\{P_i \mid i \in I\}$ is a nonempty family of prime ideals of R which is linearly ordered by inclusion, then $\bigcup_{i\in I} P_i$ and $\bigcap_{i\in I} P_i$ are prime ideals.

3. If $P_1,P_2,\dots,P_n$ are prime ideals in R and I is any ideal such that $I \not\subset P_i$ for all i, then there exists $r \in I$ such that $r \notin P_i$ for all i.

4. If R has an identity and $M_1,\dots,M_r$ are distinct maximal ideals in R, then show that $M_1 \cap M_2 \cap \cdots \cap M_r = M_1M_2\cdots M_r$. Is this true if "maximal" is replaced by "prime"?

5. If R has an identity, then the set of all zero divisors of R is a union of prime ideals.

6. Let R have an identity. A prime ideal P in R is called a **minimal prime** ideal of the ideal I if $I \subset P$ and there is no prime ideal P' such that $I \subset P' \subsetneqq P$.
   (a) If an ideal I of R is contained in a prime ideal P of R, then P contains a minimal prime ideal of I. [*Hint:* Zornify the set of all prime ideals P' such that $I \subset P' \subset P$.]
   (b) Every proper ideal possesses at least one minimal prime ideal.

7. The radical of an ideal I in a ring R with identity is the intersection of all its minimal prime ideals [see Exercise 6].

8. If R has an identity, I is an ideal and J is a finitely generated ideal such that $J \subset \operatorname{Rad} I$, then there exists a positive integer n such that $J^n \subset I$.

9. What is the radical of the zero ideal in Z_n?

10. If S is a multiplicative subset of a commutative ring R and I is an ideal of R, then $S^{-1}(\operatorname{Rad} I) = \operatorname{Rad}(S^{-1}I)$ in the ring $S^{-1}R$.

11. Let Q $(\neq R)$ be an ideal in R. Then Q is primary if and only if every zero divisor in R/Q is nilpotent (see Exercise III.1.12).

12. If F is a field, then:
    (a) the ideal (x,y) is maximal in $F[x,y]$;
    (b) $(x,y)^2 = (x^2,xy,y^2) \subsetneqq (x^2,y) \subsetneqq (x,y)$;
    (c) the ideal (x^2,y) is primary and the only proper prime ideal containing it is (x,y).

13. In the ring $\mathbf{Z}[x,y]$ the ideals $(x^2,y),(x^2,y^2),(x^2,y^3),\dots,(x^i,y^j),\dots$ are all primary ideals belonging to the prime ideal (x,y).

14. The conclusion of Theorem 2.11 is false if infinite intersections are allowed. [*Hint:* consider $\mathbf{Z}$.]

15. Let $f: R \to S$ be an epimorphism of commutative rings with identity. If J is an ideal of S, let $I = f^{-1}(J)$.
    (a) Then I is primary in R if and only if J is primary in S.
    (b) If J is primary for P, then I is primary for the prime ideal $f^{-1}(P)$.

16. Find a reduced primary decomposition for the ideal $I = (x^2,xy,2)$ in $\mathbf{Z}[x,y]$ and determine the associated primes of the primary ideals appearing in this decomposition.

17. (a) If p is prime and $n > 1$, then (p^n) is a primary, but not a prime ideal of $\mathbf{Z}$.
    (b) Obtain a reduced primary decomposition of the ideal (12600) in $\mathbf{Z}$.

18. If F is a field and I is the ideal (x^2,xy) in $F[x,y]$, then there are at least three distinct reduced primary decompositions of I; three such are:

$$\text{(i) } I = (x) \cap (x^2,y);\ \text{(ii) } I = (x) \cap (x^2,x+y);\ \text{(iii) } I = (x) \cap (x^2,xy,y^2).$$

19. (a) In the ring $\mathbf{Z}[x]$, the following are primary decompositions:

$$(4,2x,x^2) = (4,x) \cap (2,x^2);$$
$$(9,3x+3) = (3) \cap (9,x+1).$$

    (b) Are the primary decompositions of part (a) reduced?

3. PRIMARY DECOMPOSITION

We shall extend the results of Section 2 in a natural way to modules. A uniqueness statement for reduced primary decompositions (of submodules or ideals) is proved as well as the fact that every submodule [ideal] of a Noetherian module [ring] has a primary decomposition. Throughout this section *all rings are commutative with identity and all modules are unitary.*

Definition 3.1. *Let* R *be a commutative ring with identity and* B *an* R*-module. A submodule* A ($\neq$ B) *is* **primary** *provided that*

$$r \in R,\ b \notin A \text{ and } rb \in A \quad \Rightarrow \quad r^nB \subset A \text{ for some positive integer } n.$$

EXAMPLE. Consider the ring R as an R-module and let Q be a primary ideal (and hence a submodule) of R. If $rb \in Q$ with $r \in R$ and $b \notin Q$, then $r^n \in Q$ for some n. Since Q is an ideal, this implies $r^nR \subset Q$. Hence Q is a primary submodule of the module R. Conversely every primary submodule of R is a primary ideal (Exercise 1). Therefore, all results about primary submodules apply to primary ideals as well.

Theorem 3.2. *Let* R *be a commutative ring with identity and* A *a primary submodule of an* R*-module* B. *Then* $Q_A = \{r \in R \mid rB \subset A\}$ *is a primary ideal in* R.

PROOF. Since $A \neq B$, $1_R \notin Q_A$, whence $Q_A \neq R$. If $rs \in Q_A$ and $s \notin Q_A$, then $sB \not\subset A$. Consequently, for some $b \in B$, $sb \notin A$ but $r(sb) \in A$. Since A is primary $r^nB \subset A$ for some n; that is, $r^n \in Q_A$. Therefore, Q_A is primary. ■

Let R,A,B, and Q_A be as in Theorem 3.2. By Theorem 2.9 Rad $Q_A = P_1$ is a prime ideal. It is easy to see that $P_1 = \{r \in R \mid r^nB \subset A \text{ for some } n > 0\}$. A primary submodule A of a module B is said to **belong to a prime ideal** P or to be a **P-primary submodule** of B if $P = \text{Rad } Q_A = \{r \in R \mid r^nB \subset A \text{ for some } n > 0\}$. This terminology is consistent with that used for ideals. In particular, if J is a primary ideal, then $Q_J = J$.

Definition 3.3. *Let* R *be a commutative ring with identity and* B *an* R*-module. A submodule* C *of* B *has a* **primary decomposition** *if* $C = A_1 \cap A_2 \cap \cdots \cap A_n$, *with each* A_i *a* P_i*-primary submodule of* B *for some prime ideal* P_i *of* R. *If no* A_i *contains* $A_1 \cap \cdots \cap A_{i-1} \cap A_{i+1} \cap \cdots \cap A_n$ *and if the ideals* $P_1, \dots, P_n$ *are all distinct, then the primary decomposition is said to be* **reduced.**

Again the terminology here is consistent with that used for ideals. If C,A_i and P_i are as in the definition and $P_j \not\subset P_i$ for all $j \neq i$, then P_i is said to be an **isolated prime** ideal of C. In other words, P_i is isolated if it is minimal in the set $\{P_1, \dots, P_n\}$. If P_i is not isolated it is said to be **embedded.**

Theorem 3.4. *Let* R *be a commutative ring with identity and* B *an* R*-module. If a submodule* C *of* B *has a primary decomposition, then* C *has a reduced primary decomposition.*

SKETCH OF PROOF. The proof is similar to that of Theorem 2.13. Note that if $Q_A = \{r \in R \mid rB \subset A\}$, then $\bigcap_{i=1}^{n} Q_{A_i} = Q_{\cap A_i}$. Thus if $A_1, \dots, A_r$ are all P-primary submodules for the same prime ideal P, then $\bigcap_{i=1}^{r} A_i$ is also P-primary by Theorem 2.11. ■

Theorem 3.5. *Let* R *be a commutative ring with identity and* B *an* R*-module. Let* C ($\neq$ B) *be a submodule of* B *with two reduced primary decompositions,*

$$A_1 \cap A_2 \cap \cdots \cap A_k = C = A_1' \cap A_2' \cap \cdots \cap A_s',$$

where A_i *is* P_i*-primary and* A_j' *is* P_j'*-primary. Then* $k = s$ *and (after reordering if necessary)* $P_i = P_i'$ *for* $i = 1,2,\dots,k$. *Furthermore if* A_i *and* A_i' *both are* P_i*-primary and* P_i *is an isolated prime, then* $A_i = A_i'$.

PROOF. By changing notation if necessary we may assume that P_1 is maximal in the set $\{P_1, \dots, P_k, P_1', \dots, P_s'\}$. We shall first show that $P_1 = P_j'$ for some j. Suppose, on the contrary, that $P_1 \neq P_j'$ for $j = 1,2,\dots,s$. Since P_1 is maximal we have $P_1 \not\subset P_j'$ for $j = 1,2,\dots,s$. Since the first decomposition is reduced, $P_1,P_2,\dots,P_k$ are distinct, whence $P_1 \not\subset P_i$ for $i = 2,3,\dots,k$. By the contrapositive of Theorem 2.3, $P_1 \not\subset P_2 \cup \cdots \cup P_k \cup P_1' \cup \cdots \cup P_s'$. Consequently, there exists $r \in P_1$ such that $r \notin P_i$ $(i \geq 2)$ and $r \notin P_j'$ $(j \geq 1)$. Since A_1 is P_1-primary $r^nB \subset A_1$ for some positive integer n. Let C^* be the submodule $\{x \in B \mid r^nx \in C\}$. If $k = 1$, then $C = A_1$ and hence $C^* = B$. We claim that for $k \geq 1$, $C^* = C$ and for $k > 1$,

$C^* = A_2 \cap \cdots \cap A_k$. Now it is easy to see that $A_2 \cap \cdots \cap A_k \subset C^*$ and $A_1' \cap A_2' \cap \cdots \cap A_s' = C \subset C^*$ for $k > 1$. Conversely, if $x \notin A_i$ $(i \geq 2)$, then $r^n x \notin A_i$ (otherwise $r^n \in P_i$ since A_i is P_i-primary, whence $r \in P_i$ since P_i is prime). Consequently, $r^n x \notin C$, whence $x \notin C^*$. Therefore, $C^* \subset A_2 \cap \cdots \cap A_k$ for $k > 1$. A similar argument shows that $C^* \subset A_1' \cap A_2' \cap \cdots \cap A_s' = C$, so that $C^* = C$ $(k \geq 1)$ and $C^* = A_2 \cap \cdots \cap A_k$ $(k > 1)$. If $k = 1$, then as observed above $C^* = B$. Thus $C = C^* = B$, which contradicts the fact that $C \neq B$. If $k > 1$, then

$$A_2 \cap \cdots \cap A_k = C^* = C = A_1 \cap A_2 \cap \cdots \cap A_k,$$

whence $A_2 \cap \cdots \cap A_k \subset A_1$. This conclusion contradicts the fact that the first decomposition is reduced. Thus the assumption that $P_1 \neq P_j'$ for every j always leads to a contradiction. Therefore $P_1 = P_j'$ for some j, say $j = 1$.

The proof now proceeds by induction on k. If $k = 1$, then we claim $s = 1$ also. For if $s > 1$, then the argument above with $P_1 = P_1'$ and the roles of A_i, A_i' reversed) shows that $B = C^* = A_2' \cap \cdots \cap A_s'$, whence $A_j' = B$ for some $j \geq 2$. Thus the second decomposition of C is not reduced, a contradiction. Therefore, $s = 1 = k$ and $A_1 = C = A_1'$. Now assume that $k > 1$ and the theorem is true for all submodules that have a reduced primary decomposition of less than k terms. The argument of the preceding paragraph (with $P_1 = P_1'$) shows that for $k > 1$ the submodule C^* has two reduced primary decompositions:

$$A_2 \cap A_3 \cap \cdots \cap A_k = C^* = A_2' \cap \cdots \cap A_s'.$$

By induction $k = s$, and (after reindexing) $P_i = P_i'$ for all i. This completes the induction and the proof of the first part of the theorem.

Suppose A_i and A_i' are both P_i-primary and P_i is an isolated prime. For convenience of notation assume $i = 1$. Since P_1 is isolated, there exists for each $j \geq 2$, $r_j \in P_j - P_1$. Then $t = r_2 r_3 \cdots r_k \in P_j$ for $j > 1$, but $t \notin P_1$. Since A_j is P_j-primary, there exists for each $j \geq 2$ an integer n_j such that $t^{n_j} B \subset A_j$. Similarly, for each $j \geq 2$ there is an m_j such that $t^{m_j} B \subset A_j'$. Let $n = \max \{n_2, \ldots, n_k, m_2, \ldots, m_k\}$; then $t^n B \subset A_j$ and $t^n B \subset A_j'$ for all $j \geq 2$. Let D be the submodule $\{x \in B \mid t^n x \in C\}$. To complete the uniqueness proof we shall show $A_1 = D = A_1'$. If $x \in A_1$, then $t^n x \in A_1 \cap A_2 \cap \cdots \cap A_k = C$, whence $x \in D$ and $A_1 \subset D$. If $x \in D$, then $t^n x \in C \subset A_1$. Since A_1 is P_1-primary and $t \notin P_1$, we have $t^m B \not\subset A_1$, for all $m > 0$. Since A_1 is primary, we must have $x \in A_1$, (otherwise $t^n x \in A_1$ and $x \notin A_1$ imply $t^{nq} B \subset A_1$ for some positive q by Definition 2.1). Hence $D = A_1$. An identical argument shows that $A_1' = D$. Therefore, $A_1 = A_1'$. ∎

Thus far we have worked with a module that was assumed to have a primary decomposition. Now we give a partial answer to the question: which modules [ideals] have primary decompositions?

Theorem 3.6. *Let* R *be a commutative ring with identity and* B *an* R*-module satisfying the ascending chain condition on submodules. Then every submodule* A ($\neq$B) *has a reduced primary decomposition. In particular, every submodule* A ($\neq$B) *of a finitely generated module* B *over a commutative Noetherian ring* R *and every ideal* ($\neq$R) *of* R *has a reduced primary decomposition.*

PROOF OF 3.6. Let $\mathcal{S}$ be the set of all submodules of B that do *not* have a primary decomposition. Clearly no primary submodule is in $\mathcal{S}$. We must show that $\mathcal{S}$ is actually empty. If $\mathcal{S}$ is nonempty, then $\mathcal{S}$ contains a maximal element C by Theorem 1.4. Since C is not primary, there exist $r \in R$ and $b \in B - C$ such that $rb \in C$ but $r^nB \not\subset C$ for all $n > 0$. Let $B_n = \{x \in B \mid r^nx \in C\}$. Then each B_n is a submodule of B and $B_1 \subset B_2 \subset B_3 \subset \cdots$. By hypothesis there exists $k > 0$ such that $B_i = B_k$ for $i \geq k$. Let D be the submodule $\{x \in B \mid x = r^ky + c$ for some $y \in B, c \in C\}$. Clearly $C \subset B_k \cap D$. Conversely, if $x \in B_k \cap D$, then $x = r^ky + c$ and $r^kx \in C$, whence $r^{2k}y = r^k(r^ky) = r^k(x - c) = r^kx - r^kc \in C$. Therefore, $y \in B_{2k} = B_k$. Consequently, $r^ky \in C$ and hence $x = r^ky + c \in C$. Therefore $B_k \cap D \subset C$, whence $B_k \cap D = C$. Now $C \neq B_k \neq B$ and $C \neq D \neq B$ since $b \in B_k - C$ and $r^kB \not\subset C$. By the maximality of C in $\mathcal{S}$, B_k and D must have primary decompositions. Thus C has a primary decomposition, which is a contradiction. Therefore $\mathcal{S}$ is empty and every submodule has a primary decomposition. Consequently, every submodule has a reduced primary decomposition by Theorem 3.4. The last statement of the theorem is now an immediate consequence of Theorems 1.8 and 1.9. ■

EXERCISES

Note: Unless otherwise stated R is always a commutative ring with identity.

1. Consider the ring R as an R-module. If Q is a primary submodule of R, then Q is a primary ideal.

2. (a) Let $f: B \to D$ be an R-module epimorphism and $C\ (\neq D)$ a submodule of D. Then C is a primary submodule of D if and only if $f^{-1}(C)$ is a primary submodule of B.
(b) If C and $f^{-1}(C)$ are primary, then they both belong to the same prime ideal P.

3. If $A\ (\neq B)$ is a submodule of the R-module B and P is an ideal of R such that
 (i) $rx \in A$ and $x \notin A$ $(r \in R, x \in B) \Rightarrow r \in P$; and
 (ii) $r \in P \Rightarrow r^nB \subset A$ for some positive integer n,
then P is a prime ideal and A is a P-primary submodule of B.

4. If A is a P-primary submodule of an R-module B and $rx \in A$ $(r \in R, x \in B)$, then either $r \in P$ or $x \in A$.

5. If A is a P-primary submodule of an R-module B and C is any submodule of B such that $C \not\subset A$ then $\{r \in R \mid rC \subset A\}$ is a P-primary ideal. [*Hint:* Exercise 3 may be helpful.]

6. Let A be a P-primary submodule of the R-module B and let C be any submodule of B such that $C \not\subset A$. Then $A \cap C$ is a P-primary submodule of C. [*Hint:* Exercise 3 may be helpful.]

7. If B is an R-module and $x \in B$, the annihilator of x, denoted ann x, is $\{r \in R \mid rx = 0\}$. Show that ann x is an ideal.

8. If $B \neq 0$ is an R-module and P is maximal in the set of ideals $\{\text{ann}\, x \mid 0 \neq x \in B\}$ (see Exercise 7), then P is prime.

9. Let R be Noetherian and let B be an R-module. If P is a prime ideal such that $P = \text{ann}\, x$ for some nonzero $x \in B$ (see Exercise 7), then P is called an **associated prime** of B.
   (a) If $B \neq 0$, then there exists an associated prime of B. [*Hint:* use Exercise 8.]
   (b) If $B \neq 0$ and B satisfies the ascending chain condition on submodules, then there exist prime ideals $P_1, \dots, P_{r-1}$ and a sequence of submodules $B = B_1 \supset B_2 \supset \cdots \supset B_r = 0$ such that $B_i/B_{i+1} \cong R/P_i$ for each $i < r$.

10. Let R and B be as in Exercise 9(b). Then the following conditions on $r \in R$ are equivalent:
   (i) for each $x \in B$ there exists a positive integer $n(x)$ such that $r^{n(x)}x = 0$;
   (ii) r lies in every associated prime of B (see Exercises 9 and 15).

11. Let R be Noetherian, $r \in R$, and B an R-module. Then $rx = 0$ $(x \in B)$ implies $x = 0$ if and only if r does not lie in any associated prime of B (see Exercises 8 and 9).

12. Let R be Noetherian and let B be an R-module satisfying the ascending chain condition on submodules. Then the following are equivalent:
   (i) There exists exactly one associated prime of B (see Exercise 9);
   (ii) $B \neq 0$ and for each $r \in R$ one of the following is true: either $rx = 0$ implies $x = 0$ for all $x \in B$ or for each $x \in B$ there exists a positive integer $n(x)$ such that $r^{n(x)}x = 0$. [See Exercises 10 and 11.]

13. Let R and B be as in Exercise 12. Then a submodule A of B is primary if and only if B/A has exactly one associated prime P and in that case A is P-primary; (see Exercises 9 and 12).

14. Let R and B be as in Exercise 12. If A $(\neq B)$ is a submodule of B, then every associated prime of A is an associated prime of B. Every associated prime of B is an associated prime of either A or B/A; (see Exercise 9).

15. Let R and B be as in Exercise 12. Then the associated primes of B are precisely the primes $P_1, \dots, P_n$, where $0 = A_1 \cap \cdots \cap A_n$ is a reduced primary decomposition of 0 with each A_i P_i-primary. In particular, the set of associated primes of B is finite. [*Hint:* see Exercises 9, 13, 14.]

16. Let S be a multiplicative subset of R and let A be a P-primary submodule of an R-module B. If $P \cap S = \varnothing$, then $S^{-1}A$ is an $S^{-1}P$-primary submodule of the $S^{-1}R$-module $S^{-1}B$.

4. NOETHERIAN RINGS AND MODULES

This section consists of two independent parts. The first part deals primarily with Noetherian modules (that is, modules satisfying the ascending chain condition). A rather strong form of the Krull Intersection Theorem is proved. Nakayama's Lemma and several of its interesting consequences are presented. In the second part of this section, which does not depend on the first part, we prove that if R is a commutative Noetherian ring with identity, then so are the polynomial ring $R[x_1, \dots, x_n]$ and the power series ring $R[[x]]$. With few exceptions *all rings are commutative with identity.*

We begin by recalling that a *commutative* ring R is Noetherian if and only if R

satisfies the maximum condition on (two-sided) ideals (Definition 1.2 and Theorem 1.4), or equivalently if and only if every ideal of R is finitely generated (Theorem 1.9). As a matter of fact, one need only consider prime ideals of R:

Proposition 4.1. (*I. S. Cohen*). *A commutative ring* R *with identity is Noetherian if and only if every prime ideal of* R *is finitely generated.*

SKETCH OF PROOF. ($\Leftarrow$) Let $\mathcal{S}$ be the set of all ideals of R which are not finitely generated. If $\mathcal{S}$ is nonempty, then use Zorn's Lemma to find a maximal element P of $\mathcal{S}$. P is prime by Proposition 2.4 and hence finitely generated by hypothesis. This is a contradiction unless $\mathcal{S} = \varnothing$. Therefore, R is Noetherian by Theorem 1.9. ■

We now develop the preliminaries needed to prove the Krull Intersection Theorem. If B is a module over a commutative ring R, then it is easy to see that $I = \{r \in R \mid rb = 0 \text{ for all } b \in B\}$ is an ideal of R. The ideal I is called the **annihilator** of B in R.

Lemma 4.2. *Let* B *be a finitely generated module over a commutative ring* R *with identity and let* I *be the annihilator of* B *in* R. *Then* B *satisfies the ascending [resp. descending] chain condition on submodules if and only if* R/I *is a Noetherian [resp. Artinian] ring.*

SKETCH OF PROOF. Let B be generated by $b_1, \dots, b_n$ and assume B satisfies the ascending chain condition. Then $B = Rb_1 + \cdots + Rb_n$ by Theorem IV.1.5. Consequently, $I = I_1 \cap I_2 \cap \cdots \cap I_n$, where I_j is the annihilator of the submodule Rb_j. By Corollary III.2.27 there is a monomorphism of rings $\theta : R/I \to R/I_1 \times \cdots \times R/I_n$. It is easy to see that θ is also an R-module monomorphism. Verify that for each j the map $R/I_j \to Rb_j$ given by $r + I_j \mapsto rb_j$ is an isomorphism of R-modules. Since the submodule Rb_j of B necessarily satisfies the ascending chain condition, so does R/I_j. Therefore, $R/I_1 \oplus \cdots \oplus R/I_n$ satisfies the ascending chain condition on R-submodules by Corollary 1.7. Consequently its submodule $\operatorname{Im} \theta \cong R/I$ satisfies the ascending chain condition on R-submodules. But every ideal of the ring R/I is an R-submodule of R/I. Therefore, R/I is Noetherian.

Conversely suppose R/I is Noetherian. Verify that B is an R/I-module with $(r + I)b = rb$ and that the R/I submodules of B are precisely the R-submodules. Consequently, B satisfies the ascending chain condition by Theorem 1.8. ■

Recall that if I is *any* ideal in a ring R with identity and B is an R-module, then

$$IB = \left\{ \sum_{i=1}^{n} r_i b_i \mid r_i \in I;\ b_i \in B;\ n \in \mathbf{N}^* \right\}$$ is a submodule of B (Exercise IV.1.3).

Lemma 4.3. *Let* P *be a prime ideal in a commutative ring* R *with identity. If* C *is a* P*-primary submodule of the Noetherian* R*-module* A, *then there exists a positive integer* m *such that* $P^m A \subset C$.

PROOF. Let I be the annihilator of A in R and consider the ring $\overline{R} = R/I$. Denote the coset $r + I \varepsilon \overline{R}$ by $\bar{r}$. Clearly $I \subset \{r \varepsilon R \mid rA \subset C\} \subset P$, whence $\overline{P} = P/I$ is an ideal of $\overline{R}$. A and C are each $\overline{R}$-modules with $\bar{r}a = ra$ $(r \varepsilon R, a \varepsilon A)$. We claim that C is a primary $\overline{R}$-submodule of A. If $\bar{r}a \varepsilon C$ with $r \varepsilon R$ and $a \varepsilon A - C$, then $ra \varepsilon C$. Since C is a primary R-submodule, $r^nA \subset C$ for some n, whence $\bar{r}^nA \subset C$ and C is $\overline{R}$-primary. Since $\{\bar{r} \varepsilon \overline{R} \mid \bar{r}^kA \subset C$ for some $k > 0\} = \{\bar{r} \varepsilon \overline{R} \mid r^kA \subset C\}$ $= \{\bar{r} \varepsilon \overline{R} \mid r \varepsilon P\} = \overline{P}$, $\overline{P}$ is a prime ideal of $\overline{R}$ and C is a $\overline{P}$-primary $\overline{R}$-submodule of A (see Theorems 2.9 and 3.2).

Since $\overline{R}$ is Noetherian by Lemma 4.2, $\overline{P}$ is finitely generated by Theorem 1.9. Let $\bar{p}_1, \dots, \bar{p}_s$ $(p_i \varepsilon P)$ be the generators of $\overline{P}$. For each i there exists n_i such that $\bar{p}_i^{n_i}A \subset C$. If $m = n_1 + \cdots + n_s$, then it follows from Theorems III.1.2(v) and III.2.5(vi) that $\overline{P}^mA \subset C$. The facts that $\overline{P} = P/I$ and $IA = 0$ now imply that $P^mA \subset C$. ■

Theorem 4.4. *(Krull Intersection Theorem). Let* R *be a commutative ring with identity,* I *an ideal of* R *and* A *a Noetherian* R*-module. If* $B = \bigcap_{n=1}^{\infty} I^nA$, *then* $IB = B$.

Theorem 4.4 was first proved in the case where R is a Noetherian local ring with maximal ideal I. The proof we shall give depends on primary decomposition (as did the original proof). However, if one assumes that R is Noetherian, there are a number of proofs that do not use primary decomposition (Exercise 2).

PROOF OF 4.4. If $IB = A$, then $A = IB \subset B$, whence $B = A = IB$. If $IB \neq A$, then by Theorem 3.6 IB has a primary decomposition:

$$IB = A_1 \cap A_2 \cap \cdots \cap A_s,$$

where each A_i is a P_i-primary submodule of A for some prime ideal P_i of R. Since $IB \subset B$ in any case, we need only show that $B \subset A_i$ for every i in order to conclude that $B \subset IB$ and hence that $B = IB$.

Let i $(1 \leq i \leq s)$ be fixed. Suppose first that $I \subset P_i$. By Lemma 4.3 there is an integer m such that $P_i^mA \subset A_i$, whence $B = \bigcap_n I^nA \subset I^mA \subset P_i^mA \subset A_i$. Now suppose $I \not\subset P_i$. Then there exists $r \varepsilon I - P_i$. If $B \not\subset A_i$, then there exists $b \varepsilon B - A_i$. Since $rb \varepsilon IB \subset A_i$, $b \notin A_i$ and A_i is primary, $r^nA \subset A_i$ for some $n > 0$. Consequently, $r \varepsilon P_i$ since A_i is a P_i-primary submodule. This contradicts the choice of $r \varepsilon I - P_i$. Therefore $B \subset A_i$. ■

Lemma 4.5. *(Nakayama) If* J *is an ideal in a commutative ring* R *with identity, then the following conditions are equivalent.*

(i) J *is contained in every maximal ideal of* R;

(ii) $1_R - j$ *is a unit for every* $j \varepsilon J$;

(iii) *If* A *is a finitely generated* R*-module such that* $JA = A$, *then* $A = 0$;

(iv) *If* B *is a submodule of a finitely generated* R*-module* A *such that* $A = JA + B$, *then* $A = B$.

REMARK. The Lemma is true even when R is noncommutative, provided that (i) is replaced by the condition that J is contained in the Jacobson radical of R (Exercise IX.2.17).

PROOF OF 4.5. (i) $\Rightarrow$ (ii) if $j \in J$ and $1_R - j$ is not a unit, then the ideal $(1_R - j)$ is not R itself (Theorem III.3.2) and therefore is contained in a maximal ideal $M \neq R$ (Theorem III.2.18). But $1_R - j \in M$ and $j \in J \subset M$ imply that $1_R \in M$, which is a contradiction. Therefore, $1_R - j$ is a unit.

(ii) $\Rightarrow$ (iii) Since A is finitely generated, there must be a minimal generating set $X = \{a_1, \ldots, a_n\}$ of A (that is, no proper subset of X generates A). If $A \neq 0$, then $a_1 \neq 0$ by minimality. Since $JA = A$, $a_1 = j_1a_1 + j_2a_2 + \cdots + j_na_n$ $(j_i \in J)$, whence $1_Ra_1 = a_1$ so that $(1_R - j_1)a_1 = 0$ if $n = 1$ and

$$(1_R - j_1)a_1 = j_2a_2 + \cdots + j_na_n \quad \text{if} \quad n > 1.$$

Since $1_R - j_1$ is a unit in R, $a_1 = (1_R - j_1)^{-1}(1_R - j_1)a_1$. Thus if $n = 1$, then $a_1 = 0$ which is a contradiction. If $n > 1$, then a_1 is a linear combination of $a_2, \ldots, a_n$. Consequently, $\{a_2, \ldots, a_n\}$ generates A, which contradicts the choice of X.

(iii) $\Rightarrow$ (iv) Verify that the quotient module A/B is such that $J(A/B) = A/B$, whence $A/B = 0$ and $A = B$ by (iii).

(iv) $\Rightarrow$ (i) If M is any maximal ideal, then the ideal $JR + M$ contains M. But $JR + M \neq R$ (otherwise $R = M$ by (iv)). Consequently, $JR + M = M$ by maximality. Therefore $J = JR \subset M$. ■

We now give several applications of Nakayama's Lemma, beginning with a result that is the starting point of the theory of completions.

Proposition 4.6. *Let* J *be an ideal in a commutative ring* R *with identity. Then* J *is contained in every maximal ideal of* R *if and only if for every* R*-module* A *satisfying the ascending chain condition on submodules,* $\bigcap_{n=1}^{\infty} J^nA = 0$.

PROOF. ($\Rightarrow$) If $B = \bigcap_n J^nA$, then $JB = B$ by Theorem 4.4. Since B is finitely generated by Theorem 1.9, $B = 0$ by Nakayama's Lemma 4.5.

($\Leftarrow$) We may assume $R \neq 0$. If M is any maximal ideal of R, then $M \neq R$ and $A = R/M$ is a nonzero R-module that has no proper submodules (Theorem IV.1.10). Thus A trivially satisfies the ascending chain condition, whence $\bigcap_n J^nA = 0$ by hypothesis. Since JA is a submodule of A, either $JA = A$ or $JA = 0$. If $JA = A$, then $J^nA = A$ for all n. Consequently, $\bigcap_n J^nA = A \neq 0$, which is a contradiction. Hence $JA = 0$. But $0 = JA = J(R/M)$ implies that $J \subset JR \subset M$. ■

Corollary 4.7. *If* R *is a Noetherian local ring with maximal ideal* M, *then* $\bigcap_{n=1}^{\infty} M^n = 0$.

PROOF. If $J = M$ and $A = R$, then $J^nA = M^n$; apply Proposition 4.6. ■

Proposition 4.8. *If* R *is a local ring, then every finitely generated projective* R*-module is free.*

Actually a much stronger result due to I. Kaplansky [63] is true, namely: every projective module over a (not necessarily commutative) local ring is free.

PROOF OF 4.8. If P is a finitely generated projective R-module, then by Corollary IV.2.2 there exists a free R-module F with a finite basis and an epimorphism $\pi : F \to P$. Among all the free R-modules F with this property choose one with a basis $\{x_1,x_2,\dots,x_n\}$ that has a minimal number of elements. Since π is an epimorphism $\{\pi(x_1),\dots,\pi(x_n)\}$ necessarily generate P. We shall first show that $K = \operatorname{Ker}\pi$ is contained in MF, where M is the unique maximal ideal of R. If $K \not\subset MF$, then there exists $k \in K$ with $k \notin MF$. Now $k = r_1x_1 + r_2x_2 + \cdots + r_nx_n$ with $r_i \in R$ uniquely determined. Since $k \notin MF$, some r_i, say r_1, is not an element of M. By Theorem III.4.13, r_1 is a unit, whence $x_1 - r_1^{-1}k = -r_1^{-1}r_2x_2 - \cdots - r_1^{-1}r_nx_n$. Consequently, since $k \in \operatorname{Ker}\pi$, $\pi(x_1) = \pi(x_1 - r_1^{-1}k) = \pi\left(\sum_{i=2}^{n} -r_1^{-1}r_ix_i\right) = \sum_{i=2}^{n} -r_1r_i\pi(x_i)$. Therefore, $\{\pi(x_2),\dots,\pi(x_n)\}$ generates P. Thus if F' is the free submodule of F with basis $\{x_2,\dots,x_n\}$ and $\pi' : F' \to P$ the restriction of π to F', then π' is an epimorphism. This contradicts the choice of F as having a basis of minimal cardinality. Hence $K \subset MF$.

Since $0 \to K \xrightarrow{\subset} F \xrightarrow{\pi} P \to 0$ is exact and P is projective $K \oplus P \cong F$ by Theorem IV.3.4. Under this isomorphism $(k,0) \mapsto k$ for all $k \in K$ (see the proof of Theorem IV.1.18), whence F is the internal direct sum $F = K \oplus P'$ with $P' \cong P$. Thus $F = K + P' \subset MF + P'$. If $u \in F$, then $u = \sum_i m_iv_i + p_i$ with $m_i \in M$, $v_i \in F$, $p_i \in P'$. Consequently, in the R-module F/P',

$$u + P' = \sum_i m_iv_i + P' = \sum_i m_i(v_i + P') \in M(F/P'),$$

whence $M(F/P') = F/P'$. Since F is finitely generated, so is F/P'. Therefore $K \cong F/P' = 0$ by Nakayama's Lemma 4.5. Thus $P \cong P' = F$ and P is free. ∎

We close this section with two well known theorems. The proofs are independent of the preceding part of this chapter.

Theorem 4.9. (*Hilbert Basis Theorem*) *If* R *is a commutative Noetherian ring with identity, then so is* $R[x_1,\dots,x_n]$.

PROOF. Clearly it suffices to show that $R[x]$ is Noetherian. By Theorem 1.9 we need only show that every ideal J in $R[x]$ is finitely generated.

For each $n \geq 0$, let I_n be the set of all $r \in R$ such that $r = 0$ or r is the leading coefficient of a polynomial $f \in J$ of degree n. Verify that each I_n is an ideal of R. If r is a nonzero element of I_n and $f \in J$ is a polynomial of degree n with leading coefficient r, then r is also the leading coefficient of xf, which is a polynomial in J of degree $n + 1$. Hence $I_0 \subset I_1 \subset I_2 \subset \cdots$. Since R is Noetherian, there exists an integer t such that $I_n = I_t$ for all $n \geq t$; furthermore, by Theorem 1.9 each I_n $(n \geq 0)$ is finitely generated say $I_n = (r_{n1},r_{n2},\dots,r_{ni_n})$. For each r_{nj} with $0 \leq n \leq t$ and $1 \leq j \leq i_n$, let $f_{nj} \in J$ be a polynomial of degree n with leading coefficient r_{nj}. Observe that $f_{0j} = r_{0j} \in R \subset R[x]$. We shall show that the ideal J of $R[x]$ is generated by the finite set of polynomials $X = \{f_{nj} \mid 0 \leq n \leq t; 1 \leq j \leq i_n\}$.

Clearly $(X) \subset J$. Conversely, the polynomials of degree 0 in J are precisely the elements of I_0 and hence are contained in (X). Proceeding by induction assume that (X) contains all polynomials of J of degree less than k and let $g \in J$ have degree k and leading coefficient $r \neq 0$.

If $k \leq t$, then $r \in I_k$ and hence $r = s_1 r_{k1} + s_2 r_{k2} + \cdots + s_{i_k} r_{ki_k}$ for some $s_j \in R$. Therefore the polynomial $\sum_{j=1}^{i_k} s_j f_{kj} \in (X)$ has leading coefficient r and degree k. Consequently, $g - \sum_j s_j f_{kj}$ has degree at most $k - 1$. By the induction hypothesis $g - \sum_j s_j f_{kj} \in (X)$, whence $g \in (X)$.

If $k \geq t$, then $r \in I_k = I_t$ and $r = \sum_{j=1}^{i_t} s_j r_{tj}$ $(s_j \in R)$. Furthermore $\sum_{j=1}^{i_t} s_j x^{k-t} f_{tj} \in (X)$ has leading coefficient r and degree k. Thus $g - \sum_j s_j x^{k-t} f_{tj}$ has degree at most $k - 1$ and lies in (X) by the induction assumption. Consequently, $g \in (X)$ and the induction is complete. Therefore, $J = (X)$. ■

Proposition 4.10. *If* R *is a commutative Noetherian ring with identity, then so is* R[[x]].

REMARK. Our proof makes use of Proposition 4.1. Although we shall not do so, the technique used to prove Theorem 4.9 may also be used here, with nonzero coefficients of lowest degree replacing those of highest degree in the argument. However, great care must be used to insure that certain power series constructed inductively in the course of the proof are in fact validly defined. The Axiom of Choice and some version of the Recursion Theorem are necessary (this part is frequently obscured in many published proofs of Proposition 4.10).

PROOF OF 4.10. It suffices by Proposition 4.1 to prove that every prime ideal P in $R[[x]]$ is finitely generated. Define an epimorphism of rings $R[[x]] \to R$ by mapping each power series $f = \sum_{i=0}^{\infty} a_i x^i$ onto its constant term a_0. Let P^* be the image of P under this map. Then P^* is a finitely generated ideal in R (Exercise III.2.13 and Theorem 1.9), say $P^* = (r_1, \ldots, r_n)$. For each r_i choose $f_i \in P$ with constant term r_i.

If $x \in P$, we claim that P is generated by $r_1, \ldots, r_n, x$. First note that if $f_k = r_k + \sum_{i=1}^{\infty} a_i x^i$, then $r_k = f_k - x\left(\sum_{j=0}^{\infty} a_{j+1} x^j\right) \in P$. If $g = \sum_{i=0}^{\infty} b_i x^i \in P$, then $b_0 = s_1 r_1 + \cdots + s_n r_n$ for some $s_i \in R$. Consequently, $g - \sum_{i=1}^{n} s_i r_i$ has 0 constant term; that is, $g - \sum_i s_i r_i = x g_1$ $(g_1 \in R[[x]])$. Therefore $g = \sum_i s_i r_i + x g_1$ and P is generated by $r_1, \ldots, r_n, x$.

If $x \notin P$, we claim that P is generated by $f_1, \ldots, f_n \in P$. If $h = \sum_{i=0}^{\infty} c_i x^i \in P$, then $c_0 = t_1 r_1 + \cdots + t_n r_n$ for some $t_i \in R$. Consequently, $h - \sum_{i=1}^{n} t_i f_i = x h^*$ for some $h^* \in R[[x]]$. Since $x \notin P$ and $x h^* = h - \sum_i t_i f_i \in P$ and P is prime, we have $h^* \in P$. For each $h \in P$, choose $t_i \in R$ and $h^* \in P$ such that $h = \sum_{i=1}^{m} t_i f_i + x h^*$ (Axiom of

Choice). Let $\lambda : P \to P$ be the map defined by $h \mapsto h^*$. Let g be any element of P. Then by the Recursion Theorem 6.2 of the Introduction (with $\lambda = f_n$ for all n) there is a function $\phi : \mathbf{N} \to P$ such that

$$\phi(0) = g \quad \text{and} \quad \phi(k+1) = \lambda(\phi(k)) = \phi(k)^*$$

Let $\phi(k) = h_k \varepsilon R[[x]]$ and denote by t_{ki} the previously chosen elements of R such that

$$h_k = \sum_{i=1}^{n} t_{ki} f_i + x h_k^* = \sum_{i=1}^{n} t_{ki} f_i + x h_{k+1}.$$

For each i ($1 \leq i \leq n$) let $g_i = \sum_{k=0}^{\infty} t_{ki} x^k \varepsilon R[[x]]$. Then

$$g_1 f_1 + \cdots + g_n f_n = \sum_{i=1}^{n} \left(\sum_{k=0}^{\infty} t_{ki} x^k \right) f_i = \sum_{k=0}^{\infty} \left(\sum_{i=1}^{n} t_{ki} f_i \right) x^k$$
$$= \sum_{k=0}^{\infty} (h_k - x h_{k+1}) x^k.$$

Consequently, for each $m \geq 0$ the coefficient of x^m in $g_1 f_1 + \cdots + g_n f_n$ is the same as the coefficient of x^m in $\sum_{k=0}^{m} (h_k - x h_{k+1}) x^k$. Since

$$\sum_{k=0}^{m} (h_k - x h_{k+1}) x^k = h_0 - x^{m+1} h_{m+1} = g - x^{m+1} h_{m+1},$$

the coefficient of x^m in $f_1 g_1 + \cdots + f_n g_n$ is precisely the coefficient of x^m in g. Therefore, $g = g_1 f_1 + g_2 f_2 + \cdots + g_n f_n$ and $f_1, \ldots, f_n$ generate P. ■

EXERCISES

1. Let R be a commutative ring with identity and I a finitely generated ideal of R. Let C be a submodule of an R-module A. Assume that for each $r \varepsilon I$ there exists a positive integer m (depending on r) such that $r^m A \subset C$. Show that for some integer n, $I^n A \subset C$. [*Hint:* see Theorems III.1.2(v) and III.2.5(vi)].

2. Without using primary decomposition, prove this version of the Krull Intersection Theorem. If R is a commutative Noetherian ring with identity, I an ideal of R, A a finitely generated R-module, and $B = \bigcap_{n=1}^{\infty} I^n A$, then $IB = B$. [*Hints:* Let C be maximal in the set $\mathcal{S}$ of all submodules S of A such that $B \cap S = IB$. It suffices to show $I^m A \subset C$ for some m. By Exercise 1 it suffices to show that for each $r \varepsilon I$, $r^n A \subset C$ for some n (depending on r). For each k, let $D_k = \{a \varepsilon A \mid r^k a \varepsilon C\}$. $D_0 \subset D_1 \subset D_2 \subset \cdots$ is an ascending chain of R-submodules; hence for some n, $D_k = D_n$ for all $k \geq n$. Show that $(r^n A + C) \cap B = IB$. The maximality of C implies $r^n A + C = C$, that is, $r^n A \subset C$.]

3. Let R be a Noetherian local ring with maximal ideal M. If the ideal M/M^2 in R/M^2 is generated by $\{a_1 + M^2, \ldots, a_n + M^2\}$, then the ideal M is generated in R by $\{a_1, \ldots, a_n\}$.

4. (Nakayama's Lemma, second version) Let R be a commutative ring with identity, J an ideal that is contained in every maximal ideal of R, and A a finitely generated R-module. If $R/J \otimes_R A = 0$, then $A = 0$. [*Hint:* use the exact sequence $0 \to J \to R \to R/J \to 0$ and the natural isomorphism $R \otimes_R A \cong A$ to show $JA = A$.]

5. Let R and J be as in Exercise 4; let A be a finitely generated R-module and $f : C \to A$ an R-module homomorphism. Then f induces a homomorphism $\bar{f} : C/JC \to A/JA$ in the usual way (Corollary IV.1.8). Show that if $\bar{f}$ is an epimorphism, then f is an epimorphism.

6. (a) Let R be a commutative ring with identity. If every ideal of R can be generated by a finite or denumerable subset, then the same is true of $R[x]$.
(b) State and prove an analogue of part (a) for $R[[x]]$; (the answer is not quite the same here).

7. Let R be a commutative ring with identity and let $f,g \in R[[x]]$. Denote by In f, the initial degree of f (that is, the smallest n such that $a_n \neq 0$, where $f = \sum_{i=0}^{\infty} a_i x^i$). Show that
(a) $\text{In}\,(f+g) \geq \min\,(\text{In}\,f, \text{In}\,g)$.
(b) $\text{In}\,(fg) \geq \text{In}\,f + \text{In}\,g$.
(c) If R is an integral domain, $\text{In}\,(fg) = \text{In}\,f + \text{In}\,g$.

8. Let R be a commutative Noetherian ring with identity and let $Q_1 \cap \cdots \cap Q_n = 0$ be a reduced primary decomposition of the ideal 0 of R with Q_i belonging to the prime ideal P_i. Then $P_1 \cup P_2 \cup \cdots \cup P_n$ is the set of zero divisors in R.

9. Let R be a commutative ring with identity. If every maximal ideal of R is of the form (c), where $c^2 = c$, for some $c \in R$, then R is Noetherian. [*Hint:* show that every primary ideal is maximal; use Proposition 4.1.]

5. RING EXTENSIONS

In the first part of this section ring extensions are defined and the essential properties of integral extensions are developed. The last part is devoted to the study of the relations between prime ideals in rings R and S, where S is an extension ring of R. Throughout this section *all rings are commutative with identity*.

Definition 5.1. *Let* S *be a commutative ring with identity and* R *a subring of* S *containing* 1_S. *Then* S *is said to be an* **extension ring** *of* R.

EXAMPLES. Every extension field F of a field K is obviously an extension ring of K. If R is a commutative ring with identity, then $R[[x]]$ and $R[x_1, \ldots, x_n]$ are extension rings of R. The ring $\mathbf{Z}$ is *not* an extension of the subring E of even integers since E does not contain 1.

Definition 5.2. *Let S be an extension ring of R and $s \in S$. If there exists a monic polynomial $f(x) \in R[x]$ such that s is a root of f (that is, $f(s) = 0$), then s is said to be* **integral** *over R. If every element of S is integral over R, S is said to be an* **integral extension** *of R.*

The key feature of Definition 5.2 is the requirement that f be monic.

EXAMPLES. Every algebraic extension field F of a field K is an integral extension ring (see the Remarks after Definition V.1.4). The ring R is integral over itself since $r \in R$ is a root of $x - r \in R[x]$. In the extension of **Z** by the real field **R**, $1/\sqrt{3}$ is algebraic over **Z** since it is a root of $3x^2 - 1$ but $1/\sqrt{3}$ is not integral over **Z**. However, $1/\sqrt{3}$ is integral over the rational field **Q** since it is a root of $x^2 - 1/3$.

Let S be an extension ring of R and X a subset of S. Then the **subring generated by X over R** is the intersection of all subrings of S that contain $X \cup R$; it is denoted $R[X]$. The first half of Theorem V.1.3 is valid for rings and shows that $R[X]$consists of all elements $f(s_1, \dots, s_n)$ with $n \in \mathbf{N}^*$, $f \in R[x_1, \dots, x_n]$ and $s_i \in X$. In particular, for any $s_1, \dots, s_t \in S$ the subring generated by $\{s_1, \dots, s_t\}$ over R, which is denoted $R[s_1, \dots, s_t]$, consists of all elements $f(s_1, \dots, s_t)$ with $f \in R[x_1, \dots, x_t]$. An element of $R[s_1, \dots, s_t]$ is sometimes called a polynomial in $s_1, \dots, s_t$. Despite this terminology $R[s_1, \dots, s_t]$ need not be isomorphic to the polynomial ring $R[x_1, \dots, x_t]$ (for example, $f(s_1, \dots, s_t)$ may be zero even though f is a nonzero polynomial). It is easy to see that for each i $(1 \leq i \leq t)$, $R[s_1, \dots, s_{i-1}][s_i] = R[s_1, \dots, s_i]$. Since $R[s_1, \dots, s_t]$ is a ring containing R, $R[s_1, \dots, s_t]$ is an R-module in the obvious way. Likewise every module over $R[s_1, \dots, s_t]$ is obviously an R-module.

Theorem 5.3. *Let S be an extension ring of R and $s \in S$. Then the following conditions are equivalent.*

(i) *s is integral over R;*

(ii) *R[s] is a finitely generated R-module;*

(iii) *there is a subring T of S containing 1_S and R[s] which is finitely generated as an R-module;*

(iv) *there is an R[s]-submodule B of S which is finitely generated as an R-module and whose annihilator in R[s] is zero.*

SKETCH OF PROOF. (i) $\Rightarrow$ (ii) Suppose s is a root of the monic polynomial $f \in R[x]$ of degree n. We claim that $1_R = s^0, s, s^2, \dots, s^{n-1}$ generate $R[s]$ as an R-module. As observed above, every element of $R[s]$ is of the form $g(s)$ for some $g \in R[x]$. By the Division Algorithm III.6.2 $g(x) = f(x)q(x) + r(x)$ with $\deg r < \deg f$. Therefore in S, $g(s) = f(s)q(s) + r(s) = 0 + r(s) = r(s)$. Hence $g(s)$ is an R-linear combination of $1_R, s, s^2, \dots, s^m$ with $m = \deg r < \deg f = n$.

(ii) $\Rightarrow$ (iii) Let $T = R[s]$.

(iii) $\Rightarrow$ (iv) Let B be the subring T. Since $R \subset R[s] \subset T$, B is an $R[s]$-module that is finitely generated as an R-module by (iii). Since $1_S \in B$, $uB = 0$ for any $u \in S$ implies $u = u1_S = 0$; that is, the annihilator of B in $R[s]$ is 0.

(iv) $\Rightarrow$ (i) Let B be generated over R by $b_1, \dots, b_n$. Since B is an $R[s]$-module $sb_i \in B$ for each i. Therefore there exist $r_{ij} \in R$ such that

$$\begin{aligned}
sb_1 &= r_{11}b_1 + r_{12}b_2 + \cdots + r_{1n}b_n \\
sb_2 &= r_{21}b_1 + r_{22}b_2 + \cdots + r_{2n}b_n \\
&\;\;\vdots \\
sb_n &= r_{n1}b_1 + r_{n2}b_2 + \cdots + r_{nn}b_n.
\end{aligned}$$

Consequently,

$$\begin{aligned}
(r_{11} - s)b_1 + r_{12},b_2 + \cdots + r_{1n}b_n &= 0 \\
r_{21}b_1 + (r_{22} - s)b_2 + \cdots + r_{2n}b_n &= 0 \\
&\;\;\vdots \\
r_{n1}b_1 + r_{n2},b_2 + \cdots + (r_{nn} - s)b_n &= 0.
\end{aligned}$$

Let M be the $n \times n$ matrix (r_{ij}) and let $d \in R[s]$ be the determinant of the matrix $M - sI_n$. Then $db_i = 0$ for all i by Exercise VII.3.8. Since B is generated by the b_i, $dB = 0$. Since the annihilator of B in $R[s]$ is zero by (iv) we must have $d = 0$. If f is the polynomial $|M - xI_n|$ in $R[x]$, then one of f, $-f$ is monic and

$$\pm f(s) = \pm|M - sI_n| = \pm d = 0.$$

Therefore s is integral over R. ■

Corollary 5.4. *If* S *is a ring extension of* R *and* S *is finitely generated as an* R*-module, then* S *is an integral extension of* R.

PROOF. For any $s \in S$ let $S = T$ in part (iii) of Theorem 5.3. Then s is integral over R by Theorem 5.3(i). ■

The proofs of the next propositions depend on the following fact. If $R \subset S \subset T$ are rings (with $1_T \in R$) such that T is a finitely generated S-module and S is a finitely generated R-module, then T is a finitely generated R-module. The second paragraph of the proof of Theorem IV.2.16 contains a proof of this fact, *mutatis mutandis.*

Theorem 5.5. *If* S *is an extension ring of* R *and* $s_1, \ldots, s_t \in S$ *are integral over* R, *then* $R[s_1, \ldots, s_t]$ *is a finitely generated* R*-module and an integral extension ring of* R.

PROOF. We have a tower of extension rings:

$$R \subset R[s_1] \subset R[s_1,s_2] \subset \cdots \subset R[s_1, \ldots, s_t].$$

For each i, s_i is integral over R and hence integral over $R[s_1, \ldots, s_{i-1}]$. Since $R[s_1, \ldots, s_i] = R[s_1, \ldots, s_{i-1}][s_i]$, $R[s_1, \ldots, s_i]$ is a finitely generated module over $R[s_1, \ldots, s_{i-1}]$ by Theorem 5.3 (i), (ii). Repeated application of the remarks preceding the theorem shows that $R[s_1, \ldots, s_n]$ is a finitely generated R-module. Therefore, $R[s_1, \ldots, s_n]$ is an integral extension ring of R by Corollary 5.4. ■

Theorem 5.6. *If* T *is an integral extension ring of* S *and* S *is an integral extension ring of* R, *then* T *is an integral extension ring of* R.

PROOF. T is obviously an extension ring of R. If $t \varepsilon T$, then t is integral over S and therefore the root of some monic polynomial $f \varepsilon S[x]$, say $f = \sum_{i=0}^{n} s_i x^i$. Since f is also a polynomial over the ring $R[s_0,s_1, \dots, s_{n-1}]$, t is integral over $R[s_0, \dots, s_{n-1}]$. By Theorem 5.3 $R[s_0, \dots, s_{n-1}][t]$ is a finitely generated $R[s_0, \dots, s_{n-1}]$-module. But since S is integral over R, $R[s_0, \dots, s_{n-1}]$ is a finitely generated R-module by Theorem 5.5. The remarks preceding Theorem 5.5 show that

$$R[s_0, \dots, s_{n-1}][t] = R[s_0, \dots, s_{n-1},t]$$

is a finitely generated R-module. Since $R[t] \subset R[s_0, \dots, s_{n-1},t]$, t is integral over R by Theorem 5.3(iii). ■

Theorem 5.7. *Let* S *be an extension ring of* R *and let* $\hat{R}$ *be the set of all elements of* S *that are integral over* R. *Then* $\hat{R}$ *is an integral extension ring of* R *which contains every subring of* S *that is integral over* R.

PROOF. If $s,t \varepsilon \hat{R}$, then $s,t \varepsilon R[s,t]$, whence $t - s \varepsilon R[s,t]$ and $ts \varepsilon R[s,t]$. Since s and t are integral over R, so is the ring $R[s,t]$ (Theorem 5.5). Therefore $t - s \varepsilon \hat{R}$ and $ts \varepsilon \hat{R}$. Consequently, $\hat{R}$ is a subring of S (see Theorem 1.2.5). $\hat{R}$ contains R since every element of R is trivially integral over R. The definition of $\hat{R}$ insures that $\hat{R}$ is integral over R and contains all subrings of S that are integral over R. ■

If S is an extension ring of R, then the ring $\hat{R}$ of Theorem 5.7 is called the **integral closure** of R in S. If $\hat{R} = R$, then R is said to be **integrally closed in S.**

REMARKS. (i) Since $1_R \varepsilon R \subset \hat{R}$, S is an extension ring of $\hat{R}$. Theorems 5.6 and 5.7 imply that $\hat{R}$ is itself integrally closed in S. (ii) The concepts of integral closure and integrally closed rings are relative notions and refer to a given ring R and a particular extension ring S. Thus the phrase "R is integrally closed" is ambiguous unless an extension ring S is specified. There is one case, however, in which the ring S is understood without specific mention. An integral domain R is said to be **integrally closed** provided R is integrally closed in its quotient field (see p. 144).

EXAMPLE. The integral domain **Z** is integrally closed (in the rational field **Q**; Exercise 8). However, **Z** is not integrally closed in the field **C** of complex numbers since $i \varepsilon$ **C** is integral over **Z**.

EXAMPLE. More generally, every unique factorization domain is integrally closed (Exercise 8). In particular, the polynomial ring $F[x_1, \dots, x_n]$ (F a field) is integrally closed in its quotient field $F(x_1, \dots, x_n)$.

The following theorem is used only in the proof of Theorem 6.10.

Theorem 5.8. *Let* T *be a multiplicative subset of an integral domain* R *such that* $0 \notin$ T. *If* R *is integrally closed, then* $T^{-1}R$ *is an integrally closed integral domain.*

SKETCH OF PROOF. $T^{-1}R$ is an integral domain (Theorem III.4.3(ii)) and R may be identified with a subring of $T^{-1}R$ (Theorem III.4.4(ii)). Extending this identification, the quotient field $Q(R)$ of R may be considered as a subfield of the quotient field $Q(T^{-1}R)$ of $T^{-1}R$. Verify that $Q(R) = Q(T^{-1}R)$.

Let $u \in Q(T^{-1}R)$ be integral over $T^{-1}R$; then for some $r_i \in R$ and $s_i \in T$,

$$u^n + (r_{n-1}/s_{n-1})u^{n-1} + \cdots + (r_1/s_1)u + (r_0/s_0) = 0.$$

Multiply through this equation by s^n, where $s = s_0s_1\cdots s_{n-1} \in T$, and conclude that su is integral over R. Since $su \in Q(T^{-1}R) = Q(R)$ and R is integrally closed, $su \in R$. Therefore, $u = su/s \in T^{-1}R$, whence $T^{-1}R$ is integrally closed. ■

The remainder of this section is devoted to exploring the relationships between (prime) ideals in rings R and S, where S is an extension ring of R. The only point in the sequel where this material is used is in the proof of Lemma 7.3.

If S is an extension ring of R and I $(\neq S)$ is an ideal of S, it is easy to see that $I \cap R \neq R$ and $I \cap R$ is an ideal of R (Exercise 10). The ideal $J = I \cap R$ is called the **contraction** of I to R and I is said to **lie over** J.

If Q is a prime ideal in an extension ring S of a ring R, then the contraction $Q \cap R$ of Q to R is a prime ideal of R (Exercise 10). The converse problem is: given a prime ideal P in R does there exist a prime ideal Q in S that lies over P (that is, $Q \cap R = P$)? There are many examples where the answer is negative (for example, the extension of $\mathbf{Z}$ by the field $\mathbf{Q}$ of rationals). A partial solution to the problem is given by the next theorem, which is due to Cohen-Seidenberg.

Theorem 5.9. *(Lying-over Theorem) Let* S *be an integral extension ring of* R *and* P *a prime ideal of* R. *Then there exists a prime ideal* Q *in* S *which lies over* P *(that is,* Q ∩ R = P).

PROOF. Since P is prime, $R - P$ is a multiplicative subset of R (Theorem 2.1) and hence a multiplicative subset of S. Clearly $0 \notin R - P$. By Theorem 2.2 there is an ideal Q of S that is maximal in the set of all ideals I of S such that $I \cap (R - P) = \emptyset$; furthermore any such ideal Q is prime in S. Clearly $Q \cap R \subset P$. If $Q \cap R \neq P$, choose $u \in P$ such that $u \notin Q$. Then the ideal $Q + (u)$ in S properly contains Q. By maximality there exists $c \in (Q + (u)) \cap (R - P)$, say $c = q + su$ $(q \in Q; s \in S)$. Since s is integral over R, there exist $r_i \in R$ such that

$$s^n + r_{n-1}s^{n-1} + \cdots + r_1s + r_0 = 0.$$

Multiplying this equation by u^n yields

$$(su)^n + r_{n-1}u(su)^{n-1} + \cdots + r_1u^{n-1}(su) + r_0u^n = 0.$$

Since $su = c - q$ the Binomial Theorem III.1.6 implies that

$$v = c^n + r_{n-1}uc^{n-1} + \cdots + r_1u^{n-1}c + r_0u^n \in Q.$$

But $v \in R$ and hence $v \in R \cap Q \subset P$. But $u \in P$ and $v \in P$ imply $c^n \in P$. Since P is prime, c must lie in P, which is a contradiction. ■

Corollary 5.10. *(Going-up Theorem) Let* S *be an integral extension ring of* R *and* P_1, P *prime ideals in* R *such that* $P_1 \subset P$. *If* Q_1 *is a prime ideal of* S *lying over* P_1, *then there exists a prime ideal* Q *of* S *such that* $Q_1 \subset Q$ *and* Q *lies over* P.

SKETCH OF PROOF. As in the proof of Theorem 5.9, $R - P$ is a multiplicative set in S. Since $Q_1 \cap R = P_1 \subset P$, we have $Q_1 \cap (R - P) = \emptyset$. By Theorem 2.2 there is a prime ideal Q of S that contains Q_1 and is maximal in the set of all ideals I of S such that $Q_1 \subset I$ and $I \cap (R - P) = \emptyset$. The proof of Theorem 5.9 now carries over verbatim to show that $Q \cap R = P$. ■

Theorem 5.11. *Let* S *be an integral extension ring of* R *and* P *a prime ideal in* R. *If* Q *and* Q′ *are prime ideals in* S *such that* $Q \subset Q'$ *and both* Q *and* Q′ *lie over* P, *then* $Q = Q'$.

PROOF. It suffices to prove the following statement: if Q is a prime ideal in S such that $Q \cap R = P$, then Q is maximal in the set $\mathcal{S}$ of all ideals I in S with the property $I \cap (R - P) = \emptyset$.

If Q is not maximal in $\mathcal{S}$, then there is an ideal I in S with

$$Q \subsetneq I \quad \text{and} \quad I \cap (R - P) = \emptyset.$$

Consequently, $I \cap R \subset P$. Choose $u \in I - Q$. Since u is integral over R, the set of all monic polynomials $f \in R[x]$ such that $\deg f \geq 1$ and $f(u) \in Q$ is nonempty. Choose such an f of least degree, say $f = \sum_{i=0}^{n} r_i x_i$. Then

$$u^n + r_{n-1}u^{n-1} + \cdots + r_1 u + r_0 \in Q \subset I,$$

whence $r_0 \in I \cap R \subset P = Q \cap R \subset Q$. Therefore

$$u(u^{n-1} + r_{n-1}u^{n-2} + \cdots + r_2 u + r_1) \in Q.$$

By the minimality of $\deg f$, $(u^{n-1} + r_{n-1}u^{n-2} + \cdots + r_1) \notin Q$, and $u \notin Q$ by choice. This is a contradiction since Q is prime (Theorem III.2.15). Therefore Q is maximal in $\mathcal{S}$. ■

Theorem 5.12. *Let* S *be an integral extension ring of* R *and let* Q *be a prime ideal in* S *which lies over a prime ideal* P *in* R. *Then* Q *is maximal in* S *if and only if* P *is maximal in* R.

PROOF. Suppose Q is maximal in S. By Theorem III.2.18 there is a maximal ideal M of R that contains P. M is prime by Theorem III.2.19. By Corollary 5.10 there is a prime ideal Q' in S such that $Q \subset Q'$ and Q' lies over M. Since Q' is prime, $Q' \neq S$ (Definition III.2.14). The maximality of Q implies that $Q = Q'$, whence $P = Q \cap R = Q' \cap R = M$. Therefore, P is maximal in R.

Conversely suppose P is maximal in R. Since Q is prime in S, $Q \neq S$ and there is a maximal ideal N of S containing Q (Theorem III.2.18). N is prime by Theorem III.2.19, whence $1_R = 1_S \notin N$. Since $P = R \cap Q \subset R \cap N \subsetneq R$, we must have $P = R \cap N$ by maximality. Thus Q and N both lie over P and $Q \subset N$. Therefore, $Q = N$ by Theorem 5.11. ■

EXERCISES

Note: Unless otherwise specified, S is always an extension ring of R.

1. Let S be an integral extension ring of R and suppose R and S are integral domains. Then S is a field if and only if R is a field. [*Hint:* Corollary III.2.21.]

2. Let R be an integral domain. If the quotient field F of R is integral over R, then R is a field.

3. Let R be an integral domain with quotient field F. If $0 \neq a \in R$ and $1_R/a \in F$ is integral over R, then a is a unit in R.

4. (a) Let R be an integral domain with quotient field F. If $0 \neq a \in R$, then the following are equivalent:

   (i) every nonzero prime ideal of R contains a;
   (ii) every nonzero ideal of R contains some power of a;
   (iii) $F = R[1_R/a]$ (ring extension).

   An integral domain R that contains an element $a \neq 0$ satisfying (i)–(iii) is called a **Goldmann ring.**

   (b) A principal ideal domain is a Goldmann ring if and only if it has only finitely many distinct primes.
   (c) Is the homomorphic image of a Goldmann ring also a Goldmann ring?

5. If S is an integral extension ring of R and $f: S \to S$ is a ring homomorphism, such that $f(1_S) = 1_S$, then $f(S)$ is an integral extension ring of $f(R)$.

6. If S is an integral extension ring of R, then $S[x_1, \ldots, x_n]$ is an integral extension ring of $R[x_1, \ldots, x_n]$.

7. If S is an integral extension ring of R and T is a multiplicative subset of R ($0 \notin T$), then $T^{-1}S$ is an integral extension of $T^{-1}R$. [*Hint:* If $s/t \in T^{-1}S$, then $s/t = \phi_T(s)(1_R/t)$, where $\phi_T: S \to T^{-1}S$ is the canonical map (Theorem III.4.4). Show that $\phi_T(s)$ and $1_R/t$ are integral over $T^{-1}R$, whence s/t is integral over $T^{-1}R$ by Theorem 5.5.]

8. Every unique factorization domain is integrally closed. [*Hint:* Proposition III.6.8.]

9. Let T be a commutative ring with identity and $\{S_i \mid i \in I\}$, $\{R_i \mid i \in I\}$ families of subrings such that T is an extension ring of S_i and S_i is an extension ring of R_i for every i. If each R_i is integrally closed in S_i, then $\bigcap_i R_i$ is integrally closed in $\bigcap_i S_i$.

10. (a) If I ($\neq S$) is an ideal of S, then $I \cap R \neq R$ and $I \cap R$ is an ideal of R.
    (b) If Q is a prime ideal of S, then $Q \cap R$ is a prime ideal of R.

6. DEDEKIND DOMAINS

In this section we examine the class of Dedekind domains. It lies properly between the class of principal ideal domains and the class of Noetherian integral domains. Dedekind domains are important in algebraic number theory and the algebraic theory of curves. The chief result is Theorem 6.10 which characterizes Dedekind domains in several different ways.

The definition of a Dedekind domain to be given below is motivated by the following facts. Every principal ideal domain D is Noetherian (Lemma III.3.6). Consequently, every ideal ($\neq D$) has a primary decomposition (Theorem 3.6). The introduction to Section 2 shows that a particularly strong form of primary decomposition holds in a principal ideal domain, namely: every proper ideal is (uniquely) a product of prime ideals.

Definition 6.1. *A* **Dedekind domain** *is an integral domain* R *in which every ideal* ($\neq$ R) *is the product of a finite number of prime ideals.*

EXAMPLE. The preceding discussion shows that every principal ideal domain is Dedekind. The converse, however, is false. There is an example after Theorem 6.10 below of a Dedekind domain that is not a principal ideal domain.

It is not immediately evident from the definition that every Dedekind domain is in fact Noetherian. In order to prove this fact and to develop other properties of Dedekind domains we must introduce the concept of a fractional ideal.

Definition 6.2. *Let* R *be an integral domain with quotient field* K. *A* **fractional ideal** *of* R *is a nonzero* R*-submodule* I *of* K *such that* $aI \subset R$ *for some nonzero* $a \in R$.

EXAMPLE. Every ordinary nonzero ideal I in an integral domain R is an R-submodule of R and hence a fractional ideal of R. Conversely, every fractional ideal of R that is contained in R is an ordinary ideal of R.

EXAMPLE. Every nonzero finitely generated R-submodule I of K is a fractional ideal of R. For if I is generated by $b_1, \dots, b_n \in K$, then $I = Rb_1 + \cdots + Rb_n$ and for each i, $b_i = c_i/a_i$ with $0 \neq a_i$, $c_i \in R$. Let $a = a_1a_2\cdots a_n$. Then $a \neq 0$ and $aI = Ra_2\cdots a_nc_1 + \cdots + Ra_1\cdots a_{n-1}c_n \subset R$.

REMARK. If I is a fractional ideal of a domain R and $aI \subset R$ ($0 \neq a \in R$), then aI is an ordinary ideal in R and the map $I \to aI$ given by $x \mapsto ax$ is an R-module isomorphism.

Theorem 6.3. *If* R *is an integral domain with quotient field* K, *then the set of all fractional ideals of* R *forms a commutative monoid, with identity* R *and multiplication given by* $IJ = \left\{ \sum_{i=1}^{n} a_ib_i \mid a_i \in I;\ b_i \in J;\ n \in \mathbf{N}^* \right\}$.

PROOF. Exercise; note that if I and J are ideals in R, then IJ is the usual product of ideals. ■

A fractional ideal I of an integral domain R is said to be **invertible** if $IJ = R$ for some fractional ideal J of R. Thus the invertible fractional ideals[2] are precisely those that have inverses in the monoid of all fractional ideals.

[2]In the literature invertible fractional ideals are sometimes called simply invertible ideals.

REMARKS. (i) The inverse of an invertible fractional ideal I is unique and is $I^{-1} = \{a \in K \mid aI \subset R\}$. Indeed for *any* fractional ideal I the set $I^{-1} = \{a \in K \mid aI \subset R\}$ is easily seen to be a fractional ideal such that $I^{-1}I = II^{-1} \subset R$. If I is invertible and $IJ = JI = R$, then clearly $J \subset I^{-1}$. Conversely, since I^{-1} and J are R-submodules of K, $I^{-1} = RI^{-1} = (JI)I^{-1} = J(II^{-1}) \subset JR = RJ \subset J$, whence $J = I^{-1}$.

(ii) If I,A,B are fractional ideals of R such that $IA = IB$ and I is invertible, then $A = RA = (I^{-1}I)A = I^{-1}(IB) = RB = B$.

(iii) If I is an ordinary ideal in R, then $R \subset I^{-1}$.

EXAMPLE. Every nonzero principal ideal in an integral domain R is invertible. If K is the quotient field of R and $I = (b)$ with $b \neq 0$, let $J = Rc \subset K$ where $c = 1_R/b$. Then J is a fractional ideal of R such that $IJ = R$.

Invertible fractional ideals play a key role in characterizing Dedekind domains. The next five results develop some facts about them.

Lemma 6.4. *Let* $I, I_1, I_2, \ldots, I_n$ *be ideals in an integral domain* R.

(i) *The ideal* $I_1I_2\cdots I_n$ *is invertible if and only if each* I_j *is invertible.*

(ii) *If* $P_1\cdots P_m = I = Q_1\cdots Q_n$, *where the* P_i *and* Q_j *are prime ideals in* R *and every* P_i *is invertible, then* $m = n$ *and (after reindexing)* $P_i = Q_i$ *for each* $i = 1, \ldots, m$.

PROOF. (i) If J is a fractional ideal such that $J(I_1\cdots I_n) = R$, then for each $j = 1,2, \ldots, n$, $I_j(JI_1\cdots I_{j-1}I_{j+1}\cdots I_n) = R$, whence I_j is invertible. Conversely, if each I_j is invertible, then $(I_1\cdots I_n)(I_1^{-1}\cdots I_n^{-1}) = R$, whence $I_1\cdots I_n$ is invertible.

(ii) The proof is by induction on m with the case $m = 1$ being left to the reader. If $m > 1$, choose one of the P_i, say P_1, such that P_1 does not properly contain P_i for $i = 2, \ldots, m$. Since $Q_1\cdots Q_n = P_1\cdots P_m \subset P_1$ and P_1 is prime some Q_j, say Q_1, is contained in P_1 (Definition III.2.14). Similarly since $P_1\cdots P_m = Q_1\cdots Q_n \subset Q_1$, $P_i \subset Q_1$ for some i. Hence $P_i \subset Q_1 \subset P_1$. By the minimality of P_1 we must have $P_i = Q_1 = P_1$. Since $P_1 = Q_1$ is invertible, Remark (ii) after Theorem 6.3 implies

$$P_2P_3\cdots P_m = Q_2Q_3\cdots Q_n.$$

Therefore by the induction hypothesis $m = n$ and (after reindexing) $P_i = Q_i$ for $i = 1,2, \ldots, m$. ■

The example preceding Lemma 6.4 and Theorem III.3.4 show that every nonzero prime ideal in a principal ideal domain is both invertible and maximal. More generally we have

Theorem 6.5. *If* R *is a Dedekind domain, then every nonzero prime ideal of* R *is invertible and maximal.*

PROOF. We show first that every invertible prime ideal P is maximal. If $a \in R - P$, we must show that the ideal $P + Ra$ generated by P and a is R. If $P + Ra \neq R$, then since R is Dedekind, there exist prime ideals P_i and Q_j such that $P + Ra = P_1P_2\cdots P_m$ and $P + Ra^2 = Q_1Q_2\cdots Q_n$. Let $\pi : R \to R/P$ be the canonical epimorphism and consider the principal ideals in R/P generated respectively by $\pi(a)$ and $\pi(a^2)$. Clearly

$$(\pi(a)) = \pi(P_1)\cdots\pi(P_m) \quad \text{and} \quad (\pi(a^2)) = \pi(Q_1)\cdots\pi(Q_n).$$

Since $\ker \pi = P \subset P_i$ and $P \subset Q_i$ for each i, the ideals $\pi(P_i)$ and $\pi(Q_i)$ are prime in R/P (Exercise III.2.17(a)). Since R/P is an integral domain (Theorem III.2.16), every principal ideal in R/P is invertible (see the example preceding Lemma 6.4). Consequently, $\pi(P_i)$ and $\pi(Q_j)$ are invertible by Lemma 6.4(i). Since

$$\pi(Q_1)\cdots\pi(Q_n) = (\pi(a^2)) = (\pi(a))^2 = \pi(P_1)^2\cdots\pi(P_m)^2,$$

Lemma 6.4(ii) implies $n = 2m$ and (after reindexing) $\pi(P_i) = \pi(Q_{2i}) = \pi(Q_{2i-1})$ for $i = 1,2,\dots,m$. Since $\operatorname{Ker} \pi = P \subset P_i$ and $P \subset Q_j$ for all i,i,

$$P_i = \pi^{-1}(\pi(P_i)) = \pi^{-1}(\pi(Q_{2i})) = Q_{2i}$$

and similarly $P_i = Q_{2i-1}$ for $i = 1,2,\dots,m$. Consequently, $P + Ra^2 = (P + Ra)^2$ and $P \subset P + Ra^2 \subset (P + Ra)^2 \subset P^2 + Ra$. If $b = c + ra \in P$ ($c \in P^2, r \in R$), then $ra \in P$. Thus $r \in P$ since P is prime and $a \notin P$. Therefore, $P \subset P^2 + Pa \subset P$, which implies $P = P^2 + Pa = P(P + Ra)$. Since P is invertible, $R = P^{-1}P = P^{-1}P(P + Ra) = R(P + Ra) = P + Ra$. This is a contradiction. Therefore every invertible prime ideal P is maximal.

Now suppose P is any nonzero prime ideal in R and c is a nonzero element of P. Then $(c) = P_1P_2\cdots P_n$ for some prime ideals P_i. Since $P_1P_2\cdots P_n = (c) \subset P$, we have for some k, $P_k \subset P$ (Definition III.2.14). The principal ideal (c) is invertible and hence so is P_k (Lemma 6.4(i)). By the first part of the proof P_k is maximal, whence $P_k = P$. Therefore, P is maximal and invertible. ■

EXAMPLE. If F is a field, then the principal ideals (x_1) and (x_2) in the polynomial domain $F[x_1,x_2]$ are prime but not maximal (since $(x_i) \subsetneq (x_1,x_2) \subsetneq F[x_1,x_2]$). Consequently, $F[x_1,x_2]$ is not Dedekind (Theorem 6.5). Since $F[x_1,x_2]$ is Noetherian by Theorem 4.9, the class of Dedekind domains is properly contained in the class of Noetherian domains.

Lemma 6.6. *If* I *is a fractional ideal of an integral domain* R *with quotient field* K *and* $f \in Hom_R(I,R)$, *then for all* $a,b \in I$: $af(b) = bf(a)$.

PROOF. Now $a = r/s$ and $b = v/t$ ($r,s,v,t \in R$; $s,t \neq 0$) so $sa = r$ and $tb = v$. Hence $sab = rb \in I$ and $tab = va \in I$. Thus $sf(tab) = f(stab) = tf(sab)$ in R. Therefore, $af(b) = saf(b)/s = f(sab)/s = f(tab)/t = tbf(a)/t = bf(a)$. ■

Lemma 6.7. *Every invertible fractional ideal of an integral domain* R *with quotient field* K *is a finitely generated* R*-module.*

PROOF. Since $I^{-1}I = R$, there exist $a_i \in I^{-1}, b_i \in I$ such that $1_R = \sum_{i=1}^{n} a_ib_i$. If $c \in I$, then $c = \sum_{i=1}^{n} (ca_i)b_i$. Furthermore each $ca_i \in R$ since $a_i \in I^{-1} = \{a \in K \mid aI \subset R\}$. Therefore I is generated as an R-module by $b_1,\dots,b_n$ (Theorem IV.1.5(iii)). ■

We have seen that every nonzero ideal I in a principal ideal domain D is invertible. Furthermore I is isomorphic to D as a D-module (see Theorem IV.1.5(i)). Thus I is a free and hence projective D-module. This result also holds in arbitrary integral domains.

Theorem 6.8. *Let* R *be an integral domain and* I *a fractional ideal of* R. *Then* I *is invertible if and only if* I *is a projective* R-*module.*

PROOF. ($\Rightarrow$) By Lemma 6.7 and Theorem IV.1.5, $I = Rb_1 + \cdots + Rb_n$ with $b_i \in I$ and $1_R = \sum_{i=1}^{n} a_i b_i$ $(a_i \in I^{-1})$. Let F be a free R-module with a basis of n elements $e_1, \ldots, e_n$. Then the map $\pi : F \to I$ defined by $e_i \mapsto b_i$ is an R-module epimorphism (see Theorem IV.2.1), and there is a short exact sequence: $0 \to \operatorname{Ker} \pi \to F \xrightarrow{\pi} I \to 0$. Define $\zeta : I \to F$ by $\zeta(c) = ca_1e_1 + \cdots + ca_ne_n$ $(c \in I)$ and verify that ζ is an R-module homomorphism such that $\pi\zeta = 1_I$; (note that $ca_i \in R$ for each i since $a_i \in I^{-1}$). Consequently the exact sequence splits and I is a direct summand of a free R-module (Theorem IV.1.18). Therefore, I is projective by Theorem IV.3.4.

($\Leftarrow$) Let $X = \{b_j \mid j \in J\}$ be a (possibly infinite) set of nonzero generators of the projective R-module I. Let b_0 be a fixed element of X. Let F be a free R-module with basis $\{e_j \mid j \in J\}$ and let $\phi : F \to I$ be the R-module epimorphism defined by $e_i \mapsto b_i$ (Theorem IV.2.1). Since I is projective there is an R-module homomorphism $\psi : I \to F$ such that $\phi\psi = 1_I$. For each $j \in J$ let $\pi_j : F \to Re_j \cong R$ be the canonical projection that maps $\sum_i r_ie_i \in F$ onto $r_j \in R$ (see Theorem IV.2.1). Then for each j the map $\theta_j = \pi_j\psi : I \to R$ is an R-module homomorphism. Let $c_j = \theta_j(b_0)$. For any $c \in I$, $cc_j = c\theta_j(b_0) = b_0\theta_j(c)$ by Lemma 6.6, whence in the quotient field K of R, $c(c_j/b_0) = cc_j/b_0 = b_0\theta_j(c)/b_0 = \theta_j(c) \in R$. Therefore

$$c_j/b_0 \in I^{-1} = \{a \in K \mid aI \subset R\}.$$

Consequently, for any $c \in I$

$$\psi(c) = \sum_{j \in J_1} \theta_j(c)e_j = \sum_{j \in J_1} c(c_j/b_0)e_j,$$

where J_1 is the finite subset $\{j \in J \mid \theta_j(c) \neq 0\}$. Therefore, for any nonzero $c \in I$,

$$c = \phi\psi(c) = \phi(\sum_{j \in J_1} c(c_j/b_0)e_j) = \sum_{j \in J_1} c(c_j/b_0)b_j = c(\sum_{j \in J_1} (c_j/b_0)b_j),$$

whence $1_R = \sum_{j \in J_1} (c_j/b_0)b_j$ with $c_j/b_0 \in I^{-1}$. It follows that $R \subset I^{-1}I$. Since $I^{-1}I \subset R$ is always true, $R = I^{-1}I$. Therefore I is invertible. ∎

The characterization of Dedekind domains to be given below requires us to introduce another concept. A **discrete valuation ring** is a principal ideal domain that has exactly one *nonzero* prime ideal; (the zero ideal is prime in any integral domain).

Lemma 6.9. *If* R *is a Noetherian, integrally closed integral domain and* R *has a unique nonzero prime ideal* P, *then* R *is a discrete valuation ring.*

PROOF. We need only show that every proper ideal in R is principal. This requires the following facts, which are proved below:

(i) Let K be the quotient field of R. For every fractional ideal I of R the set $\bar{I} = \{a \in K \mid aI \subset I\}$ is precisely R;

(ii) $R \subsetneq P^{-1}$;

(iii) P is invertible;

(iv) $\bigcap_{n\in N^*} P^n = 0$;

(v) P is principal.

Assuming (i)–(v) for now, let I be any proper ideal of R. Then I is contained in a nonzero maximal ideal M of R (Theorem III.2.18), which is necessarily prime (Theorem III.2.19). By uniqueness $M = P$, whence $I \subset P$. Since $\bigcap_{n\in N^*} P^n = 0$ by (iv), there is a largest integer m such that $I \subset P^m$ and $I \not\subset P^{m+1}$. Choose $b \in I - P^{m+1}$. Since $P = (a)$ for some $a \in R$ by (v), $P^m = (a)^m = (a^m)$. Since $b \in P^m$, $b = ua^m$. Furthermore, $u \notin P = (a)$ (otherwise $b \in P^{m+1} = (a^{m+1})$). Consequently, u is a unit in R; (otherwise (u) would be a proper ideal by Theorem III.3.2 and hence contained in P by the argument used above). Therefore by Theorem III.3.2 $P^m = (a^m) = (ua^m) = (b) \subset I$, whence I is the principal ideal $P^m = (a^m)$.

Statements (i)–(v) are justified as follows.

(i) Clearly $R \subset \bar{I}$. It is easy to see that $\bar{I}$ is a subring of K and a fractional ideal of R, whence $\bar{I}$ is isomorphic (as an R-module) to an ideal of R (Remark preceding Theorem 6.3). Thus since R is Noetherian, $\bar{I}$ is finitely generated (Theorem 1.9). Theorem 5.3 (with $T = \bar{I}$) implies that every element of $\bar{I}$ is integral over R. Therefore, $\bar{I} \subset R$ since R is integrally closed. Hence $\bar{I} = R$.

(ii) Recall that $R \subset J^{-1}$ for every *ideal* J in R. Let $\mathfrak{F}$ be the set of all ideals J in R such that $R \subsetneq J^{-1}$. Since P is a proper ideal (Definition III.2.14), every nonzero element of P is a nonunit by Theorem III.3.2. If $J = (a)$, $(0 \neq a \in P)$, then $1_R/a \in J^{-1}$, but $1_R/a \notin R$, whence $R \subsetneq J^{-1}$. Therefore, $\mathfrak{F}$ is nonempty. Since R is Noetherian, $\mathfrak{F}$ contains a maximal element M (Theorem 1.4). We claim M is a prime ideal of R. If $ab \in M$ with $a,b \in R$ and $a \notin M$, choose $c \in M^{-1} - R$. Then $c(ab) \in R$, whence $bc(aR + M) \subset R$ and $bc \in (aR + M)^{-1}$. Therefore, $bc \in R$ (otherwise, $aR + M \in \mathfrak{F}$, contradicting the maximality of M). Consequently, $c(bR + M) \subset R$, and thus $c \in (bR + M)^{-1}$. Since $c \notin R$ the maximality of M implies that $bR + M = M$, whence $b \in M$. Therefore M is prime by Theorem III.2.15. Since $M \neq 0$, we must have $P = M$ by uniqueness. Thus $R \subsetneq M^{-1} = P^{-1}$.

(iii) Clearly $P \subset PP^{-1} \subset R$. The argument in the first paragraph of the proof shows that P is the unique maximal ideal in R, whence $P = PP^{-1}$ or $PP^{-1} = R$. But if $P = PP^{-1}$, then $P^{-1} \subset \bar{P}$ and by (i) and (ii), $R \subsetneq P^{-1} \subset \bar{P} = R$, which is a contradiction. Therefore $PP^{-1} = R$ and P is invertible.

(iv) If $\bigcap_{n\in N^*} P^n \neq 0$, then $\bigcap_{n\in N^*} P^n$ is a fractional ideal of R. Verify that $P^{-1} \subset \overline{\bigcap_{n\in N^*} P^n}$. Then by (i) and (ii) $R \subsetneq P^{-1} \subset \overline{\bigcap_{n\in N^*} P^n} = R$, which is a contradiction.

(v) There exists $a \in P$ such that $a \notin P^2$; (otherwise $P = P^2$, whence $\bigcap_{n\in N^*} P^n = P \neq 0$ contradicting (iv)). Then aP^{-1} is a nonzero ideal in R such that $aP^{-1} \not\subset P$ (otherwise, $a \in aR = aP^{-1}P \subset P^2$). The first paragraph of the proof shows that every proper ideal in R is contained in P, whence $aP^{-1} = R$. Therefore by (iii), $(a) = (a)R = (a)P^{-1}P = (aP^{-1})P = RP = P$. ■

Theorem 6.10. *The following conditions on an integral domain* R *are equivalent.*

(i) R *is a Dedekind domain;*

(ii) *every proper ideal in* R *is uniquely a product of a finite number of prime ideals;*

(iii) *every nonzero ideal in* R *is invertible;*

(iv) *every fractional ideal of* R *is invertible;*

(v) *the set of all fractional ideals of* R *is a group under multiplication;*

(vi) *every ideal in* R *is projective;*

(vii) *every fractional ideal of* R *is projective;*

(viii) R *is Noetherian, integrally closed and every nonzero prime ideal is maximal;*

(ix) R *is Noetherian and for every nonzero prime ideal* P *of* R, *the localization* R_P *of* R *at* P *is a discrete valuation ring.*

PROOF. The equivalence (iv) $\Leftrightarrow$ (v) is trivial (see Theorem 6.3). (i) $\Rightarrow$ (ii) and (ii) $\Rightarrow$ (iii) follow from Lemma 6.4 and Theorem 6.5. (iii) $\Leftrightarrow$ (vi) and (vii) $\Leftrightarrow$ (iv) are immediate consequences of Theorem 6.8. (vi) $\Rightarrow$ (vii) follows from the Remark preceding Theorem 6.3. In order to complete the proof we need only prove the implications (iv) $\Rightarrow$ (viii), (viii) $\Rightarrow$ (ix) and (ix) $\Rightarrow$ (i).

(iv) $\Rightarrow$ (viii) Every ideal of R is invertible by (iv) and hence finitely generated by Lemma 6.7. Therefore R is Noetherian by Theorem 1.9. Let K be the quotient field of R. If $u \in K$ is integral over R, then $R[u]$ is a finitely generated R-submodule of K by Theorem 5.3. Consequently, the second example after Definition 6.2 shows that $R[u]$ is a fractional ideal of R. Therefore, $R[u]$ is invertible by (iv). Thus since $R[u]R[u] = R[u]$, $R[u] = RR[u] = (R[u]^{-1}R[u])R[u] = R[u]^{-1}R[u] = R$, whence $u \in R$. Therefore R is integrally closed. Finally if P is a nonzero prime ideal in R, then there is a maximal ideal M of R that contains P (Theorem III.2.18). M is invertible by (iv). Consequently $M^{-1}P$ is a fractional ideal of R with $M^{-1}P \subset M^{-1}M = R$, whence $M^{-1}P$ is an ideal in R. Since $M(M^{-1}P) = RP = P$ and P is prime; either $M \subset P$ or $M^{-1}P \subset P$. But if $M^{-1}P \subset P$, then $R \subset M^{-1} = M^{-1}R = M^{-1}PP^{-1} \subset PP^{-1} \subset R$, whence $M^{-1} = R$. Thus $R = MM^{-1} = MR = M$, which contradicts the fact that M is maximal. Therefore $M \subset P$ and hence $M = P$. Therefore, P is maximal.

(viii) $\Rightarrow$ (ix) R_P is an integrally closed integral domain by Theorem 5.8. By Lemma III.4.9 every ideal in R_P is of the form $I_P = \{i/s \mid i \in I; s \notin P\}$, where I is an ideal of R. Since every ideal of R is finitely generated by (viii) and Theorem 1.9, it follows that every ideal of R_P is finitely generated. Therefore, R_P is Noetherian by Theorem 1.9. By Theorem III.4.11 every nonzero prime ideal of R_P is of the form I_P, where I is a nonzero prime ideal of R that is contained in P. Since every nonzero prime ideal of R is maximal by (viii), P_P must be the unique nonzero prime ideal in R_P. Therefore, R_P is a discrete valuation ring by Lemma 6.9.

(ix) $\Rightarrow$ (i) We first show that every ideal I ($\neq 0$) is invertible. II^{-1} is a fractional ideal of R contained in R (Remark (i) after Theorem 6.3), whence II^{-1} is an ideal in R. If $II^{-1} \neq R$, then there is a maximal ideal M containing II^{-1} (Theorem III.2.18). Since M is prime (Theorem III.2.19), the ideal I_M in R_M is principal by (ix); say $I_M = (a/s)$ with $a \in I$ and $s \in R - M$. Since R is Noetherian, I is finitely generated, say $I = (b_1, \dots, b_n)$, by Theorem 1.9. For each i, $b_i/1_R \in I_M$, whence in R_M, $b_i/1_R = (r_i/s_i)(a/s)$ for some $r_i \in R$, $s_i \in R - M$. Therefore $s_isb_i = r_ia \in I$. Let $t = ss_1s_2\cdots s_n$. Since $R - M$ is multiplicative, $t \in R - M$. In the quotient field of R we have for every t, $(t/a)b_i = tb_i/a = s_1\cdots s_{i-1}s_{i+1}\cdots s_nr_i \in R$, whence $t/a \in I^{-1}$. Consequently $t = (t/a)a \in I^{-1}I \subset M$, which contradicts the fact that $t \in R - M$. Therefore $II^{-1} = R$ and I is invertible.

For each ideal I ($\neq R$) of R choose a maximal ideal M_I of R such that

$I \subset M_I \subsetneq R$ (Theorem III.2.18; Axiom of Choice). If $I = R$, let $M_R = R$. Then IM_I^{-1} is a fractional ideal of R with $IM_I^{-1} \subset M_I M_I^{-1} \subset R$. Therefore, IM_I^{-1} is an ideal of R that clearly contains I. Also, if I is proper, then $I \subsetneq IM_I^{-1}$ (otherwise since I and M_I are invertible, $R = RR = (I^{-1}I)(M_I^{-1}M_I) = I^{-1}(IM_I^{-1})M_I = I^{-1}IM_I = RM_I = M_I$, which contradicts the choice of M_I). Let S be the set of all ideals of R and define a function $f: S \to S$ by $I \mapsto IM_I^{-1}$. Given a proper ideal J, there exists by the Recursion Theorem 6.2 of the Introduction (with $f_n = f$ for all n) a function $\phi: \mathbf{N} \to S$ such that $\phi(0) = J$ and $\phi(n+1) = f(\phi(n))$. If we denote $\phi(n)$ by J_n and M_{J_n} by M_n, then we have an ascending chain of ideals $J = J_0 \subset J_1 \subset J_2 \subset \cdots$ such that $J = J_0$ and $J_{n+1} = f(J_n) = J_n M_n^{-1}$. Since R is Noetherian and J is proper, there is a least integer k such that

$$J = J_0 \subsetneq J_1 \subsetneq \cdots \subsetneq J_{k-1} \subsetneq J_k = J_{k+1}.$$

Thus $J_k = J_{k+1} = f(J_k) = J_k M_k^{-1}$. The remarks above show that this can occur only if $J_k = R$. Consequently, $R = J_k = f(J_{k-1}) = J_{k-1}M_{k-1}^{-1}$, whence

$$J_{k-1} = J_{k-1}R = J_{k-1}M_{k-1}^{-1}\, M_{k-1} = RM_{k-1} = M_{k-1}.$$

Since $M_{k-1} = J_{k-1} \subsetneq J_k = R$, M_{k-1} is a maximal ideal. The minimality of k insures that each of $M_0, \ldots, M_{k-2}$ is also maximal (otherwise $M_j = R$, whence $J_{j+1} = J_j M_j^{-1} = J_j R^{-1} = J_j R = J_j$). It is easy to verify that

$$M_{k-1} = J_{k-1} = J_{k-2}M_{k-2}^{-1} = J_{k-3}M_{k-3}^{-1}M_{k-2}^{-1} = \cdots = JM_0^{-1}M_1^{-1}\cdots M_{k-2}^{-1}.$$

Consequently, since each M_i is invertible,

$$M_{k-1}(M_0\cdots M_{k-2}) = JM_0^{-1}\cdots M_{k-2}^{-1}(M_0\cdots M_{k-2}) = J.$$

Thus J is the product maximal (hence prime) ideals. Therefore R is Dedekind. ∎

We close with an example showing that the class of principal ideal domains is properly contained in the class of Dedekind domains.

EXAMPLE. The integral domain $\mathbf{Z}[\sqrt{10}] = \{a + b\sqrt{10} \mid a,b \in \mathbf{Z}\}$ has quotient field $\mathbf{Q}(\sqrt{10}) = \{r + s\sqrt{10} \mid r,s \in \mathbf{Q}\}$. A tedious calculation and elementary number theory show that $\mathbf{Z}[\sqrt{10}]$ is integrally closed (Exercise 14). Since the evaluation map $\mathbf{Z}[x] \to \mathbf{Z}[\sqrt{10}]$ given by $f(x) \mapsto f(\sqrt{10})$ is an epimorphism and $\mathbf{Z}[x]$ is Noetherian (Theorem 4.9), $\mathbf{Z}[\sqrt{10}]$ is also Noetherian (Exercise 1.5). Finally it is not difficult to prove that every nonzero prime ideal of $\mathbf{Z}[\sqrt{10}]$ is maximal (Exercise 15). Therefore $\mathbf{Z}[\sqrt{10}]$ is a Dedekind domain by Theorem 6.10(viii). However $\mathbf{Z}[\sqrt{10}]$ is not a principal ideal domain (Theorem III.3.7 and Exercise III.3.4).

EXERCISES

1. The ideal generated by 3 and $1 + \sqrt{5}i$ in the subdomain $\mathbf{Z}[\sqrt{5}i]$ of $\mathbf{C}$ is invertible.

2. An invertible ideal in an integral domain that is a local ring is principal.

3. If I is an invertible ideal in an integral domain R and S is a multiplicative set in R with $0 \notin S$, then $S^{-1}I$ is invertible in $S^{-1}R$.

4. Let R be any ring with identity and P an R-module. Then P is projective if and only if there exist sets $\{a_i \mid i \varepsilon I\} \subset P$ and $\{f_i \mid i \varepsilon I\} \subset \text{Hom}_R(P,R)$ such that for all $a \varepsilon P$, $a = \sum_{i \varepsilon I} f_i(a)a_i$. [See the proof of Theorem 6.8.]

5. (Converse of Lemma 6.9) A discrete valuation ring R is Noetherian and integrally closed. [*Hint:* Exercise 5.8.]

6. (a) If every prime ideal in an integral domain R is invertible, then R is Dedekind.
(b) If R is a Noetherian integral domain in which every maximal ideal is invertible, then R is Dedekind.

7. If S is a multiplicative subset of a Dedekind domain R (with $1_R \varepsilon S, 0 \notin S$), then $S^{-1}R$ is a Dedekind domain.

8. If R is an integral domain and P a prime ideal in $R[x]$ such that $P \cap R = 0$, then $R[x]_P$ is a discrete valuation ring.

9. If a Dedekind domain R has only a finite number of nonzero prime ideals $P_1, \ldots, P_n$, then R is a principal ideal domain. [*Hint:* There exists $a_i \varepsilon P_i - P_i^2$ and by the Chinese Remainder Theorem III.2.25 there exists $b_i \varepsilon P_i$ such that $b_i \equiv a_i \pmod{P_i}$ and $b_i \equiv 1_R \pmod{P_j}$ for $j \neq i$. Show that $P_i = (b_i)$, which implies that every ideal is principal.]

10. If I is a nonzero ideal in a Dedekind domain R, then R/I is an Artinian ring.

11. Every proper ideal in a Dedekind domain may be generated by at most two elements.

12. An R-module A is *divisible* if $rA = A$ for all nonzero $r \varepsilon R$. If R is a Dedekind domain, every divisible R-module is injective. [N.B. the converse is also true, but harder.]

13. (Nontrivial) If R is a Dedekind domain with quotient field K, F is a finite dimensional extension field of K and S is the integral closure of R in F (that is, the ring of all elements of F that are integral over R), then S is a Dedekind domain.

14. (a) Prove that the integral domain $\mathbf{Z}[\sqrt{10}]$ is an integral extension ring of $\mathbf{Z}$ with quotient field $\mathbf{Q}(\sqrt{10})$.
(b) Let $u \varepsilon \mathbf{Q}(\sqrt{10})$ be integral over $\mathbf{Z}[\sqrt{10}]$. Then u is integral over $\mathbf{Z}$ (Theorem 5.6). Furthermore if $u \varepsilon \mathbf{Q}$, then $u \varepsilon \mathbf{Z}$ (Exercise 5.8). Prove that if $u \varepsilon \mathbf{Q}(\sqrt{10})$ and $u \notin \mathbf{Q}$, then u is the root of an irreducible monic polynomial of degree 2 in $\mathbf{Z}[x]$. [*Hint:* Corollary III.6.13 and Theorem V.1.6.]
(c) Prove that if $u = r + s\sqrt{10} \varepsilon \mathbf{Q}(\sqrt{10})$ and u is a root of $x^2 + ax + b \varepsilon \mathbf{Z}[x]$, then $a = -2r$ and $b = r^2 - 10s^2$. [*Hint:* note that $u^2 - 2ru + (r^2 - 10s^2) = 0$; if $u \notin \mathbf{Q}$ use Theorem V.1.6.]
(d) Prove that $\mathbf{Z}[\sqrt{10}]$ is integrally closed. [*Hint:* if $u = r + s\sqrt{10} \varepsilon \mathbf{Q}(\sqrt{10})$ is a root of $x^2 + ax + b \varepsilon \mathbf{Z}[x]$ and a is even, then $r \varepsilon \mathbf{Z}$ by (c); it follows that $s \varepsilon \mathbf{Z}$. The assumption that a is odd leads to a contradiction.]

15. (a) If P is a nonzero prime ideal of the ring $\mathbf{Z}[\sqrt{10}]$, then $P \cap \mathbf{Z}$ is a nonzero prime ideal of $\mathbf{Z}$. [*Hint:* if $0 \neq u \varepsilon P$, then u is a root of $x^2 + ax + b \varepsilon \mathbf{Z}[x]$ by Exercise 14. Show that one of a,b is nonzero and lies in P.]
(b) Every nonzero prime ideal of $\mathbf{Z}[\sqrt{10}]$ is maximal. [Use (a), Theorem III.3.4 and either an easy direct argument or Theorem 5.12.]

16. A **valuation domain** is an integral domain R such that for all $a,b \in R$ either $a \mid b$ or $b \mid a$. (Clearly a discrete valuation ring is a valuation domain.) A **Prüfer domain** is an integral domain in which every finitely generated ideal is invertible.

(a) The following are equivalent: (i) R is a Prüfer domain; (ii) for every prime ideal P in R, R_P is a valuation domain; (iii) for every maximal ideal M in R, R_M is a valuation domain.

(b) A Prüfer domain is Dedekind if and only if it is Noetherian.

(c) If R is a Prüfer domain with quotient field K, then any domain S such that $R \subset S \subset K$ is Prüfer.

7. THE HILBERT NULLSTELLENSATZ

The results of Section VI.1 and Section 5 are used to prove a famous result of classical algebraic geometry, the Nullstellensatz (Zeros Theorem) of Hilbert. Along the way we also prove the Noether Normalization Lemma. We begin with a very brief sketch of the geometric background (this discussion is continued at the end of the section).

Classical algebraic geometry is the study of simultaneous solutions of systems of polynomial equations:

$$f(x_1,x_2,\dots,x_n) = 0 \qquad (f \in S)$$

where K is a field and $S \subset K[x_1,\dots,x_n]$. A solution of this system is an n-tuple, $(a_1,\dots,a_n) \in F^n = F \times F \times \cdots \times F$ (n factors), where F is an algebraically closed extension field of K and $f(a_1,\dots,a_n) = 0$ for all $f \in S$. Such a solution is called a **zero of S in $\mathbf{F}^n$**. The set of all zeros of S is called the **affine K-variety** (or **algebraic set**) in F^n defined by S and is denoted $V(S)$. Thus

$$V(S) = \{(a_1,\dots,a_n) \in F^n \mid f(a_1,\dots,a_n) = 0 \quad \text{for all } f \in S\}.$$

Note that if I is the ideal of $K[x_1,\dots,x_n]$ generated by S, then $V(I) = V(S)$.

The assignment $S \mapsto V(S)$ defines a function from the set of all subsets of $K[x_1,\dots,x_n]$ to the set of all subsets of F^n. Conversely, define a function from the set of subsets of F^n to the set of subsets of $K[x_1,\dots,x_n]$ by $Y \mapsto J(Y)$, where $Y \subset F^n$ and

$$J(Y) = \{f \in K[x_1,\dots,x_n] \mid f(a_1,\dots,a_n) = 0 \quad \text{for all} \quad (a_1,\dots,a_n) \in Y\}.$$

Note that $J(Y)$ is actually an ideal of $K[x_1,\dots,x_n]$. The correspondence given by V and J has the same formal properties as does the Galois correspondence (priming operations) between intermediate fields of an extension and subgroups of the Galois group. In other words we have the following analogue of Lemma V.2.6.

Lemma 7.1. *Let* F *be an algebraically closed extension field of* K *and let* S, T *be subsets of* $K[x_1,\dots,x_n]$ *and* X,Y *subsets of* F^n. *Then*

(i) $V(K[x_1,\dots,x_n]) = \varnothing$; $J(F^n) = \varnothing$; $J(\varnothing) = K[x_1,\dots,x_n]$;

(ii) $S \subset T \Rightarrow V(T) \subset V(S)$ *and* $X \subset Y \Rightarrow J(Y) \subset J(X)$;

(iii) $S \subset J(V(S))$ *and* $Y \subset V(J(Y))$;

(iv) $V(S) = V(J(V(S)))$ *and* $J(Y) = J(V(J(Y)))$.

PROOF. Exercise. ■

It is natural to ask which objects are **closed** under this correspondence, that is, which S and Y satisfy $J(V(S)) = S$ and $V(J(Y)) = Y$. Closed subsets of F^n are easily described (Exercise 1), but the characterization of closed subsets of $K[x_1,\ldots,x_n]$ requires the Nullstellensatz, which states that $J(V(I)) = \text{Rad } I$ for every proper ideal I of $K[x_1,\ldots,x_n]$. In order to prove the Nullstellensatz we need two preliminary results, the first of which is of interest in its own right.

Theorem 7.2. (*Noether Normalization Lemma*) *Let* R *be an integral domain which is a finitely generated extension ring of a field* K *and let* r *be the transcendence degree over* K *of the quotient field* F *of* R. *Then there exists an algebraically independent subset* $\{t_1,t_2,\ldots,t_r\}$ *of* R *such that* R *is integral over* $K[t_1,\ldots,t_r]$.

PROOF. Let $R = K[u_1,\ldots,u_n]$; then $F = K(u_1,\ldots,u_n)$. If $\{u_1,\ldots,u_n\}$ is algebraically independent over K, $\{u_1,\ldots,u_n\}$ is a transcendence base of F over K by Corollary VI.1.6, whence $r = n$ and the theorem is trivially true. If $\{u_1,\ldots,u_n\}$ is algebraically dependent over K, then $r \leq n - 1$ (Corollary VI.1.7) and

$$\sum_{(i_1,\ldots,i_n)\in I} k_{i_1\ldots i_n}u_1^{i_1}u_2^{i_2}\cdots u_n^{i_n} = 0,$$

where I is a finite set of distinct n-tuples of nonnegative integers and $k_{i_1\ldots i_n}$ is a nonzero element of K for every $(i_1,\ldots,i_n) \in I$. Let c be a positive integer that is greater than every component i_s of every element $(i_1,\ldots,i_n)$ of I. If $(i_1,\ldots,i_n)$, $(j_1,\ldots,j_n) \in I$ are such that

$$i_1 + ci_2 + c^2i_3 + \cdots + c^{n-1}i_n = j_1 + cj_2 + c^2j_3 + \cdots + c^{n-1}j_n,$$

then $c \mid i_1 - j_1$ which is impossible unless $i_1 = j_1$ (since $c > i_1 \geq 0$ and $c > j_1 \geq 0$ imply $c > |i_1 - j_1|$). Consequently, $i_2 + ci_3 + \cdots + c^{n-2}i_n = j_2 + cj_3 + \cdots + c^{n-2}j_n$. As before $c \mid i_2 - j_2$, whence $i_2 = j_2$. Repetition of this argument shows that $(i_1,\ldots,i_n) = (j_1,\ldots,j_n)$. Therefore, the set

$$\{i_1 + ci_2 + c^2i_3 + \cdots + c^{n-1}i_n \mid (i_1,\ldots,i_n) \in I\}$$

consists of $|I|$ distinct nonnegative integers; in particular, it has a *unique* maximum element $j_1 + cj_2 + \cdots + c^{n-1}j_n$ for some $(j_1,\ldots,j_n) \in I$. Let

$$v_2 = u_2 - u_1^{c},\ v_3 = u_3 - u_1^{c^2},\ldots,v_n = u_n - u_1^{c^{n-1}}.$$

If we expand the algebraic dependence relation above, after making the substitutions $u_i = v_i + u_1^{c^{i-1}}$ $(2 \leq i \leq n)$, we obtain

$$k_{j_1\ldots j_n}u_1^{j_1+cj_2+c^2j_3+\cdots+c^{n-1}j_n} + f(u_1,v_2,v_3,\ldots,v_n) = 0,$$

where the degree of $f \in K[x_1,\ldots,x_n]$ in x_1 is strictly less than $j_1 + cj_2 + \cdots + c^{n-1}j_n$. Therefore, u_1 is a root of the monic polynomial

$$x^{j_1+cj_2+\cdots+c^{n-1}j_n} + k_{j_1\ldots j_n}^{-1}f(x,v_2,\ldots,v_n) \in K[v_2,\ldots,v_n][x].$$

Consequently, u_1 is integral over $K[v_2,\ldots,v_n]$. By Theorem 5.5 $K[u_1,v_2,\ldots,v_n] = K[v_2,\ldots,v_n][u_1]$ is integral over $K[v_2,\ldots,v_n]$. Since each u_i $(2 \leq i \leq n)$ is obviously integral over $K[u_1,v_2,\ldots,v_n]$, Theorems 5.5 and 5.6 imply that

$$R = K[u_1,\ldots,u_n]$$

is integral over $K[v_2, \ldots, v_n]$ (whence F is algebraic over $K(v_2, \ldots, v_n)$). If $\{v_2, \ldots, v_n\}$ is algebraically independent, then $r = n - 1$ by Corollary VI.1.6 and the theorem is proved. If not, the preceding argument with $K[v_2, \ldots, v_n]$ in place of R shows that for some $w_3, \ldots, w_n \in R$, $K[v_2, \ldots, v_n]$ is integral over $K[w_3, \ldots, w_n]$. By Theorem 5.6 R is integral over $K[w_3, \ldots, w_n]$ (whence F is algebraic over $K(w_3, \ldots, w_n)$ and $r \leq n - 2$). If $\{w_3, \ldots, w_n\}$ is algebraically independent, we are finished. If not, the preceding process may be repeated and an inductive argument will yield an algebraically independent subset $\{z_{n-r+1}, \ldots, z_n\}$ of r elements of R such that R is integral over $K[z_{n-r+1}, \ldots, z_n]$. ■

Now let K be a field and F an algebraically closed extension field of K. If a proper ideal I of $K[x_1, \ldots, x_n]$ is finitely generated, say $I = (g_1, \ldots, g_k)$, then the affine variety $V(I)$ clearly consists of every $(a_1, \ldots, a_n) \in F^n$ that is a common root of $g_1, \ldots, g_k$ (see Exercise 4). If $n = 1$, $K[x_1]$ is a principal ideal domain and it is obvious that $V(I)$ is nonempty. More generally (and somewhat surprisingly) we have:

Lemma 7.3. *If* F *is an algebraically closed extension field of a field* K *and* I *is a proper ideal of* $K[x_1, \ldots, x_n]$, *then the affine variety* V(I) *defined by* I *in* F^n *is nonempty.*

PROOF. By Theorems III.2.18 and III.2.19 I is contained in a proper prime ideal P, whence $V(P) \subset V(I)$. Consequently, it suffices to prove that $V(P)$ is nonempty for every proper prime ideal P of $K[x_1, \ldots, x_n]$. Observe that $P \cap K = 0$; (otherwise $0 \neq a \in P \cap K$, whence $1_K = a^{-1}a \in P$, contradicting the fact that P is proper).

Let R be the integral domain $K[x_1, \ldots, x_n]/P$ (see Theorem III.2.16) and let $\pi : K[x_1, \ldots, x_n] \to R$ be the canonical epimorphism. If we denote $\pi(x_i) \in R$ by u_i, then $R = \pi(K)[u_1, \ldots, u_n]$. Furthermore since $K \cap P = 0$, π maps K isomorphically onto $\pi(K)$; in particular, $\pi(K)$ is a field. By the Noether Normalization Lemma there exists a subset $\{t_1, \ldots, t_r\}$ of R such that $\{t_1, \ldots, t_r\}$ is algebraically independent over $\pi(K)$ and R is integral over $S = \pi(K)[t_1, \ldots, t_r]$. If M is the ideal of S generated by $t_1, \ldots, t_r$, then the map $\pi(K) \to S/M$ given by $\pi(a) \mapsto \pi(a) + M$ is an isomorphism (see Theorem VI.1.2). Consequently M is a maximal ideal of S by Theorem III.2.20. Therefore, there is a maximal ideal N of R such that $N \cap S = M$ (Theorems 5.9 and 5.12). Let $\tau : R \to R/N$ be the canonical epimorphism. Then $\tau(R) = R/N$ is a field by Theorem III.2.20. The Second Isomorphism Theorem III.2.12 together with the maps defined above now yields an isomorphism

$$K \cong \pi(K) \cong S/M = S/(N \cap S) \cong (S + N)/N = \tau(S),$$

which is given by $a \mapsto \pi(a) \mapsto \pi(a) + M \mapsto \pi(a) + N = \tau(\pi(a))$. Let $\overline{\tau(R)}$ be an algebraic closure of $\tau(R)$. Since R is integral over S, $\tau(R)$ is an algebraic field extension of $\tau(S)$, whence $\overline{\tau(R)}$ is also an algebraic closure of $\tau(S)$ (Theorem V.3.4). Now F contains an algebraic closure $\overline{K}$ of K (Exercise V.3.7). By Theorem V.3.8 the isomorphism $K \cong \tau(S)$ extends to an isomorphism $\overline{K} \cong \overline{\tau(R)}$. Restriction of the *inverse* of this isomorphism yields a monomorphism $\sigma : \tau(R) \to \overline{K} \subset F$. Let ϕ be the composition $K[x_1, \ldots, x_n] \xrightarrow{\pi} R \xrightarrow{\tau} \tau(R) \xrightarrow{\sigma} F$ and verify that $\phi \mid K = 1_K$ and $\phi \mid P = 0$. Consequently, for any $f(x_1, \ldots, x_n) \in P \subset K[x_1, \ldots, x_n]$, $f(\phi(x_1), \ldots, \phi(x_n)) =$

$\phi(f(x_1, \ldots, x_n)) = 0$, whence $(\phi(x_1), \ldots, \phi(x_n))$ is a zero of P in F^n. Therefore, $V(P)$ is nonempty. ■

Proposition 7.4. *(Hilbert Nullstellensatz) Let* F *be an algebraically closed extension field of a field* K *and* I *a proper ideal of* $K[x_1, \ldots, x_n]$. *Let* $V(I) = \{(a_1, \ldots, a_n) \in F^n \mid g(a_1, \ldots, a_n) = 0$ *for all* $g \in I\}$. *Then*

$$\begin{aligned} Rad\, I &= J(V(I)) \\ &= \{f \in K[x_1, \ldots, x_n] \mid f(a_1, \ldots, a_n) = 0 \quad \textit{for all} \quad (a_1, \ldots, a_n) \in V(I)\}. \end{aligned}$$

In other words, $f(a_1, \ldots, a_n) = 0$ *for every zero* $(a_1, \ldots, a_n)$ *of* I *in* F^n *if and only if* $f^m \in I$ *for some* $m \geq 1$.

REMARK. We shall use Lemma 7.3 to prove the theorem. Since the theorem implies the lemma (Exercise 6), the two are actually equivalent.

PROOF OF 7.4. If $f \in \text{Rad } I$, then $f^m \in I$ for some $m \geq 1$ (Theorem 2.6). If $(a_1, \ldots, a_n)$ is a zero of I in F^n, then $0 = f^m(a_1, \ldots, a_n) = (f(a_1, \ldots, a_n))^m$. Consequently, since F is a field, $f(a_1, \ldots, a_n) = 0$. Therefore, Rad $I \subset JV(I)$.

Conversely, suppose $f \in JV(I)$. We may assume $f \neq 0$ since $0 \in \text{Rad } I$. Consider $K[x_1, \ldots, x_n]$ as a subring of the ring $K[x_1, \ldots, x_n, y]$ of polynomials in $n + 1$ indeterminates over K. Let L be the nonzero ideal of $K[x_1, \ldots, x_n, y]$ generated by I and $yf - 1_F$. Clearly if $(a_1, \ldots, a_n, b)$ is a zero of L in F^{n+1} then $(a_1, \ldots, a_n)$ must be a zero of I in F^n. But $(yf - 1_F)(a_1, \ldots, a_n, b) = bf(a_1, \ldots, a_n) - 1_F = -1_F$ for all zeros $(a_1, \ldots, a_n)$ of I in F^n. Therefore, L has no zeros in F^{n+1}; that is, $V(L)$ is empty. Consequently, $L = K[x_1, \ldots, x_n, y]$ by Lemma 7.3, whence $1_F \in L$. Thus

$$1_F = \sum_{i=1}^{t-1} g_i f_i + g_t(yf - 1_F),$$

where $f_i \in I$ $(1 \leq i \leq t - 1)$ and $g_i \in K[x_1, \ldots, x_n, y]$. Define an evaluation homomorphism $K[x_1, \ldots, x_n, y] \to K(x_1, \ldots, x_n)$ by $x_i \mapsto x_i$ and $y \mapsto f^{-1} = 1_K/f(x_1, \ldots, x_n)$ (Corollary III.5.6). Then in the field $K(x_1, \ldots, x_n)$

$$1_F = \sum_{i=1}^{t-1} g_i(x_1, \ldots, x_n, f^{-1}) f_i(x_1, \ldots, x_n).$$

Let m be a positive integer larger than the degree of g_i in y for every i $(1 \leq i \leq t - 1)$. Then for each i, $f^m(x_1, \ldots, x_n)g_i(x_1, \ldots, x_n, f^{-1})$ lies in $K[x_1, \ldots, x_n]$, whence $f^m = f^m 1_F = \sum_{i=1}^{t-1} f^m(x_1, \ldots, x_n)g_i(x_1, \ldots, x_n, f^{-1}) f_i(x_1, \ldots, x_n) \in I$. Therefore $f \in \text{Rad } I$ and hence $JV(I) \subset \text{Rad } I$. ■

The determination of closed objects as mentioned in the introduction of this section is now straightforward (Exercises 1–3).

We close this section with an informal attempt to establish the connection between geometry and algebra which characterizes the classical approach to algebraic geometry. Let K be a field. Every polynomial $f \in K[x_1, \ldots, x_n]$ determines a function $F^n \to F$ by substitution: $(a_1, \ldots, a_n) \mapsto f(a_1, \ldots, a_n)$. If $V = V(I)$ is an affine variety contained in F^n, the restriction of this function to V is called a **regular function** on V. The regular functions $V \to F$ form a ring $\Gamma(V)$ which is isomorphic to

$$K[x_1, \ldots, x_n]/J(V(I))$$

(Exercise 10). This ring is called the **coordinate ring** of V. Since $I \subset J(V(I)) = \text{Rad } I$ the ring $\Gamma(V)$ has no nonzero nilpotent elements. Furthermore $\Gamma(V)$ is a finitely generated algebra over K (since $K[x_1, \ldots, x_n]$ and the ideal $J(V(I))$ are; see Section IV.7). Conversely it can be proved that every finitely generated K-algebra with no nonzero nilpotent elements is the coordinate ring of some affine variety. Therefore, there is a one-to-one correspondence between affine varieties and a rather special class of commutative rings. With a suitable definition of morphisms the affine varieties form a category as do the commutative rings in question and this correspondence is actually an "equivalence" of categories. Thus statements about affine varieties are equivalent to certain statements of commutative algebra. For further information see W. Fulton [53] and I. G. MacDonald [55].

EXERCISES

Note: F is always an algebraically closed extension field of a field K; J, V, and F^n are as above.

1. A subset Y of F^n is closed (that is, $V(J(Y)) = Y$) if and only if Y is an affine K-variety determined by some subset S of $K[x_1, \ldots, x_n]$.
2. A subset S of $K[x_1, \ldots, x_n]$ is closed (that is, $J(V(S)) = S$) if and only if S is a radical ideal (that is, S is an ideal and $S = \text{Rad } S$).
3. There is a one-to-one inclusion reversing correspondence between the set of affine K-varieties in F^n and the set of radical ideals of $K[x_1, \ldots, x_n]$. [See Exercises 1, 2.]
4. Every affine K-variety in F^n is of the form $V(S)$ where S is a finite subset of $K[x_1, \ldots, x_n]$. [*Hint:* Theorems 1.9 and 4.9 and Exercise 3.]
5. If $V_1 \supset V_2 \supset \cdots$ is a descending chain of K-varieties in F^n, then $V_m = V_{m+1} = \cdots$ for some m. [*Hint:* Theorem 4.9 and Exercise 3.]
6. Show that the Nullstellensatz implies Lemma 7.3.
7. If $I_1, \ldots, I_k$ are ideals of $K[x_1, \ldots, x_n]$, then $V(I_1 \cap I_2 \cap \cdots \cap I_k) = V(I_1) \cup V(I_2) \cup \cdots \cup V(I_k)$ and $V(I_1 I_2 \cdots I_k) = V(I_1) \cap V(I_2) \cap \cdots \cap V(I_k)$.
8. A K-variety V in F^n is **irreducible** provided that whenever $V = W_1 \cup W_2$ with each W_i a K-variety in F^n, either $V = W_1$ or $V = W_2$.
   (a) Prove that V is irreducible if and only if $J(V)$ is a prime ideal in $K[x_1, \ldots, x_n]$.
   (b) Let $F = \mathbf{C}$ and $S = \{x_1^2 - 2x_2^2\}$. Then $V(S)$ is irreducible as a $\mathbf{Q}$-variety but not as an $\mathbf{R}$-variety.
9. Every nonempty K-variety in F^n may be written uniquely as a finite union $V_1 \cup V_2 \cup \cdots \cup V_k$ of affine K-varieties in F^n such that $V_j \not\subset V_i$ for $i \neq j$ and each V_i is irreducible (Exercise 8).
10. The coordinate ring of an affine K-variety $V(I)$ is isomorphic to $K[x_1, \ldots, x_n]/J(V(I))$.

CHAPTER IX

THE STRUCTURE OF RINGS

In the first part of this chapter a general structure theory for rings is presented. Although the concepts and techniques introduced have widespread application, complete structure theorems are available only for certain classes of rings. The basic method for determining such a class of rings might be described intuitively as follows. One singles out an "undesirable" property P that satisfies certain conditions, in particular, that every ring has an ideal which is maximal with respect to having property P. This ideal is called the P-radical of the ring. One then attempts to find structure theorems for the class of rings with zero P-radical. Frequently one must include additional hypotheses (such as appropriate chain conditions) in order to obtain really strong structure theorems. These ideas are discussed in full detail in the introductions to Sections 1 and 2 below. The reader would do well to read both these discussions before beginning serious study of the chapter.

We shall investigate two different radicals, the Jacobson radical (Section 2) and the prime radical (Section 4). Very deep and useful structure theorems are obtained for left Artinian semisimple rings (that is, left Artinian rings with zero Jacobson radical) in Section 3. Goldie's Theorem is discussed in Section 4. It includes a characterization of left Noetherian semiprime rings (that is, left Noetherian rings with zero prime radical). The basic building blocks for all of these structure theorems are the endomorphism rings of vector spaces over division rings and certain "dense" subrings of such rings (Section 1).

The last two sections of the chapter deal with algebras over a commutative ring with identity. The Jacobson radical and related concepts and results are carried over to algebras (Section 5). Division algebras are studied in Section 6.

A theme that occurs continually in this chapter is the close interconnection between the structure of a ring and the structure of modules over the ring. The use of modules in the study of rings has resulted in a host of new insights and deep theorems.

The interdependence of the sections of this chapter is as follows:

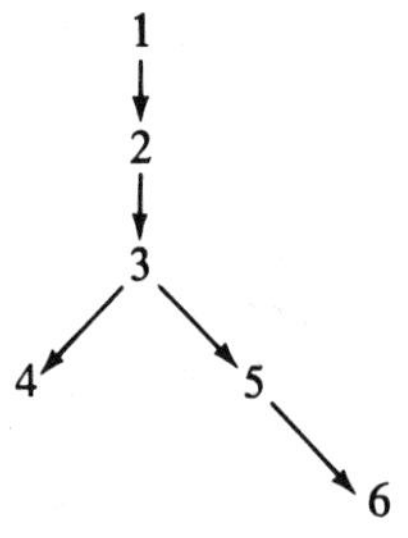

Much of the discussion here depends on the results of Section VIII.1 (Chain conditions).

1. SIMPLE AND PRIMITIVE RINGS

In this section we study those rings that will be used as the basic building blocks in the structure theory of rings.

We begin by recalling several facts that motivate a large part of this chapter.

(i) If V is a vector space over a division ring D, then $\mathrm{Hom}_D(V,V)$ is a ring (Exercise IV.1.7), called the **endomorphism ring** of V.

(ii) The endomorphism ring of a finite dimensional vector space over a division ring is isomorphic to the ring of all $n \times n$ matrices over a (possibly different) division ring (Theorem VII.1.4).

(iii) If D is a division ring, then $\mathrm{Mat}_n D$ is simple (that is, has no proper ideals; Exercise III.2.9) and is both left and right Artinian (Corollary VIII.1.12). Consequently by (ii) every endomorphism ring of a finite dimensional vector space over a division ring is both simple and Artinian.

(iv) The endomorphism ring of an infinite dimensional vector space over a division ring is neither simple nor Artinian (Exercise 3). However, such a ring is primitive, in a sense to be defined below.

Matrix rings and endomorphism rings of vector spaces over division rings arise naturally in many different contexts. They are extremely useful mathematical concepts. Consequently it seems reasonable to take such rings, or at least rings that closely resemble them, as the basis of a structure theory and to attempt to describe arbitrary rings in terms of these basic rings.

With the advantage of hindsight we single out two fundamental properties of the endomorphism ring of a vector space V: simplicity (Definition 1.1) and primitivity (Definition 1.5). As noted above these two concepts roughly correspond to the cases when V is finite or infinite dimensional respectively. In this section we shall analyze simple and primitive rings and show that in several important cases they coincide with endomorphism rings. In other cases they come as close to being endomorphism rings as is reasonably possible.

More precisely, an arbitrary primitive ring R is shown to be isomorphic to a particular kind of subring (called a dense subring) of the endomorphism ring of a vector space V over a division ring D (Theorem 1.12). R is left Artinian if and only if $\dim_D V$

is finite (Theorem 1.9). In this classical case, simple and primitive rings coincide and R is actually isomorphic to the complete endomorphism ring of V (Theorem 1.14). Furthermore in this situation $\dim_D V$ is uniquely determined and V is determined up to isomorphism (Proposition 1.17). These results amply justify the designation of simplicity and primitivity as fundamental concepts.

As noted in the introduction to this chapter modules play a crucial role in ring theory. Consequently we begin by defining and developing the elementary properties of simplicity for both rings and modules.

Definition 1.1. *A (left) module* A *over a ring* R *is* **simple** *(or* **irreducible***) provided* $RA \neq 0$ *and* A *has no proper submodules. A ring* R *is* **simple** *if* $R^2 \neq 0$ *and* R *has no proper (two-sided) ideals.*

REMARKS. (i) Every simple module [ring] is nonzero.

(ii) Every simple module over a ring with identity is unitary (Exercise IV.1.17). A unitary module A over a ring R with identity has $RA \neq 0$, whence A is simple if and only if A has no proper submodules.

(iii) Every simple module A is cyclic; in fact, $A = Ra$ for every nonzero $a \in A$. [Proof: both Ra ($a \in A$) and $B = \{c \in A \mid Rc = 0\}$ are submodules of A, whence each is either 0 or A by simplicity. But $RA \neq 0$ implies $B \neq A$. Consequently $B = 0$, whence $Ra = A$ for all nonzero $a \in A$.] However a cyclic module need not be simple (for example, the cyclic **Z**-module Z_6).

(iv) The definitions of "simple" for groups, modules, and rings can be subsumed into one general definition, which might be roughly stated as: an algebraic object C that is nontrivial in some reasonable sense (for example, $RA \neq 0$ or $R^2 \neq 0$) is *simple*, provided that every homomorphism with domain C has kernel 0 or C. The point here is that the absence of nontrivial kernels is equivalent to the absence of proper normal subgroups of a group or proper submodules of a module or proper ideals of a ring as the case may be.

EXAMPLE. Every division ring is a simple ring and a simple D-module (see the Remarks preceding Theorem III.2.2).

EXAMPLE. Let D be a division ring and let $R = \text{Mat}_n D$ ($n > 1$). For each k ($1 \leq k \leq n$), $I_k = \{(a_{ij}) \in R \mid a_{ij} = 0 \text{ for } j \neq k\}$ is a simple left R-module (see the proof of Corollary VIII.1.12).

EXAMPLE. The preceding example shows that $\text{Mat}_n D$ (D a division ring) is not a simple left *module* over itself if $n > 1$. However, the *ring* $\text{Mat}_n D$ ($n \geq 1$) is simple by Exercise III.2.9. Thus by Theorem VII.1.4 the endomorphism ring of any finite dimensional vector space over a division ring is a simple ring.

EXAMPLE. A left ideal I of a ring R is said to be a **minimal left ideal** if $I \neq 0$ and for every left ideal J such that $0 \subset J \subset I$, either $J = 0$ or $J = I$. A left ideal I of R such that $RI \neq 0$ is a simple left R-module if and only if I is a minimal left ideal.

EXAMPLE. Let F be a field of characteristic zero and R the *additive* group of polynomials $F[x,y]$. Define multiplication in R by requiring that multiplication be distributive and that $xy = yx + 1$ and $ax = xa$, $ay = ya$ for $a \in F$. Then R is a well-defined simple ring that has no zero divisors and is *not* a division ring (Exercise 1).

Let $A = Ra$ be a cyclic R-module. The map $\theta : R \to A$ defined by $r \mapsto ra$ is an R-module epimorphism whose kernel I is a left ideal (submodule) of R (Theorem IV.1.5). By the First Isomorphism Theorem IV.1.7 R/I is isomorphic to A. By Theorem IV.1.10 every submodule of R/I is of the form J/I, where J is a left ideal of R that contains I. Consequently R/I (and hence A) has no proper submodules if and only if I is a maximal left ideal of R. Since every simple R-module is cyclic by Remark (iii) above, every simple R-module is isomorphic to R/I for some maximal left ideal I. Conversely, if I is a maximal left ideal of R, R/I will be simple provided $R(R/I) \neq 0$. A condition that guarantees that $R(R/I) \neq 0$ is given by

Definition 1.2. *A left ideal* I *in a ring* R *is* **regular** (*or* **modular**) *if there exists* $e \in R$ *such that* $r - re \in I$ *for every* $r \in R$. *Similarly, a right ideal* J *is* **regular** *if there exists* $e \in R$ *such that* $r - er \in J$ *for every* $r \in R$.

REMARK. Every left ideal in a ring R with identity is regular (let $e = 1_R$).

Theorem 1.3. *A left module* A *over a ring* R *is simple if and only if* A *is isomorphic to* R/I *for some regular maximal left ideal* I.

REMARKS. If R has an identity, the theorem is an immediate consequence of the discussion above. The theorem is true if "left" is replaced by "right" throughout.

PROOF OF 1.3. The discussion preceding Definition 1.2 shows that if A is simple, then $A = Ra \cong R/I$ where the maximal left ideal I is the kernel of θ. Since $A = Ra$, $a = ea$ for some $e \in R$. Consequently, for any $r \in R$, $ra = rea$ or $(r - re)a = 0$, whence $r - re \in \operatorname{Ker} \theta = I$. Therefore I is regular.

Conversely let I be a regular maximal left ideal of R such that $A \cong R/I$. In view of the discussion preceding Definition 1.2 it suffices to prove that $R(R/I) \neq 0$. If this is not the case, then for all $r \in R$ $r(e + I) = I$, whence $re \in I$. Since $r - re \in I$, we have $r \in I$. Thus $R = I$, contradicting the maximality of I. ■

Having developed the necessary facts about simplicity we now turn to primitivity. In order to define primitive rings we need:

Theorem 1.4. *Let* B *be a subset of a left module* A *over a ring* R. *Then* $\mathcal{A}(B) = \{r \in R \mid rb = 0$ *for all* $b \in B\}$ *is a left ideal of* R. *If* B *is a submodule of* A, *then* $\mathcal{A}(B)$ *is an ideal.*

$\mathcal{A}(B)$ is called the **(left) annihilator** of B. The right annihilator of a right module is defined analogously.

SKETCH OF PROOF OF 1.4. It is easy to verify that $\mathcal{A}(B)$ is a left ideal. Let B be a submodule. If $r \in R$ and $s \in \mathcal{A}(B)$, then for every $b \in B$ $(sr)b = s(rb) = 0$ since $rb \in B$. Consequently, $sr \in \mathcal{A}(B)$, whence $\mathcal{A}(B)$ is also a right ideal. ■

Definition 1.5. *A (left) module* A *is* **faithful** *if its (left) annihilator* $\mathcal{A}(A)$ *is* 0. *A ring* R *is* **(left) primitive** *if there exists a simple faithful left* R*-module.*

Right primitive rings are defined analogously. There do exist right primitive rings that are not left primitive (see G. Bergman [58]). Hereafter "primitive" will always mean "left primitive." However, all results proved for left primitive rings are true, *mutatis mutandis*, for right primitive rings.

EXAMPLE. Let V be a (possibly infinite dimensional) vector space over a division ring D and let R be the endomorphism ring $\mathrm{Hom}_D(V,V)$ of V. Recall that V is a left R-module with $\theta v = \theta(v)$ for $v \varepsilon V$, $\theta \varepsilon R$ (Exercise IV.1.7). If u is a nonzero vector in V, then there is a basis of V that contains u (Theorem IV.2.4). If $v \varepsilon V$, then there exists $\theta_v \varepsilon R$ such that $\theta_v u = v$ (just define $\theta_v(u) = v$ and $\theta_v(w) = 0$ for all other basis elements w; then $\theta_v \varepsilon R$ by Theorems IV.2.1 and IV.2.4). Therefore $Ru = V$ for any nonzero $u \varepsilon V$, whence V has no proper R-submodules. Since R has an identity, $RV \neq 0$. Thus V is a simple R-module. If $\theta V = 0$ ($\theta \varepsilon R$), then clearly $\theta = 0$, whence $\mathcal{A}(V) = 0$ and V is a faithful R-module. Therefore, R is primitive. If V is finite dimensional over D, then R is simple by Exercise III.2.9 and Theorem VII.1.4. But if V is infinite dimensional over D, then R is not simple: the set of all $\theta \varepsilon R$ such that $\mathrm{Im}\,\theta$ is finite dimensional subspace of V is a proper ideal of R (Exercise 3).

The next two results provide other examples of primitive rings.

Proposition 1.6. *A simple ring* R *with identity is primitive.*

PROOF. R contains a maximal left ideal I by Theorem III.2.18. Since R has an identity I is regular, whence R/I is a simple R-module by Theorem 1.3. Since $\mathcal{A}(R/I)$ is an ideal of R that does not contain 1_R, $\mathcal{A}(R/I) = 0$ by simplicity. Therefore R/I is faithful. ■

Proposition 1.7. *A commutative ring* R *is primitive if and only if* R *is a field.*

PROOF. A field is primitive by Proposition 1.6. Conversely, let A be a faithful simple left R-module. Then $A \cong R/I$ for some regular maximal left ideal I of R. Since R is commutative, I is in fact an ideal and $I \subset \mathcal{A}(R/I) = \mathcal{A}(A) = 0$. Since $I = 0$ is regular, there is an $e \varepsilon R$ such that $r = re$ $(= er)$ for all $r \varepsilon R$. Thus R is a commutative ring with identity. Since $I = 0$ is maximal, R is a field by Corollary III.2.21. ■

In order to characterize noncommutative primitive rings we need the concept of density.

Definition 1.8. *Let* V *be a (left) vector space over a division ring* D. *A subring* R *of the endomorphism ring* $\mathrm{Hom}_D(V,V)$ *is called a* **dense ring of endomorphisms** *of* V *(or a* **dense subring** *of* $\mathrm{Hom}_D(V,V)$*) if for every positive integer* n, *every linearly independent subset* $\{u_1, \ldots, u_n\}$ *of* V *and every arbitrary subset* $\{v_1, \ldots, v_n\}$ *of* V, *there exists* $\theta \varepsilon R$ *such that* $\theta(u_i) = v_i$ $(i = 1,2,\ldots,n)$.

EXAMPLE. $\text{Hom}_D(V,V)$ is a dense subring of itself. For if $\{u_1, \ldots, u_n\}$ is a linearly independent subset of V, then there is a basis U of V that contains $u_1, \ldots, u_n$ by Theorem IV.2.4. If $v_1, \ldots, v_m \in V$, then the map $\theta : V \to V$ defined by $\theta(u_i) = v_i$ and $\theta(u) = 0$ for $u \in U - \{u_1, \ldots, u_n\}$ is a well-defined element of $\text{Hom}_D(V,V)$ by Theorems IV.2.1 and IV.2.4. In the finite dimensional case, $\text{Hom}_D(V,V)$ is the only dense subring as we see in

Theorem 1.9. *Let* R *be a dense ring of endomorphisms of a vector space* V *over a division ring* D. *Then* R *is left [resp. right] Artinian if and only if* $\dim_D V$ *is finite, in which case* $R = \text{Hom}_D(V,V)$.

PROOF. If R is left Artinian and $\dim_D V$ is infinite, then there exists an infinite linearly independent subset $\{u_1,u_2, \ldots\}$ of V. By Exercise IV.1.7 V is a left $\text{Hom}_D(V,V)$-module and hence a left R-module. For each n let I_n be the left annihilator in R of the set $\{u_1, \ldots, u_n\}$. By Theorem 1.4, $I_1 \supset I_2 \supset \cdots$ is a descending chain of left ideals of R. Let w be any nonzero element of V. Since $\{u_1, \ldots, u_{n+1}\}$ is linearly independent for each n and R is dense, there exists $\theta \in R$ such that

$$\theta u_i = 0 \quad \text{for} \quad i = 1,2, \ldots, n \quad \text{and} \quad \theta u_{n+1} = w \neq 0.$$

Consequently $\theta \in I_n$ but $\theta \notin I_{n+1}$. Therefore $I_1 \underset{\neq}{\supset} I_2 \underset{\neq}{\supset} \cdots$ is a properly descending chain, which is a contradiction. Hence $\dim_D V$ is finite.

Conversely if $\dim_D V$ is finite, then V has a finite basis $\{v_1, \ldots, v_m\}$. If f is any element of $\text{Hom}_D(V,V)$, then f is completely determined by its action on $v_1, \ldots, v_m$ by Theorems IV.2.1 and IV.2.4. Since R is dense, there exists $\theta \in R$ such that

$$\theta(v_i) = f(v_i) \quad \text{for} \quad i = 1,2, \ldots, m,$$

whence $f = \theta \in R$. Therefore $\text{Hom}_D(V,V) = R$. But $\text{Hom}_D(V,V)$ is Artinian by Theorem VII.1.4 and Corollary VIII.1.12. ■

In order to prove that an arbitrary primitive ring is isomorphic to a dense ring of endomorphisms of a suitable vector space we need two lemmas.

Lemma 1.10. (*Schur*) *Let* A *be a simple module over a ring* R *and let* B *be any* R*-module.*

(i) *Every nonzero* R*-module homomorphism* $f : A \to B$ *is a monomorphism;*
(ii) *every nonzero* R*-module homomorphism* $g : B \to A$ *is an epimorphism;*
(iii) *the endomorphism ring* $D = \text{Hom}_R(A,A)$ *is a division ring.*

PROOF. (i) Ker f is a submodule of A and Ker $f \neq A$ since $f \neq 0$. Therefore Ker $f = 0$ by simplicity. (ii) Im g is a nonzero submodule of A since $g \neq 0$, whence Im $g = A$ by simplicity. (iii) If $h \in D$ and $h \neq 0$, then h is an isomorphism by (i) and (ii). Thus f has a two-sided inverse $f^{-1} \in \text{Hom}_R(A,A) = D$ (see the paragraph after Definition IV.1.2). Consequently every nonzero element of D is a unit, whence D is a division ring. ■

REMARK. If A is a simple R-module, then A is a vector space over the division ring $\mathrm{Hom}_R(A,A)$ with $fa = f(a)$ (Exercise IV.1.7 and Lemma 1.10).

Lemma 1.11. *Let* A *be a simple module over a ring* R. *Consider* A *as a vector space over the division ring* D = *Hom*$_R$(A,A). *If* V *is a finite dimensional* D*-subspace of the* D*-vector space* A *and* a ε A − V, *then there exists* r ε R *such that* ra ≠ 0 *and* rV = 0.

PROOF. The proof is by induction on $n = \dim_D V$. If $n = 0$, then $V = 0$ and $a \neq 0$. Since A is simple, $A = Ra$ by Remark (iii) after Definition 1.1. Consequently, there exists $r \in R$ such that $ra = a \neq 0$ and $rV = r0 = 0$. Suppose $\dim_D V = n > 0$ and the theorem is true for dimensions less than n. Let $\{u_1, \ldots, u_{n-1}, u\}$ be a D-basis of V and let W be the $(n-1)$-dimensional D-subspace spanned by $\{u_1, \ldots, u_{n-1}\}$ ($W = 0$ if $n = 1$). Then $V = W \oplus Du$ (vector space direct sum). Now W may not be an R-submodule of A, but in any case the left annihilator $I = \mathfrak{a}(W)$ in R of W is a left ideal of R by Theorem 1.4. Consequently, Iu is an R-submodule of A (Exercise IV.1.3). Since $u \in A - W$, the induction hypothesis implies that there exists $r \in R$ such that $ru \neq 0$ and $rW = 0$ (that is, $r \in I = \mathfrak{a}(W)$). Consequently $0 \neq ru \in Iu$, whence $Iu \neq 0$. Therefore $A = Iu$ by simplicity.

[*Note:* The contrapositive of the inductive argument used above shows that if $v \in A$ and $rv = 0$ for all $r \in I$, then $v \in W$.]

We must find $r \in R$ such that $ra \neq 0$ and $rV = 0$. If no such r exists, then we can define a map $\theta : A \to A$ as follows. For $ru \in Iu = A$ let $\theta(ru) = ra \in A$. We claim that θ is well defined. If $r_1u = r_2u$ ($r_i \in I = \mathfrak{a}(W)$), then $(r_1 - r_2)u = 0$, whence $(r_1 - r_2)V = (r_1 - r_2)(W \oplus Du) = 0$. Consequently by hypothesis $(r_1 - r_2)a = 0$. Therefore, $\theta(r_1u) = r_1a = r_2a = \theta(r_2u)$. Verify that $\theta \in \mathrm{Hom}_R(A,A) = D$. Then for every $r \in I$,

$$0 = \theta(ru) - ra = r\theta(u) - ra = r(\theta(u) - a).$$

Therefore $\theta(u) - a \in W$ by the parenthetical Note above. Consequently

$$a = \theta u - (\theta u - a) \in Du + W = V,$$

which contradicts the fact that $a \notin V$. Therefore, there exists $r \in R$ such that $ra \neq 0$ and $rV = 0$. ■

Theorem 1.12. (*Jacobson Density Theorem*) *Let* R *be a primitive ring and* A *a faithful simple* R*-module. Consider* A *as a vector space over the division ring Hom*$_R$(A,A) = D. *Then* R *is isomorphic to a dense ring of endomorphisms of the* D*-vector space* A.

REMARK. A converse of Theorem 1.12 is also true, in fact in a much stronger form (Exercise 4).

PROOF OF 1.12. For each $r \in R$ the map $\alpha_r : A \to A$ given by $\alpha_r(a) = ra$ is easily seen to be a D-endomorphism of A: that is, $\alpha_r \in \mathrm{Hom}_D(A,A)$. Furthermore for all $r,s \in R$

$$\alpha_{(r+s)} = \alpha_r + \alpha_s \quad \text{and} \quad \alpha_{rs} = \alpha_r\alpha_s.$$

Consequently the map $\alpha : R \to \mathrm{Hom}_D(A,A)$ defined by $\alpha(r) = \alpha_r$ is a well-defined homomorphism of rings. Since A is a faithful R-module, $\alpha_r = 0$ if and only if

$r \in \mathcal{A}(A) = 0$. Therefore α is a monomorphism, whence R is isomorphic to the subring $\operatorname{Im}\alpha$ of $\operatorname{Hom}_D(A,A)$.

To complete the proof we must show that $\operatorname{Im}\alpha$ is a dense subring of $\operatorname{Hom}_D(A,A)$. Given a D-linearly independent subset $U = \{u_1, \ldots, u_n\}$ of A and an arbitrary subset $\{v_1, \ldots, v_n\}$ of A we must find $\alpha_r \in \operatorname{Im}\alpha$ such that $\alpha_r(u_i) = v_i$ for $i = 1,2, \ldots, n$. For each i let V_i be the D-subspace of A spanned by $\{u_1, \ldots, u_{i-1}, u_{i+1}, \ldots, u_n\}$. Since U is D-linearly independent, $u_i \notin V_i$. Consequently, by Lemma 1.11 there exists $r_i \in R$ such that $r_iu_i \neq 0$ and $r_iV_i = 0$. We next apply Lemma 1.11 to the zero subspace and the nonzero element r_iu_i: there exists $s_i \in R$ such that $s_ir_iu_i \neq 0$ and $s_i0 = 0$. Since $s_ir_iu_i \neq 0$, the R-submodule Rr_iu_i of A is nonzero, whence $Rr_iu_i = A$ by simplicity. Therefore exists $t_i \in R$ such that $t_ir_iu_i = v_i$. Let

$$r = t_1r_1 + t_2r_2 + \cdots + t_nr_n \in R.$$

Recall that for $i \neq j$, $u_i \in V_j$, whence $t_jr_ju_i \in t_j(r_jV_j) = t_j0 = 0$. Consequently for each $i = 1,2, \ldots, n$

$$\alpha_r(u_i) = (t_1r_1 + \cdots + t_nr_n)u_i = t_ir_iu_i = v_i.$$

Therefore $\operatorname{Im}\alpha$ is a dense ring of endomorphisms of the D-vector space A. ■

REMARK. The only point in the proof of Theorem 1.12 at which the faithfulness of A is used is to show that α is a monomorphism. Consequently the proof shows that any ring that has a simple module A also has a homomorphic image that is a dense ring of endomorphisms of the D-vector space A.

Corollary 1.13. *If* R *is a primitive ring, then for some division ring* D *either* R *is isomorphic to the endomorphism ring of a finite dimensional vector space over* D *or for every positive integer* m *there is a subring* R_m *of* R *and an epimorphism of rings* $R_m \to Hom_D(V_m,V_m)$, *where* V_m *is an* m*-dimensional vector space over* D.

REMARK. The Corollary may also be phrased in terms of matrix rings over a division ring via Theorem VII.1.4.

SKETCH OF PROOF OF 1.13. In the notation of Theorem 1.12,

$$\alpha : R \to \operatorname{Hom}_D(A,A)$$

is a monomorphism such that $R \cong \operatorname{Im}\alpha$ and $\operatorname{Im}\alpha$ is dense in $\operatorname{Hom}_D(A,A)$. If $\dim_D A = n$ is finite, then $\operatorname{Im}\alpha = \operatorname{Hom}_D(A,A)$ by Theorem 1.9. If $\dim_D A$ is infinite and $\{u_1,u_2, \ldots\}$ is an infinite linearly independent set, let V_m be the m-dimensional D-subspace of A spanned by $\{u_1, \ldots, u_m\}$. Verify that $R_m = \{r \in R \mid rV_m \subset V_m\}$ is a subring of R. Use the density of $R \cong \operatorname{Im}\alpha$ in $\operatorname{Hom}_D(A,A)$ to show that the map $R_m \to \operatorname{Hom}_D(V_m,V_m)$ given by $r \mapsto \alpha_r \mid V_m$ is a well-defined ring epimorphism. ■

Theorem 1.14. (*Wedderburn-Artin*) *The following conditions on a left Artinian ring* R *are equivalent.*

(i) R *is simple;*
(ii) R *is primitive;*

(iii) R *is isomorphic to the endomorphism ring of a nonzero finite dimensional vector space* V *over a division ring* D;

(iv) *for some positive integer* n, R *is isomorphic to the ring of all* n × n *matrices over a division ring.*

PROOF. (i) ⇒ (ii) We first observe that $I = \{r \in R \mid Rr = 0\}$ is an ideal of R, whence $I = R$ or $I = 0$. Since $R^2 \neq 0$, we must have $I = 0$. Since R is left Artinian the set of all nonzero left ideals of R contains a minimal left ideal J. J has no proper R-submodules, (an R-submodule of J is a left ideal of R). We claim that the left annihilator $\mathcal{Q}(J)$ of J in R is zero. Otherwise $\mathcal{Q}(J) = R$ by simplicity and $Ru = 0$ for every nonzero $u \in J$. Consequently, each such nonzero u is contained in $I = 0$, which is a contradiction. Therefore $\mathcal{Q}(J) = 0$ and $RJ \neq 0$. Thus J is a faithful simple R-module, whence R is primitive.

(ii) ⇒ (iii) By Theorem 1.12 R is isomorphic to a dense ring T of endomorphisms of a vector space V over a division ring D. Since R is left Artinian, $R \cong T = \text{Hom}_D(V,V)$ by Theorem 1.9.

(iii) ⇔ (iv) Theorem VII.1.4.

(iv) ⇒ (i) Exercise III.2.9. ■

We close this section by proving that for a simple left Artinian ring R the integers $\dim_D V$ and n in Theorem 1.14 are uniquely determined and the division rings in Theorem 1.14 (iii) and (iv) are determined up to isomorphism. We need two lemmas.

Lemma 1.15. *Let* V *be a finite dimensional vector space over a division ring* D. *If* A *and* B *are simple faithful modules over the endomorphism ring* $R = Hom_D(V,V)$, *then* A *and* B *are isomorphic* R*-modules.*

PROOF. By Theorems VII.1.4, VIII.1.4 and Corollary VIII.1.12, the ring R contains a (nonzero) minimal left ideal I. Since A is faithful, there exists $a \in A$ such that $Ia \neq 0$. Thus Ia is a nonzero submodule of A (Exercise IV.1.3), whence $Ia = A$ by simplicity. The map $\theta : I \to Ia = A$ given by $i \mapsto ia$ is a nonzero R-module epimorphism. By Lemma 1.10 θ is an isomorphism. Similarly $I \cong B$. ■

Lemma 1.16. *Let* V *be a nonzero vector space over a division ring* D *and let* R *be the endomorphism ring* $Hom_D(V,V)$. *If* $g : V \to V$ *is a homomorphism of additive groups such that* $gr = rg$ *for all* $r \in R$, *then there exists* $d \in D$ *such that* $g(v) = dv$ *for all* $v \in V$.

PROOF. Let u be a nonzero element of V. We claim that u and $g(u)$ are linearly dependent over D. If $\dim_D V = 1$, this is trivial. Suppose $\dim_D V \geq 2$ and $\{u, g(u)\}$ is linearly independent. Since R is dense in itself (Example after Definition 1.8), there exists $r \in R$ such that $r(u) = 0$ and $r(g(u)) \neq 0$. But by hypothesis

$$r(g(u)) = rg(u) = gr(u) = g(r(u)) = g(0) = 0,$$

which is a contradiction. Therefore for some $d \in D$, $g(u) = du$. If $v \in V$, then there exists $s \in R$ such that $s(u) = v$ by density. Consequently, since $s \in R = \text{Hom}_D(V,V)$, $g(v) = g(s(u)) = gs(u) = sg(u) = s(du) = ds(u) = dv$. ■

Proposition 1.17. *For* i = 1,2 *let* V_i *be a vector space of finite dimension* n_i *over the division ring* D_i.

(i) *If there is an isomorphism of rings* $Hom_{D_1}(V_1,V_1) \cong Hom_{D_2}(V_2,V_2)$, *then* $dim_{D_1}V_1 = dim_{D_2}V_2$ *and* D_1 *is isomorphic to* D_2.

(ii) *If there is an isomorphism of rings* $Mat_{n_1}D_1 \cong Mat_{n_2}D_2$, *then* $n_1 = n_2$ *and* D_1 *is isomorphic to* D_2.

SKETCH OF PROOF. (i) For $i = 1,2$ the example after Definition 1.5 shows that V_i is a faithful simple $\mathrm{Hom}_{D_i}(V_i,V_i)$-module. Let $R = \mathrm{Hom}_{D_1}(V_1,V_1)$ and let

$$\sigma : R \to \mathrm{Hom}_{D_2}(V_2,V_2)$$

be an isomorphism. Then V_2 is a faithful simple R-module by pullback along σ (that is, $rv = \sigma(r)v$ for $r \in R$, $v \in V_2$). By Lemma 1.15 there is an R-module isomorphism $\phi : V_1 \to V_2$. For each $v \in V_1$ and $f \in R$,

$$\phi[f(v)] = f\phi(v) = (\sigma f)[\phi(v)],$$

whence

$$\phi f \phi^{-1} = \sigma(f)$$

as a homomorphism of additive groups $V_2 \to V_2$. For each $d \in D_i$ let $\alpha_d : V_i \to V_i$ be the homomorphism of additive groups defined by $x \mapsto dx$. Clearly $\alpha_d = 0$ if and only if $d = 0$. For every $f \in R = \mathrm{Hom}_{D_1}(V_1,V_1)$ and every $d \in D_1$, $f\alpha_d = \alpha_d f$. Consequently,

$$\begin{aligned}[\phi\alpha_d\phi^{-1}](\sigma f) &= \phi\alpha_d\phi^{-1}\phi f\phi^{-1} = \phi\alpha_d f\phi^{-1} = \phi f\alpha_d\phi^{-1} \\ &= \phi f\phi^{-1}\phi\alpha_d\phi^{-1} = (\sigma f)[\phi\alpha_d\phi^{-1}].\end{aligned}$$

Since σ is surjective, Lemma 1.16 (with $V = V_2$, $g = \phi\alpha_d\phi^{-1}$) implies that there exists $d^* \in D_2$ such that $\phi\alpha_d\phi^{-1} = \alpha_{d^*}$. Let $\tau : D_1 \to D_2$ be the map given by $\tau(d) = d^*$. Then for every $d \in D_1$,

$$\phi\alpha_d\phi^{-1} = \alpha_{\tau(d)}.$$

Verify that τ is a monomorphism of rings. Reversing the roles of D_1 and D_2 in the preceding argument (and replacing ϕ,σ by ϕ^{-1},σ^{-1}) yields for every $k \in D_2$ an element $d \in D$ such that

$$\phi^{-1}\alpha_k\phi = \alpha_d : V_1 \to V_1,$$

whence $\alpha_k = \phi\alpha_d\phi^{-1} = \alpha_{\tau(d)}$. Consequently $k = \tau(d)$ and hence τ is surjective. Therefore τ is an isomorphism. Furthermore for every $d \in D_1$ and $v \in V_1$,

$$\phi(dv) = \phi\alpha_d(v) = \alpha_{\tau(d)}\phi(v) = \tau(d)\phi(v).$$

Use this fact to show that $\{u_1, \ldots, u_k\}$ is D_1-linearly independent in V_1 if and only if $\{\phi(u_1), \ldots, \phi(u_k)\}$ is D_2-linearly independent in V_2. It follows that $\dim_{D_1}V_1 = \dim_{D_2}V_2$.

(ii) Use (i), Exercise III.1.17(e) and Theorem VII.1.4. ■

EXERCISES

1. Let F be a field of characteristic 0 and $R = F[x,y]$ the additive group of polynomials in two indeterminates. Define multiplication in R by requiring that multiplication be distributive, that $ax = xa$, $ay = ya$ for all $a \in F$, that the product of x and y (in that order) be the polynomial xy as usual, but that the product of y and x be the polynomial $xy + 1$.

(a) R is a ring.

(b) $yx^k = x^ky + kx^{k-1}$ and $y^kx = xy^k + ky^{k-1}$.

(c) R is simple. (*Hint:* Let f be a nonzero element in an ideal I of R; then either f has no terms involving y or $g = xf - fx$ is a nonzero element of I that has lower degree in y than does f. In the latter case, consider $xg - gx$. Eventually, find a nonzero $h \in I$, which is free of y. If h is nonconstant, consider $hy - yh$. In a finite number of steps, obtain a nonzero constant element of I; hence $I = R$.)

(d) R has no zero divisors.

(e) R is not a division ring.

2. (a) If A is an R-module, then A is also a well-defined $R/\mathfrak{A}(A)$-module with $(r + \mathfrak{A}(A))a = ra$ $(a \in A)$.
(b) If A is a simple left R-module, then $R/\mathfrak{A}(A)$ is a primitive ring.

3. Let V be an infinite dimensional vector space over a division ring D.
(a) If F is the set of all $\theta \in \text{Hom}_D(V,V)$ such that Im θ is finite dimensional, then F is a proper ideal of $\text{Hom}_D(V,V)$. Therefore $\text{Hom}_D(V,V)$ is not simple.
(b) F is itself a simple ring.
(c) F is contained in every nonzero ideal of $\text{Hom}_D(V,V)$.
(d) $\text{Hom}_D(V,V)$ is not (left) Artinian.

4. Let V be a vector space over a division ring D. A subring R of $\text{Hom}_D(V,V)$ is said to be **n-fold transitive** if for every k $(1 \leq k \leq n)$ and every linearly independent subset $\{u_1, \dots, u_k\}$ of V and every arbitrary subset $\{v_1, \dots, v_k\}$ of V, there exists $\theta \in R$ such that $\theta(u_i) = v_i$ for $i = 1,2, \dots, k$.
(a) If R is one-fold transitive, then R is primitive. [*Hint:* examine the example after Definition 1.5.]
(b) If R is two-fold transitive, then R is dense in $\text{Hom}_D(V,V)$. [*Hints:* Use (a) to show that R is a dense subring of $\text{Hom}_\Delta(V,V)$, where $\Delta = \text{Hom}_R(V,V)$. Use two-fold transitivity to show that $\Delta = \{\beta_d \mid d \in D\}$, where $\beta_d : V \to V$ is given by $x \mapsto dx$. Consequently $\text{Hom}_\Delta(V,V) = \text{Hom}_D(V,V)$.]

5. If R is a primitive ring such that for all $a,b \in R$, $a(ab - ba) = (ab - ba)a$, then R is a division ring. [*Hint:* show that R is isomorphic to a dense ring of endomorphisms of a vector space V over a division ring D with $\dim_D V = 1$, whence $R \cong D$.]

6. If R is a primitive ring with identity and $e \in R$ is such that $e^2 = e \neq 0$, then
(a) eRe is a subring of R, with identity e.
(b) eRe is primitive. [*Hint:* if R is isomorphic to a dense ring of endomorphisms of the vector space V over a division ring D, then Ve is a D-vector space and eRe is isomorphic to a dense ring of endomorphisms of Ve.]

7. If R is a dense ring of endomorphisms of a vector space V and K is a nonzero ideal of R, then K is also a dense ring of endomorphisms of V.

2. THE JACOBSON RADICAL

The Jacobson radical is defined (Theorem 2.3) and its basic properties are developed (Theorems 2.12–2.16). The interrelationships of simple, primitive, and semisimple rings are examined (Theorem 2.10) and numerous examples are given.

Before pursuing further our study of the structure of rings, we summarize the general technique that we shall use. There is little hope at present of classifying all rings up to isomorphism. Consequently we shall attempt to discover classes of rings for which some reasonable structure theorems are obtainable. Here is a classic method of determining such a class. Single out some "bad" or "undesirable" property of rings and study only those rings that do not have this property. In order to make this method workable in practice one must make some additional assumptions.

Let P be a property of rings and call an ideal [ring] I a P-ideal [P-ring] if I has property P. Assume that

(i) the homomorphic image of a P-ring is a P-ring;
(ii) every ring R (or at least every ring in some specified class $\mathfrak{C}$) contains a P-ideal $P(R)$ (called the **P-radical** of R) that contains all other P-ideals of R;
(iii) the P-radical of the quotient ring $R/P(R)$ is zero;
(iv) the P-radical of the ring $P(R)$ is $P(R)$.

A property P that satisfies (i)–(iv) is called a **radical property.**

The P-radical may be thought of as measuring the degree to which a given ring possesses the "undesirable" property P. If we have chosen a radical property P, we then attempt to find structure theorems for those "nice" rings whose P-radical is zero. Such a ring is said to be P-radical free or P-semisimple. In actual practice we are usually more concerned with the P-radical itself rather than the radical property P from which it arises. By condition (iii) every ring that has a P-radical has a P-semisimple quotient ring. Thus the larger P-radical is, the more one discards (or factors out) when studying P-semisimple rings. The basic problem is to find radicals that enable us to discard as little as possible and yet to obtain reasonably deep structure theorems.

Wedderburn first introduced a radical in the study of finite dimensional algebras. His results were later extended to (left) Artinian rings. However, the radical of Wedderburn (namely the maximal nilpotent ideal) and the remarkably strong structure theorems that resulted applied only to (left) Artinian rings. In subsequent years many other radicals were introduced. Generally speaking each of these coincided with the radical of Wedderburn in the left Artinian case, but were also defined for non-Artinian rings.

The chief purpose of this section is to study one such radical, the Jacobson radical. Another radical, the prime radical, is discussed in Section 4; see also Exercise 4.11. For an extensive treatment of radicals see N. J. Divinsky [22] or M. Gray [23]. The host of striking theorems that have resulted from its use provide ample justification for studying the Jacobson radical in some detail. Indeed Section 1 was developed with the Jacobson radical in mind. Rings that are Jacobson semisimple (that is, have zero Jacobson radical) can be described in terms of simple and primitive rings (Section 3).

Two preliminaries are needed before we define the Jacobson radical.

Definition 2.1. *An ideal* P *of a ring* R *is said to be* **left** *[resp.* **right***]* **primitive** *if the quotient ring* R/P *is a left [resp. right] primitive ring.*

REMARK. Since the zero ring has no simple modules and hence is not primitive, R itself is not a left (or right) primitive ideal.

Definition 2.2. *An element* a *in a ring* R *is said to be* **left quasi-regular** *if there exists* $r \in R$ *such that* $r + a + ra = 0$. *The element* r *is called a* **left quasi-inverse** *of* a. *A (right, left or two-sided) ideal* I *of* R *is said to be* **left quasi-regular** *if every element of* I *is left quasi-regular. Similarly,* $a \in R$ *is said to be* **right quasi-regular** *if there exists* $r \in R$ *such that* $a + r + ar = 0$. *Right quasi-inverses and right quasi-regular ideals are defined analogously.*

REMARKS. It is sometimes convenient to write $r \circ a$ for $r + a + ra$. If R has an identity, then a is left [resp. right] quasi-regular if and only if $1_R + a$ is left [resp. right] invertible (Exercise 1).

In order to simplify the statement of several results, we shall adopt the following convention (which is actually a theorem of axiomatic set theory).

If the class $\mathcal{C}$ *of those subsets of a ring* R *that satisfy a given property is empty, then* $\bigcap_{I \in \mathcal{C}} I$ *is defined to be* R.

Theorem 2.3. *If* R *is a ring, then there is an ideal* J(R) *of* R *such that:*

(i) J(R) *is the intersection of all the left annihilators of simple left* R*-modules;*
(ii) J(R) *is the intersection of all the regular maximal left ideals of* R;
(iii) J(R) *is the intersection of all the left primitive ideals of* R;
(iv) J(R) *is a left quasi-regular left ideal which contains every left quasi-regular left ideal of* R;
(v) *Statements* (i)–(iv) *are also true if "left" is replaced by "right".*

Theorem 2.3 is proved below (p. 428). The ideal $J(R)$ is called the **Jacobson radical** of the ring R. Historically it was first defined in terms of quasi-regularity (Theorem 2.3 (iv)), which turns out to be a radical property as defined in the introductory remarks above (see p. 431). As the importance of the role of modules in the study of rings became clearer the other descriptions of $J(R)$ were developed (Theorem 2.3 (i)–(iii)).

REMARKS. According to Theorem 2.3 (i) and the convention adopted above, $J(R) = R$ if R has no simple left R-modules (and hence no annihilators of same). If R has an identity, then every ideal is regular and maximal left ideals always exist (Theorem III.2.18), whence $J(R) \neq R$ by Theorem 2.3(ii). Theorem 2.3(iv) does *not* imply that $J(R)$ contains every left quasi-regular *element* of R; see Exercise 4.

The proof of Theorem 2.3 (which begins on p. 428) requires five preliminary lemmas. The lemmas are stated and proved for left ideals. However, *each of Lemmas* 2.4–2.8 *is valid with "left" replaced by "right" throughout.* Examples are given after the proof of Theorem 2.3.

Lemma 2.4. *If* I ($\neq$R) *is a regular left ideal of a ring* R, *then* I *is contained in a maximal left ideal which is regular.*

SKETCH OF PROOF. Since I is regular, there exists $e \in R$ such that $r - re \in I$ for all $r \in R$. Thus any left ideal J containing I is also regular (with the same element

$e \in R$). If $I \subset J$ and $e \in J$, then $r - re \in I \subset J$ implies $r \in J$ for every $r \in R$, whence $R = J$. Use this fact to verify that Zorn's Lemma is applicable to the set $\mathcal{S}$ of all left ideals L such that $I \subset L \subsetneq R$, partially ordered by inclusion. A maximal element of $\mathcal{S}$ is a regular maximal left ideal containing I. ■

Lemma 2.5. *Let* R *be a ring and let* K *be the intersection of all regular maximal left ideals of* R. *Then* K *is a left quasi-regular left ideal of* R.

PROOF. K is obviously a left ideal. If $a \in K$ let $T = \{r + ra \mid r \in R\}$. If $T = R$, then there exists $r \in R$ such that $r + ra = -a$. Consequently $r + a + ra = 0$ and hence a is left quasi-regular. Thus it suffices to show that $T = R$.

Verify that T is a regular left ideal of R (with $e = -a$). If $T \neq R$, then T is contained in a regular maximal left ideal I_0 by Lemma 2.4. (Thus $T \neq R$ is impossible if R has no regular maximal left ideals.) Since $a \in K \subset I_0$, $ra \in I_0$ for all $r \in R$. Thus since $r + ra \in T \subset I_0$, we must have $r \in I_0$ for all $r \in R$. Consequently, $R = I_0$, which contradicts the maximality of I_0. Therefore $T = R$. ■

Lemma 2.6. *Let* R *be a ring that has a simple left* R*-module. If* I *is a left quasi-regular left ideal of* R, *then* I *is contained in the intersection of all the left annihilators of simple left* R*-modules.*

PROOF. If $I \not\subset \bigcap \mathfrak{a}(A)$, where the intersection is taken over all simple left R-modules A, then $IB \neq 0$ for some simple left R-module B, whence $Ib \neq 0$ for some nonzero $b \in B$. Since I is a left ideal, Ib is a nonzero submodule of B. Consequently $B = Ib$ by simplicity and hence $ab = -b$ for some $a \in I$. Since I is left quasi-regular, there exists $r \in R$ such that $r + a + ra = 0$. Therefore, $0 = 0b = (r + a + ra)b = rb + ab + rab = rb - b - rb = -b$. Since this conclusion contradicts the fact that $b \neq 0$, we must have $I \subset \bigcap \mathfrak{a}(A)$. ■

Lemma 2.7. *An ideal* P *of a ring* R *is left primitive if and only if* P *is the left annihilator of a simple left* R*-module.*

PROOF. If P is a left primitive ideal, let A be a simple faithful R/P-module. Verify that A is an R-module, with ra ($r \in R, a \in A$) defined to be $(r + P)a$. Then $RA = (R/P)A \neq 0$ and every R-submodule of A is an R/P-submodule of A, whence A is a simple R-module. If $r \in R$, then $rA = 0$ if and only if $(r + P)A = 0$. But $(r + P)A = 0$ if and only if $r \in P$ since A is a faithful R/P-module. Therefore P is the left annihilator of the simple R-module A.

Conversely suppose that P is the left annihilator of a simple R-module B. Verify that B is a simple R/P-module with $(r + P)b = rb$ for $r \in R, b \in B$. Furthermore if $(r + P)B = 0$, then $rB = 0$, whence $r \in \mathfrak{a}(B) = P$ and $r + P = 0$ in R/P. Consequently, B is a faithful R/P-module. Therefore R/P is a left primitive ring, whence P is a left primitive ideal of R. ■

Lemma 2.8. *Let* I *be a left ideal of a ring* R. *If* I *is left quasi-regular, then* I *is right quasi-regular.*

PROOF. If I is left quasi-regular and $a \varepsilon I$, then there exists $r \varepsilon R$ such that $r \circ a = r + a + ra = 0$. Since $r = -a - ra \varepsilon I$, there exists $s \varepsilon R$ such that $s \circ r = s + r + sr = 0$, whence s is right quasi-regular. The operation $\circ$ is easily seen to be associative. Consequently

$$a = 0 \circ a = (s \circ r) \circ a = s \circ (r \circ a) = s \circ 0 = s.$$

Therefore a, and hence I, is right quasi-regular. ■

PROOF OF THEOREM 2.3. Let $J(R)$ be the intersection of all the *left* annilators of simple *left* R-modules. If R has no simple left R-modules, then $J(R) = R$ by the convention adopted above. $J(R)$ is an ideal by Theorem 1.4. We now show that statements (ii)–(iv) are true for all *left* ideals.

We first observe that R itself cannot be the annihilator of a simple left R-module A (otherwise $RA = 0$). This fact together with Theorem 1.3 and Lemma 2.7 implies that the following conditions are equivalent:

(a) $J(R) = R$;
(b) R has no simple left R-modules;
(c) R has no regular maximal left ideals;
(d) R has no left primitive ideals.

Therefore by the convention adopted above, (ii), (iii), and (iv) are true if $J(R) = R$.

(ii) Assume $J(R) \neq R$ and let K be the intersection of all the regular maximal left ideals of R. Then $K \subset J(R)$ by Lemmas 2.5 and 2.6. Conversely suppose $c \varepsilon J(R)$. By Theorem 1.3, $J(R)$ is the intersection of the left annihilators of the quotients R/I, where I runs over all regular maximal left ideals of R. For each regular maximal ideal I there exists $e \varepsilon R$ such that $c - ce \varepsilon I$. Since $c \varepsilon \mathfrak{A}(R/I)$, $cr \varepsilon I$ for all $r \varepsilon R$; in particular, $ce \varepsilon I$. Consequently, $c \varepsilon I$ for every regular maximal ideal I. Thus $J(R) \subset \bigcap I = K$. Therefore $J(R) = K$.

(iii) is an immediate consequence of Lemma 2.7.

(iv) $J(R)$ is a left quasi-regular left ideal by (ii) and Lemma 2.5. $J(R)$ contains every left quasi-regular left ideal by Lemma 2.6.

To complete the proof we must show that (i)–(iv) are true with "right" in place of "left." Let $J_1(R)$ be the intersection of the *right* annihilators of all simple *right* R-modules. Then the preceding proof is valid with "right" in place of "left," whence (i)–(iv) hold for the ideal $J_1(R)$. Since $J(R)$ is right quasi-regular by (iv) and Lemma 2.8, $J(R) \subset J_1(R)$ by (iv). Similarly $J_1(R)$ is left quasi-regular, whence $J_1(R) \subset J(R)$. Therefore, $J(R) = J_1(R)$. ■

EXAMPLE. Let R be a local ring with unique maximal ideal M (consisting of all nonunits of R; see Theorem III.4.13). We shall show that $J(R) = M$. Since R has an identity, $J(R) \neq R$. Since a proper ideal contains only nonunits by Theorem III.3.2, $J(R) \subset M$. On the other hand if $r \varepsilon M$, then $1_R + r \notin M$ (otherwise $1_R \varepsilon M$). Consequently, $1_R + r$ is a unit, whence r is left quasi-regular (Exercise 1). Thus $M \subset J(R)$ by Theorem 2.3 (iv). Therefore $J(R) = M$. Here are two special cases:

EXAMPLE. The power series ring $F[[x]]$ over a field F is a local ring with principal maximal ideal (x) by Corollary III.5.10. Therefore $J(F[[x]]) = (x)$.

EXAMPLE. If p is prime, then Z_{p^n} $(n \geq 2)$ is a local ring with principal maximal ideal (p), which is isomorphic *as an abelian group* to $Z_{p^{n-1}}$. Therefore $J(Z_{p^n}) = (p)$. The radical of Z_m (m arbitrary) is considered in Exercise 10.

Definition 2.9. *A ring* R *is said to be (Jacobson)* **semisimple** *if its Jacobson radical* J(R) *is zero.* R *is said to be a* **radical ring** *if* J(R) = R.

REMARK. Throughout this book "radical" always means "Jacobson radical" and "semisimple" always means "Jacobson semisimple." When reading the literature in ring theory, one must determine which notion of radical and semisimplicity is being used in a particular theorem. A number of definitions of radical (and semisimplicity) require that the ring be (left) Artinian. This is not the case with the Jacobson radical, which is defined for every ring.

EXAMPLE. Every division ring is semisimple by Theorem 2.3 (ii) since the only regular maximal left ideal is the zero ideal.

EXAMPLE. Every maximal ideal in **Z** is of the form (p) with p prime by Theorem III.3.4. Consequently, $J(\mathbf{Z}) = \bigcap_p (p) = 0$, whence **Z** is Jacobson semisimple. For a generalization, see Exercise 9.

EXAMPLE. If D is a division ring, then the polynomial ring

$$R = D[x_1,x_2, \dots , x_m]$$

is semisimple. For if $f \in J(R)$, then f is both right and left quasi-regular by Theorem 2.3 (iv). Consequently $1_R + f = 1_D + f$ is a unit in R by Exercise 1. Since the only units in R are the nonzero elements of D (see Theorem III.6.1), it follows that $f \in D$. Thus $J(R)$ is an ideal of D, whence $J(R) = 0$ or $J(R) = D$ by the simplicity of D. Since -1_D is not left quasi-regular (verify!), $-1_D \notin J(R)$. Therefore $J(R) = 0$ and R is semisimple.

Theorem 2.10. *Let* R *be a ring.*

(i) *If* R *is primitive, then* R *is semisimple.*
(ii) *If* R *is simple and semisimple, then* R *is primitive.*
(iii) *If* R *is simple, then* R *is either a primitive semisimple or a radical ring.*

PROOF. (i) R has a faithful simple left R-module A, whence $J(R) \subset \mathcal{a}(A) = 0$.

(ii) $R \neq 0$ by simplicity. There must exist a simple left R-module A; (otherwise by Theorem 2.3 (i) $J(R) = R \neq 0$, contradicting semisimplicity). The left annihilator $\mathcal{a}(A)$ is an ideal of R by Theorem 1.4 and $\mathcal{a}(A) \neq R$ (since $RA \neq 0$). Consequently $\mathcal{a}(A) = 0$ by simplicity, whence A is a simple faithful R-module. Therefore R is primitive.

(iii) If R is simple then the ideal $J(R)$ is either R or zero. In the former case R is a radical ring and in the latter R is semisimple and primitive by (ii). ■

EXAMPLES. The endomorphism ring of a (left) vector space over a division ring is semisimple by Theorem 2.10 (i) and the example after Definition 1.5. Consequently by Theorem VII.1.4 the ring of all $n \times n$ matrices over a division ring is semisimple.

EXAMPLE. An example of a simple radical ring is given in E. Sasiada and P. M. Cohn [66].

The classical radical of Wedderburn (in a left Artinian ring) is the maximal nilpotent ideal. We now explore the connection between this radical and the Jacobson radical.

Definition 2.11. *An element* a *of a ring* R *is* **nilpotent** *if* $a^n = 0$ *for some positive integer* n. *A (left, right, two-sided) ideal* I *of* R *is* **nil** *if every element of* I *is nilpotent;* I *is* **nilpotent** *if* $I^n = 0$ *for some integer* n.

Every nilpotent ideal is nil since $I^n = 0$ implies $a^n = 0$ for all $a \varepsilon I$. It is possible, however, to have a nil ideal that is not nilpotent (Exercise 11).

Theorem 2.12. *If* R *is a ring, then every nil right or left ideal is contained in the radical* J(R).

REMARK. The theorem immediately implies that every nil ring is a radical ring.

PROOF OF 2.12. If $a^n = 0$, let $r = -a + a^2 - a^3 + \cdots + (-1)^{n-1}a^{n-1}$. Verify that $r + a + ra = 0 = a + r + ar$, whence a is both left and right quasi-regular. Therefore every nil left [right] ideal is left [right] quasi-regular and hence is contained in $J(R)$ by Theorem 2.3 (iv). ■

Proposition 2.13. *If* R *is a left [resp. right] Artinian ring, then the radical* J(R) *is a nilpotent ideal. Consequently every nil left or right ideal of* R *is nilpotent and* J(R) *is the unique maximal nilpotent left (or right) ideal of* R.

REMARK. If R is left [resp. right] Noetherian, then every nil left or right ideal is nilpotent (Exercise 16).

PROOF OF 2.13. Let $J = J(R)$ and consider the chain of (left) ideals $J \supset J^2 \supset J^3 \supset \cdots$. By hypothesis there exists k such that $J^i = J^k$ for all $i \geq k$. We claim that $J^k = 0$. If $J^k \neq 0$, then the set S of all left ideals I such that $J^kI \neq 0$ is nonempty (since $J^kJ^k = J^{2k} = J^k \neq 0$). By Theorem VIII.1.4 S has a minimal element I_0. Since $J^kI_0 \neq 0$, there is a nonzero $a \varepsilon I_0$ such that $J^ka \neq 0$. Clearly J^ka is a left ideal of R that is contained in I_0. Furthermore $J^ka \varepsilon S$ since $J^k(J^ka) = J^{2k}a = J^ka \neq 0$. Consequently $J^k a = I_0$ by minimality. Thus for some nonzero $r \in J^k$, $ra = a$. Since $-r \in J^k \subset J(R)$, $-r$ is left quasi-regular, whence $s - r - sr = 0$ for some $s \in R$. Consequently,

$$\begin{aligned} a = ra &= -[-ra] = -[-ra + 0] = -[-ra + sa - sa] \\ &= -[-ra + sa - s(ra)] = -[-r + s - sr]a = -0a = 0. \end{aligned}$$

This contradicts the fact that $a \neq 0$. Therefore $J^k = 0$. The last statement of the theorem is now an immediate consequence of Theorem 2.12. ■

Finally we wish to show that left quasi-regularity is a radical property as defined in the introduction to this section. By Theorem 2.3 (iv) its associated radical is clearly the Jacobson radical and a left quasi-regular ring is precisely a radical ring (Definition 2.9). Since a ring homomorphism necessarily maps left quasi-regular elements onto left quasi-regular elements, the homomorphic image of a radical ring is also a radical ring. To complete the discussion we must show that $R/J(R)$ is semisimple and that $J(R)$ is a radical ring.

Theorem 2.14. *If* R *is a ring, then the quotient ring* R/J(R) *is semisimple.*

PROOF. Let $\pi : R \to R/J(R)$ be the canonical epimorphism and denote $\pi(r)$ by $\bar{r}$ ($r \in R$). Let $\mathcal{C}$ be the set of all regular maximal left ideals of R. If $I \in \mathcal{C}$, then $J(R) \subset I$ by Theorem 2.3 (ii) and $\pi(I) = I/J(R)$ is a maximal left ideal of $R/J(R)$ by Theorem IV.1.10. If $e \in R$ is such that $r - re \in I$ for all $r \in R$, then $\bar{r} - \bar{r}\bar{e} \in \pi(I)$ for all $\bar{r} \in R/J(R)$. Therefore, $\pi(I)$ is regular for every I in $\mathcal{C}$. Since $J(R) = \bigcap_{I \in \mathcal{C}} I$ it is easy to verify that if $\bar{r} \in \bigcap_{I \in \mathcal{C}} \pi(I) = \bigcap_{I \in \mathcal{C}} I/J(R)$, then $r \in J(R)$. Consequently, by Theorem 2.3 (ii) (applied to $R/J(R)$)

$$J(R/J(R)) \subset \bigcap_{I \in \mathcal{C}} \pi(I) \subset \pi(J(R)) = 0,$$

whence $R/J(R)$ is semisimple. ■

Lemma 2.15. *Let* R *be a ring and* $a \in R$.

(i) *If* $-a^2$ *is left quasi-regular, then so is* a.
(ii) $a \in J(R)$ *if and only if* Ra *is a left quasi-regular left ideal.*

PROOF. (i) If $r + (-a^2) + r(-a^2) = 0$, let $s = r - a - ra$. Verify that $s + a + sa = 0$, whence a is left quasi-regular.

(ii) If $a \in J(R)$, then $Ra \subset J(R)$. Therefore, Ra is left quasi-regular since $J(R)$ is. Conversely suppose Ra is left quasi-regular. Verify that $K = \{ra + na \mid r \in R, n \in \mathbf{Z}\}$ is a left ideal of R that contains a and Ra. If $s = ra + na$, then $-s^2 \in Ra$. By hypothesis $-s^2$ is left quasi-regular and hence so is s by (i). Thus K is a left quasi-regular left ideal. Therefore $a \in K \subset J(R)$ by Theorem 2.3 (iv). ■

Theorem 2.16. (i) *If an ideal* I *of a ring* R *is itself considered as a ring, then* $J(I) = I \cap J(R)$.

(ii) *If* R *is semisimple, then so is every ideal of* R.
(iii) J(R) *is a radical ring.*

PROOF. (i) $I \cap J(R)$ is clearly an ideal of I. If $a \varepsilon I \cap J(R)$, then a is left quasi-regular in R, whence $r + a + ra = 0$ for some $r \varepsilon R$. But $r = -a - ra \varepsilon I$. Thus every element of $I \cap J(R)$ is left quasi-regular in I. Therefore $I \cap J(R) \subset J(I)$ by Theorem 2.3 (iv) (applied to I).

Suppose $a \varepsilon J(I)$. For any $r \varepsilon R$, $-(ra)^2 = -(rar)a \varepsilon IJ(I) \subset J(I)$, whence $-(ra)^2$ is left quasi-regular in I by Theorem 2.3 (iv). Consequently by Lemma 2.15 (i) ra is left quasi-regular in I and hence in R. Thus Ra is a left quasi-regular left ideal of R, whence $a \varepsilon J(R)$ by Lemma 2.15 (ii). Therefore $a \varepsilon J(I) \cap J(R) \subset I \cap J(R)$. Consequently $J(I) \subset I \cap J(R)$, which completes the proof that $J(I) = I \cap J(R)$. Statements (ii) and (iii) are now immediate consequences of (i). ■

Theorem 2.17. *If* $\{R_i \mid i \varepsilon I\}$ *is a family of rings, then* $J(\prod_{i \varepsilon I} R_i) = \prod_{i \varepsilon I} J(R_i)$.

SKETCH OF PROOF. Verify that an element $\{a_i\} \varepsilon \prod R_i$ is left quasi-regular in $\prod R_i$ if and only if a_i is left quasi-regular in R_i for each i. Consequently $\prod J(R_i)$ is a left quasi-regular ideal of $\prod R_i$, whence $\prod J(R_i) \subset J(\prod R_i)$ by Theorem 2.3 (iv).

For each $k \varepsilon I$, let $\pi_k : \prod R_i \to R_k$ be the canonical projection. Verify that $I_k = \pi_k(J(\prod R_i))$ is a left quasi-regular ideal of R_k. It follows that $I_k \subset J(R_k)$ and therefore that $J(\prod R_i) \subset \prod J(R_i)$. ■

EXERCISES

Note: R is always a ring.

1. For each $a,b \varepsilon R$ let $a \circ b = a + b + ab$.
   (a) $\circ$ is an associative binary operation with identity element $0 \varepsilon R$.
   (b) The set G of all elements of R that are *both* left and right quasi-regular forms a group under $\circ$.
   (c) If R has an identity, then $a \varepsilon R$ is left [resp. right] quasi-regular if and only if $1_R + a$ is left [resp. right] invertible. [*Hint:* $(1_R + r)(1_R + a) = 1_R + r \circ a$ and $r(1_R + a) - 1_R = (r - 1_R) \circ a$.]

2. (Kaplansky) R is a division ring if and only if every element of R except one is left quasi-regular. [Note that the only element in a division ring D that is *not* left quasi-regular is -1_D; also see Exercise 1.]

3. Let I be a left ideal of R and let $(I : R) = \{r \varepsilon R \mid rR \subset I\}$.
   (a) $(I : R)$ is an ideal of R. If I is regular, then $(I : R)$ is the largest ideal of R that is contained in I.
   (b) If I is a regular maximal left ideal of R and $A \cong R/I$, then $\mathcal{A}(A) = (I : R)$. Therefore $J(R) = \bigcap (I : R)$, where I runs over all the regular maximal left ideals of R.

4. The radical $J(R)$ contains no nonzero idempotents. However, a nonzero idempotent may be left quasi-regular. [*Hint:* Exercises 1 and 2].

5. If R has an identity, then
   (a) $J(R) = \{r \in R \mid 1_R + sr$ is left invertible for all $s \in R\}$.
   (b) $J(R)$ is the largest ideal K such that for all $r \in K$, $1_R + r$ is a unit.

6. (a) The homomorphic image of a semisimple ring need not be semisimple.
   (b) If $f : R \to S$ is a ring epimorphism, then $f(J(R)) \subset J(S)$.

7. If R is the ring of all rational numbers with odd denominators, then $J(R)$ consists of all rational numbers with odd denominator and even numerator.

8. Let R be the ring of all upper triangular $n \times n$ matrices over a division ring D (see Exercise VII.1.2). Find $J(R)$ and prove that $R/J(R)$ is isomorphic to the direct product $D \times D \times \cdots \times D$ (n factors). [*Hint:* show that a strictly triangular matrix is nilpotent.]

9. A principal ideal domain R is semisimple if and only if R is a field or R contains an infinite number of distinct nonassociate irreducible elements.

10. Let D be a principal ideal domain and d a nonzero nonunit element of D. Let R be the quotient ring $D/(d)$.
    (a) R is semisimple if and only if d is the product of distinct nonassociate irreducible elements of D. [*Hint:* Exercise VIII.1.2.]
    (b) What is $J(R)$?

11. If p is a prime, let R be the subring $\sum_{n\geq 1} Z_{p^n}$ of $\prod_{n\geq 1} Z_{p^n}$. The ideal $I = \sum_{n\geq 1} I_n$, where I_n is the ideal of Z_{p^n} generated by $p \in Z_{p^n}$, is a nil ideal of R that is not nilpotent.

12. Let R be a ring without identity. Embed R in a ring S with identity which has characteristic zero, as in Theorem III.1.10. Prove that $J(R) = J(S)$. Consequently every semisimple ring may be embedded in a semisimple ring with identity.

13. $J(\text{Mat}_n R) = \text{Mat}_n J(R)$. Here is an outline of a proof:
    (a) If A is a left R-module, consider the elements of $A^n = A \oplus A \oplus \cdots \oplus A$ (n summands) as column vectors; then A^n is a left $(\text{Mat}_n R)$-module (under ordinary matrix multiplication).
    (b) If A is a simple R-module, A^n is a simple $(\text{Mat}_n R)$-module.
    (c) $J(\text{Mat}_n R) \subset \text{Mat}_n J(R)$.
    (d) $\text{Mat}_n J(R) \subset J(\text{Mat}_n R)$. [*Hint:* prove that $\text{Mat}_n J(R)$ is a left quasi-regular ideal of $\text{Mat}_n R$ as follows. For each $k = 1,2,\dots,n$ let K_k consist of all matrices (a_{ij}) such that $a_{ij} \in J(R)$ and $a_{ij} = 0$ if $j \neq k$. Show that K_k is a left quasi-regular left ideal of $\text{Mat}_n R$ and observe that $K_1 + K_2 + \cdots + K_n = \text{Mat}_n J(R)$.]

14. (a) Let I be a nonzero ideal of $R[x]$ and $p(x)$ a nonzero polynomial of least degree in I with leading coefficient a. If $f(x) \in R[x]$ and $a^m f(x) = 0$, then $a^{m-1}p(x)f(x) = 0$.
    (b) If a ring R has no nonzero nil ideals (in particular, if R is semisimple), then $R[x]$ is semisimple. [*Hint:* Let M be the set of nonzero polynomials of least degree in $J(R[x])$. Let N be the set consisting of 0 and the leading coefficients of polynomials in M. Use (a) to show that N is a nil ideal of R, whence $J(R[x]) = 0$.]
    (c) There exist rings R such that $R[x]$ is semisimple, but R is not. [*Hint,* consider $R = F[[x]]$, with F a field.]

15. Let L be a left ideal and K a right ideal of R. Let $M(R)$ be the ideal generated by *all* nilpotent ideals of R.
   (a) $L + LR$ is an ideal such that $(L + LR)^n \subset L^n + L^nR$ for all $n \geq 1$.
   (b) $K + RK$ is an ideal such that $(K + RK)^n \subset K^n + RK^n$ for all $n \geq 1$.
   (c) If L [resp. K] is nilpotent, so is the ideal $L + LR$ [resp. $K + RK$], whence $L \subset M(R)$ [resp. $K \subset M(R)$].
   (d) If N is a maximal nilpotent ideal of R, then R/N has no nonzero nilpotent left or right ideals. [*Hint:* first show that R/N has no nonzero nilpotent ideals; then apply (*c*) to the ring R/N.]
   (e) If K [resp. L] is nil, but not nilpotent and $\pi : R \to R/N$ is the canonical epimorphism, then $\pi(K)$ [resp. $\pi(L)$] is a nil right [resp. left] ideal of R/N which is not nilpotent.

16. (Levitsky) Every nil left or right ideal I in a left Noetherian ring R is nilpotent. [*Sketch of Proof.* It suffices by Exercise 15 to assume that R has no nonzero nilpotent left or right ideals. Suppose I is a left or a right ideal which is not nilpotent and $0 \neq a \in I$. Show that aR is a nil right ideal (even though I may be a left ideal), whence the left ideal $\mathcal{a}(u)$ is nonzero for all $u \in aR$. There exists a nonzero $u_0 \in aR$ with $\mathcal{a}(u_0)$ maximal, whence $\mathcal{a}(u_0) = \mathcal{a}(u_0x)$ for all $x \in R$ such that $u_0x \neq 0$. Show that $(u_0y)u_0 = 0$ for all $y \in R$, so that $(Ru_0)^2 = 0$. Therefore $Ru_0 = 0$, which implies that $\{r \in R \mid Rr = 0\}$ is a nonzero nilpotent right ideal of R; contradiction.]

17. Show that Nakayama's Lemma VIII.4.5 is valid for any ring R with identity, provided condition (i) is replaced by the condition

   (i′) J is contained in the Jacobson radical of R.

   [*Hint:* Use Theorem 2.3(iv) and Exercise 1 (c) to show (i′) $\Rightarrow$ (ii).]

3. SEMISIMPLE RINGS

In accordance with the theory of radicals outlined in the first part of Section 2 we now restrict our study to rings that are Jacobson semisimple. Arbitrary semisimple rings are characterized as particular kinds of subrings of direct products of primitive rings (Proposition 3.2). Much stronger results are proved for semisimple (left) Artinian rings. Such rings are actually finite direct products of simple rings (Theorem 3.3). They may also be characterized in numerous ways in terms of modules (Theorem 3.7). Along the way semisimple modules over arbitrary rings are defined and their basic properties developed (Theorem 3.6).

Definition 3.1. *A ring* R *is said to be a* **subdirect product** *of the family of rings* $\{R_i \mid i \in I\}$ *if* R *is a subring of the direct product* $\prod_{i \in I} R_i$ *such that* $\pi_k(R) = R_k$ *for every* $k \in I$, *where* $\pi_k : \prod_{i \in I} R_i \to R_k$ *is the canonical epimorphism.*

REMARK. A ring S is isomorphic to a subdirect product of the family of rings $\{R_i \mid i \in I\}$ if and only if there is a monomorphism of rings $\phi : S \to \prod_{i \in I} R_i$ such that $\pi_k\phi(S) = R_k$ for every $k \in I$.

EXAMPLE. Let P be the set of prime integers. For each $k \in \mathbf{Z}$ and $p \in P$ let $k_p \in Z_p$ be the image of k under the canonical epimorphism $\mathbf{Z} \to Z_p$. Then the map $\phi : \mathbf{Z} \to \prod_{p \in P} Z_p$ given by $k \mapsto \{k_p\}_{p \in P}$ is a monomorphism of rings such that $\pi_p\phi(\mathbf{Z}) = Z_p$ for every $p \in P$. Therefore $\mathbf{Z}$ is isomorphic to a subdirect product of the family of fields $\{Z_p \mid p \in P\}$. More generally we have:

Proposition 3.2. *A nonzero ring R is semisimple if and only if R is isomorphic to a subdirect product of primitive rings.*

REMARK. Propositions 1.7 and 3.2 imply that a nonzero commutative semisimple ring is a subdirect product of fields.

SKETCH OF PROOF OF 3.2. Suppose R is nonzero semisimple and let $\mathcal{P}$ be the set of all left primitive ideals of R. Then for each $P \in \mathcal{P}$, R/P is a primitive ring (Definition 2.1). By Theorem 2.3 (iii), $0 = J(R) = \bigcap_{P \in \mathcal{P}} P$. For each P let $\lambda_P : R \to R/P$ and $\pi_P : \prod_{Q \in \mathcal{P}} R/Q \to R/P$ be the respective canonical epimorphisms. The map $\phi : R \to \prod_{P \in \mathcal{P}} R/P$ given by $r \mapsto \{\lambda_P(r)\}_{P \in \mathcal{P}} = \{r + P\}_{P \in \mathcal{P}}$ is a monomorphism of rings such that $\pi_P\phi(R) = R/P$ for every $P \in \mathcal{P}$.

Conversely suppose there is a family of primitive rings $\{R_i \mid i \in I\}$ and a monomorphism of rings $\phi : R \to \prod_{i \in I} R_i$ such that $\pi_k\phi(R) = R_k$ for each $k \in I$. Let ψ_k be the epimorphism $\pi_k\phi$. Then $R/\text{Ker } \psi_k$ is isomorphic to the primitive ring R_k (Corollary III.2.10), whence Ker ψ_k is a left primitive ideal of R (Definition 2.1). Therefore $J(R) \subset \bigcap_{k \in I} \text{Ker } \psi_k$ by Theorem 2.3 (iii). However, if $r \in R$ and $\psi_k(r) = 0$, then the kth component of $\phi(r)$ in $\prod R_i$ is zero. Thus if $r \in \bigcap_{k \in I} \text{Ker } \psi_k$, we must have $\phi(r) = 0$. Since ϕ is a monomorphism $r = 0$. Therefore $J(R) \subset \bigcap_{k \in I} \text{Ker } \psi_k = 0$, whence R is semisimple. ■

In view of the results on primitive rings in Section 1, we can now characterize semisimple rings as those rings that are isomorphic to subdirect products of families of rings, each of which is a dense ring of endomorphisms of a vector space over a division ring. Unfortunately subdirect products (and dense rings of endomorphisms) are not always the most tractable objects with which to deal. But in the absence of further restrictions this is probably the best one can do. In the case of (left) Artinian rings, however, these results can be considerably sharpened.

Theorem 3.3. (*Wedderburn-Artin*). *The following conditions on a ring R are equivalent.*

(i) R *is a nonzero semisimple left Artinian ring;*

(ii) R *is a direct product of a finite number of simple ideals each of which is isomorphic to the endomorphism ring of a finite dimensional vector space over a division ring;*

(iii) *there exist division rings* $D_1, \ldots, D_t$ *and positive integers* $n_1, \ldots, n_t$ *such that* R *is isomorphic to the ring* $Mat_{n_1}D_1 \times Mat_{n_2}D_2 \times \cdots \times Mat_{n_t}D_t$.

REMARK. By a simple ideal of R we mean an ideal that is itself a simple ring.

PROOF OF 3.3. (ii) $\Leftrightarrow$ (iii) Exercise III.2.9 and Theorem VII.1.4.

(ii) $\Rightarrow$ (i) By hypothesis $R \cong \prod_{i=1}^{t} R_i$ with each R_i the endomorphism ring of a vector space. The example after Definition 1.5 shows that each R_i is primitive, whence $J(R_i) = 0$ by Theorem 2.10 (i). Consequently by Theorem 2.17

$$J(R) \cong \prod_{i=1}^{t} J(R_i) = 0.$$

Therefore R is semisimple. R is left Artinian by Theorem VII.1.4 and Corollaries VIII.1.7 and VIII.1.12.

(i) $\Rightarrow$ (ii) Since $R \neq 0$ and $J(R) = 0$, R has left primitive ideals by Theorem 2.3 (iii). Suppose that R has only finitely many distinct left primitive ideals: $P_1, P_2, \ldots, P_t$. Then each R/P_i is a primitive ring (Definition 2.1) that is left Artinian (Corollary VIII.1.6). Consequently, by Theorem 1.14 each R/P_i is a simple ring isomorphic to an endomorphism ring of a finite dimensional left vector space over a division ring. Since R/P_i is simple, each P_i is a maximal ideal of R (Theorem III.2.13). Furthermore $R^2 \not\subset P_i$ (otherwise $(R/P_i)^2 = 0$), whence $R^2 + P_i = R$ by maximality. Likewise if $i \neq j$, then $P_i + P_j = R$ by maximality. Consequently by Corollary III.2.27 (of the Chinese Remainder Theorem) and Theorem 2.3 (iii) there is an isomorphism of rings:

$$R = R/0 = R/J(R) = R/\bigcap_{i=1}^{t} P_i \cong R/P_1 \times \cdots \times R/P_t.$$

If $\iota_k : R/P_k \to \prod_{i=1}^{t} R/P_i$ is the canonical monomorphism (Theorem III.2.22), then each $\iota_k(R/P_k)$ is a simple ideal of $\prod_{i=1}^{t} R/P_i$. Under the isomorphism $\prod_{i=1}^{t} R/P_i \cong R$, the images of the $\iota_k(R/P_k)$ are simple ideals of R. Clearly R is the (internal) direct product of these ideals.

To complete the proof we need only show that R cannot have an infinite number of distinct left primitive ideals. Suppose, on the contrary, that $P_1, P_2, P_3, \ldots$ is a sequence of distinct left primitive ideals of R. Since

$$P_1 \supset P_1 \cap P_2 \supset P_1 \cap P_2 \cap P_3 \supset \cdots$$

is a descending chain of (left) ideals there is an integer n such that $P_1 \cap \cdots \cap P_n = P_1 \cap \cdots \cap P_n \cap P_{n+1}$, whence $P_1 \cap \cdots \cap P_n \subset P_{n+1}$. The previous paragraph shows that $R^2 + P_i = R$ and $P_i + P_j = R$ $(i \neq j)$ for $i,j = 1,2,\ldots,n+1$. The proof of Theorem III.2.25 shows that $P_{n+1} + (P_1 \cap \cdots \cap P_n) = R$. Consequently $P_{n+1} = R$, which contradicts the fact that P_{n+1} is left primitive (see the Remark after Definition 2.1). Therefore R has only finitely many distinct primitive ideals and the proof is complete. ■

Corollary 3.4. (i) *A semisimple left Artinian ring has an identity.*

(ii) *A semisimple ring is left Artinian if and only if it is right Artinian.*

(iii) *A semisimple left Artinian ring is both left and right Noetherian.*

REMARK. Somewhat more is actually true: any left Artinian ring with identity is left Noetherian (Exercise 13).

SKETCH OF PROOF OF 3.4. (i) Theorem 3.3. (ii) Theorem 3.3 is valid with "left" replaced by "right" throughout. Consequently the equivalence of conditions (i) and (iii) of Theorem 3.3 implies that R is left Artinian if and only if R is right Artinian.

(iii) Corollaries VIII.1.7 and VIII.1.12 and Theorem 3.3 (iii). ■

The following corollary is not needed in the sequel. Recall that an element e of a ring R is said to be **idempotent** if $e^2 = e$.

Corollary 3.5. *If* I *is an ideal in a semisimple left Artinian ring* R, *then* I = Re, *where* e *is an idempotent which is in the center of* R.

SKETCH OF PROOF. By Theorem 3.3 R is a (ring) direct product of simple ideals, $R = I_1 \times \cdots \times I_n$. For each j, $I \cap I_j$ is either 0 or I_j by simplicity. After reindexing if necessary we may assume that $I \cap I_j = I_j$ for $j = 1,2,\ldots,t$ and $I \cap I_j = 0$ for $j = t+1,\ldots,n$. Since R has an identity by Corollary 3.4, there exist $e_j \in I_j$ such that $1_R = e_1 + e_2 + \cdots + e_n$. Since $I_jI_k = 0$ for $j \neq k$ we have

$$e_1 + e_2 + \cdots + e_n = 1_R = (1_R)^2 = e_1^2 + e_2^2 + \cdots + e_n^2,$$

whence $e_j^2 = e_j$ for each j. It is easy to verify that each e_i lies in the center of R and that $e = e_1 + e_2 + \cdots + e_t$ is an idempotent in I which is in the center of R. Since I is an ideal, $Re \subset I$. Conversely if $u \in I$, then $u = u1_R = ue_1 + \cdots + ue_n$. But for $j > t$, $ue_j \in I \cap I_j = 0$. Thus $u = ue_1 + \cdots + ue_t = ue$. Therefore $I \subset Re$. ■

Theorem 3.3 is a characterization of semisimple left Artinian rings in ring theoretic terms. As one might suspect from the close interrelationship of rings and modules, such rings can also be characterized strictly in terms of modules. In order to obtain these characterizations we need a theorem that is valid for modules over an arbitrary ring.

Theorem 3.6. *The following conditions on a nonzero module* A *over a ring* R *are equivalent.*

(i) A *is the sum of a family of simple submodules.*
(ii) A *is the (internal) direct sum of a family of simple submodules.*
(iii) *For every nonzero element* a *of* A, Ra $\neq$ 0; *and every submodule* B *of* A *is a direct summand (that is,* A = B $\oplus$ C *for some submodule* C).

A module that satisfies the equivalent conditions of Theorem 3.6 is said to be **semisimple** or **completely reducible.** The terminology semisimple is motivated by Theorem 3.3 (ii) and the fact (to be proved below) that every module over a (left) Artinian semisimple ring is semisimple.

SKETCH OF PROOF OF 3.6. (i) $\Rightarrow$ (ii) Suppose A is the sum of the family $\{B_i \mid i \varepsilon I\}$ of simple submodules (that is, A is generated by $\bigcup_{i \varepsilon I} B_i$). Use Zorn's Lemma to show that there is a nonempty subset J of I which is maximal with respect to the property: the submodule generated by $\{B_j \mid j \varepsilon J\}$ is in fact a direct sum $\sum_{j \varepsilon J} B_j$. We claim that $A = \sum_{j \varepsilon J} B_j$. To prove this we need only show that $B_i \subset \sum_{j \varepsilon J} B_j$ for every $i \varepsilon I$. Since B_i is simple and $B_i \cap (\sum B_j)$ is a submodule of B_i, either $B_i \cap (\sum B_j) = B_i$, which implies $B_i \subset \sum B_j$, or $B_i \cap (\sum B_j) = 0$. The second case cannot occur. For if it did, $K = \{i\} \cup J$ would be a set such that the submodule generated by $\{B_k \mid k \varepsilon K\}$ is a direct sum (Theorem IV.1.15), which contradicts the maximality of J.

(ii) $\Rightarrow$ (iii) Suppose A is the direct sum $\sum_{i \varepsilon I} B_i$ with each B_i a simple submodule. If a is a nonzero element of A, then $a = b_{i_1} + \cdots + b_{i_k}$ with $0 \neq b_{i_k} \varepsilon B_{i_k}$ $(i_1, \ldots, i_k \varepsilon I)$. Clearly $Ra = 0$ if and only if $Rb_{i_k} = 0$ for each i_k. But Remark (iii) after Definition 1.1 shows that $Rb_{i_k} = B_{i_k} \neq 0$. Therefore $Ra \neq 0$.

Let B be a nonzero submodule of A. By simplicity $B \cap B_i$ is either 0 or B_i. If $B \cap B_i = B_i$ for all i, then $A = B$ and B is trivially a direct summand, $A = B \oplus 0$. Otherwise $B \cap B_i = 0$ for some i. Use Zorn's Lemma to find a subset J of I which is maximal with respect to the property: $B \cap (\sum_{j \varepsilon J} B_j) = 0$. We claim that $A = B \oplus (\sum_{j \varepsilon J} B_j)$. It suffices by Theorem IV.1.15 to show that $B_i \subset B \oplus (\sum_{j \varepsilon J} B_j)$ for each i. If $i \varepsilon J$, then $B_i \subset \sum_{j \varepsilon J} B_j$ and we are done. If $i \notin J$ and $B_i \not\subset B \oplus \sum_{j \varepsilon J} B_j$, then $B_i \cap (B \oplus \sum_{j \varepsilon J} B_j) = 0$ by the simplicity of B_i. It follows that $J \cup \{i\}$ is a set that contradicts the maximality of J. Therefore $B_i \subset B \oplus \sum_{j \varepsilon J} B_j$.

(iii) $\Rightarrow$ (i) We first observe that if N is any submodule of A, then every submodule K of N is a direct summand of N. For by hypothesis K is a direct summand of A, say $A = K \oplus L$. Verify that $N = N \cap A = (N \cap K) \oplus (N \cap L) = K \oplus (N \cap L)$.

Next we show that A has simple submodules. Since $A \neq 0$, there exists a nonzero element a of A. Use Zorn's Lemma to find a submodule B of A that is maximal with respect to the property that $a \notin B$. By hypothesis $A = B \oplus C$ for some nonzero submodule C and $RC \neq 0$. We claim that C is simple. If it were not, then C would have a *proper* submodule D, which would be a direct summand of C by the previous paragraph. Consequently $C = D \oplus E$ with $E \neq 0$, whence $A = B \oplus C = B \oplus D \oplus E$, with $D \neq 0$ and $E \neq 0$. Now $B \oplus D$ and $B \oplus E$ both contain B properly. Therefore by the maximality of B we must have $a \varepsilon B \oplus D$ and $a \varepsilon B \oplus E$. Thus $b + d = a = b' + e$ $(b,b' \varepsilon B; d \varepsilon D; e \varepsilon E)$. Now $0 = a - a = (b - b') + d - e \varepsilon B \oplus D \oplus E$ implies that $d = 0$, $e = 0$, and $b - b' = 0$. Consequently, $a = b \varepsilon B$ which is a contradiction. Therefore C is simple.

Let A_0 be the submodule of A generated by all the simple submodules of A. Then $A = A_0 \oplus N$ for some submodule N. N satisfies the same hypotheses as A by the paragraph before last. If $N \neq 0$, then the argument in the immediately preceding paragraph shows that N contains a nonzero simple submodule T. Since T is a simple submodule of A, $T \subset A_0$. Thus $T \subset A_0 \cap N = 0$, which is a contradiction. Therefore $N = 0$, whence $A = A_0$ is the sum of a family of simple submodules. ■

We are now able to give numerous characterizations of semisimple left Artinian rings in terms of modules. Since the submodules of a ring R (considered as a left

R-module) are precisely the left ideals of R, some of these characterizations are stated in terms of left ideals. A subset $\{e_1, \ldots, e_m\}$ of R is a set of **orthogonal idempotents** if $e_i^2 = e_i$ for all i and $e_ie_j = 0$ for all $i \neq j$.

Theorem 3.7. *The following conditions on a nonzero ring* R *with identity are equivalent.*

(i) R *is semisimple left Artinian;*
(ii) *every unitary left* R*-module is projective;*
(iii) *every unitary left* R*-module is injective;*
(iv) *every short exact sequence of unitary left* R*-modules is split exact;*
(v) *every nonzero unitary left* R*-module is semisimple;*
(vi) R *is itself a unitary semisimple left* R*-module;*
(vii) *every left ideal of* R *is of the form* Re *with* e *idempotent;*
(viii) R *is the (internal) direct sum (as a left* R*-module) of minimal left ideals* $K_1, \ldots, K_m$ *such that* $K_i = Re_i$ $(e_i \in R)$ *for* $i = 1,2, \ldots, m$ *and* $\{e_1, \ldots, e_m\}$ *is a set of orthogonal idempotents with* $e_1 + e_2 + \cdots + e_m = 1_R$.

REMARKS. Since a semisimple ring is left Artinian if and only if it is right Artinian (Corollary 3.4), each condition in Theorem 3.7 is equivalent to its obvious analogue for right modules or right ideals. There is no loss of generality in assuming R has an identity, since every semisimple left Artinian ring necessarily has one by Corollary 3.4. The theorem is false if the word "unitary" is omitted (Exercise 10).

SKETCH OF PROOF OF 3.7. (ii) $\Leftrightarrow$ (iii) $\Leftrightarrow$ (iv) is Exercise IV.3.1. To complete the proof we shall prove the implications (iv) $\Leftrightarrow$ (v) and (v) $\Rightarrow$ (vii) $\Rightarrow$ (vi) $\Rightarrow$ (i) $\Rightarrow$ (viii) $\Rightarrow$ (v).

(iv) $\Rightarrow$ (v) If B is a submodule of a nonzero unitary R-module A, then

$$0 \rightarrow B \xrightarrow{\subset} A \rightarrow A/B \rightarrow 0$$

is a short exact sequence, which splits by hypothesis. The proof of Theorem IV.1.18 shows that $A = B \oplus C$ with $C \cong A/B$. Since A is unitary, $Ra \neq 0$ for every nonzero $a \in A$. Therefore A is semisimple by Theorem 3.6.

(v) $\Rightarrow$ (iv) Let $0 \rightarrow A \xrightarrow{f} B \xrightarrow{g} C \rightarrow 0$ be a short exact sequence of unitary R-modules. Then $f : A \rightarrow f(A)$ is an isomorphism. Since B is semisimple by (v), $f(A)$ is a direct summand of B by Theorem 3.6. If $\pi : B \rightarrow f(A)$ is the canonical epimorphism, then $\pi f = f$ and $f^{-1}\pi : B \rightarrow A$ is an R-module homomorphism such that $(f^{-1}\pi)f = 1_A$. Therefore the sequence splits by Theorem IV.1.18.

(v) $\Rightarrow$ (vii) The left ideals of R are precisely its submodules. If L is a left ideal, then $R = L \oplus I$ for some left ideal I by (v) and Theorem 3.6. Consequently, there are $e_1 \in L$ and $e_2 \in I$ such that $1_R = e_1 + e_2$. Since $e_1 \in L$, $Re_1 \subset L$. If $r \in L$, then $r = re_1 + re_2$, whence $re_2 = r - re_1 \in L \cap I = 0$. Thus $r = re_1$ for every $r \in L$; in particular, $e_1e_1 = e_1$ and $L \subset Re_1$. Therefore, $L = Re_1$ with e_1 idempotent.

(vii) $\Rightarrow$ (vi) A submodule L of R is a left ideal, whence $L = Re$ with e idempotent. Verify that $R(1_R - e)$ is a left ideal of R such that $R = Re \oplus R(1_R - e)$. Therefore, R is semisimple by Theorem 3.6.

(vi) $\Rightarrow$ (i) By hypothesis R is a direct sum $\sum_{i \in I} B_i$, with each B_i a simple submodule (left ideal) of R. Consequently there is a finite subset I_0 of I (whose elements will be labeled 1,2, . . . , k for convenience) such that $1_R = e_1 + e_2 + \cdots + e_k$ $(e_i \varepsilon B_i)$. Thus for every $r \varepsilon R$, $r = re_1 + re_2 + \cdots + re_k \varepsilon \sum_{i=1}^{k} B_i$, whence $R = \sum_{i=1}^{k} B_i$. If $r \varepsilon J(R)$, then $rB_i = 0$ for all i by Theorem 2.3 (i). Consequently,

$$r = r1_R = re_1 + re_2 + \cdots + re_k = 0.$$

Therefore, $J(R) = 0$ and R is semisimple. Since B_i is simple and

$$(B_1 \oplus \cdots \oplus B_i)/(B_1 \oplus \cdots \oplus B_{i-1}) \cong B_i,$$

the series

$$R = B_1 \oplus \cdots \oplus B_k \supset B_1 \oplus \cdots \oplus B_{k-1} \supset \cdots \supset B_1 \oplus B_2 \supset B_1 \supset 0$$

is a composition series for R. Therefore, R is left Artinian by Theorem VIII.1.11.

(i) $\Rightarrow$ (viii) In view of Theorem 3.3 it suffices to assume that $R = \prod_{i=1}^{t} \text{Mat}_{n_i} D_i$ with each $n_i > 0$ and each D_i a division ring. For each fixed i and each $j = 1,2, \ldots, n_i$ let e_{ij} be the matrix in $\text{Mat}_{n_i} D_i$ with 1_{D_i} in position (j,j) and 0 elsewhere. Then $\{e_{i1}, \ldots, e_{in_i}\}$ is a set of orthogonal idempotents in $\text{Mat}_{n_i} D_i = R_i$ whose sum is the identity matrix. The proof of Corollary VIII.1.12 shows that each $R_i e_{ij}$ is a minimal left ideal of R_i and $R_i = R_i e_{i1} \oplus \cdots \oplus R_i e_{in_i}$. Since R is the ring direct product $R_1 \times \cdots \times R_t$, it follows that $R_i R_j = 0$ for $i \neq j$; that $Re_{ij} = R_i e_{ij}$; that Re_{ij} is a minimal left ideal of R; and that $\{e_{ij} \mid 1 \leq i \leq t; 1 \leq j \leq n_i\}$ is a set of orthogonal idempotents in R whose sum is $\sum_{i=1}^{t} (\sum_j e_{ij}) = \sum_{i=1}^{t} 1_{R_i} = 1_R$. Clearly $R = \sum_{i=1}^{t} \sum_{j=1}^{n_i} Re_{ij}$.

(viii) $\Rightarrow$ (v) Let A be a unitary R-module. For each $a \varepsilon A$ and each i, $K_i a$ is a submodule of A (Exercise IV.1.3) and $a = 1_R a = e_1 a + \cdots + e_m a \varepsilon K_1 a + \cdots + K_m a$. Consequently the submodules $K_i a$ $(a \varepsilon A, 1 \leq i \leq m)$ generate A. For each $a \varepsilon A$ and each i, the map $f : K_i \to K_i a$ given by $k \mapsto ka$ is an R-module epimorphism. Since K_i is a minimal left ideal of a ring with identity, K_i is a simple R-module. Consequently if $K_i a \neq 0$, then f is an isomorphism by Schur's Lemma 1.10. Thus $\{K_i a \mid 1 \leq i \leq m; a \varepsilon A; K_i a \neq 0\}$ is a family of simple submodules whose sum is A. Therefore A is semisimple by Theorem 3.6. ■

Theorems 3.3 and 3.7 show that a semisimple left Artinian ring may be decomposed as a direct product [resp. sum] of simple ideals [resp. minimal left ideals]. We turn now to the question of the uniqueness of these decompositions.

Proposition 3.8. *Let* R *be a semisimple left Artinian ring.*

(i) $R = I_1 \times \cdots \times I_n$ *where each* I_j *is a simple ideal of* R.

(ii) *If* J *is any simple ideal of* R, *then* $J = I_k$ *for some* k.

(iii) *If* $R = J_1 \times \cdots \times J_m$ *with each* J_k *a simple ideal of* R, *then* $n = m$ *and* (*after reindexing*) $I_k = J_k$ *for* $k = 1,2, \ldots, n$.

REMARKS. The conclusion $J = I_j$ [resp. $J_k = I_k$] is considerably stronger than the statement "J [resp. J_k] is isomorphic to I_k." The uniquely determined simple ideals $I_1, \ldots, I_n$ in Proposition 3.8 are called the **simple components** of R.

PROOF OF 3.8. (i) is true by Theorem 3.3. (ii) If J is a simple ideal of R, then $RJ \neq 0$, whence $I_kJ \neq 0$ for some k. Since I_kJ is a nonzero ideal that is contained in both I_k and J, the simplicity of I_k and J implies $I_k = I_kJ = J$. (iii) The ideals $I_1, \dots, I_n$ [resp. $J_1, \dots, J_m$] are nonzero and mutually disjoint by hypothesis. Define a map θ from the m element set $\{J_1, \dots, J_m\}$ to the n element set $\{I_1, \dots, I_n\}$ by $J_k \mapsto I_k$, where $J_k = I_k$. θ is well defined and injective by (ii), whence $m \leq n$. The same argument with the roles of J_k and I_k reversed shows that $n \leq m$. Therefore $n = m$ and θ is a bijection. ∎

A semisimple left Artinian ring R is a direct sum of minimal left ideals by Theorem 3.7 (viii). The uniqueness (up to isomorphism) of this decomposition will be an immediate consequence of the following proposition. For R is a semisimple R-module (Theorem 3.7 (vi)) and the minimal left ideals of R are precisely its simple submodules.

Proposition 3.9. *Let* A *be a semisimple module over a ring* R. *If there are direct sum decompositions*

$$A = B_1 \oplus \cdots \oplus B_m \quad \textit{and} \quad A = C_1 \oplus \cdots \oplus C_n,$$

where each B_i, C_j *is a simple submodule of* A, *then* $m = n$ *and (after reindexing)* $B_i \cong C_i$ *for* $i = 1,2, \dots, m$.

REMARK. The uniqueness statement here is weaker than the one in Proposition 3.8. Proposition 3.9 is false if "$B_i \cong C_i$" is replaced by "$B_i = C_i$" (Exercise 11).

PROOF OF 3.9. The series

$$A = B_1 \oplus \cdots \oplus B_m \supset B_2 \oplus \cdots \oplus B_m \supset \cdots \supset B_m \supset 0$$

is a composition series for A with simple factors $B_1, B_2, \dots, B_m$ (see p. 375). Similarly $A = C_1 \oplus \cdots \oplus C_n \supset C_2 \oplus \cdots \oplus C_m \supset \cdots \supset C_m \supset 0$ is a composition series for A with simple factors $C_1, \dots, C_n$. The Jordan-Hölder Theorem VIII.1.10 implies that $m = n$ and (after reindexing) $B_i \cong C_i$ for $i = 1,2, \dots, m$. ∎

The following theorem will be used only in the proof of Theorem 6.7.

Theorem 3.10. *Let* R *be a semisimple left Artinian ring.*

(i) *Every simple left [resp. right]* R*-module is isomorphic to a minimal left [resp. right] ideal of* R.

(ii) *The number of nonisomorphic simple left [resp. right]* R*-modules is the same as the number of simple components of* R.

PROOF. R is right Artinian by Corollary 3.4. Since the preceding results are left-right symmetric, it suffices to prove the theorem for left modules.

(i) By Theorem 3.7, $R = K_1 \oplus \cdots \oplus K_m$ with each K_i a nonzero minimal left ideal (simple submodule) of R. R has an identity (Corollary 3.4) and every simple R-module A is unitary by Remark (ii) after Definition 1.1. The proof of (viii) $\Rightarrow$ (v) of Theorem 3.7 shows that for some i $(1 \leq i \leq m)$ and $a \varepsilon A$, A contains a nonzero submodule K_ia such that $K_ia \cong K_i$. The simplicity of A implies that $A = K_ia \cong K_i$.

(ii) The simple components of R are the unique simple ideals I_j of R such that $R = I_1 \times \cdots \times I_n$ (Proposition 3.8). In view of (i) it suffices to prove:

(a) each K_i is contained in some I_t;

(b) each I_t contains some K_i;

(c) $K_i \cong K_j$ as R-modules if and only if K_i and K_j are contained in the same simple component I_t.

These statements are proved as follows.

(a) Since R has an identity, $K_i = RK_i = I_1K_i \times \cdots \times I_nK_i$. Since each I_jK_i is a left ideal of R contained in K_i, we must have $I_tK_i = K_i$ for some t and $I_jK_i = 0$ for $j \neq t$ by minimality. Therefore $K_i = I_tK_i \subset I_t$.

(b) If I_t contains no K_i, then $R = \sum K_i$ is contained in

$$I_1 \times \cdots \times I_{t-1} \times I_{t+1} \times \cdots \times I_n$$

by (a). Since $I_t \neq 0$ by simplicity and $R = \prod I_j$,

$$0 \neq I_t = I_t \cap R = I_t \cap (I_1 \times \cdots \times I_{t-1} \times I_{t+1} \times \cdots \times I_n) = 0,$$

which is a contradiction.

(c) If $K_i \subset I_{t_1}$ and $K_j \subset I_{t_2}$ with $t_1 \neq t_2$, then by (a), $0 \neq K_i = I_{t_1}K_i$ and $0 \neq K_j = I_{t_2}K_j$. Since $R = \prod I_j$, $I_{t_1}I_{t_2} = 0 = I_{t_2}I_{t_1}$. Consequently, there can be no R-module isomorphism $K_i \cong K_j$. Conversely suppose $K_i \subset I_t$ and $K_j \subset I_t$. Then K_i and K_j are I_t-modules. Since I_t is simple and $0 \neq K_i = I_tK_i$ by (a), the left annihilator ideal of K_i in I_t must be zero. Consequently, $K_jK_i \neq 0$ since $0 \neq K_j \subset I_t$. Thus for some $a \varepsilon K_i$, $K_ja \neq 0$. Since K_i and K_j are left ideals of R, K_ja is a nonzero left ideal of R and $K_ja \subset K_i$. Therefore $K_ja = K_i$ by minimality. The proof (viii) $\Rightarrow$ (v) of Theorem 3.7 shows that $K_ja \cong K_j$, whence $K_i \cong K_j$. ■

EXERCISES

1. A ring R is isomorphic to a subdirect product of the family of rings $\{R_i \mid i \varepsilon I\}$ if and only if there exists for each $i \varepsilon I$ an ideal K_i of R such that $R/K_i \cong R_i$ and $\bigcap_{i \varepsilon I} K_i = 0$.

2. A ring R is **subdirectly irreducible** if the intersection of all nonzero ideals of R is nonzero.

   (a) R is subdirectly irreducible if and only if whenever R is isomorphic to a subdirect product of $\{R_i | i \varepsilon I\}$, $R \cong R_i$ for some $i \varepsilon I$ [see Exercise 1].

   (b) (Birkhoff) Every ring is isomorphic to a subdirect product of a family of subdirectly irreducible rings.

   (c) The zero divisors in a commutative subdirectly irreducible ring (together with 0) form an ideal.

3. A commutative semisimple left Artinian ring is a direct product of fields.

4. Determine up to isomorphism all semisimple rings of order 1008. How many of them are commutative? [*Hint:* Exercise V.8.10.]

5. An element a of a ring R is **regular** (in the sense of Von Neumann) if there exists $x \varepsilon R$ such that $axa = a$. If every element of R is regular, then R is said to be a **regular ring.**

(a) Every division ring is regular.

(b) A finite direct product of regular rings is regular.

(c) Every regular ring is semisimple. [The converse is false (for example, **Z**).]

(d) The ring of all linear transformations on a vector space (not necessarily finite dimensional) over a division ring is regular.

(e) A semisimple left Artinian ring is regular.

(f) R is regular if and only if every principal left [resp. right] ideal of R is generated by an idempotent element.

(g) A nonzero regular ring R with identity is a division ring if and only if its only idempotents are 0 and 1_R.

6. (a) Every nonzero homomorphic image and every nonzero submodule of a semisimple module is semisimple.
   (b) The intersection of two semisimple submodules is 0 or semisimple.

7. The following conditions on a semisimple module A are equivalent:
   (a) A is finitely generated.
   (b) A is a direct sum of a finite number of simple submodules.
   (c) A has a composition series (see p. 375).
   (d) A satisfies both the ascending and descending chain conditions on submodules (see Theorem VIII.1.11).

8. Let A be a module over a left Artinian ring R such that $Ra \neq 0$ for all nonzero $a \in A$ and let $J = J(R)$. Then $JA = 0$ if and only if A is semisimple. [*Hints:* if $JA = 0$, then A is an R/J-module, with R/J semisimple left Artinian; see Exercise IV.1.17.]

9. Let R be a ring that (as a left R-module) is the sum of its minimal left ideals. Assume that $\{r \in R \mid Rr = 0\} = 0$. If A is an R-module such that $RA = A$, then A is semisimple. [*Hint:* if I is a minimal left ideal and $a \in A$, show that Ia is either zero or a simple submodule of A.]

10. Show that a nonzero R-module A such that $RA = 0$ is not semisimple, but may be projective. Consequently Theorem 3.7 may be false if the word "unitary" is omitted. [See Exercise IV.2.2, Theorem IV.3.2 and Proposition IV.3.5.]

11. Let R be the ring of 2×2 matrices over an infinite field.
    (a) R has an infinite number of distinct proper left ideals, any two of which are isomorphic as left R-modules.
    (b) There are infinitely many distinct pairs (B,C) such that B and C are minimal left ideals of R and $R = B \oplus C$.

12. A left Artinian ring R has the same number of nonisomorphic simple left R-modules as nonisomorphic simple right R-modules. [*Hint:* Show that A is a simple R-module if and only if A is a simple $R/J(R)$-module; use Theorem 2.14 and Theorem 3.10.]

13. (a) (Hopkins) If R is a left Artinian ring with identity, then R is left Noetherian. [*Hints:* Let n be the least positive integer such that $J^n = 0$ (Proposition 2.13). Let $J^0 = R$. Since $J(J^i/J^{i+1}) = 0$ and R is left Artinian each J^i/J^{i+1} $(0 \leq i \leq n-1)$ has a composition series by Exercises 7 and 8. Use these and Theorem IV.1.10 to construct a composition series for R; apply Theorem VIII.1.11.]

Remark. Hopkins' Theorem is valid even if the hypothesis "R has an identity" is replaced by the much weaker hypothesis that $\{r \in R \mid rR = 0 \text{ and } Rr = 0\} = 0$; see L. Fuchs [13; pp. 283–286].

(b) The converse of Hopkins' Theorem is false.

4. THE PRIME RADICAL; PRIME AND SEMIPRIME RINGS

We now introduce the prime radical of a ring and call a ring semiprime if it has zero prime radical (Definition 4.1). We then develop the analogues of the results proved in Sections 2 and 3 for the Jacobson radical and semisimple rings (Propositions 4.2–4.4). There is a strong analogy between the prime radical, prime ideals, semiprime rings, prime rings, and the Jacobson radical, left primitive ideals, semisimple rings, and primitive rings respectively.

The remainder of the section is devoted to a discussion of Goldie's Theorem 4.8, which is a structure theorem for semiprime rings satisfying the ascending chain condition on certain types of left ideals. Goldie's Theorem plays the same role here as do the Wedderburn-Artin Theorems 1.14 and 3.3 for rings with the descending chain condition on left ideals. In fact Goldie's Theorem may be considered as an extension of the Wedderburn-Artin Theorems to a wider class of rings. A fuller explanation of these statements is contained in discussions after Proposition 4.4, preceding Theorem 4.8 and after Corollary 4.9.

This section is not needed in the sequel.

Definition 4.1. *The* **prime radical** $\mathrm{P(R)}$ *of a ring* R *is the intersection of all prime ideals of* R. *If* R *has no prime ideals, then* $\mathrm{P(R) = R}$. *A ring* R *such that* $\mathrm{P(R)} = 0$ *is said to be* **semiprime.**

REMARKS. The prime radical (also called the **Baer lower radical** or the **McCoy radical**) is the radical with respect to a certain radical property, as defined in the introduction to Section 2; for details, see Exercises 1 and 2. A semiprime ring is one that is semisimple with respect to the prime radical (see the introduction to Section 2). We use the term "semiprime" to avoid both awkward phrasing and confusion with Jacobson semisimplicity. The relationship of the prime radical with the Jacobson radical is discussed in Exercise 3.

Just as in the case of the Jacobson radical, there is a close connection between the prime radical of a ring R and the nilpotent ideals of R. In order to prove one such result, we must recall some terminology.

Let S be a subset of a ring R. By Theorem 1.4 the set $\{r \in R \mid rS = 0\}$ is a left ideal of R, which is actually an ideal if S is a left ideal. The set $\{r \in R \mid rS = 0\}$ is called the **left annihilator** of S and is denoted $\mathcal{A}(S)$. Similarly the set

$$\mathcal{A}_r(S) = \{r \in R \mid Sr = 0\}$$

is a right ideal of R that is an ideal if S is a right ideal. $\mathcal{A}_r(S)$ is called the **right annihilator** of S. A left [resp. right] ideal I of R is said to be a left [resp. right] annihilator if $I = \mathcal{A}(S)$ [resp. $I = \mathcal{A}_r(S)$] for some subset S of R.

REMARK. The intersection of two left [resp. right] annihilators is also a left [resp. right] annihilator since $\mathfrak{A}(S) \cap \mathfrak{A}(T) = \mathfrak{A}(S \cup T)$. If S and T are actually left ideals, then $\mathfrak{A}(S) \cap \mathfrak{A}(T) = \mathfrak{A}(S \cup T) = \mathfrak{A}(S + T)$.

Proposition 4.2. *A ring* R *is semiprime if and only if* R *has no nonzero nilpotent ideals.*

SKETCH OF PROOF. ($\Rightarrow$) If I is a nilpotent ideal and K is *any* prime ideal, then for some n, $I^n = 0 \in K$, whence $I \subset K$. Therefore $I \subset P(R)$. Consequently, if R is semiprime, so that $P(R) = 0$, then the only nilpotent ideal is the zero ideal.

($\Leftarrow$) Conversely suppose that R has no nonzero nilpotent ideals. We must show that $P(R) = 0$. It suffices to prove that for every nonzero element a of R there is a prime ideal K such that $a \notin K$, whence $a \notin P(R)$. We first observe that $\mathfrak{A}(R) \cap R$ is a nilpotent ideal of R since

$$(\mathfrak{A}(R) \cap R)(\mathfrak{A}(R) \cap R) \subset \mathfrak{A}(R)R = 0.$$

Consequently, $\mathfrak{A}(R) = \mathfrak{A}(R) \cap R = 0$. Similarly $\mathfrak{A}_r(R) = 0$. If b is any nonzero element of R, we claim that $RbR \neq 0$. Otherwise $Rb \subset \mathfrak{A}(R) = 0$, whence $Rb = 0$. Thus $b \in \mathfrak{A}_r(R) = 0$, which is a contradiction. Therefore RbR is a nonzero ideal of R and hence not nilpotent. Consequently $bRb \neq 0$ (otherwise $(RbR)^2 \subset RbRbR = 0$). For each nonzero $b \in R$ choose $f(b) \in bRb$ such that $f(b) \neq 0$. Then by the Recursion Theorem 6.2 of the Introduction there is a function $\varphi : \mathbf{N} \to R$ such that

$$\varphi(0) = a \quad \text{and} \quad \varphi(n + 1) = f(\varphi(n)).$$

Let $a_n = \varphi(n)$ so that $a_{n+1} = f(a_n) \neq 0$. Let $S = \{a_i \mid i \geq 0\}$. Use Zorn's Lemma to find an ideal K that is maximal with respect to the property $K \cap S = \emptyset$ (since $0 \notin S$ there is at least one ideal disjoint from S).

Since $a = a_0 \in S$, $a \notin K$ and $K \neq R$. To complete the proof we need only show that K is prime. If A and B are ideals of R such that $A \not\subset K$ and $B \not\subset K$, then $(A + K) \cap S \neq \emptyset$ and $(B + K) \cap S \neq \emptyset$ by maximality. Consequently for some i,j, $a_i \in A + K$ and $a_j \in B + K$. Choose $m > \max\{i,j\}$. Since $a_{n+1} = f(a_n) \in a_nRa_n$ for each n, it follows that $a_m \in (a_iRa_i) \cap (a_jRa_j) \subset (A + K) \cap (B + K)$. Consequently,

$$a_{m+1} = f(a_m) \in a_mRa_m \subset (A + K)(B + K) \subset AB + K.$$

Since $a_{m+1} \notin K$, we must have $AB \not\subset K$. Therefore K is a prime ideal. ∎

A ring R is said to be a **prime ring** if the zero ideal is a prime ideal (that is, if I, J are ideals such that $IJ = 0$, then $I = 0$ or $J = 0$). The relationships among prime ideals, prime rings, and semiprime rings are analogous to the relationships between left primitive ideals, primitive rings, and semisimple rings. In particular, we note the following:

(i) The prime [resp. Jacobson] radical is the intersection of all prime [resp. primitive] ideals (see Theorem 2.3(iii)).

(ii) Every prime ring is semiprime since 0 is a prime ideal. This corresponds to the fact that every primitive ring is semisimple (Theorem 2.10(i)).

Proposition 4.3. *K is a prime ideal of a ring R if and only if R/K is a prime ring.*

REMARK. This is the analogue of Definition 2.1 (left primitive ideals).

SKETCH OF PROOF OF 4.3. If R/K is prime, let $\pi : R \to R/K$ be the canonical epimorphism. If I and J are ideals of R such that $IJ \subset K$, then $\pi(I)$, $\pi(J)$ are ideals of R/K (Exercise III.2.13(b)) such that $\pi(I)\pi(J) = \pi(IJ) = 0$. Since R/K is prime, either $\pi(I) = 0$ or $\pi(J) = 0$; that is, $I \subset K$ or $J \subset K$. Therefore, K is a prime ideal (Definition III.2.14). The converse is an easy consequence of Theorem III.2.13 and Definition III.2.14. ■

The final part of the semiprime-semisimple analogy is given by

Proposition 4.4. *A ring R is semiprime if and only if R is isomorphic to a subdirect product of prime rings.*

SKETCH OF PROOF. Proposition 4.4 is simply Proposition 3.2 with the words "semisimple" and "primitive" changed to "semiprime" and "prime" respectively. With this change and the use of Proposition 4.3 in place of Definition 2.1, the proof of Proposition 3.2 carries over verbatim to the present case. ■

We have seen that primitive rings are the basic building blocks for semisimple rings. Proposition 4.4 shows that the basic building blocks for semiprime rings are the prime rings. At this point the analogy between primitive and prime rings fails. Primitive rings may be characterized in terms of familiar matrix rings and endomorphism rings of vector spaces (Section 1). There are no comparable results for prime rings. But the situation is not completely hopeless. We have obtained very striking results for primitive and semisimple left Artinian rings (Sections 1 and 3). Consequently it seems plausible that one could obtain useful characterizations of prime and semiprime rings that satisfy certain chain conditions. We shall now do precisely that.

We first observe that in a left Artinian ring the prime radical coincides with the Jacobson radical (Exercise 3(c)). Consequently, left Artinian semiprime rings are also semisimple, whence their structure is determined by the Wedderburn-Artin Theorem 3.3. Since every semiprime (semisimple) left Artinian ring is also left Noetherian by Corollary 3.4, the next obvious candidate to consider is the class of semiprime left Noetherian rings (that is, semiprime rings that satisfy the ascending chain condition on left ideals). Note that there are semiprime left Noetherian rings that are not left Artinian (for example, **Z**). Consequently, a characterization of semiprime left Noetherian rings would be a genuine extension of our previous results.

We shall actually characterize a wider class of rings that properly includes the class of all semiprime left Noetherian rings. The class in question is the class of all semiprime left Goldie rings, which we now define.

A family of left ideals of R $\{I_j \mid j \in J\}$ is said to be **independent** provided that for each $k \in J$, $I_k \cap I_k^* = 0$, where I_k^* is the left ideal generated by $\{I_j \mid j \neq k\}$. In other words, $\{I_j \mid j \in J\}$ is independent if and only if the left ideal I generated by $\{I_j \mid j \in J\}$ is actually the internal direct sum $I = \sum_{j \in J} I_j$ (see Theorem IV.1.15).

Definition 4.5. *A ring* R *is said to be a* **(left) Goldie ring** *if*

(i) R *satisfies the ascending chain condition on left annihilators;*
(ii) *every independent set of left ideals of* R *is finite.*

REMARKS. (i) Condition (i) of Definition 4.5 means that given any chain of left annihilators $\mathcal{A}(S_1) \subset \mathcal{A}(S_2) \subset \cdots$, there exists an n such that $\mathcal{A}(S_i) = \mathcal{A}(S_n)$ for all $i \geq n$. This condition is equivalent to the condition

(i′) R *satisfies the maximum condition on left annihilators (that is, every nonempty set of left annihilators contains a maximal element with respect to set theoretic inclusion).*

To see this one need only observe that the proof of Theorem VIII.1.4 carries over to the present situation, *mutatis mutandis.*

(ii) Right Goldie rings are defined in the obvious way. A right Goldie ring need not be a left Goldie ring; see A. W. Goldie [62].

EXAMPLE. Every left Noetherian ring R is a left Goldie ring. Condition (i) is obviously satisfied. If $\{I_j \mid j \in J\}$ were an infinite independent set of left ideals, then there would exist $I_1, I_2, \ldots$ such that $I_1 \subsetneqq I_1 \times I_2 \subsetneqq I_1 \times I_2 \times I_3 \subsetneqq \cdots$, which contradicts the ascending chain condition. Therefore (ii) is satisfied and R is a Goldie ring. There do exist left Goldie rings that are *not* left Noetherian rings.

The preceding example shows that the class of semiprime left Goldie rings contains the class of semiprime left Noetherian rings. Our characterization of semiprime left Goldie rings will be given in terms of their left quotient rings, in the sense of the following definitions.

Definition 4.6. *A nonzero element* a *in a ring* R *is said to be* **regular** *if* a *is neither a left nor right zero divisor.*

Definition 4.7. *A ring* Q(R) *with identity is said to be a* **left quotient ring** *of a ring* R *if*

(i) $R \subset Q(R)$;
(ii) *every regular element in* R *is a unit in* Q(R);
(iii) *every element* c *of* Q(R) *is of the form* $c = a^{-1}b$, *where* $a,b \in R$ *and* a *is regular.*

REMARKS. (i) A ring R need not have a left quotient ring. If it does, however, it is easy to see that $Q(R)$ is determined up to isomorphism by Definition 4.7.

(ii) A right quotient ring of R is defined in the same way, except that "$c = a^{-1}b$" is replaced by "$c = ba^{-1}$" in condition (iii). A ring may have a right quotient ring, but no left quotient ring (see N. J. Divinsky [22; p. 71]).

(iii) If R is a ring that has a left quotient ring $Q(R) = T$, then R is said to be a **left order** in T.

EXAMPLE. Let R be a commutative ring that has at least one regular element. Let S be the set of all regular elements of R. Then the complete ring of quotients $S^{-1}R$

is a ring with identity (Theorem III.4.3) that contains an isomorphic copy $\varphi_S(R)$ of R (Theorem III.4.4(ii)). If we identify R and $\varphi_S(R)$ as usual, then $R \subset S^{-1}R$, every regular element of R is a unit in $S^{-1}R$ (Theorem III.4.4(i)) and every element of $S^{-1}R$ is of the form $s^{-1}r$ ($r \in R$, $s \in S \subset R$). Therefore $S^{-1}R$ is a left quotient ring of R. Special case: the rational field **Q** is a left quotient ring of the left Noetherian ring **Z**.

EXAMPLE. Every semisimple left Artinian ring is its own left quotient ring (Exercise 6).

It is clear from Definition 4.7 that the structure of a left quotient ring $Q(R)$ is intimately connected with the structure of the ring R. Consequently, if one cannot explicitly describe the ring R in terms of well-known rings, the next best thing is to show that R has a left quotient ring that can be explicitly described in such terms. This is precisely what Goldie's Theorem does.

Theorem 4.8. *(Goldie)* R *is a semiprime [resp. prime] left Goldie ring if and only if* R *has a left quotient ring* Q(R) *which is semisimple [resp. simple] left Artinian.*

Theorem 4.8 will not be proved here for reasons of space. One of the best proofs is due to C. Procesi and L. Small [65]; a slightly expanded version appears in I. Herstein [24]. Although long, this proof is no more difficult than many proofs presented earlier in this chapter. It does use Ore's Theorem, a proof of which is sketched in I. N. Herstein [24; p. 170] and given in detail in N. J. Divinsky [22; p. 66].

Since the structure of semisimple left Artinian rings has been completely determined, Theorem 4.8 gives as good a description as we are likely to get of semiprime left Goldie rings (special case: semiprime left Noetherian rings). The "distance" between the rings R and $Q(R)$ is the price that must be paid for replacing the descending chain condition with the ascending chain condition. For as we observed in the discussion after Proposition 4.4 and in Exercise 3.13, the latter is a considerably weaker condition than the former.

Corollary 4.9. R *is a semiprime [resp. prime] left Goldie ring if and only if* R *has a quotient ring* Q(R) *such that* $Q(R) \cong Mat_{n_1}D_1 \times \cdots \times Mat_{n_k}D_k$, *[resp.* $Q(R) \cong Mat_{n_1}D_1$*], where* $n_1, \ldots, n_k$ *are positive integers and* $D_1, \ldots, D_n$ *are division rings.*

PROOF. Theorems 1.14, 3.3, and 4.8. ■

Goldie's Theorem, as rephrased in Corollary 4.9, may be thought of as an extension of the Wedderburn-Artin Theorems 1.14 and 3.3 to a wider class of rings. For instance, Theorem 3.3 states that a semisimple left Artinian ring is a direct product of matrix rings over division rings. Goldie's Theorem states that every semiprime left Goldie ring has a quotient ring that is a direct product of matrix rings over division rings. But every semisimple left Artinian ring is a semiprime left Goldie ring (Corollary 3.4, Exercise 3(a), and the Example after Definition 4.5). Furthermore every semisimple left Artinian ring is its own quotient ring (Exercise 6). Thus Goldie's Theorem reduces to the Wedderburn-Artin Theorem in this case. An analogous argument holds for simple left Artinian rings and Theorem 1.14.

EXERCISES

Note: R is always a ring.

1. A subset T of R is said to be an **m-system** (generalized multiplicative system) if

$$c,d \in T \quad \Rightarrow \quad cxd \in T \quad \text{for some } x \in R.$$

(a) P is a prime ideal of R if and only if $R - P$ is an m-system. [*Hint:* Exercise III.2.14.]
(b) Let I be an ideal of R that is disjoint from an m-system T. Show that I is contained in an ideal Q which is maximal respect to the property that $Q \cap T = \emptyset$. Then show that Q is a prime ideal. [*Hint:* Adapt the proof of Theorem VIII.2.2.]
(c) An element r of R is said to have the *zero property* if every m-system that contains r also contains 0. Show that the prime radical $P(R)$ is the set M of all elements of R that have the zero property. [*Hint:* use (a) to show $M \subset P(R)$ and (b) to show $P(R) \subset M$.]
(d) Every element c of $P(R)$ is nilpotent. [*Hint:* $\{c^i \mid i \geq 1\}$ is an m-system.] If R is commutative, $P(R)$ consists of all nilpotent elements of R.

2. (a) If I is an ideal of R, then $P(I) = I \cap P(R)$. In particular, $P(P(R)) = P(R)$. [*Hint:* Exercise 1(c).]
(b) $P(R)$ is the smallest ideal K of R such that $P(R/K) = 0$. In particular, $P(R/P(R)) = 0$, whence $R/P(R)$ is semiprime. [*Hint:* Exercise III.2.17(d).]
(c) An ideal I is said to have the *zero property* if every element of I has the zero property (Exercise 1(c)). Show that the zero property is a radical property (as defined in the introduction to Section 2), whose radical is precisely $P(R)$.

3. (a) Every semisimple ring is semiprime.
(b) $P(R) \subset J(R)$. [*Hint:* Exercise 1(d); or (a) and Exercise 2(b).]
(c) If R is left Artinian, $P(R) = J(R)$. [*Hint:* Proposition 2.13.]

4. R is semiprime if and only if for all ideals A, B

$$AB = 0 \quad \Rightarrow \quad A \cap B = 0.$$

5. (a) Let R be a ring with identity. The matrix ring $\text{Mat}_n R$ is prime if and only if R is prime.
(b) If R is any ring, then $P(\text{Mat}_n R) = \text{Mat}_n P(R)$. [*Hint:* Use Exercise 2 and part (a) if R has an identity. In the general case, embed R in a ring S with identity via Theorem III.1.10; then $P(R) = R \cap P(S)$ by Exercise 2.]

6. If R is semisimple left Artinian, then R is its own quotient ring. [*Hint:* Since R has an identity by Theorem 3.3, it suffices to show that every regular element of R is actually a unit. By Theorem 3.3 and a direct argument it suffices to assume $R = \text{Mat}_n D$ for some division ring D. Theorem VII.2.6 and Proposition VII.2.12 may be helpful.]

7. The following are equivalent:
(a) R is prime;
(b) $a,b \in R$ and $aRb = 0$ imply $a = 0$ or $b = 0$;
(c) the right annihilator of every nonzero right ideal of R is 0;
(d) the left annihilator of every nonzero left ideal of R is 0.

8. Every primitive ring is prime [see Exercise 7].

9. The center of a prime ring with identity is an integral domain. [See Exercise 7; for the converse see Exercise 10.]

10. Let J be an integral domain and let F be the complete field of quotients of J. Let R be the set of all infinite matrices (row, columns indexed by $\mathbf{N}^*$) of the form

$$\begin{pmatrix} A_n & & & & \\ & d & & 0 & \\ & & d & & \\ & & & d & \\ & 0 & & & \ddots \end{pmatrix}$$

where $A_n \in \mathrm{Mat}_n(F)$ and $d \in J \subset F$.

(a) R is a ring.

(b) The center of R is the set of all matrices of the form

$$\begin{pmatrix} d & & & \\ & d & & 0 \\ & & d & \\ & 0 & & \ddots \end{pmatrix}$$

with $d \in J$ and hence is isomorphic to J.

(c) R is primitive (and hence prime by Exercise 8).

11. The **nil radical** $N(R)$ of R is the ideal generated by the set of all nil ideals of R.

(a) $N(R)$ is a nil ideal.

(b) $N(N(R)) = N(R)$.

(c) $N(R/N(R)) = 0$.

(d) $P(R) \subset N(R) \subset J(R)$.

(e) If R is left Artinian, $P(R) = N(R) = J(R)$.

(f) If R is commutative $P(R) = N(R)$.

5. ALGEBRAS

The concepts and results of Sections 1–3 are carried over to algebras over a commutative ring K with identity. In particular, the Wedderburn-Artin Theorem is proved for K-algebras (Theorem 5.4). The latter part of the section deals with algebras over a field, including algebraic algebras and the group algebra of a finite group. Throughout this section *K is always a commutative ring with identity.*

The first step in carrying over the results of Sections 1–3 to K-algebras is to review the definitions of a K-algebra, a homomorphism of K-algebras, a subalgebra and an algebra ideal (Section IV.7). We recall that *if a K-algebra A has an identity, then (left, right, two-sided) algebra ideals coincide with (left, right, two-sided) ideals of the ring A* (see the Remarks after Definition IV.7.3). This fact will be used frequently without explicit mention.

A **left Artinian K-algebra** is a K-algebra that satisfies the descending chain condition on left algebra ideals. A left Artinian K-algebra may not be a left Artinian ring (Exercise 1).

EXAMPLE. If D is a division algebra over K, then $\text{Mat}_n D$ is a K-algebra (p. 227) which is left Artinian by Corollary VIII.1.12.

Definition 5.1. *Let* A *be an algebra over a commutatuve ring* K *with identity.*

(i) *A* **left (algebra) A-module** *is a unitary left* K*-module* M *such that* M *is a left module over the ring* A *and* $k(rc) = (kr)c = r(kc)$ *for all* $k \varepsilon K, r \varepsilon A, c \varepsilon M$.

(ii) *An* **A-submodule** *of an* A*-module* M *is a subset of* M *which is itself an algebra* A*-module (under the operations in* M*).*

(iii) *An algebra* A*-module* M *is* **simple** *(or* **irreducible***) if* $AM \neq 0$ *and* M *has no proper* A*-submodules.*

(iv) *A* **homomorphism** $f : M \to N$ *of algebra* A*-modules is a map that is both a* K*-module and an* A*-module homomorphism.*

REMARKS. If A is a K-algebra the term "A-module" will always indicate an algebra A-module. Modules over the *ring A* will be so labeled. A right A-module N is defined analogously and satisfies $k(cr) = (kc)r = c(kr)$ for all $k \varepsilon K, r \varepsilon A, c \varepsilon N$.

Simple K-algebras, primitive K-algebras, the Jacobson radical of a K-algebra, semisimple K-algebras, etc. are now defined in the same way the corresponding concepts for rings were defined, with algebra ideals, modules, homomorphisms, etc. in place of ring ideals, modules, and homomorphisms. In order to carry over the results of Sections 1–3 to K-algebras (in particular, the Wedderburn-Artin Theorems) the following two theorems are helpful.

Theorem 5.2. *Let* A *be a* K*-algebra.*

(i) *A subset* I *of* A *is a regular maximal left algebra ideal if and only if* I *is a regular maximal left ideal of the ring* A.

(ii) *The Jacobson radical of the ring* A *coincides with the Jacobson radical of the algebra* A. *In particular* A *is a semisimple ring if and only if* A *is a semisimple algebra.*

REMARK. Theorem 5.2 is trivial if A has an identity since algebra ideals and ring ideals coincide in this case.

PROOF OF 5.2. (i) If I is a regular maximal left ideal of the ring A, it suffices to show that $kI \subset I$ for all $k \varepsilon K$. Suppose $kI \not\subset I$ for some $k \varepsilon K$. Since $r(kI) = k(rI)$ by Definition 5.1(i), $I + kI$ is a left ideal of A that properly contains I. Therefore, $A = I + kI$ by maximality. By hypothesis there exists $e \varepsilon A$ such that $r - re \varepsilon I$ for all $r \varepsilon A$. Let $e = a + kb$ $(a,b \varepsilon I)$. Then

$$e^2 = e(a + kb) = ea + e(kb) = ea + (ke)b \varepsilon I.$$

Since $e - e^2 \varepsilon I$ and $e^2 \varepsilon I$, we must have $e \varepsilon I$. Consequently, the fact that $r - re \varepsilon I$ for all $r \varepsilon A$ implies $A = I$. This contradicts the maximality of I. Therefore, $kI \subset I$ for all $k \varepsilon K$.

Conversely let I be a regular maximal left algebra ideal and hence a regular left ideal of the ring A. By Lemma 2.4 I is contained in a regular maximal left ideal I_1 of the ring A. The previous paragraph shows that I_1 is actually a regular left algebra ideal, whence $I = I_1$ by maximality.

(ii) follows from (i) and Theorem 2.3(ii). ■

Theorem 5.3. *Let* A *be a* K*-algebra. Every simple algebra* A*-module is a simple module over the ring* A. *Every simple module* M *over the ring* A *can be given a unique* K*-module structure in such a way that* M *is a simple algebra* A*-module.*

PROOF. Let N be a simple algebra A-module, whence $AN \neq 0$. If N_1 is a submodule of N, then AN_1 is an algebra submodule of N, whence $AN_1 = N$ or $AN_1 = 0$. If $AN_1 = N$, then $N_1 = N$. If $AN_1 = 0$, then $N_1 \subset D = \{c \in N \mid Ac = 0\}$. But D is an algebra submodule of N and $D \neq N$ since $AN \neq 0$. Therefore $D = 0$ by simplicity, whence $N_1 = 0$. Consequently, N has no proper submodules and hence is a simple module over the ring A.

If M is a simple module over the ring A, then M is cyclic, say $M = Ac$ $(c \in M)$, by Remark (iii) after Definition 1.1. Define a K-module structure on $M = Ac$ by

$$k(rc) = (kr)c, \quad (k \in K, r \in A).$$

Since $kr \in A$, $(kr)c$ is an element of $Ac = M$. In order to show that the action of K on M is well defined we must show that

$$rc = r_1c \quad \Rightarrow \quad (kr)c = (kr_1)c, \quad (k \in K; r,r_1 \in A).$$

Clearly it will suffice to prove

$$rc = 0 \quad \Rightarrow \quad (kr)c = 0, \quad (k \in K, r \in A).$$

Now by the proof of Theorem 1.3, $M \cong A/I$ where the regular maximal left ideal I is the kernel of the map $A \to Ac = M$ given by $x \mapsto xc$. Consequently, $rc = 0$ implies $r \in I$. But I is an algebra ideal by Theorem 5.4, whence $kr \in I$. Therefore $(kr)c = 0$ and the action of K on M is well defined. It is now easy to verify that M is a K-module and an algebra A-module. The K-module structure of M is uniquely determined since any K-module structure on M that makes $M = Ac$ an A-module necessarily satisfies $k(rc) = (kr)c$ for all $k \in K$, $r \in A$. ■

Theorem 5.4. A *is a semisimple left Artinian* K*-algebra if and only if there is an isomorphism of* K*-algebras*

$$A \cong Mat_{n_1}D_1 \times Mat_{n_2}D_2 \times \cdots \times Mat_{n_t}D_t,$$

where each n_i *is a positive integer and each* D_i *a division algebra over* K.

REMARK. Theorem 5.4 is valid for any semisimple finite dimensional algebra A over a field K since any such A is left Artinian (Exercise 2).

SKETCH OF PROOF OF 5.4. Use Theorems 5.2 and 5.3 and Exercises 3 and 4 to carry over the proof of the Wedderburn-Artin Theorem 3.3 to K-algebras. ■

The remainder of this section deals with selected topics involving algebras over a field. We first obtain a sharper version of Theorem 5.4 in case K is an algebraically closed field and finally we consider group algebras over a field.

If A is a nonzero algebra with identity over a field K, then the map $\alpha : K \to A$, de-defined by $k \mapsto k1_A$, is easily seen to be a homomorphism of K-algebras. Since $\alpha(1_K) = 1_A \neq 0$, ker $\alpha \neq K$. But the field K has no proper ideals, whence Ker $\alpha = 0$. Thus α is a monomorphism. Furthermore the image of α lies in the center of A since for all $k \in K$, $r \in A$:

$$\alpha(k)r = (k1_A)r = k(1_A r)1_A = (1_A r)(k1_A) = r\alpha(k).$$

Consequently we adopt the following convention:

If A *is a nonzero algebra with identity over a field* K, *then* K *is to be identified with Im* α *and considered to be a subalgebra of the center of* A.

Under this identification the K-module action of K on A coincides with multiplication by elements of the subalgebra K in A since $ka = (k1_A)a = \alpha(k)a$.

Definition 5.5. *An element* a *of an algebra* A *over a field* K *is said to be* **algebraic** *over* K *if* a *is the root of some polynomial in* K[x]. A *is said to be an* **algebraic algebra** *over* K *if every element of* A *is algebraic over* K.

EXAMPLE. If A is finite dimensional then A is an algebraic algebra. For if $\dim_K A = n$ and $a \in A$, then the $n + 1$ elements $a, a^2, a^3, \ldots, a^{n+1}$ must be linearly dependent. Thus $k_1 a + k_2 a^2 + \cdots + k_{n+1}a^{n+1} = 0$ for some $k_i \in K$, not all zero. Thus $f(a) = 0$ where f is the nonzero polynomial $k_1 x + k_2 x^2 + \cdots + k_{n+1}x^{n+1} \in K[x]$.

EXAMPLE. The algebra of countably infinite matrices over a field K with only a finite number of nonzero entries is an infinite dimensional simple algebraic algebra (Exercise 5).

REMARK. The radical of an algebraic algebra is nil (Exercise 6).

Lemma 5.6. *If* D *is an algebraic division algebra over an algebraically closed field* K, *then* D = K.

PROOF. K is contained in the center of D by the convention adopted above. If $a \in D$, then $f(a) = 0$ for some $f \in K[x]$. Since K is algebraically closed $f(x) = k(x - k_1)(x - k_2)\cdots(x - k_n)$ $(k, k_i \in K;\ k \neq 0)$, whence

$$0 = f(a) = k(a - k_1)(a - k_2)\cdots(a - k_n).$$

Since D is a division ring, $a - k_i = 0$, for some i. Therefore $a = k_i \in K$ and thus $D \subset K$. ■

Theorem 5.7. *Let* A *be a finite dimensional semisimple algebra over an algebraically closed field* K. *Then there are positive integers* $n_1, \ldots, n_t$ *and an isomorphism of* K-*algebras*

$$A \cong Mat_{n_1}K \times \cdots \times Mat_{n_t}K.$$

PROOF. By Theorem 5.4 (and the subsequent Remark) $A \cong \text{Mat}_{n_1}D_1 \times \text{Mat}_{n_2}D_2 \times \cdots \times \text{Mat}_{n_t}D_t$ where each D_i is a division algebra over K. Each D_i is necessarily finite dimensional over K; (otherwise $\text{Mat}_{n_i}D_i$ and hence A would be infinite dimensional). Therefore $D_i = K$ for every i by Lemma 5.6. ■

A great deal of research over the years has been devoted to group algebras over a field (see p. 227). They are useful, among other reasons, because they make it possible to exploit ring-theoretic techniques in the study of groups.

Proposition 5.8. *(Maschke) Let* K(G) *be the group algebra of a finite group* G *over a field* K. *If* K *has characteristic* 0, *then* K(G) *is semisimple. If* K *has prime characteristic* p, *then* K(G) *is semisimple if and only if* p *does not divide* $|G|$.

SKETCH OF PROOF. Suppose char $K = 0$ or p, where $p \nmid |G|$. If B is any K-algebra with identity (in particular $K(G)$), verify that there is a well-defined monomorphism of K-algebras $\alpha : B \to \text{Hom}_K(B,B)$ given as follows: $\alpha(b)$ is defined to be the map $\alpha_b : B \to B$, where $\alpha_b(x) = bx$.

If $g \in G$, we denote the element 1_Kg of $K(G)$ simply by g. By definition $K(G)$ is a K-vector space with basis $X = \{g \mid g \in G\}$ and finite dimension $n = |G|$. For each $u \in K(G)$ let M_u be the matrix of α_u relative to the basis X. Let $g \in G$ with $g \neq e$. Then for all $g_1 \in G$, $\alpha_g(g_1) = gg_1 \neq g_1$ (since G is a group). Thus α_g simply permutes the elements of the basis X and leaves no basis element fixed. Consequently, the matrix M_g of α_g relative to the basis X may be obtained from the identity matrix I_n by an appropriate permutation of the rows that leaves no row fixed (see Theorem VII.1.2). Recall that the trace, Tr M_u, is the sum of the main diagonal entries of M_u (see p. 369). It is easy to see that

(i) $\text{Tr } M_g = 0$ for $g \in G$, $g \neq e$;
(ii) $M_e = I_n$, whence $\text{Tr } M_e = n1_K$;
(iii) if $u = k_1g_1 + \cdots + k_ng_n \in K(G)$, then

$$\alpha_u = \sum_{i=1}^{n} k_i\alpha_{g_i} \quad \text{and} \quad \text{Tr } M_u = \sum_{i=1}^{n} k_i \text{ Tr } M_{g_i}.$$

If the radical J of $K(G)$ is nonzero, then there is a nonzero element $v \in J$ with $v = k_1g_1 + \cdots + k_ng_n$. We may assume $g_1 = e$ and $k_1 = 1_K$ (if not, replace v by $k_i^{-1}g_i^{-1}v$, where $k_i \neq 0$, and relabel). Since $K(G)$ is finite dimensional over K, $K(G)$ is left Artinian (Exercise 2). Consequently J is nilpotent by Proposition 2.13 (for algebras). Therefore $v \in J$ is nilpotent, whence α_v is nilpotent. Thus by Theorem VII.1.3 M_v is a nilpotent matrix. Therefore $\text{Tr } M_v = 0$ (Exercise VII.5.10). On the other hand (i)–(iii) above imply

$$\begin{aligned}\text{Tr } M_v &= \sum_{i=1}^{n} k_i \text{ Tr } M_{g_i} = 1_K \text{ Tr } M_e + \sum_{i=2}^{n} k_i \text{ Tr } M_{g_i}\\ &= \text{Tr } M_e + 0 = n1_K.\end{aligned}$$

But $n1_K \neq 0$ since char $K = 0$ or char $K = p$ and p does not divide $|G| = n$. This is a contradiction. Therefore $J = 0$ and $K(G)$ is semisimple.

Conversely suppose char $K = p$ and $p \mid n$. Let w be the sum in $K(G)$ of all the elements of the basis X; that is, $w = g_1 + g_2 + \cdots + g_n \in K(G)$. Clearly $w \neq 0$. Verify that $wg = gw$ for all $g \in G$, which implies that w is in the center of $K(G)$. Show that $w^2 = nw = (n1_K)w$, whence $w^2 = 0$ (since $p \mid n$). Thus $(K(G)w)(K(G)w) = 0$ so that the nonzero left ideal $K(G)w$ is nilpotent. Since $K(G)w \subset J$ by Theorem 2.12, $J \neq 0$. Therefore $K(G)$ is not semisimple. ■

The following corollary (with K the field of complex numbers) is quite useful in the study of representations and characters of finite groups.

Corollary 5.9. *Let* K(G) *be the group algebra of a finite group* G *over an algebraically closed field* K. *If char* K = 0 *or char* K = p *and* $p \nmid |G|$, *then there exist positive integers* $n_1, \ldots, n_t$ *and an isomorphism of* K*-algebras*

$$K(G) \cong Mat_{n_1}K \times \cdots \times Mat_{n_t}K.$$

PROOF. Since G is finite, $K(G)$ is a finite dimensional K-algebra and hence left Artinian (Exercise 2). Apply Theorem 5.7 and Proposition 5.8. ■

EXERCISES

Note: K is always a commutative ring with identity and A a K-algebra.

1. The **Q**-algebra A of Exercise IV.7.4 is a left Artinian **Q**-algebra that is *not* a left Artinian ring.

2. A finite dimensional algebra over a field K satisfies both the ascending and descending chain conditions on left and right algebra ideals.

3. (a) If M is a left algebra A-module, then $\mathcal{a}(M) = \{r \in A \mid rc = 0 \text{ for all } c \in M\}$ is an algebra ideal of A.
(b) An algebra ideal P of A is said to be **primitive** if the quotient algebra R/P is primitive (that is, has a faithful simple algebra R/P-module). Show that every primitive algebra ideal is a primitive ideal of the ring A and vice versa.

4. Let M be a simple algebra A-module.
(a) $D = \mathrm{Hom}_A(M,M)$ is a division algebra over K, where $\mathrm{Hom}_A(M,M)$ denotes all endomorphisms of the algebra A-module M.
(b) M is a left algebra D-module.
(c) The ring $\mathrm{Hom}_D(M,M)$ of all D-algebra endomorphisms of M is a K-algebra.
(d) The map $A \to \mathrm{Hom}_D(M,M)$ given by $r \mapsto \alpha_r$ (where $\alpha_r(x) = rx$) is a K-algebra homomorphism.

5. Let A be the set of all denumerably infinite matrices over a field K (that is, matrices with rows and columns indexed by $\mathbf{N}^*$) which have only a finite number of nonzero entries.
(a) A is a simple K-algebra.
(b) A is an infinite dimensional algebraic K-algebra.

6. The radical J of an algebraic algebra A over a field K is nil. [*Hint:* if $r \in J$ and $k_nr^n + k_{n-1}r^{n-1} + \cdots + k_tr^t = 0$ ($k_t \neq 0$), then $r^t = r^tu$ with $u = -k_t^{-1}k_nr^{n-t} - \cdots - k_t^{-1}k_{t+1}r$, whence $-u$ is right quasi-regular, say $-u + v - uv = 0$. Show that $0 = r^t(-u + v - uv) = -r^t$.]

7. Let A be a K-algebra and C the center of the ring A.

   (a) C is a K-subalgebra of A.

   (b) If K is an algebraically closed field and A is finite dimensional semisimple, then the number t of simple components of A (as in Theorem 5.7) is precisely $\dim_K C$.

6. DIVISION ALGEBRAS

We first consider certain simple algebras over a field and then turn to the special case of division algebras over a field. We show that the structure of a division algebra is greatly influenced by its maximal subfields. Finally the Noether-Skolem Theorem (6.7) is proved. It has as corollaries two famous theorems due to Frobenius and Wedderburn respectively (Corollaries 6.8 and 6.9). The tensor product of algebras (Section IV.7) is used extensively throughout this section.

Definition 6.1. *An algebra* A *with identity over a field* K *is said to be* **central simple** *if* A *is a simple* K*-algebra and the center of* A *is precisely* K.

EXAMPLE. Let D be a division ring and let K be the center of D. It is easy to verify that if d is a nonzero element of K, then $d^{-1} \varepsilon K$. Consequently K is a field. Clearly D is an algebra over K (with K acting by ordinary multiplication in D). Furthermore since D is a simple ring with identity, it is also simple as an algebra. Thus D is a central simple algebra over K.

Recall that if A and B are K-algebras with identities, then so is their tensor product $A \otimes_K B$ (Theorem IV.7.4). The product of $a \otimes b$ and $a_1 \otimes b_1$ is $aa_1 \otimes bb_1$. Here and below we shall denote the set $\{1_A \otimes b \mid b \varepsilon B\}$ by $1_A \otimes_K B$ and $\{a \otimes 1_B \mid a \varepsilon A\}$ by $A \otimes_K 1_B$. Note that $A \otimes_K B = (A \otimes_K 1_B)(1_A \otimes_K B)$; see p. 124.

Theorem 6.2. *If* A *is a central simple algebra over a field* K *and* B *is a simple* K*-algebra with identity, then* $A \otimes_K B$ *is a simple* K*-algebra.*

PROOF. Since B is a vector space over K, it has a basis Y and by Theorem IV.5.11 every element u of $A \otimes_K B$ can be written $\sum_{i=1}^{n} a_i \otimes y_i$, with $y_i \varepsilon Y$ and the a_i unique. If U is any nonzero ideal of $A \otimes_K B$, choose a nonzero $u \varepsilon U$ such that $u = \sum_{i=1}^{n} a_i \otimes y_i$, with all $a_i \neq 0$ and n minimal. Since A is simple with identity and Aa_1A is a nonzero ideal, $Aa_1A = A$. Consequently there are elements $r_1, \ldots, r_t, s_1, \ldots, s_t \varepsilon A$ such that $1_A = \sum_{j=1}^{t} r_j a_1 s_j$. Since U is an ideal, the element $v = \sum_{j=1}^{t} (r_j \otimes 1_B)u(s_j \otimes 1_B)$ is in U. Now

$$v = \sum_j (r_j \otimes 1_B)(\sum_i a_i \otimes y_i)(s_j \otimes 1_B) = \sum_i (\sum_j r_j a_i s_j) \otimes y_i$$

$$= \sum_j r_j a_1 s_j \otimes y_1 + \sum_{i=2}^{n} (\sum_j r_j a_i s_j) \otimes y_i = 1_A \otimes y_1 + \sum_{i=2}^{n} \bar{a}_i \otimes y_i,$$

where $\bar{a}_i = \sum_{j=1}^{t} r_j a_i s_j$. By the minimality of n, $\bar{a}_i \neq 0$ for all $i \geq 2$. If $a \varepsilon A$, then the element $w = (a \otimes 1_B)v - v(a \otimes 1_B)$ is in U and

$$w = \left(a \otimes y_1 + \sum_{i=2}^{n} a\bar{a}_i \otimes y_i\right) - \left(a \otimes y_1 + \sum_{i=2}^{n} \bar{a}_i a \otimes y_i\right)$$

$$= \sum_{i=2}^{n} (a\bar{a}_i - \bar{a}_i a) \otimes y_i.$$

By the minimality of n, $w = 0$ and $a\bar{a}_i - \bar{a}_i a = 0$ for all $i \geq 2$. Thus $a\bar{a}_i = \bar{a}_i a$ for all $a \varepsilon A$ and each $\bar{a}_i$ is in the center of A, which by assumption is precisely K. Therefore

$$v = 1_A \otimes y_1 + \sum_{i=2}^{n} \bar{a}_i \otimes y_i = 1_A \otimes y_1 + \sum_{i=2}^{n} 1_A \otimes \bar{a}_i y_i = 1_A \otimes b,$$

where $b = y_1 + \bar{a}_2 y_2 + \cdots + \bar{a}_n y_n \varepsilon B$. Since each $\bar{a}_i \neq 0$ and the y_i are linearly independent over K, $b \neq 0$. Thus, since B has an identity, the ideal BbB is precisely B by simplicity. Therefore,

$$1_A \otimes_K B = 1_A \otimes BbB = (1_A \otimes_K B)(1_A \otimes b)(1_A \otimes_K B)$$
$$= (1_A \otimes_K B)v(1_A \otimes_K B) \subset U.$$

Consequently,

$$A \otimes_K B = (A \otimes_K 1_B)(1_A \otimes_K B) \subset (A \otimes_K 1_B)U \subset U.$$

Therefore $U = A \otimes_K B$ and there is only one nonzero ideal of $A \otimes_K B$. Since $A \otimes_K B$ has an identity $1_A \otimes 1_B, (A \otimes_K B)^2 \neq 0$, whence $A \otimes_K B$ is simple. ∎

We now consider division rings. If D is a division ring and F is a subring of D containing 1_D that is a field, F is called a **subfield** of D. Clearly D is a vector space over any subfield F. A subfield F of D is said to be a **maximal subfield** if it is not properly contained in any other subfield of D. Maximal subfields always exist (Exercise 4). Every maximal subfield F of D contains the center K of D (otherwise F and K would generate a subfield of D properly containing F; Exercise 3). It is easy to see that F is actually a simple K-algebra. The maximal subfields of a division ring strongly influence the structure of the division ring itself, as the following theorems indicate.

Theorem 6.3. *Let* D *be a division ring with center* K *and let* F *be a maximal subfield of* D. *Then* $D \otimes_K F$ *is isomorphic (as a* K*-algebra) to a dense subalgebra of* $Hom_F(D,D)$, *where* D *is considered as a vector space over* F.

PROOF. $\text{Hom}_F(D,D)$ is an F-algebra (third example after Definition IV.7.1) and hence a K-algebra. For each $a \varepsilon D$ let $\alpha_a : D \to D$ be defined by $\alpha_a(x) = xa$. For each $c \varepsilon F$ let $\beta_c : D \to D$ be defined by $\beta_c(x) = cx$. Verify that $\alpha_a, \beta_c \varepsilon \text{Hom}_F(D,D)$

and that $\alpha_a\beta_c = \beta_c\alpha_a$ for all $a \in D$, $c \in F$. Verify that the map $D \times F \to \mathrm{Hom}_F(D,D)$ given by $(a,c) \mapsto \alpha_a\beta_c$ is K-bilinear. By Theorem IV.5.6 this map induces a K-module homomorphism $\theta : D \otimes_K F \to \mathrm{Hom}_F(D,D)$ such that

$$\theta\left(\sum_{i=1}^{n} a_i \otimes c_i\right) = \sum_{i=1}^{n} \alpha_{a_i}\beta_{c_i} \quad (a_i \in D,\ c_i \in F).$$

Verify that θ is a K-algebra homomorphism, which is not zero (since $\theta(1_D \otimes 1_D)$ is the identity map on D). Since D is a central simple and F a simple K-algebra, $D \otimes_K F$ is simple by Theorem 6.2. Since $\theta \neq 0$ and Ker θ is an algebra ideal, Ker $\theta = 0$, whence θ is a monomorphism. Therefore $D \otimes_K F$ is isomorphic to the K-subalgebra Im θ of $\mathrm{Hom}_F(D,D)$. We must show that $A = \mathrm{Im}\,\theta$ is dense in $\mathrm{Hom}_F(D,D)$.

D is clearly a left module over $\mathrm{Hom}_F(D,D)$ with $fd = f(d)$ ($f \in \mathrm{Hom}_F(D,D)$, $d \in D$). Consequently D is a left module over $A = \mathrm{Im}\,\theta$. If d is a nonzero element of D, then since D is a division ring,

$$Ad = \{\theta(u)(d) \mid u \in D \otimes_K F\} = \{\sum_i c_i d a_i \mid i \in \mathbf{N}^*;\ c_i \in F;\ a_i \in D\} = D.$$

Consequently, D has no nontrivial A-submodules, whence D is a simple A-module. Furthermore D is a faithful A-module since the zero map is the only element f of $\mathrm{Hom}_F(D,D)$ such that $fD = 0$. Therefore by the Density Theorem 1.12 A is isomorphic to a dense subring of $\mathrm{Hom}_\Delta(D,D)$, where Δ is the division ring $\mathrm{Hom}_A(D,D)$ and D is a lett Δ-vector space. Under the monomorphism $A \to \mathrm{Hom}_\Delta(D,D)$ the image of $f \in A$ is f considered as an element of $\mathrm{Hom}_\Delta(D,D)$.

We now construct an isomorphism of rings $F \cong \Delta$. Let $\beta : F \to \Delta = \mathrm{Hom}_A(D,D)$ be given by $c \mapsto \beta_c$ (notation as above). Verify that $\beta_c \in \Delta$ and that β is a monomorphism of rings. If $f \in \Delta$ and $x \in D$, then $\alpha_x = \theta(x \otimes 1_D) \in A$ and

$$f(x) = f(1_D x) = f[\alpha_x(1_D)] = \alpha_x(f(1_D)) = f(1_D)x = \beta_c(x),$$

where $c = f(1_D)$. In order to show that β is an epimorphism it suffices to prove that $c \in F$; for in that case $f(x) = cx = \beta_c(x)$ for all $x \in D$, whence $f = \beta_c = \beta(c)$. If $y \in F$, then $\beta_y = \theta(1_D \otimes y) \in A$ and $\alpha_y = \theta(y \otimes 1_D) \in A$ and

$$\begin{aligned} cy = f(1_D)y &= \alpha_y(f(1_D)) = f(\alpha_y(1_D)) = f(1_D y) = f(y1_D) \\ &= f(\beta_y(1_D)) = \beta_y f(1_D) = \beta_y(c) = yc. \end{aligned}$$

Therefore c commutes with every element of F. If $c \notin F$, then c and F generate a subfield of D that properly contains the maximal subfield F (Exercise 3). Since this would be a contradiction, we must have $c \in F$. Therefore β: F $\cong \Delta$.

To complete the proof, let $v_1, \dots, v_n \in D$ and let $\{u_1, \dots, u_n\}$ be a subset of D that is linearly independent over F. We claim that $\{u_1, \dots, u_n\}$ is also linearly independent over Δ. If $\sum_{i=1}^{n} g_i u_i = 0$, ($g_i \in \Delta$), then

$$0 = \sum g_i u_i = \sum \beta_{c_i}(u_i) = \sum c_i u_i,$$

where $c_i \in F$ and $g_i = \beta(c_i) = \beta_{c_i}$. The F-linear independence of $\{u_1, \dots, u_n\}$ implies that every $c_i = 0$, whence $g_i = \beta(0) = 0$ for all i. Therefore $\{u_1, \dots, u_n\}$ is linearly independent over Δ. By the density of A in $\mathrm{Hom}_\Delta(D,D)$ (Definition 1.7), there exists $h \in A$ such that $h(u_i) = v_i$ for every i. Therefore A is dense in $\mathrm{Hom}_F(D,D)$. ∎

Theorem 6.3 has an interesting corollary that requires two preliminary lemmas.

Lemma 6.4. *Let* A *be an algebra with identity over a field* K *and* F *a field containing* K; *then* $A \otimes_K F$ *is an* F*-algebra such that* $\dim_K A = \dim_F(A \otimes_K F)$.

SKETCH OF PROOF. Since F is commutative and a K-F bimodule, $A \otimes_K F$ is a vector space over F with $b(a \otimes b_1) = (a \otimes b_1)b = a \otimes b_1 b$ $(a \in A; b,b_1 \in F$; see Theorem IV.5.5 and the subsequent Remark). $A \otimes_K F$ is a K-algebra by Theorem IV.7.4 and is easily seen to be an F-algebra as well. If X is a basis of A over K, then by (the obvious analogue of) Theorem IV.5.11 every element of $A \otimes_K F$ can be written

$$\sum_i x_i \otimes c_i = \sum_i (x_i \otimes 1_F)c_i = \sum_i c_i(x_i \otimes 1_F) \; (x_i \in X; c_i \in F),$$

with the elements x_i and c_i uniquely determined. It follows that

$$X \otimes_K 1_F = \{x \otimes 1_F \mid x \in X\}$$

is a basis of $A \otimes_K F$ over F. Clearly $\dim_K A = |X| = |X \otimes_K 1_F| = \dim_F(A \otimes_K F)$ ∎

Lemma 6.5. *Let* D *be a division algebra over a field* K *and* A *a finite dimensional* K*-algebra with identity. Then* $D \otimes_K A$ *is a left Artinian* K*-algebra.*

SKETCH OF PROOF. $D \otimes_K A$ is a vector space over D with the action of $d \in D$ on a generator $d_1 \otimes a$ of $D \otimes_K A$ given by $d(d_1 \otimes a) = dd_1 \otimes a = (d \otimes 1_A)(d_1 \otimes a)$ (Theorem IV.5.5). Consequently every left ideal of $D \otimes_K A$ is also a D-subspace of $D \otimes_K A$. The proof of Lemma 6.4 is valid here, *mutatis mutandis*, and shows that $\dim_D(D \otimes_K A) = \dim_K A$. Since $\dim_K A$ is finite, a routine dimension argument shows that $D \otimes_K A$ is left Artinian. ∎

Theorem 6.6. *Let* D *be a division ring with center* K *and maximal subfield* F. *Then* $\dim_K D$ *is finite if and only if* $\dim_K F$ *is finite, in which case* $\dim_F D = \dim_K F$ *and* $\dim_K D = (\dim_K F)^2$.

PROOF. If $\dim_K F$ is infinite, so is $\dim_K D$. If $\dim_K F$ is finite, then $D \otimes_K F$ is a left Artinian K-algebra by Lemma 6.5. Thus $D \otimes_K F$ is isomorphic to a dense left Artinian subalgebra of $\mathrm{Hom}_F(D,D)$ by Theorem 6.3. The proof of Theorem 6.3 shows that this isomorphism is actually an F-algebra isomorphism. Consequently, there is an F-algebra isomorphism $D \otimes_K F \cong \mathrm{Hom}_F(D,D)$ and $n = \dim_F D$ is finite by Theorem 1.9. Therefore $D \otimes_K F \cong \mathrm{Hom}_F(D,D) \cong \mathrm{Mat}_n F$ by Theorem VII.1.4 (and the subsequent Remark). Lemma 6.4 now implies

$$\dim_K D = \dim_F(D \otimes_K F) = \dim_F(\mathrm{Mat}_n F) = n^2 = (\dim_F D)^2.$$

On the other hand $\dim_K D = (\dim_F D)(\dim_K F)$ by Theorem IV.2.16. Therefore $\dim_K F = \dim_F D$. ∎

Recall that if u is a unit in a ring R with identity, then the map $R \to R$ given by $r \mapsto uru^{-1}$ is an automorphism of the ring R. It is called the **inner automorphism** induced by u.

Theorem 6.7. *(Noether-Skolem) Let* R *be a simple left Artinian ring and let* K *be the center of* R *(so that* R *is a* K*-algebra). Let* A *and* B *be finite dimensional simple* K*-subalgebras of* R *that contain* K. *If* $\alpha : \mathrm{A} \to \mathrm{B}$ *is a* K*-algebra isomorphism that leaves* K *fixed elementwise, then* α *extends to an inner automorphism of* R.

PROOF. It suffices by the Wedderburn-Artin Theorem 1.14 to assume $R = \mathrm{Hom}_D(V,V)$, where V is an n-dimensional vector space over the division ring D. The remarks after Theorem VII.1.3 show that there is an anti-isomorphism of rings $R = \mathrm{Hom}_D(V,V) \to \mathrm{Mat}_n D$. Under this map the center K of R is necessarily mapped *isomorphically* onto the center of $\mathrm{Mat}_n D$. But the center of $\mathrm{Mat}_n D$ is isomorphic to the center of D by Exercise VII.1.3. Consequently we shall identify K with the center of D so that D is a central simple K-algebra.

Observe that V is a left R-module with $rv = r(v)$ $(v \in V; r \in R = \mathrm{Hom}_D(V,V))$. Since V is a left D-vector space, it follows that V is a left algebra module over the K-algebra $D \otimes_K R$, with the action of a generator $d \otimes r$ of $D \otimes_K R$ on $v \in V$ given by

$$(d \otimes r)v = d(rv) = d(r(v)) = r(dv). \tag{i}$$

If $\bar{A}$ is the subalgebra $D \otimes_K A$ of $D \otimes_K R$, then V is clearly a left $\bar{A}$-module. Similarly if $\bar{B} = D \otimes_K B$, then V is a left $\bar{B}$-module. Now the map $\bar{\alpha} = 1_D \otimes \alpha : \bar{A} \to \bar{B}$ is an isomorphism of K-algebras. Consequently, V has a second $\bar{A}$-module structure given by pullback along $\bar{\alpha}$; (that is, $\bar{a}v$ is defined to be $\bar{\alpha}(\bar{a})v$ for $v \in V$, $\bar{a} \in \bar{A}$; see p. 170). Under this second $\bar{A}$-module structure the action of a generator $d \otimes r$ of $\bar{A} = D \otimes_K A$ on $v \in V$ is given by

$$(d \otimes r)v = \bar{\alpha}(d \otimes r)v = (d \otimes \alpha(r))v = d(\alpha(r)(v)) = \alpha(r)(dv). \tag{ii}$$

By Theorem 6.2 and Lemma 6.5 $\bar{A}$ is a simple left Artinian K-algebra. Consequently by Theorem 3.10 there is (up to isomorphism) only one simple $\bar{A}$-module. Now V with either the $\bar{A}$-module structure (i) or (ii) is semisimple by Theorem 3.7. Consequently there are $\bar{A}$-module isomorphisms

$$V = \sum_{i \in I} U_i \quad \text{(corresponding to structure (i)) and} \tag{iii}$$

$$V = \sum_{j \in J} W_j \quad \text{(corresponding to structure (ii)),} \tag{iv}$$

with each U_i, W_j a simple $\bar{A}$-module and $U_i \cong W_j$ for all i,j. Since $dv = (d \otimes 1_R)v$ $(d \in D, v \in V)$, every $\bar{A}$-submodule of V is a D-subspace of V and every $\bar{A}$-module isomorphism is an isomorphism of D-vector spaces. Since $\dim_D V = n$ is finite, each U_i, W_j has finite dimension t over D and the index sets I, J are finite, say

$$I = \{1,2,\ldots,m\} \quad \text{and} \quad J = \{1,2,\ldots,s\}.$$

Therefore

$$\dim_D V = \dim_D\left(\sum_{i=1}^{m} U_i\right) = \sum_{i=1}^{m} \dim_D U_i = mt, \quad \text{and}$$

$$\dim_D V = \dim_D\left(\sum_{j=1}^{s} W_j\right) = \sum_{j=1}^{s} \dim_D W_j = st,$$

whence $m = s$. Since $U_i \cong W_j$ for all i,j, $\sum_{i=1}^{m} U_i \cong \sum_{j=1}^{m} W_j$. This isomorphism combined with the isomorphisms (iii) and (iv) above yields an $\bar{A}$-module isomorphism β of V (with the $\bar{A}$-module structure (i)) and V (with the $\bar{A}$-module structure (ii)). Thus for all $\bar{a} \in \bar{A}$ and $v \in V$

$$\beta(\bar{a}v) = \bar{\alpha}(\bar{a})(\beta(v)).$$

In particular, for $d \in D$ and $\bar{d} = d \otimes 1_A \in \bar{A}$,

$$\beta(dv) = \beta(\bar{d}v) = \bar{\alpha}(\bar{d})(\beta(v)) = (d \otimes 1_B)\beta(v) = d\beta(v),$$

whence $\beta \in \mathrm{Hom}_D(V,V) = R$. Since β is an isomorphism, β is a unit in R. Furthermore for $r \in A$ and $\bar{r} = 1_D \otimes r \in \bar{A}$,

$$\begin{aligned}\beta r(v) &= \beta[r(v)] = \beta[\bar{r}v] = \bar{\alpha}(\bar{r})\beta(v)\\ &= (1_D \otimes \alpha(r))\beta(v) = \alpha(r)[\beta(v)] = [\alpha(r)\beta](v),\end{aligned}$$

whence $\beta r = \alpha(r)\beta$ in $R = \mathrm{Hom}_D(V,V)$. In other words,

$$\beta r \beta^{-1} = \alpha(r) \quad \text{for all} \quad r \in A.$$

Therefore the inner automorphism of R induced by β extends the map $\alpha : A \to B$. ■

The division algebra of real quaternions, which is mentioned in the following corollary is defined on pages 117 and 227.

Corollary 6.8. *(Frobenius) Let* D *be an algebraic division algebra over the field* **R** *of real numbers. Then* D *is isomorphic to either* **R** *or the field* **C** *of complex numbers or the division algebra* T *of real quaternions.*

SKETCH OF PROOF. Let K be the center of D and F a maximal subfield. We have $\mathbf{R} \subset K \subset F \subset D$, with F an algebraic field extension of $\mathbf{R}$. Consequently $\dim_K F \leq \dim_{\mathbf{R}} F \leq 2$ by Corollary V.3.20. By Theorem 6.6 $\dim_F D = \dim_K F$ and $\dim_K D = (\dim_K F)^2$. Thus the only possibilities are $\dim_K D = 1$ and $\dim_K D = 4$. If $\dim_K D = 1$, then $D = F$, and D is isomorphic to $\mathbf{R}$ or $\mathbf{C}$ by Corollary V.3.20.

If $\dim_K D = 4$, then $\dim_K F = 2 = \dim_F D$, whence $K = \mathbf{R}$ and F is isomorphic to $\mathbf{C}$ by Corollary V.3.20. Furthermore D is noncommutative; otherwise D would be a proper algebraic extension field of the algebraically closed field $\mathbf{C}$. Since F is isomorphic to $\mathbf{C}$, $F = \mathbf{R}(i)$ for some $i \in F$ such that $i^2 = -1$. The map $F \to F$ given by $a + bi \mapsto a - bi$ is a nonidentity automorphism of F that fixes $\mathbf{R}$ elementwise. By Theorem 6.7 it extends to an inner automorphism β of D, given by $\beta(x) = dxd^{-1}$ for some nonzero $d \in D$.

Since $-i = \beta(i) = did^{-1}$, $-id = di$ and hence $id^2 = d^2 i$. Consequently $d^2 \in D$ commutes with every element of $F = \mathbf{R}(i)$. Therefore $d^2 \in F$; otherwise d^2 and F would generate a subfield of D that properly contained the maximal subfield F. Since the only elements of F that are fixed by β are the elements of $\mathbf{R}$ and $\beta(d^2) = dd^2d^{-1} = d^2$, we have $d^2 \in \mathbf{R}$. If $d^2 > 0$, then $d \in \mathbf{R}$. This is impossible since $d \in \mathbf{R}$ implies β is the identity map. Thus $d^2 = -r^2$ for some nonzero $r \in \mathbf{R}$, whence $(d/r)^2 = -1$. Let $j = d/r$ and $k = ij$. Verify that $\{1,i,j,k\}$ is a basis of D over $\mathbf{R}$ and that there is an $\mathbf{R}$-algebra isomorphism $D \cong T$. ■

Corollary 6.9. *(Wedderburn) Every finite division ring* D *is a field.*

REMARK. An elementary proof of this fact, via cyclotomic polynomials, is given in Exercise V.8.10.

PROOF OF 6.9. Let K be the center of D and F any maximal subfield. By Theorem 6.6 $\dim_K D = n^2$, where $\dim_K F = n$. Thus every maximal subfield is a finite field of order q^n, where $q = |K|$. Hence any two maximal subfields F and F' are isomorphic under an isomorphism $\beta : F \to F'$ that fixes K elementwise (Corollary V.5.8). By Theorem 6.7, β is given by an inner automorphism of D. Thus $F' = aFa^{-1}$ for some nonzero $a \varepsilon D$.

If $u \varepsilon D$, then $K(u)$ is a subfield of D (Exercise 3). $K(u)$ is contained in some maximal subfield that is of the form aFa^{-1} (for some $a \varepsilon D$). Thus $D = \bigcup_{0 \neq a \varepsilon D} aFa^{-1}$ and $D^* = \bigcup_{a \varepsilon D^*} aF^*a^{-1}$ (where D^*,F^* are the multiplicative groups of nonzero elements of D, F respectively). This is impossible unless $F = D$ according to Lemma 6.10 below. ∎

Lemma 6.10. *If* G *is a finite (multiplicative) group and* H *is a proper subgroup, then*

$$\bigcup_{x \varepsilon G} xHx^{-1} \subsetneq G.$$

PROOF. The number of distinct conjugates of H is $[G : N]$, where N is the normalizer of H in G (Corollary II.4.4). Since $H < N < G$ and $H \neq G$, $[G : N] \leq [G : H]$ and $[G : H] > 1$. If r is the number of distinct elements in $\bigcup_{x \varepsilon G} xHx^{-1}$, then

$$\begin{aligned} r &\leq 1 + (|H| - 1)[G : N] \leq 1 + (|H| - 1)[G : H] \\ &= 1 + |H|[G : H] - [G : H] = 1 + |G| - [G : H] < |G|, \end{aligned}$$

since $[G : H] > 1$. ∎

EXERCISES

1. If A is a finite dimensional central simple algebra over the field K, then $A \otimes_K A^{op} \cong \mathrm{Mat}_n K$, where $n = \dim_K A$ and A^{op} is defined in Exercise III.1.17.
2. If A and B are central simple algebras over a field K, then so is $A \otimes_K B$.
3. Let D be a division ring and F a subfield. If $d \varepsilon D$ commutes with every element of F, then the subdivision ring $F(d)$ generated by F and d (the intersection of all subdivision rings of D containing F and d) is a subfield. [See Theorem V.1.3.]
4. If D is a division ring, then D contains a maximal subfield.
5. If A is a finite dimensional central simple algebra over a field K, then $\dim_K A$ is a perfect square.
6. If A and B are left Artinian algebras over a field K, then $A \otimes_K B$ need not be left Artinian. [*Hint:* let A be a division algebra with center K and maximal subfield B such that $\dim_B A$ is infinite.]

7. If D is finite dimensional division algebra over its center K and F is a maximal subfield of D, then there is a K-algebra isomorphism $D \otimes_K F \cong \mathrm{Mat}_n F$, where $n = \dim_F D$.

8. If A is a simple algebra finite dimensional over its center, then any automorphism of A that leaves the center fixed elementwise is an inner automorphism.

9. (Dickson) Let D be a division ring with center K. If $a,b \in D$ are algebraic over the field K and have the same minimal polynomial, then $b = dad^{-1}$ for some $d \in D$.

CHAPTER X

CATEGORIES

This chapter completes the introduction to the theory of categories, which was begun in Section I.7. Categories and functors first appeared in the work of Eilenberg-MacLane in algebraic topology in the 1940s. It was soon apparent that these concepts had far wider applications. Many different mathematical topics may be interpreted in terms of categories so that the techniques and theorems of the theory of categories may be applied to these topics. For example, two proofs in disparate areas frequently use "similar" methods. Categorical algebra provides a means of precisely expressing these similarities. Consequently it is frequently possible to provide a proof in a categorical setting, which has as special cases the previously known results from two different areas. This unification process provides a means of comprehending wider areas of mathematics as well as new topics whose fundamentals are expressible in categorical terms.

In this book category theory is used primarily in the manner just described — as a convenient language of unification. In recent years, however, category theory has begun to emerge as a mathematical discipline in its own right. Frequently the source of inspiration for advances in category theory now comes to a considerable extent from within the theory itself. This wider development of category theory is only hinted at in this chapter.

The basic notions of functor and natural transformation are thoroughly discussed in Section 1. Two especially important types of functors are representable functors (Section 1) and adjoint pairs of functors (Section 2). Section 3 is devoted to carrying over to arbitrary categories as many concepts as possible from well-known categories, such as the category of modules over a ring.

This chapter depends on Section I.7, but is independent of the rest of this book, except for certain examples. Sections 1 and 3 are essentially independent. Section 1 is a prerequisite for Section 2.

1. FUNCTORS AND NATURAL TRANSFORMATIONS

As we have observed frequently in previous chapters the study of any mathematical object necessarily requires consideration of the "maps" of such objects. In the present case the mathematical objects in question are categories (Section I.7). A functor may be roughly described as a "map" from one category to another which preserves the appropriate structure. A natural transformation, in turn, is a "map" from one functor to another.

We begin with the definition of covariant and contravariant functors and numerous examples. Natural transformations are then introduced and more examples given. The last part of the section is devoted to some important functors in the theory of categories, the representable functors.

The reader should review the basic properties of categories (Section I.7), particularly the notion of universal object (which is needed in the study of representable functors). We shall frequently be dealing with several categories simultaneously. Consequently, if A and B are objects of a category $\mathcal{C}$, the set of all morphisms in $\mathcal{C}$ from A to B will sometimes be denoted by $\hom_{\mathcal{C}}(A,B)$ rather than $\hom(A,B)$ as previously.

Definition 1.1. *Let $\mathcal{C}$ and $\mathfrak{D}$ be categories. A* **covariant functor** T *from $\mathcal{C}$ to $\mathfrak{D}$ (denoted $\mathrm{T} : \mathcal{C} \to \mathfrak{D}$) is a pair of functions (both denoted by T), an object function that assigns to each object C of $\mathcal{C}$ an object $\mathrm{T(C)}$ of $\mathfrak{D}$ and a morphism function which assigns to each morphism $\mathrm{f : C \to C'}$ of $\mathcal{C}$ a morphism*

$$\mathrm{T(f) : T(C) \to T(C')}$$

of $\mathfrak{D}$, such that

(i) $\mathrm{T(1_C) = 1_{T(C)}}$ *for every identity morphism* $\mathrm{1_C}$ *of* $\mathcal{C}$;

(ii) $\mathrm{T(g \circ f) = T(g) \circ T(f)}$ *for any two morphisms* $\mathrm{f, g}$ *of* $\mathcal{C}$ *whose composite* $\mathrm{g \circ f}$ *is defined.*

EXAMPLE. The (covariant) **identity functor** $I_{\mathcal{C}} : \mathcal{C} \to \mathcal{C}$ assigns each object and each morphism of the category $\mathcal{C}$ to itself.

EXAMPLE. Let R be a ring and A a fixed left R-module. For each R-module C, let $T(C) = \mathrm{Hom}_R(A,C)$. For each R-module homomorphism $f : C \to C'$, let $T(f)$ be the usual induced map $\bar{f} : \mathrm{Hom}_R(A,C) \to \mathrm{Hom}_R(A,C')$ (see the remarks after Theorem IV.4.1). Then T is a covariant functor from the category of left R-modules to the category of abelian groups.

EXAMPLE. More generally, let A be a fixed object in a category $\mathcal{C}$. Define a covariant functor h_A from $\mathcal{C}$ to the category $\mathcal{S}$ of sets by assigning to an object C of $\mathcal{C}$ the set $h_A(C) = \hom(A,C)$ of all morphisms in $\mathcal{C}$ from A to C. If $f : C \to C'$ is a morphism of $\mathcal{C}$, let $h_A(f) : \hom(A,C) \to \hom(A,C')$ be the function given by $g \mapsto f \circ g$ ($g \in \hom(A,C)$). The functor h_A, which will be discussed in some detail below, is called the **covariant hom functor.**

EXAMPLE. Let F be the following covariant functor from the category of sets to the category of left modules over a ring R with identity. For each set X, $F(X)$ is the free R-module on X (see the Remarks after Theorem IV.2.1). If $f: X \to X'$ is a function, let $F(f): F(X) \to F(X')$ be the unique module homomorphism $\bar{f}: F(X) \to F(X')$ such that $\bar{f}i = f$, where i is the inclusion map $X \to F(X)$ (Theorem IV.2.1).

EXAMPLE. Let $\mathcal{C}$ be a concrete category (Definition I.7.6), such as the category of left R-modules or groups or rings. The (covariant) **forgetful functor** from $\mathcal{C}$ to the category $\mathcal{S}$ of sets assigns to each object A its underlying set (also denoted A) and to each morphism $f: A \to A'$ the function $f: A \to A'$ (see Definition I.7.6).

Definition 1.2. *Let $\mathcal{C}$ and $\mathcal{D}$ be categories. A* **contravariant functor** S *from $\mathcal{C}$ to $\mathcal{D}$ (denoted $S: \mathcal{C} \to \mathcal{D}$) is a pair of functions (both denoted by S), an object function which assigns to each object C of $\mathcal{C}$ an object $S(C)$ of $\mathcal{D}$ and a morphism function which assigns to each morphism $f: C \to C'$ of $\mathcal{C}$ a morphism*

$$S(f): S(C') \to S(C)$$

of $\mathcal{D}$ such that

(i) $S(1_C) = 1_{S(C)}$ *for every identity morphism 1_C of $\mathcal{C}$;*

(ii) $S(g \circ f) = S(f) \circ S(g)$ *for any two morphisms f, g of $\mathcal{C}$ whose composite $g \circ f$ is defined.*

Thus the morphism function of a contravariant functor $S: \mathcal{C} \to \mathcal{D}$ reverses the direction of morphisms.

EXAMPLE. Let R be a ring and B a fixed left R-module. Define a contravariant functor S from the category of left R-modules to the category of abelian groups by defining $S(C) = \mathrm{Hom}_R(C,B)$ for each R-module C. If $f: C \to C'$ is an R-module homomorphism, then $S(f)$ is the induced map $\bar{f}: \mathrm{Hom}_R(C',B) \to \mathrm{Hom}_R(C,B)$ (see the Remarks after Theorem IV.4.1).

EXAMPLE. More generally, let B be a fixed object in a category $\mathcal{C}$. Define a contravariant functor h^B from $\mathcal{C}$ to the category $\mathcal{S}$ of sets by assigning to each object C of $\mathcal{C}$ the set $h^B(C) = \hom(C,B)$ of all morphisms in $\mathcal{C}$ from C to B. If $f: C \to C'$ is a morphism of $\mathcal{C}$, let $h^B(f): \hom(C',B) \to \hom(C,B)$ be the function given by $g \mapsto g \circ f$ ($g \in \hom(C',B)$). The functor h^B is called the **contravariant hom functor.**

The following method may be used to reduce the study of contravariant functors to the study of covariant functors. If $\mathcal{C}$ is a category, then the **opposite** (or **dual**) **category** of $\mathcal{C}$, denoted $\mathcal{C}^{op}$, is defined as follows. The objects of $\mathcal{C}^{op}$ are the same as the objects of $\mathcal{C}$. The set $\hom_{\mathcal{C}^{op}}(A,B)$ of morphisms in $\mathcal{C}^{op}$ from A to B is defined to be the set $\hom_{\mathcal{C}}(B,A)$ of morphisms in $\mathcal{C}$ from B to A. When a morphism $f \in \hom_{\mathcal{C}}(B,A)$ is considered as a morphism in $\hom_{\mathcal{C}^{op}}(A,B)$, we denote it by f^{op}. Composition of morphisms in $\mathcal{C}^{op}$ is defined by

$$g^{op} \circ f^{op} = (f \circ g)^{op}.$$

If $S: \mathcal{C} \to \mathcal{D}$ is a contravariant functor, let $\bar{S}: \mathcal{C}^{op} \to \mathcal{D}$ be the unique *covariant* functor defined by

$$\bar{S}(A) = S(A) \quad \text{and} \quad \bar{S}(f^{op}) = S(f)$$

for each object A and morphism f of $\mathcal{C}^{op}$. Conversely, it is easy to verify that every covariant functor on $\mathcal{C}^{op}$ arises in this way from a contravariant functor on $\mathcal{C}$.

Recall that every statement involving objects and morphisms in a category has a dual statement obtained by reversing the direction of the morphisms (see p. 54). It follows readily that a statement is true in a category $\mathcal{C}$ if and only if the dual statement is true in $\mathcal{C}^{op}$. Consequently a statement involving objects, morphisms and a contravariant functor S on $\mathcal{C}$ is true provided the dual statement is true for the covariant functor $\bar{S}$ on $\mathcal{C}^{op}$. For this reason many results in the sequel will be proved only for covariant functors, the contravariant case being easily proved by dualization.

In order to define functors of several variables, it is convenient to introduce the concept of a **product category.** If $\mathcal{C}$ and $\mathcal{D}$ are categories, their product is the category $\mathcal{C} \times \mathcal{D}$ whose objects are all pairs (C,D), where C and D are objects of $\mathcal{C}$ and $\mathcal{D}$ respectively. A morphism $(C,D) \to (C',D')$ of $\mathcal{C} \times \mathcal{D}$ is a pair (f,g), where $f : C \to C'$ is a morphism of $\mathcal{C}$ and $g : D \to D'$ is a morphism of $\mathcal{D}$. Composition is given by $(f',g') \circ (f,g) = (f' \circ f, g' \circ g)$. The axioms for a category are readily verified. The product of more than two categories is defined similarly.

Functors of several variables are defined on an appropriate product category. Such a functor may be covariant in some variables and contravariant in others. For example, if $\mathcal{C},\mathcal{D},\mathcal{E}$ are categories, a functor T of two variables (contravariant in the first and covariant in the second variable) from $\mathcal{C} \times \mathcal{D}$ to $\mathcal{E}$ consists of an object function, which assigns to each pair of objects (C,D) in $\mathcal{C} \times \mathcal{D}$ an object $T(C,D)$ of $\mathcal{E}$, and a morphism function, which assigns to each pair of morphisms $f : C \to C'$, $g : D \to D'$ of $\mathcal{C} \times \mathcal{D}$ a morphism of $\mathcal{E}$:

$$T(f,g) : T(C',D) \to T(C,D'),$$

subject to the conditions:

(i) $T(1_C,1_D) = 1_{T(C,D)}$ for all (C,D) in $\mathcal{C} \times \mathcal{D}$;

(ii) $T(f' \circ f, g' \circ g) = T(f,g') \circ T(f',g)$, whenever the compositions $f' \circ f$, $g' \circ g$ are defined in $\mathcal{C}$ and $\mathcal{D}$ respectively. The second condition implies that for each fixed object C of $\mathcal{C}$ the object function $T(C,-)$ and the morphism function $T(1_C,-)$ constitute a covariant functor $\mathcal{D} \to \mathcal{E}$. Similarly for each fixed object D of $\mathcal{D}$, $T(-,D)$ and $T(-,1_D)$ constitute a contravariant functor $\mathcal{C} \to \mathcal{E}$.

EXAMPLE. $\mathrm{Hom}_R(-,-)$ is a functor of two variables, contravariant in the first and covariant in the second, from the category $\mathfrak{M}$ of left R-modules[1] to the category of abelian groups.

EXAMPLE. More generally let $\mathcal{C}$ be any category. Consider the functor that assigns to each pair (A,B) of objects of $\mathcal{C}$ the set $\hom_{\mathcal{C}}(A,B)$ and to each pair of morphisms $f : A \to A'$, $g : B \to B'$ the function

$$\hom(f,g) : \hom_{\mathcal{C}}(A',B) \to \hom_{\mathcal{C}}(A,B')$$

given by $h \mapsto g \circ h \circ f$. Then $\hom_{\mathcal{C}}(-,-)$ is a functor of two variables from $\mathcal{C}$ to the category $\mathcal{S}$ of sets, contravariant in the first variable and covariant in the second. Note that for a fixed object A, $\hom_{\mathcal{C}}(A,-)$ is just the covariant hom functor h_A and $h_A(g) = \hom(1_A,g)$. Similarly for fixed B $\hom_{\mathcal{C}}(-,B)$ is the contravariant hom functor h^B and $h^B(f) = \hom(f,1_B)$.

[1]Strictly speaking $\mathrm{Hom}_R(-,-)$ is a functor on $\mathfrak{M} \times \mathfrak{M}$, but this abuse of language is common and causes no confusion.

EXAMPLE. Let K be a commutative ring with identity. Then the functor given by

$$T(A_1, \ldots, A_n) = A_1 \otimes_K \cdots \otimes_K A_n$$
$$T(f_1, \ldots, f_n) = f_1 \otimes \cdots \otimes f_n$$

is a functor of n covariant variables from the category of K-modules to itself.

If $T_1 : \mathcal{C} \to \mathcal{D}$ and $T_2 : \mathcal{D} \to \mathcal{E}$ are functors, then their **composite** (denoted T_2T_1) is the functor from $\mathcal{C}$ to $\mathcal{E}$ with object and morphism functions given by

$$C \to T_2(T_1(C));$$
$$f \to T_2(T_1(f)).$$

T_2T_1 is covariant if T_1 and T_2 are both covariant or both contravariant. T_2T_1 is contravariant if one T_i is covariant and the other is contravariant.

Definition 1.3. *Let $\mathcal{C}$ and $\mathcal{D}$ be categories and $S : \mathcal{C} \to \mathcal{D}$, $T : \mathcal{C} \to \mathcal{D}$ covariant functors. A* **natural transformation** $\alpha : S \to T$ *is a function that assigns to each object C of $\mathcal{C}$ a morphism $\alpha_C : S(C) \to T(C)$ of $\mathcal{D}$ in such a way that for every morphism $f : C \to C'$ of $\mathcal{C}$, the diagram*

$$\begin{array}{ccc} S(C) & \xrightarrow{\alpha_C} & T(C) \\ \downarrow{\scriptstyle S(f)} & & \downarrow{\scriptstyle T(f)} \\ S(C') & \xrightarrow[\alpha_{C'}]{} & T(C') \end{array}$$

in $\mathcal{D}$ is commutative. If α_C is an equivalence for every C in $\mathcal{C}$, then α is a **natural isomorphism** (*or* **natural equivalence**) *of the functors S and T.*

A natural transformation [isomorphism] $\beta : S \to T$ of *contravariant* functors $S,T : \mathcal{C} \to \mathcal{D}$ is defined in the same way, except that the required commutative diagram is:

$$\begin{array}{ccc} S(C) & \xrightarrow{\beta_C} & T(C) \\ \uparrow{\scriptstyle S(f)} & & \uparrow{\scriptstyle T(f)} \\ S(C') & \xrightarrow[\beta_{C'}]{} & T(C'), \end{array}$$

for each morphism $f : C \to C'$ of $\mathcal{C}$.

REMARKS. The composition of two natural transformations is clearly a natural transformation. Natural transformations of functors of several variables are defined analogously.

EXAMPLE. If $T : \mathcal{C} \to \mathcal{C}$ is any functor, then the assignment $C \mapsto 1_{T(C)}$ defines a natural isomorphism $I_T : T \to T$, called the **identity natural isomorphism.**

EXAMPLE. Let $\mathfrak{M}$ be the category of left modules over a ring R and $T:\mathfrak{M}\to\mathfrak{M}$ the double dual functor, which assigns to each module A its double dual module $A^{**}=\text{Hom}_R(\text{Hom}_R(A,R),R)$. For each module A let $\theta_A : A\to A^{**}$ be the homomorphism of Theorem IV.4.12. Then the assignment $A\mapsto\theta_A$ defines a natural transformation from the identity functor $I_{\mathfrak{M}}$ to the functor T (Exercise IV.4.9). If the category $\mathfrak{M}$ is replaced by the category $\mathcal{V}$ of all finite dimensional left vector spaces over a division ring and T considered as a functor $\mathcal{V}\to\mathcal{V}$, then the assignment $A\mapsto\theta_A$ ($A \in \mathcal{V}$) defines a natural *isomorphism* from $I_{\mathcal{V}}$ to T by Theorem IV.4.12 (iii). Also see Exercise 5.

Natural transformations frequently appear in disguised form in specific categories. For example, in the category of R-modules (and similarly for groups, rings, etc.), a statement may be made that a certain homomorphism is natural, without any mention of functors. This is usually a shorthand statement that means: there are two (reasonably obvious) functors and a natural transformation between them.

EXAMPLE. If B is a unitary left module over a ring R with identity, then there is a natural isomorphism of modules $\alpha_B : R\otimes_R B\cong B$ (see Theorem IV.5.7). It is easy to verify that for any module homomorphism $f: B\to C$, the diagram

$$\begin{array}{ccc} R\otimes_R B & \xrightarrow{\alpha_B} & B \\ {\scriptstyle 1_R\otimes f}\downarrow & & \downarrow{\scriptstyle f} \\ R\otimes_R C & \xrightarrow[\alpha_C]{} & C \end{array}$$

is commutative. Thus the phrase "natural isomorphism" means that the assignment $B\mapsto\alpha_B$ defines a natural isomorphism $\alpha : T\to I_{\mathfrak{M}}$, where $\mathfrak{M}$ is the category of unitary left R-modules and $T:\mathfrak{M}\to\mathfrak{M}$ is given by $B\mapsto R\otimes_R B$ and $f\mapsto 1_R\otimes f$.

EXAMPLE. If A,B,C are left modules over a ring R, then the isomorphism of abelian groups

$$\phi : \text{Hom}_R(A\oplus B,C)\cong\text{Hom}_R(A,C)\oplus\text{Hom}_R(B,C)$$

of Theorem IV.4.7 is natural. One may interpret the word "natural" here by fixing any two variables, say A and C, and observing that for each module homomorphism $f: B\to B'$ the diagram

$$\begin{array}{ccc} \text{Hom}_R(A\oplus B',C) & \xrightarrow{\phi} & \text{Hom}_R(A,C)\oplus\text{Hom}_R(B',C) \\ {\scriptstyle \text{Hom}(1_A\oplus f,1_C)}\downarrow & & \downarrow{\scriptstyle \text{Hom}(1_A,1_C)\oplus\text{Hom}(f,1_C)} \\ \text{Hom}_R(A\oplus B,C) & \xrightarrow{\phi} & \text{Hom}_R(A,C)\oplus\text{Hom}_R(B,C) \end{array}$$

is commutative, where $1_A\oplus f: A\oplus B\to A\oplus B'$ is given by $(a,b)\mapsto(a,f(b))$. Thus ϕ defines a natural isomorphism of the contravariant functors S and T, where

$$S(B)=\text{Hom}_R(A\oplus B,C)\quad\text{and}\quad T(B)=\text{Hom}_R(A,C)\oplus\text{Hom}_R(B,C).$$

One says that the isomorphism ϕ is **natural in B.** A similar argument shows that ϕ is natural in A and C as well.

Other examples are given in Exercise 4.

Definition 1.4. *Let* T *be a covariant functor from a category* $\mathcal{C}$ *to the category* $\mathcal{S}$ *of sets.* T *is said to be a* **representable functor** *if there is an object* A *in* $\mathcal{C}$ *and a natural isomorphism* α *from the covariant hom functor* $h_A = hom_{\mathcal{C}}(A,-)$ *to the functor* T. *The pair* (A,α) *is called a* **representation** *of* T *and* T *is said to be represented by the object* A.

Similarly a contravariant functor $S : \mathcal{C} \to \mathcal{S}$ *is said to be* **representable** *if there is an object* B *of* $\mathcal{C}$ *and a natural isomorphism* $\beta : h^B \to S$, *where* $h^B = hom_{\mathcal{C}}(-,B)$. *The pair* (B,β) *is said to be a* **representation** *of* S.

EXAMPLE. Let A and B be unitary modules over a commutative ring K with identity and for each K-module C let $T(C)$ be the set of all K-bilinear maps $A \times B \to C$. If $f : C \to C'$ is a K-module homomorphism, let $T(f) : T(C) \to T(C')$ be the function that sends a bilinear map $g : A \times B \to C$ to the bilinear map $fg : A \times B \to C'$. Then T is a covariant functor from the category $\mathfrak{M}$ of K-modules to the category $\mathcal{S}$ of sets. We claim that T is represented by the K-module $A \otimes_K B$. To see this, define for each K-module C a function

$$\alpha_C : \mathrm{Hom}_K(A \otimes_K B,C) \to T(C)$$

by $\alpha_C(f) = fi$, where $i : A \times B \to A \otimes_K B$ is the canonical bilinear map (see p. 211). Now $\alpha_C(f) : A \times B \to C$ is obviously bilinear for each $f \in \mathrm{Hom}_K(A \otimes_K B,C)$. By Theorem IV.5.6 every bilinear map $g : A \times B \to C$ is of the form $\bar{g}i$ for a unique K-module homomorphism $\bar{g} : A \otimes_K B \to C$. Therefore α_C is a bijection of sets (that is, an equivalence in the category $\mathcal{S}$). It is easy to verify that the assignment $C \mapsto \alpha_C$ defines a natural isomorphism from $h_{A\otimes_K B}$ to T, whence $(A \otimes_K B,\alpha)$ is a representation of T. It is not just coincidence that $A \otimes_K B$ is a universal object in an appropriate category (Theorem IV.5.6). We shall now show that a similar fact is true for any representable functor.

Let (A,α) be a representation of a covariant functor $T : \mathcal{C} \to \mathcal{S}$. Let $\mathcal{C}_T$ be the category with objects all pairs (C,s), where C is an object of $\mathcal{C}$ and $s \in T(C)$. A morphism in $\mathcal{C}_T$ from (C,s) to (D,t) is defined to be a morphism $f : C \to D$ of $\mathcal{C}$ such that $T(f)(s) = t \in T(D)$. Note that f is an equivalence in $\mathcal{C}_T$ if and only if f is an equivalence in $\mathcal{C}$. A universal object in the category $\mathcal{C}_T$ (see Definition I.7.9) is called a **universal element** of the functor T.

EXAMPLE. In the example after Definition 1.4 the statement that $(A \otimes_K B,\alpha)$ is a representation of the functor $T : \mathfrak{M} \to \mathcal{S}$ clearly implies that for each K-module C and bilinear map $f : A \times B \to C$ (that is, for each pair (C,f) with $f \in T(C)$), there is a unique K-module homomorphism $\bar{f} : A \otimes_K B \to C$ such that $\bar{f}i = f$ (that is, such that $T(\bar{f})(i) = f$ with $i = \alpha_{A\otimes_K B}(1_{A\otimes_K B}) \in T(A \otimes_K B)$). Consequently the pair $(A \otimes_K B,i) = (A \otimes_K B,\alpha_{A\otimes_K B}(1_{A\otimes_K B}))$ is a universal object in the category $\mathfrak{M}_T$, that is, a universal element of T.

With the preceding example as motivation we shall now show that representations of a functor $T : \mathcal{C} \to \mathcal{S}$ are essentially equivalent to universal elements of T. We shall need

Lemma 1.5. *Let* $T : \mathcal{C} \to \mathcal{S}$ *be a covariant functor from a category* $\mathcal{C}$ *to the category* $\mathcal{S}$ *of sets and let* A *be an object of* $\mathcal{C}$.

(i) *If* $\alpha : \mathrm{h}_{\mathrm{A}} \to \mathrm{T}$ *is a natural transformation from the covariant hom functor* h_{A} *to* T *and* $\mathrm{u} = \alpha_{\mathrm{A}}(1_{\mathrm{A}}) \in \mathrm{T(A)}$, *then for any object* C *of* $\mathcal{C}$ *and* $\mathrm{g} \in hom_{\mathcal{C}}(\mathrm{A,C})$

$$\alpha_{\mathrm{C}}(\mathrm{g}) = \mathrm{T(g)(u)}.$$

(ii) *If* $\mathrm{u} \in \mathrm{T(A)}$ *and for each object* C *of* $\mathcal{C}$ $\beta_{\mathrm{C}} : hom_{\mathcal{C}}(\mathrm{A,C}) \to \mathrm{T(C)}$ *is the map defined by* $\mathrm{g} \mapsto \mathrm{T(g)(u)}$, *then* $\beta : \mathrm{h}_{\mathrm{A}} \to \mathrm{T}$ *is a natural transformation such that* $\beta_{\mathrm{A}}(1_{\mathrm{A}}) = u$.

PROOF. (i) Let C be an object of $\mathcal{C}$ and $g \in \hom_{\mathcal{C}}(A,C)$. By hypothesis the diagram

$$\begin{array}{ccc} h_A(A) = \hom_{\mathcal{C}}(A,A) & \xrightarrow{\alpha_A} & T(A) \\ {\scriptstyle h_A(g)}\downarrow & & \downarrow{\scriptstyle T(g)} \\ h_A(C) = \hom_{\mathcal{C}}(A,C) & \xrightarrow[\alpha_C]{} & T(C) \end{array}$$

is commutative. Consequently,

$$\begin{aligned} \alpha_C(g) &= \alpha_C(g \circ 1_A) = \alpha_C[h_A(g)(1_A)] \\ &= [\alpha_C h_A(g)]\,(1_A) = (T(g)\alpha_A)(1_A) = T(g)[\alpha_A(1_A)] \\ &= T(g)(u). \end{aligned}$$

(ii) We must show that for every morphism $k : B \to C$ of $\mathcal{C}$ the diagram

$$\begin{array}{ccc} h_A(B) = \hom_{\mathcal{C}}(A,B) & \xrightarrow{\beta_B} & T(B) \\ {\scriptstyle h_A(k)}\downarrow & & \downarrow{\scriptstyle T(k)} \\ h_A(C) = \hom_{\mathcal{C}}(A,C) & \xrightarrow[\beta_C]{} & T(C) \end{array}$$

is commutative. This fact follows immediately since for any $f \in \hom_{\mathcal{C}}(A,B)$

$$\begin{aligned} [\beta_C h_A(k)](f) &= \beta_C(k \circ f) = T(k \circ f)(u) = [T(k)T(f)](u) \\ &= T(k)[T(f)(u)] = T(k)[\beta_B(f)] \\ &= [T(k)\beta_B](f). \end{aligned}$$

Therefore β is a natural transformation. Finally,

$$\beta_A(1_A) = T(1_A)(u) = 1_{T(A)}(u) = u. \quad \blacksquare$$

Theorem 1.6. *Let* $\mathrm{T} : \mathcal{C} \to \mathcal{S}$ *be a covariant functor from a category* $\mathcal{C}$ *to the category* $\mathcal{S}$ *of sets. There is a one-to-one correspondence between the class* X *of all representations of* T *and the class* Y *of all universal elements of* T, *given by* $(\mathrm{A},\alpha) \mapsto (\mathrm{A},\alpha_{\mathrm{A}}(1_{\mathrm{A}}))$.

REMARK. Since $\alpha_A : \hom_{\mathcal{C}}(A,A) \to T(A)$, $\alpha_A(1_A)$ is an element of $T(A)$.

PROOF OF 1.6. Let (A,α) be a representation of T and let $\alpha_A(1_A) = u \in T(A)$. Suppose (B,s) is an object of $\mathcal{C}_T$. By hypothesis $\alpha_B : h_A(B) = \hom_{\mathcal{C}}(A,B) \to T(B)$ is a bijection, whence $s = \alpha_B(f)$ for a unique morphism $f : A \to B$. By Lemma 1.5, $T(f)(u) = \alpha_B(f) = s$. Therefore, f is a morphism in $\mathcal{C}_T$ from (A,u) to (B,s). If g is another morphism in $\mathcal{C}_T$ from (A,u) to (B,s) then $g \in \hom_{\mathcal{C}}(A,B)$ and $T(g)(u) = s$.

Consequently, by Lemma 1.5 $\alpha_B(g) = T(g)(u) = s = \alpha_B(f)$. Since α_B is a bijection, $f = g$. Therefore, f is the unique morphism in $\mathcal{C}_T$ from (A,u) to (B,s), whence (A,u) is universal in $\mathcal{C}_T$. Thus (A,u) is a universal element of T.

Conversely suppose (A,u) is a universal element of T. Let $\beta : h_A \to T$ be the natural transformation of Lemma 1.5 (ii) such that for any object C of $\mathcal{C}$, $\beta_C : \hom_{\mathcal{C}}(A,C) \to T(C)$ is given by $\beta_C(f) = T(f)(u)$. If $s \in T(C)$, then (C,s) is in $\mathcal{C}_T$. Since (A,u) is universal in $\mathcal{C}_T$, there exists $f \in \hom_{\mathcal{C}}(A,C)$ such that $s = T(f)(u) = \beta_C(f)$. Therefore β_C is surjective. If $\beta_C(f_1) = \beta_C(f_2)$, then $T(f_1)(u) = \beta_C(f_1) = \beta_C(f_2) = T(f_2)(u)$, whence f_1 and f_2 are both morphisms in $\mathcal{C}_T$ from (A,u) to $(C,T(f_1)(u)) = (C,T(f_2)(u))$. Consequently, $f_1 = f_2$ by universality. Therefore each β_C is injective and hence a bijection (equivalence in $\mathcal{S}$). Thus β is a natural isomorphism, whence (A,β) is a representation of T.

To complete the proof use Lemma 1.5 to verify that $\phi\psi = 1_Y$ and $\psi\phi = 1_X$, where $\phi : X \to Y$ is given by $(A,\alpha) \mapsto (A,\alpha_A(1_A))$ and $\psi : Y \to X$ is given by $(A,u) \mapsto (A,\beta)$ (β as in the previous paragraph). Therefore ϕ is a bijection. ■

Corollary 1.7. *Let* T : $\mathcal{C} \to \mathcal{S}$ *be a covariant functor from a category* $\mathcal{C}$ *to the category* $\mathcal{S}$ *of sets. If* (A,α) *and* (B,β) *are representations of* T, *then there is a unique equivalence* f : A → B *such that the following diagram is commutative for all objects* C *of* $\mathcal{C}$:

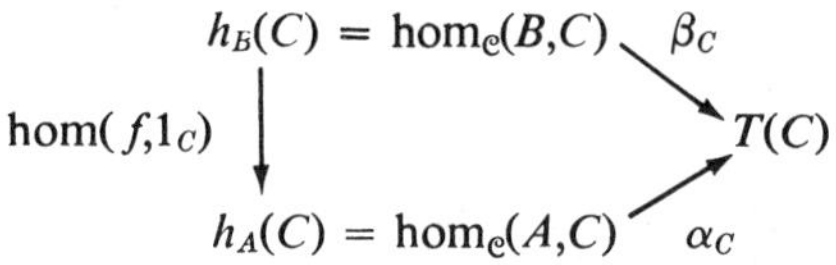

PROOF. Let $u = \alpha_A(1_A)$ and $v = \beta_B(1_B)$. By Theorem 1.6 (A,u) and (B,v) are universal elements of T, whence by Lemma I.7.10 there is a unique equivalence $f : A \to B$ in $\mathcal{C}$ such that $T(f)(u) = v$. Lemma 1.5 (i) implies that for any object C of $\mathcal{C}$ and $g \in \hom_{\mathcal{C}}(B,C)$

$$\begin{aligned}[\alpha_C \hom(f,1_C)](g) &= \alpha_C(g \circ f) = T(g \circ f)(u)\\ &= [T(g)T(f)](u) = T(g)[T(f)(u)] = T(g)(v)\\ &= \beta_C(g),\end{aligned}$$

so that the required diagram is commutative. Furthermore if $f_1 : A \to B$ also makes the diagram commutative, then for $C = B$ and $g = 1_B$,

$$T(f_1)(u) = \alpha_B(f_1) = \alpha_B(1_B \circ f_1) = \alpha_B[\hom(f_1,1_B)(1_B)] = \beta_B(1_B) = v.$$

Therefore $f_1 = f$ by uniqueness. ■

Corollary 1.8. (*Yoneda*) *Let* T : $\mathcal{C} \to \mathcal{S}$ *be a covariant functor from a category* $\mathcal{C}$ *to the category* $\mathcal{S}$ *of sets and let* A *be an object of* $\mathcal{C}$. *Then there is a one-to-one correspondence between the set* T(A) *and the set Nat*(h$_A$,T) *of all natural transformations from the covariant hom functor* h$_A$ *to the functor* T. *This bijection is natural in* A *and* T.

SKETCH OF PROOF. Define a function $\psi = \psi_A : \text{Nat}(h_A,T) \to T(A)$ by

$$\alpha \mapsto \alpha_A(1_A) \in T(A)$$

and a function $\phi : T(A) \to \mathrm{Nat}(h_A,T)$ by

$$u \mapsto \beta,$$

where β is given by Lemma 1.5 (ii). Verify that $\phi\psi$ and $\psi\phi$ are the respective identity maps. Therefore ψ_A is a bijection.

The naturality statement of the corollary means that the diagrams

$$\begin{array}{ccc} \mathrm{Nat}(h_A,T) & \xrightarrow{\psi_A} & T(A) \\ {\scriptstyle N^*(f)}\downarrow & & \downarrow{\scriptstyle T(f)} \\ \mathrm{Nat}(h_B,T) & \xrightarrow[\psi_B]{} & T(B) \end{array} \quad ,$$

$$\begin{array}{ccc} \mathrm{Nat}(h_A,T) & \xrightarrow{\psi_A} & T(A) \\ {\scriptstyle N_*(\alpha)}\downarrow & & \downarrow{\scriptstyle \alpha_A} \\ \mathrm{Nat}(h_A,S) & \xrightarrow[\psi_A]{} & S(A) \end{array}$$

are commutative, where $f : A \to B$ is any morphism of $\mathcal{C}$, $\alpha : T \to S$ is any natural transformation of functors and $N^*(f)$, $N_*(\alpha)$ are defined as follows. For each object C of $\mathcal{C}$ and $\beta \in \mathrm{Nat}(h_A,T)$,

$$N^*(f)(\beta)_C : h_B(C) = \hom_{\mathcal{C}}(B,C) \to T(C)$$

is given by $g \mapsto \beta_C(g \circ f)$. The map $N_*(\alpha) : \mathrm{Nat}(h_A,T) \to \mathrm{Nat}(h_A,S)$ is given by $\beta \mapsto \alpha\beta$. ■

A representable functor is a functor of one variable that is naturally isomorphic to the covariant (or contravariant) hom functor. But for a given category $\mathcal{D}$, $\hom_{\mathcal{D}}(-,-)$ is a functor of two variables. We now investigate conditions under which a functor T of two variables is naturally isomorphic to $\hom_{\mathcal{D}}(-,-)$.

We shall deal with the following somewhat more general situation. Let $\mathcal{C}$ and $\mathcal{D}$ be categories and $T : \mathcal{C} \times \mathcal{D} \to \mathcal{S}$ a functor that is contravariant in the first variable and covariant in the second. If $S : \mathcal{C} \to \mathcal{D}$ is a covariant functor, then it is easy to verify that the assignments $(C,D) \mapsto \hom_{\mathcal{D}}(S(C),D)$ and $(f,g) \mapsto \hom_{\mathcal{D}}(S(f),g)$ define a functor $\mathcal{C} \times \mathcal{D} \to \mathcal{S}$ that is contravariant in the first variable and covariant in the second.

Theorem 1.9. *Let $\mathcal{C}$ and $\mathcal{D}$ be categories and T a functor from the product category $\mathcal{C} \times \mathcal{D}$ to the category $\mathcal{S}$ of sets, contravariant in the first variable and covariant in the second, such that for each object C of $\mathcal{C}$, the covariant functor $T(C,-) : \mathcal{D} \to \mathcal{S}$ has a representation (A_C,α^C). Then there is a unique covariant functor $S : \mathcal{C} \to \mathcal{D}$ such that $S(C) = A_C$ and there is a natural isomorphism from $\hom_{\mathcal{D}}(S(-),-)$ to T, given by*

$$\alpha^C{}_D : \hom_{\mathcal{D}}(S(C),D) \to T(C,D).$$

REMARK ON NOTATION. For each object C of $\mathcal{C}$, A_C is an object in $\mathcal{D}$ and α^C is a natural isomorphism from $\hom_{\mathcal{D}}(A_C,-)$ to $T(C,-)$. Thus for each D in $\mathcal{D}$ there is an equivalence $\alpha^C{}_D : \hom_{\mathcal{D}}(A_C,D) \to T(C,D)$.

PROOF OF 1.9. The object function of the functor S is defined by $S(C) = A_C$ for each object C of $\mathcal{C}$. The morphism function of S is defined as follows. For each object C of $\mathcal{C}$ $\alpha^C{}_{A_C} : \hom_{\mathfrak{D}}(A_C,A_C) \to T(C,A_C)$ and $u_C = \alpha^C{}_{A_C}(1_{A_C}) \in T(C,A_C)$. By Theorem 1.6 (A_C,u_C) is a universal element of the functor $T(C,-)$. If $f : C \to C'$ is a morphism of $\mathcal{C}$, let $v = T(f,1_{A_{C'}})(u_{C'}) \in T(C,A_{C'})$. By the universality of (A_C,u_C) in $\mathcal{D}$ there exists a unique morphism $\bar{f}: A_C \to A_{C'}$ in $\mathcal{D}$ such that

$$T(1_C,\bar{f})(u_C) = v = T(f,1_{A_{C'}})(u_{C'}).$$

Define $S(f)$ to be the morphism $\bar{f}$.

Clearly $S(1_C) = 1_{A_C} = 1_{S(C)}$. If $C \xrightarrow{f} C' \xrightarrow{g} C''$ are morphisms of $\mathcal{C}$, then by definition $S(g)$ is the unique morphism $\bar{g} : A_{C'} \to A_{C''}$ such that

$$T(1_{C'},\bar{g})(u_{C'}) = T(g,1_{A_{C''}})(u_{C''}).$$

Similarly $S(g \circ f)$ is the *unique* morphism $\bar{h} : A_C \to A_{C''}$ such that

$$T(1_C,\bar{h})(u_C) = T(g \circ f,1_{A_{C''}})(u_{C''}).$$

Consequently $S(g) \circ S(f) = \bar{g} \circ \bar{f}$ is a morphism $A_C \to A_{C''}$ such that

$$\begin{aligned} T(1_C,\bar{g} \circ \bar{f})(u_C) &= T(1_C,\bar{g})T(1_C,\bar{f})(u_C) = T(1_C,\bar{g})T(f,1_{A_{C'}})(u_{C'}) \\ &= T(f,\bar{g})(u_{C'}) = T(f,1_{A_{C''}})T(1_{C'},\bar{g})(u_{C}') \\ &= T(f,1_{A_{C''}})T(g,1_{A_{C''}})(u_{C''}) = T(g \circ f,1_{A_{C''}})(u_{C''}) \\ &= T(1_C,\bar{h})(u_C). \end{aligned}$$

Therefore by the uniqueness property of universal objects in $\mathfrak{D}_{T(C,-)}$ we must have

$$S(g) \circ S(f) = \bar{g} \circ \bar{f} = \bar{h} = S(g \circ f).$$

Thus $S : \mathcal{C} \to \mathfrak{D}$ is a covariant functor.

In order to show that $\alpha : \hom_{\mathfrak{D}}(S(-),-) \to T$ is a natural isomorphism we need only show that for morphisms $f : C \to C'$ in $\mathcal{C}$ and $g : D \to D'$ in $\mathfrak{D}$ the diagram

$$\begin{array}{ccc} \hom_{\mathfrak{D}}(A_{C'},D) & \xrightarrow{\alpha^{C'}{}_D} & T(C',D) \\ {\scriptstyle \hom(S(f),1_D)} \downarrow & & \downarrow {\scriptstyle T(f,1_D)} \\ \hom_{\mathfrak{D}}(A_C,D) & \xrightarrow{\alpha^C{}_D} & T(C,D) \\ {\scriptstyle \hom(1_{A_C},g)} \downarrow & & \downarrow {\scriptstyle T(1_C,g)} \\ \hom_{\mathfrak{D}}(A_C,D') & \xrightarrow[\alpha^C{}_{D'}]{} & T(C,D') \end{array}$$

is commutative. The lower square is commutative since for fixed C,

$$\alpha^C : \hom_{\mathfrak{D}}(A_C,-) \to T(C,-)$$

is a natural isomorphism by hypothesis. As for the upper square let $k \in \hom_{\mathfrak{D}}(A_{C'},D)$. Then by Lemma 1.5 (i):

$$\begin{aligned} T(f,1_D)\alpha^{C'}{}_D(k) &= T(f,1_D)T(1_{C'},k)(u_{C'}) = T(f,k)(u_C') \\ &= T(1_C,k)T(f,1_{AC'})(u_{C'}) = T(1_C,k)T(1_C,\bar{f})(u_C) \\ &= T(1_C,k \circ \bar{f})(u_C) = T(1_C,k \circ S(f))(u_C) \\ &= \alpha^C{}_{D'}(k \circ S(f)) \\ &= \alpha^C{}_{D'}\text{hom}(S(f),1_D)(k). \quad \blacksquare \end{aligned}$$

EXERCISES

Note: In these exercises $\mathcal{S}$ is the category of sets and functions; $\mathcal{R}$ is the category of rings and ring homomorphisms; R is a ring; $\mathfrak{M}$ is the category of left R-modules and R-module homomorphisms; $\mathcal{G}$ is the category of groups and group homomorphisms.

1. Construct functors as follows:
   (a) A covariant functor $\mathcal{G} \to \mathcal{S}$ that assigns to each group the set of all its subgroups.
   (b) A covariant functor $\mathcal{R} \to \mathcal{R}$ that assigns to each ring N the polynomial ring $N[x]$.
   (c) A functor, covariant in both variables $\mathfrak{M} \times \mathfrak{M} \to \mathfrak{M}$ such that
   $$(A,B) \mapsto A \oplus B.$$
   (d) A covariant functor $\mathcal{G} \to \mathcal{G}$ that assigns to each group G its commutator subgroup G' (Definition II.7.7).

2. (a) If $T : \mathcal{C} \to \mathcal{D}$ is a covariant functor, let Im T consist of the objects $\{T(C) \mid C \in \mathcal{C}\}$ and the morphisms $\{T(f) : T(C) \to T(C') \mid f : C \to C'$ a morphism in $\mathcal{C}\}$. Then show that Im T need *not* be a category.
   (b) If the object function of T is injective, then show that Im T is a category.

3. (a) If $S : \mathcal{C} \to \mathcal{D}$ is a functor, let $\sigma(S) = 1$ if S is covariant and -1 if S is contravariant. If $T : \mathcal{D} \to \mathcal{E}$ is another functor, show that TS is a functor from $\mathcal{C}$ to $\mathcal{E}$ whose variance is given by $\sigma(TS) = \sigma(T)\sigma(S)$.
   (b) Generalize part (a) to any finite number of functors, $S_1 : \mathcal{C}_1 \to \mathcal{C}_2, S_2 : \mathcal{C}_2 \to \mathcal{C}_3, \dots, S_n : \mathcal{C}_n \to \mathcal{C}_{n+1}$.

4. (a) If A,B,C are sets, then there are natural bijections: $A \times B \to B \times A$ and $(A \times B) \times C \to A \times (B \times C)$.
   (b) Prove that the isomorphisms of Theorems IV.4.9, IV.5.8, IV.5.9, and IV.5.10 are all natural.

5. Let $\mathcal{V}$ be the category whose objects are all finite dimensional vector spaces over a field F (of characteristic $\neq 2,3$) and whose morphisms are all vector-space *isomorphisms*. Consider the dual space V^* of a left vector space V as a left vector space (see the Remark after Proposition VII.1.10).
   (a) If $\phi : V \to V_1$ is a vector-space isomorphism (morphism of $\mathcal{V}$), then so is the dual map $\bar{\phi} : V_1^* \to V^*$ (see Theorem IV.4.10). Hence $\bar{\phi}^{-1} : V^* \to V_1^*$ is also a morphism of $\mathcal{V}$.
   (b) $D : \mathcal{V} \to \mathcal{V}$ is a covariant functor, where $D(V) = V^*$ and $D(\phi) = \bar{\phi}^{-1}$.
   (c) For each V in $\mathcal{V}$ choose a basis $\{x_1, \dots, x_n\}$ and let $\{f_{x_1}, \dots, f_{x_n}\}$ be the dual bases of V^* (Theorem IV.4.11). Then the map $\alpha_V : V \to V^*$ defined by $x_i \mapsto f_{x_i}$ is an isomorphism. Thus $\alpha_V : V \cong D(V)$.

(d) The isomorphism α_V is *not* natural; that is, the assignment $V \mapsto \alpha_V$ is *not* a natural isomorphism from the identity functor $I_{\mathcal{V}}$ to D. [*Hint:* consider a one dimensional space with basis $\{x\}$ and let $\phi(x) = cx$ with $c \neq 0, \pm 1_F$.]

6. (a) Let $S : \mathcal{C} \to \mathcal{D}$ and $T : \mathcal{C} \to \mathcal{D}$ be covariant functors and $\alpha : S \to T$ a natural isomorphism. Then there is a natural isomorphism $\beta : T \to S$ such that $\beta\alpha = I_S$ and $\alpha\beta = I_T$, where $I_S : S \to S$ is the identity natural isomorphism and similarly for I_T. [*Hint:* for each C of $\mathcal{C}$, $\alpha_C : S(C) \to T(C)$ is an equivalence and hence has an inverse morphism $\beta_C : T(C) \to S(C)$.]
(b) Extend (a) to functors of several variables.

7. Covariant representable functors from $\mathcal{S}$ to $\mathcal{S}$ preserve surjective maps.

8. (a) The forgetful functor $\mathcal{M} \to \mathcal{S}$ (see the Example preceding Definition 1.2) is representable.
(b) The forgetful functor $\mathcal{G} \to \mathcal{S}$ is representable.

9. (a) Let $P : \mathcal{S} \to \mathcal{S}$ be the functor that assigns to each set X its power set (set of all subsets) $P(X)$ and to each function $f: A \to B$ the map $P(f) : P(B) \to P(A)$ that sends a subset X of B onto $f^{-1}(X) \subset A$. Then P is a representable contravariant functor.
(b) Let the object function of $Q : \mathcal{S} \to \mathcal{S}$ be defined by $Q(A) = P(A)$. If $f: A \to B$, let $Q(f) : Q(A) \to Q(B)$ be given by $X \mapsto f(X)$. Then Q is a covariant functor. Is Q representable?

10. Let (A,α) and (B,β) be representations of the covariant functors $S : \mathcal{C} \to \mathcal{S}$ and $T : \mathcal{C} \to \mathcal{S}$ respectively. If $\tau : S \to T$ is a natural transformation, then there is a unique morphism $f : A \to A$ in $\mathcal{C}$ such that the following diagram is commutative for every object C of $\mathcal{C}$:

$$\begin{array}{ccc} \hom_{\mathcal{C}}(A,C) & \xrightarrow{\alpha_C} & S(C) \\ {\scriptstyle \hom(f,1_C)} \big\downarrow & & \big\downarrow {\scriptstyle \tau_C} \\ \hom_{\mathcal{C}}(B,C) & \xrightarrow[\beta_C]{} & T(C) \end{array}$$

2. ADJOINT FUNCTORS

Adjoint pairs of functors are defined and discussed. Although they occur in many branches of mathematics formal descriptions of them are relatively recent.

Let $S : \mathcal{C} \to \mathcal{D}$ and $T : \mathcal{D} \to \mathcal{C}$ be covariant functors. As observed in the discussion preceding Theorem 1.9, the assignments $(C,D) \mapsto \hom_{\mathcal{D}}(S(C),D)$ and $(f,g) \mapsto \hom_{\mathcal{D}}(S(f),g)$ define a functor $\mathcal{C} \times \mathcal{D} \to \mathcal{S}$ which is contravariant in the first variable and covariant in the second. We denote this functor by $\hom_{\mathcal{D}}(S(-),-)$. Similarly the functor $\hom_{\mathcal{C}}(-,T(-)) : \mathcal{C} \times \mathcal{D} \to \mathcal{S}$ is defined by

$$(C,D) \mapsto \hom_{\mathcal{C}}(C,T(D)) \quad \text{and} \quad (f,g) \mapsto \hom_{\mathcal{C}}(f,T(g)).$$

Definition 2.1. *Let* $\mathrm{S} : \mathcal{C} \to \mathfrak{D}$ *and* $\mathrm{T} : \mathfrak{D} \to \mathcal{C}$ *be covariant functors.* S *is said to be a* **left adjoint** *of* T (*or* T *a* **right adjoint** *of* S, *or* (S,T) *an* **adjoint pair**) *if there is a natural isomorphism from the functor* $hom_{\mathfrak{D}}(\mathrm{S}(-),-)$ *to the functor* $hom_{\mathcal{C}}(-,\mathrm{T}(-))$.

Thus if S is a left adjoint of T, there is for each C of $\mathcal{C}$ and D of $\mathfrak{D}$ a bijection

$$\alpha_{C,D} : \hom_{\mathfrak{D}}(S(C),D) \to \hom_{\mathcal{C}}(C,T(D)),$$

which is natural in C and D. The theory of adjoint functors was first suggested by the following example.

EXAMPLE. Let R, S be rings and A_R, ${}_RB_S$, C_S (bi)modules as indicated. By Theorem IV.5.10 there is an isomorphism of abelian groups

$$\mathrm{Hom}_S(A \otimes_R B,C) \cong \mathrm{Hom}_R(A, \mathrm{Hom}_S(B,C)),$$

which is easily shown to be natural in A and C (also in B). Note that $A \otimes_R B$ is a right S-module by Theorem IV.5.5 (iii) and $\mathrm{Hom}_S(B,C)$ a right R-module by Exercise IV.4.4 (c). Let B be a fixed R-S bimodule. Let $\mathcal{C}$ be the category of right R-modules and $\mathfrak{D}$ the category of right S-modules so that $\hom_{\mathcal{C}}(X,Y) = \mathrm{Hom}_R(X,Y)$ and $\hom_{\mathfrak{D}}(U,V) = \mathrm{Hom}_S(U,V)$. Then the isomorphism above simply states that the functor $-\otimes_R B$ from $\mathcal{C}$ to $\mathfrak{D}$ is a left adjoint of the functor $\hom_S(B,-)$ from $\mathfrak{D}$ to $\mathcal{C}$.

EXAMPLE. Let R be a ring with identity and $\mathfrak{M}$ the category of unitary left R-modules. Let $T : \mathfrak{M} \to \mathcal{S}$ be the forgetful functor, which assigns to each module its underlying set. Then for each set X and module A, $\hom_{\mathcal{S}}(X,T(A))$ is just the set of all functions $X \to A$. Let $F : \mathcal{S} \to \mathfrak{M}$ be the functor that assigns to each X the free R-module $F(X)$ on the set X (see p. 182). Let $i_X : X \to F(X)$ be the canonical map. For each set X and module A, the map

$$\alpha_{X,A} : \mathrm{Hom}_R(F(X),A) \to \hom_{\mathcal{S}}(X,T(A))$$

defined by $g \mapsto gi_X$ is easily seen to be natural in X and A. Since $F(X)$ is free on X, $\alpha_{X,A}$ is injective (Theorem IV.2.1 (iv)). Furthermore every function $f : X \to T(A)$ is of the form $f = \bar{f}i_X$ for a unique homomorphism $\bar{f} : F(X) \to A$ (Theorem IV.2.1 (iv)). Consequently $\alpha_{X,A}$ is surjective and hence a bijection. Therefore F is a left adjoint of T.

Other examples are given in the exercises.

There is a close connection between adjoint pairs of functors and representable functors.

Proposition 2.2. *A covariant functor* $\mathrm{T} : \mathfrak{D} \to \mathcal{C}$ *has a left adjoint if and only if for each object* C *in* $\mathcal{C}$ *the functor* $hom_{\mathcal{C}}(\mathrm{C},\mathrm{T}(-)) : \mathfrak{D} \to \mathcal{S}$ *is representable.*

PROOF. If $S : \mathcal{C} \to \mathfrak{D}$ is a left adjoint of T, then there is for each object C of $\mathcal{C}$ and D of $\mathfrak{D}$ a bijection

$$\alpha_{C,D} : \hom_{\mathfrak{D}}(S(C),D) \to \hom_{\mathcal{C}}(C,T(D)),$$

which is natural in C and D. Thus for a fixed C, $(S(C),\alpha_{C,-})$ is a representation of the functor $\hom_{\mathcal{C}}(C,T(-))$.

Conversely suppose that for each C, A_C is an object of $\mathcal{D}$ that represents $\hom_{\mathcal{C}}(C,T(-))$. By Theorem 1.9 there is a covariant functor $S:\mathcal{C}\to\mathcal{D}$ such that $S(C) = A_C$ and there is a natural isomorphism of functors

$$\hom_{\mathcal{D}}(S(-),-)\to\hom_{\mathcal{C}}(-,T(-)).$$

Therefore S is a left adjoint of T. ■

Corollary 2.3. *A covariant functor* $T:\mathcal{D}\to\mathcal{C}$ *has a left adjoint if and only if there exists for each object* C *of* $\mathcal{C}$ *an object* $S(C)$ *of* $\mathcal{D}$ *and a morphism* $u_C : C\to T(S(C))$ *such that* $(S(C),u_C)$ *is a universal element of the functor* $\hom_{\mathcal{C}}(C,T(-)):\mathcal{D}\to\mathcal{S}$.

PROOF. Exercise; see Theorem 1.6. ■

Corollary 2.4. *Any two left adjoints of a covariant functor* $T:\mathcal{D}\to\mathcal{C}$ *are naturally isomorphic.*

PROOF. If $S_1:\mathcal{C}\to\mathcal{D}$ and $S_2:\mathcal{C}\to\mathcal{D}$ are left adjoints of T, then there are natural isomorphisms

$$\alpha:\hom_{\mathcal{D}}(S_1(-),-)\to\hom_{\mathcal{C}}(-,T(-)),$$
$$\beta:\hom_{\mathcal{D}}(S_2(-),-)\to\hom_{\mathcal{C}}(-,T(-)).$$

For each object C of $\mathcal{C}$ the objects $S_1(C)$ and $S_2(C)$ both represent the functor $\hom_{\mathcal{C}}(C,T(-))$ by the first part of the proof of Proposition 2.2. Consequently for each object C of $\mathcal{C}$ there is by Corollary 1.7 an equivalence $f_C:S_1(C)\to S_2(C)$. We need only show that f_C is natural in C; that is, given a morphism $g:C\to C'$ of $\mathcal{C}$ we must prove that

$$\begin{array}{ccc} S_1(C) & \xrightarrow{f_C} & S_2(C) \\ {\scriptstyle S_1(g)}\downarrow & & \downarrow{\scriptstyle S_2(g)} \\ S_1(C') & \xrightarrow[f_{C'}]{} & S_2(C') \end{array}$$

is commutative. We claim that it suffices to prove that

$$\begin{array}{ccc} \hom_{\mathcal{D}}(S_1(C'),S_2(C')) & \xleftarrow{\hom(f_{C'},1)} & \hom_{\mathcal{D}}(S_2(C'),S_2(C')) \\ {\scriptstyle \hom(S_1(g),1)}\downarrow & & \downarrow{\scriptstyle \hom(S_2(g),1)} \\ \hom_{\mathcal{D}}(S_1(C),S_2(C')) & \xleftarrow[\hom(f_C,1)]{} & \hom_{\mathcal{D}}(S_2(C),S_2(C')) \end{array}$$

is commutative (where $1 = 1_{S_2(C')}$). For the image of $1_{S_2(C')}$ in one direction is $S_2(g)\circ f_C$ and in the other direction $f_{C'}\circ S_1(g)$.

Consider the following three-dimensional diagram (in which $1 = 1_{S_2(C')}$, $\alpha_X = \alpha_{X,S_2(C')}$ and the induced map $\hom(k,1)$ is denoted $\bar{k}$ for simplicity):

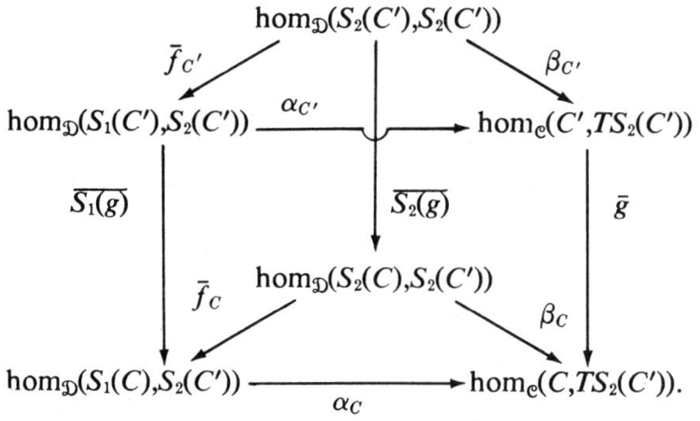

We must prove that the left rear rectangle is commutative. The top and bottom triangles are commutative by Corollary 1.7. The front and right rear rectangles are commutative since α and β respectively are natural. Consequently

$$\alpha_C\overline{S_1(g)}\,\bar{f}_{C'} = \bar{g}\alpha_{C'}\bar{f}_{C'} = \bar{g}\beta_{C'} = \beta_C\overline{S_2(g)} = \alpha_C\,\bar{f}_C\overline{S_2(g)}.$$

Since $\alpha_C = \alpha_{C,S_2(C')}$ is injective by hypothesis, we must have $\overline{S_1(g)}\,\bar{f}_{C'} = \bar{f}_C\overline{S_2(g)}$. Therefore the left rear rectangle is commutative. ■

EXERCISES

Note: $\mathcal{S}$ denotes the category of sets.

1. If $T : \mathfrak{C} \to \mathcal{S}$ is a covariant functor that has a left adjoint, then T is representable.

2. Let $\mathfrak{C}$ be a concrete category and $T : \mathfrak{C} \to \mathcal{S}$ the forgetful functor. If T has a left adjoint $F : \mathcal{S} \to \mathfrak{C}$, then F is called a **free-object functor** and $F(X)$ ($X \in \mathcal{S}$) is called a free F-object on X.
   (a) The category of groups has a free-object functor.
   (b) The category of commutative rings with identity and identity preserving homomorphisms has a free-object functor. [If X is finite, use Exercise III.5.11 to define $F(X)$.]

3. Let X be a fixed set and define a functor $S : \mathcal{S} \to \mathcal{S}$ by $Y \mapsto X \times Y$. Then S is a left adjoint of the covariant hom functor $h_X = \hom_{\mathcal{S}}(X,-)$.

4. Let $\mathcal{G}$ be the category of groups, $\mathfrak{A}$ the category of abelian groups, $\mathfrak{F}$ the category of fields, $\mathfrak{D}$ the category of integral domains, $\mathfrak{M}$ the category of unitary left K-modules, and $\mathfrak{B}$ the category of unitary K-K bimodules (K,R rings with identity).
   In each of the following cases let T be the appropriate forgetful functor (for example, $T : \mathfrak{F} \to \mathfrak{D}$ sends each field F to itself, considered as an integral domain). Show that (S,T) is an adjoint pair.
   (a) $T : \mathfrak{A} \to \mathcal{G}$, $S : \mathcal{G} \to \mathfrak{A}$, where $S(G) = G/G'$ with G' the commutator subgroup of G (Definition II.7.7).
   (b) $T : \mathfrak{F} \to \mathfrak{D}$, $S : \mathfrak{D} \to \mathfrak{F}$, where $S(D)$ is the field of quotients of D (Section III.4).
   (c) $T : \mathfrak{M} \to \mathfrak{A}$, $S : \mathfrak{A} \to \mathfrak{M}$, where $S(A) = K \otimes_{\mathbf{Z}} A$ (see Theorem IV.5.5).
   (d) $T : \mathfrak{B} \to \mathfrak{M}$, $S : \mathfrak{M} \to \mathfrak{B}$, where $S(M) = M \otimes_{\mathbf{Z}} R$.

3. MORPHISMS

A significant part of the elementary theory of categories is the attempt to generalize as many concepts as possible from well-known categories (for example, sets or modules) to arbitrary categories. In this section we extend to (more or less) arbitrary categories the concepts of monomorphisms, epimorphisms, kernels and cokernels of morphisms.

NOTATION. Hereafter we shall usually denote the composite of two morphisms of a category by gf instead of $g \circ f$ as previously.

We begin by recalling that a morphism $f : C \to D$ in a category is an equivalence if and only if there is a morphism $g : D \to C$ such that $gf = 1_C$ and $fg = 1_D$. This definition is simply a reflection of the fact that a homomorphism in the category of groups (or rings, or modules, etc.) is an isomorphism if and only if it has a two sided inverse (see Theorem I.2.3). In a similar fashion we may extend the concepts of monomorphisms and epimorphisms to arbitrary categories as follows.

Definition 3.1. *A morphism* $f : C \to D$ *of a category* $\mathcal{C}$ *is* **monic** (*or a* **monomorphism**) *if*

$$fh = fg \quad \Rightarrow \quad h = g$$

for all objects B *and morphisms* $g,h \in hom(B,C)$. *The morphism* f *is* **epic** (*or an* **epimorphism**) *if*

$$kf = tf \quad \Rightarrow \quad k = t$$

for all objects E *and morphisms* $k, t \in hom(D,E)$.

EXAMPLE. A morphism in the category of sets is monic [resp. epic] if and only if it is injective [resp. surjective] (Exercise 1).

EXAMPLES. Let $\mathcal{C}$ be any one of the following categories: groups, rings, left modules over a ring. If $f : C \to D$ and $g,h : B \to C$ are homomorphisms (that is, morphisms of $\mathcal{C}$), then by Exercise IV.1.2(a), $fh = fg$ implies $h = g$ if and only if f is an injective homomorphism (that is, a monomorphism in the usual sense).[2] Thus the categorical definition of monomorphism agrees with the previous definition in these familiar categories.

EXAMPLES. Exercise IV.1.2(b) shows that a morphism f in the category of left modules over a ring R is epic if and only if f is a surjective homomorphism (that is, an epimorphism in the usual sense). The same fact is true in the category of groups, but the proof is more difficult (Exercise 2). Thus the categorical definition of epimorphism agrees with the previous definition in these two categories.

EXAMPLES. In the category of rings every surjective homomorphism is easily seen to be epic. However, if $f,g : \mathbf{Q} \to R$ are homomorphisms of rings such that

[2]The Exercise deals only with modules, but the same argument is valid for groups and rings.

$f \mid \mathbf{Z} = g \mid \mathbf{Z}$, then $f = g$ by Exercise III.1.18. Consequently the inclusion map $\mathbf{Z} \to \mathbf{Q}$ is epic in the category of rings. But this map is obviously not surjective.

EXAMPLE. In the category of divisible abelian groups (p. 195) and group homomorphisms the canonical map $\pi : \mathbf{Q} \to \mathbf{Q}/\mathbf{Z}$ is monic, but clearly not injective. To see this, suppose $g,h : A \to \mathbf{Q}$ are homomorphisms with A divisible and $\pi g = \pi h$. If $g \neq h$, then there exist $a \in A, r,s \in \mathbf{Z}$ $(s \neq \pm 1)$ such that $g(a) - h(a) = r/s \neq 0$. By hypothesis $rb = a$ for some $b \in A$. Consequently, $r(g(b) - h(b)) = g(a) - h(a) = r(1/s)$, whence $g(b) - h(b) = 1/s$. Therefore $0 = \pi g(b) - \pi h(b) = \pi(g(b) - h(b)) = \pi(1/s)$. Thus $1/s \in \text{Ker } \pi = \mathbf{Z}$, which is a contradiction since $s \neq \pm 1$. Therefore $g = h$ and hence π is monic.

Proposition 3.2. *Let* $f : B \to C$ *and* $g : C \to D$ *be morphisms of a category* $\mathcal{C}$.

(i) f *and* g *monic* $\Rightarrow$ gf *monic;*
(ii) gf *monic* $\Rightarrow$ f *monic;*
(iii) f *and* g *epic* $\Rightarrow$ gf *epic;*
(iv) gf *epic* $\Rightarrow$ g *epic;*
(v) f *is an equivalence* $\Rightarrow$ f *is monic and epic.*

PROOF. Exercise. ∎

REMARK. The two examples preceding Proposition 3.2 show that the converse of (v) is false.

An object 0 in a category $\mathcal{C}$ is said to be a **zero object** if 0 is both universal and couniversal in $\mathcal{C}$ (see Definition I.7.9). Thus for any object C of $\mathcal{C}$ there is a unique morphism $0 \to C$ and a unique morphism $C \to 0$.

EXAMPLE. The zero module is a zero object in the category of left modules over a ring; similarly for groups and rings. The category of sets has no zero objects.

Proposition 3.3. *Let* $\mathcal{C}$ *be a category and* C *an object of* $\mathcal{C}$.

(i) *Any two zero objects of* $\mathcal{C}$ *are equivalent.*

(ii) *If* 0 *is a zero object, then the unique morphism* $0 \to C$ *is monic and the unique morphism* $C \to 0$ *is epic.*

SKETCH OF PROOF. (i) Theorem I.7.10. (ii) If $0_C \circ f = 0_C \circ g$, where $0_C : 0 \to C$, then $f = g$ by the couniversality of 0. Therefore 0_C is monic. ∎

Proposition 3.4. *Let* $\mathcal{C}$ *be a category which has a zero object* 0. *Then for each pair* C,D *of objects of* $\mathcal{C}$ *there is a unique morphism* $0_{C,D} : C \to D$ *such that*

$$f \circ 0_{C,D} = 0_{C,E} \quad \textit{and} \quad 0_{C,D} \circ g = 0_{B,D}$$

for all morphisms $f \in hom(D,E)$, $g \in hom(B,C)$.

REMARK. $0_{C,D}$ is called a **zero morphism.**

PROOF OF 3.4. (Uniqueness) If $\{0'_{C,D}\}$ and $\{0_{C,D}\}$ are two families of morphisms with the stated properties, then for each pair C,D

$$0_{C,D} = 0'_{D,D}0_{C,D} = 0'_{C,D}.$$

(Existence) For each object A of $\mathcal{C}$ let $\iota_A : 0 \to A$ and $\pi_A : A \to 0$ be the unique morphisms. For any $f \in \hom(D,E)$, $f\iota_D = \iota_E : 0 \to E$ by universality. For any $g \in \hom(B,C)$ $\pi_C g = \pi_B : B \to 0$ by couniversality. Define $0_{C,D}$ to be the composition $C \xrightarrow{\pi_C} 0 \xrightarrow{\iota_D} D$. Then for $f \in \hom(D,E)$, $f \circ 0_{C,D} = f\iota_D\pi_C = \iota_E\pi_C = 0_{C,E}$ and similarly in the other case. ■

The final step in extending properties of morphisms in familiar categories to morphisms in arbitrary categories is to develop reasonable definitions of kernels and cokernels of morphisms. We begin in a somewhat more general setting.

Definition 3.5. *Let* $f : C \to D$ *and* $g : C \to D$ *be morphisms of a category* $\mathcal{C}$. *A* **difference kernel** (*or* **equalizer**) *for the pair* (f,g) *is a morphism* $i : B \to C$ *such that:*

(i) $fi = gi$;

(ii) *if* $h : A \to C$ *is a morphism with* $fh = gh$, *then there exists a unique morphism* $\bar{h} : A \to B$ *such that* $i\bar{h} = h$.

A **difference cokernel** (*or* **coequalizer**) *for the pair* (f,g) *is a morphism* $j : D \to E$ *such that:*

(iii) $jf = jg$;

(iv) *if* $k : D \to F$ *is a morphism with* $kf = kg$, *then there exists a unique morphism* $\bar{k} : E \to F$ *such that* $\bar{k}j = k$.

EXAMPLES. In the category $\mathcal{S}$ of sets a difference kernel of $f : C \to D$ and $g : C \to D$ is the inclusion map $B \to C$, where $B = \{c \in C \mid f(c) = g(c)\}$. The same construction shows that every pair of morphisms has a difference kernel in the categories of groups, rings, and modules respectively.

EXAMPLE. Let $f : G \to H$ and $g : G \to H$ be homomorphisms of groups. Let N be the smallest normal subgroup of H containing $\{f(a)g(a)^{-1} \mid a \in G\}$. Then the canonical epimorphism $H \to H/N$ is a difference cokernel of (f,g) by Theorem I.5.6.

Proposition 3.6. *Let* $f : C \to D$ *and* $g : C \to D$ *be morphisms of a category* $\mathcal{C}$.

(i) *If* $i : B \to C$ *is a difference kernel of* (f,g), *then* i *is a monomorphism.*

(ii) *If* $i : B \to C$ *and* $j : A \to C$ *are difference kernels of* (f,g), *then there is a unique equivalence* $h : A \to B$ *such that* $ih = j$.

PROOF. (i) Let $h, k : F \to B$ be morphisms such that $ih = ik$. Then $f(ih) = (fi)h = (gi)h = g(ih)$. Since i is a difference kernel of (f,g), there is a unique morphism $t : F \to B$ such that $it = ih$. But both $t = h$ and $t = k$ satisfy this condition, whence $h = k$ by uniqueness. Therefore i is monic.

(ii) By hypothesis there exist unique morphisms $h : A \to B$ and $k : B \to A$ such that $ih = j$ and $jk = i$ respectively. Consequently $ihk = jk = i = i \circ 1_B$ and $jkh = ih = j = j \circ 1_A$. Since i and j are monomorphisms by (i), $hk = 1_B$ and $kh = 1_A$. Therefore h is an equivalence. ∎

REMARK. Difference cokernels are epimorphisms and the dual of Proposition 3.6 (ii) holds for difference cokernels.

Suppose that $\mathcal{C}$ is a category with a zero object 0 and hence zero morphisms (Proposition 3.4). A **kernel** of a morphism $f: C \to D$ (if one exists) is defined to be any difference kernel of the pair $(f, 0_{C,D})$; it is sometimes denoted Ker f. Definition 3.5 and Propositions 3.4 and 3.6 show that $k : K \to C$ is a kernel of $f : C \to D$ if and only if

(i) k is a monomorphism with $fk = 0_{K,D}$; and
(ii) if $h : B \to C$ is a morphism such that $fh = 0_{B,D}$, then there is a unique morphism $\bar{h} : B \to K$ such that $k\bar{h} = h$.
By Proposition 3.6 K is unique up to equivalence.

A **cokernel** $t : D \to E$ of a morphism $f : C \to D$ is defined dually as a difference cokernel of the pair $(f, 0_{C,D})$; it is sometimes denoted Coker f. As above t is characterized by the conditions:

(iii) t is an epimorphism with $tf = 0_{C,E}$; and
(iv) if $g : D \to F$ is a morphism such that $gf = 0_{C,F}$, then there is a unique morphism $\bar{g} : E \to F$ such that $\bar{g}t = g$.

EXAMPLES. In the categories of groups, rings and modules, a kernel of the morphism $f : C \to D$ is the inclusion map $K \to C$, where K is the usual kernel, $K = \{c \in C \mid f(c) = 0\}$. In the category of modules, the canonical epimorphism $D \to D/\text{Im } f$ is a cokernel of f.

EXERCISES

1. A morphism in the category of sets is monic [resp. epic] if and only if it is injective [resp. surjective].

2. A morphism $f : G \to H$ in the category of groups is epic if and only if f is a surjective homomorphism (that is, an epimorphism in the usual sense). [*Hint:* If f is epic, $K = \text{Im } f$, and $j : K \to H$ is the inclusion map, then j is epic by Proposition 3.2. Show that f is surjective (that is, $K = H$) as follows. Let S be the set of left cosets of K in H; let $T = S \cup \{u\}$ with $u \notin S$. Let A be the group of all permutations of T. Let $t : H \to A$ be given by $t(h)(h'K) = hh'K$ and $t(h)(u) = u$. Let $s : H \to A$ be given by $\sigma t(h)\sigma$, where $\sigma \in A$ is the transposition interchanging u and K. Show that s and t are homomorphisms such that $sj = tj$, whence $s = t$. Show that $hK = K$ for all $h \in H$; therefore $K = H$.]

3. A commutative diagram

$$\begin{array}{ccc} B & \xrightarrow{g_1} & C_1 \\ {\scriptstyle g_2}\downarrow & & \downarrow{\scriptstyle f_1} \\ C_2 & \xrightarrow[f_2]{} & D \end{array}$$

of morphisms of a category $\mathcal{C}$ is called a **pullback** for f_1 and f_2 if for every pair of morphisms $h_1 : B' \to C_1$, $h_2 : B' \to C_2$ such that $f_1h_1 = f_2h_2$ there exists a unique morphism $t : B' \to B$ such that $h_1 = g_1t$ and $h_2 = g_2t$.

(a) If there is another pullback diagram for f_1,f_2 with B_1 in the upper left-hand corner, then B and B_1 are equivalent.

(b) In the pullback diagram above, if f_2 is a monomorphism, then so is g_1.

(c) Every pair of functions $f_1 : C_1 \to D, f_2 : C_2 \to D$ in the category of sets has a pullback.

4. Show that every pair of functions $f, g : C \to D$ has a difference cokernel in the category of sets.

5. Let $f, g : C \to D$ be morphisms of a category $\mathcal{C}$. For each X in $\mathcal{C}$ let

$$\operatorname{Eq}(X,f,g) = \{h \varepsilon \hom(X,C) \mid fh = gh\}.$$

(a) $\operatorname{Eq}(-,f,g)$ is a contravariant functor from $\mathcal{C}$ to the category of sets.

(b) A morphism $i : K \to C$ is a difference kernel of (f,g) if and only if $\operatorname{Eq}(-,f,g)$ is representable with representing object K (that is, there is a natural isomorphism $\tau : \hom_{\mathcal{C}}(-,K) \to \operatorname{Eq}(-,f,g)$). [*Hint:* show that for $h : X \to K$, $\tau_X(h) = ih$, where $i = \tau_K(1_K)$.]

6. If each square in the following diagram is a pullback and $B' \to B$ is a monomorphism, then the outer rectangle is a pullback. [*Hint:* See Exercise 3.]

$$\begin{array}{ccccc} P & \longrightarrow & Q & \longrightarrow & B' \\ \downarrow & & \downarrow & & \downarrow \\ A & \longrightarrow & I & \longrightarrow & B. \end{array}$$

7. In a category with a zero object, the kernel of a monomorphism is a zero morphism.

List of Symbols

SYMBOL	MEANING	PAGE REFERENCE
$\mathbf{Q}$	field of rational numbers	1
$\mathbf{R}$	field of real numbers	1
$\mathbf{C}$	field of complex numbers	1
$\Rightarrow$	implies	1
$\Leftrightarrow$	if and only if	1
$\in$	is an element of	2
$\notin$	is not an element of	2
$\{x \mid P(x)\}$	the class of all x such that $P(x)$ is true	2
$\subset$	is a subclass (or subset) of	2
$\varnothing$	empty set	3
$P(A)$ or 2^A	power set of A	3
$\bigcup_{i \in I} A_i$	union of the sets A_i	3
$\bigcap_{i \in I} A_i$	intersection of the sets A_i	3
$B - A$	relative complement of A in B	3
A'	complement of A	3
$f: A \to B$	f is a function from A to B	3
$a \mapsto f(a)$	the function f maps a to $f(a)$	3
$f \mid S$	restriction of the function f to S	4
1_A	identity function on the set A	4
	identity element of the ring A	115
$g \circ f$ or gf	composite function of f and g	4
	composite morphism of f and g	52
$\operatorname{Im} f$	image of the function f	4, 31
$f^{-1}(T)$	inverse image of the set T	4

SYMBOL	MEANING	PAGE REFERENCE
$A \times B$	Cartesian product of sets A and B	6
	direct product of groups A and B	26
$\sim$	is equivalent to	6
	is equipollent with	15
$\bar{a}$	equivalence class of a	6
$\prod_{i \in I} A_i$	(Cartesian) product of the sets A_i;	7
	product of the family of objects $\{A_i \mid i \in I\}$	53
	direct product of the family of groups [or rings or modules] $\{A_i \mid i \in I\}$	59, 130, 173
$\mathbf{Z}$	set of integers	9
$\mathbf{N}$	set of nonnegative integers (natural numbers)	9
$\mathbf{N}^*$	set of positive integers	9
$a \mid b$	a divides b	11, 135
$a \nmid b$	a does not divide b	11, 135
$(a_1, a_2, \dots, a_n)$	greatest common divisor of $a_1, \dots, a_n$	11
	ideal generated by $a_1, \dots, a_n$	123
$a \equiv b \pmod{m}$	a is congruent to b modulo m	12
$\|A\|$	cardinal number of the set A	16
	order of the group A	24
	determinant of the matrix A	351
$\aleph_0$	aleph-naught	16
D_4^*	group of symmetries of the square	26
S_n	symmetric group on n letters	26
$G \oplus H$	direct sum of additive groups G and H	26
Z_m	integers modulo m	27
$\mathbf{Q}/\mathbf{Z}$	group of rationals modulo one	27
$Z(p^\infty)$	Sylow p-subgroup of $\mathbf{Q}/\mathbf{Z}$	30, 37
$\cong$	is isomorphic to	30
$\operatorname{Ker} f$	kernel of the homomorphism f	31, 119, 170
$H < G$	H is a subgroup of G	31
$<X>$	subgroup generated by the set X	32
$<a>$	cyclic (sub)group generated by a	32
$H \vee K$, $H + K$	the join of subgroups H and K	33
Q_8	quaternion group	33
$\|a\|$	order of the element a	35
$a \equiv_r b \pmod{H}$	$ab^{-1} \in H$	37
$a \equiv_l b \pmod{H}$	$a^{-1}b \in H$	37
Ha, aH	right and left cosets of a	38
$[G:H]$	index of a subgroup H in a group G	38
HK	$\{ab \mid a \in H, b \in K\}$	39
$N \lhd G$	N is a normal subgroup of G	41
G/N	factor group of G by N	42

SYMBOL	MEANING	PAGE REFERENCE
sgn τ	sign of the permutation τ	48
A_n	alternating group on n letters	49
D_n	dihedral group of degree n	50
$\dot{\bigcup}_{i\in I} A_i$	disjoint union of the sets A_i	58
$\prod^w_{i\in I} G_i$	weak direct product of the groups G_i	60
$\sum_{i\in I} G_i$	direct sum of the groups (or modules) G_i	60, 173
$\prod^*_{i\in I} G_i$	free product of the groups G_i	68
$G[m]$	$\{u \in G \mid mu = 0\}$	77, 224
$G(p)$	$\{u \in G \mid u$ has order a power of $p\}$	77, 222
G_t	torsion subgroup [submodule] of G	78, 220
G_x	stabilizer of x	89
$C_H(x)$	centralizer of x in H	89
$N_H(K)$	normalizer of K in H	89
$C(G)$	center of G	91
$C_n(G)$	n-th term of ascending central series	100
G'	commutator subgroup of G	102
$G^{(n)}$	n-th derived subgroup of G	102
End A	endomorphism ring of A	116
$\binom{n}{k}$	binomial coefficient	118
char R	characteristic of the ring R	119
R^{op}	opposite ring of R	122, 330
(X)	ideal generated by the set X	123
(a)	principal ideal generated by a	123
$S^{-1}R$	ring of quotients of R by S	143
R_P	localization of R at P	147
$R[x]$	ring of polynomials over R	149
$R[x_1, \ldots, x_n]$	ring of polynomials in n indeterminates over R	151
$R[[x]]$	ring of formal power series over R	154
deg f	degree of the polynomial f	157, 158
$C(f)$	content of the polynomial f	162
$\mathrm{Hom}_R(A,B)$	set of all R-module homomorphisms $A \to B$	174
$\dim_D V$	dimension of the D-vector space V	185
${}_RA_S$	R-S bimodule A	202
${}_RA,[A_R]$	left [resp. right] R-module A	202
A^*	dual module of A	203
$\langle a,f \rangle$	$f(a)$	204
δ_{ij}	Kronecker delta	204
$\mathfrak{M}(A,B)$	category of middle linear maps on $A \times B$	207

SYMBOL	MEANING	PAGE REFERENCE
$A \otimes_R B$	tensor product of modules A and B	208
$f \otimes g$	induced map on the tensor product	209
$\mathcal{O}_a$	order ideal of a	220
$[F:K]$	dimension of field F as a K-vector space	231
$K[u_1, \ldots, u_n]$, [resp. $K[X]$]	subring generated by K and $u_1, \ldots, u_n$ [resp. X]	232
$K(u_1, \ldots, u_n)$ [resp. $K(X)$]	subfield generated by K and $u_1, \ldots, u_n$ [resp. X]	232
$K(x_1, \ldots, x_n)$	field of rational functions in n indeterminates	233
$\mathrm{Aut}_K F$	Galois group of F over K	243
Δ	discriminant of a polynomial	270
F^{p^n}	$\{u^{p^n} \mid u \in F; \text{char } F = p\}$	285
$[F:K]_s$	separable degree of F over K	285
$[F:K]_i$	inseparable degree of F over K	285
$N_K{}^F(u)$	norm of u	289
$T_K{}^F(u)$	trace of u	289
$g_n(x)$	n-th cyclotomic polynomial	298
tr.d. F/K	transcendence degree of F over K	316
K^{1/p^n}	$\{u \in C \mid u^{p^n} \in K\}$	320
K^{1/p^∞}	$\{u \in C \mid u^{p^n} \in K \text{ for some } n \geq 0\}$	320
I_n	$n \times n$ identity matrix	328
$\mathrm{Mat}_n R$	ring of $n \times n$ matrices over R	328
A^t	transpose of the matrix A	328
A^{-1}	inverse of the invertible matrix A	331
$E_r^{n,m}$	a certain matrix	337
A^a	classical adjoint of the matrix A	353
$q_\phi(x)$, $q_A(x)$	minimal polynomial of ϕ [resp. A]	356
Tr A	trace of the matrix A	369
Rad I	radical of the ideal I	379
$V(S)$	affine variety determined by S	409
$\mathcal{A}(B)$	left annihilator of B	417
$r \circ a$	$r + a + ra$	426
$J(R)$	Jacobson radical of R	426
$P(R)$	prime radical of R	444
hom(A,B) or hom$_{\mathcal{C}}(A,B)$	set of morphisms $A \to B$ in a category $\mathcal{C}$	52, 465
h_A	covariant hom functor	465
h^B	contravariant hom functor	466
$\mathcal{C}_T$	category formed from $\mathcal{C}$ and T	470
$0_{C,D}$	zero morphism from C to D	482

Bibliography

All books and articles actually referred to in the text are listed below. The list also includes a number of other books that may prove to be useful references. No attempt has been made to make the list complete. It contains only a reasonable selection of English language books in algebra and related areas. Almost all of these books are accessible to anyone able to read this text, though in some cases certain parts of this text are prerequisites.

For the reader's convenience the books are classified by topic. However, this classification is not a rigid one. For instance, several books classified as "general" contain a fairly complete treatment of group theory as well as fields and Galois theory. Other books, such as [26] and [39], readily fit into more than one classification.

BOOKS

GENERAL

1. Chevalley, C., *Fundamental Concepts of Algebra.* NewYork: Academic Press, Inc., 1956.
2. Faith, C., *Algebra: Rings, Modules and Categories I.* Berlin: Springer-Verlag, 1973.
3. Goldhaber, J. and G. Ehrlich, *Algebra.* New York: The Macmillan Company, 1970.
4. Herstein, I., *Topics in Algebra.* Waltham, Mass.: Blaisdell Publishing Company, 1964.
5. Lang, S., *Algebra.* Reading, Mass.: Addison-Wesley, Publishing Company, Inc., 1965.
6. MacLane, S. and G. Birkhoff, *Algebra.* New York: The Macmillan Company, 1967.
7. Van der Waerden, B. L., *Algebra.* (7th ed., 2 vols.), New York: Frederick Ungar Publishing Co., 1970.

SET THEORY

8. Eisenberg, M., *Axiomatic Theory of Sets and Classes.* New York: Holt, Rinehart and Winston, Inc., 1971.
9. Halmos, P., *Naive Set Theory*, Princeton, N. J.: D. Van Nostrand Company Inc., 1960.
10. Suppes, P., *Axiomatic Set Theory.* Princeton, N. J.: D. Van Nostrand Company, Inc., 1960.

GROUPS

11. Curtis, C. W. and I. Reiner, *Representation Theory of Finite Groups and Associative Algebras*. New York: Interscience Publishers, 1962.
12. Dixon, J., *Problems in Group Theory*. Waltham, Mass.: Blaisdell Publishing Company, 1967.
13. Fuchs, L., *Infinite Abelian Groups*. New York: Academic Press, Inc., 1970.
14. Gorenstein, D., *Finite Groups*. New York: Harper and Row, Publishers, 1968.
15. Hall, M., *The Theory of Groups*. New York: The Macmillan Company, 1959.
16. Hall, M. and J. K. Senior, *The Groups of Order* $2^n (n \leq 6)$. New York: The Macmillan Company, 1964.
17. Kaplansky, I., *Infinite Abelian Groups* (2d ed.), Ann Arbor, Mich.: University of Michigan Press, 1969.
18. Kurosh, A. G., *The Theory of Groups* (2 vols.), New York: Chelsea Publishing Company, 1960.
19. Rotman, J., *The Theory of Groups* (2d ed.). Boston: Allyn and Bacon, Inc., 1973.
20. Scott, W. R., *Group Theory*. Englewood Cliffs, N. J.: Prentice-Hall, Inc., 1964.
21. Zassenhaus, H., *The Theory of Groups*. New York: Chelsea Publishing Company, 1958.

RINGS AND MODULES

22. Divinsky, N. J., *Rings and Radicals*. Toronto: University of Toronto Press, 1965.
23. Gray, M., *A Radical Approach to Algebra*. Reading, Mass.: Addison-Wesley Publishing Company, Inc., 1970.
24. Herstein, I. N., *Noncommutative Rings*. Math. Assoc. of America, distributed by J. Wiley, 1968.
25. Jacobson N., *Structure of Rings*. Amer. Math. Soc. Colloq. Publ., vol. 37, 1964.
26. Jans, J., *Rings and Homology*. New York: Holt, Rinehart and Winston, Inc., 1964.
27. Lambek, J., *Lectures on Rings and Modules*. Waltham, Mass.: Blaisdell Publishing Company, 1966.
28. McCoy, N., *Theory of Rings*. New York: The Macmillan Company, 1964.
29. Northcott, D. G., *Lessons on Rings, Modules and Multiplicity*. New York: Cambridge University Press, 1968.

COMMUTATIVE ALGEBRA

30. Atiyah, M. F. and I. G. MacDonald, *Introduction to Commutative Algebra*. Reading, Mass.: Addison-Wesley Publishing Company, Inc., 1969.
31. Kaplansky, I., *Commutative Rings*. Boston: Allyn and Bacon, Inc., 1970.
32. Larsen, M. and P. J. McCarthy, *Multiplicative Theory of Ideals*. New York: Academic Press, Inc., 1971.
33. Zariski, O. and P. Samuel, *Commutative Algebra* (vols. I and II). Princeton N. J., D. Van Nostrand Company, Inc., 1958. 1960.

HOMOLOGICAL ALGEBRA

34. Hilton, P. J. and U. Stammbach, *A Course in Homological Algebra*. Berlin: Springer-Verlag, 1971.
35. S. MacLane, *Homology*. Berlin: Springer-Verlag, 1963.

FIELDS

36. Artin, E., *Galois Theory*. Notre Dame, Ind.: Notre Dame Mathematical Lectures No. 2 (2d ed.), 1944.
37. Gaal, L., *Classical Galois Theory with Examples*. Chicago: Markham, 1971.
38. Jacobson, N., *Lectures in Abstract Algebra* (vol. III). Princeton, N. J.: D. Van Nostrand Company, Inc., 1964.
39. Kaplansky, I., *Fields and Rings* (2d ed.). Chicago: University of Chicago Press, 1972.
40. McCarthy, P. J., *Algebraic Extensions of Fields*. Waltham, Mass.: Blaisdell Publishing Company, 1966.

LINEAR AND MULTILINEAR ALGEBRA

41. Greub, W., *Linear Algebra* (3rd ed.). Berlin: Springer-Verlag, 1967.
42. Greub, W., *Multilinear Algebra*. Berlin: Springer-Verlag, 1967.
43. Halmos, P. R., *Finite Dimensional Vector Spaces* (2d ed.). Princeton, N. J.: D. Van Nostrand Company, Inc., 1958.
44. Jacobson, N., *Lectures in Abstract Algebra* (vol. II). Princeton, N. J.: D. Van Nostrand Company, Inc., 1953.

CATEGORIES

45. MacLane, S., *Categories for the Working Mathematician*. Berlin: Springer-Verlag, 1972.
46. Mitchell, B., *Theory of Categories*. New York: Academic Press, Inc., 1965.
47. Pareigis, B., *Categories and Functors*. New York: Academic Press, Inc., 1970.

NUMBER THEORY

48. Artin E., *Algebraic Numbers and Algebraic Functions*. New York: Gordon and Breach, 1967.
49. Lang, S., *Algebraic Number Theory*. Reading Mass.: Addison-Wesley Publishing Company, Inc., 1970.
50. O'Meara, O. T., *Introduction to Quadratic Forms*. Berlin: Springer-Verlag, 1963.
51. Shockley, J. E., *Introduction to Number Theory*. New York: Holt, Rinehart and Winston, Inc., 1967.
52. Weiss E., *Algebraic Number Theory*. New York: McGraw-Hill Inc., 1963.

ALGEBRAIC GEOMETRY

53. Fulton, W., *An Introduction to Algebraic Geometry*. New York: W. A. Benjamin Inc., 1969.
54. Lang, S., *Introduction to Algebraic Geometry*. New York: Interscience Publishers, 1959.
55. MacDonald, I. G., *Algebraic Geometry: Introduction to Schemes*. New York: W. A. Benjamin Inc., 1968.

ANALYSIS

56. Burrill, C. W., *Foundations of Real Numbers*. New York: McGraw-Hill, Inc., 1967.

57. Hewitt, E. and K. Stromberg, *Real and Abstract Analysis.* Berlin: Springer-Verlag, 1969.

ARTICLES AND NOTES

58. Bergman, G., "A Ring Primitive on the Right but Not on the Left," *Proc. Amer. Math. Soc.*, 15 (1964), pp. 473–475; correction, pg. 1000.
59. Cohen, P. J., *Set Theory and the Continuum Hypothesis*, New York: W. A. Benjamin Inc., 1966.
60. Corner, A. L. S., "On a Conjecture of Pierce Concerning Direct Decomposition of Abelian Groups," *Proc. Colloq. on Abelian Groups*, Budapest, 1964, pp. 43–48.
61. Feit, W. and J. Thompson, "Solvability of Groups of Odd Order," *Pac. Jour. Math.*, 13 (1963), pp. 775–1029.
62. Goldie, A. W., "Semiprime rings with maximum condition," *Proc. Lond. Math. Soc.*, 10 (1960), pp. 201–220.
63. Kaplansky, I., "Projective Modules," *Math. Ann.*, 68 (1958), pp. 372–377.
64. Krull, W., "Galoissche Theorie der unendlichen algebraischen Erweiterungen," *Math. Ann.*, 100 (1928), pp. 687–698.
65. Procesi, C. and L. Small, "On a theorem of Goldie," *Jour. of Algebra*, 2 (1965), pp. 80–84.
66. Sasiada, E. and P. M. Cohn, "An Example of a Simple Radical Ring," *Jour. of Algebra*, 5 (1967), pp. 373–377.

Index

Graduate Texts in Mathematics

(continued from page ii)

68 WEIDMANN. Linear Operators in Hilbert Spaces.
69 LANG. Cyclotomic Fields II.
70 MASSEY. Singular Homology Theory.
71 FARKAS/KRA. Riemann Surfaces. 2nd ed.
72 STILLWELL. Classical Topology and Combinatorial Group Theory. 2nd ed.
73 HUNGERFORD. Algebra.
74 DAVENPORT. Multiplicative Number Theory. 2nd ed.
75 HOCHSCHILD. Basic Theory of Algebraic Groups and Lie Algebras.
76 IITAKA. Algebraic Geometry.
77 HECKE. Lectures on the Theory of Algebraic Numbers.
78 BURRIS/SANKAPPANAVAR. A Course in Universal Algebra.
79 WALTERS. An Introduction to Ergodic Theory.
80 ROBINSON. A Course in the Theory of Groups. 2nd ed.
81 FORSTER. Lectures on Riemann Surfaces.
82 BOTT/TU. Differential Forms in Algebraic Topology.
83 WASHINGTON. Introduction to Cyclotomic Fields. 2nd ed.
84 IRELAND/ROSEN. A Classical Introduction to Modern Number Theory. 2nd ed.
85 EDWARDS. Fourier Series. Vol. II. 2nd ed.
86 VAN LINT. Introduction to Coding Theory. 2nd ed.
87 BROWN. Cohomology of Groups.
88 PIERCE. Associative Algebras.
89 LANG. Introduction to Algebraic and Abelian Functions. 2nd ed.
90 BRØNDSTED. An Introduction to Convex Polytopes.
91 BEARDON. On the Geometry of Discrete Groups.
92 DIESTEL. Sequences and Series in Banach Spaces.
93 DUBROVIN/FOMENKO/NOVIKOV. Modern Geometry—Methods and Applications. Part I. 2nd ed.
94 WARNER. Foundations of Differentiable Manifolds and Lie Groups.
95 SHIRYAEV. Probability. 2nd ed.
96 CONWAY. A Course in Functional Analysis. 2nd ed.
97 KOBLITZ. Introduction to Elliptic Curves and Modular Forms. 2nd ed.
98 BRÖCKER/TOM DIECK. Representations of Compact Lie Groups.
99 GROVE/BENSON. Finite Reflection Groups. 2nd ed.
100 BERG/CHRISTENSEN/RESSEL. Harmonic Analysis on Semigroups: Theory of Positive Definite and Related Functions.
101 EDWARDS. Galois Theory.
102 VARADARAJAN. Lie Groups, Lie Algebras and Their Representations.
103 LANG. Complex Analysis. 3rd ed.
104 DUBROVIN/FOMENKO/NOVIKOV. Modern Geometry—Methods and Applications. Part II.
105 LANG. $SL_2(\mathbf{R})$.
106 SILVERMAN. The Arithmetic of Elliptic Curves.
107 OLVER. Applications of Lie Groups to Differential Equations. 2nd ed.
108 RANGE. Holomorphic Functions and Integral Representations in Several Complex Variables.
109 LEHTO. Univalent Functions and Teichmüller Spaces.
110 LANG. Algebraic Number Theory.
111 HUSEMÖLLER. Elliptic Curves.
112 LANG. Elliptic Functions.
113 KARATZAS/SHREVE. Brownian Motion and Stochastic Calculus. 2nd ed.
114 KOBLITZ. A Course in Number Theory and Cryptography. 2nd ed.
115 BERGER/GOSTIAUX. Differential Geometry: Manifolds, Curves, and Surfaces.
116 KELLEY/SRINIVASAN. Measure and Integral. Vol. I.
117 SERRE. Algebraic Groups and Class Fields.
118 PEDERSEN. Analysis Now.
119 ROTMAN. An Introduction to Algebraic Topology.
120 ZIEMER. Weakly Differentiable Functions: Sobolev Spaces and Functions of Bounded Variation.
121 LANG. Cyclotomic Fields I and II. Combined 2nd ed.
122 REMMERT. Theory of Complex Functions. *Readings in Mathematics*
123 EBBINGHAUS/HERMES et al. Numbers. *Readings in Mathematics*
124 DUBROVIN/FOMENKO/NOVIKOV. Modern Geometry—Methods and Applications. Part III.
125 BERENSTEIN/GAY. Complex Variables: An Introduction.
126 BOREL. Linear Algebraic Groups. 2nd ed.
127 MASSEY. A Basic Course in Algebraic Topology.
128 RAUCH. Partial Differential Equations.

129 FULTON/HARRIS. Representation Theory: A First Course.
Readings in Mathematics
130 DODSON/POSTON. Tensor Geometry.
131 LAM. A First Course in Noncommutative Rings.
132 BEARDON. Iteration of Rational Functions.
133 HARRIS. Algebraic Geometry: A First Course.
134 ROMAN. Coding and Information Theory.
135 ROMAN. Advanced Linear Algebra.
136 ADKINS/WEINTRAUB. Algebra: An Approach via Module Theory.
137 AXLER/BOURDON/RAMEY. Harmonic Function Theory.
138 COHEN. A Course in Computational Algebraic Number Theory.
139 BREDON. Topology and Geometry.
140 AUBIN. Optima and Equilibria. An Introduction to Nonlinear Analysis.
141 BECKER/WEISPFENNING/KREDEL. Gröbner Bases. A Computational Approach to Commutative Algebra.
142 LANG. Real and Functional Analysis. 3rd ed.
143 DOOB. Measure Theory.
144 DENNIS/FARB. Noncommutative Algebra.
145 VICK. Homology Theory. An Introduction to Algebraic Topology. 2nd ed.
146 BRIDGES. Computability: A Mathematical Sketchbook.
147 ROSENBERG. Algebraic K-Theory and Its Applications.
148 ROTMAN. An Introduction to the Theory of Groups. 4th ed.
149 RATCLIFFE. Foundations of Hyperbolic Manifolds.
150 EISENBUD. Commutative Algebra with a View Toward Algebraic Geometry.
151 SILVERMAN. Advanced Topics in the Arithmetic of Elliptic Curves.
152 ZIEGLER. Lectures on Polytopes.
153 FULTON. Algebraic Topology: A First Course.
154 BROWN/PEARCY. An Introduction to Analysis.
155 KASSEL. Quantum Groups.
156 KECHRIS. Classical Descriptive Set Theory.
157 MALLIAVIN. Integration and Probability.
158 ROMAN. Field Theory.
159 CONWAY. Functions of One Complex Variable II.
160 LANG. Differential and Riemannian Manifolds.
161 BORWEIN/ERDÉLYI. Polynomials and Polynomial Inequalities.
162 ALPERIN/BELL. Groups and Representations.
163 DIXON/MORTIMER. Permutation Groups.
164 NATHANSON. Additive Number Theory: The Classical Bases.
165 NATHANSON. Additive Number Theory: Inverse Problems and the Geometry of Sumsets.
166 SHARPE. Differential Geometry: Cartan's Generalization of Klein's Erlangen Program.
167 MORANDI. Field and Galois Theory.
168 EWALD. Combinatorial Convexity and Algebraic Geometry.
169 BHATIA. Matrix Analysis.
170 BREDON. Sheaf Theory. 2nd ed.
171 PETERSEN. Riemannian Geometry.
172 REMMERT. Classical Topics in Complex Function Theory.
173 DIESTEL. Graph Theory.
174 BRIDGES. Foundations of Real and Abstract Analysis.
175 LICKORISH. An Introduction to Knot Theory.
176 LEE. Riemannian Manifolds.
177 NEWMAN. Analytic Number Theory.
178 CLARKE/LEDYAEV/STERN/WOLENSKI. Nonsmooth Analysis and Control Theory.
179 DOUGLAS. Banach Algebra Techniques in Operator Theory. 2nd ed.
180 SRIVASTAVA. A Course on Borel Sets.
181 KRESS. Numerical Analysis.
182 WALTER. Ordinary Differential Equations.
183 MEGGINSON. An Introduction to Banach Space Theory.
184 BOLLOBAS. Modern Graph Theory.
185 COX/LITTLE/O'SHEA. Using Algebraic Geometry.
186 RAMAKRISHNAN/VALENZA. Fourier Analysis on Number Fields.
187 HARRIS/MORRISON. Moduli of Curves.
188 GOLDBLATT. Lectures on the Hyperreals: An Introduction to Nonstandard Analysis.
189 LAM. Lectures on Modules and Rings.
190 ESMONDE/MURTY. Problems in Algebraic Number Theory.